International Conference on Engineering, Technology and Innovative Researches (ICETIR 2020)

IOP Conference Series: Materials Science and Engineering
Volume 982

Purwokerto, Indonesia
2 – 3 September 2020

Editors:

Maulana Rizki Aditama Aini Hanifa
Yogiek Indra Kurniawan
FX Anjar Trilaksono
Januar Aziz Zaenurrochman
Huzaely Latief Sunan
Ahmad Khalil Gibran
M. Syaiful A'alim
Ari Fadli
Rani Aulia I.

ISBN: 978-1-7138-4796-0
ISSN: 1757-8981

Printed from e-media with permission by:

Curran Associates, Inc.
57 Morehouse Lane
Red Hook, NY 12571

Some format issues inherent in the e-media version may also appear in this print version.

Institute of Physics
Dirac House, Temple Back
Bristol BS1 6BE UK

Phone: 44 1 17 929 7481
Fax: 44 1 17 920 0979

techtracking@iop.org

Additional copies of this publication are available from:

Curran Associates, Inc.
57 Morehouse Lane
Red Hook, NY 12571 USA
Phone: 845-758-0400
Fax: 845-758-2633
Email: curran@proceedings.com
Web: www.proceedings.com

TABLE OF CONTENTS

IOP Conf. Series: Materials Science and Engineering **982** (2020) 011001 doi:10.1088/1757-899X/982/1/011001

PREFACE

The International Conference on Engineering, Technology and Innovative Researches (ICETIR) 2020 is held in September 2nd 2020, organized by the Faculty of Engineering, Universitas Jenderal Soedirman. This conference is aimed to communicate the results of research in the field of engineering and to jointly contribute to society through innovative research. In 2020, we highlight the theme "**Sustainable Engineering and Technology for Reshaping our Community**".

However, due to the covid19 and travel restriction, we held the conference virtually through zoom meeting, inviting the keynote speakers from the various fields of engineering, along with the accepted papers, which consist of some tracks, i.e., information system, software engineering, computer engineering, mechanical engineering, geological engineering, industrial engineering, electrical power, telecommunication, control systems, agriculture engineering, and civil engineering. On this occasion, we have received submissions from authors originated from 5 countries (Indonesia, Taiwan, Malaysia, Saudi Arabia, Japan), with the attendance number were around 100 participants. We had plenary and parallel session for each presenter with 10 minutes talk, followed by 4 minutes of discussion (question answer session).

We would like to acknowledge all supports for the success of this conference, especially, the board of reviewers, the board of editors, the organizing committee, the sponsors, and all conference participants. We hope that the 2nd ICETIR 2020 has been an excellent discussion forum in the field of engineering.

Sugeng Waluyo

The Chairman of ICETIR 2020

IOP Conf. Series: Materials Science and Engineering **982** (2020) 011001 doi:10.1088/1757-899X/982/1/011001

Scientific Committee

1. Prof. Silvio Ranise (Fondazione Bruno Kessler, Italy)
2. Prof. Hitoshi Habe (Kindai University, Japan)
3. Prof. Toshihiko Noguchi (Shizuoka University, Japan)
4. Dr. Albert Causo (Nanyang Technological University, Singapore)
5. Prof. Imam Sadisun (Bandung Institute of Technology, Indonesia)
6. Prof. Budi Santosa (Sepuluh Nopember Institute of Technology, Indonesia)
7. Dr.-Ing. Supriyanto (Advantech Europe, Jerman)
8. Dr. Adie Wahyudie(United Arab Emirates University, United Arab Emirates)
9. Dr. Suroso (Universitas Jenderal Soedirman, Indonesia)
10. Prof. Retno Supriyanti (Universitas Jenderal Soedirman, Indonesia)
11. Dr. Gito Sugiyanto (Universitas Jenderal Soedirman, Indonesia)
12. Dr. Eng. Agus Maryoto (Universitas Jenderal Soedirman, Indonesia)

Organizing Committee

1. Sugeng Waluyo (Universitas Jenderal Soedirman, Indonesia)
2. Asmoro Widagdo (Universitas Jenderal Soedirman, Indonesia)
3. Rani Aulia Imran (Universitas Jenderal Soedirman, Indonesia)
4. FX Anjar Trilaksono (Universitas Jenderal Soedirman, Indonesia)

Editors

Editor in Chief : Maulana Rizki Aditama
Aini Hanifa

Editor :

1. Yogiek Indra Kurniawan
2. FX Anjar Trilaksono
3. Januar Aziz Zaenurrochman
4. Huzaely Latief Sunan
5. Ahmad Khalil Gibran
6. M. Syaiful A'alim
7. Ari Fadli
8. Rani Aulia I.

9. Tigar Dian P. A.
10. Amanda Sofiana
11. Katon Muhammad
12. Indro Prakoso
13. Ayu Anggraeni
14. Yudi Syahrullah
15. Dani Nugroho
16. Redityo Januardi

Peer review declaration

All papers published in this volume of *Material Science and Engineering: Conference Series* have been peer-reviewed through processes administered by the proceedings Editors. All papers have been processed using the double-blind system, considering the topics, contents and the plagiarism indexes through Turnitin. All papers which passed the editing process is then given to the two different reviewers, to meet the standard of quality for an international publication at this level. Expert referees conducted reviews to the professional and scientific standards expected of a proceedings journal published by IOP Publishing.

- **Type of peer review: Double blind**
- **Conference submission management system: Submission is done through the system**
- **Number of submissions received: 83**
- **Number of submissions sent for review: 83**
- **Number of submissions accepted: 67**
- **Acceptance Rate (Number of Submissions Accepted / Number of Submissions Received X 100): 80.7**
- **Average number of reviews per paper: 5**
- **Total number of reviewers involved: 17**
- **Any additional info on review process: No**
- **Contact person for queries: maulanarizkia@unsoed.ac.id and sugeng.waluyo@unsoed.ac.id**

IOP Conf. Series: Materials Science and Engineering **982** (2020) 012001 doi:10.1088/1757-899X/982/1/012001

Computer Aided System for Gambung Tea Identification using Convolutional Neural Network

Nur Ibrahim[1], YN Fu'adah[2], NK Caecar Pratiwi[3], Syamsul Rizal[4], Koredianto Usman[5]

[1,2,3,4,5] Electrical Engineering Department - Telkom University, Bandung, Indonesia

e-mail: nuribrahim@telkomuniversity.ac.id

Abstract. Tea commodity has a very strategic role for Indonesian economy. In 2012 the tea commodity was able to generate foreign exchange of US $ 156.74 million. Nationally, tea industry contributes a Gross Domestic Product (GDP) of around Rp. 1.2 trillion. The total tea plantation area in Indonesia entirely covering 123 938 hectares. The conception of sustainable tea production includes three aspects,economic development, social development and environmental protection. One of the steps towards sustainable tea production is survey and identification. The process continues with the selection of suitable planting material (seedlings). Gambung series, are planting material seeds that have been recommended by the Indonesia Ministry of Agriculture. The Gambung series has a potential yield of 4.000 - 5.800 kg/hectare of dried tea. The morphological similarity level of Gambung series is very high, because the elders of the clones are from the same crossing parents. Experts who are able to identify Gambung clone are very limited. This process is susceptible to errors and is very dependent on the presence of experts. If an error occurs in the process of identifying the type of clone, it will interferes with the breeding process. Errors in the selection of recommended clones will be detrimental to the process for long time period, due to the economic age of the tea plant can reach 50 years. From this issue, it is very necessary to design a system that is able to identify the planting material of Gambung series clones. The system is designed to classify Gambung and Non Gambung tea series using Convolutional Neural Network (CNN) with a high accuracy and low loss rate.

1. Introduction

Indonesia has been known as one of foremost tea producers in the world. Tea is a commodity from the plantation sector which has experienced a triumph over the past decades. But from year to year, Indonesia's tea ranking in the international market continues to decline [1]. The geographical location of Indonesia which is on the equatorial trajectory and the presence of hundreds volcanoes are the main factors causing Indonesia's fertile soil. The fertile soil are suitable for the conditions for growing tea and has great potential to expand the land and increase the quantity and quality of tea Indonesia [2]. In addition to the increasingly export opportunities, the domestic tea market is still quite large even though it has not been optimally explored yet. Tea plantations in Indonesia are distinguished according to their operations became Perkebunan Besar (PB) and Perkebunan Rakyat (PR). Perkebunan Besar consist of Perkebunan Besar Negara (PBN) and Perkebunan Besar Swasta (PBS). In 2016, the area of PBN was recorded experiencing decreased by 5.29% and continued to decline until 2018 decreased by 15.39%. While the total area of PBS in 2016 decreased by 1.92% and in 2018 decreased again by 6.71 % [3]. This land reduction are caused by several things, one of is due to the favorable business prospects of oil palm, so that tea plantations have been converted into oil palm plantations. It was also caused by some tea plantations was stop the production and switch to more profitable commodities. Mistakes in selecting superior seedlings are one causes of decreased productivity of tea plants.

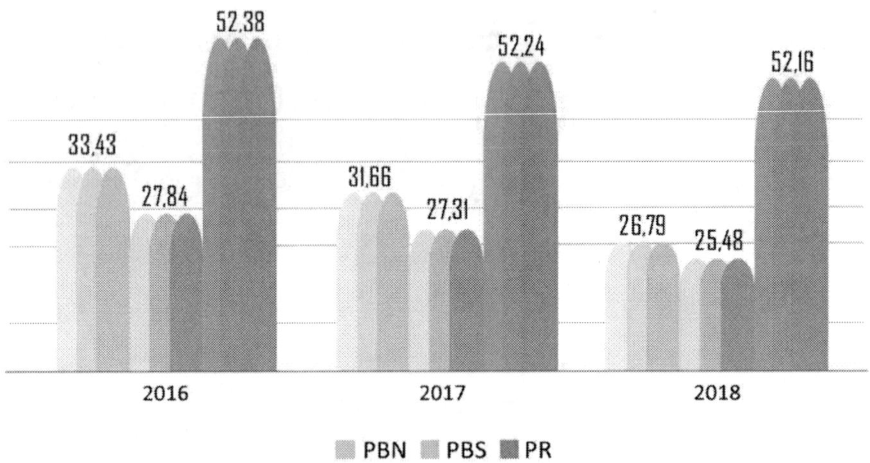

Figure 1. Development of Indonesia's Tea Plantation Area in the 2016 - 2018 Period [3]

One of the steps towards sustainable tea production is survey and identification. The survey was conducted to determine the condition related to: soil, biodiversity, social and human resources. After knowing the survey results, identification of deficiencies can be more easily done. This is useful to determine the direction and priority of improvement towards sustainable tea plantations, one of which is election suitable planting material (seeds). Based on the Decree of the Indonesian Minister of Agriculture, the Gambung series superior clones have been released. Gambung series has a potential yield of 4000 - 5800 kg / hectare of dry tea. During this time, the process of identifying gambung series is done manually using visual eyes of experts at PPTK Gambung. This process is susceptible to errors in the reading of clone types, and is very dependent on the presence of experts. If an error occurs in the process of identifying the type of Gambung series clone, would certainly would be detrimental to the production process for many years. Errors in the selection of Gambung series clone plant material will have a long-term loss due to the economic life of the tea plant which can reach 50 years. The potential loss of production in an area of 127357 hectare due to misuse of plant material can reach 1200 kg/hectare/year or equivalent to loss of potential income of Rp. 30 billion/year (assuming the selling price of dry tea is Rp. 20000/kg). From this issue, it is very necessary to design a system that is able to identify Gambung series clones. The system is designed to classify Gambung and Non Gambung tea series using Convolutional Neural Network (CNN).

There are several identification systems using leaf recognition that have been developed by several previous researchers. However, most of the leaves data-set are originating from abroad plants. So, it is not necessarily suitable when used to detect leaves plants from Indonesia. Hossain and Amin [4] developed a system that was able to detect 30 types of leaves from different plants. The system consists of 5 main processes, images are taken from a digital camera or scanner, the selection of leaf base points and several other reference points, separating leaves from the background and converting the RGB into a binary images, equalizing the leaf with the base point, and extraction of leaf morphological features. After that, the results of feature extraction are used as input in classification processes using a Probabilistic Neural Network (PNN). From the testing results of 1200 samples (30 different plants), system was able to identify the type of leaf with an accuracy of 91.41%. Kumar, et al [5] proposed a smartphone application to identify plant types by using leaves visual recognition. The methods are separating non-leaf imagery, segmentation of leaves from non-textured backgrounds, extraction of curvature features, and classification using the K-Nearest Neighbor. The system uses data sets of 184 plants in Northeastern United States. Unfortunately there are no reports on how accurate the testing of this system is. Aakif and Khan [6] created a system of identifying 14 different fruit plants. The method used consists of 3 main steps, pre-processing, feature extraction and classification. Features extracted including morphological features, fourier descriptors, and shapes. The classification used is artificial neural network (ANN). From the testing/ validating process of 817 samples, the results give an accuracy of 96%. Research related to the classification of leaves-based plants using CNN has been carried out by Lee, et al [7] and Syamsul et al [8]. Lee developed a leaf identification system using the Convolutional Neural Network (CNN) classification method, with feature selection made using the Deconvolutional Network (DN) approach. This system was tested on MalayaKew (MK) Leaf dataset consisting of 44 different plant species. The accuracy of the system validating is 98.1%. Syamsul has been built a system using CNN with 3 hidden layers plus a fully connected layer with softmax activation to determine its classification [8]. After doing the training and validation process for 50 iterations, obtained 86% accuracy and loss of 0.087. Suyanto and Munte [9] proposed a system that could identify the type of rubber plant from the leaves. The methods used in the system are data acquisition, image processing, edge detection using Sobel, and identification using template matching. System testing was carried out using 14 rubber clones in the White Sungei Rubber Research Center. From the test results it can be seen that the accuracy of this system is 91.79%. Turkoglu and Hanbay [10] developed a plant identification system using leaf recognition. Feature extraction used is color features, leaf bone, Fourier Descriptor (FD), and Gray-Level-Co-occurance (GLCM). Next, the classification method used is Extreme Learning Machines (ELM). This system was tested on Flavia dataset, the test results found that this system has an accuracy of 99.10%.

Reviewing some of the previous research above, then we propose a system that is able to identify Indonesia's superior tea plants, Gambung tea series. We propose a system using Convolutional Neural Network (CNN) modeling. Nowadays, CNN has been widely used in various applications, especially for images and videos processing. CNN will continue to practice learning filters by itself without explicit mentioning it.

2. Convolutional Neural Network (CNN)

Convolutional Neural Network (CNN) is one of the deep learning methods that designed to process images (two-dimensional) data [11]. The Convolutional Neural Network is a development of back-propagation method that composed of several layers namely input layer, followed by several hidden layers and output layers [12].

IOP Conf. Series: Materials Science and Engineering **982** (2020) 012001 doi:10.1088/1757-899X/982/1/012001

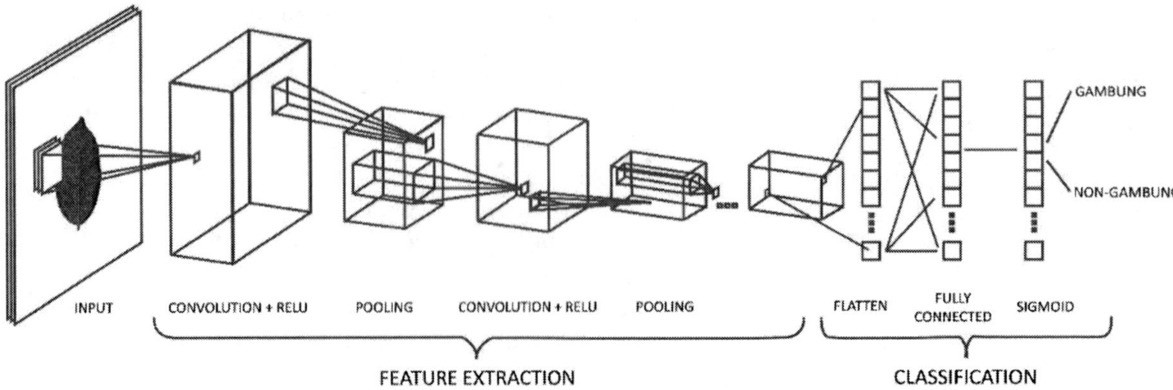

Figure 2. The Architecture of CNN

At each layer is composed of several neurons that are connected to all neurons in the previous layer and the layer afterwards. In order that, CNN imitates the work of visual cortex in the brain to process and to recognize image [12]. The architecture of CNN is shown in Figure 2 as follows [13]. Based on Figure 2, the architecture of CNN consists of two main process namely feature extraction and classification that are connected respectively.

2.1. Feature Extraction

Feature extraction neural network consists of the convolutional layer and the pooling layer. CNN works hierarchically, so the output at the first convolution layer is used as input at the next convolution layer. The convolutional layer uses filter to generate feature maps from input image. The number of feature maps generated depends on the numbers of filters used. Therefore, for instance, if the convolutional contains three filters, it will generate three feature maps [12]. There are parameters that influence the convolutional layer, namely filter size, stride and padding. Stride is useful for determining the shift in the number of filters. Padding or zero padding is a parameter that determines the number of pixels (containing the value 0) to be added to each side of the image with the aim of accurate information on the edges of the image. An illustration of the convolution process is shown in Figure 3.

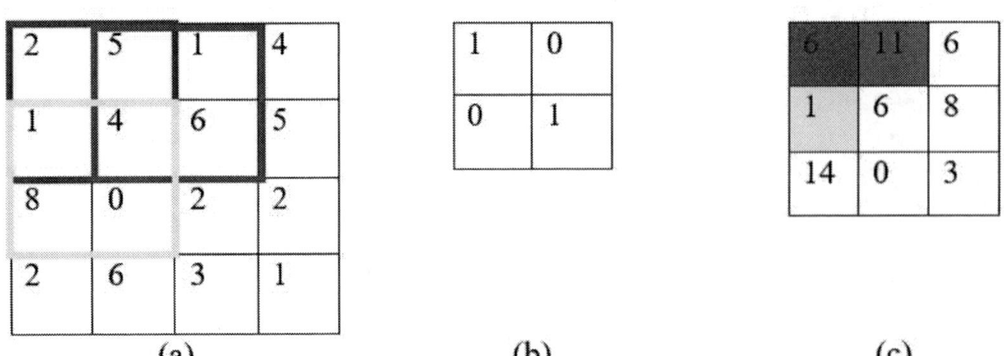

Figure 3. Convolutional process of input data with size 4 and stride 1(a), filter size (2) (b), and result (c)

Rectified Linear Unit (ReLU) is an important element in the convolutional layer to change the negative pixel value on the feature map to zero. The ReLU activation function is given in Equation 1 as follows [14].

$$f(x) = max(0, x) \tag{1}$$

The output value of a neuron can be expressed as 0 if the input is negative. If the input value is positive, then the output of the neuron is the activation input value itself [15].

Pooling or sub-sampling is a reduction in matrix size. There are two types of pooling that are often used, namely average pooling and max pooling [2] The value taken at average pooling is the average value while at max pooling is the maximum value [12]. Figure 4 is shown the illustrate of average pooling and maximum pooling.

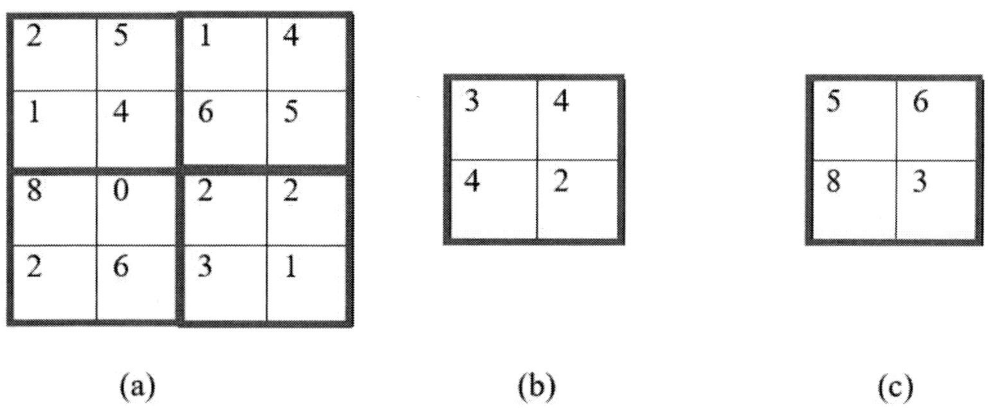

Figure 4. Illustrate of Pooling Layer, (a). The four by four pixel input image ; (b). Mean Pooling and (c). Maximum Pooling

2.2. Classification

Feature maps as a result of feature learning layer are multidimensional arrays. The flatten process convert feature maps from multidimensional arrays to one dimensional array so after that it can be processed at the fully connected layer. This layer gets input from the previous process to determine which features are most correlated with a particular class [16]. The sigmoid activation function is used to get the classification results. The sigmoid function is used for binary classification in the Logistic Regression model, it transforms the range of values from input x to be between 0 and 1 with the distribution form of the function as in Figure 5 and shown by Equation 2 as follows [17]:

$$S(x) = \frac{1}{1 + e^{-x}} \tag{2}$$

IOP Conf. Series: Materials Science and Engineering **982** (2020) 012001 doi:10.1088/1757-899X/982/1/012001

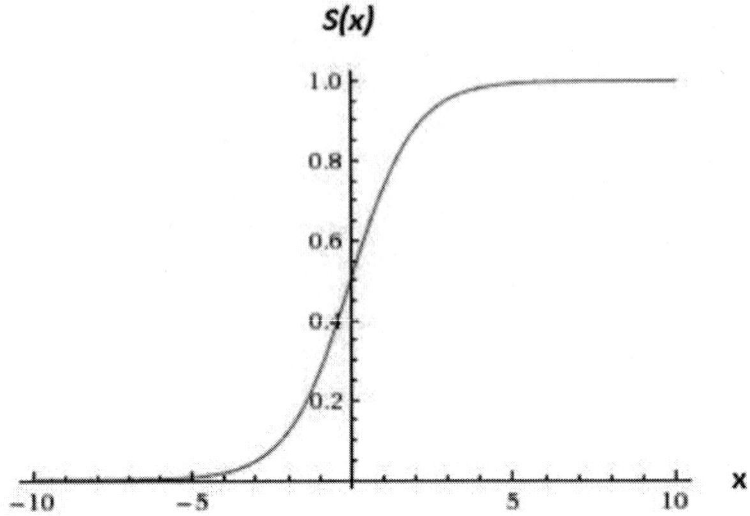

Figure 5. Sigmoid Activation Graph

3. System Design

In this research, the system is built using CNN with two hidden convolutional layer and a fully connected later. Figure 6 and Table 1 show the layer configuration in our CNN model.

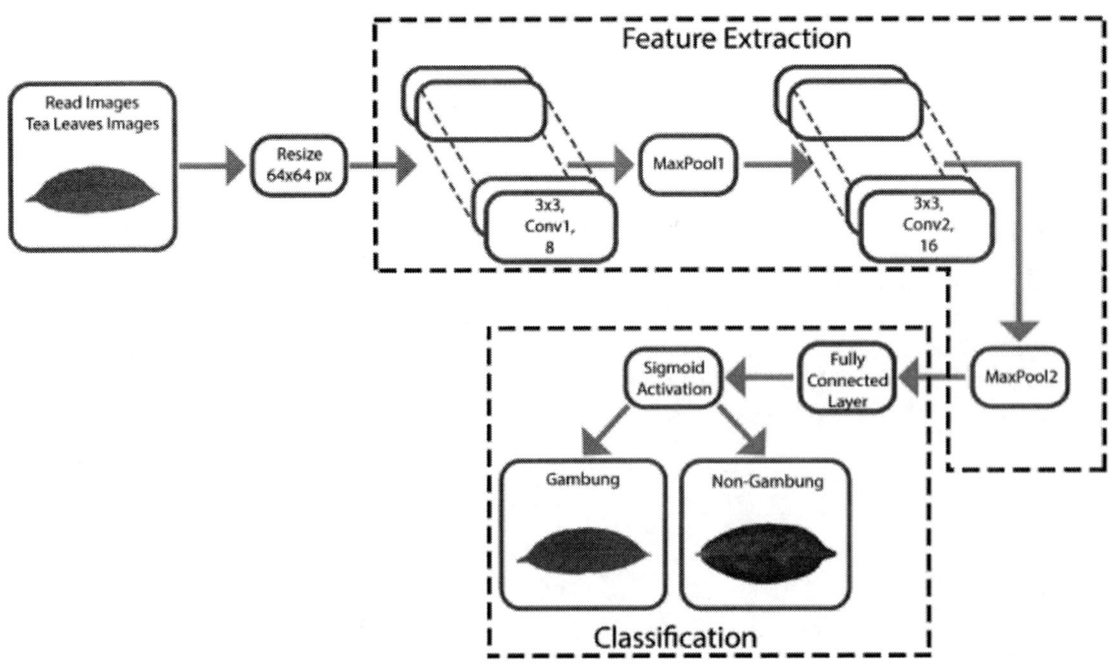

Figure 6. Proposed Model of CNN

IOP Conf. Series: Materials Science and Engineering **982** (2020) 012001 doi:10.1088/1757-899X/982/1/012001

Table 1. Summary of CNN Configuration

Layer (type)	Output Shape	Parameter
conv2d_1 (Conv2D)	(None, 64, 64, 8)	224
activation_1 (Activation)	(None, 64, 64, 8)	0
max_pooling2d_1 (MaxPooling2)	(None, 32, 32, 8)	0
conv2d_2 (Conv2D)	(None, 32, 32, 16)	1168
activation_2 (Activation)	(None, 32, 32, 16)	0
max_pooling2d_2 (MaxPooling2	(None, 16, 16, 16)	0
flatten_1 (Flatten)	(None, 4096)	0
dense_1 (Dense)	(None, 2)	8194
activation_3 (Activation)	(None, 2)	0
Total parameters: 9,586		
Trainable parameters: 9,586		
Non-trainable parameters: 0		

The Dataset of tea laves that used in this research is 212 image of training image (consist of 36 image non Gambung tea leaves images and 176 Gambung tea leaves images) 71 of testing images (consist of 27 images for non Gambung tea leaves images and 44 Gambung tea leaves images). Based on Figure 6 and Table 1, the resolution of tea leaves images is changed to 64×64 pixel and it inserted into CNN. In hidden layer 1, we use filter with size 3×3 and channel output 8. In hidden layer 2, we use filter with size 3×3 and channel output 16. After that, we do the flatten to change the image featured from 3 dimension became 1 dimension so that we can do the classification into 2 class (Gambung Tea Class and Non Gambung Tea Class). Activation that used to do the classification is sigmoid.

In the first convolution layer, named conv2d_1 layer, the data images that has been resized to 64×64 pixels in three layers (Red, Green, and Blue), will be convoluted by filter with 2×2 size, and the output is the 32×32 pixels images size with 8 layers. The output from first convolution layer illustrated in figure 7 below.

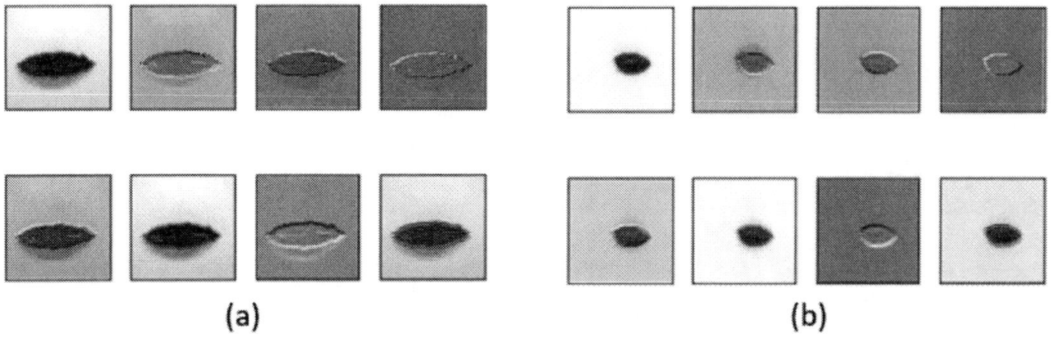

Figure 7. Output Image of the first convolutional layer with (a). GMB image input and (b). nonGMB image input

These images will go through the second convolutional layer, named conv2d_2 layer, and it will be convoluted by filter with 2×2 size, so the output is the 16×16 pixels images size with 16 layers. The output from first convolution layer illustrated in figure 8 below.

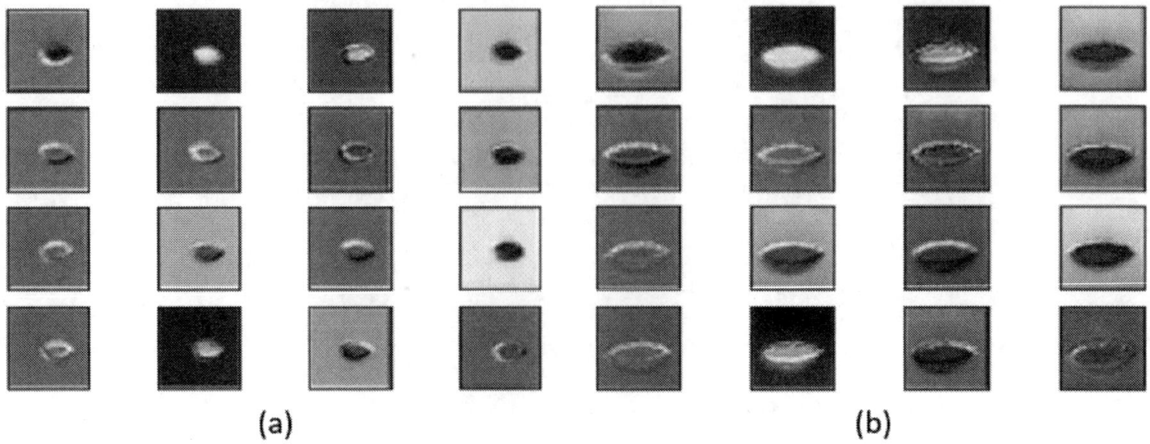

Figure 8. Output Image of the second convolutional layer with (a). GMB image input and (b). nonGMB image input

4. Result and Analysis

Adam with learning rate 0.001 is used for Optimizer, and Binary Crossentropy for loss. The measured parameters is accuracy and loss. Total iteration for training data is 50 times (50 epoch). After do the training, we can see the result of accuracy and loss from Figure 7 and Figure 8 below.

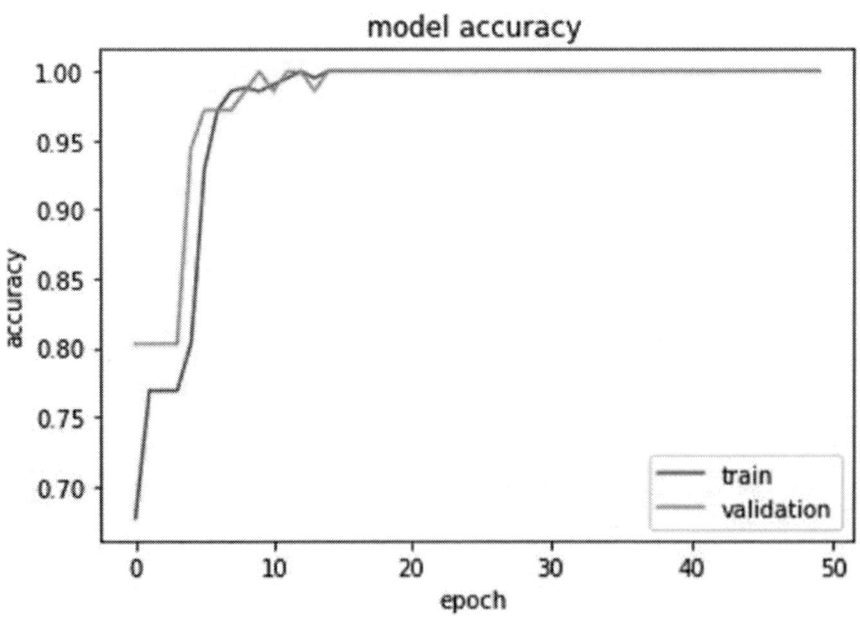

Figure 9. The Increase of accuracy value in training process and validation process

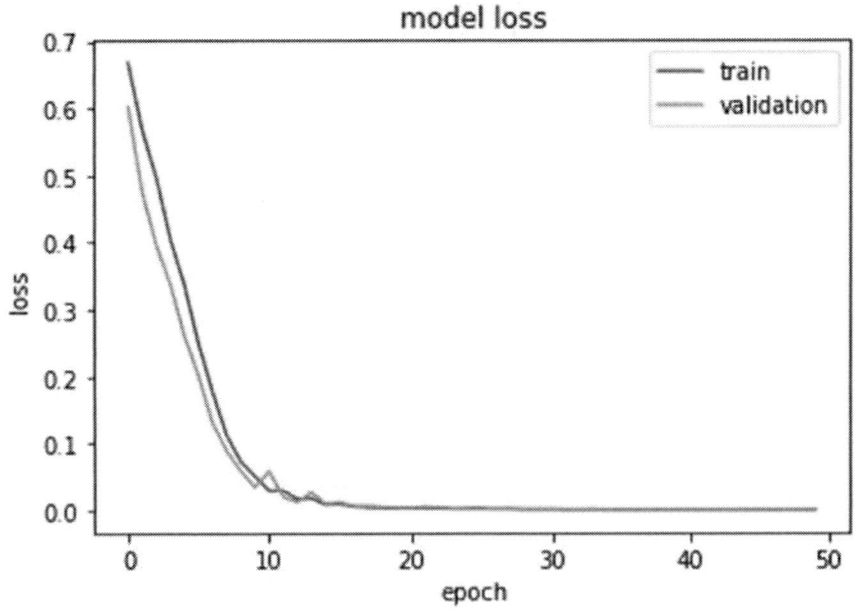

Figure 10. The decrease of loss value in training process and validation process

Figure 9 shows the increase of accuracy value for each iteration (epoch). The model is not over-fitting that we can see from the difference between training accuracy and validation accuracy is not too far. Figure 10 shows the decrease of loss of value for each iteration. The difference between training loss and validation loss is not too far, so we can conclude that the model system can be used to classify between Gambung Tea and non Gambung Tea with the accuracy 100% and loss 0.0007.

5. Conclusion

To classify the Gambung and Non Gambung Series tea leaves, a system has been built using CNN with two hidden layers plus one fully connected layer with sigmoid activation to determine its classification. After doing the 50 iterations for training and validation process, obtained an accuracy value of 100% with 0.0007 loss. So it can be concluded that this system does not occur over-fitting, which means the system can recognize the type of Gambung and Non Gambung tea leaves with new data.

References

[1] Ramadhani, Fadhilah. "Daya Saing Teh Indonesia di Pasar Internasional." Economics Development Analysis Journal 2, no. 4. 2013.
[2] Zakariyah, Mochamad Yuzi, Ratya Anindita, and Nur Baladina. "Analisis daya saing teh Indonesia di pasar internasional." Jurnal Agrimeta 4, no. 08. 2014.
[3] BPS, Indonesian Statistics. "Indonesian Tea Statistics 2018." Indonesian Statistics. 2018.
[4] Hossain, Javed, and M. Ashraful Amin. "Leaf shape identification based plant biometrics." In 2010 13th International conference on computer and information technology (ICCIT), pp. 458-463. IEEE. 2010.
[5] Kumar, Neeraj, Peter N. Belhumeur, Arijit Biswas, David W. Jacobs, W. John Kress, Ida C. Lopez, and Joo VB Soares. "Leafsnap: A computer vision system for automatic plant species identification." In European conference on computer vision, pp. 502-516. Springer, Berlin, Heidelberg. 2012.
[6] Aakif, Aimen, and Muhammad Faisal Khan. "Automatic classification of plants based on their leaves." Biosystems Engineering 139: 66-75. 2015.

[7] Lee, Sue Han, Chee Seng Chan, Simon Joseph Mayo, and Paolo Remagnino. "How deep learning extracts and learns leaf features for plant classification." Pattern Recognition 71: 1-13. 2017.
[8] Rizal, S., Pratiwi, N.K.C., Ibrahim, N., Vidya, H., Saidah, S. and Fu'adah, R.Y.N,. "Tea Leaves GMB Series Classification using Convolutional Neural Network. JESCE-Journal of Electrical and System Contorl Engineering, 3(2), pp.108-115. 2020.
[9] Munte, J. "Implementation of sobel method to detect the seed rubber plant leaves." In Journal of Physics: Conference Series, vol. 983, no. 1, p. 012128. IOP Publishing, 2018.
[10] Turkoglu, M., Alcin, O.F., Ustundag, M., Hanbay, D. "Leaf recognition and Identification based on geometric and statistical features." Proceedings of the International Conference on Natural Science and Engineering (ICNASE16). 2016.
[11] Phil Kim. "Matlab Deep Learning with Machine Learning, Neural Networks and Artificial Intelligence." Apress. 2017.
[12] O'Shea, Keiron, and Ryan Nash. "An introduction to convolutional neural networks." arXiv preprint arXiv:1511.08458. 2015.
[13] Krizhevsky, A., Sutskever, I., and Hinton, G. E. "ImageNet Classification with Deep Convolutional Neural Networks." Proceedings of the Twenty-Sixth Annual Conference on Neural Information Processing Systems. Lake Tahoe, NY, USA, 38, 10971105. 2012.
[14] Heaton, J. "Artificial Intelligence for Humans: Deep learning and neural networks of Artificial Intelligence for Humans Series." Createspace Independent Publishing Platform. 2015.
[15] Kim, J., Sangjun, O., Kim, Y., & Lee, M. Convolutional Neural Network with Biologically Inspired Retinal Structure. Procedia Computer Science, 88, 145154. 2016.
[16] Erlyna Nour Arrofiqoh and Harintaka. "The Implementation of Convolutional Neural Network Method for Agricultural Plant Classification in High Resolution Imagery." Geomatika Volume 24 No.2 November:61-68. 2018.
[17] Yanhua Guo. "Algorithm Research on Improving Activation Function of Convolutional Neural Networks." The 31th Chinese Control and Decision Conference. 2019.

Automatic Diabetic Detection System based on Retina Using an artificial intelligent system

Zayd Assyarif Alaydrus[1], Winda Astuti[1,2], and Sofyan Tan[1]

[1]Automotive and Robotics Program, Computer Engineering Department, BINUS ASO School of Engineering, Bina Nusantara University, Jakarta, Indonesia 11480

[2]wastuti@binus.edu

Abstract. The Diabetes measurement level usually uses the blood sugar test method, where patients have to fast for eight hours before the test. Glucose tests generally only require a small sample of blood using a sterile needle from the glucometer. The sufferer may feel pain due to a pin prick. However, not all diabetics can do glucose tests regarding having a great fear of syringes, especially in older patients. Therefore, in this work, the diabetic detection system with the retina is an alternative for early checking of patients without having to take a sample of the blood patient. The Contrast Limited Adaptive Histogram Equalization (CLAHE) is used to extract the signal, in which the extracted the retina image is later used as input to the artificial neural network (ANN) based identifier. The retina image parameters are compared and classified to identify the human retina that are intended to be performed. The results of the computer simulation show that this technique produces good accuracy 100% and 96.6%, for training and testing phases, respectively.

1. Introduction

Diabetes is one of several non-communicable diseases (PTM) which is the leading cause of death in Indonesia[1]. Diabetes checks usually use the blood sugar test method, where patients must fast eight hours before the test. Glucose tests generally only require a small sample of blood using a sterile needle from the glucometer. The sufferer may feel pain due to a pin prick. However, not all diabetics can do glucose tests regarding having a great fear of syringes, especially in older patients. This diabetic detection system with the retina is an alternative for the early checking of patients without having to take a sample of the sick blood to be taken[2].

An automatic diabetic detection based on a retina recognition system has been developed to detect the diabetes of the patients. The morphology method od retina processing has been developed to detect the diabetes level based on countras[3]. By using Gray Level Co-occurrence Matrix as a feature extraction method and Support Vector Machines as a method for recognition of brain signal activity[4]. The accuracy of the training and the accuracy of the proposed system testing are 95.93% and 91.07%, respectively [4]. Other experiments that recognize the diabetes recognition system by applied Neural Network to identify the level of diabetes which have accuracy of 96%[5].

Based on the previous study, in this work, An automatic identification of diabetes based on the retina is developed. The system uses CLAHE for feature extraction techniques. Binary classification is based on a Neural network algorithm as applied as a recognition system of the brain activity patterns. The significance of this work is observed the better intelligent identification system suitable for the brain recognition system. The effectiveness of the proposed system based on binary classification of Neural Network ANN is evaluated experimentally.

2. Proposed System

This paper considers diabetic detection based on the Retina identification system. The other technique is an artificial neural network (Ann) which is a very popular pattern machine for the fruit classification system. The multilayer perceptron (MLP) network with a backpropagation learning algorithm is the most well-known of ANN [1]. A block diagram of the conventional Retina identification system is shown in Figure 1. There are two important phases on the diabetic identification system, namely training and testing phases respectively. In the training phase, the system is trained to classify the retina data of the eyes, which are taken by the camera and processed in the Personal Computer (PC). The retina image is digitized and some of them are carried out to create a template for the retina pattern of the normal and diabetic condition and stored in the memory. In the testing phases, the system is tested with the new data, the data which are not applied to train the system.

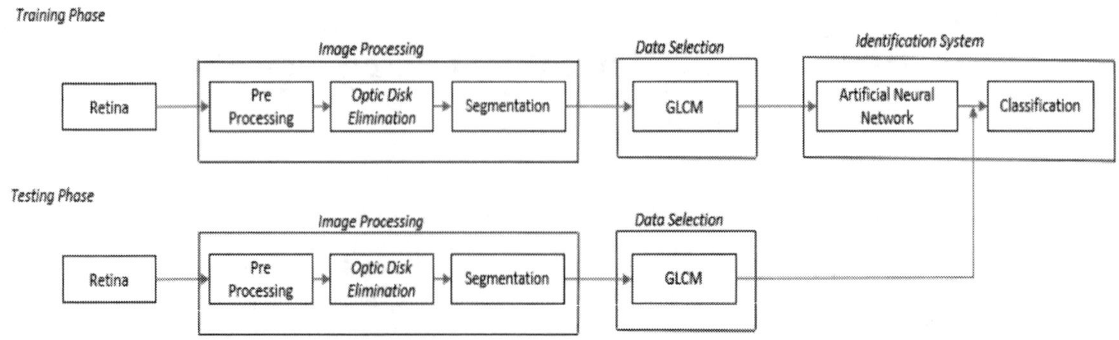

Figure 1. Block diagram of conventional speaker identification

The overall model of the proposed system for the ANN-based retina identification system as depicted in Figure 1. The training phase consists of three important steps, capturing the data, feature extraction, and identification system. Capturing the data is a process of taking the data using a special camera called Fundus. The data is capturing and convert to the matrix form to use as an input to the identification system. The digitized retina image signal, that has the process of image processing and data selection, passed through the processing stages from which Contrast Limited Adaptive Histogram Equalization (CLAHE) features are extracted and passed through the ANN learning algorithm for both training and testing . Image processing and data selection process called feature extraction is used to perform a feature vector of the retina image that will be used as input to the system. In this work, the feature extraction method which applied are consist of image processing and data selection. In image processing, data applied to the three important steps, namely Pre-processing, optic disk elimination, and segmentation, respectively. There is three important parts classification system, namely image preprocessing, feature extraction, and pattern matching [7].

Image preprocessing is the process of converting a continuous image signal to a discrete form. It consists of uploading images into the PC, convert the color image into the gray image, and determine the edge of the image (segmentation of the image). Feature Extraction is to extract a sequence of feature vector of the image signal that will be used as input to the system. Many techniques are used to extract the image such as wavelet transform [8], Principle Component Analysis (PCA) [9], Discrete Wavelet Transform (DWT) [10] and Fast Fourier Transform (FFT) [11]. In this research, a Fast Fourier Transform (FFT) is used to extract the fruit image.

Pattern matching is to measure the similarity between unknown feature vector and feature vector that is saved on the memory of the system. The model of the fruit image is constructed from the extracted feature of the fruit image signal. The pattern matching algorithm compares the incoming fruit image to the reference model and scores their difference called as distance. The distance is used to determine the unknown pattern. There are two types ofmodels: stochastic models and template models. In stochastic models, the pattern matching is a probabilistic result, such as HMM, Ann, and SVM. For template models, pattern matching is deterministic such as dynamic time warping (DTW) with K-nearest neighbors (KNN).

2.1 Convert the image into the grayscale

A grayscale image as shown in Figure 2, is an image that has one channel value at each pixel, the grayscale image consists of three kinds of colors namely red, green, and blue. All images will turn gray and white black, an example for 8-bit grayscale is the number of possible values 28 = 256, the color and the maximum value is 28-I = 255, the following possible colors of 0-255 are the values for each red, green, and blue.

Figure 2. *Grayscale*

2.2 Fungsi Imadjust

The adjust function is a function used to transform the intensity of an image that has a grayscale. This function provides the gamma parameter option which defines the shape of the function between coordinates (a, c) and (b, d). if the gamma value is more than 1, then a concave rises, if less than 1, a concave falls, as shown in Figure 3.

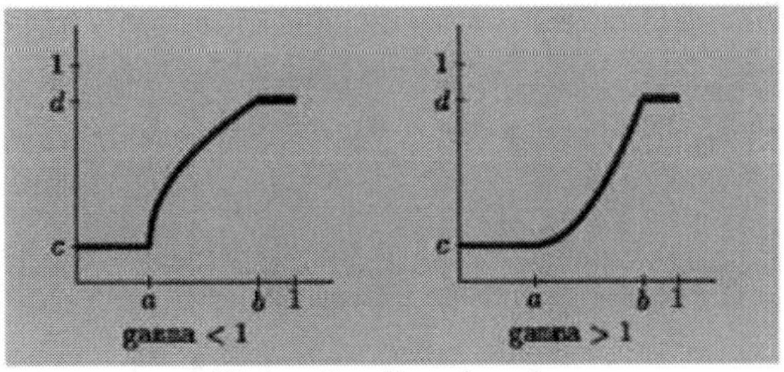

Figure 3. adjust function

2.3 CLAHE

A histogram is a diagram that displays the number of pixel sizes and the number of frequency sizes with gray levels in an image. The histogram has 2 axes namely the X-axis and the Y-axis. The Y-axis is the pixel size value while the X-axis is the frequency value. CLAHE is used to enhance and modify contrast in an image, as shown in Figure 4.

Figure 4. Contrast Limited Adaptive Histogram Equalization (CLAHE)

IOP Conf. Series: Materials Science and Engineering **982** (2020) 012002 doi:10.1088/1757-899X/982/1/012002

2.4 Median Filter

A median filter is a filter used to refine images and reduce noise. The median filter is a median that has nonlinear properties, the median filter is sought by sorting the number of pixels with a predetermined mask area, as shown in Figure 5.

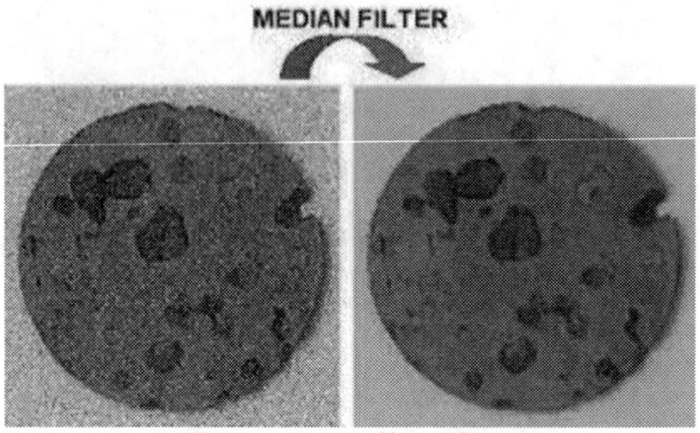

Figure 5. Median Filter

2.5 Eliminasi Optical Disk

Optical Disk is the part of the retina, which is the optical center of the conscious place as the meeting point of the nerve of the eye. Optical disk elimination is a step in analyzing retinal images for disease diagnosis, as shown in Figure 6. This optical disk erase must ensure the right position of the circle, optical disk has an important role to be erased because it has a high contrast value, otherwise this circle will be detected along with the exudate.

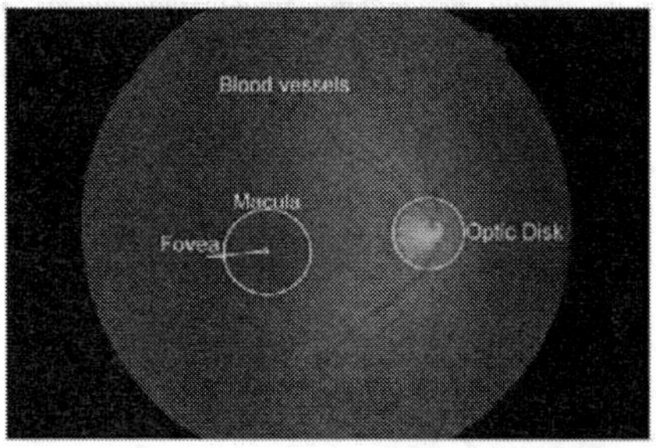

Figure 6. Optic Disk Location

2.6 Segmentation

Segmentation is one of the important processes in image processing. The results of this segmentation will be used for further high-level processes such as the classification process and object identification process. Segmentation itself serves to simplify and / or change the presentation of an image to something easier to analyze, as shown in Figure 7.

Figure 7. Segmentation

2.7 Artificial Neural Network (ANN) Approach

Artificial Neural Network (ANN) is a system that resembles the workings of the brain and can also be trained like the brain in humans. After training or commonly called neural network training, ANN can provide output in the form of identification results[12]. By looking at the amount of input-output in ANN, ANN can find quite accurate results. ANN has also been used in the engineering control field for its identification system and also for its control system[13].

In Figure 8 above, ANN has three main layers, namely the input layer, hidden layer, and the output layer. The input layer functions to enter object data input and carry it for further processing in the hidden layer. In the hidden layer, data will be processed again using artificial neurons. The output layer functions to receive data from the hidden layer and output it as output data in this case in the form of identification of fish species.

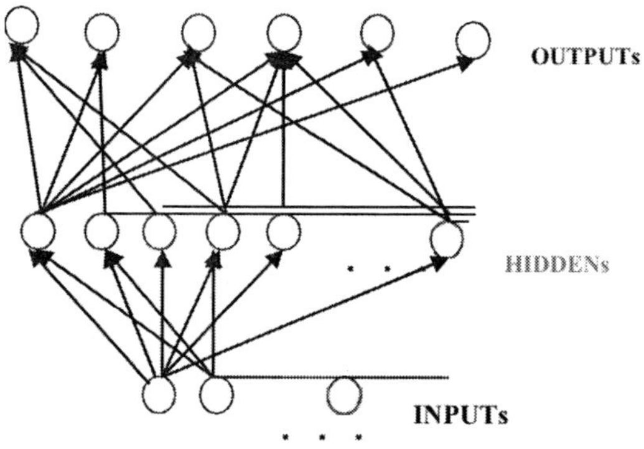

Figure 8. MLP neural network [14]

3. Experimental Result and Discussion

3.1. Experiment setup

Figure 9 shows the proposed system description, the system basically consists of two main components, namely camera (Fundus) capturing for the retina, laptop for processing the retina image processing and identification system. A Fundus camera is commonly used for capturing the retina image. the computer system is used as an image processing system and recognition of the person's retina image. The capturing image is processed in the image-based recognition system which will recognize the retina where the person has a diabetic or normal condition. A personal computer (PC) of 2.1 MHz Intel Core i3 processor equipped with an image card is used for speaker recognition system implementation. In this system, all of the image data processing and diabetics recognition algorithms are implemented in the PC using MATLAB and its toolboxes [15].

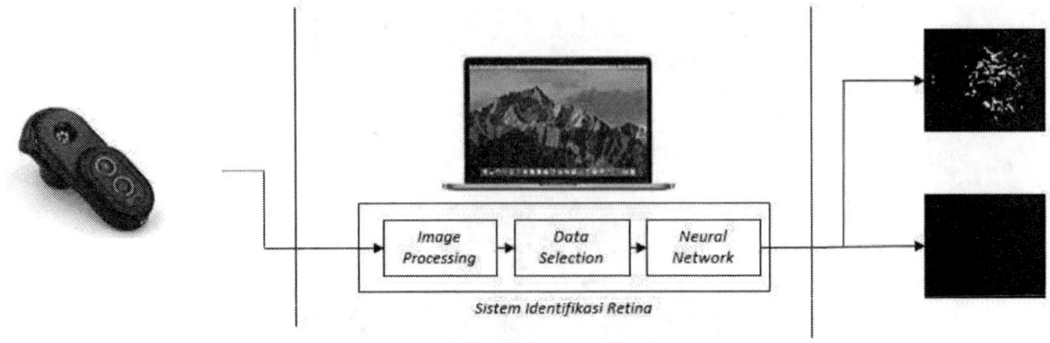

Figure 9. identification system of the R*etinopathy Diabetic*

The overall process of the proposed system for diabetic detection based on retina recognition system is shown in Figure 9. A digital image processing is passed through a preprocessing procedure. The preprocessing step is used to spectrally flatten the signal and making less susceptible to finite precision effects at a later stage. The preprocessing image is then used as input to the feature extraction steps. Feature extraction is the process of converting the raw image into the feature vector which will be used as input for the classification system. There are two important phases in the proposed system, namely training phases, and testing phases. In the training phase, the system is trained to develop a retina based model of the diabetic and normal retina, respectively, a template for the model pattern stored in the memory. In the testing phase, the input data will match with the model to recognize the retina.

In this work, the data used are provided by the STARE (Structured Analysis And of The Retina Dataset) & HRF (High-Resolution Fundus) database. The data consists of two different conditions; Diabetic and normal condition [16]. To evaluate the effectiveness of the proposed diabetes and normal retina recognition system, the diabetics data is applied. The database consists of two different images, which are diabetics and normal image retina, which are illustrated in Figure 10 and Figure 11, respectively.

The retina and normal image convert the green part of the image to the grayscale, as shown in Figure 12 and Figure 13, for diabetics and normal retina, respectively. The CLAHE method is then applied to improve the contrast of the image, as shown in Figure 14 and Figure 15, respectively. The median filter is applied to reduce the noise, as shown in Figure 16 and Figure 17, for diabetics and normal retina, respectively. The segmentation is applied to the filtered image to detect the exudate which indicates the retina involves the diabetic disease, as shown in Figure 18 and Figure 19 for both diabetic and Normal Retina, the results of the excudate is in then applied as 8 by 8 matrics using GLCM function. To obtain the CLAHE coefficients, the 8 by 8 coefficients are used. These methods improving the contrast of the image segments in which resulting in the matrix 8 by 8 matrix from the extracting method.

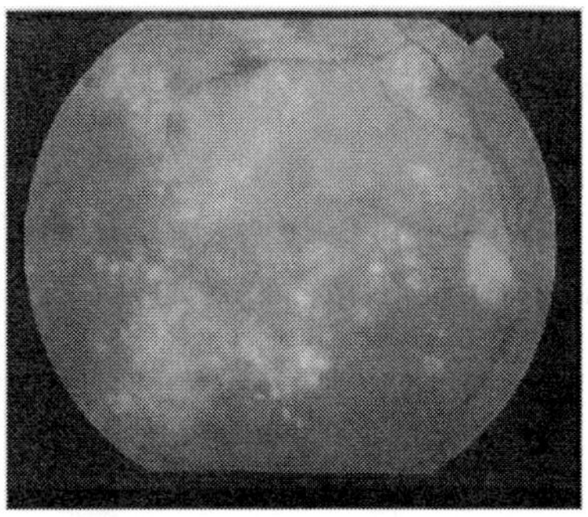

Figure 10. Retina with Diabetic

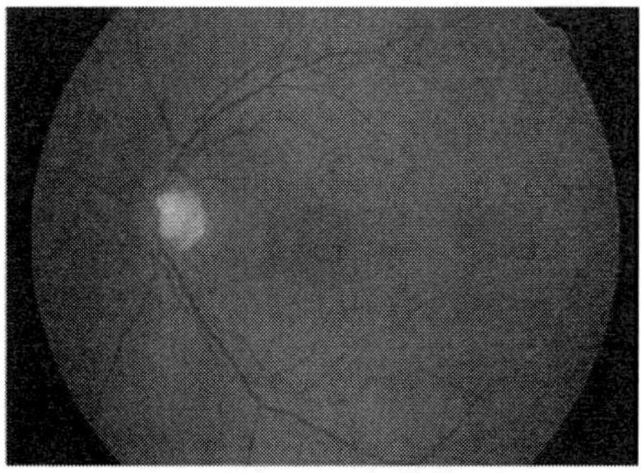

Figure 11. Normal Retina

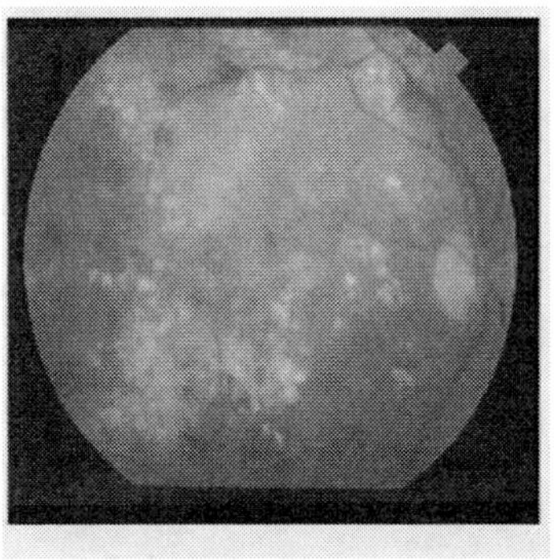

Figure 12. Gray diabetic retina

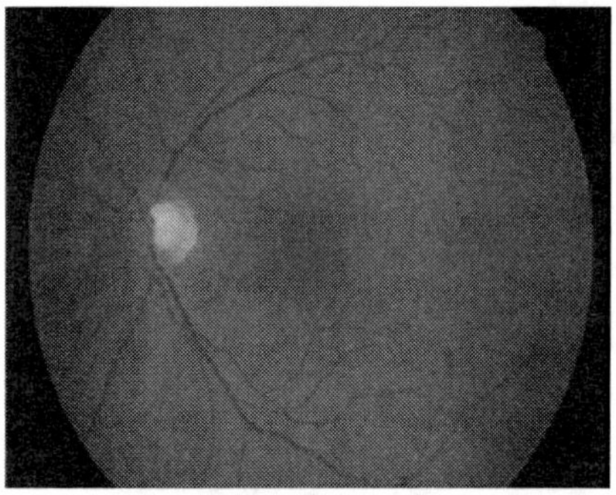

Figure 13. Gray normal retina

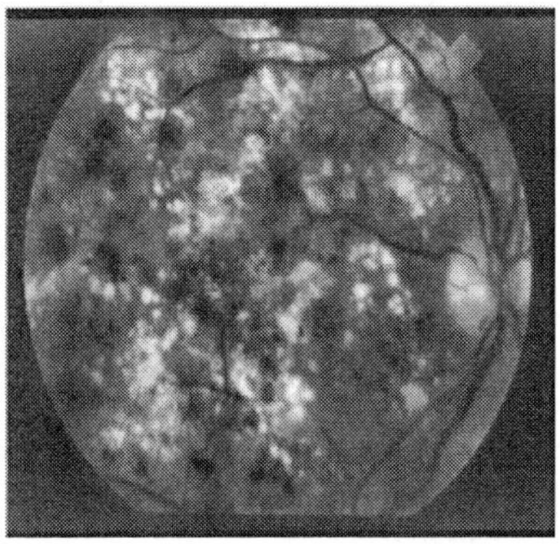

Figure 14. CLAHE method for retina image

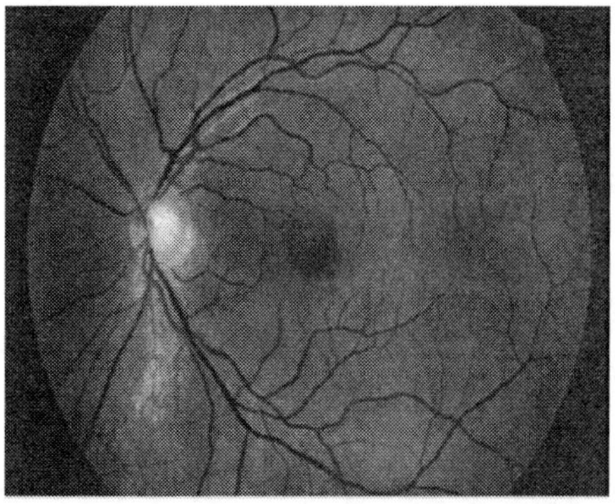

Figure 15. CLAHE method for normal image

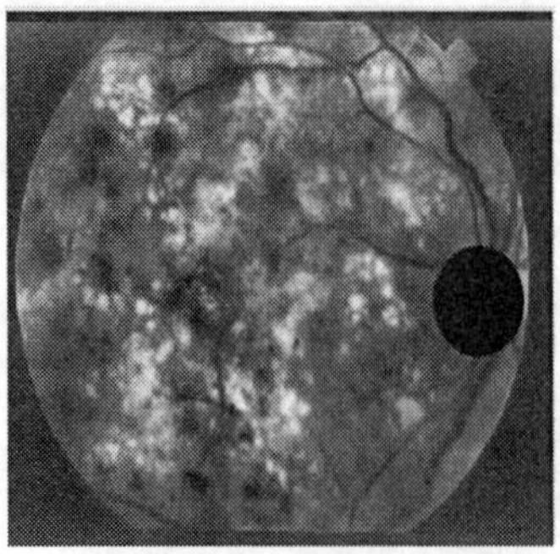

Figure 16. Retina diabetic image with applied median filter.

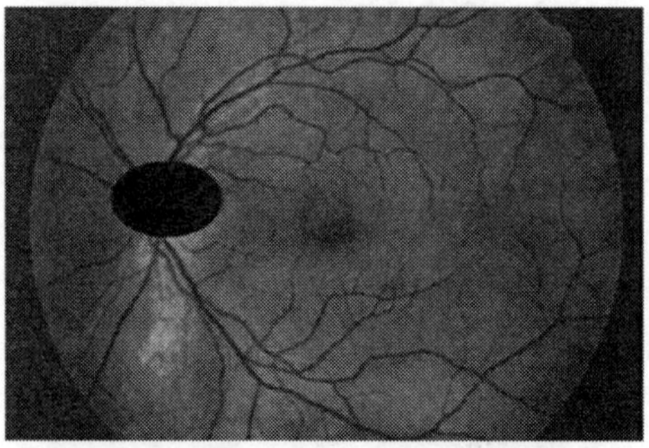

Figure 17. Normal diabetic image with applied median filter.

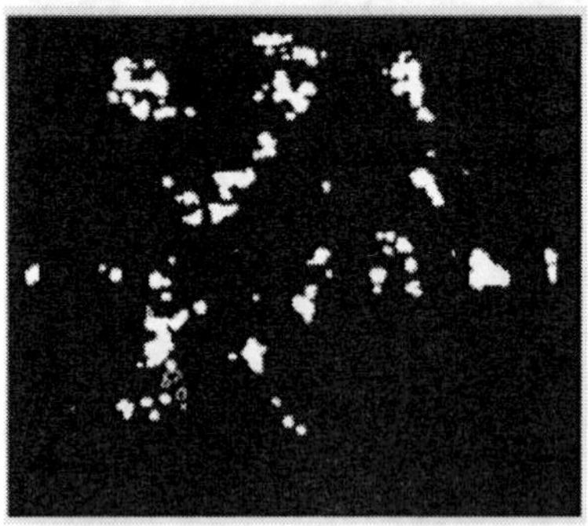

Figure 18. Retina diabetic image with applied segmentation.

Figure 19. Retina diabetic image with applied segmentation.

3.2. Training and classification

To evaluate the effectiveness of the proposed system, the 21 data of each of the diabetic and normal retina which a total of 42 data respectively are performed to train the system. as discussed above the training phase is applied in order of getting the model each of the cases, diabetes, and normal retina respectively. The 30 other data for normal and diabetes retina are evaluated. Each of the images of the retina is modeled based on the CLAHE technique into eight by eight CLAHE parameters. The training data involve 21 retina data each of the normal and diabetes condition. Therefore, the training data involve 248 parameters for all two kinds of normal and diabetes retina. The other brain signal activity, 248 parameters are applied as testing data.

The ANN-based classification is applied to perform identification of processes using the binary classification method. The ANN-based multilayer perceptron (MLP) network with the backpropagation learning algorithm has been used in this paper and it contained a hidden layer. The numbers of neurons in the input, hidden and output layers are 11,11,11 respectively. The ANN gives 100% accuracy for training phased. This means the ANN has perfectly classified all of the two kinds of retina normal and diabetes, respectively. Furthermore, the results of analysis with another retina data testing stage produce 96.67% classification rate which is similar to the result obtained with ANN.

The testing classification rate results have shown that the performance of the ANN-based pattern matching technique has better performance in terms of training and testing accuracy.

4. Conclusion

An ANN-based diabetic recognition system has been proposed and discussed in this paper. This technique is developed based on image processing and intelligent system, the method based on ANN. The accuracy of the ANN-based retina image recognition was found good result. It was also clear, from our computer simulation results, that the proposed method requires much less training time than the existing one. These results identify whether normal or diabetic based on the retina image is considered. In the future, it is needed to clarify the performance and complexity with increasing number of human retina images to prove its usefulness in the real system.

5. References

[1] W. D. Foundation, "Diabetes Indonesia Project," *World Diabetes Foundation*, 2013. [Online]. Available: https://www.worlddiabetesfoundation.org/projects/indonesia-wdf13-828. [Accessed: 05-Jul-2020].

[2] Mayo Clinic Staff, "Blood sugar testing: Why, when and how," *Mayo Clinic Staff*, 2020. [Online]. Available: https://www.mayoclinic.org/diseases-conditions/diabetes/in-depth/blood-sugar/art-20046628. [Accessed: 05-Jul-2020].

[3] Lukman Hermawan, "Deteksi Dini Retinopati Diabetik dengan Pengolahan Citra Berbasis Morfologi Matematika," *IJCCS*, vol. 11, no. 2, pp. 209–218, 2017.

[4] T. Galang, A. Putranto, and I. Candradewi, "Sistem Klasifikasi Tingkat Keparahan Retinopati Diabetik Menggunakan Support Vector Machine," *Indones. J. Electron. Instrum. Syst.*, vol. 8, no. 1, pp. 37–48, 2018.

[5] H. Saiyar, "Klasifikasi Retinopati diabets dengan metode Neural Network," *Paradigma*, vol. 19, no. 2, pp. 92–101, 2017.

[6] D. Buvaneash and M. R. Stalin John, "Brain robot interface using artificial neural network," *IOP Conf. Ser. Mater. Sci. Eng.*, vol. 402, no. 1, 2018.

[7] E. Niedermeyer and F. H. L. da Silva, "Electroencephalography: Basic Principles, Clinical Applications, and Related Fields," vol. Lippincott, pp. 24–28, 2005.

[8] Z. Xizhi, "The Application of Wavelet Transform in Digital Image Processing," 2008.

[9] G. Kumar, "A Detailed Review of Feature Extraction in Image Processing Systems," pp. 5–12, 2014.

[10] I. I. Journal, "IMAGE PROCESSING USING DISCRETE," vol. 3, no. 1, pp. 53–59, 2015.

[11] A. Kadir and A. Susanto, *Teori Dan Aplikasi Pengolahan Citra*. Yogyakarta: Andi, 2013.

[12] A. M. Aibinu, M. J. E. Salami, and A. A. Shafie, "A novel signal diagnosis technique using pseudo complex-valued autoregressive technique," *Expert Syst. Appl.*, vol. 38, no. 8, 2011.

[13] K. J. N. Wilfred, S. Sreeraj, and B. Vijay, "System Identification Using Artificial Neural Network," 2015.

[14] J. Uglov, L. Jakaite, V. Schetinin, and C. Maple, "Comparing Robustness of Pairwise and Multiclass Neural-Network Systems for Face Recognition NOISE PROBLEMS," *EURASIP J. Adv. Signal Process.*, vol. 2008, 2008.

[15] I. Mporas, T. Ganchev, M. Siafarikas, and N. Fakotakis, "Comparison of speech features on the speech recognition task," *J. Comput. Sci.*, vol. 3, no. 8, pp. 608–616, 2007.

[16] Berlin, Brain-Computer, and Interface, "BCI Competition III," 2005. [Online]. Available: http://bbci.de/competition/iii/#download. [Accessed: 16-Jul-2019].

IOP Conf. Series: Materials Science and Engineering **982** (2020) 012003 doi:10.1088/1757-899X/982/1/012003

Fast and Accurate Fish Classification from Underwater Video using You Only Look Once

Hasna Maudi Lathifah[1], Ledya Novamizanti[1,2], Syamsul Rizal[1]

[1]School of Electrical Engineering, Telkom University
Jl. Telekomunikasi Terusan Buah Batu Bandung 40257 Indonesia

[2]ledyaldn@telkomuniversity.ac.id

Abstract. Indonesia is a maritime country and one of the largest archipelago countries in the world. Indonesian fisheries have many types of fish in stock, this causes difficulties in introducing fish species directly. This study designed a fish species classification system using the You Only Look Once (YOLO) architecture. YOLO is an object detection method using a convolutional network that will only be just once. Unlike the convolutional networks in general that spend thousands of networks to obtain an image with computing that is long enough. The architecture of this work using YOLO9000. The dataset consists of 6 classes, that is banded butterflyfish, blue tang surgeonfish, barred hamlet, black side hawkfish, Arabian Picasso triggerfish, dan black margate grunt. System testing produces an accuracy of 92%, IoS 0.75, and 2.223 FPS using Adam optimizer. The proposed system model has good accuracy and fast detection time.

1. Introduction

Indonesia is a maritime country and one of the largest archipelago countries in the world. The vast area of Indonesian fisheries can increase the potential to improve the Indonesian economy, especially in the fisheries sector. In Indonesia, there are many species of fish. Different types have thousands of different species. The classification of fish species is not only seen based on the analysis of color patterns, but also from the shape characteristics of the fins, and so on [1]. After using the characteristics, various features are matched according to the reference characteristics. The process of classifying fish manually requires a long time and a level of human accuracy [2]. Therefore, we need a system that can recognize fish species automatically, and can shorten the time.

Fish Recognition processes the framework of recognition and feature extraction recognition in an informative manner. Feature extraction is divided into two methods, namely, supervised and unsupervised. In the supervised method, one fish has been determined beforehand to be used as a contour. Fish descriptors are labeled as appearance like texture. In the unsupervised method, mathematically, this algorithm will group features or fish that are almost the same in a certain area [3].

Along with the increasing ocean observation, scientists began to research the introduction of fish species with automatic [2]. Ranju Mandal proposed the classification of fish underwater video use deep learning [4]. The method used is Faster R-CNN with additional architectures. So it takes longer to process fish introduction. Sushil Kumar Mahapatra investigated algorithm Fuzzy C-Mean Multithreading for detecting fish underwater. This algorithm resulted in a high detection rate and

IOP Conf. Series: Materials Science and Engineering **982** (2020) 012003 doi:10.1088/1757-899X/982/1/012003

showed an accuracy rate of 98.35% [5]. However, in this study, it is difficult to track objects moving underwater. Elhoseny proposed technology and applications using Machine Learning. These studies use a robot for automated immersion tools that try to take a potential underwater video with high-resolution video [6]. The data obtained is the only video with high resolution. Anak Agung Putra Adi Kesava Esa performs the classification of marine fish underwater video with methods Faster R-CNN. The study using deep learning to identify four types of fish with VGG-16 architecture and AlexNet and searching the best combination [7]. However, in that study, just classify four types of fish with an accuracy of 87.25%.

This research applies You Only Look Once (YOLO) in the classification of types of fish from the underwater video. There are six different types of fish. YOLO is an object detection method that simplifies the task generalize object recognition task more quickly because the convolutional network is only passed once in detecting objects [8]. The way YOLO works is dividing the image input into grids of S size where the S value is 7 with the image input size of 488×488. To calculate bounding boxes, YOLO implements two main process steps, namely, Intersect over Union (IoU) and Non-Maximum Suppression (NMS) [8].

2. Methods

2.1. You Only Look Once

YOLO is an object detection method that is suitable for realtime processing [9]. The object detection method is tasked to determine the location where the object is in an image and determine the object class [10]. The R-CNN method uses several steps to perform the task, so that performance is slow. This is so because each component needs to be trained separately, unlike YOLO, which uses a single neural network. YOLO will divide the input image into 7 × 7 grid cells to match the use of the PASCAL VOC dataset [9]. Each grid cell is responsible for providing predicted bounding boxes. Predicted bounding boxes will provide an object existence score and a prediction score for the class of objects detected.

The YOLO9000 architecture is a development of the YOLOv1 architecture, which has 24 convolutional layers and two fully connected layers (FC) [9]. In YOLO9000 softmax layer is added as the last output layer, which functions to do multi-class classification. YOLO9000 also adds an average pooling layer to its architecture. YOLO9000 uses only the convolutional layer and the pooling layer. Each layer has its own role. Figure 1 is the Architecture of the YOLO network [11].

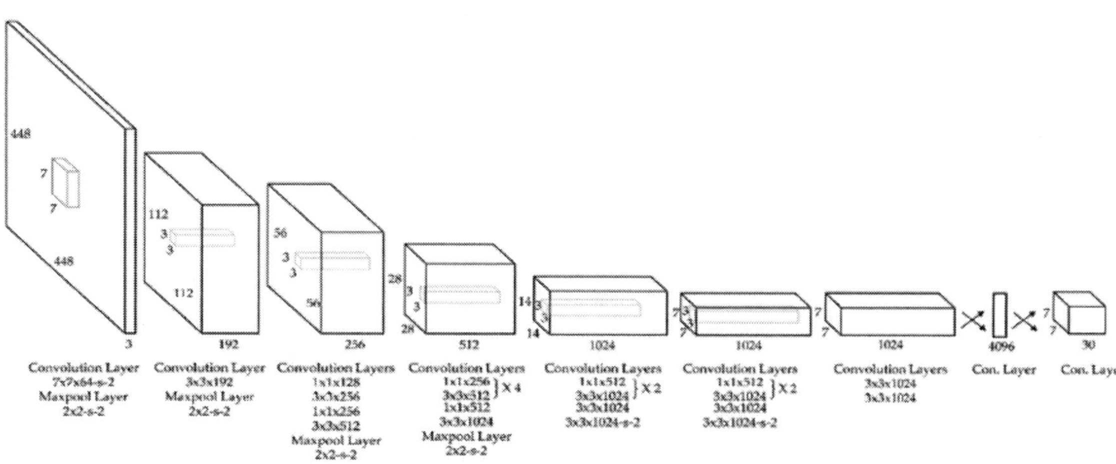

Figure 1. The architecture of the YOLO network.

2.2. Convolutional Layers

Convolutional layers are convolution layers that function as filters for the detection and extraction of image features. In other words, convolutional layers are the main layer on CNN. The extracted image features are then processed by convolution operations. The results obtained are edge detection, corner detection, and so on. The formula of Convolution as follows

$$h(x) = f(x) * g(x) \tag{1}$$

with $f(x)$ and $g(x)$ is an input image and filter/ kernel windows [12],[13].

2.3. Pooling Layer

Pooling layers are used to reduce the size of features. It takes important information by comparing the values of the matrix elements in one window. Pooling layers use the pooling function, which is max pooling, and average pooling. The concept of max pooling is to create a new matrix with a smaller size by taking the strongest features available on a layer. In its implementation, max pooling will take the largest pixel to be arranged into a new matrix. The average pooling will take the average value on a layer [12].

2.4. Transfer Learning

Transfer learning is a method that utilizes models that have been trained on a dataset to solve similar problems. The solution is to use it as a starting point, modify and update its parameters so that it matches the new dataset [14],[15]. For example, in real life, when we can recognize a type of banded fish, we can also recognize other types of fish from the characteristics of the type of banded or previous fish. This can be done by searching a dataset about the types of fish in the process of use. Transfer learning is encouraged because there is prior knowledge to solve new problems quickly.

2.5. Optimizer

The optimizer is a parameter that is used to improve learning on the system by using the value of learning rate or a certain level of learning in object detection. It optimizes the process of deep learning CNN via layer deep enough. The learning rate value determines the system's ability to detect objects quickly or slowly. The size of the learning rate depends on the default value of each optimizer. Optimizer functions that are commonly used are Stochastic Gradient Descent (SGD) and Adaptive with momentum (Adam) [16]. SGD is an optimizer function with a good level of effectiveness in the process of machine learning to do deep learning. This optimizer function is stochastic or able to measure an event with unstable data. SGD in the machine learning process uses a learning rate or learning rate of 0.01. Adam is an optimizer function, which is an optimization of the SGD optimizer function. Adam has an advantage in the computational process that is better than SGD and is widely used for large amounts of data. The learning rate value used by Adam is 0.001 [16].

3. System Design

This section designed a system for classifying fish species from underwater videos using YOLO architecture. Figure 2 shows a flow diagram of a fish species classification system.

3.1 Dataset

The dataset is in the form of 6 types of fish, namely banded butterflyfish, blue tang surgeonfish, barred hamlet, blackside hawkfish, Arabian Picasso triggerfish, and black margate grunt. The dataset was obtained from fishdb with a total dataset of 220 images, with training data for 120 images and test data for 100 images.

3.2 YOLO Process

This research uses a pre train model that has been trained using the COCO and PASCAL VOC dataset. In YOLO training, the system performs predictions or performs functions of the YOLO model with the trained dataset. The training process in YOLO has three stages: (1) Resize the image by resizing the image to 448 × 488, (2) Run a convolutional network that is one way or single past on the image,

(3) Perform non-max suppression, i.e., limit the detection produced, with the number of bounding boxes to output only one bounding box based on the specified threshold. From the dataset that has been trained, a new image is obtained with a bounding box along with the image's json file. In the bounding box, there is also the class name of each object, which is the type of fish. The json file is the confidence score and coordinates of the bounding box. Weight parameters have been trained with the COCO 20 class dataset, which is then carried out transfer learning into 6 classes. Update weight is a pre-training weight that must be in accordance with the basic model used. If the pre-training weights do not match the basic model used, then the train cannot be done. This study uses the config model YOLO 6 class, with the basic model yolo.cfg. Figure 3 is an example of the results of the test image detection with bounding boxes.

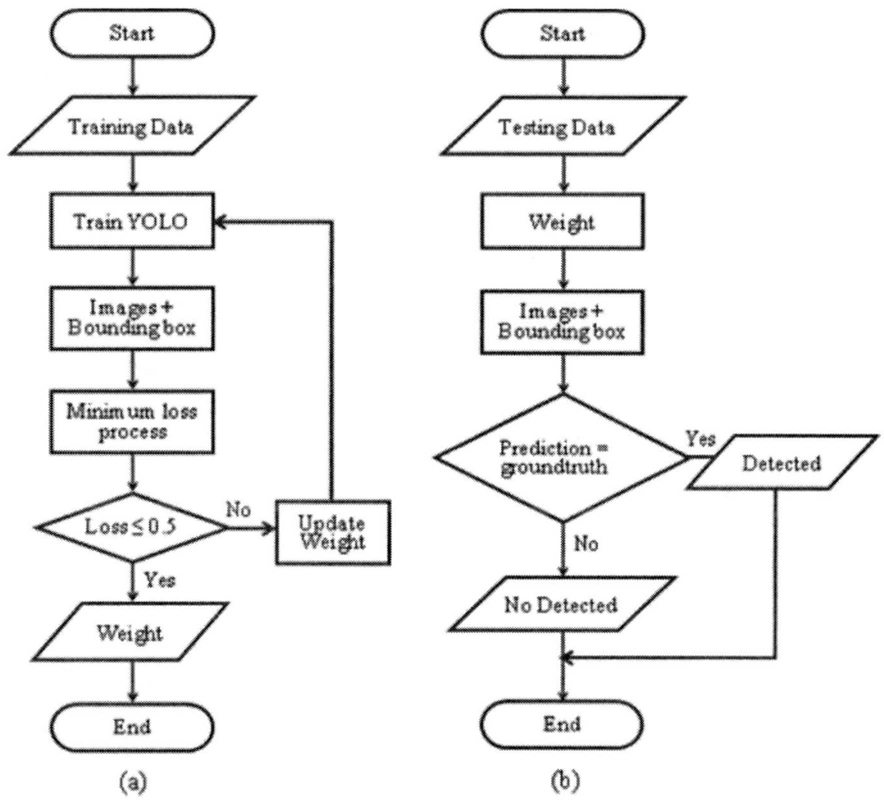

Figure 2. Flow diagram of a fish classification using YOLO (a) Training, and (b) Testing.

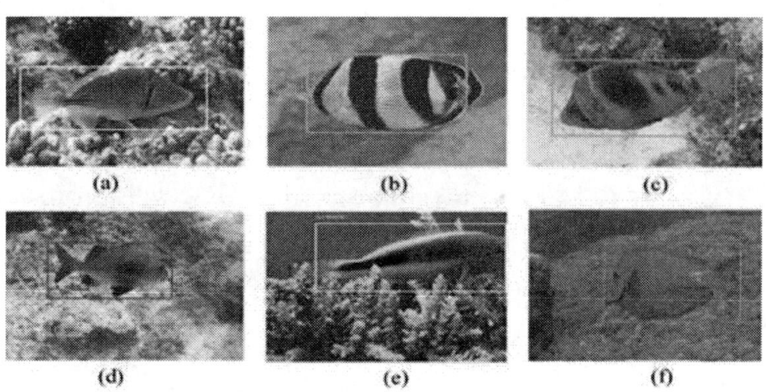

Figure 3. Fish images with bounding box (a) arabian picasso, (b) banded, (c) barred hamlet, (d) blackside, (e) black margate, dan (f) blue tang.

3.3 Performance Parameter
In this study, the parameters used to measure system performance, namely accuracy, precision, and Intersect of Union (IoU).

- Accuracy is used to measure the truth in recognizing fish species. Mathematically, accuracy is calculated as follows

$$\eta = \frac{B}{C} \times 100 \tag{2}$$

  with η is the percentage accuracy, B is an accurate type of fish, and C is the total data [15-17].
- Precision is the calculation of the distance between the centroid ground-truth and the actual centroid bounding box. Precision is calculated as follows

$$E = \sqrt[2]{(C_{x_{GT}}) - (C_{x_{ac}})^2 + (C_{y_{GT}}) - (C_{y_{ac}})^2} \tag{3}$$

  with E is the precision parameter, $C_{x_{GT}}$ is the x coordinate for the centroid groundtruth, $C_{x_{ac}}$ is the x coordinate for the actual centroid, $C_{y_{GT}}$ is the groundtruth, and $C_{y_{ac}}$ is the y coordinate for the actual centroid [2].
- Intersect of Union (IoU) is a parameter to calculate how much prediction is done to produce overlapping with groundtruth. IoU is calculated using a formula

$$IoU = \frac{Area\ Of\ Overlap}{Area\ Of\ Union} \tag{4}$$

  IoU requires two areas to be intersected and unionized. The two areas are the groundtruth bounding box area, which is the actual bounding box, and the area detected from the model built [2].

4. Result and Discussion
The results of the detection are in the form of bounding boxes in the fish object area. From the bounding box, calculations can be done to get system performance. In this study, it is implemented and compared two types of optimizers, namely Adam and Stochastic Gradient Descent Optimizer (SGD). The test scenario is carried out to find the best optimizer and threshold value for the performance parameters.

4.1. System performance on IoU
This section tests the parameters of the IoU to measure the accuracy of the method in detecting fish objects. The IoU parameter will play an important role in the accuracy parameter. The measurement is based on two areas, namely between groundtruth bounding boxes and predicted bounding boxes. Groundtruth boxes with predicted bounding boxes get better when IoU approaches 1. The system can detect well if IoU ≥ 0.5. Figure 4 shows the performance of YOLO with Adam and SGD optimizers. Based on Fig. 4, in Adam and SGD optimizers, the best IoU is at the 0.1 thresholds. The IoU value will get better if it gets closer to 1. In Adam's optimizer, when the 0.1 threshold is obtained, the IoU value is 0.75. In the SGD optimizer, when the threshold is 0.1, the IoU value is 0.69. The resulting IoU value is optimizer Adam, 0.06 points greater than the SGD optimizer. System performance continues to decline if the threshold value is enlarged to 0.9. The greater the threshold, the bounding box is thinning with confidence score close to 0. The smaller the threshold, the bounding box is thickening

with a confidence score close to 1. However, when tested with a threshold of less than 0.1, the bounding box overlaps and does not match the groundtruth bounding box.

4.2. System performance on precision

This section tests the parameters of precision to test the level of centroid groundtruth proximity with centroid predicted bounding boxes. A system with good precision if precision approaches 0. The closer the centroid groundtruth is to the centroid predicted bounding boxes, the smaller the precision. Precision approaching 0 means the system is getting better. Figure 5 shows the performance of Adam and SGD optimizers on precision.

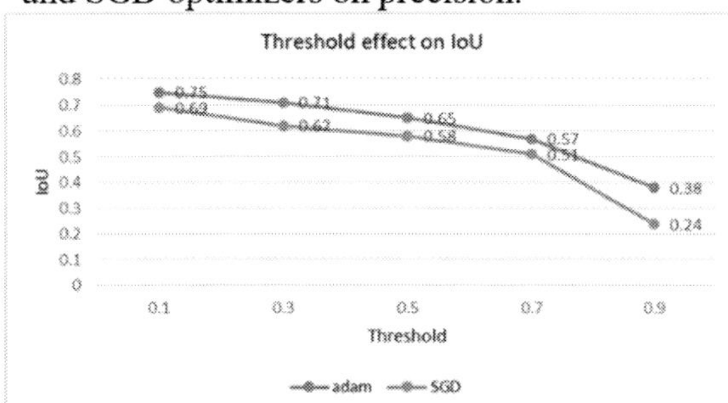

Figure 4. The results of the optimizer on IoU.

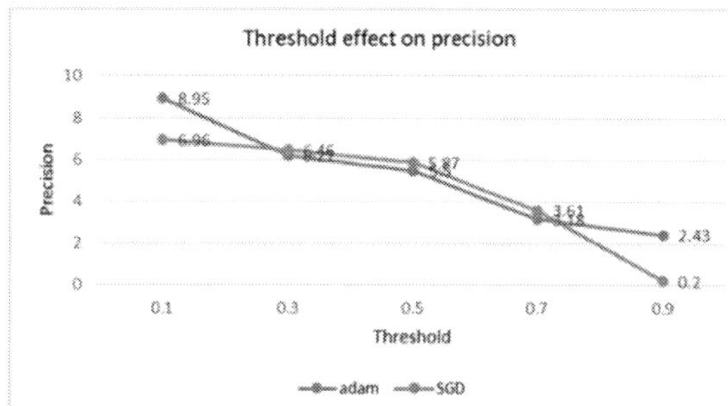

Figure 5. The results of the optimizer on precision.

Based on Fig. 5, in the Adam and SGD optimizers, the best precision is obtained at the threshold 0.9. The greater the threshold, the two centroids predicted bounding boxes will be closer to the centroid bounding boxes groundtruth. The smaller the threshold, the two centroids predicted bounding boxes are increasingly away from the groundtruth centroid bounding boxes. However, when tested with a threshold of more than 0.9, centroid predicted bounding boxes cannot be seen whether approaching or moving away from centroid bounding boxes groundtruth because bounding boxes are very thin.

4.3. System performance on accuracy

This section examines system performance in recognizing fish species against accuracy parameters. The output network that has been trained with the training dataset is then tested in recognizing an object test image. Figure 6 shows the performance of YOLO with Adam and SGD optimizer to parameter accuracy.

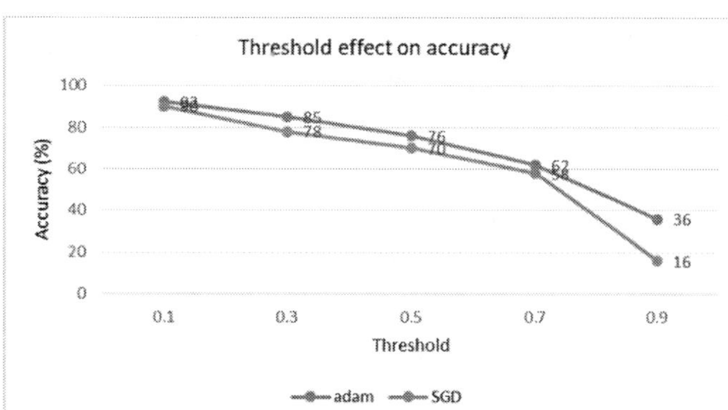

Figure 6. The results of the optimizer on accuracy.

Based on Fig. 6, the best accuracy is obtained when the threshold is 0.1. The accuracy obtained by Adam Optimizer is 2% better than the accuracy obtained by the SGD optimizer. In the Adam and SGD optimizers, the accuracy parameter is directly proportional to the IoU parameter. When the predicted bounding boxes fail to approach the groundtruth bounding boxes, the system cannot detect the object. So the system cannot classify fish. IoU parameters for Adam optimization and SGD optimization, get results that are directly proportional to accuracy. A system with high accuracy does not guarantee the best precision. The system is more intelligent in recognizing an object if the accuracy is higher. But the proximity between centroid groundtruth and centroid predicted bounding boxes is not necessarily close. The precision value between centroid groundtruth and centroid predicted bounding boxes will still be included in the calculation, even though the results of the precision are quite large.

5. Conclusion

This study proposes the classification of fish species from underwater videos using YOLO architecture. The system successfully classified six fish classes with IoU 0.75, 92% accuracy, 0.24 precision, and 2.223 FPS. The best configuration is obtained at threshold 0.1 with Adam Optimizer. Adam's optimization uses the adaptive learning rate calculation, so it is good to be applied to the classification of fish species by the YOLO method. The system precisely detects fish objects from a short distance compared to a distance.

6. Reference

[1] Andayani U, Wijaya A, Rahmat R F, Siregar B, Syahputra M F 2019 Fish Species Classification Using Probabilistic Neural Network *IOP Conf. Series: Journal of Physics: Conf. Series* **1235** 012094

[2] Li X, Shang M, Hao J and Yang Z 2016 Accelerating fish detection and recognition by sharing CNNs with objectness learning *Proc. Ocean* (Shanghai) IEEE

[3] Chuang M.-C, Hwang J.-N and Williams K A 2016 Feature Learning and Object Recognition Framework for Underwater Fish Images *IEEE Trans. Image Process.* **25**(4) 1-1

[4] Mandal R, Connolly R M, Schlacher T A and Stanticv B 2018 Assessing fish abundance from underwater video using deep neural networks *Proc. Int. Jt. Conf. Neural Networks* (Brazil) IEEE

[5] Mahapatra S K, Mohapatra S K, Mahapatra S and Tripathy S K 2016 A Proposed Multithreading Fuzzy C-Mean Algorithm for Detecting Underwater Fishes *2nd International Conference on Computational Intelligence and Networks* (India) IEEE

[6] Elhoseny H, Elhoseny M and Riad A M 2018 *The International Conference on Advanced Machine Learning Technologies and Applications (AMLTA2018)* **723** Springer international publishing

[7] Kesava A A E , Raharjo J and Wibowo S A 2019 Klasifikasi Ikan Laut Menggunakan Video Bawah Laut dengan Metode Faster RCNN (Region Convolutional Neural Network) Skripsi (Bandung: Universitas Telkom)

[8] Redmon J, Divvala S, Girshick R and Farhadi 2016 A You only look once: Unified, realtime object detection *Proceedings of the IEEE Computer Society Conference on Computer Vision and Pattern Recognition* 779

[9] Redmon J, Divvala S, Girshick R and Farhadi 2016 A You only look once: Unified, realtime object detection *Proc. IEEE Comput. Soc. Conf. Comput. Vis. Pattern Recognit.* pp 779–788

[10] Phadnis R, Mishra J and Bendale S 2018 Objects Talk - Object Detection and Pattern Tracking Using TensorFlow *Proc. Int. Conf. Inven. Commun. Comput. Technol.* pp 1216–1219

[11] Redmon, Divvala, Girshick R and Farhadi A 2016 You only look once: Unified, realtime object detection *In Proceedings of the IEEE Conference on Computer Vision and Pattern Recognition* pp 779–788

[12] Wulandari C D R, Wibowo S A and Novamizanti L 2019 Classification of diabetic retinopathy using statistical region merging and convolutional neural network *IEEE Asia Pacific Conference on Wireless and Mobile*

[13] Priatama M B, Novamizanti L, Aulia S, Candrasari E B 2020 Hand gesture recognition using discrete wavelet transform and convolutional neural network *Bulletin of Electrical Engineering and Informatics* **9**(3)

[14] Yang Q and Pan S J 2010 A Survey on Transfer Learning *IEEE Trans.Knowl. Data Eng.* **22(**10) pp 1345–1359

[15] Radiuk P M 2018 Impact of Training Set Batch Size on the Performance of Convolutional Neural Networks for Diverse Datasets *Inf. Technol. Manag. Sci* **20(**1) pp 20–24

[16] Kingma D P and Ba J L 2015 Adam: A method for stochastic optimization *3rd Int. Conf. Learn. Represent.* pp 1–15

IOP Conf. Series: Materials Science and Engineering **982** (2020) 012004 doi:10.1088/1757-899X/982/1/012004

Computer-Aided Detection (CAD) for COVID-19 based on Chest X-Ray Images using Convolutional Neural Network

NK Caecar Pratiwi[1], Nur Ibrahim[1], Yunendah Nur Fu'adah[1], Kholidiyah Masykuroh[2]

[1] Electrical Engineering Faculty - Telkom University, Bandung, Indonesia
[2] Telecommunication and Electrical Engineering Faculty, IT Telkom Purwokerto

[3] caecarnkcp@telkomuniversity.ac.id

Abstract. Covid-19 has spread throughout the world and has been declared as a pandemic by the World Health Organization (WHO). The disease was discovered by the end of 2019 in Wuhan- China. The number of deaths continues to surge sharply and spread to many countries. Covid-19 has sent billions of people on earth into lock-down when health services struggle to cope. A swift and reliable Covid-19 diagnosis system is needed, to direct the patient to the appropriate treatment and prevent the disease dissemination. During this time, we are familiar with rapid tests and Real-Time Polymerase Chain Reaction (RT-PCR) as the procedure of Covid-19 detection. Both of these procedures tend to be impractical and require specialized laboratories that are arranged in such away. It can also take several hours to wait for the amplification process until the results are known. In this study, we introduce a Covid-19 detection system based on Chest C-Ray images using Convolutional Neural Network (CNN). The dataset consists of 1000 images, 500 images each for positive Covid-19, and Pneumonia. The CNN model that was designed consisted of three hidden layers, a fully connected layer with sigmoid activation. The evaluation was conducted to determine the performance of the proposed model using matrices of precision, recall, F1, and accuracy. The experimental results show that the proposed method provides precision, recall, F1 was 1 and 100% accuracy, respectively. This research is expected to be tested in field validation, to help the medical authorities for clinical diagnosis.

1. Introduction

It is estimated that around the end of December 2019, a new and unknown coronavirus was discovered, the first case appeared in Huanan Seafood Wholesale Market, Wuhan (Hubei Province) - China. An early indication of the Plague pneumonia is characterized by fever, dry cough, fatigue, and sometimes gastrointestinal symptoms [1]. This virus then carries the disease that we know by the name of Corona Virus Disease 2019, briefly names as Covid-19. Covid-19 has spread throughout the world and has been declared as a pandemic by the World Health Organization (WHO). As shown in figure 1, At least, until 22 June 2020, WHO stated that there had been 8.844.171 confirmed cases, 465.460 confirm deaths, and 216 countries were expose [2]. The number of deaths continues to surge sharply and spread to many countries. Covid-19 has sent billions of people on earth into lock-down when health services struggle to cope.

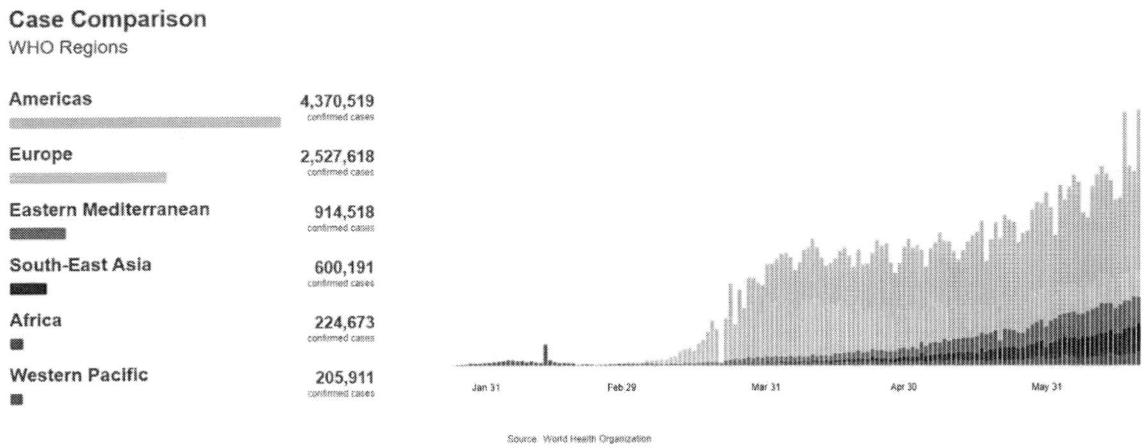

Figure 1. Case Comparison of Covid-19 in WHO Regions by 22 June 2020 [2]

The emergence of Covid-19 had an impact on all aspects, millions of employees lost their jobs, lost revenue in the air service sector, the decline in foreign exchange tourism, rising inflation, and lead to the risk of worsening economic conditions. This prompted scientists in all corners of the world, from various branches of science continue to work with optimism for research related to this outbreak, including researchers in the field of biomedical image processing. As we all realize, one of the biggest challenges when dealing with COVID-19 is the necessity for fast and reliable diagnostic identification that can serve as an alternative to reverse-transcriptase testing polymerase chain reaction (RT-PCR), bearing in mind that this epidemic is spreading rapidly and globally. Togacar and colleagues conducted a study related to COVID-19 detection with MobileNetV2 and SqueezeNet modeling, each modeling used Support Vector Machine (SVM) as a classifier [3]. The results of this study are a system that can 100% detect COVID-19 images, and 99.27% for the detection of Normal and pneumonia images. Chest radiograph (CXR) plays an essential role in triage for COVID-19 detection, so as Murphy and colleagues conducted a study to evaluate the performance of artificial intelligence (AI) for detecting COVID-19 and pneumonia based on CRX [4]. The study concluded that the AI system correctly classified CXR images as COVID-19 pneumonia with an Area under the ROC curve (AUC) of 0.81. The Coronavirus novel also motivated Asif Khan and his colleagues to propose CoroNet, a Deep Convolutional Neural Network-based modeling that can detect COVID-19 infection from chest X-ray images [5]. This study concluded that CoroNet model substantially can be a very helpful tool for clinical practitioners and radiologists to aid them in the diagnosis of COVID-19. A system that can detect whether a person is exposed by Covid-19 or not based on chest CT images has been designed by using multi-objective differential evolution (MODE) and convolutional neural networks (CNN) [6]. The results showed that the CNN model yielded accuracy, F-measure, sensitivity, specificity, and Kappa statistics by 1.9789%, 2.0928%, 1.8262%, 1.6827%, and 1.9276%, respectively. Research related to the detection of COVID-19 based on chest X-rays images has been carried out using Alexnet, GoogleGene, and Restnet18 [7]. The results showed that the Googlenet architecture produced the best accuracy of 80.6%, this result was obtained when the system was designed for 4 class classifications (Covid-19, Normal, Pneumonia Bacteria, and Pneumonia Bacteria Virus). Alexnet was selected to be the best model as it achieved 85.2% accuracy when the system was designed for 3 class classifications (Covid-19, Normal, and Pneumonia Bacteria). In the third scenario, the system is designed for the 2 classes classification (COVID-19 and normal), Googlenet was chosen as the best architecture, it reaches 100% in testing accuracy and 99.9% in validation accuracy.

In this study, we propose a small CNN architecture which can conduct early detection whether someone is exposed to Covid-19 or not based on Chest X-ray images. The model designed is expected to have a simple complexity level, but produces high accuracy and short computing time.

2. Convolutional Neural Network (CNN)

In recent years, CNN has been very widely used to identify visual forms of images, allegedly using the least amount of calculations [8]. CNN has been widely used in medical image detection midwives, such as the retina, fundus, chest X-ray, chest CT, breast cancer, cardiac, abdominal, and musculoskeletal image analysis. Thus, convolutional neural networks have almost unlimited applications. Convolutional neural networks are also called ConvNets, built of three main layers; convolutional layer, pooling layer, and fully connected layer.

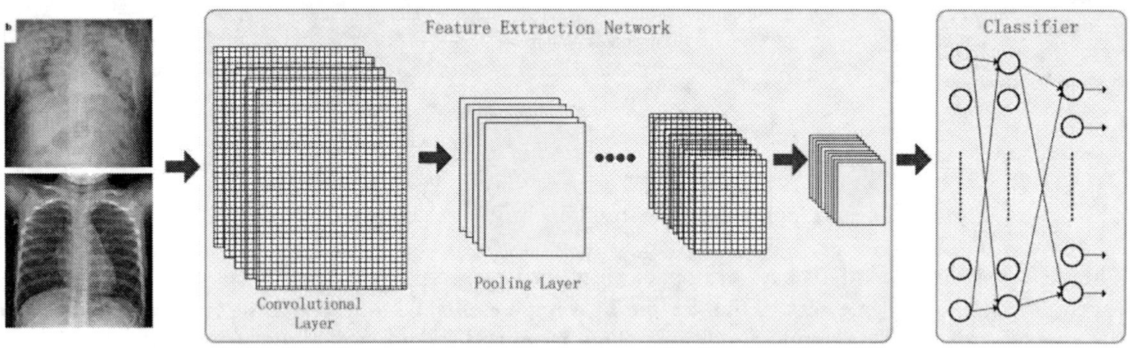

Figure 2. The Convoluted feature Matrix created

As shown in Figure 2 The convolutional layer has the task of carrying out mathematical operations, convoluted the image with a filter to extract important features [9]. That is the first operation on CNN. The input image will be entered into a filter, then produces an activation map or feature map. Figure 3 below illustrates the 5 × 5 input image convolution process with a 3 × 3 kernels/ filter size, and a 1 × 1 stride, producing a 3 × 3 feature map.

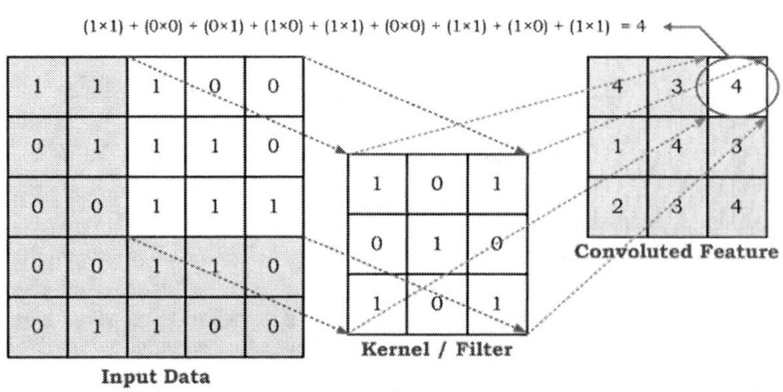

Figure 3. The Convoluted feature Matrix created

Rectified Linear Unit (ReLU) is a very simple activation function, and lately, it is very popular to use because it helps to overcome some optimization problems in the sigmoid activation function [10]. ReLU is simply defined as $f(x) = max(0, x)$, the derivate of ReLU:

IOP Conf. Series: Materials Science and Engineering **982** (2020) 012004 doi:10.1088/1757-899X/982/1/012004

$$f'(x) = \begin{cases} 1, if\ x > 0 \\ 0,\ otherwise \end{cases} \quad (1)$$

Take a look at the following RelU Activation function graph as shown in figure 4, it appears that the function will make the value zero for negative values and grow linearly for positive values.

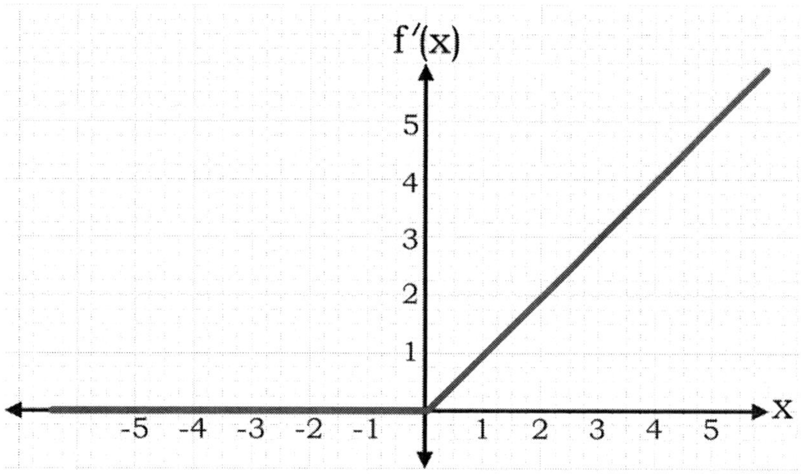

Figure 4. The ReLU (REctified Linear Unit) Activation Function

After the convolution, the second fundamental operation on CNN is pooling. This operation is much easier to understand than convolution [11]. Pooling operation is used to down-sampling or reduced the complexity process for the next layer [12]. The reduction in complexity is achieved by reducing the size of the feature map matrix. The way to reduce the size of a matrix is by comparing neighboring pixels and making them into one value. The representative value is set as the average or maximum value. To understand this, let's look at a concrete example of max-pooling in figure 5

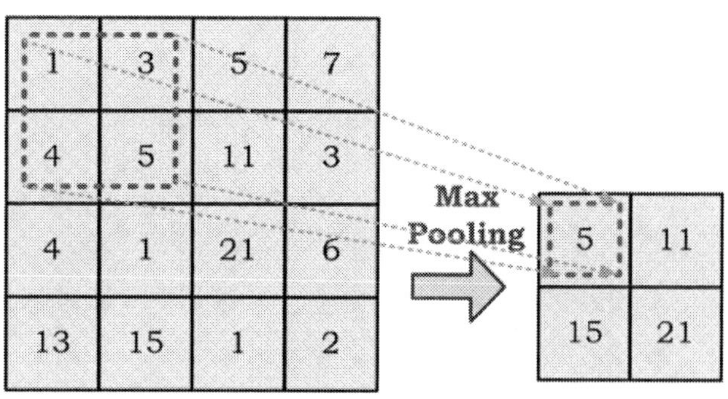

Figure 5. The Resultant Cases of Max Pooling

Sigmoid activation returns the output value which falls in the range of 0 or 1, mainly used for binary data classification [14]. The equation for the Sigmoid classifier produces the curve in S shape.

$$f(x) = \frac{1}{1 + e^{-x}} \quad (2)$$

IOP Conf. Series: Materials Science and Engineering **982** (2020) 012004 doi:10.1088/1757-899X/982/1/012004

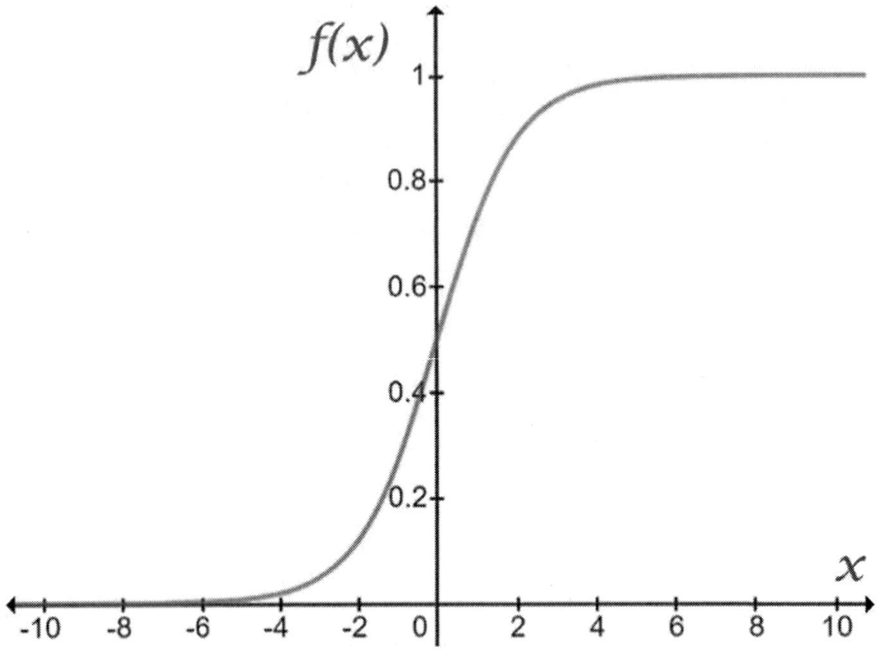

Figure 6. Sigmoid Curve

Adaptive Moment Estimation (ADAM) optimizer, is a combination of momentum and RMSprop [15]. In terms of computing, ADAM tends to be more efficient and requires less memory [15]. ADAM is an algorithm that serves to update the network weights in the training data, upgrading of weights is done to minimize the loss value [16]. ADAM will store the exponential average of the square decay gradient v_t, and calculate the average of the second moment of the gradient m_t [17].

$$m_t = \beta_1 m_{t-1} + (1 - \beta_1)gt \tag{3}$$

$$v_t = \beta_2 v_{t-1} + (1 - \beta_2)gt^2 \tag{4}$$

Whereas m_t and v_t are values of the mean and uncentered variance. Adam updates the exponential moving averages of the gradient and the squared gradient where the $\beta_1, \beta_2 \in [0, 1]$.

$$\widehat{m}_t = \frac{m_t}{1 - \beta_1^t} \tag{5}$$

$$\hat{v}_t = \frac{v_t}{1 - \beta_2^t} \tag{6}$$

The final update weight is given as:

$$w_{t+1} = w_t - \frac{\mu}{\sqrt{\hat{v}_t}} . \widehat{m}_t \tag{7}$$

3. System Design

In this study, the proposed model consists of three hidden convolutional layers, with RelU activation, Max Pooling, and sigmoid classifier. The learning rate of 0.001 using in Adam Optimizer with Binary Crossentropy for loss. CNN Architecture or the designed model can be seen in figure 7 below. The dataset consists of 1000 Chest X-Ray images, 500 images each for positive Covid-19, and Pneumonia. The dataset is taken from Kaggle Chest X-ray Images, they are collected from various publicly available resources [18] [19] [20]. Total data will be divided into 75% (750 images) as training data and 25% (250 images) as validation data. This dataset will serve as input for the designed CNN model. Based on Figure 7 and Table 1, the chest x-ray images are resized into 64 × 64 pixels. In hidden layer 1, we use a filter with size 3 × 3 and channel output 16. In hidden layer 2, we use the filter with size 3 × 3 channel output 32. And last, in hidden layer 3, we use a filter with size 3 × 3 and channel output 64. Then, we do the flatten to change the image featured into one layer/dimension so that we can classify it into two classes (whether Covid-19 or Pneumonia). Activation that is used to do the classification is sigmoid.

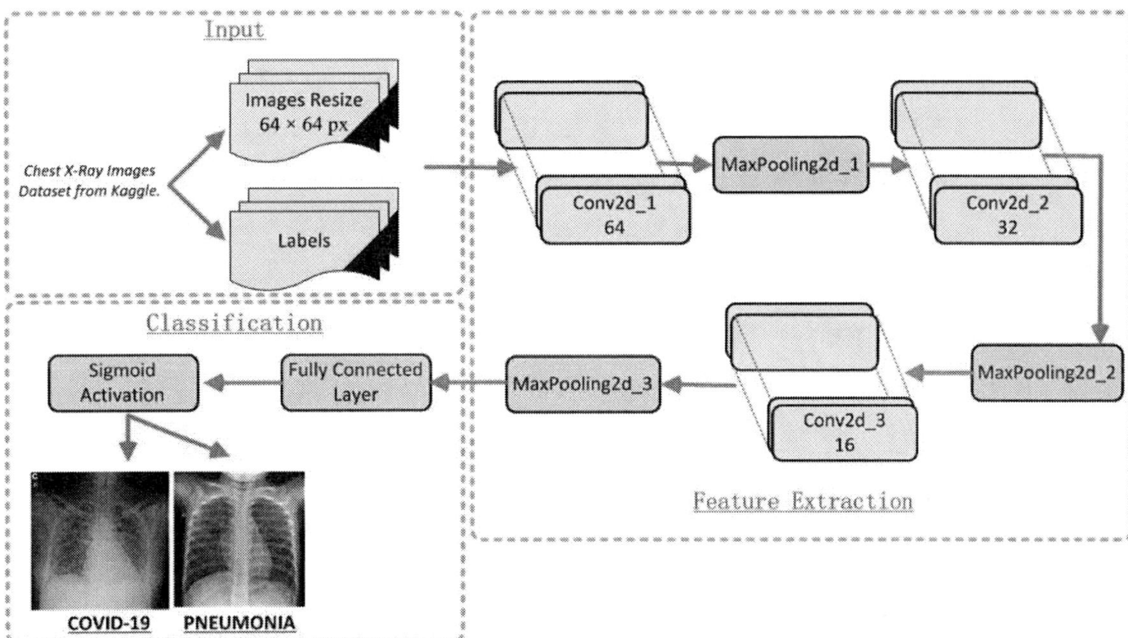

Figure 7. Proposed Model for COVID-19 Detection based on Chest X-Ray Images

Table 1. Summary of CNN Configuration

Layer (type)	Output Shape	Parameter
conv2d 1 (Conv2D)	(None, 64, 64, 16)	448
activation 1 (Activation)	(None, 64, 64, 16)	0
max pooling2d 1 (MaxPooling2)	(None, 32, 32, 16)	0
conv2d 2 (Conv2D)	(None, 32, 32, 32)	4640
activation 2 (Activation)	(None, 32, 32, 32)	0
max pooling2d 2 (MaxPooling2	(None, 16, 16, 32)	0
conv2d 3 (Conv2D)	(None, 16, 16, 64)	18496

IOP Conf. Series: Materials Science and Engineering **982** (2020) 012004 doi:10.1088/1757-899X/982/1/012004

activation_3 (Activation)	(None, 16, 16, 64)	0
max_pooling2d_3 (MaxPooling2	(None, 8, 8, 64)	0
dropout_1 (Dropout)	(None, 8, 8, 64)	0
flatten_1 (Flatten)	(None, 4096)	0
dense_1 (Dense)	(None, 2)	8194
Total parameters: 31,778		
Trainable parameters: 31,778		
Non-trainable parameters: 0		

4. Result and Analysis

Evaluating the model aims to check whether the predicted value is equal to the actual value during the testing process. In this study, the evaluation was carried out by using a confusion matrix. The parameters that become the reference related to system performance are Accuracy, Precision, Recall, and F1-score.

$$Accuracy = \frac{TP+TN}{TP+TN+FP+FN} \tag{8}$$

$$Precision = \frac{TP}{TP+FP} \tag{9}$$

$$Recall = \frac{TP}{TP+FN} \tag{10}$$

$$F1 - Score = 2\,x\,\frac{Precision\ x\ Recall}{Precision - Recall} \tag{11}$$

True Positive (TP) is likened to a case where both the prediction and the actual are TRUE. True Negative (TN) is the case when both the prediction and the actual are FALSE. False Positive (FP) is a case when something is estimated as positive (TRUE) but is negative. False Negative (FN) when predicted as FALSE, but in reality, it is TRUE [8]. F1-score is the harmonic mean of precision and recall.

The experimental results show that the proposed method provides the precision, recall, F1- score of 1, and accuracy of 100%. The total iteration for training data is 100 times (100 epoch). The accuracy increases for each iteration, and the difference between training accuracy and validation accuracy is not much different. It can be concluded that the designed model does not experience overfitting.

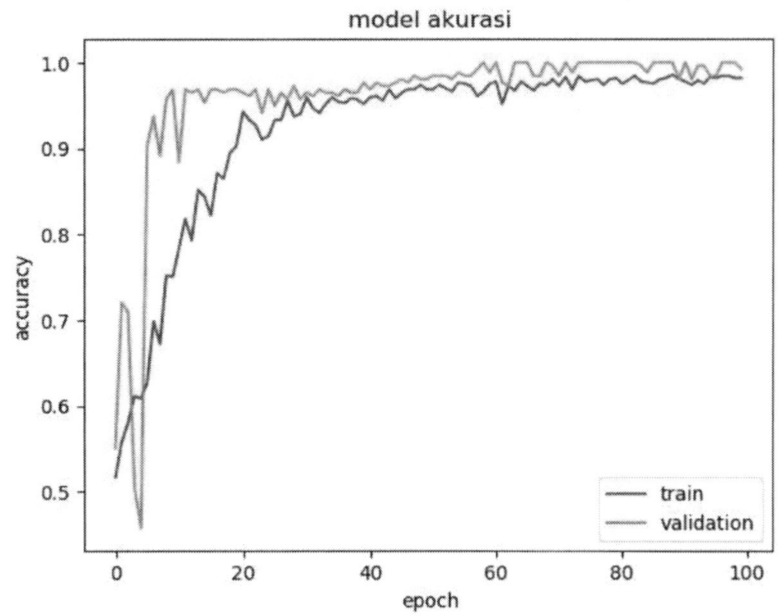

Figure 8. The Accuracy of the Proposed Model for COVID-19 Detection

Through the Confusion Matrix in Figure 9, from a total of 128 validation data, all the images detected correctly to their class.

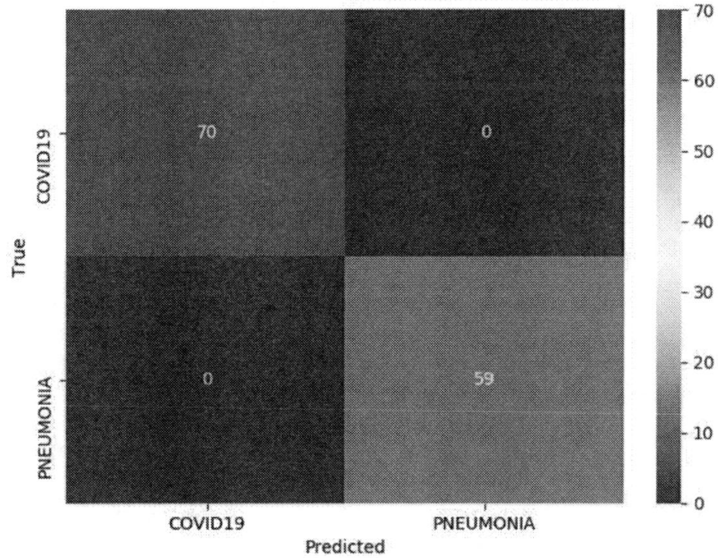

Figure 9. The Accuracy of the Proposed Model for COVID-19 Detection

In 100 iterations, the value of precision, recall, F1-score can be seen in Table 2 below:

Table 2. Precision, Recall and F1-Score of Proposed Model

	Precision	**Recall**	**F1-Score**
COVID-19	1.00	1.00	1.00
PNEUMONIA	1.00	1.00	1.00
Average	1.00	1.00	1.00

5. Conclusion

To classify Covid-19 and Pneumonia from chest x-ray images, a system has been built using CNN with three hidden layers plus one fully connected layer with sigmoid activation to determine its classification. After doing the 100 iterations for the training and validation process, obtained an accuracy value of 100% with precision, recall, F1-score is 1.00. So it can be concluded that this system does not occur over-fitting, which means the system can recognize the type of Covid-19 and Pneumonia with new chest x-ray images. This research is expected to be tested in field validation so that it can help the medical authorities for clinical diagnosis.

References

[1] Wu, Yi-Chi, Ching-Sung Chen, and Yu-Jiun Chan 2020 The outbreak of COVID-19: An overview /text journal of the Chinese Medical Association 83 no. 3: 217.
[2] WHO 2020 *Coronavirus Disease (COVID-19)* at https://covid19.who.int/.
[3] Toğacar, Mesut, Burhan E, and Zafer C 2020 *COVID-19 detection using deep learning models to exploit Social Mimic Optimization and structured chest X-ray images using fuzzy color and stacking approaches* Computers in Biology and Medicine: 103805.
[4] Murphy, Keelin, Henk Smits, Arnoud JG Knoops, Mike BJM Korst, Tijs Samson, Ernst T. Scholten, Steven Schalekamp et al 2020 *COVID-19 on the Chest Radiograph: A Multi-Reader Evaluation of an AI System* Radiology: 201874.
[5] Khan, Asif Iqbal, Junaid Latief Shah, and Mohammad Mudasir Bhat. 2020 *Coronet: A deep neural network for detection and diagnosis of COVID-19 from chest x-ray images* Computer Methods and Programs in Biomedicine: 105581.
[6] Singh, Dilbag, Vijay Kumar, and Manjit Kaur 2020 *Classification of COVID-19 patients from chest CT images using multi-objective differential evolution–based convolutional neural networks* European Journal of Clinical Microbiology & Infectious Diseases: 1-11
[7] Loey, Mohamed, Florentin Smarandache, and Nour Eldeen M Khalifa 2020 *Within the Lack of Chest COVID- 19 X-ray Dataset: A Novel Detection Model Based on GAN and Deep Transfer Learning.* Symmetry 12 no. 4: 651.
[8] Balaji, K. and K. Lavanya 2019 *Medical Image Analysis With Deep Neural Networks* (Academic Press: In Deep Learning and Parallel Computing Environment for Bioengineering Systems) pp 75-97.
[9] Ciaburro, Giuseppe, and Balaji Venkateswaran 2017 *Neural Networks with R: Smart models using CNN, RNN, deep learning, and artificial intelligence principles* Packt Publishing Ltd.
[10] Gulli, Antonio, Amita Kapoor, and Sujit Pal 2019 *Deep Learning with TensorFlow 2 and Keras: Regression, ConvNets, GANs, RNNs, NLP, and More with TensorFlow 2 and the Keras API* Packt PublishingLimited.
[11] Michelucci and Umberto 2019 *Advanced applied deep learning: convolutional neural networks and object detection* Apress.
[12] Albawi, Saad, Tareq Abed Mohammed and Saad Al-Zawi 2017 *Understanding of a convolutional neural network." In 2017 International Conference on Engineering and Technology (ICET)* IEEE pp. 1-6.
[13] Sangaiah, Arun Kumar et al 2019 *Deep Learning and Parallel Computing Environment for Bioengineering Systems* Academic Press.

IOP Conf. Series: Materials Science and Engineering **982** (2020) 012004 doi:10.1088/1757-899X/982/1/012004

[14] Chauhan, Karan, and Shrwan Ram 2018 *Image classification with deep learning and comparison between different convolutional neural network structures using TensorFlow and Keras* International Journal of Advance Engineering and Research Development 5 no. 02: 533-538.
[15] El-Amir, Hisham and Mahmoud Hamdy 2020 *Deep Learning Pipeline.*
[16] Kumar, Avinash, Sobhangi Sarkar, and Chittaranjan Pradhan 2020 *Malaria Disease Detection Using CNN Technique with SGD, RMSprop and ADAM Optimizers* In Deep Learning Techniques for Biomedical and Health Informatics pp. 211-230 Springer, Cham,
[17] Wani, M. Arif, Farooq Ahmad Bhat, Saduf Afzal, and Asif Iqbal Khan 2020 *Advances in deep learning. Vol.57. Berlin: Springer.*
[18] https://github.com/ieee8023/covid-chestxray-dataset
[19] https://www.kaggle.com/paultimothymooney/chest-xray-pneumonia
[20] https://github.com/agchung

Convolutional Neural Network (CNN) for Automatic Skin Cancer Classification System

Yunendah Nur Fu'adah[1,2], **NK Caecar Pratiwi**[1], **Muhammad Adnan Pramudito**[1], **Nur Ibrahim**[1]

[1] Electrical Engineering Department - Telkom University, Bandung, Indonesia
[2] yunendah@telkomuniversity.ac.id

Abstract. Skin cancer is a type of cancer that grows in the skin tissue, which can cause damage to the surrounding tissue, disability, and even death. In Indonesia, skin cancer is the third leading for most cancer cases after cervical and breast cancer. The accuracy of diagnosis and the early proper treatment can minimize and control the harmful effects of skin cancer. Due to the similar shape of the lesion between skin cancer and benign tumor lesions, physicians consuming much more time in diagnosing these lesions. The system was developed in this study could identify skin cancer and benign tumor lesions automatically using the Convolutional Neural Network (CNN). The proposed model consists of three hidden layers with an output channel of 16,32, and 64 for each layer respectively. The proposed model uses several optimizers such as SGD, RMSprop, Adam, and Nadam with a learning rate of 0.001. Adam optimizer provides the best performance with an accuracy value of 99% in identifying the skin lesions from the ISIC dataset into 4 classes, namely dermatofibroma, nevus pigmentosus, squamous cell carcinoma, and melanoma. The results obtained outperform the performance of the existing skin cancer classification system.

1. Introduction

Skin cancer is a disease caused by changes in the properties of normal skin cells to become malignant, in which cells will continue to divide into abnormal shapes that are uncontrolled due to DNA damage. Based on histopathology view, skin cancer has an irregular structure with cell differentiation in various levels of chromatin, nucleus, and cytoplasm [1]. Skin cancer is a malignancy disease that is often found in Indonesia in addition to cervical cancer and breast cancer. Skin cancer found 5.9 to 7.8% for all types of skin cancer per year. The most common skin cancers in Indonesia are basal cell carcinoma (65.5%), followed by squamous cell carcinoma (23%), malignant melanoma (7.9%), and other skin cancers [2]. Even though the number of Malignant Melanoma incidents is smaller than Basal Cell Carcinoma and Squamous Cell Carcinoma, the death rate tends to be greater, which causes 75% of deaths from skin cancer [1]. The most invasive skin cancer is melanoma, which has a high mortality rate, especially if it is not early detected. Non-melanoma skin cancers, such as basal cell carcinoma and squamous cell carcinoma are more common but less metastatic, and only partially leads to disability or death. Accurate diagnosis and early detection of skin cancer can help the healing process, proper medical treatment, and avoid the worst effects of skin cancer. Therefore, an early detection system is needed that can facilitate and increase public awareness in identifying types of skin cancer or other skin disorders such as a benign tumor on the skin that look very similar to skin cancer.

The automatic skin disorders classification can help people in identifying skin disorders that occur and immediately consult with medical personnel to get appropriate medical treatment. Several related studies based on digital image processing for the detection and classification of skin cancers were developed as a tool for medical personnel to diagnose skin disorders more accurately with fast computing time. The previous research [3] was developed using an edge detection method with K-NN and C-NN algorithms provide an accuracy of 75% and 75.6% respectively to classify skin disorders that potentially benign cancer and skin disorders that have the potential to be malignant cancer by using the International Skin Imaging dataset Collaboration (ISIC). In research [4], using the ISIC dataset for the conditions of skin cancer and skin benign tumors, an automatic skin disease classification system was developed based on deep learning with PNASNet-5-Large architecture which gives the best performance accuracy of 76%. Furthermore, other studies using CNN for the detection of skin diseases [5] [6] [7] provide performance accuracy of 80.52%, 86.21%, 87.25% respectively. To increase the amount of data and to improve the performance of the skin cancer detection system, the ISIC data augmentation process was performed [8]. The best accuracy was obtained by 95.91% using Alexnet. In the study [9], the CNN method with random regulators gave performance accuracy of 97.49% to distinguish some lesions of skin disorders such as nevus lesions, carcinomas, and melanomas. In this study, augmentation data from the ISIC dataset will be used to recognize the condition of skin cancer lesions or benign tumor lesions that look similar to cancer. The proposed method in this study uses CNN with various optimizer methods such as SGD, RMSprop, Adam, and Nadam optimizer to find out which optimizer provides the best performance.

2. Skin Cancer ISIC Dataset

This study uses the International Skin Imaging Collaboration (ISIC) data set shown in Figure 1 [10]. The data set consists of squamous cell carcinoma and melanoma for skin cancer, while for tumor conditions, there are dermatofibroma and nevus pigmentosus.

2.1. Dermatofibroma

Dermatofibromas include the category of benign tumors caused by an overgrowth of a mixture of various types of cells in the dermis layer of the skin. The skin growth that causes dermatofibroma usually occurs after experiencing several types of minor trauma to the skin, such as stab wounds caused by glass splinters or insect bites. Characteristics of dermatofibromas measuring around 2-3 mm, purplish brown, hard structure, and painful when pressed [11].

2.2. Nevus Pigmentosus

Nevus pigmentosus is a benign tumor originating from melanocytes, dendritic cells that produce pigment, normally found between keratinocytes in the basal layer of the epidermis. The developing Nevus Pigmentosus is very dangerous and quite difficult to handle. Nevus Pigmentosus has characteristics such as birthmarks or moles, if not early detected and exposure to pollution, ultraviolet light, and harmful chemicals, its potential to develop into melanoma which is very deadly skin cancer. Other effects of this disease for patients affected by complications will experience nerve disorders, such as seizures, fainting, and vomiting [12].

2.3. Squamous Cell Carcinoma

Squamous Cell Carcinoma is a type of skin cancer that attacks parts of the body that are often exposed to sunlight, such as legs, arms, lips, ears, face, neck, and head [13]. This disease is not too aggressive as like as other skin cancers. This disease tends to grow slowly and it can be treated easily through non-surgical therapy if early diagnosed. The benign tumor can continue to grow to be cancer which can spread to bones, tissues, and even lymph nodes due to the lateness of treatment. The more widespread the spread, the more difficult the cancer is to be treated.

2.4. *Melanoma*
Melanoma is a type of skin cancer that is very dangerous. This condition starts with human skin and can spread to other organs in the body. This disease is a type of skin cancer originating from melanocyte cells, melanin-producing cells that are usually found in the skin. Melanoma has an irregular shape and more than one color. Moles affected by melanoma can feel itchy and can bleed, also, their size can exceed normal moles [14].

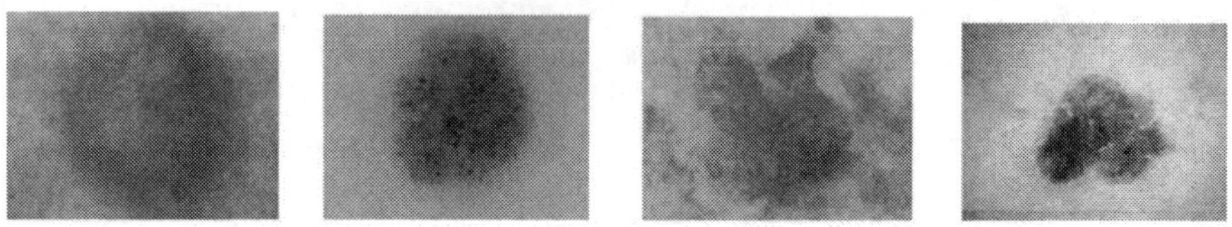

Dermatofibroma Nevus Pigmentosus Squamous Cell Carcinoma Melanoma

Figure 1. ISIC Dataset

3. Convolutional Neural Network (CNN)

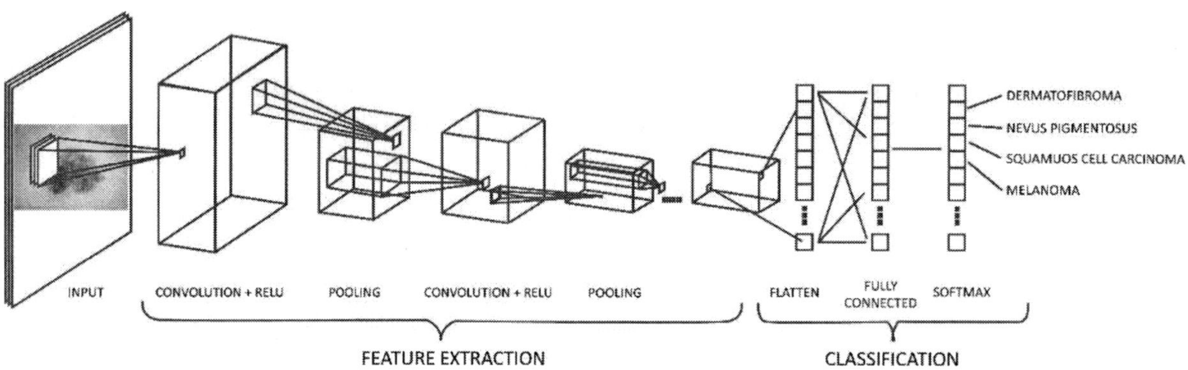

Figure 2. The Architecture of CNN

Convolutional Neural Network (CNN) is a development of the Multilayer Perceptron (MLP) which is designed to process two-dimensional data. CNN is included in the type of Deep Neural Network because it has a high network depth and has been widely applied to image data [15]. CNN has an architecture as like as neural networks in general, neurons in CNN have a weight, bias, and activation function. CNN architecture as shown in Figure 2, which consists of the convolution layer with ReLU activation, pooling layer as feature extraction layer, and fully connected layer with softmax activation as classification layer.

3.1. *Convolution Layer*
In the Convolution layer, the convolution process is the main process that underlies CNN. Convolution layer is the first layer that will process the image as an input system model. The image will be convoluted with a filter to extract features from the input image that is called the feature map. Figure 3 shows an illustration of the convolution process.

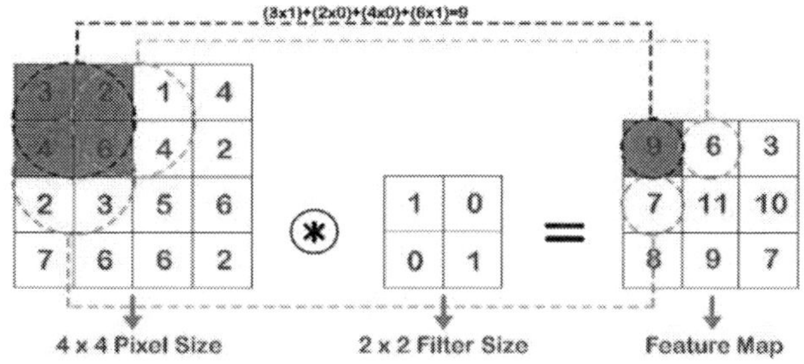

Figure 3. Illustration of the convolution process

3.2. *Activation Rel-U*

ReLU (Rectified Linear Unit) is an activation layer in CNN to increase the training stage on neural networks that have advantages to minimize errors. Rel-U activation makes all pixel values to be zero when a pixel image has a value of less than zero [16].

$$f(x) = \begin{cases} x, x > 0 \\ 0, x \leq 0 \end{cases} \quad (1)$$

3.3. *Activation Rel-U*

Polling layers in the CNN method usually will be inserted regularly after several convolution layers, There are several advantages of the pooling layer, which can progressively reduce the size of the output volume on the Feature Map so that it can control over-fitting [15]. Pooling Layer is used to reduce data using max-pooling or mean Pooling. The max-pooling will select the maximum value, whereas the main pooling finds the average value. The pooling process illustration using the four by four-pixel input image is shown in Figure 4.

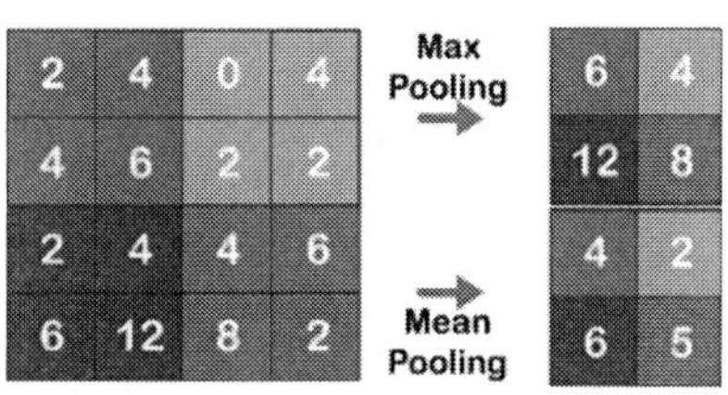

Figure 4. Pooling Process Illustration

3.4. *Fully Connected Layer*

The Fully-Connected Layer is the layer at the end of the architecture used in the multi-layer perceptron. This layer will connect all the neurons of the previous activation layer. In this stage, all neurons in the input layer need to be transformed into one-dimensional data (flatten process) [17]. After that softmax activation as another form of the logistic regression algorithm can be used to classify more than two classes.

3.5. *Hyperparameter*

Hyperparameter has variable values that remain during the model training process and can affect the performance of the model trains. In this study, the hyperparameter that used is an optimizer, such as Stochastic Gradient Descent (SGD), Root Mean Square Propagation (RMSprop), Adaptive Moment Estimation (Adam), and Nesterov-accelerated Adaptive Moment Estimation (Nadam). Stochastic Gradient Descent (SGD) is a repetitive optimization method that functions to optimize the model using better functions such as differential or subdifferential [18]. SGD uses each training sample as a new parameter. Root Mean Square Propagation (RMSprop) is widely used in the design

IOP Conf. Series: Materials Science and Engineering **982** (2020) 012005 doi:10.1088/1757-899X/982/1/012005

of deep learning models [19]. This optimizer is an improvisation from Root Propagation (Rprop). Initially, Rprop cannot be used on files with large amounts of data. The essence of RMSprop is moving the average gradient at the time of the model. Adam optimizer is a combination of RMSprop and momentum. This optimizer also uses an average gradient of weight [20]. The advantage of Adam over other optimizers is efficient in computing time, consume less memory, and can handle sparse gradients on noisy problems. Nadam (Nesterov-accelerated Adaptive Moment Estimation) thus combines Adam and NAG (Nesterov accelerated gradient).

4. System Design

4.1. Proposed System Model

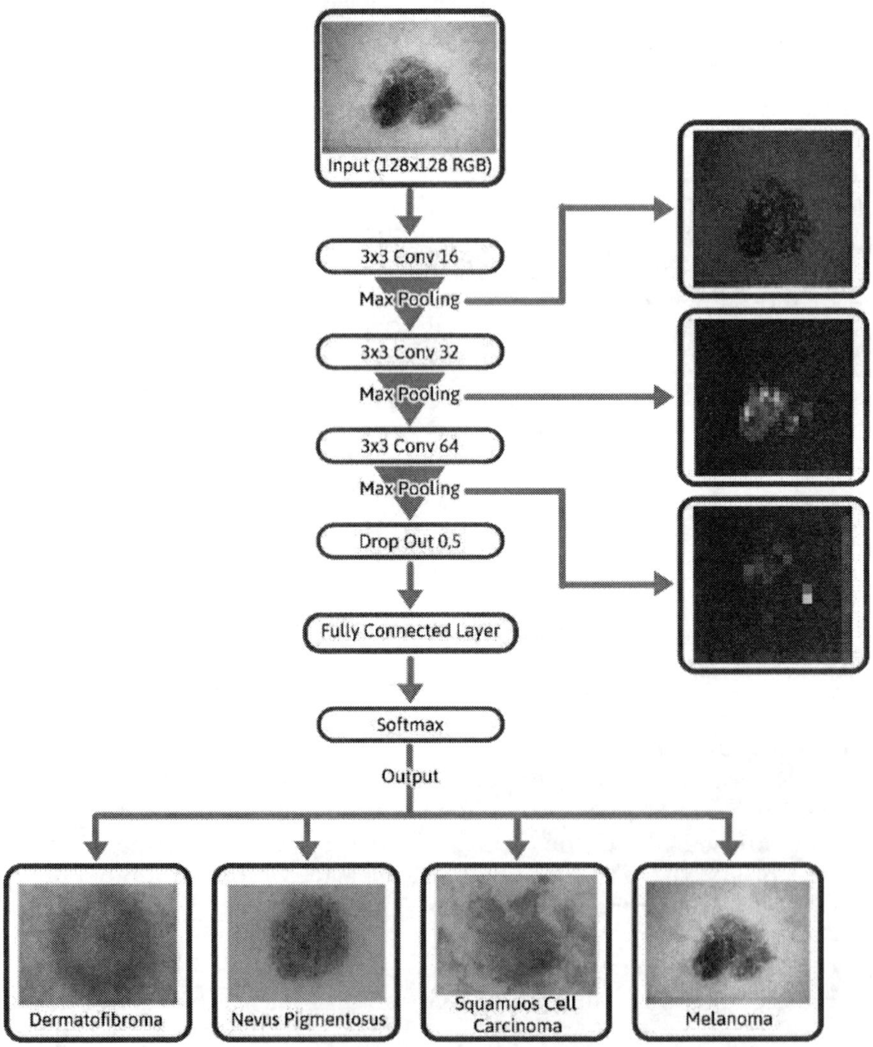

Figure 5. The Proposed System Model

In this study, the dataset used is an augmentation of the ISIC dataset for the conditions of dermatofibroma, nevus pigmentosus, squamous cell carcinoma, and melanoma. Dermatofibroma and nevus pigmentosus are benign tumors, while squamous cell carcinoma and melanoma are skin cancers. The total amount of augmentation data is 4000 images consisting of 1000 images for each class. The distribution of training data and validation data were 75% and 25% so that the training data used were 3000 images and the validation data used were 1000 images.

Based on Figure 5 and Table 1, the resolution of skin images is changed to 128 × 128 pixels as an input of the CNN model which consists of 3 hidden layers. The image is convoluted using 3 × 3 filters on each hidden layer with the number of output channels on each layer is 16, 32, 64 respectively. At each layer, the activation process uses Rel-U activation and Max pooling. The result of Maxpooling reduces the size of the image as can be seen in Figure 5 and Table 1. After that, the flatten process will change image features from 3 dimensions became 1 dimension. The last, softmax activation function will be used to classify the condition of skin image into four classes namely, dermatofibroma, nevus pigmentosus, squamous cell carcinoma, and melanoma.

Table 1. Details of CNN Model Proposed

Layer (type)	Output Shape	Parameter
Input Image	128,128,3	0
Convolution	128,128,16	448
ReLU	128,128,16	0
Max-Pooling	64,64,16	0
Convolution	64,64,32	4640
ReLU	64,64,32	0
Max Pooling	32,32,32	0
Convolution	32,32,64	18496
ReLU	32,32,64	0
Max Pooling	16,16,64	0
Dropout	16,16,64	0
Flatten	16384	0
Dense	4	65540
Softmax	4	0

4.2. *System Performance*

System performance in classifying the conditions of dermatofibroma, nevus pigmentosus, squamous cell carcinoma, and melanoma was measured using a confusion matrix to obtain accuracy, recall, precision, and F1 scores. The equation is used to determine the accuracy of the system in classifying skin cancer lesions and benign tumor lesions.

$$Accuracy = \frac{TP+TN}{TP+TN+FP+FN} \tag{2}$$

$$Recall = \frac{TP}{TP+FN} \tag{3}$$

$$Precision = \frac{TP}{TP+FP} \tag{4}$$

$$F1 - Score = 2\,x\,\frac{Precision\ x\ Recall}{Precision+Recall} \tag{5}$$

True Positive (TP) indicates the condition while the data is positive and correctly predicted as positive. True Negative (TN) indicates the condition where the data is negative and correctly predicted as negative. False Positive (FP) indicates the conditions where the data is negative but incorrectly detected as positive. Whereas False Negative (FN) indicates the conditions where the data is positive, but incorrectly detected as negative.

5. Result and Discussion

In this study, 3000 training images and 1000 validation images were used in the training model. The images obtained from the ISIC dataset, consist of four classes; dermatofibroma, nevus pigmentosus, squamous cell carcinoma, and melanoma. These images were trained using the CNN model with various optimizer methods such as SGD, RMSprop, Adam, and Nadam optimizer with a learning rate of 0.001, and use loss categorical cross-entropy. The performance parameters measured in this study are accuracy, recall, precision, F1 scores, and loss. After training with 100 iterations (epoch), it can be seen the comparison of the accuracy and loss performance of the proposed model for each optimizer used in Figure 6.

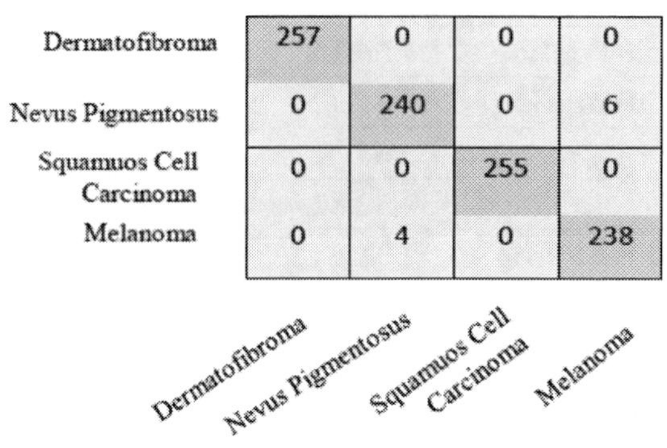

Figure 6. Confusion Matrix

Table 2. Details Performance of Model Proposed

Class	Precision	Recall	F1-Score	No of Images
Dermatofibroma	1.00	1.00	1.00	257
Nevus Pigmentosus	0.98	0.98	0.98	246
Squamous Cell Carcinoma	1.00	1.00	1.00	255
Melanoma	0.98	0.98	0.98	242
Total	0.91	0.91	0.91	114

Based on the results shown in Figure 7, it shows the performance comparison of each optimizer used, Adam optimizer provides the best accuracy performance and loss performance when compared to other optimizers. The performance accuracy using SGD, RMSprop, and Nadam optimizer for training and validation data can decrease suddenly so that the loss also rises at certain epochs. This is indicated by the appearance of many spikes on the accuracy and the loss graph. Whereas, the system model that uses Adam optimizer

IOP Conf. Series: Materials Science and Engineering **982** (2020) 012005 doi:10.1088/1757-899X/982/1/012005

continues to show an increase in accuracy at each iteration (epoch) and the difference in accuracy between the training data and the validation data is not much different, as well as for the system loss continues to decrease at each iteration. This condition shows that the proposed model is not overfitting and the system model used can recognize the condition of dermatofibroma, nevus pigmentosus, squamous cell carcinoma, and melanoma with the best accuracy performance of 99% and loss of0.0346.

Confusion Matrix for system models with Adam optimizer can be seen in Figure 6. It can be seen that from 1000 validation images used, 990 images were successfully classified according to their class. The error occurs in four images on nevus pigmentosus detected as melanoma and six images on melanoma detected as nevus pigmentosus. Other parameters used to evaluate system

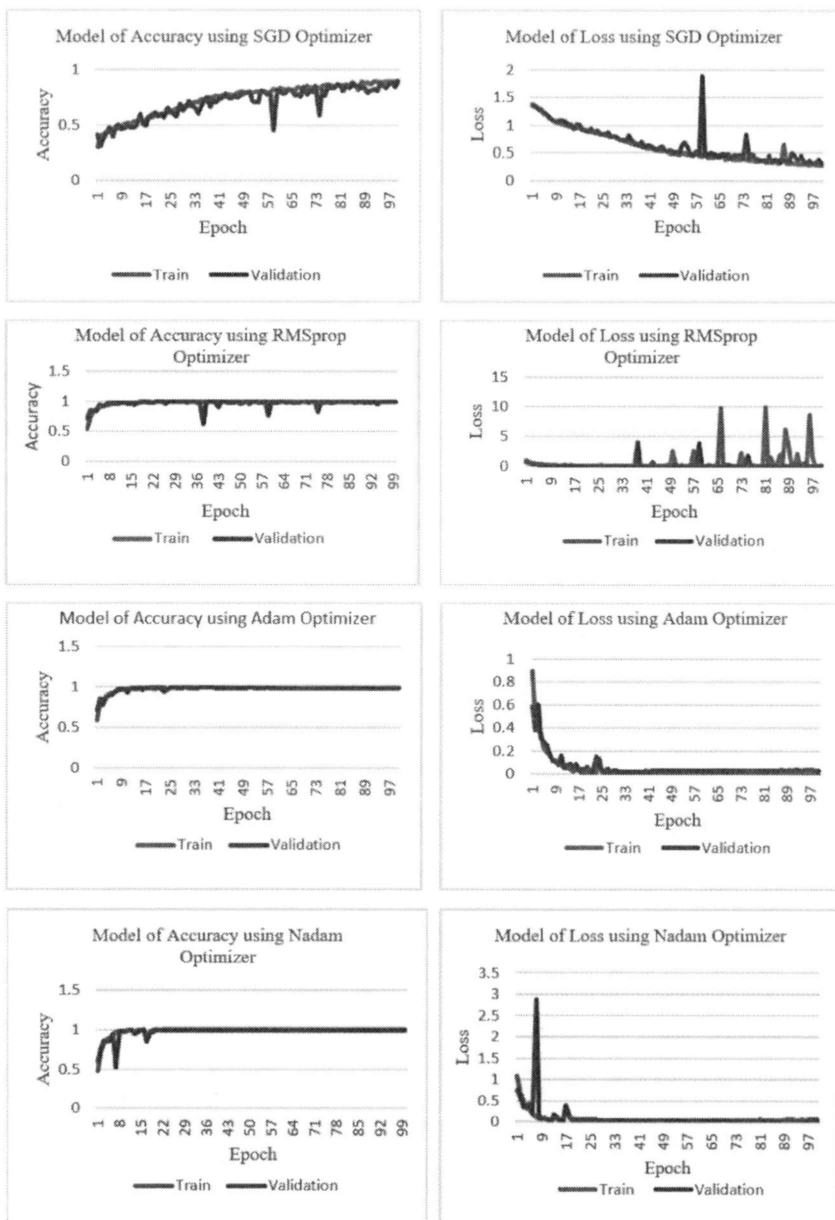

Figure 7. System Performance Comparison

performance are Precision, Recall, and F1-score, which has a range of values from 0 to 1 (a value of 1 indicates no error). Based on the data shown in Table 2, the value of system performance parameters obtained is close to 1, so it can be concluded that the CNN model proposed can classify the conditions of dermatofibroma, nevus pigmentosus, squamous cell carcinoma and melanoma with high accuracy and provides a minimum error.

6. Conclusion

In this study, an automatic system was designed to classify the conditions of dermatofibroma, nevus pigmentosus, squamous cell carcinoma, and melanoma based on digital image processing. The CNN model used in this study consists of 3 hidden layers, using 3 × 3 filter sizes with 16, 32, and 64 channel outputs in sequence, a fully connected layer, and softmax activation. The optimization is performed on the proposed model using SGD, RMSprop, Adam, and Nadam optimizer. Based on testing that has been carried out the CNN model proposed with Adam optimizer provides the best performance in classifying the dataset of skin cancer lesions and benign tumor lesions with 99% accuracy, loss of 0.0346 and the value of precision, recall, F1- score is almost 1. Based on the performance results, the system shows that the proposed model is promising to use as an existing tool for medical personnel in determining the diagnosis of skin cancer or benign tumors. In further research, systems can be developed to classify the various types of skin cancer and other skin diseases.

References

[1] Buljan M, Bulana V and Sandra S 2008 *Variation in Clinical Presentation of Basal Cell Carcinoma* (Kroasia: University Department of Dermatology and Venereology Zagreb Croatia) p 25-30.

[2] Cipto H, Suriadiredja AS. Tumor kulit. Dalam: Menaldi SL, Bramono K, Indriatmi W, editor. Ilmu penyakit kulit dan kelamin. Edisi ketujuh. Jakarta: Badan Penerbit FKUI; 2016. h.262-276.

[3] Teresia R, Savera, Winsya H, Suryawan and Agung WS 2020 Deteksi Dini Kanker Kulit Menggunakan K-NN dan Convolutional Neural Network *J. JTIIK.* **7** **2** p 373-378.

[4] Md Ashraful Alam Milton 2018 *Automated Skin Lesion Classification Using Ensemble of Deep Neural Networks in ISIC: Skin Lesion Analysis Towards Melanoma Detection Challenge.*

[5] Serban Radu SJ, Loretta Ichim, et al 2019 *Automatic Diagnosis of Skin Cancer Using Neural Networks* (Bucharest, Romania: The XIth International Symposium on Advanced Topics in Electrical Engineering March 28-30).

[6] Nazia Hameed, et al *Multi-Class Skin Diseases Classification Using Deep Convolutional Neural Networkand Support Vector Machine.*

[7] Xinyuan Zhang, et al 2018 *Towards Improving Diagnosis of Skin Diseases by Combining Deep Neural Network and Human Knowledge BMC Medical Informatics and Decision Making* **18(Suppl 2)** p 59

[8] Khalid M. Hosny, et al 2019 *Classification of skin lesions using transfer learning and augmentation with Alex-net* PLOS ONE.

[9] Marwan AA 2019 *Skin Lesion Classification Using Convolutional Neural Network With Novel Regularizer* IEEE Access.

[10] The International Skin Imaging Collaboration (ISIC).Accessed: June 2020. [Online]. Available: https: //www.isicarchive.com/#!/topWithHeader/onlyHeaderTop/gallery

[11] Mousumi Roy, et al 2016 Dermatofibroma: Atypical Presentations Indian *J. Dermatology.*

[12] R. Delila Tsaniyah, Aspitriani and Fatmawati "Prevalensi dan Gambaran Histopatologi Nevus Pigmentosus di Bagian Patologi Anatomi Rumah Sakit Dr. Mohammad Hoesin Palembang," Periode 1 Januari 2009-31 Desember 2013.

[13] Syril Keena T. Cutaneous squamous cell carcinoma. Journal of The American Academy of

Dermatology. Volume 78, Issue 2, p237-432, e33-e55.
[14] Marco Rastrelli, et al 2014 *Melanoma: Epidemiology, Risk Factors, Pathogenesis, Diagnosis, and Classification* 28 no. 6 p 1005-1011
[15] P. Kim, 2017 *MATLAB Deep Learning: With Machine Learning, Neural Networks, and Artificial Intelligence.*
[16] A. F. Agarap 2008 *Deep Learning using Rectified Linear Units (ReLU)* **1** p 2–8 [Online]. Available: http://arxiv.org/abs/1803.08375.
[17] S. Khan, H. Rahmani, S. A. A. Shah, M. Bennamoun, G. Medioni and S. Dickinson 2018 A Guide to Convolutional Neural Networks forComputer Vision. Morgan Claypool [Online] Available: https: //ieeexplore.ieee.org/document/8295029
[18] H. Robbins and S. Monro 1985 A Stochastic Approximation Method" in Springer
[19] J. Duchi, E. Hasan and Y. Singer 2011 Subgradient Methods for Online Learning and Stochastic Optimization J. of Machine Learning Research
[20] Y. Yunlong, L. Fuxian. 2019 Effective Neural Network Training with a New Weighting Mechanism-Based Optimition Algorithm IEEE Access.

IOP Conf. Series: Materials Science and Engineering **982** (2020) 012006 doi:10.1088/1757-899X/982/1/012006

Unsupervised Method for Calculating Diameter and Number of Leukocyte Cells

Retno Supriyanti[1,3], Ahmad Haeromi[1], Yogi Ramadhani[1], Wahyu Siswandari[2]

[1]Electrical Engineering Department, Universitas Jenderal Soedirman, Purwokerto, Indonesia
[2]Medical Department, Universitas Jenderal Soedirman, Purwokerto, Indonesia

[3]retno_supriyanti@unsoed.ac.id

Abstract. Leukocyte has an essential role in the body. Leukocyte has functioned as the body's immune system function to protect the body from viruses, bacteria, and pathogens. Under normal circumstances, leukocyte has a relatively constant amount. If the number of leukocytes exceeds or is less than usual, health problems can occur. So that the leukocyte cell counts as an indicator of abnormalities and diseases. Digital image processing is an alternative method to identify leukocyte in addition to a hematology analyzer or manually by medical personnel by processing blood images so that the desired information is obtained. Blood images have a variety of lighting conditions, staining time, and cell thickness that cause the various attributes of the blood image. This experiment segmented leukocyte using unsupervised method especially the K-Means clustering method on the hue component of the image after being transformed into HSV color coordinates by grouping leukocyte cell images (great hue value) into one cluster. The results show that the performance of our system as follows, sensitivity 84%, specificity 94%, and accuracy 97%.

1. Introduction

Blood is a liquid in the body that functions to carry the oxygen needed by cells throughout the body. Blood also supplies body tissues with nutrients, transports metabolic waste, and contains various ingredients that make up the immune system that aims to defend the body from various diseases. Blood cells consist of red blood cells erythrocytes, leukocytes, and platelets. Leukocytes are blood cells that function as the body's immune system. Increasing or decreasing the number of leukocytes in the blood, indicating various disorders and diseases. The number and diameter of leukocytes in the body could apply for the classification of types of leukocyte cells as well as indicators of abnormalities that can help medical staff diagnose diseases. On the other hand, the development of information technology, especially in terms of digital image processing, is also proliferating. This field is also widely used in the health sector, namely the field of medical image processing. One of the applications that have been widely developed is microscopic image processing, which includes processing and analyzing leukocytes based on image processing. According to this condition, currently a lot of research about leukocyte in the world. Nizam [1] researched intending to evaluate the efflux of leukocytes and platelets fbrin (L-PRF) and combining with bovine bone mineral in the treatment of maxillary sinus. Moshavash [2] researched texture-based leukocyte segmentation for leukemia identification. Li [3] developed a holographic imaging system to determine the differences in three parts of leukocytes. Wang [4] used the use of machine learning in classifying leukocytes, in particular, the use of the Single Shot Multibox Detector method. The results obtained have a pretty good precision of around 90%. Choi [5] researched the automatic counting system of leukocytes cells from the bone marrow using a two-stage artificial neural network. After that, classify myeloid and erythroid. Reta [6] researched identifying leukocyte morphology by using the contextual analysis method in acute leukemia. Blood cell data came from the bone marrow. From this analysis, he can produce an automatic leukemia detection system with an accuracy of around 90%. Hegde [7] researched leukocyte image processing to detect cell nuclei and leukocyte classification based on cell core features using the TissueQuant method. The primary variable used in the identification and classification of this research is color. Abdulhay [8] researched leukocyte segmentation by focusing on identifying the region of interest (ROI) accurately. The input image used is a static microscopic image. Kelley [9] In his research, he uses a protocol used to explain multichannel in confocal images in living cells. The method he uses includes all stages of research, such as preparation, image synchronization, image acquisition, and quantitative analysis of leukocyte images. Loddo [10] reviewed a variety of research into the operation of blood cell morphology in the classification of various diseases that have relevance to blood cells. Our research also discusses the application of digital image processing in the health sector. The long-term goal of our research is the development of computer-aided diagnosis (CAD) in early detection of leukemia. Based on the fact that in developing countries like Indonesia, the number of specialist doctors is minimal. So that the presence of CAD can help improve the quality of health, especially optimal health services in the community. We have conducted several experiments in the context of developing CAD [11] [12][13][14][15][16][17]. In this paper, we focus on calculating the diameter and number of leukocytes using the K-means clustering method. The main objective is to obtain one of the leukocyte morphology variables precisely and accurately as apart of our CAD leukemia identification.

2. Methods

2.1. Input images

This experiment uses a digital image of blood cells taken from the Acute lymphoblastic Leukemia Image Database for Image Processing (ALL-IDB) of the University of Degli Studi in Milano, Italy, and is available at http://crema.di.unimi.it/~fscotti/ all /. Image data available on ALL-IDB was taken with a laboratory optical microscope plus a Canon PowerShot G5 camera with different microscope magnifications ranging from 300 to 500. All images are in JPG format with 24-bit color depth and a resolution of 2592 x 1944. Figure 1 shows examples of input images which is used in this experiment. In this paper, we will analyze the results of the segmentation of white blood cell objects from the image of blood cells of Acute Lymphoblastic Leukemia (ALL) patients, which then count the number and diameter. There are two types of leukocytes cells in the image data, namely ordinary leukocyte cell objects that are dark blue in solid and round shape. Furthermore, abnormal leukocyte cell objects are purplish in blue and have a non-solid and round shape. Normal leukocyte cell objects to be segmented and counted in number and diameter.

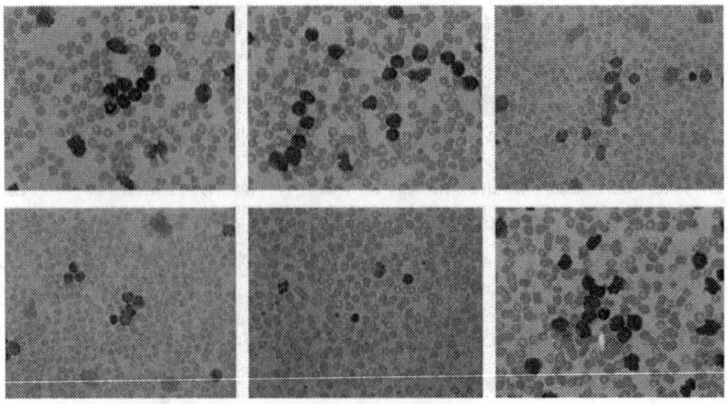

Figure 1. Examples of input images

2.2. Pre-Processing Image

At the stage of image pre-processing, we performed contrast stretching operations and the conversion of RGB color space to HSV color space. Contrast stretching techniques have a function to improve image contrast so that the values of the image pixels are distributed evenly. It refers to the input image, as shown in Figure 1, each component of the image of a blood cell has low contrast; that is, the image pixels grouped in a particular area. The initial image is in the form of an RGB image that has been stretched out before contrast and then transformed into the HSV color space. Figure 2 shows an example of an input image that has been converted into the HSV color space and is ready for use in this research, especially for the image in the H component.

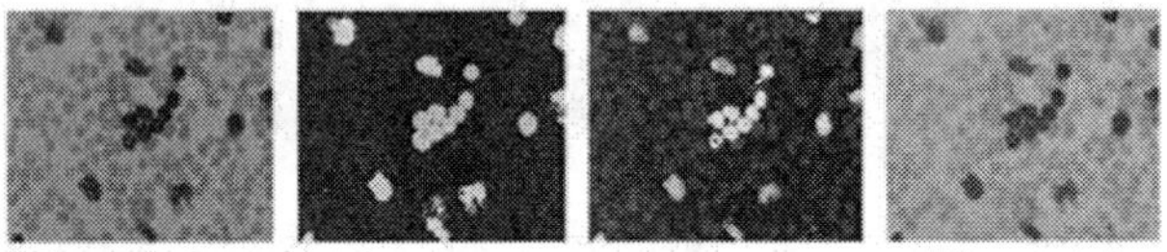

Figure 2. An example of an RGB image conversion to HSV color space

2.3. K-Means Clustering

K-means clustering is one type of clustering algorithm that is commonly used by grouping data according to similar characteristics. This data group is called a cluster. Data in a cluster has similar characteristics and is not similar to data in another cluster. The input received is the data or object and the desired group (cluster). This algorithm will group data or objects into these groups. In each cluster, there is a center point (centroid) that represents the cluster. The process of grouping data into a cluster could be done by calculating the closest distance from a data to a centroid point. The distance of each data to the centroid added, and the smallest total distance obtained. Minkowski distance calculation could use to calculate the distance between 2 pieces of data, as shown in Equation 1.

$$d(X_i,X_j) = (|X_{i1} - X_{j1}|^g + |X_{i2} - X_{j2}|^g + \ldots + |X_{ip} - X_{jp}|^g)^{1/g} \quad (1)$$

which is:

$g = 1$, to calculate Manhattan distance

$g = 2$, to calculate Euclidian distance

$g = \infty$, to calculate Chebychev distance

x_i, x_j are the two data to be calculated the distance

p = dimension of a data

If the total distance of each data is already obtained, then the centroid will be updated to the midpoint of the data in the cluster. The renewal of a centroid can be done by Equation 2.

$$-\sum \qquad (2)$$

Which is μ_k is centroid point of the K-cluster, N_k is the amount of data in the K cluster, and x_q is q data in the K-cluster.

2.4. Number of Objects Calculation

Calculation of leukocyte objects number by tracing the edge of pixels of each object or hole in the object. The results of the calculation would be stored in the matrix M x 2, where the length M is the number of detected leukocyte cell objects.

2.5. Object Diameter Calculation

Diameter is the longest distance between two points of a shape. In digital image processing can use the maximum distance between each pair of points that form an object. The calculation is quickly done by testing the points on the edges of the object, ie, the distance between each pair of edge points. This experiment will calculate the diameter of each object one by one —the result obtained by calculating the distance of the one-pixel pair and counting the other pixel pairs. Then the most significant distance between pixels is obtained and taken as the diameter of the object.

3. Results and Discussions

As already explained in Subsection 2, the image to be analyzed in this experiment is the image in the HSV color space, precisely the image of the H component only. In the hue component, leukocyte cell objects have a high intensity value used to segment leukocyte cell objects using the K-Means method. The hue component which is the matrix M x N will be changed to 2 x (M * N). This experiment has two clusters, cluster 1 for high-value data, and cluster 2 for low-value data. The system makes two centroids that represent each cluster. Data of each pixel that has been grouped is then labeled based on the cluster and stored for later in comparison with the original image so that the K-Means clustering segmentation image can be displayed. Figure 3 shows some examples of the clustering result.

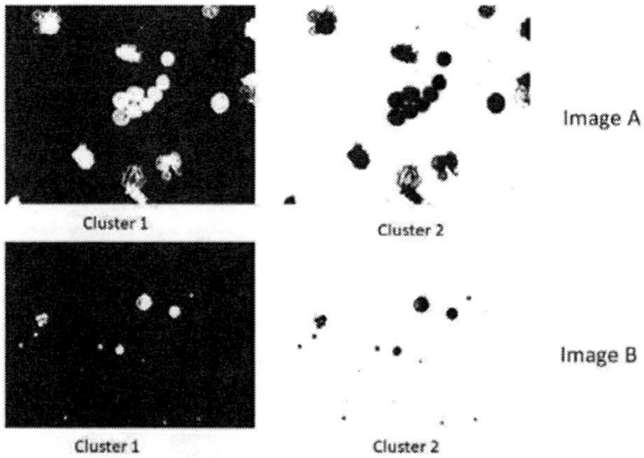

Figure 3. Examples of clustering

However, image segmentation results not only produce leukocyte cell images, but there is also much noise. If not removed, noise can interfere with subsequent processes. Our algorithm will delete objects with a value of fewer than 500 pixels. After the noise is removed, the image of leukocytes will be obtained. Figure 4 shows an example of noise removal in image B.

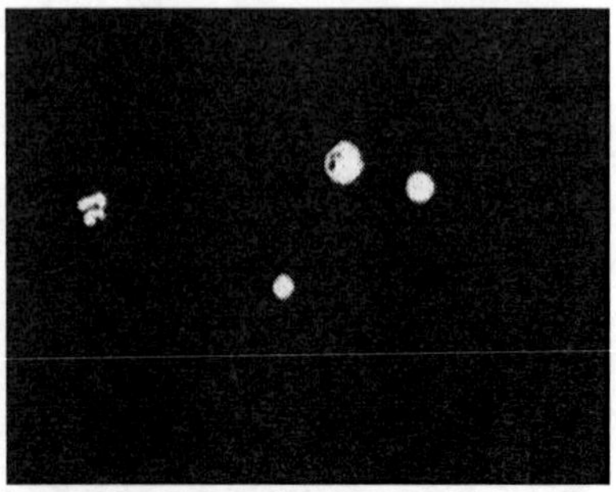

Figure 4. Image B after noise removal

According to Figure 3, there are several round objects. The first is a single leukocyte cell objects. The second is an irregularly shaped. The others are non-round objects are abnormal leukocyte cell objects, and almost round-shaped objects, which are several leukocyte cell objects which are attached to produce one only object. In order to get the number and diameter of leukocytes cells accurately, these leukocyte cell objects must be separated or cut. Before the object is almost rounded and not rounded, it must be grouped first in order to facilitate the process of cutting the leukocyte cell objects together. Figure 5 shows some examples of the results of measuring the degree of roundness of each object.

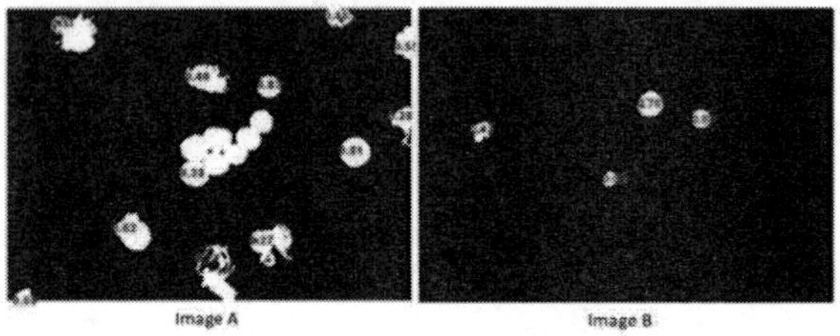

Figure 5. Examples of roundness degree measurement of each Object

Objects that have a degree of roundness approaching value 1 are increasingly rounded the object according to some of the experiments we did. Then the value of the degree of roundness to separate a single object that approaches 1 with a coincident object and unusual object is 0.77. If an object has a degree of roundness less than the boundary value, then the object will be grouped into coinciding and unusual objects, which will then be separated as shown in Figure 6.

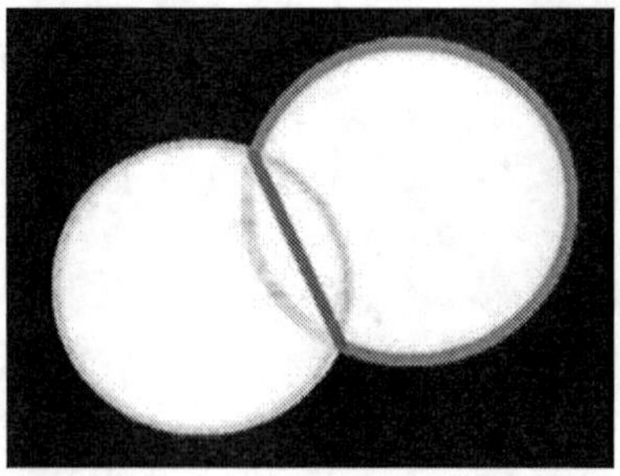

Figure 6. Cutting two Circles Into Each Other

In the object that coincide will be detected several forms of intersecting circles. The area between the two intersecting circle points will then be cut as in Figure 6 so that the adjacent leukocyte cell object can be separated. System accuracy calculation uses sensitivity and specificity methods as we discussed in [17]. In the test using the sensitivity and specificity methods, then there are four variables used, namely True Positive (TP), True Negative (TN), False Positive (FP) and False Negative (FN). Calculation results are shown in Figure 7.

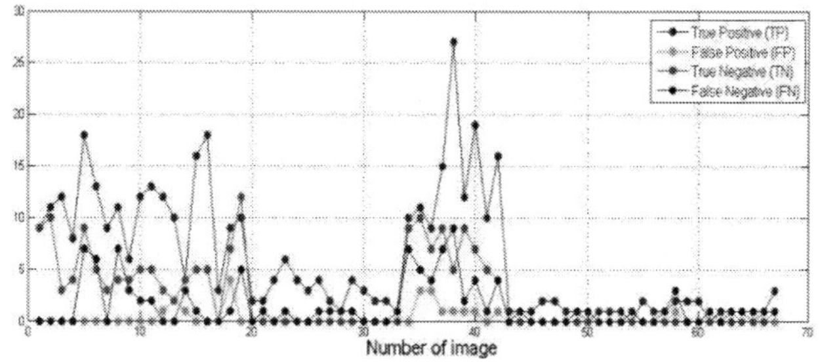

Figure 7. Graph of Calculation Results of Leukocyte Cell Count

Based on the results in Figure 7, the performance results that we obtain in this experiment are as shown in Table 1.

Table 1. Performance of Our System

No.	Performance	%
1.	Sensitivity	84%
2.	Specificity	94%
3.	Accuracy	97%

4. Conclusions

Based on the results we have presented in the previous sub-section, our system can segment 67 cell leukocyte images using the K-Means clustering method. The first step is to improve the input images using contrast stretching operation and the conversion to the HSV color space. There are 2 clusters used, where cluster 1 succeeds in segmenting leukocytes cells, and cluster 2 succeeds in segmenting red blood cells. Although some noise should remove in the input image due to the degree of roundness that is not appropriate, our system can get a sensitivity performance of 84%, a specificity of 94%, and an accuracy of 97%. These results are quite promising for the further development of computer-aided rediagnosis.

Acknowledgements

We want to thank the Ministry of Research and Technology - BRIN for funding support through the Fiscal Year 2020 "Penelitian Terapan" Grant.

References

[1] N. Nizam, G. Eren, A. Akcalı, and N. Donos, —Maxillary sinus augmentation with leukocyte and platelet-rich fibrin and deproteinized bovine bone mineral: A split-mouth histological

and histomorphometric study,‖ Clin. Oral Implants Res., vol. 29, no. 1, pp. 67–75, 2018, DOI: 10.1111/clr.13044.

[2] Z. Moshavash, H. Danyali, and M. S. Helfroush, —An Automatic and Robust Decision Support System for Accurate Acute Leukemia Diagnosis from Blood Microscopic Images,‖ J. Digit. Imaging, vol. 31, no. 5, pp. 702–717, 2018, DOI: 10.1007/s10278-018-0074-y.

[3] Y. Li et al., —Accurate label-free 3-part leukocyte recognition with single cell lens-free imaging flow cytometry,‖ Comput. Biol. Med., vol. 96, no. March, pp. 147–156, 2018, DOI: 10.1016/j.compbiomed.2018.03.008.

[4] Q. Wang, S. Bi, M. Sun, Y. Wang, D. Wang, and S. Yang, —Deep learning approach to peripheral leukocyte recognition,‖ PLoS One, vol. 14, no. 6, 2019, DOI: 10.1371/journal.pone.0218808.

[5] J. W. Choi et al., —White blood cell differential count of maturation stages in bone marrow smear using dual-stage convolutional neural networks,‖ PLoS One, vol. 12, no. 12, pp. 1–16, 2017, DOI: 10.1371/journal.pone.0189259.

[6] C. Reta et al., —Segmentation and classification of bone marrow cells images using contextual information for medical diagnosis of acute leukemias,‖ PLoS One, vol. 10, no. 6, pp. 1–19, 2015, DOI: 10.1371/journal.pone.0130805.

[7] R. B. Hegde, K. Prasad, H. Hebbar, and B. M. K. Singh, —Development of a robust algorithm for detection of nuclei of white blood cells in peripheral blood smear images,‖ Multimed. Tools Appl., vol. 78, no. 13, pp. 17879–17898, 2019, DOI: 10.1007/s11042-018-7107-x.

[8] E. Abdulhay, M. A. Mohammed, D. A. Ibrahim, N. Arunkumar, and V. Venkatraman, —Computer Aided Solution for Automatic Segmenting and Measurements of Blood Leucocytes Using Static Microscope Images,‖ J. Med. Syst., vol. 42, no. 4, 2018, DOI: 10.1007/s10916-018-0912-y.

[9] L. C. Kelley et al., —Live-cell confocal microscopy and quantitative 4D image analysis of anchor-cell invasion through the basement membrane in Caenorhabditis elegans,‖ Nat. Protoc., vol. 12, no. 10, pp. 2081–2096, 2017, DOI: 10.1038/nprot.2017.093.

[10] A. Loddo, C. Di Ruberto, and M. Kocher, —Recent advances of malaria parasites detection systems based on mathematical morphology,‖ Sensors (Switzerland), vol. 18, no. 2, pp. 1– 22, 2018, DOI: 10.3390/s18020513.

[11] R. Supriyanti, A. Chrisanty, Y. Ramadhani, and W. Siswandari, —Computer aided diagnosis for screening the shape and size of leukocyte cell nucleus based on morphological image,‖ Int. J. Electr. Comput. Eng., vol. 8, no. 1, pp. 150–158, 2018, DOI: 10.11591/ijece.v8i1.pp150-158.

[12] R. Supriyanti, A. Chrisanty, Y. Ramadhani, and W. Siswandari, —Computer Aided Diagnosis for Screening the Shape and Size of Leukocyte Cell Nucleus based on Morphological Image,‖ Int. J. Electrical Comput. Eng., vol. 8, no. 1, pp. 150–158, 2018, DOI: 10.11591/ijece.v8i1.pp150-158.

[13] R. Supriyanti, -Technology supporting health services for rural areas based on image processing Technology supporting health services for rural areas based on image processing,‖ J. Phy, vol. 1367, 2019, DOI: 10.1088/1742-6596/1367/1/012090.

[14] R. Supriyanti, G. Satrio, Y. Ramadhani, and W. Siswandari, —Contour Detection of Leukocyte Cell Nucleus Using Morphological Image,‖ J. Phys. Conf. Ser., vol. 824, no. 1, 2017.

[15] R. Supriyanti, B. L. Nababan, Y. Ramadhani, and W. Siswandari, —A Simple and Easy-to-Use Tool for Detecting Outer Contour of Leukocytes Based on Image Processing Techniques,‖ in (ICBAB), 19th International Conference on Biomedical Applications and Bioinformatics, 2017.

[16] R. Supriyanti, A. Rifai, Y. Ramadhani, and W. Siswandari, —Characteristics identification of myeloblast cell using K-means clustering for uncontrolled images,‖ Int. J. Mach. Learn. Comput., vol. 9, no. 3, 2019, DOI: 10.18178/ijmlc.2019.9.3.809.

[17] R. Supriyanti, M. M. Afif, I. T. Hasan, Y. Ramadhani, and W. Siswandari, —a Simple Tool for Identifying Outer Shape of White Blood Cell Based on Image Processing Techniques in

Order To Develop Health Facilities in Developing Countries,‖ PONTE Int. Sci. Res. J., vol. 73, no. 12, pp. 314–325, 2017, DOI: 10.21506/j.ponte.2017.12.21.

IOP Conf. Series: Materials Science and Engineering **982** (2020) 012007 doi:10.1088/1757-899X/982/1/012007

Simple Classification of the Alzheimer's Severity in Supporting Strengthening the Diagnosis of Patients based on ROC Diagram

Retno Supriyanti1[1,3], Ays Rahmadhani Subhi1[1], Egi Julian Asharil[1], Fathoni Ahmad1[1], Yogi Ramadhani1[1], Haris B. Widodo[2]

[1]Electrical Engineering Department, Jenderal Soedirman University
[2]Medical Department, Jenderal Soedirman University

[3]retno_supriyanti@unsoed.ac.id

Abstract. Alzheimer's disease is the most common cause of dementia and accounts for 60-80 percent of all cases of dementia. Dementia is a brain disorder that results in the loss of one's intellectual and social abilities. Progressive disease is one characteristic of this disease that interferes with a person's mental functions, such as memory and behavior. Experts believe that Alzheimer's disease interferes with part of the cell factory so that it does not go well. These scientists are not sure how this problem started, but like a real factory, backups, and disruptions in one system cause problems in other areas. When damage spreads, cells lose the ability to do work and eventually die, causing changes in the brain that cannot be changed. At present, the Conventional diagnosis of the severity of Alzheimer's is still carrying. The main objective of our research is to develop a Computer-Aided Diagnosis tool in the classification of the severity of Alzheimer's. This paper will discuss optimizing the use of the Receiver Operating Characteristics (ROC) parameter in identifying the severity of Alzheimer's. The main objective of this research is the development of automated Alzheimer's diagnoses in developing countries and rural areas that have limited health facilities and human resources. So even though the method used is relatively simple, it is proven to have a high level of accuracy above 90%. These results indicate that the use of this method has very promising results to be applied in developing countries and rural areas.

1. Introduction

Alzheimer's disease is a disease that attacks brain function, resulting in memory loss and other cognitive functions. This disease develops progressively, which is getting worse over time. Intellectual and social abilities in people living with Alzheimer's will continue to decline due to brain cells in people with Alzheimer's deteriorating and eventually dying. One part of what doctors do in diagnosing Alzheimer's is by scanning the brain. This examination is done to detect abnormalities or changes in the brain. Also, to ascertain the cause of the symptoms that appear. The brain scan method can be done with a CT scan or MRI. Based on the MRI image, the doctor can analyze the severity of a patient's Alzheimer's. However, unfortunately, to get a certain level of analysis, the MRI image that is used as a reference must be a high-resolution image. In reality, for many areas in developing countries such as Indonesia, hospitals may not necessarily have MRI machines, even if they have an MRI machine, then it is usually only MRI with low resolution. Another problem is, in developing countries, the neurologist is usually very limited in number. Also, in the analysis of radiological images, it is possible for human error in the analysis. To overcome these problems, we need a technique that can optimize the image of low-resolution MRI results while simultaneously analyzing the characteristics of determining the severity of Alzheimer's automatically.

Alzheimer's disease is a disease that attacks brain function, resulting in memory loss and other cognitive functions. This disease develops progressively, which is getting worse over time. Intellectual and social abilities in people living with Alzheimer's will continue to decline due to brain cells in people with Alzheimer's deteriorating and eventually dying. One part of what doctors do in diagnosing Alzheimer's is by scanning the brain. This examination is done to detect abnormalities or changes in the brain. Also, to ascertain the cause of the symptoms that appear. The brain scan method can be done with a CT scan or MRI. Based on the MRI image, the doctor can analyze the severity of a patient's Alzheimer's. However, unfortunately, to get a certain level of analysis, the MRI image that is used as a reference must be a high-resolution image. In reality, for many areas in developing countries such as Indonesia, hospitals may not necessarily have MRI machines, even if they have an MRI machine, then it is usually only MRI with low resolution. Another problem is, in developing countries, the neurologist is usually very limited in number. Also, in the analysis of radiological images, it is possible for human error in the analysis. To overcome these problems, we need a technique that can optimize the image of low-resolution MRI results while simultaneously analyzing the characteristics of determining the severity of Alzheimer's automatically.

There is research on the diagnosis of Alzheimer's, which has been done by several people, including the following. Mikhno [1] developed the Voxels of Interest (VOI) template to control the subject treatment of people with AD. The developed VOI template can be used for both automatic and manual treatment. Boenink [2] analyzed diagnostic guidelines that could be used as a way to assess the existence of new technology in diagnosing Alzheimer's. Platero [3]introduced the analysis of longitudinal images, hippocampal segmentation, and classification of longitudinal boundaries in the diagnosis of Alzheimer's. Hajipour [4] conducted a review of nanotechnology for Alzheimer's diagnosis and therapy and discussed his accuracy so far. Heung-Il [5] proposed a multi-task learning method for feature selection in the classification of computer-based Alzheimer's diagnosis. In his research, he uses a multipeak class distribution. In this case, he uses the grouping method to get the characteristics of the multipeak distribution and defines the subclasses based on the results of the grouping. Balamurugan [6] he proposed reducing the KNN Algorithm-based dimension to analyze and clarify the severity of Alzheimer's. Baiying [7] proposed an analysis of discriminatory features and an analysis of canonical correlations in Alzheimer's disease. This method is proposed to improve performance in exploring data from various modalities. Baig [8] discussed peptide compounds for initial diagnosis and treatment of Alzheimer's and their development in a variety of ways regarding applications in Alzheimer's disease. Liu [9] proposed the construction of multilevel convolutional neural networks to analyze the features of various modalities in MRI and PET brain images to classify the severity of Alzheimer's. Zhu [10] proposed a method for converting the original features of various modalities into a comparable general form for canonical correlation analysis in Alzheimer's classification. Our research focuses on the development of Computer Aided Diagnosis in helping to speed up the diagnosis of Alzheimer's based on MRI images that have a low resolution as is the case in many developing countries, although this system is also possible to be applied in developed countries. In our previous papers [11] [12] [13][14] [15] we discussed the segmentation of the hippocampus and ventricular regions which are important areas in diagnosing the severity of Alzheimer's. But in our previous paper, we have not yet discussed the Alzheimer's classification method based on the results of segmentation. Referring to this, then in this paper we discuss a simple method of classifying the severity of Alzheimer's by using a ROC curve based on Clinical Dementia Rating (CDR) values.

2. Research Methods

2.1. Input Images

All data used in this research came from the Open Access Series of Imaging Studies (OASIS) databases [16][17][18][19]. Images from OASIS have been grouped into CDR 0, CDR 1 and CDR 2. In previous research we also pre-processed the image so that the input image is ready to be analyzed, for example by doing image enhancement and segmentation of the hippocampus and ventricular areas using the active contour and watershed segmentation methods [11] [12] [13][14][15]. In this paper, we conducted an experiment by developing a very simple classification method, in classifying the severity of Alzheimer's based on CDR values. Figure 1 shows an example of the input image of the three types of axial, sagittal and coronal slices.

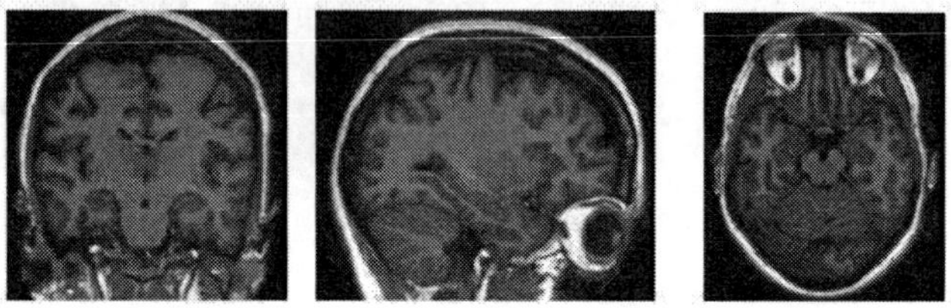

Fig.1. Examples of the input image

2.2. Receiver Operating Characteristics (ROC) for Determination of Pixel Characteristics in Segmentation

The ROC curve was first used by electrical engineers and radar technicians during the Second World War to detect enemy objects on the battlefield, then this is known as signal detection theory. ROC analysis has subsequently evolved and has been used in medicine, radiology, as well as in several other fields for decades. Further ROC analysis has been introduced in relatively new fields such as machine learning and data mining [20]. ROC is a measurement in diagnostic tests, in the medical world, these measurements are used for evaluating medical tests, for example, to compare a new device with standard medical equipment that is standard. A segmentation application must have sufficient accuracy, to meet these requirements, researchers use the ROC measurement method, which is to calculate the ratio of False Positive (FP) and False Negative Ratio (FN) on the segmented image by comparing the results of the experimental image segmentation on the original image. Suppose there is a classification problem with two classes. We can consider each pair of data I to map a set of elements {p, n} as positive class labels and negative class labels. The classification model maps the pair of data to the predicted class. To distinguish the actual class from the predicted class, the predicted class is symbolized by {Y, N}. In terms of identifying sick and healthy images, the mapping will produce four outputs, namely TP, TN, FP, and FN. In contrast to the evaluation parameters for pixel segmentation, TP in identification is image data that is diagnosed by doctors as a sick image, correctly detected pain after going through hypothesis testing, whereas FP is image data with a health diagnosis but is indicated incorrectly as a sick image after testing, TN is image data that is diagnosed by doctors as being healthy, correctly detected healthy. After going through hypothesis testing, and FN is image data with a diagnosis of pain but is indicated incorrectly as a healthy image after testing. The parameters used as performance indicators in this research are those shown by Equation 1 through equation 5.

$$Sensitivity\ (Se) = \frac{TP}{TP+FN} = \frac{TP}{nD} \tag{1}$$

$$Specificity\ (Sp) = \frac{TN}{FP+TN} = \frac{TN}{nC} \tag{2}$$

$$Accuracy = \frac{TP+TN}{TP+TN+FP+fN} \quad (3)$$

$$PPV = \frac{TP}{(TP+FP)} = \frac{TP}{nP} \quad (4)$$

$$NPV = \frac{TN}{TN+FN} = \frac{TN}{nN} \quad (5)$$

Positive Predictive Value (PPV) is the possibility that people with positive test results (images declared sick) will have the conditions tested. NPV is the possibility that people with negative test results (otherwise healthy images) do not have that condition. The ROC curve shows the tradeoff between the level at which a model can recognize positive data accurately and the level at which the model incorrectly recognizes negative data as positive data.

3. Result and Discussion

As explained in the above paragraph, that in our previous research, we have succeeded in segmenting the three sagittal, axial, and coronal slices in localizing the ventricular and hippocampal areas. In this paper, the emphasis is on identifying the severity of Alzheimer's. Because the ultimate goal of this research is the development of simple technology, it can be used as an accelerator for diagnosing the severity of Alzheimer's especially in developing countries and rural areas to pursue industrial technology 4.0 especially in health, so we use identification based on the ROC curve only. In the sagittal slice, three areas are classified as the left hippocampus, right hippocampus, and ventricle. Table 1, Table 2, and Table 3 show the average values of the right hippocampus, left hippocampus, and ventricle areas according to sagittal slice.

Table 1. The average value of the right hippocampus left hippocampus and ventricle area based on the sagittal slice.

	Right Hippocampus		**Left Hippocampus**		**Ventricle**	
	Alzheimer	Healthy	Alzheimer	Healthy	Alzheimer	Healthy
TP	108.5	100.7	85.8	50.1	1259.6	946.9
TN	2981.1	2900.2	2960.7	2848.7	7846.7	8112.8
FP	92.8	173.5	113.3	225.2	315.2	49.10
FN	103.5	108.8	126.1	154.1	154.3	469.7
Se	51.1	48.5	40.5	27.0	89.0	66.8
Sp	96.9	94.3	96.3	92.6	96.1	99.3
Ac	94.0	91.4	92.7	88.4	95.0	94.5
RFP	0.0302	0.0564	0.0368	0.0732	0.038	0.006
RFN	0.4882	0.5141	0.5949	0.7297	0.109	0.331

According to Table 1, the FP values in the Alzheimer's image have an average value less than the FP values in the healthy image. This shows that the object with Alzheimer's has a smaller number of pixels than the reference object. However, this is the opposite of the results obtained in the measurement of the ventricle area as shown in Table 3. Referring to our previous research [12][13][14][15], for the hippocampus area we use pixels for reference image is 212, while for the ventricle area is 1557. Using Equation 1 to 5, we obtained the accurate identification of the severity of

IOP Conf. Series: Materials Science and Engineering **982** (2020) 012007 doi:10.1088/1757-899X/982/1/012007

Alzheimer's by 96% for the right hippocampus, 98% for the left hippocampus, and 91% for ventricle areas based on sagittal slices.
In the axial slice, the only area that can be analyzed is the hippocampus area. Table 4 shows the average value of the results of the segmentation of the hippocampus area based on axial slices. Referring to Table 2, RFP values increase and RFN decreases along with the sequence based on the smallest pixel area until the largest pixel in the healthy test image data. RFP is the ratio between pixels that should not be vessels in the reference image but segmented as vessel pixels. In images with Alzheimer's, the RFN shows a greater value than the RFP, and the FN ratio graph increases as the number of pixels decrease from the test image. This shows that objects with Alzheimer's diagnosis have fewer pixel counts than reference objects so that the RFP is small. Conversely, objects with healthy diagnoses have more pixels than reference objects so that the RFP is high. Identification results through the ROC parameters stated in Equations 1 to 5 give an accuracy value of 98%.

Table 2. The average value of the hippocampus area based on an axial slice

	Hippocampus	
	Alzheimer	Healthy
TP	303.25	435.15
TN	1968.083	1851.5
FP	56.666	211.25
FN	272.25	30.5
Se	53.951	93.277
Sp	97.087	89.737
Ac	87.206	90.333
RFP	0.029	0.102
RFN	0.460	0.067

In the coronal slice, the area analyzed is the ventricle and the hippocampus but does not distinguish between the right and left hippocampus. Table 3 shows the average value of ventricle and hippocampus areas based on coronal slices.

Table 3. Average value of ventricle and hippocampus area based on coronal slice

	Ventricle		**Hippocampus**	
	Alzheimer	Healthy	Alzheimer	Healthy
TP	439.2	274.4	80.7	274.4
TN	30327.06	30458.8	614.9	30458.8
FP	175.9	44.1	30.0	44.3
FN	33.8	198.5	57.2	198.5
Se	92.03	58.02	54.01	89.47
Sp	99.54	99.85	95.07	84.49
RFP	0.005	0.001	0.04	0.15
RFN	0.001	0.006	0.07	0.01

According to Table 3, RFP and RFN values represent the pixel characteristics of the segmented results with the reference image. The ratio value of 0 to 1, if the image has a lot of segmentation areas following the reference image, the ratio will be of high value. From the test results, it can be concluded that Alzheimer's image for ventricular objects has a higher number of pixels than the healthy image with an average value of RFP 0.005 and RFN 0.001. According to Table 3, In images with Alzheimer's, the RFN shows a greater value than the RFP. This shows that objects with Alzheimer's have fewer pixel counts than reference objects so that the RFN is large. Vice versa, the object with a health diagnosis has more pixels than the reference object so that the RFN is of low value. Based on the test results, it can be concluded that Alzheimer's image for hippocampus objects has a lower pixel count than in healthy images with RFP values of 0.04 and RFN of 0.07. Identification results through the ROC parameters provide an accuracy value of 86.3% for the ventricle area and 98.4% for the hippocampus area.

4. Conclusion

The experimental results show that in almost all types of slices, implementation using ROC parameters shows a performance value of more than 90%. So if the system that we develop can be applied to health service units in developing countries and rural areas, it will support the implementation of the industrial revolution 4.0, especially in the field of health services

5. Acknowledgment

It is supported by the Ministry of Education and Culture through "Fundamental Research" Scheme fiscal year 2020. We also thank the Open Access Series of Imaging Studies (OASIS) datasets for all MRI data used in this research.

.

6. Reference

[1] Mikhno A, Devanand D, Pelton G, Cuasay K, Gunn R, and Al E 2008 Voxel-Based Analysis of sup 11 C-PIB Scans for Diagnosing Alzheimer ' s Disease *J Nucl Med.* **49** (8).

[2] Boenink M 2018 Gatekeeping and trailblazing: The role of biomarkers in novel guidelines for diagnosing Alzheimer's disease *Biosocieties* **13** (1), 213–231.

[3] Platero C, Lin L, and Tobar M C 2019 Longitudinal Neuroimaging Hippocampal Markers for Diagnosing Alzheimer's Disease *Neuroinformatics.* **17**(1), 43–61.

[4] Hajipour M J, Santoso M R, Rezaee F, Aghaverdi H, Mahmoudi M, and Perry G 2017 Advances in Alzheimer's Diagnosis and Therapy: The Implications of Nanotechnology *Trends Biotechnol.* **35** (10) 9.

[5] Suk H I and Shen D 2014 Subclass-based multi-task learning for Alzheimer's disease diagnosis *Front. Aging Neurosci.* **6** no. JUN 1–20.

[6] Balamurugan M, Nancy A, and Vijaykumar S 2017 Alzheimer's disease diagnosis by using dimensionality reduction based on KNN Classifier, *Biomed Pharmacol J* 10 (4) 1823–1830.

[7] Lei B, Chen S, Ni D, and Wang T 2016 Discriminative learning for Alzheimer's disease diagnosis via canonical correlation analysis and multimodal fusion Front. Aging Neurosci. **8** no. MAY, 1–18.

[8] Baig M H, Ahmad K, Rabbani G, and Choi I 2018 Use of peptides for the management of Alzheimer's disease: Diagnosis and inhibition *Front Aging Neurosci.* 10, no. FEB, 1–7, 2018.

[9] Liu M, Cheng D, Wang K, and Wang Y, 2018 Multi-Modality Cascaded Convolutional Neural Networks for Alzheimer's Disease Diagnosis *Neuroinformatics*, **16** (3–4).

[10] Zhu X, Suk H I, Lee S W, and Shen D 2016 Canonical feature selection for joint regression and multi-class identification in Alzheimer's disease diagnosis *Brain Imaging Behav.***10**, (3) 818–828.

[11] Supriyanti R, Subhi A R, Ramadhani Y, and Widodo H B 2019 Coronal slices segmentation of mri images using active contour method on initial identification of alzheimer severity level based on clinical dementia rating (CDR) *J. Eng. Sci. Technol.* **14** (3) 1672–1686.

[12] Supriyanti R, Fanthoni A, Ashari E J,Ramadhani Y, and Widodo H B 2019 Active Contour Method for Identifying Area of Sagittal and Axial Slices to Identify the Severity of Alzheimer's disease, *International Conference on Recent Advances in Engineering and*

Technology (ICRAET).

[13] Supriyanti R, Subhi A R, Ramadhani Y, and Widodo H B, 2019 CORONAL SLICES SEGMENTATION OF MRI IMAGES USING ACTIVE CONTOUR METHOD ON INITIAL IDENTIFICATION OF ALZHEIMER SEVERITY LEVEL BASED ON CLINICAL DEMENTIA RATING (CDR) J. Eng. Sci. Technol. **14** (3)1672–1686.

[14] Supriyanti R, Subhi A R, Ramadhani Y, and Widodo H B 2018 A Simple Tool For Identifying The Severity of Alzheimer's Based on Hippocampal and Ventricular Size Using A ROC Curve on a Coronal Slice Image *Asian J. Inf. Technol.* **17** (7) 321–327,.

[15] Supriyanti R, Subhi A R, Ramadhani Y, and Widodo H B 2018 Calculating Ventricle's Area based on Clinical Dementia Rating (CDR) Values on Coronal MRI image *20th International Conference on Computer-Based Patient Records and Error Prevention (ICCPREP)* 2018.

[16] Buckner R 2004 A unified approach for morphometric and functional data analysis in young, old, and demented adults using automated atlas-based head size normalization: reliability and validation against manual measurement of total intracranial volume. *Neuroimage.* **23** 724–738.

[17] Rubin E 1998 A prospective study of cognitive function and onset of dementia in cognitively healthy elders *Arch Neurol.* **55** 395–401.

[18] Zhang Y, Brady M, and Smith S 2001 Segmentation of brain MR images through a hidden Markov random field model and the expectation maximization algorithm *IEEE Trans. Med. Imaging* **20** (1) 45–57.

[19] Fotenos A, Snyder A, Girton L, Morris J, and Buckner R 2005 Normative estimates of cross-sectional and longitudinal brain volume decline in aging and AD *Neurology.* **64** 1032–1039.

[20] Fawcett T 2006 An introduction to ROC analysis *Pattern Recognit* **27** 861–874.

IOP Conf. Series: Materials Science and Engineering **982** (2020) 012008 doi:10.1088/1757-899X/982/1/012008

Artificial Neural Networks For Pattern Recognize Handwritten

Arief Kelik Nugroho[1,2], Ipung Permadi[1], Aini Hanifa[1], Dian Tri Wiyanti[2]

[1]Departement of Informatics, Engineering Faculty, Universitas Jenderal Soedirman, Indonesia,

[2]Departement of Mathematics, Faculty of Mathematics and Natural Science, Universitas Negeri Semarang

[1]arief.nugroho@unsoed.ac.id

Abstract. Backpropagation is a supervised learning method on artificial neural networks. The process of training artificial neural networks that are built will be tested several structures of artificial neural networks, with the hope that the best structure is obtained to produce the output with the smallest error. The evaluation purposes use the same data set to form a neural model with a hidden level and the neuron model with two hidden levels then compare these three models both in terms of classification accuracy and training time. A neural network can be concluded that for each architecture, the neural network architecture used to recognize type 3 is 82,42%, type 2 77,5%,type 1 98,1%.

1. Introduction

Backpropagation is a supervised learning method on artificial neural networks. Usually used in the multilayer perceptron (MLP) to change the weights associated with neurons that exist in the hidden layer[1,2]. This algorithm tries to make a gradient graph that shows the error value during the iteration at the time of training. Then the backpropagation algorithm can be accelerated significantly after finding the gradient component, and the weight is modified along the gradient direction until the minimum value. This process can be carried out without the need for intensive gradient computation calculations at each step. The new gradient components are calculated shortly after a minimum value in the direction of the previous gradient is reached. This process is only possible for cumulative weighting. A method for finding the minimum value along the gradient direction is a three-step process of finding errors for three points along the gradient direction and then using a parabolic approach, jumping directly to the minimum[1,2].

The process of training artificial neural networks that are built will be tested several structures of artificial neural networks, with the hope that the best structure is obtained in order to produce the output with the smallest error. Some parameters considered in designing neural networks in this study include are weight (w), bias (b), learning rate (α), MSE (e), momentum, number of layers, number of nodes, and speed of process[7]. Other algorithms or methods in pattern recognition are using artificial neural networks. This pattern may originate from the results of the extraction process from the pattern to be recognized. For example, the pattern of an image will be recognized, then the process of extracting the features of the image will be carried out. The extraction results then become input for artificial neural networks to be recognized, one of which is through the classification process of the pattern of input provided. Pattern recognition by artificial neural networks is done by first training it with certain patterns[3].

2. Method

The goal of this study is to create software that recognizes handwritten characters. To achieve the goal, we have developed a classification module that uses a Neural network model. Form the neural network models with a series of images of data of handwritten characters. For evaluation purposes, we also use the same data set to form a neural model with a hidden level and the neuron model with two hidden levels. We compare these three models, both in terms of classification accuracy and training time. This section describes the research methods used in this document. The methodology consists of gathering data for characters written by hand, constructing the neural network model, modeling and testing models, and classifying handwritten characters. Each part of the methodology will be described in the following subsections

2.1. Thinning Algorithm

To obtain an image skeleton by reducing the object pixels in the image to produce the simplest form of the object contained in the image[9]. Although the thinning method is similar to the erosion or opening method, the difference between the morphological operations methods is the result of the operation. The difference is that the thinning method maintains the continuity of the object's pixels in the image, thus maintaining the shape and structure of the object. The thinning method used to process the images obtained is the Zhang-Suen thinning algorithm. The first check removes the pixels from the lower right corner (southeast) of the object[9]. Meanwhile, the second examination removes the pixels in the upper left corner (northwestward) of the object. The processes that have been carried out in this procedure will produce images that are ready to be converted into data sets used for procedures related to artificial neural networks (procedures for forming and training neural networks and testing procedures for artificial neural networks). However, before it is used to form, train, and test artificial neural networks, the resulting image matrix must be converted to a row matrix.

2.2. Neural Network

An artificial neural network is one of the branches of discussion in the field of Artificial Intelligence studies (Artificial Intelligence). An artificial neural network is an information processing system that has characteristics such as biological neural networks and is formed as a generalization of mathematical models of artificial neural networks[10]. An artificial neural network is built consisting of an input layer, a hidden layer, and also the output layer.

Artificial neural network model Back Propagation is one of the models of artificial neural networks that are included in supervised learning. Like other artificial neural network models, Back Propagation trains networks to strike a balance between the ability of the network to recognize patterns used during training and the ability of networks to provide correct responses to input patterns similar to those used during the training process[10].

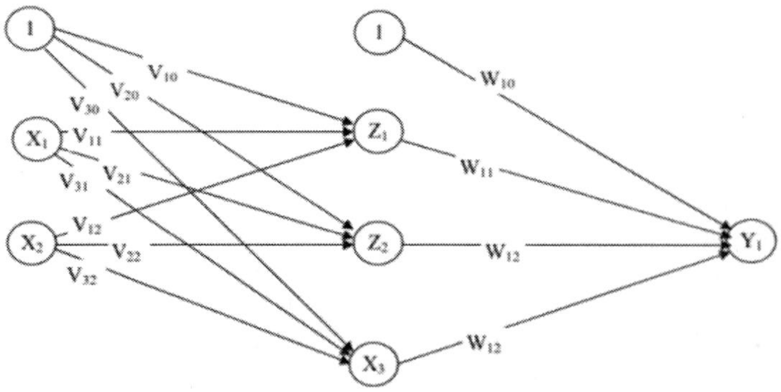

Figure 1. Model Back Propagation

3. Results and Discussion

The images were collected in the form of two kinds, namely images from the results of computer writing and images from the results of handwriting. 20 images were collected for each character, with 5 pieces of the computer writing and 15 pieces of handwriting. The scripts produced from word processing are arranged so that the five characters formed have different thicknesses. The thickness of

IOP Conf. Series: Materials Science and Engineering **982** (2020) 012008 doi:10.1088/1757-899X/982/1/012008

the letters formed is divided into normal (thickest), brighter 5%, brighter 15%, brighter 25%, and brighter 35% (thinnest).

3.1. Data Acquisition

Each instance of the dataset is presented to the classification algorithm using a representation system. The quality of the presentation system is a key factor in most classification algorithms. The most common representation system is characteristic [6]. In this system, each case is represented by a vector form value. The dataset (containing both the exercise set and the test set) is usually represented in a tabular form, with each sequence being an example and each column representing different problem areas. The quality of the action space is defined in the problem area, which usually includes the amount of action and their importance for the task in question, which has a significant impact on the performance of the classification algorithm.

The main database of handwritten characters was manually collected by some, as shown in Figure 1. Each character class consists of 100 training data. Each data set in the dataset is an 8-bit grayscale image that is 20 x 20 pixels in size.

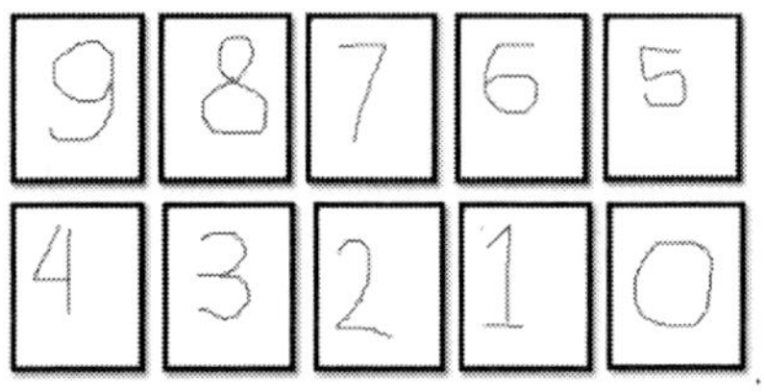

Figure 2. Type 1

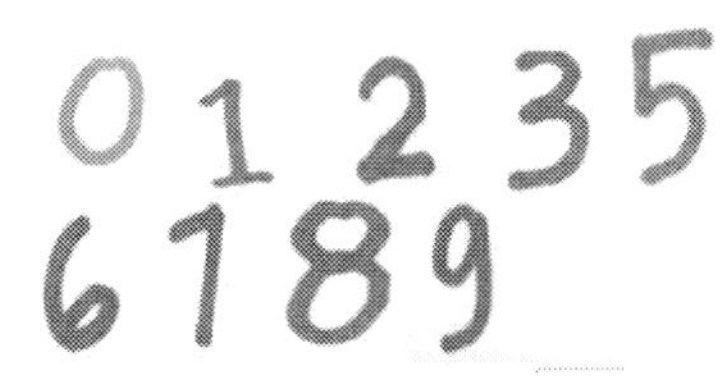

Figure 3. Type 2

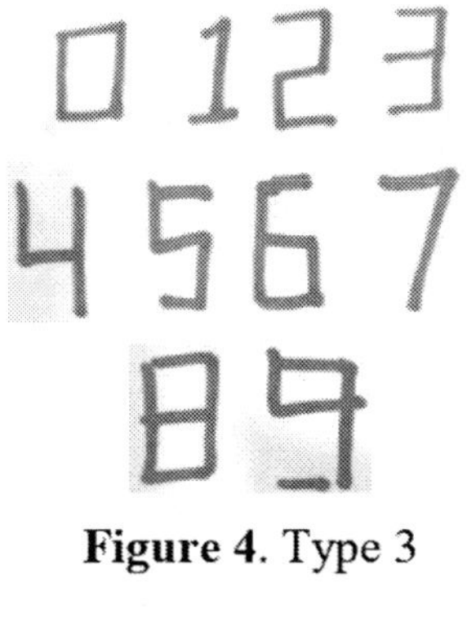

Figure 4. Type 3

3.2. Preprocessing Data

The scripted image data that has been collected will go through sequential processes so that the data obtained will be used in the next procedure. These processes aim to make image data that will be used in neural network training to be clean from noise or interference so that the results obtained are accurate. This normalization must be carried out to uniform the size of the training image. When this process is not carried out, subsequent processes will be hampered due to the unequal size of the characters used.

The next process is extracting the shape characteristics of the image. This process is intended to take the characteristics possessed by the image that will be used as training data in artificial neural networks. This characteristic is used as a differentiator between images. The characteristic shape that can be extracted from the image is by using the skeleton method. Skeleton is a unique form of an object that resembles the framework of an object and is used for handwriting recognition and recognition, fingerprint patterns, biological cell structures, and other such things. This skeleton has three characteristics, its thickness is 1 mm; passes through the center of the object; and states the object's topology.

3.3. Formation and Learning of Artificial Neural Networks

This procedure is carried out to form and train artificial neural networks. This procedure will produce an artificial neural network that can recognize the types. This procedure begins by compiling data sets (datasets) that will be used, determining the stages that need to be done in learning in artificial neural networks, determining the value of the parameters needed to build an artificial neural network, and conducting neural network learning. imitation. The learning data used is divided into two parts. The two parts are training data and testing data. Training data is used in the training stage of artificial neural networks. Meanwhile, the test data is the data used in the testing stage of artificial neural networks. Of the 20 literacy image data that have been collected and preprocessed, 80% of the illiterate sample data are used as training data. Meanwhile, the remaining 20% of sample image data is used as test data.

Much of the training data and testing data used in each artificial neural network architecture are listed in table 1

Table 1. Amount of training and testing data

Type	Data(Learning)	Data (Training)	Data (Testing)
Type 1	920	736	184
Type 2	400	320	80
Type 3	100	80	20

3.4. Learning ANN

The first step taken in the procedure this time is to compile data sets (datasets). The data sets that will be used in the learning process of artificial neural networks are taken from the pixel values of images that have been processed in such a way as to produce the shape. Then, the image matrices will be converted into a row matrix (matrix with dimensions of 1 x dimensional image of the beginning of [the number of rows multiplied by the number of columns]).

The parameters used are the number of layers/layer on each architecture, the number of neurons in each layer/layers neural network, the pace of understanding/learning rate, momentum, value fault tolerance (in this case is the value of the mean square error) learning, and coefficients maximum iteration. These parameters determine the results of the artificial neural network training process.

3.5. Number of Layers

The number of layers/layers used in artificial neural network architecture is 3 layers, namely 1 input layer, 1 hidden layer, and 1 output layer. The number of neurons for each layer/layer of artificial neural networks is determined as follows. Neurons in the lining of the input/input layer amounted to 400 neurons. This number adjusts the size of the input image 20 x 20 pixels or in other words 400 pixels. Neurons in the hidden layer / hidden layers totaling 50 pieces of neurons. Neurons in the output layer vary in number. The number of neurons in this layer is determined based on the type of character. The number of neurons in the output layer of artificial neural networks to recognize the type of goujon script is the number of neurons. Meanwhile, the number of neurons in the output layer of artificial neural networks to recognize the type of dotaon is 20 neurons and the number of neurons in the layer of an artificial neural network to recognize the type 1 type of alphabet is 5 neurons. The value of the coefficient of the learning rate, momentum, errors tolerance, and maximum iteration is shown in the following table 2.

Table 2. The Coefficient Value

(learning rate)	0.005; 0.003; 0.001
Momentum	0.1; 0.05; 0.01
Error tolerance Learning	1×10^{-20}
Maks iteration	10000

The training phase is carried out using predetermined learning parameters. Training of the training data sets results in the data listed in table 3 and table 4.

Table 3. The Second Stage Of Learning Parameters

Type	A	B	*Hidden layer*	Iteration	Average (Training)	Avg (Testing)
Type 1	0.005	0.1	50	10000	0,8673	0,7656
	0.003	0.05			0,9014	0,7703
	0.001	0.01			0,8322	0,7311
Type 2	0.005	0.1			0,7063	0,6733
	0.003	0.05			0,6368	0,5651
	0.001	0.01			0,6918	0,6458
Type 3	0.005	0.1			0,7060	0,6882
	0.003	0.05			0,8360	0,8429
	0.001	0.01			0,7658	0,7586

Table 4. Learning parameters for the second stage

Architecture	Hidden Layer	A	B	Error tolerance	Max iteration
Type 1		0.003	0.05		
Type 2	50	0.005	0.1	1 x 10-20	25000
Type 3		0.003	0.05		

The second learning parameters have been set. The second stage of learning about neural network architecture is ready to do. After the second stage of learning, the learning score is shown in table 5.

Table 5. The second stage of learning neural network architecture

Architecture	AVG (training Score)	AVG (testing score)
Type 1	0,8242	0,8252
Type 2	0,6520	0,5505
Type 3	0,9080	0,7420

This training score does not indicate whether artificial neural networks can identify characters completely or not. The results of these calculations are shown in table 6. These calculations are performed on each learning data used, both training data and testing data.

Table 6. Percentage of characters recognized by each architecture in the second experiment

Architecture	Data(training) %	Data (testing)%
Type 3	82.42%	8,152%
Type 2	77,5%	46,25%
Type 1	98.1%	82,5%

4. Conclusion

Learning that has been done shows the scores generated by the parameters used in the learning phase. Based on the learning results that have been shown in table 5 and 6, it can be concluded that for each architecture the neural network architecture used to recognize Type 3 picture characters has the most optimal performance score on the second learning parameter (learning rate $\alpha = 0.003$ and momentum $\beta = 0.05$). The artificial neural network architecture used to recognize type 1 characters have the most optimal performance score on the first learning parameter (learning rate $\alpha = 0.005$ and momentum $\beta = 0.1$). The artificial neural network architecture to recognize the type 1 has the most optimal performance score on the second learning parameter (learning rate $\alpha = 0.003$ and momentum $\beta = 0.05$).

References

[1]. Wilamowski, B.M.. *Comparison of Training Algorithms and Network Architectures*. 2013, p.4799.
[2]. Wilamowski, B.M. *Neural Network Architectures and Learning*. International Conference on Industrial Technology, 2003.pp.1–12
[3]. Rojas, R., 1996. *Neural Networks A Systematic Introduction*, Berlin: SpringerVerlag.
[4]. Cun LY.,et al. Handwritten digit recognition with a back-propagation network. In *Advances in neural information processing systems*, 1990.
[5]. Y. LeCun, F.J. Huang, and L. Bottou. *Learning methods for generic object recognition with invariance to pose and lighting In Computer Vision and Pattern Recognition. CVPR 2004. Proceedings of the 2004 IEEE Computer Society Conference on.*,2004.volume 2, pages II–97. IEEE.

[6]. Nugroho,A.K.,Permadi,I.*Composite Image with a Geographic Information System Approach*, IOP Conference Series: Earth and Environmental Science, 2020.Volume 406
[7]. Dash, T., Ranjan, S., Nayak, T. & Mishra, G., 2014. *Neural Network Approach to Control Wall Following Robot Navigation*. In IEEE International Conference on Advanced Communication Control and Computing Technologies (ICACCCT). pp. 1072–1076
[8]. Rajashekararadhya, S. V, and Ranjan, P. V. *Efficient Zone Based Feature Extraction Algorithm For Handwritten Nuemarl Recognition of Four Popular South Indian Scripts*. Journal of Theoretical and Applied Information Technology, 2008.Volume IV, pp. 1171-1181
[9]. Fisher, R., Perkins, S., Walker, A. & Wolfart, E., 2003. Morphology – Thinning
[10]. Siang, JJ., *Jaringan Syaraf Tiruan dan Pemrogramannya menggunakan MATLAB*. 2009.Yogyakarta: Penerbit Andi

Study of Linear Discriminant Analysis to Identify Baby Cry Based on DWT and MFCC

Ledya Novamizanti[1,2], Anggunmeka Luhur Prasasti[1], Bangun Satria Utama[1]
[1]School of Electrical Engineering, Telkom University
Jl. Telekomunikasi Terusan Buah Batu Bandung 40257 Indonesia
[2]ledyaldn@telkomuniversity.ac.id

Abstract. Baby crying is a common behavior among babies and is a means of verbal communication for babies who express their needs and desires. A baby's cry identification system is needed because it makes it easy for adults to find out the meaning of a baby's cry. This study proposes a system for classifying infant crying sounds using Linear Discriminant Analysis (LDA) with Discrete Wavelet Transform (DWT) and Mel-frequency Cepstral Coefficient (MFCC) as a feature extraction method. Based on experiments, the system can identify the sound of crying babies grouped into 5 (five) classes, namely discomfort, hunger, colds, burp, and drowsiness. The system achieves an accuracy of 94% and an average computing time of 1.5506 seconds. The performance of Linear Discriminant Analysis (LDA) outperformed Principal Component Analysis (LDA) in the identification of crying babies

1. Introduction

In newborns, communication is done through crying, movement or certain cues. Infants over the age of 9 months will develop the ability to mimic the sounds of adults [1]. After the baby is born, the baby cry has a function of responses such as changes in temperature, pain, hunger, or discomfort [2]. The research carried out by Dunstan Baby Language (DBL) explains that the sound of a baby cry has meanings. For example, "HEH" means feeling uncomfortable, usually occurs when a baby wants her diaper or clothes to be replaced because she feels hot and dirty. Other examples, there are "NEH" that means hungry, "EAIRH" that means the baby is sick and depressed in the abdomen or caught cold, "EH" that means to belch, and "OWH" that means tired and sleepy [3].

Related research conducted by [1], identified baby cry using MFCC and the Linear Predictive Coding classification. The number of baby cry data is 25 sound samples through which there are 5 classes of data identified such as uncomfortableness, hunger, cold, belching and drowsiness. It reported achieving an accuracy of 76%. Welly et al. [4] using MFCC and K-Nearest Neighbor classification. The number of their baby cry data is 139 samples through which there are 4 classes of data identified such as 'Eairh', 'Heh', 'Neh', 'Owh'. They reported yielding an accuracy of 96.57%. Another study is also carried out by Bhagatpatil and Sardar [5] using LFCC and VQ codebook classification. The number of baby crying data is 120 samples. The system is divided into 5 classes: discomfort, hunger, cold, belching and drowsiness, and reach an accuracy of 92.45%. Bhagatpatil also tested MFCC performance and the code book method. Their system yielded an accuracy of 83.75%. Meanwhile, [6] used MFCC and Reservoir Network (RN). The number of their baby cry data is 132 samples through which there are 5 classes of data identified such as uncomfortableness, hunger, cold, belching and drowsiness, with an accuracy of 94.5%. Another study was reported by [7]. They use DWT and extreme as their method.

The MFCC method is a feature extraction method using a mel calculation that can equate with the human voice. Hence, MFCC will easily recognize the sound of crying babies by extracting traits through the mel scale [6-8]. Meanwhile, DWT is a feature extraction method of decomposing a signal that separates between a low signal and a high signal. This is very useful and makes it efficient when a signal can be separated according to its frequency and can distinguish the desired frequency signal [9-10]. The LDA is a classification method to discriminate data according to linear analysis. LDA maximizes the covariance matrix between classes and minimizes the covariance matrix with classes. More class members spread and ultimately, can increase the success of the introduction [11].

2. Methodology

2.1. Discrete Wavelet Transform (DWT)

DWT is a signal decomposition at a frequency, and a sub-band wavelet component is produced by decreasing the decomposition level. Implementation of DWT can be done by passing the signal into two DWT filters, i.e. Low Pass Filter (LPF) and High Pass Filter (HPF) [10]. DWT serves as pre-processing aim for decomposing the sound signals so that the signals can conform to required specifications. The general formula of LDA is represented by the following equation [12-14],

$$x[n] = \sum_{k=-\sim}^{\sim} \left(y_{high}[k].g[-n+2k]\right) + \left(y_{low}[k].h[-n+2k]\right) \quad (1)$$

2.2. MFCC

MFCC is a signal feature extraction method with a Mel frequency scale which is known to approach the human hearing response of 20-20,000 Hz [6]. MFCC adheres to the workings of the human ear where the human ear is a linear filter at low frequencies and works logarithmically at high frequencies. Before entering the MFCC process, a pre-processing process will be carried out aiming for incoming sound signals by the required specifications. The pre-processing process is DC removal and pre-emphasize filtering [15-16].

2.3. Linear Discriminant Analysis (LDA)

LDA is a method with a combination of arithmetic operations and statistics that perform separate statistical tests for each object. The general formula of LDA is represented as follows [17],

$$SW = \frac{1}{S}\sum_{S-1}^{S}\frac{1}{T_s}\sum_{T-1}^{T_s}(X_s + \emptyset_s)(X_1 + \emptyset_s)^T \quad (2)$$

The *SW* covariance matrix is minimized, and SB is maximized, then eigenvalue and equation are searched. The formula is defined as follows.

$$SB = SW.V \quad (3)$$

$$cov = SB.SW \quad (4)$$

3. Proposed Method

This system is divided into three stages of the process, namely the acquisition of baby crying sound, feature extraction with DWT and MFCC, and classification with LDA. The system identifies the sound of crying babies into 5 classes, namely discomfort, hunger, colds, belching, and drowsiness. Figure 1 shows the block diagram of system design—total records 200 sound of crying babies, consisting of 150 training data and test data 50 [12]. Figure 2 is an example of a baby crying signal with five classes.

IOP Conf. Series: Materials Science and Engineering **982** (2020) 012009 doi:10.1088/1757-899X/982/1/012009

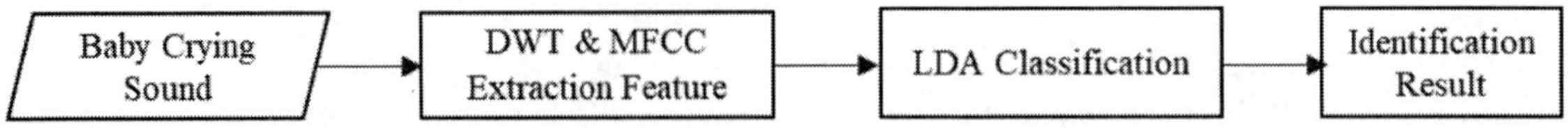

Figure 1. Block Diagram of System Design

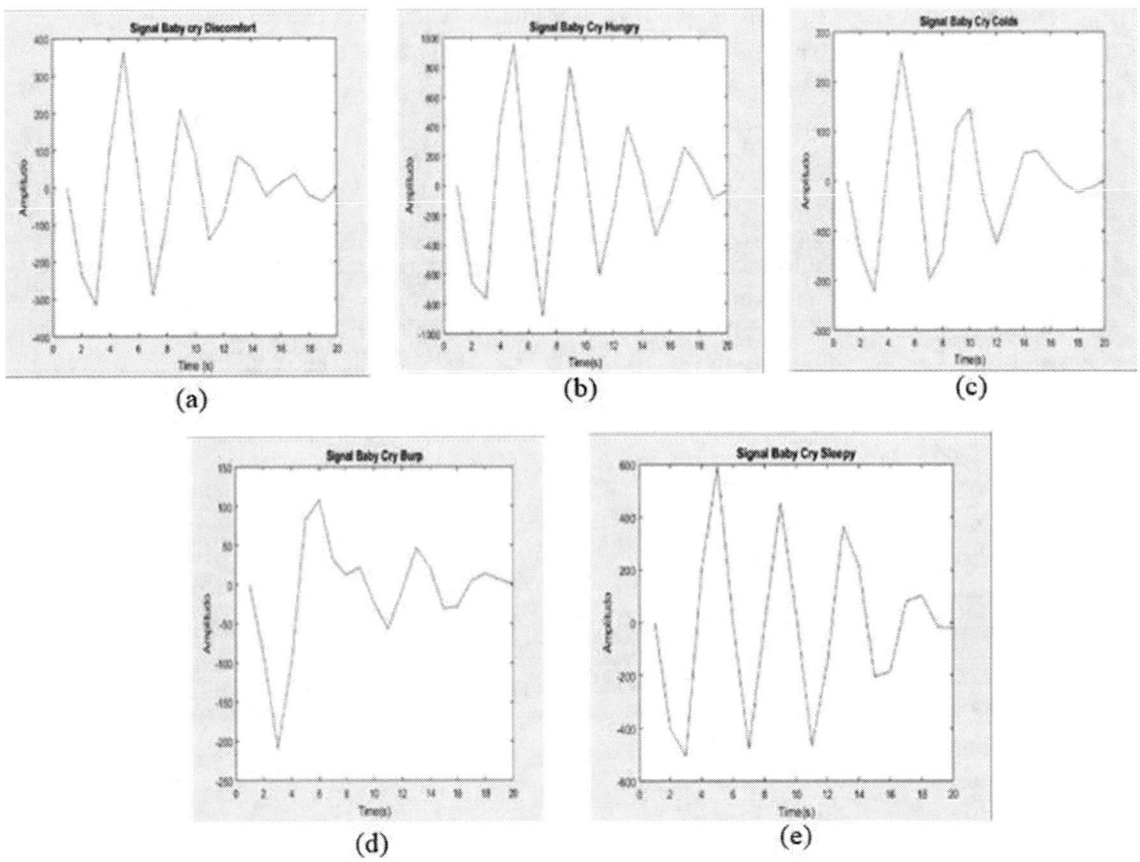

Figure 2. Signal of baby cry (a) Discomfort, (b) Hungry, (c) Colds, (d) Burp, (e) Sleepy

3.1. Pre Processing

Pre-processing is conducted so that the incoming sound matches the required specifications. DC removal is done by reducing the value of each sampled sound with the average value of the data. The function of pre-emphasis filtering is to balance or stabilize the signal spectrum. Figure 3 shows the block diagram of Pre-Processing.

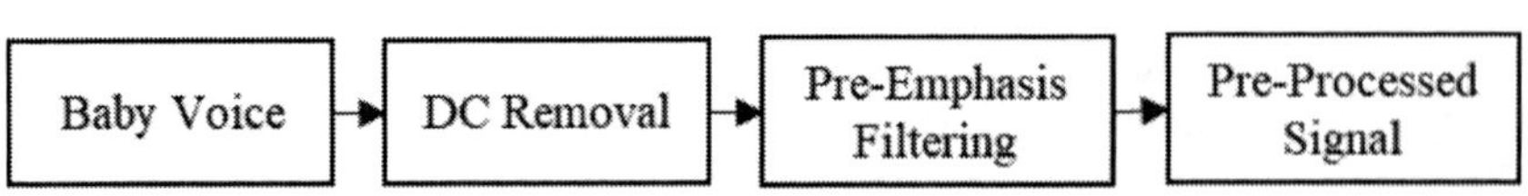

Figure 3. Block Diagram of Pre-Processing

3.2. Feature Extraction

The filtering process is done until it finds the desired level. Feature extraction with Mel-frequency Cepstral Coefficient (MFCC) is performed to find out the special traits that are in the sound signal taken.

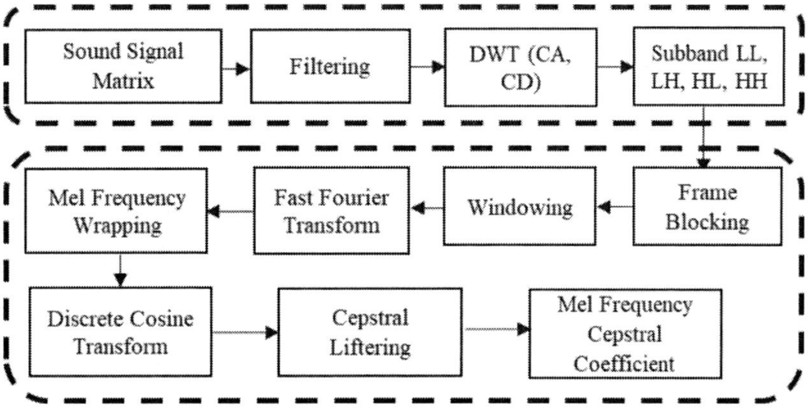

Figure 4. Block Diagram of DWT+MFCC Feature Extraction

3.3. Classification

The classification process aims to classify data based on class. Feature vectors that have been obtained through the feature extraction process are classified using LDA according to their baby's cry identification class. The type of class, which is uncomfortable, hungry, catching cold, burping and sleepy. Figure 5 displays the flowchart of classification using LDA.

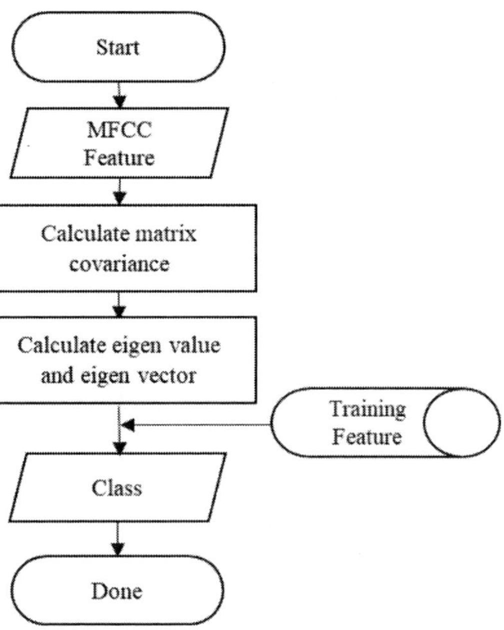

Figure 5. Flowchart of LDA Classification

4. Result and Discussion

4.1. MFCC coefficients and DWT level decomposition

This scenario evaluates the effect of changing the number of MFCC coefficients and the DWT level decomposition on accuracy. The accuracy parameter is calculated based on the amount of data classified correctly divided by the total data. Figure 6 displays the effect of the number of coefficients and the level of DWT decomposition on accuracy. The system shows 100% accuracy when the MFCC coefficient value is 20, and the DWT Level 1. The more MFCC coefficient numbers are tested, the more features are produced for testing by the system. The MFCC coefficient is greater the degree of impact of decomposition on system accuracy. This occurs because an indication that the level of additional decomposition causes information to appear more detailed. More information can contribute to better model performance, but we suspect it can reduce system stability. In the end, we decided that the baby cry detection system used level 1 DWT.

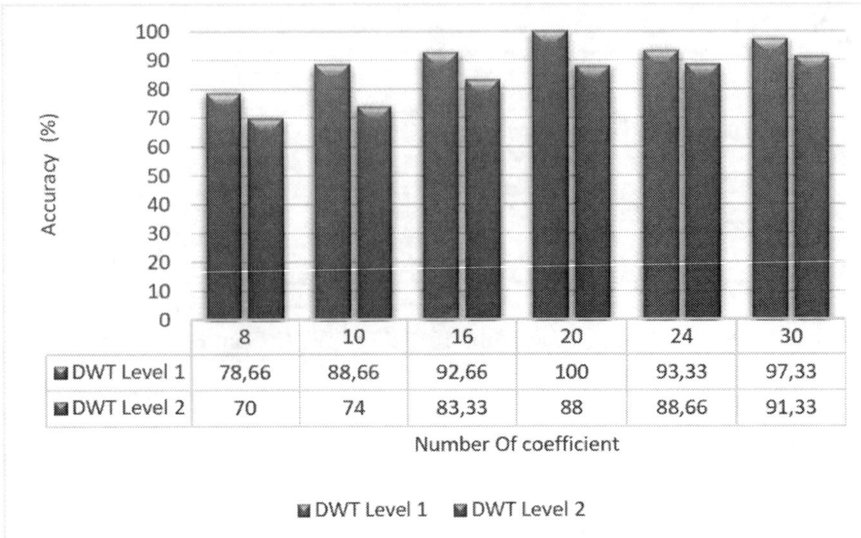

Figure 6. The effect of MFCC coefficient and DWT level on accuracy

4.2. Frame Numbers

This scenario observes the effect of the number of frames on system accuracy. Figure 7 displays the effect of the number of frames on system accuracy.

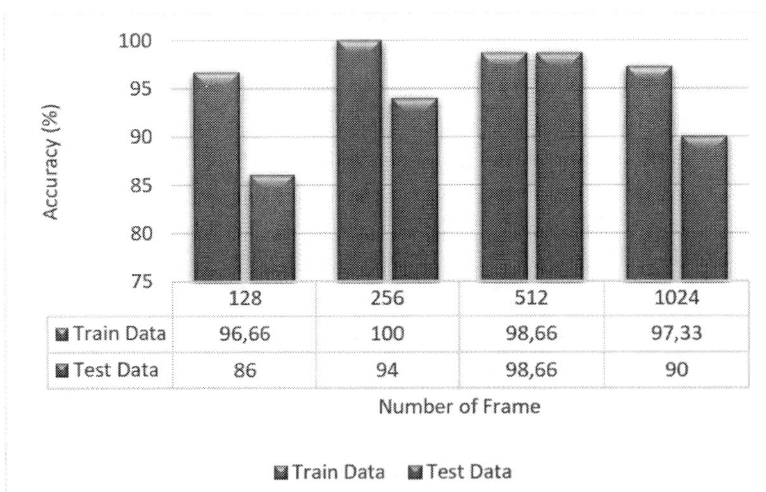

Figure 7. Effect of frame number on accuracy

The best results are achieved when the number of frames is 256, with 100% system accuracy. Signal cuts per frame into smaller parts, making the signal diameter more stable. It was observed that the number of frames tested resulted in fluctuating system accuracy. The results of our system show the number of frames 256 to be the best barometer.

5. Conclusion

In this study, the proposed baby cry classification system uses the DWT, MFCC and LDA methods. The baby's crying is separated into 5 classes, namely discomfort, hungry, catching a cold, burp and drowsiness. This system has an accuracy of 94% and a computing time of around 1.5506 seconds. The best parameters obtained during training are DWT level 1, DB 8, C 20, and the number of frames is 256. The more coefficient numbers are used, the more features are produced from the crying baby. When compared with previous studies, with the same feature method (MFCC) but different classification methods using PCA [12], this LDA method has better system accuracy. When using LFCC feature extraction, and the KNN classification method [18], which only reaches 90% accuracy, this system still has better accuracy.

6. References

[1] Dewi I A, Adriana Z and Lestari A A 2018 Identifikasi Suara Tangisan Bayi menggunakan Metode LPC dan Euclidean Distance *Jurnal ELKOMIKA* **6** 153

[2] Brazelton T B 1962 Crying in infancy *Pediatrics* **29** 579.

[3] P. Dunstan, "Open Up and Discover Your Baby's Language," 2012. [Online]. Available: www.babytaal.nl/media/PDF/ComprehensiveBooklet(2).pdf. [Accessed: 24-July-2020]

[4] Welly S L, Chastine F, Umi Y L 2016 Rancang Bangun Aplikasi Pendeteksi Suara Tangisan Bayi *Thesis* (Surabaya: Institut Teknologi Sepuluh Nopember)

[5] Bhagatpatil V V, Sardar V M 2015, An Automatic Infant's Cry Detection Using Linear Frequency Cepstrum Coefficients (LFCC) *International Journal of Scientific & Engineering Research* **3** 29

[6] Renanyi M D, Buono A, Kusuma W A 2013 Infant Cries Identification By Using Codebook As Feature Matching and MFCC as Feature Extraction Thesis (Bogor: Institut Pertanian Bogor)

[7] Putra D K, Tritoasmoro I I and Atmaja R D 2017 Simulasi Dan Analisis Speaker Recognition Menggunakan Metode Mel Frequency Cepstrum Coefficient (MFCC) Dan Gaussian Mixture Model (GMM) *e-Proceeding of Engineering* **4** 1766

[8] Ikhsan I, Novamizanti L, Ramatryana I N A 2014 Automatic musical genre classification of audio using Hidden Markov Model *2nd International Conference on Information and Communication Technology* (Indonesia: IEEE)

[9] Stavros N 2015 Audio Pattern Recognition of Baby Crying Sound Events *Journal of the Audio Engineering Society* **63** 358

[10] Chaiwachiragompol A, Suwannata N 2016 The Features Extraction of Infants Cries by Using Discrete Wavelet Transform Techniques *Procedia Computer Science* **86** 285

[11] Borade S N, Deshmukh R R and Ramu S 2016 Face recognition using fusion of PCA and LDA: Borda count approach *Proceedings of the 2016 24th Mediterranean Conference on Control and Automation* (Greece: IEEE) 1164

[12] Prasasti A L, Novamizanti L, Razik M I 2019 Identification of baby cry with Discrete Wavelet Transform, Mel Frequency Cepstral Coefficient and Principal Component Analysis *Journal of Physics: Conference Series* **1367** 012061

[13] Duzenli T and Ozkurt N 2011 Comparison OF Wavelet Based Feature Extraction Methods for Speech/Music Discrimination *National Conference on Electrical, Electronics and Computer Engineering* (Turkey: IEEE)

[14] Arabjamaloei R, et al. 2011 Exact Lithologic Boundary Detection Based on Wavelet Transform Analysis and Real-Time Investigation of Facies Discontinuities Using Drilling Data *Petroleum Science and Technology* **29** 569

[15] Tiwari V 2010 MFCC and its applications in speaker recognition *International Journal on Emerging Technologies* **1** 19

[16] Permana I S, Indrawaty Y and Zulkarnain A 2018 Implementasi metode MFCC dan DTW untuk jenis suara pria dan wanita *Mind Journal* **4** 49

[17] Wang, Y. and Y. Chen, 2017. A new feature extraction algorithm based on fisher linear discriminant analysis. Proceedings of the 2017 3rd International Conference on Control, Automation and Robotics (Japan: IEEE) 414

[18] Dewi S P, Prasasti A L, Irawan B 2019 Analysis of LFCC Feature Extraction in Baby Crying Classification using KNN *International Conference on Internet of Things and Intelligence System (IoTaIS)* (Indonesia: IEEE)

IOP Conf. Series: Materials Science and Engineering **982** (2020) 012010 doi:10.1088/1757-899X/982/1/012010

Cholesterol Level Detection Through Eye Image Using Fractal and Decision Tree

Dian Sari[1], Jangkung Raharjo[1,2], Ledya Novamizanti[1,3]

[1]School of Electrical Engineering, Telkom University
Jl. Telekomunikasi Terusan Buah Batu Bandung 40257 Indonesia

[2]jangkungraharjo@telkomuniversity.ac.id, [3]ledyaldn@telkomuniversity.ac.id

Abstract. Cholesterol is a natural substance with physical properties of fat but has a steroid group. High cholesterol levels will cause hypertension, coronary heart disease. Cholesterol levels can now be detected through eye images. This research produces a cholesterol level detection system with input in the form of iris images. First, the image is resized, converted to grayscale, and cropped by the system. Then feature extraction is done by the fractal method, which has characteristics that can explain dimensions in non-integers. The last stage is a classification using the decision tree method because it can simplify a complex decision-making process to be more specific. Eye images are classified into three, namely cholesterol, cholesterol risk, and no cholesterol. In total, there are 105 images, consisting of 63 training data images and 42 test data images. The result is 95.23% accuracy, 90.47% precision, 100% recall, and 40 ms computing time.

1. Introduction

Cholesterol is a natural substance with physical properties, but it has a high-grade steroid sum of steroids that can create hypertension and clogs in brain veins, heart, causing limestone [1]. Cholesterol checks can usually be done in hospitals and clinics by fasting for 9-12 hours. Data collection on total cholesterol levels can be measured from peripheral blood samples with a cholesterol test kit [2]. But with the development of technology, cholesterol levels can be detected through the iris, commonly called iridology. Iridology is the study of the pattern of arrangement of the contents of the eye's fiber by observing the model that will be known to one's health problem [3]. Cholesterol detection studies have been carried out as in [4], which uses the iris recognition method to divide the iris region, the normalization process, and finally, determine the presence of cholesterol using the thresholding and OTSU histogram. Andana et al. [5] investigated FLBP for feature extraction with sampling point 8, radius 4, and F = 7. Linear regression is used to measure cholesterol values with the computation time of each image is 11 seconds.

Mukti et al. [6] using LBPH for feature extraction and then use linear regression to obtain cholesterol level value with a standard value error is 2.5111 and computation time 9.196 seconds. Raharjo et al. [7] found that detection cholesterol using GLCM and linear regression can inform cholesterol levels with a result of the accuracy of 88.52%, computation time 0.0365 seconds, and standard deviation 6.9595. One study of fractals [8] by analyzing the Higuchi fractal dimension and the Katz fractal dimension. The best accuracy results obtained at the Katz fractal dimension were 90.33%. Safitri et al. [9] investigated the Fractal dimension is analyzed for the classification of diabetic retinopathy, with an accuracy of 89.17%.

In this study, the authors propose the application of fractals for the detection of cholesterol levels classified using a decision tree. Fractal was introduced by Benoit Mandelbrot for the first time. It represented the shape of objects in nature and had self-similarity, self transformability [10]. First, the image is resized, convert to grayscale, and cropped by the system. Then feature extraction is carried out by the fractal method, which has characteristics that can explain dimension in non-integer. The final stage is classification using the decision tree method because it can simplify the complex decision process becomes more specific, where there are three classifications, namely the risk of cholesterol, cholesterol, and no cholesterol.

2. Basic Theory

2.1. Fractal

The history of fractals can be traced from Benoit Mandelbrot's book, The Fractal Geometry of Nature. Euclidean geometry is used to represent shapes created by humans (rectangles, circles, balls, etc.). In contrast, fractal geometry is a natural way to represent the shapes of objects in nature. The main attraction of fractal geometry comes from its strong ability to describe disorder or fragmented nature shapes and other complex objects [10].
The key parameter in fractal geometry is Dimension Fractal (D), which can offer a systematic approach to measuring irregular patterns. The fractal dimension itself has dimensions with non-integer numbers, such as dimension 1.9, 4.6. Among all the computational method of fractal dimension, the box-counting method is the most popular method for calculating fractal dimensions because it is simpler, but has limited accuracy [11]. The box-counting technique is used to get a 2-D fractal object scaling by closing the 2-D image with a range of box size (s) and the number of boxes that cover the object (N) [12].

$$D = \frac{\log(N)}{\log(^1/_s)} \tag{1}$$

2.2. Decision Tree

The decision tree is one of the classification methods that use tree representations, some nodes represent attributes, leaves that represent classes, and braches represent the values of those classes. The method is used to further visualize decision trees and display detailed information using complicated graph nodes and links [13]. The root node is the node located at the top of the tree. The internal node is a fork node, where on this node, there is only one input and has a minimum of two outputs. The leaf is the end node, it has only one input, and it has no output. In the decision tree, each leaf node marks the class label. In the decision tree, each branch of the stated conditions must be met. Each end of the tree states the value of the data class [14].

1. Find GINI Index
   GINI Index is used to find out positive or negative objects. The resulting values are squared in each class. Here is a formula to find the square value of the probability of each class.
   $$I_{GINI} = 1 - \Sigma(P_i)^2 \tag{2}$$

2. Find GINI Splitting Index
   The GINI splitting index calculating of each GINI index is done to find the value of each class feature. The smallest splitting index GINI value will be used as the root node.
   $$I_{GINI} = 1 - \sum_{i=1}^{n}(p_i)^2 \; x \; (1 - P_i) \tag{3}$$

3. System Design

In this study, the first step is collecting some images that are objects. The next is preprocessing and then proceed to the extraction stage feature with the fractal method to get the data feature of the

IOP Conf. Series: Materials Science and Engineering **982** (2020) 012010 doi:10.1088/1757-899X/982/1/012010

whole data. Then the data of the feature will be classified using the decision tree method. Figure 1 is a block diagram of system design.

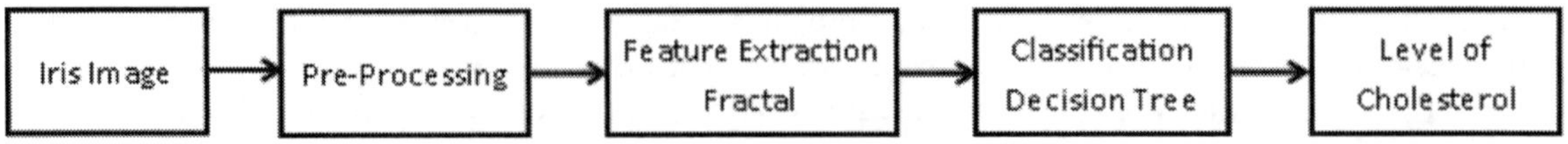

Figure 1. System Design

3.1 Dataset

This iris image used amounted to 105 images; data has a format of *.bmp. The image is derived from the result of a person's blood test when the cholesterol check is performed by the previous researcher. The imagery used for the training data will be stored on the database in each class, which is the risk of cholesterol, cholesterol, and non-cholesterol. Figure 2 shows the risk of cholesterol, cholesterol, and non-cholesterol [5], [6], [7], [15].

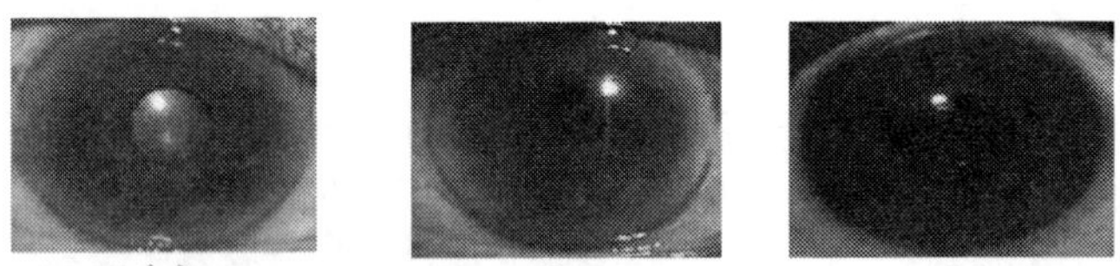

(a) (b) (c)

Figure 2. Image (a) risk of cholesterol (b) cholesterol (c) non-cholesterol

3.2 Pre-processing

In this process, the iris image will be done resizing with a size of 768×768, then convert to grayscale. Then cropping is done, this process aims to take the area that contains information only in the edges of the iris that is commonly called arcus senilis. Firstly iris will be determined its radius and also specified the angel of the area containing the information after obtaining the area. Then, the cropping process begins. This process is done up to an angle of 360° or circle of the iris image. Figure 3 is the result of cropping converted into a rectangular form.

Figure 3. Cropping result image

3.3 Feature extraction

In this study, the features extraction process using the Fractal Dimension. The first step is to retrieve the objects that will be extracted. Then specify the size of the boxes to cover the entire image section after that specifies the iteration to be used. Then calculate the number of square N (s) on the image, the value of N (s) depends on s. the last step is to calculate the fractal dimension with the equation (1) so that the vector is obtained. Figure 4 shows a block diagram of the Fractal Dimension.

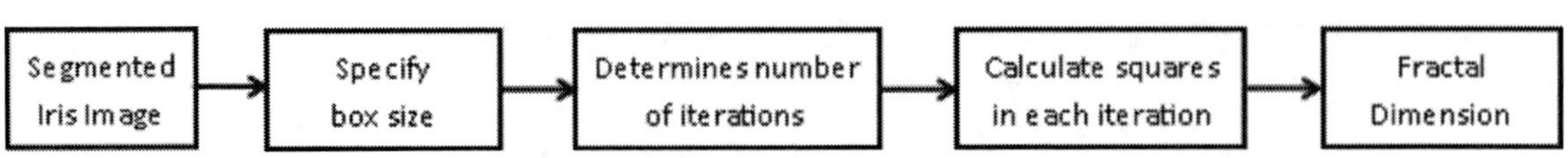

Figure 4. Block diagram of Fractal Dimension

If the parameter s used is 8 so that the box is covering the iris as much as 32 squares. Figure 5 displays an illustration of the iris image using the Fractal Dimension.

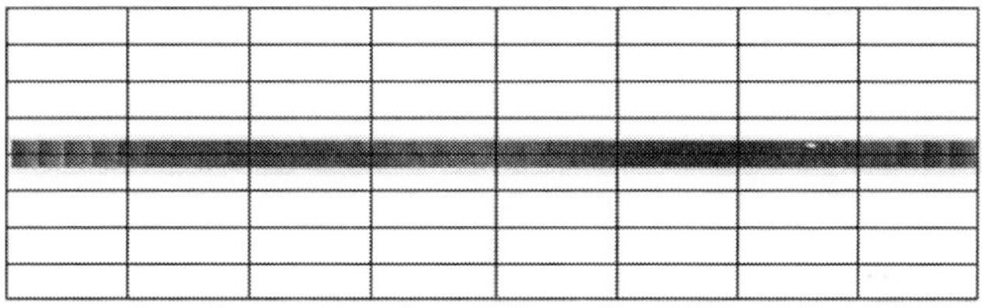

Figure 5. Illustration of the iris image using the Fractal Dimension

3.4 Classification

The classification process will be carried out a vector feature retrieval of imagery, which is done by two processes to separate the image into namely the trainer image and test image. The classifying process is done by the decision tree classification method. This training phase creates a classification pattern for class prediction references on the data to be tested. In each process consists of a practice input process that has been extracted in the form of a vector trainer feature, then determine the root of the tree is to make the class risk cholesterol, cholesterol, and non-cholesterol. After that, all data gets the class corresponding data feature that has been extracted using the fractal method. The process of rules all the information recorded will be grouped to know the GINI index of each class. In each process, the testing phase consists of a test image input process in the form of a vector test feature.

Furthermore, the rule decision tree is to classify the unclassified data. Data that has been classified next will be labeled, whether the data belongs to risk categories of cholesterol, cholesterol, non-cholesterol in the data labeling process. Figure 6 shows the block diagram of the Decision Tree.

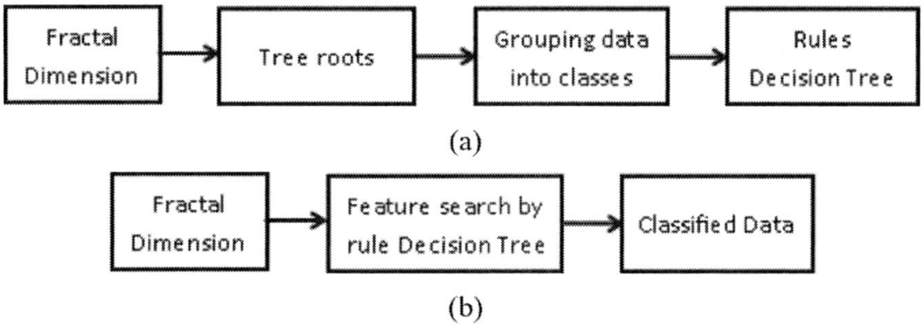

Figure 6. Block diagram of Decision Tree (a) Training, (b) Testing

4. Result and Discussion

The result tests several test scenarios. The scenario is to resize the box in the fractal dimension method, change the fork of the decision tree classification, and change the amount of training data and test data. Test simulation using MATLAB 2018b.

4.1. Effect of fractal dimension

In this test, a fractal dimension change can be calculated using equation (1). The fractal dimension is affected by the parameter s, where the s value changes from 1 to 2^k. These dimension changes are 2, 4, 8, 16, 32, and 64. The image is used when the image resizes 768 × 768 pixels. Figure 7 shows the effect of the range of box size on accuracy. Based on Figure 7, optimal results are obtained when the fractal dimension is 32 with an accuracy result of 95.23% and a computing time of 45.3 ms. This is due to the greater value of the fractal dimension, the greater the accuracy value because the boxes that cover the image are more and more precise with the input image. In contrast, the lowest accuracy value is the 64 range of box size with an accuracy value of 83.33% and 65.5 ms because the fractal dimension itself has a limit until the size of the input image.

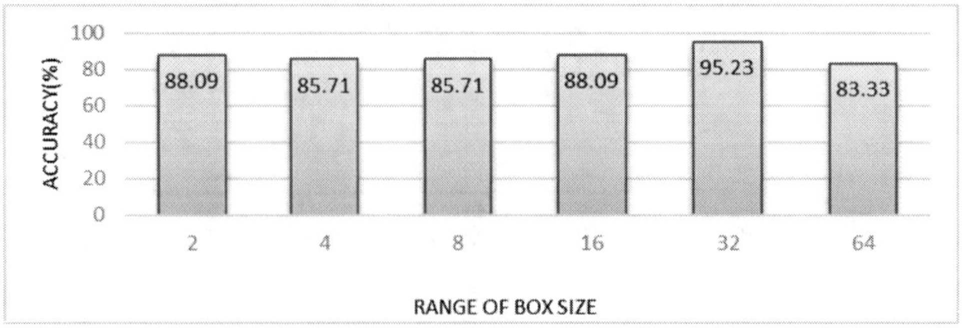

Figure 7. Effect of range of box size on accuracy

4.2. Effect of the number of branches

In this scenario, a test of the number of branches is made to make the tree more shallow or reduce the branch. It can reduce the complexity and time of system computing. Figure 8 and Figure 9 show the effect of the number of branches on accuracy and computing time.

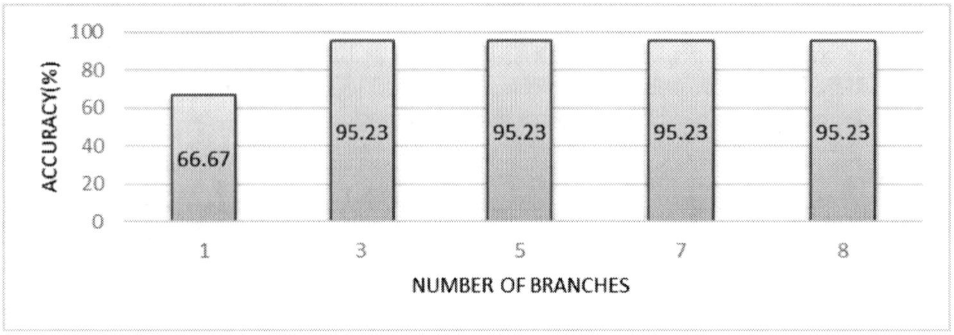

Figure 8. Effect of the number of branches on accuracy

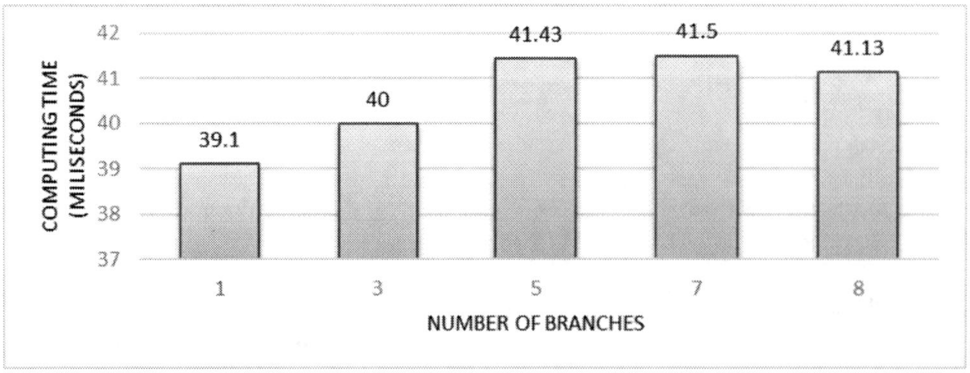

Figure 9. Effect of the number of branches on computing time

Based on Figure 8 and Figure 9, the change in this branch has no significant effect on the accuracy value, but it greatly affects the computing time. The optimal result is the branches at number 3, with 95.23% accuracy value and 40 ms compute time. In contrast, the lowest accuracy value is when the branch amounts to 1 with a value of 66.67% accuracy and computational time 39.1 ms. It is due to the absence of other attributes used to divide the sample further. The more the number of branches, the computation time will be long because the system takes a long time to calculate the probability of existing branch options.

5. Conclusion

The study proposed a cholesterol level detection system in a person by using fractals as the feature extraction and decision tree method as the classification, with the best results on the resizing 768 × 768 pixels, 32 fractal dimensions, and the number of branches on the decision tree 3 which gives the accuracy 95.23%, precision 90.47%, recall 100% and process time 40 ms. The fractal dimension of 32 (at a resize condition 768 × 768 pixels) provides the highest accuracy 95.23% as the larger the fractal dimension value, the greater the accuracy because the boxes that cover the image are more and more precise with the input image. The number of branches in the decision tree of 3 or more does not affect the outcome of both accuracies due to the absence of other attributes used to divide the sample further. The more the number of branches, the computation time will be long because the system takes a long time to calculate the probability of existing branch options.

6. Reference

[1] Ingrid D, Fathrah L 2018 Activity Test of Suji Leaf Extract (Dracaena angustifolia Roxb.) on invitro cholesterol lowering *J Kim Sains dan Ap*l. **21**(2) 54–8
[2] Maryati H 2017 Hubungan Kadar Kolesterol dengan Tekanan Darah Penderita Hipertensi Di Dusun Sidomulyo Desa Rejoagung Kecamatan Ploso Kabupaten Jombang *J. Keperawatan* **8** 128–137
[3] Yohandy DH, Wiranata IMN, Ferbia TQ 2018 Identifikasi Pola Penyakit Pada Citra Iris Mata dengan RBF Neural Network *J Inform.* **5**(2) 195–201
[4] Songire S, Joshi M 2016 Automated Detection of Cholesterol Presence using Iris Recognition Algorithm *Int J Comput Appl.* **133**(6) 41–5
[5] Andana SN, Novamizanti L, Apraz Ramatryana IN 2019 Measurement of Cholesterol Conditions of Eye Image using Fuzzy Local Binary Pattern (FLBP) and Linear Regression *IEEE Int Conf Signals Syst.* 79–84
[6] Mukti GBY, Novamizanti L, Fadhila PA 2020 Measurement of Cholesterol Level through Eye Image using Local Binary Pattern Histogram Test Eng Manag. **83** 12553–12560
[7] Raharjo J, Novamizanti L, Ramatryana INA 2019 Cholesterol Level Measurement Through Iris image Using Gray Level Co-Occurrence Matrix and Linear Regression *ARPN J Eng Appl Sci.* **14**(21) 3757–3763
[8] Jacob JE, Gopakumar K 2018 Automated diagnosis of encephalopathy using fractal dimensions of EEG sub-bands *IEEE Recent Adv Intell Comput Syst.* 94–97.
[9] Safitri DW, Juniati D 2017 Classification of diabetic retinopathy using fractal dimension analysis of eye fundus image *AIP Conference Proceedings* **1867** 020011
[10] Peitgen H-O, Jurgens H, Saupe D 1992 *Fractals for the Classroom Part One Introduction to Fractals and Chaos* p 452
[11] Xue S, Jiang X, Duan J 2018 A New Box-Counting Method for Image Fractal Dimension Estimation *IEEE Int Conf Comput Commun.* 1786-1791
[12] Liu K, Ostadhassan M 2017 Multi-Scale Fractal Analysis of Pores in Shale Rocks *J Appl Geophys.* **140** 1–10
[13] Mrva J, Neupauer S, Hudec L, Sevcech J, Kapec P 2019 *Decision support in medical data using 3D decision tree visualisationi* E-Health Bioeng Conf EHB pp 21-23
[14] Alpaydin E 2004 *Introduction to Machine Learning* (Massachusetts: The MIT Press Cambridge) p 407
[15] Daniel M, Raharjo J, Usman K 2020 Iris-based Image Processing for Cholesterol Level Detection Using Gray Level Co-Occurrence Matrix and Support Vector Machine *Engineering J* **24** 135–144

Identification of Foot Posture using *Foot Posture Index-6* (FPI-6) based on Image Processing

Hasya Azqia Hanifan[1,2], **Ledya Novamizanti**[1,3], **Husnaeni Mukhtar**[1,4]
[1]School of Electrical Engineering, Telkom University Jl. Telekomunikasi Terusan Buah Batu Bandung 40257 Indonesia
[2]hasyaazqia@student.telkomuniversity.ac.id, [3]ledyaldn@telkomuniversity.ac.id, [4]husneni.mukhtar@telkomuniversity.ac.id

Abstract. Legs and ankles have an important role in daily life for activities. Unconsciously often the leg posture becomes abnormal due to the bad habits, so the risk of injury. Assessment of the type of leg posture can be done by various methods, both uni-planar and multi-planar. This study developed the FPI-6 method as a multi-planar tool based on image processing. The data used are 30 reference data and 20 test data from the age of 19-23 years. The results of this study are image processing systems that can identify the type of leg posture automatically quickly and accurately.

1. Introduction

The comfort of activities using legs such as walking and running is very much influenced by the type of posture of one's feet. Abnormal leg posture will reduce the efficiency of leg movement and increase the risk of injury [1]. Diagnosis and rehabilitation of abnormal leg posture are increasingly needed to maintain leg posture. However, the amount of money that must be spent to diagnose this type of foot posture by experts becomes an obstacle for the community. Therefore we need test equipment that can diagnose this type of leg posture effectively and efficiently without having to pay a large fee.

In general, there are two assessment tools, namely uniplanar and multi-planar. There are two studies on uniplanar devices, namely radiography and anthropometrics. Radiographically, studies of the normal range of joint motion have been carried out using computed tomography (CT) [2]. Whereas anthropometrically, studies of leg posture types have been carried out with Rear Foot Arch (RFA) measurements using goniometer [3], Medial-Longitudinal Arch Angle (MLAA) measurements [4], and Navicular Position Test (NPT) which evaluates the position of the navicular bone [5] In the latest research, the RFA assessment was carried out using image processing technology and was considered capable of increasing efficiency [1], [6]. In multi-planar tools, Foot Posture Index-6 (FPI-6) and FPI-8 methods have been found [7], [8]. FPI-6 assesses the type of leg posture based on six criteria [7]. Multi-planar assessment using FPI-6 is considered the most reliable and can validate leg posture simply and quickly [9]. Therefore, there is an opportunity to develop FPI-6 as the most reliable multi-planar tool, using image processing technology that can improve efficiency.

This study proposes a GUI system that can identify foot posture using the FPI-6 method based on image processing. The identification process uses Otsu's thresholding segmentation to recognize the characteristics of each of the six criteria. The output of the FPI-6 system made is in the form of three types of foot posture, namely normal, pronation, or supination.

2. Methodology

2.1. Foot Posture Index-6 (FPI-6)
FPI-6 is a multi-planar tool that is judged to be fast and reliable for determining human foot posture. The FPI-6 assessment is based on six observation criteria. Before observing the FPI-6, patients are required to stand in a relaxed position. Then various foot regions will be observed based on six criteria. Figure 1 shows an illustration of the FPI-6 criteria [10].

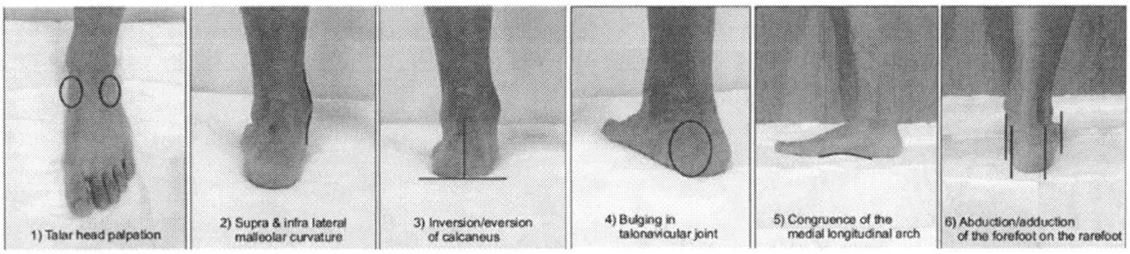

Figure 1. Six Criteria of FPI-6.

In the six observations, each criteria will be given an index value according to its position. For each criteria, values +1 or +2 are given for pronation positions, values -1 or -2 are given for supination positions, and a zero score is given for normal or neutral positions. Furthermore, the six indexes of each criteria are added up to produce a total index score as the basis for the classification of foot posture types. A score of 0 to 5 can be classified as normal. More than five are classified as pronated, and less than 0 are classified as supination.

2.2. Experimental Setup
In this study, data were taken in the form of right foot images from 50 Telkom University students ranging in age from 19-23 years. Of the 50 students, 3-foot images were taken each from a different perspective. So that the total number of images used in the system is 150-foot images with details of 50 rear foot images when the lateral position is straightened to a 90-degree angle, 50 rear right foot images when the standing position is relaxed, and 50 side foot images in the inside.

2.3. System Specification
In designing a foot posture identification system, hardware, and software that support the system are needed to work properly and optimally. The software needed is the MATLAB R2018a program, while the hardware used is a device with specifications, namely Processor Intel® Core™ i3-5005U CPU @ 2.00GHz, Memory: 4GB (RAM), Operating system: Windows 10 Pro 64-bit. Also, some equipment that is used for primary data collection is needed, including the following Camera Canon D3100, Tripod, two light stands, two lamps 45W, two white umbrellas, and Black Box 40 × 45 × 40 cm.

3. System Design
In general, the design and implementation stages of the system consisted of the stages of image acquisition, image pre-processing, quantification using FPI-6, and classification based on the index obtained from the quantization stage. Figure 2 displays a block diagram of a foot posture identification system.

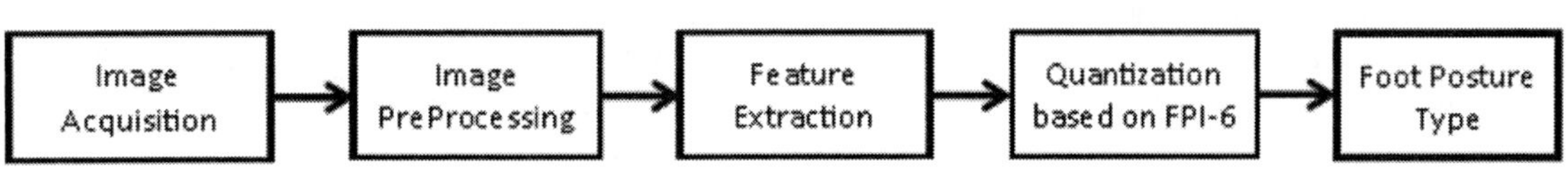

Figure 2. Block Diagram of Foot Posture Identification System.

3.1. Image Acquisition

The acquisition process uses an image object in the form of a photo of the back right foot and the right side of the inside using a digital camera with dimensions of 1708 × 1139 pixels and a resolution of 96 dpi. The image is taken by inserting the foot object into a black box measuring 40 × 40 × 45 cm. Subjects were asked to stand up straight for foot photographs. The camera is placed parallel to the object at the height of 50 cm above the ground. The distance of the object to the lens is set at 25 cm because it is considered quite clear and not too far away. Two light stands with 45W lamps are used, each on the right and left side of the camera. The light from the two lights is filtered using a white umbrella, so the results are even and do not cause disturbing shadows. How to capture the right rear foot image is shown in Figure 3. Three points of view of the foot image can be seen sequentially in Figure 4.

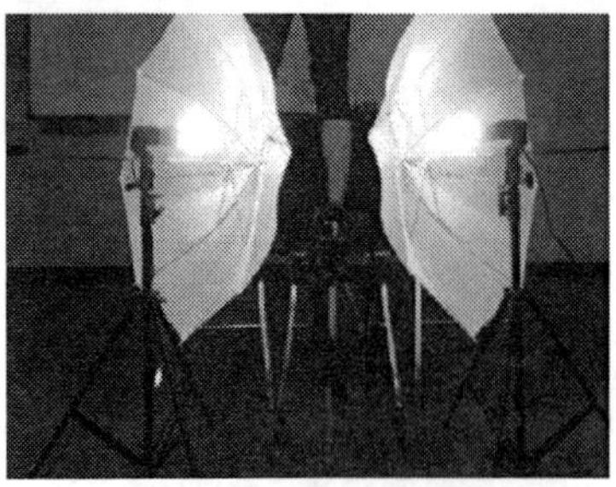

Figure 3. Image Acquisition.

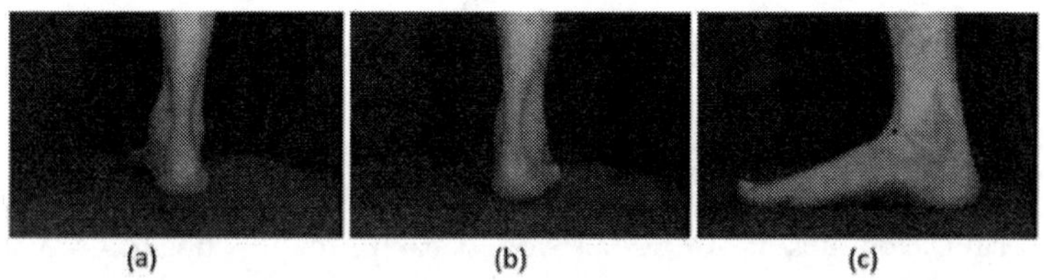

Figure 4. Three Foot Viewpoints Used in the System.

Each subject was only taken the image of his right foot, with three angles of the field, namely when the back foot in the lateral position was straightened to a 90-degree angle, when the right rear foot in a relaxed standing position, and the right inner foot. Figure 4 shows the image (a) needed to see the 2nd criteria in FPI-6, image (b) for the 3rd and 6th number criteria, and image (c) for the 5th criteria. The pictures (a) and (b) are the images of the back foot. Still, the difference is that in picture (a) the lateral part of the foot is positioned to be straight at an angle of 90 degrees to the camera position. The aim is to get clearer information about supra and infra lateral curvature. Whereas in picture (b) is the position of the feet when standing relaxed.

3.2. Image Preprocessing

Digital imagery can be expressed as a two-dimensional function $f(x, y)$ with x or y being the coordinate position and f is the amplitude at position (x, y) which is often known as grayscale intensity [11], [12]. This research used several image pre-processing stages, including resizing, grayscaling, Otsu thresholding [13], [14], noise removing [15], and filling holes [14]. This stage is used for each criteria before feature extraction is performed in the next process. Figure 5 displays the flow diagram of the FPI-6 image pre-processing system stage.

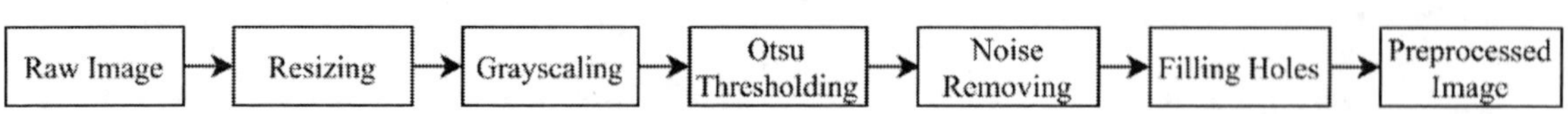

Figure 5. Block Diagram of the FPI-6 Image Preprocessing System.

The image size of 1708 × 1139 pixels is resized by 25% to 427 × 285 pixels. Then the image is converted into an RGB color space grayscale. Each image is converted to binary using the Otsu thresholding method [15]. In the separation of background images and grayscale objects with thresholding, there is still a background that is read as an object so that it forms certain areas that are undesirable or referred to as noise. Therefore, to eliminate noise levels, a threshold value for noise removal is performed for each criteria. The noise removal stage is removing the background pixel area that is read as an object. Therefore, the hole filling process is carried out to fill the black hole with a pixel of 1 value.

3.3. Feature Extraction

3.3.1. Criteria 2 (Supra and Infra Lateral Malleolar Curvature. The amount of curvature can be seen from the position of the pixels at the maximum infra and supra lateral curvature. The pixel position can be compared by finding the distance between the two. The output of this first step is the two coordinates (x_1, y_1) represent the maximum supra lateral arc position, and (x_2, y_2) the infra lateral maximum arc position. Then, to find out the distance of x coordinate between the two maximum arches (SI) obtained, the reduction of x infra lateral coordinate (x_2) is reduced by supra lateral x coordinate (x_1). The output of this stage is the feature value of criteria 2 in the form of the x-coordinate distance between the two maximum arches (SI).

3.3.2. Criteria 3 (Inversion or Eversion Calcaneus). In this criteria, the three angular slopes formed by the center of the calcaneus are calculated against the surface of the plane. Figure 6 shows the steps to get the Calcaneus feature [16].

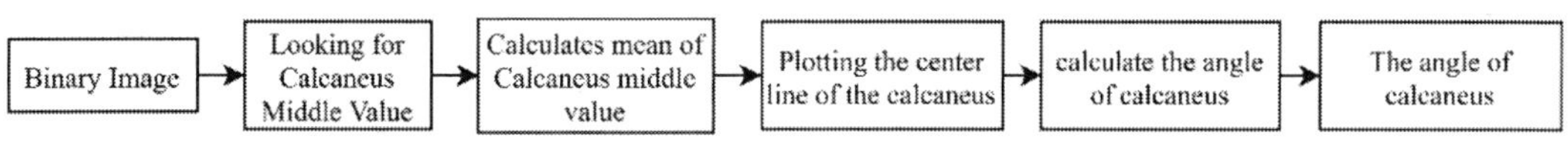

Figure 6. Feature Extraction Block diagram in the calcaneus.

This stage starts by storing information in the form of pixel positions on the right and left side of the calcaneus object. Then from the pixel position of the right side (x_2, y_2) and the left side (x_1, y_1), the pixel midpoint (x, y) is searched by calculating the mean coordinates of x from the right and left sides of the calcaneus (Figure 7(a)). Figure 7(b) shows the results of the plotting line marked with a red line with (x_1, y_1) as the top point and (x_2, y_2) as the bottom point. The angle is calculated by first looking for x as the difference $x_2 - x_1$, as well as y as the difference $y_2 - y_1$. Then the values of x and y that have been obtained as follows

$$\tan \alpha = \frac{y}{x} \quad \alpha = tan^{-1}\left(\frac{y}{x}\right) \tag{1}$$

Furthermore, the angle α in units of degrees becomes the value of the system output results in criteria 3, which will then be quantized. Figure 7(c) illustrates the process of obtaining an angle (α).

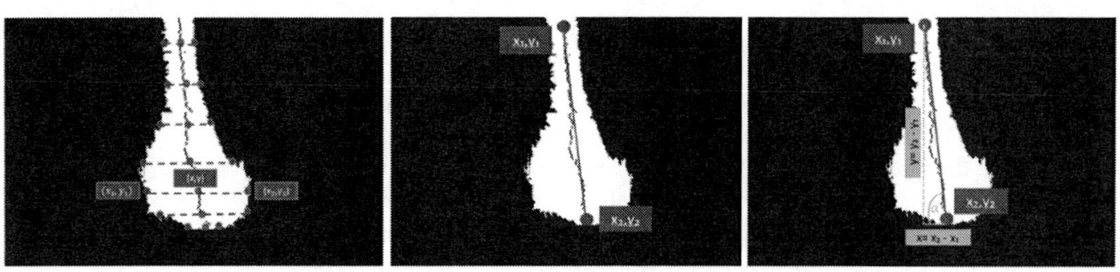

(a) (b) (c)

Figure 7. Feature Extraction of Calcaneus.

3.3.3. Criteria 5 (Congruence of the Medial Longitudinal Arch). The medial elongated arch height is obtained by storing pixel information at the bottom of the foot object and then detecting pixels. The pixel position of the leg curve is obtained by finding the upper, lower, right, and left boundaries first. From this boundary, the lower boundary is drawn for the plot. After that, we can see the maximum point formed by the MLA chart. The maximum MLA curve is calculated from the distance to the surface of the foot. High MLA is the value of the system output feature in criteria 5.

3.3.4. Criteria 6 (Abduction or Adduction of the Forefoot on the Rarefoot). Abduction is the appearance of the forefoot on the lateral side of the foot (outer side), and adduction is the appearance of the front foot on the medial side of the foot (the inside). Both will be sought the difference by reducing abduction by adduction. If the result is positive, the feet are more abducted, so pronation tends. And conversely, if more foot adduction will tend to supination. The result is a feature value in the form of the difference in the magnitude of the abduction and induction distance.

3.4. Quantization Using Scoring Foot Posture Index-6 (FPI-6) Rules

In this study, image processing was not carried out for criteria 1 and 4 due to image limitations that can only detect two dimensions. Quantization using the FPI-6 rule is performed on the value of features that have been obtained in the extraction process. There are five indexes of -2 to +2 given to the six criteria of qualitative standards that are in line with FPI. The six indexes of each FPI criteria are then added up to get a total value. Based on this total value, the classification is done. Total values for normal legs range from 0 to +5. If more than +5, prone foot posture. The posture of fewer than 0 feet can be called supination.

4. Result and Discussion

System validation is carried out by bringing ten subjects to the expert for manual FPI-6 assessment. Because FPI-6 assessment is qualitative, the mol is repeated by three experts in each subject so that the assessment can be more objective. Each criteria will have three index values based on the assessment of each expert. From the total index value, the foot posture is classified based on the FPI-6 assessment rules. In this study, testing was carried out on 50 subjects. Another 40 subjects were validated based on the foot image taken. However, validation is only on criteria 2, 3, 5, and 6, considering criteria 1 and 4 must be palpated (touched) directly.

4.1. Otsu Threshold Testing

In criteria 2 and 6, the information to be extracted is the amount of curvature of the supra and infra lateral. Otsu thresholding can separate objects and backgrounds more perfectly using the multiplier coefficient (k). The value k = 0.7 is chosen so that the object does not lose much information. Calcaneus information can be seen starting at k = 1.4, but at this value, calcaneus still joins with the rest of the foot. The best threshold is obtained when Otsu thresholding is multiplied by the value of k = 1.7 in Criteria 3 and 6. In criteria 5, the threshold value is automatically obtained when the multiplier coefficient is one. The MLA arch is clear and doesn't lose information. Figure 8 displayed the best threshold value in criteria 2, 3, and 6.

4.2. Noise Removal Testing

The results of thresholding in the previous process still leave noise in the background in the form of white dots that have a certain area. In this test, the threshold value of noise (N) removal for the noise in the image will be chosen. A value of N = 10000 is used for each criteria in the FPI-6 system. Figure 9 displays the results of the noise removal process with N=10000 in criteria 2 and 3.

4.3. Testing the Results of the FPI-6 System

A total of 20 test data will be tested on a system that has previously been set parameters based on reference data. Table 1 is the result of an assessment of 20 test data tested on the FPI-6 criteria 2 system.

(a) (b) (c)

Figure 8. Threshold testing (a) k = 0.7 in criteria 2, and 6, (b) k = 1.7 in criteria 3, and 6, (c) k = 1 in criteria 5.

(a) (b)

Figure 9. The Noise removal threshold N=10000 (a) criteria 2, (b) criteria 3.

Table 1. Accuracy of FPI-6 systems on test data.

Verification by	Foot Posture	Number of valid indexes	Index Verification		Accuracy (%)
			Valid	Not Valid	
6 Criteria of FPI-6	Normal	17	16	1	95
	Pronation	3	3	0	
	Total	20	19	1	
4 Criteria of FPI-6	Normal	17	15	2	85
	Pronation	3	2	1	
	Total	20	17	3	

A comparison is performed on testing the total accuracy of the FPI-6 system by six criteria and 4 FPI-6 criteria. A total of 20 results of test data validation were tested on testing six criteria (including criteria 1 and 4), and then the index of each criteria was added up. Based on the range of values classified type of foot posture, likewise, in testing four criteria (without criteria 1 and 4), but classified with a different range of values. Table 1 shows a comparison of the results of the verification of foot posture classification using six criteria and 4 FPI-6 criteria. Based on the verification of 20 verified test data, the accuracy of the 6 FPI-6 assessment values is greater than the 4 FPI-6 criteria, which is 95%. In comparison, the accuracy of the assessment by four criteria is 85%.

5. Conclusion

This study proposes a system to identify the type of leg posture using FPI-6 automatically based on image processing. Pre-processing stage in the form of resizing, grayscaling, Otsu thresholding, noise removal, and filling holes. Based on the experiment obtained 95% system accuracy in classification based on six criteria (4 criteria automatically and two criteria by palpation by a doctor), and 85% accuracy in classification based on four criteria.

6. References

[1] Lin C H, Qiu Z H and Yeh C C 2018 Image processing for rear foot image evaluating leg and foot angles *Meas. J. Int. Meas. Confed.* **126** 168
[2] Beimers L and Amsterdam S Z 2015 Subtalar Joint Kinematics in Healthy Individuals Using Computed *Univ. Hosp. AMC* (Netherlands: Dept. Orthopedics, University Hospital AMC)
[3] Calhoun J H 2011 *Foot and ankle* **4**
[4] Caravaggi P, Matias A B, Taddei U T, Ortolani M, Leardini A and Sacco I C N 2019 Reliability of medial-longitudinal-arch measures for skin-markers based kinematic analysis *J. Biomech.*, **88** 180
[5] Spörndly-Nees S, Dåsberg B, Nielsen R O, Boesen M I and Langberg H 2011 The navicular position test - a reliable measure of the navicular bone position during rest and loading *Int. J. Sports Phys. Ther.* **6** 199
[6] Langley B, Cramp M and Morrison S C 2016 Clinical measures of static foot posture do not agree *J. Foot Ankle Res.* 1
[7] Redmond A C, Crane Y Z and Menz H B 2008 Normative values for the Foot Posture Index J Foot Ankle Res. **1** 6

[8] Menz H and Munteanu S E 2005 Validity of 3 Clinical Techniques for the Measurement of Static Foot Posture *Journal of Orthopaedic and Sports Physical Therapy* **3**
[9] Abel E W, Unger A, Fletcher R and Jain A S 2002 Development of clinical measurement of the axes of rotation of the ankle and subtalar joints *Annu. Int. Conf. IEEE Eng. Med. Biol. - Proc.* **3** 2455
[10] Lee J S, Kim K B, Jeong J O, Kwon N Y and Jeong S M 2015 Correlation of Foot Posture Index With Plantar Pressure and Radiographic Measurements in Pediatric Flatfoot *Ann. Rehabil. Med.* **39** 10
[11] Gonzales R C, Woods R E 2002 Digital Image Processing (New Jersey: Prentice-Hall, Inc.)
[12] Tan J J L 2013 Image Processing Basics Digital Signal Processing (Second Edition)
[13] Senthilkumaran N and Vaithegi S 2016 Image Segmentation By Using Thresholding Techniques For Medical Images *Comput. Sci. Eng. An Int. J.* **6** 1
[15] Putri E I, Magdalena R and Novamizanti L 2015 The detection of cervical cancer disease using an adaptive thresholding method through digital image processing *Journal of Advances in Health and Medical Sciences* **1** 30
[16] Atmaja R D, Novamizanti L, Halomoan J and Murty M A 2017 An image processing algorithm to find edges in grayscale image *Far East Journal of Electronics and Communications* **17** 781
[17] Mukhtar H, Rahmawati D, Silalahi D K, Novamizanti L, Ghifari M R, Dzikri A A A, Rachman F F, Khatami A A 2020 Fast Algorithm to Measure the Types of Foot Postures with Anthropometric Tests Using Image Processing *IJEEMI* **2** 48

IOP Conf. Series: Materials Science and Engineering **982** (2020) 012012 doi:10.1088/1757-899X/982/1/012012

Preprocessing Using Correlation Based Features Selection on Naive Bayes Classification

Yogiek Indra Kurniawan[1,4], Teguh Cahyono[1], Nofiyati[1], Eddy Maryanto[1], Ari Fadli[2] and Naisha Rahma Indraswari[3]

[1]Informatics Department, Universitas Jenderal Soedirman, Indonesia
[2]Electrical Engineering Department, Universitas Jenderal Soedirman, Indonesia
[3]Informatics Department, Universitas Muhammadiyah Surakarta, Indonesia

[4]yogiek@unsoed.ac.id

Abstract. The birth of a healthy baby is generally around 38-42 weeks of pregnancy. However, there are many babies born at an inadequate age of birth and the age of birth that is past its time. This study aims to predict the age of the birth of a patient. The method used is a classification with the Naïve Bayes algorithm with input variable (X), the factors experienced by pregnant women in the form of 8 variables X and Y variable in the form of Birth Age. Problems that arise are too many attributes that affect the results of accuracy. To overcome this, preprocessing is used with the Correlation Based Features Selection (CBFS) method. CBFS chose the X variables which had the highest correlation with the Y variable (Birth Age) but had the least correlation between the X variables. From the CBFS that had been done, produced 4 X variables, namely: blood pressure, number of babies, congenital diseases before pregnancy, and problems during pregnancy. The results of the test showed an increase in Precision, recall, and accuracy in the Naïve Bayes classification when implemented CBFS. The highest value of accuracy after preprocessing is 67% with an increase of 2 percent compared to before preprocessing.

1. Introduction

The birth rate of babies in Indonesia tends to increase every year. This baby's birth rate is accompanied by a large newborn mortality rate. Many things cause newborns to die, some of them are nutritional deficiencies when in the womb, fetuses that have defects in the body, premature age birth factors, postmature / postdate age birth factors, and others. Death due to premature birth is certainly a pretty serious problem. According to research conducted by WHO in 2010, Indonesia is the fifth-highest country with 675,700 premature births in one year. This figure is still quite low compared to India which reaches 3.5 million premature births per year. This causes India to rank first in the highest number of premature births among 184 other countries. In second place is China with 1.1 million premature births followed by Nigeria and Pakistan. Indonesia ranks 9th in the highest average number of premature births with birth rates of 15.5 per 100 births[1].

More than 15 million premature births occur in the world and continue to increase every year. More than 1 million children die each year due to complications due to premature birth. Premature birth is the cause of death of newborns age less than 4 weeks and is the second leading cause of death in children under 5 years other than pneumonia. Meanwhile, babies who successfully survive experience disabilities have experience throughout their lives including learning and visual impairments and hearing problems[2].

Based on the problems mentioned above, we must know what factors influence the age of birth in infants so that nothing untoward happens in the future. One solution that can solve that problem is to classify data mining. In the classification process, it is found a set of objects with the same characteristics in a database, then classifies them into some classes that have been determined through the process of building the model. The purpose of classification is to find a model of training data or training data that distinguishes objects into appropriate categories or classes, the model is then used to classify objects whose class has not been previously known. There are several classification methods in Data Mining, including Decision Tree Classifier, Bayesian Classifier, Rule-Based Classifier, etc[3]. This research aims to predict the age of birth into 3 categories based on factors experienced by pregnant women using the classification algorithm. The categories of age at birth in this study include premature, normal, or term and postdate.

The method used to predict the age of birth in this study is classification using the Naïve Bayes algorithm. Naïve Bayes is a simple probabilistic based prediction technique based on the application of Bayes theorem or Bayes rules with strong (naïve) independence assumptions[4]. The Bayes Method is a statistical approach to induction inference on classification problems.

In this study, data taken from pregnant women patients from RSUD. Dr. Moewardi Central Java Province and An-Nisa Primary Clinic as much as 550 data. 8 attributes are used as variable X in the initial data, namely: age, blood pressure, number of babies, history of childbirth, history of abortion, malnutrition, other diseases, and current pregnancy problems. The attribute used as the Y variable is the age of birth which contains normal, premature, and postdate data. However, many attributes can make the accuracy of the Naïve Bayes algorithm worse. Therefore, it takes preprocessing data to eliminate the attributes that affect the Y variable.

The Correlation-based feature selection or abbreviated as CBFS is a simple filter algorithm that ranks the subset based on a correlation-based heuristic evaluation function. Based on the hypothesis that a good attribute is an attribute that has a high correlation with the class and does not correlate to each other. The high correlation between attributes indicates that the attribute is redundant. The attributes which have a low correlation to the class is said to be an irrelevant attribute. The irrelevant attribute and redundant then must be eliminated[5].

Correlation Based Features Selection (CBFS) has been implemented on several classification algorithms, such as K-Nearest Neighbor (K-NN), Support Vector Machine (SVM), Artificial Neural Network and Decision Tree which show increased accuracy in the classification[6]. This research[6] shows that CBFS can be applied to classification and can improve accuracy. However, in previous studies, the CBFS method was not applied to the Naive Bayes algorithm.

In this research, Preprocessing was implemented using the Correlation Based Features Selection (CBFS) method on Naïve Bayes Algorithm. CBFS chooses X variables which have the highest correlation with the Y variable but has the least correlation between X variables[7]. After that, measurements are made between the accuracy before using preprocessing and after implementing preprocessing to find out how much effect the use of CBFS on the Naïve Bayes algorithm.

2. Method

This part of the method will explain the stages carried out in this research from beginning to end. The Method from this research divided into some stages in Figure 1:

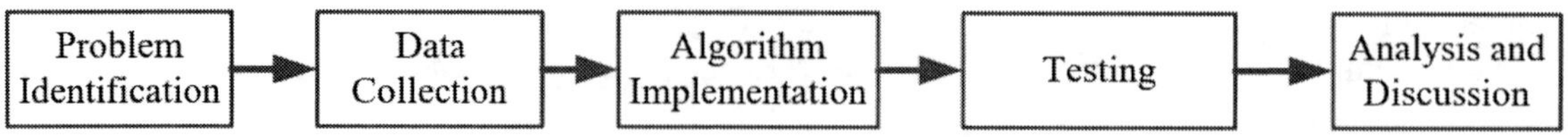

Figure 1. Stages of Research Method

IOP Conf. Series: Materials Science and Engineering **982** (2020) 012012 doi:10.1088/1757-899X/982/1/012012

2.1. Problem Identification
Problem identification is an initial stage that aims to determine what problems happen[8], especially present in the birth of a baby. Based on the problems in the field, a method for prediction is needed to predict birth ages into 3 age categories : normal, premature, and postdate.

2.2. Data Collection
The data obtained in this study are data obtained from medical records of childbirth patients in hospitals. Dr. Moewardi (Central Java Province, Indonesia) and An-Nisa Primary Clinic(Indonesia), with the amount of data reaching 550 rows.

Data is divided into 2 parts, namely training data and testing data. The amount of testing data is fixed at 100 rows. While the training data is divided into several parts again, namely training data of 100 rows, 200 rows, 300 rows, 400 rows, and finally 500 rows. This is intended to determine the effect of the amount of training data on the value of accuracy.

Explanation of the data taken in the patient's medical record is in table 1.

Table 1. Data Variables

Variable	Attribute	Type	Information
X1	Mother's age	Polynomial	a. Less (<20 years) b. Sufficient (20-35 years) c. More (> 35 years)
X2	Blood pressure	Polynomial	a. Low (<90/70 mmHg) b. Normal (90/70 - 140/90 mmHg) c. Height (> 140/90 mmHg)
X3	Number of Babies	Binomials	{1, 2}
X4	Childbirth History	Polynomial	a. Premature history b. Postmature / postdate history c. History of childbirth of normal age d. Short distance from previous pregnancy (<2 years) e. First pregnancy
X5	Abortion history	Binomials	{Yes, No}
X6	Malnutrition	Polynomial	{Less, Normal, More}
X7	Other Diseases	Polynomial	a. Heart b. Asthma c. Hypertension d. Anemia e. Diabetes mellitus f. HIV g. There is no
X8	Problems During This Pregnancy	Polynomial	a. Mild pre-eclampsia b. Severe pre-eclampsia c. Gestational hypertension d. Bleeding e. There is no
Y	Age of Birth	Label	a. Premature (<37 weeks) b. Normal (38 - 42 weeks) c. Postmature / postdate (> 42 weeks)

2.3. Algorithm Implementation

At this stage, it is divided into 2 phases. The first phase is applying Naive Bayes to classify cases the of age of birth. The second phase is implementing CBFS as preprocessing before implementing the Naive Bayes algorithm.

For the first phase, the method used to predict the age of birth in this study is classification using the Naïve Bayes algorithm. Naïve Bayes is a simple probabilistic based prediction technique based on the application of Bayes theorem or Bayes rules with strong (naïve) independence assumptions[4]. The Bayes Method is a statistical approach to induction inference on classification problems.

The advantage of using the Naïve Bayes classification is this method only requires very little training data to create the (mean and variance of variables) needed for classification. Because the independent variable is assumed, only the variance of each class variable must be determined and not the whole covariance matrix[9].

Bayes' prediction is based on the Bayes theorem with the general formula contained in the equation 1:[10]

$$P(H|E) = \frac{P(E|H) \times P(H)}{P(E)} \quad (1)$$

Information :
P (H | E) : The conditional probability of a hypothesis H occurs if the evidence is given.
P (E | H) : Probability of evidence E will affect the hypothesis H.
P (H) : The initial probability of hypothesis H occurs regardless of any evidence.
P (E) : The initial probability of proof E occurs regardless of the hypothesis / other evidence.

For the second phase, CBFS was implemented as a preprocessing stage before applying the Naive Bayes algorithm. Following steps are to calculate the correlation value on CBFS :

1. Calculate the correlation coefficient with Symmetrical Uncertainty by using the following equation :

$$\text{Gain}(S, A) = \text{Entropy}(S) - \sum_{t=1}^{n} \frac{|Si|}{|S|} * \text{Entropy}(Si) \quad (2)$$

Where S is a special set, A is an attribute, n is a total number of a partition of A attribute, $|Si|$ is a total number of cases i-th partition and $|S|$ is a total number of cases in S.

$$SU = 2.0 * \left(\frac{Gain}{H(X)+H(Y)}\right) \quad (3)$$

Where H is entropy attribute, X is Attribute 1 and Y is Attribute 2.

2. Calculate the value of the merits by using the equation (4) :

$$\text{Merits} = \left(\frac{k\overline{r_{cf}}}{\sqrt{k+k)k-1)\overline{r_{ff}}}}\right) \quad (4)$$

Where K is the total number of attributes, $\overline{_{cf}}$ is the correlation between class and attribute and $\overline{r_{ff}}$ is intercorrelation between attribute and attribute.

3. Perform the merits calculation for all possible attribute selection combinations to find the highest Merits

CBFS is done to eliminate the variable X which is less influential on the variable Y. After elimination, then the Naive Bayes algorithm is applied to the remaining variable X.

2.4. Testing

Testing is done to find out whether the system or algorithm is running[11], to find out the performance of the algorithm that has been implemented and to test calculations in training data by using data testing[12]. During this testing phase, calculation of the level of precision, recall, and accuracy is also carried out to test the accuracy of the application being built.

A Precision is a calculation of the estimated proportion of true positive cases and is formulated in equation 5[13]:

$$Precision = \frac{TP}{TP+FP} \quad (5)$$

A Recall is a calculation of the estimated proportion of positive cases that are correctly identified and formulated in equation 6:

$$Recall = \frac{TP}{TP+FN} \quad (6)$$

An Accuracy is a calculation of the proportion of the total number of correct predictions and is formulated in equation 7:

$$Accuracy = \frac{TP+TN}{TP+FP+TN+FN} \quad (7)$$

Where
TP : True Positive
TN : True Negative
FP : False Positive
FN : False Negative

2.5. Analysis and Discussion
At this stage, analysis and discussion of the test results are carried out. The first analysis is about the effect of the amount of training data on the accuracy value of the Naive Bayes algorithm. The next analysis is about a comparison between the accuracy value of the pure Naive Bayes Algorithm and the accuracy value of the Naive Bayes Algorithm that has been applied by CBFS before. After that, conclusions are drawn from the research that has been done.

3. Result and Discussion
This section contains the results of the test as well as analysis and discussion of the results of the tests that have been carried out.

The amount of data in this research is 550 data, divided into 2 parts, namely training data and testing data. The amount of testing data is fixed at 100 rows. While the training data is divided into several parts again, namely training data of 100 rows, 200 rows, 300 rows, 400 rows and finally 500 rows. This is intended to determine the effect of the amount of training data on the value of accuracy.

3.1. Precision, Recall and Accuracy Testing on Naïve Bayes Algorithm
The first testing has been done for the precision, recall, and accuracy of the Naive Bayes algorithm on each training data. The results of the test can be seen in table 2.

Table 2. Precision, Recall and Accuracy Testing on Naïve Bayes Algorithm

Training Data	Precision	Recall	Accuracy
100	0.567	0.58	58%
200	0.62	0.62	62%
300	0.63	0.63	63%
400	0.68	0.64	64%
500	0.703	0.65	65%

3.2. Precision, Recall and Accuracy Testing on CBFS before Applying Naïve Bayes Algorithm
Before the second testing, CBFS has been implemented on the training data to get optimum variables. From the CBFS that had been done, produced 4 X variables, namely: blood pressure, number of babies, congenital diseases before pregnancy, and problems during pregnancy. After CBFS has been implemented as preprocessing before applied Naïve Bayes, testing has been done for precision, recall, and accuracy. The results of the test can be seen in table 3.

Table 3. Precision, Recall and Accuracy Testing on CBFS before applying Naïve Bayes Algorithm

Training Data	Precision	Recall	Accuracy
100	0.484	0.55	55%
200	0.495	0.59	59%
300	0.64	0.63	63%
400	0.70	0.65	65%
500	0.72	0.67	67%

3.3. Discussion
From table 2, the result of the test about Precision and Recall on Naïve Bayes algorithm can be processed into figure 2.

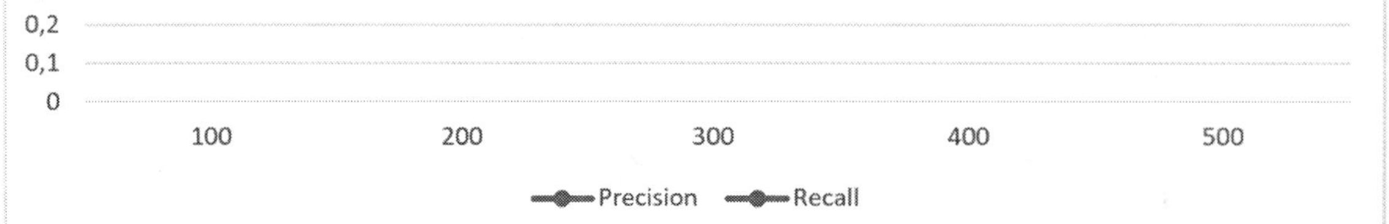

Figure 2. Precision and Recall Trends on Naïve Bayes Algorithm

From table 3, the result of the test about Precision and Recall on CBFS before Naïve Bayes can be processed into figure 3.

From figures 2 and 3, it can be seen that the value of Precision and recall from both testings tends to increase when the amount of training data increases. This is because the more the amount of training data, then the possibility of similarity of data between training data with testing data is also increasing, so it can increase the value of precision and recall.

Higher levels of precision and recall indicate that the data in the application have a collective level of data high enough to measure the closeness between the actual value and the predicted results.
From Tables 2 and 3, the result of the test about accuracy on Naïve Bayes and CBFS before Naïve Bayes can be processed into figure 4.

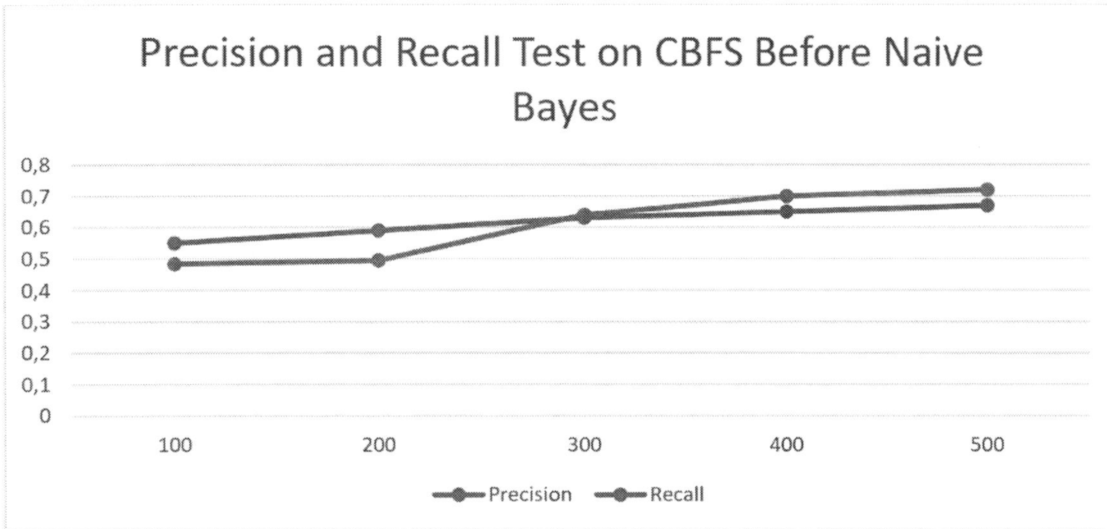

Figure 3. Precision and Recall Trends on CBFS before Naïve Bayes Algorithm

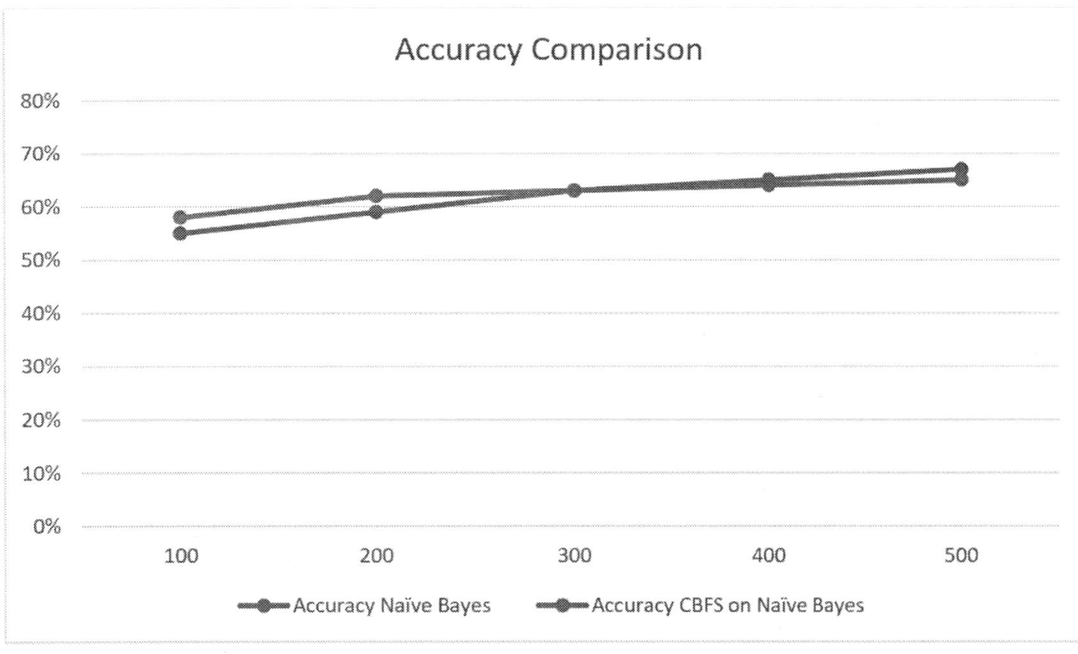

Figure 4. Accuracy Comparison from Naïve Bayes and CBFS on Naïve Bayes

Based on Figure 4, it can be seen that the value of Accuracy increases with an increasing amount of training data. Also, when the amount of training data is small (100 lines and 200 lines), the accuracy in CBFS Naive Bayes is lower than the accuracy of Naive Bayes, but when the amount of training data increases, starting at 400 lines up to 500 lines of training data, the accuracy value at CBFS Naive Bayes increases and exceeds Naive Bayes accuracy. This is because, in CBFS Naive Bayes, the number of X attributes that affect the value of the Y attribute (gestational age) is getting smaller, so the more training data provided, each attribute X can have more effect for the Y attribute. With the least attribute X, then the probability of calculation is increasing which can increase the accuracy value of Naive Bayes.

4. Conclusion

Based on the results of tests that have been done, it can be concluded that the greater the amount of training data, it can increase the value of Precision, Recall, and Accuracy on Naive Bayes and Correlation Based Feature Selection (CBFS) as preprocessing before Naive Bayes. Also, accuracy values on CBFS Naive Bayes tend to increase beyond accuracy in Naive Bayes when the amount of training data is large.

References

[1] WHO, "Preterm birth," 2018. https://www.who.int/en/news-room/fact-sheets/detail/preterm-birth (accessed Jul. 14, 2020).

[2] W. March of Dimes, PMNCH, Save the Children, *Born to Soon : The Global Action Report on Preterm Birth*. Geneva: Eds CP Howson, MV Kinney, JE Lawn. World Health Organization., 2012.

[3] T. M. Tran, X. M. T. Le, H. T. Nguyen, and V. N. Huynh, "A novel non-parametric method for time series classification based on k-Nearest Neighbors and Dynamic Time Warping Barycenter Averaging," *Eng. Appl. Artif. Intell.*, vol. 78, no. October 2017, pp. 173–185, 2019, doi: 10.1016/j.engappai.2018.11.009.

[4] L. Hou, Y. Guo, and J. Cao, "The Improved Joint Bayesian Method for Person Re-identification Across Different Camera," vol. 15, no. 4, pp. 785–796, 2019.

[5] S. Adi, Y. Pristyanto, and A. Sunyoto, "The best features selection method and relevance variable for web phishing classification," in *2019 International Conference on Information and Communications Technology, ICOIACT 2019*, 2019, no. June 2020, pp. 578–583, doi: 10.1109/ICOIACT46704.2019.8938566.

[6] Y. Pristyanto, S. Adi, and A. Sunyoto, "The Effect of Feature Selection on Classification Algorithms in Credit Approval," in *2019 International Conference on Information and Communications Technology (ICOIACT)*, 2019, doi: 10.1109/ICOIACT46704.2019.8938523.

[7] T. Astuti, R. Wahyudi, U. Hasanah, B. P. Hartato, and Z. Rifai, "The Impact of Features Selection on Performance of Artificial Neural Network in Diagnosis of Diabetic Retinopathy," in *2019 4th International Conference on Information Technology, Information Systems and Electrical Engineering (ICITISEE)*, 2019, doi: 10.1109/ICITISEE48480.2019.9003856.

[8] Y. I. Kurniawan, A. Rahmawati, N. Chasanah, and A. Hanifa, "Application for determining the modality preference of student learning," in *Journal of Physics: Conference Series*, 2019, vol. 1367, no. 1, pp. 1–11, doi: 10.1088/1742-6596/1367/1/012011.

[9] V. S and D. S, "Data Mining Classification Algorithms for Kidney Disease Prediction," *Int. J. Cybern. Informatics*, vol. 4, no. 4, pp. 13–25, 2015, doi: 10.5121/ijci.2015.4402.

[10] Y. I. Kurniawan, "Perbandingan Algoritma Naive Bayes dan C.45 Dalam Klasifikasi Data Mining," vol. 5, no. 4, pp. 455–463, 2018, doi: http://dx.doi.org/10.25126/jtiik.201854803.

[11] F. Y. Al Irsyadi, Supriyadi, and Y. I. Kurniawan, "Interactive educational animal identification game for primary schoolchildren with intellectual disability," *Int. J. Adv. Trends Comput. Sci. Eng.*, vol. 8, no. 6, pp. 3058–3064, 2019, doi: 10.30534/ijatcse/2019/64862019.

[12] Y. I. Kurniawan, E. Soviana, and I. Yuliana, "Merging Pearson Correlation and TAN-ELR algorithm in recommender system," in *AIP Conference Proceedings*, 2018, vol. 1977, doi: 10.1063/1.5042998.

[13] T. Vafeiadis, K. I. Diamantaras, G. Sarigiannidis, and K. C. Chatzisavvas, "A comparison of machine learning techniques for customer churn prediction," *Simul. Model. Pract. Theory*, vol. 55, pp. 1–9, 2015, doi: 10.1016/j.simpat.2015.03.003.

IOP Conf. Series: Materials Science and Engineering **982** (2020) 012013 doi:10.1088/1757-899X/982/1/012013

Thermal Hotspots Distribution of MODIS Aqua/Terra Satellite in Humbang Hasudutan Regency, Northern Sumatra

Yahya Darmawan [1,4], Lamtupa Nainggolan [2], Thahir Daniel Hutapea [1], Erwin Eka Syahputra Makmur [1], Ivan Mambaul Munir [3]

[1]Research and Development Center, The Agency for Meteorology, Climatology, and Geophysics of the Republic of Indonesia, Jl. Angkasa I No.2, Kemayoran, Jakarta Pusat 10720, Indonesia.
[2]Education and Culture Board, Medan City Government, Sidorame Barat II, Medan Perjuangan, Medan City, North Sumatra 20233, Indonesia.
[3]Banten Assessment Institute for Agricultural Technology (AIAT), Indonesian Agency for Agricultural Research and Development, Ministry of Agriculture Republic of Indonesia, JL. Ciptayasa, KM. 01, Singamerta, Ciruas, Serang, Banten, 42182, Indonesia.

[4]yahya.darmawan@bmkg.go.id

Abstract. The hotspot of MODIS Aqua/Terra represents the high risk of land/forest fire due to an extreme temperature over the hotspot location. Spatial-temporal analysis of the hotspot can be used to map the regions with a high vulnerability of forest/land fires. This study has an objective to use hotspot data from MODIS Aqua/Terra to map the forest/land fires over the Humbang Hasudutan Regency from 2001 to 2019. In spatial, the hotspot mostly occurs in the eastern part of the study area (Doloksanggul and Pollung), which this area has assigned as the peatland area. Based on land cover, the hotspot often detected over the dryland agriculture, dryland with shrubs, and bare soil. In temporal, the hotspot mostly increases during the dry seasons such as February, June, and July. The hotspot decreases during the wet season, such as January, October, November, and December. Besides, there was an inversely proportional between the number of detected hotspots and rainfall over the study area.

1. Introduction

Indonesia is one of the tropical countries and is included as the third-largest peatland in the world. The peatland area is around 265,500 km2, spread on the islands of Sumatra, Kalimantan, and Papua. Based on the breadth, the Indonesian peatland area is half of the tropics' peat area [1]. According to the 2008 reports, Indonesia's peatland stores are the third-largest carbon stock in the world (after Canada and Russia), which is around 54,016 Mton. One of the areas in North Sumatra that have peatlands is the Humbang Hasundutan Regency (starting now referred to as Humbahas), which has an estimated peatland of 2,358 ha [2]. Dolok Sanggul Sub-district, Pollung Sub-district, and Lintong Ni Huta Sub-districts are sub-districts with a peat area in Humbang Hasundutan Regency. The type of peatland in the Humbang Hasundutan Regency is a tophogenic peat type or highland peat type with many benefits such as carbon stock storage and a Lake Toba catchment area. Peat utilization activities that have occurred in Humbang Hasundutan Regency have become a problem given the number of emissions that will be generated from these activities and affect their carbon storage capacity [2]. Peatland is a land rich in organic matter formed from the remains of plants that have not been weathered entirely due to environmental conditions saturated with water and nutrient-poor. Peatlands whose primary function as carbon storage can change into carbon source areas and other greenhouse gases, the greenhouse gases emitted by peatlands are CO2, CH4, and N2O; among these three gases, CO2 is the most critical greenhouse gas due to its relatively large amount.

Hotspots are hotspots that are indicated as locations of forest and land fires. This parameter has been used extensively in various countries to monitor forest and land fires from satellites. The way to detect forest and land fires is by observing hotspots. Hotspots (hotspots) can be detected with NOAA (National Oceanic and Atmospheric Administration) satellites equipped with AVHRR (Advanced Very High-Resolution Radiometer) sensors. In detecting forest fires, NOAA satellites do not directly detect fires (temperature), but those detected are hotspots. Hotspots can be detected with NOAA satellites equipped with AVHRR sensors that work based on the thermal energy emitted from objects observed from an area with a temperature of 42^0 C.

This satellite is often used to detect the region because one of its sensors can distinguish surface temperatures on land or sea. Another advantage is that the satellites often visit the same place twice a day and night; another advantage is the low price. A hotspot (hotspot) can reflect an area that may be partially or completely burned because it does not indicate precisely how large the area burned. The number of hotspots can vary significantly from a subsequent measurement depending on the measurement time of the day (fire activity decreases at night and highest in the afternoon), weather (the sensor used cannot penetrate clouds and smoke) and what organization which provides such data (there is no temperature or temperature threshold standard to identify hot spots) [3].

Hotspots provide little information if they are not supported by further analysis and interpretation. A group of hotspots which detected in a large place continuously is a good indicator of fires (hotspots). Hotspot data is useful when combined with information such as land use, crop cover, animal habitat, or other maps. The bias or geographic error of a hotspot can be as far as 3 km [3]. This hotspot data is free downloadable by the public, but the use of this data for study the distribution of hotspots is rarely used by the public. This study's objective is to use MODIS Aqua / Terra satellite to assess the risk of forest and land fires in the Humbang Hasudutan Regency (Humbahas) based on the distribution of hotspot from MODIS satellite data. The result of this study can be used as a reference for Humbang Hasudutan District to anticipate forest fires over the study area.

2. Material and Methods

2.1. Study area

The study area is located in the central part of North Sumatra Province. Humbang Hasundutan Regency is located on the lines 20 1' – 20 28 'LU and 980 10' – 980 58 'BT. The area of Humbang Hasundutan Regency is 251,765.93 Ha and consists of 10 sub-districts (Pakkat sub-district, Onanganjang sub-district, Sijamapolang sub-district, Lintongnihuta sub-district, Paranginan sub-district, Doloksanggul sub-district, Pollung sub-district, Parlilitan sub-district, Tarabintang sub-district, and Baktiraja sub-district).

2.2. Data used

We used several data, including land use and land cover data, precipitation, Humbahas Regency administration map data. Hotspot data can be downloaded from MODIS aqua/terra. MODIS (Moderate Resolution Imaging Spectroradiometer) is the main sensor on the Terra satellite and Aqua satellites that orbit the earth polarly (north to south) at an altitude of 705 Kilometers and crosses the equator at 10:30 and at 22:30 local time [4]. The width of land coverage on the surface of the earth each rotation is around 2330 Kilometers. Reflections of electromagnetic waves received by MODIS (Moderate Resolution Imaging Spectroradiometer) sensors are 36 channels (36 wavelength intervals), ranging from 0.620 μm to 14,385 μm (1 μm = 1 / 1,000,000 meters). MODIS (Moderate Resolution Imaging Spectroradiometer) data using the brightness temperature data for channel 21 or channel 22 (T4) and channel 31 (T11). Canal 32 for cloud masking, the brightness temperature for this channel is denoted by (T12) [4] can be used to detect the location and distribution of hotspots. The hotspot data can easily download by this site: https://firms.modaps.eosdis.nasa.gov/. The output will be provided in shapefile (shp).

The precipitation data relied on satellite processing, the Climate Vulnerabilities Group InfraRed Precipitation with Station data (CHIRPS) from January 2001 to December 2019 with a spatial resolution of 5 km. CHIRPS is a new 30+ year quasi-global groundwater level dataset. It has area scanning from 50°S to 50°N in all longitudes, which is the daily data that has duration from 1981 to present. For spatial resolution, CHIRPS incorporates 0.05° resolution satellite imagery, which validates by station data to create gridded groundwater level time series for many applications. Since February 12th, 2015, version 2.0 of CHIRPS is finished and opened to the public. Compared to other satellite data (GSMaP, IMERG, and CHIRPS,) CHIRPS has a better correlation to the observation data in Indonesia, especially for monthly periods [5]. We also used the Humbahas Regency administration map data from BPN (2010) and Land cover from the Indonesian Forestry Planning (2011).

2.3. Methods

The steps taken in this study are as follows: the MODIS satellite data processing process includes downloading, extracting, and displaying data in spatial form. The format of the data obtained from the website is in the form of shapefile so that it can be directly plotted using ArcGIS software. Second, the TRMM satellite data processing process includes downloading, extracting, and displaying data in spatial form. Data processing is performed using GrADS. Third, Overlay between the MODIS satellite and Humbahas versus administration map and MODIS Satellite versus Land Cover. Spatial analysis is carried out to determine the pattern of hotspots distribution both by village / sub-district and land cover. Fourth, Comparison of MODIS satellite data vs. TRMM satellite has been done to find out the relationship between hotspots and average rainfall in Humbahas regency.

3. Results

3.1. Hotspot spatial distribution

The map was based on hotspot data from the MODIS Aqua / Terra satellite from January 1, 2001 to 31 December 2019. Based on available MODIS data, 575 were received in the Humbahas Regency (Figure 1).

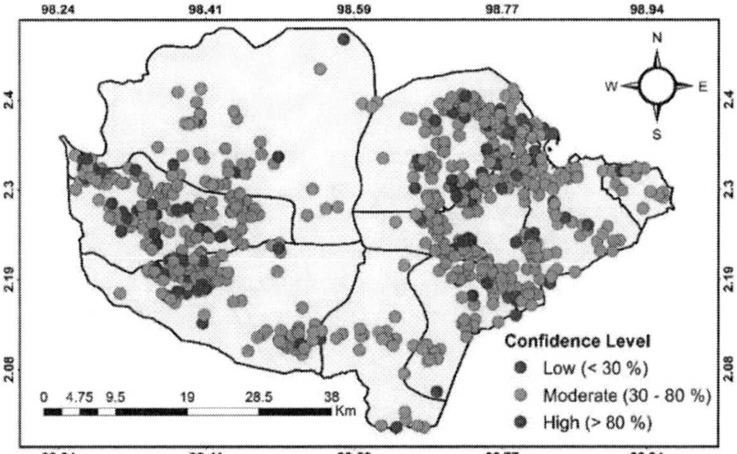

Figure 1. Hotspots distribution in Humbang Hasudutan Regency from 1 Jan 2001 to 31 December 2019

Basically, the data are being classified into three classes based on the confidence level of MODIS Aqua/Terra. The total of hotspots is about 744 hotspots from 2001 to 2019. As many as 33 points were included as the "Low" of confidence level. For the moderate confidence level, we found 602 hotspots in the "Medium" category. There were 109 hotspots which assigned as "high" of the confidence level.

In general, the distribution of in Humbahas is spreading only in the western part, eastern part, and southern part.

In this study, a hotspot distribution overlay was carried out on the administrative map of the Humbahas Regency, where the Humbahas Regency consisted of 10 districts and 91 villages. The overlay distribution of and Humbahas Regency administration map show that the distribution mostly occurs in Pollung Sub-district, Doloksanggul Sub-district, and followed by other sub-districts (Figure 2).

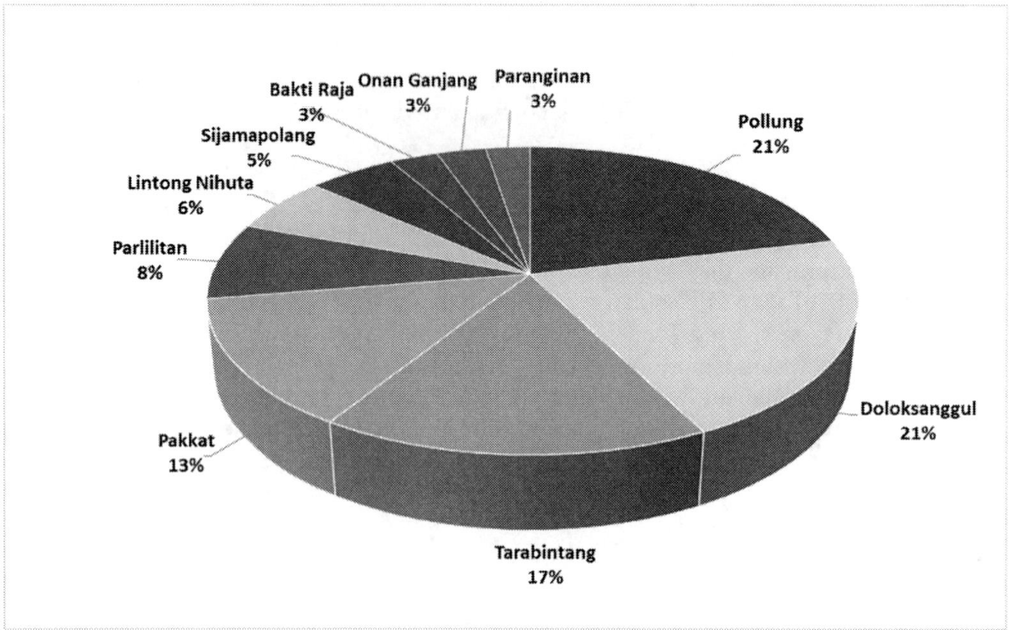

Figure 2. Distribution of the hotspot based on the district in the study area

3.2. Spatial pattern based on land cover

In addition to overlaying the administrative map of the Humbahas Regency, this study also analysed the relationship between distribution and land cover. It will be useful and vital to investigate the relationship between land cover and in Humbahas Regency. Broadly speaking, as many as 40% of the 744 detected were on land cover in the form of dryland agriculture (Pt / 20091). Based on the classification, all agricultural activities on this dry land include dry fields, mixed gardens, and fields. All agricultural activities on dry land such as dry fields, mixed gardens, and fields. Based on SNI 7645-2010 concerning Land Cover Classification by Bakosurtanal, Dryland Agriculture is included in the classification of agricultural land carried out on dry land. Dryland is characterized by low rainfall (< 250 - 300 mm / year). Drought index (ratio/ratio between rainfall and evapotranspiration is less than 0.2), plant variation is minimal (only shrubs, grasses, and small trees in certain areas), very high temperatures (+ 49 degrees Celsius in summer), the texture of the soil is sand and has high salination of the soil and groundwater caused by high evaporation and infiltration.

Typically, the types of plants on dry land are corn, sweet potatoes, and beans. In addition to Dryland Agriculture, we detected in dryland agriculture with Shrubs (Pc / 20092), as many as 18% of the 744 were detected. This type of land cover includes all types of dryland agriculture intermittent with shrubs, shrubs, and logged forests. This type of land cover often appears in areas of shifting cultivation and rotation of planting karst land. The third type of land cover where are detected is open land (T / 2014) as much as 17% of the 744 detected. Open land is the entire appearance of open land without vegetation (outcrops of mountain peaks, volcanic craters, sandbanks, beach sand), open land after the fire, and open land covered with grass/grass. The appearance of open land for mining is explained by mining, while open land used for land clearing is classified as agriculture, plantation, or plantation forest.

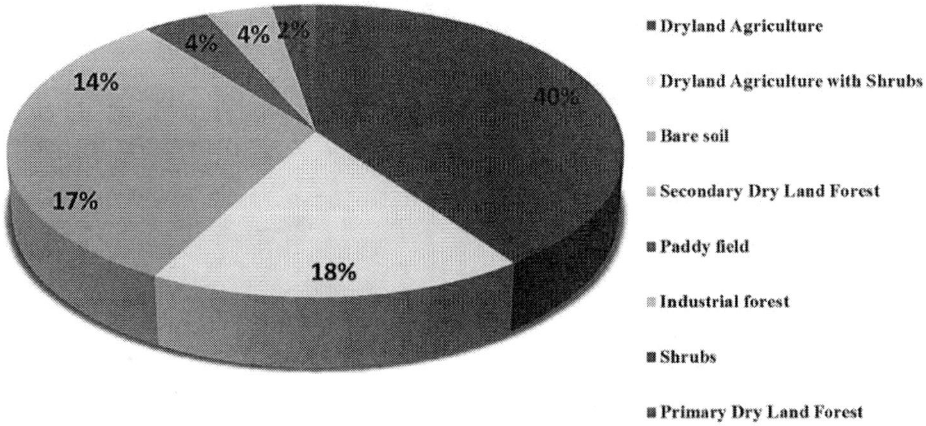

Figure 3. Hotspot based on land cover in Humbahas Regency

3.3. Hotspots distribution based on time series

In addition to conducting spatial analysis, a temporal analysis is based on annual and monthly time series to determine the general pattern of during the last 19 years in Humbahas Regency. By using linear regression, it was found that there were variations in the form of fluctuations in the number of detected each year in the 19 years of existing data. However, based on linear regression analysis, a positive trend has been detected in the last 19 years. The highest peak of occurrence was in 2016 (98 hotspots detected) and followed in 2003 (68 hotspots detected) (Figure 4).

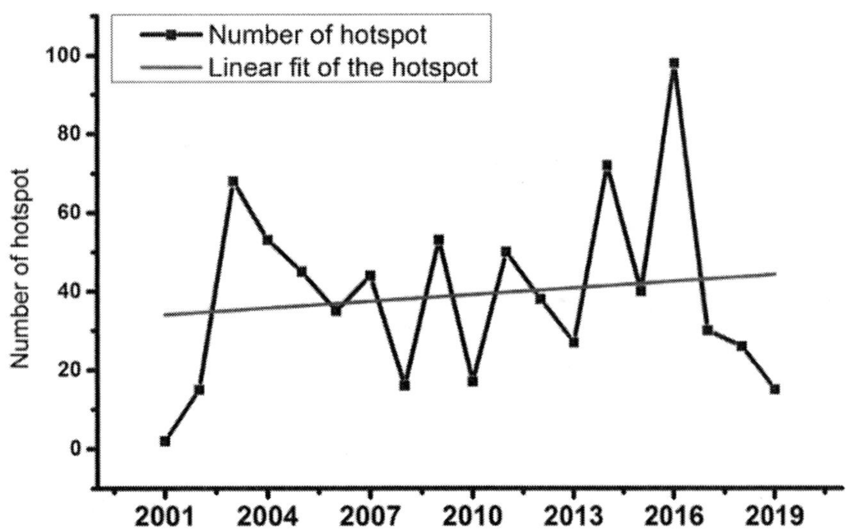

Figure 4. Annual Hotspot based on land cover over Humbahas Regency

For the number of monthly hotspots analysis, it is seen that the spread of hotspots was mostly detected in February, June, and July (Figure 5). In February there were 67 hotspots. Whereas for June, the number of hotspots detected was 161 hotspots. A total of 147 hotspots were detected in July. Hotspots decrease in intensity in January (11 hotspots), April (4 hotspots), October (7 hotspots), November (1 hotspot), and December (5 hotspots) (Figure 5).

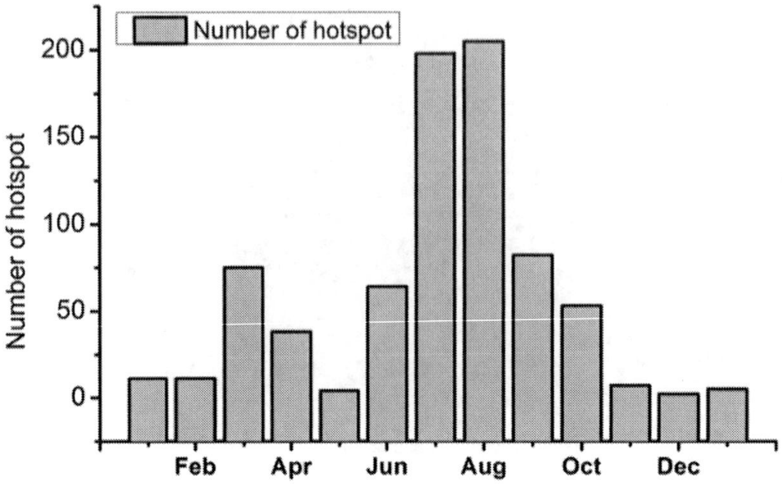

Figure 5. Hotspot based on month over the Humbahas Regency

Referring to the results of the analysis of the number of hotspots per month in Humbahas, a comparison was made between the number of hotspots and the average monthly rainfall for 19 years based on the TRMM rain satellite in 3 districts namely Pollung, Doloksanggul, and Lintong Nihuta. The analysis shows that the detected hotspots have the same pattern as the average rainfall in Humbahas Regency. When rainfall is low, the more hotspots are detected and vice versa when the rainfall is high, the hotspots tend to decrease (Figure 6).

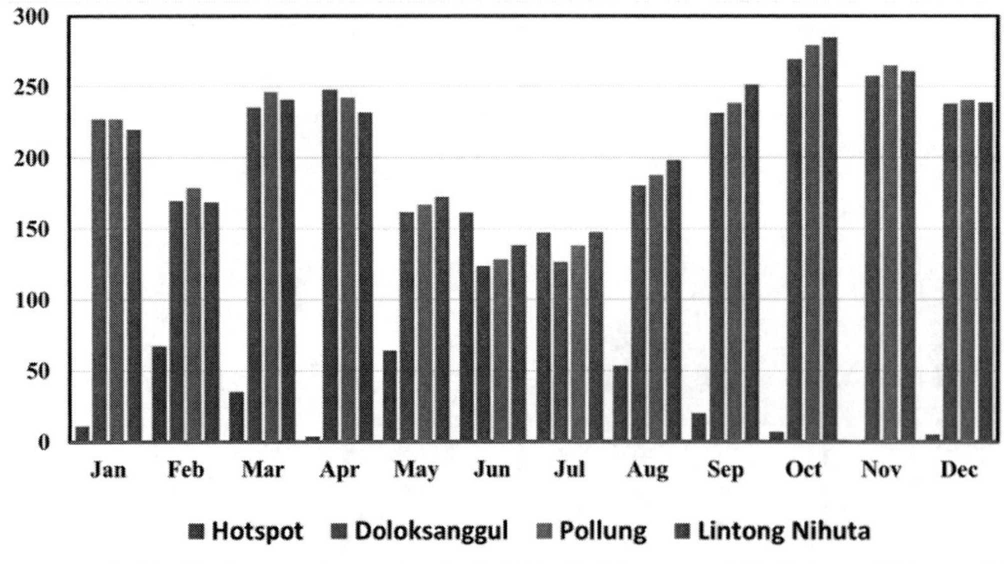

Figure 6. Relationship between hotspot and rainfall in Doloksangggul, Pollung and Lintong Nihuta

4. Discussion

Spatial analysis of hotspots has been carried out for Humbahas Regency. The spatial pattern of hotspots in Humbahas is spatially centered in two districts, namely Pollung and Doloksanggul. Based on Sitanggang et al. (2013), these two districts have natural resource potential in the form of peatlands [7]. In addition to the Lintong Nihuta sub-district. From the spatial map of the distribution of hotspots (figure 2), hotspots in Humbahas District were detected a lot, especially in sub-districts classified as peatlands such as in Pollung and Doloksanggul sub-districts. For the Doloksanggul sub-district, several villages have quite several hotspots such as Hutaraja (23 points), Saitnihuta (15 points), and Simangaronsang (14 points). As for the Pollung sub-district, three villages have the highest records for hotspots, namely Parsingguran I (16 points), Parsingguran II (31 points), and Ria Ria (22 points).

Based on the overlay results with the 2011 land cover map from Bakosurtanal, some hotspots detected in Humbahas District indicated a lot to appear in dryland agriculture, dryland farming mixed with bushes and open land. For Humbahas, conditions with a high fire risk can be seen from the number of hotspots detected, especially in areas with potential peatlands. A hotspot is a pixel in a satellite portrait that is a representation of an actual area on earth covering an area of 1.1 km2 and which has a temperature above 42° C [8]. When a hotspot is detected at a location, the surface temperature at that location reaches 42^0 C and is prone to burning.

Usually, peatlands are not flammable because of their sponge-like nature, which absorbs and holds water optimally so that in the rainy and dry seasons, there is no difference in extreme conditions. However, if the condition of the peatland has begun to be disrupted as a result of land conversion or canal making, the ecological balance will be disrupted. In the dry season, peatlands will be very dry to a certain depth and flammable. Peat contains fuel (plant residues) below the surface, so that fire in peatlands slowly spreads beneath the surface of the soil and is difficult to detect and causes thick smoke. Fire on peatlands is challenging to extinguish so that it can last for months (months).

Temporal analysis of hotspots in Humbahas District shows a positive Trend that indicates an increase in the number of hotspots detected each year. Based on temporal analysis of the number of monthly hotspots, many hotspots were detected in February, June, and July. Comparing the monthly pattern with average rainfall in Doloksanggul, Pollung, and Lintong Nihuta sub-districts shows a correlation between decreasing rainfall and an increasing number of hotspots detected. In general, hotspots will be detected if the surface temperature reaches 42^0C. In this case, it is clearly seen that more hotspots will be detected due to rising surface temperatures during the dry season. In this case, the real cause of the temperature rises at a location identified by the hot spot must be studied more deeply. The increase in the number of hotspots, which is inversely proportional to the average rainfall, only gives the idea that the dry season (dry season) indicates an increase in the number of hotspots.

The increase in the number of hotspots can be used as reference information to conduct a direct inspection in the field to see whether the temperature rise is related to forest fires or only because it is purely an increase in temperature. The cause of rising temperatures in a location can only be known by direct observation in the field. There are two factors related to the changes in temperature, namely natural factors and human factors (anthropogenic). Natural factors include changes in local microclimates, such as drought. Human factors include human intervention in the process of increasing ambient temperature, such as the burning of forest/land, factory roof, and burning of waste. The land burn system can be one of the scapegoats of rising hot spots in the Humbahas during the dry season because, during the dry season, farmers are encouraged to burn land.

5. Conclusions

Spatially, most hotspots distribution in Humbahas Regency occurred in the eastern part of the Humbahas Regency, such as the Doloksanggul and Pollung sub-district. Besides, areas with peatland potential tend to have more hotspots. Hotspots in Humbahas Regency were detected on land cover in the form of dryland agriculture, dryland agriculture with bushes, and open land. Temporally, many hotspots are detected, especially in the dry months such as February, June, and July. Conversely, hotspots will decrease in intensity when entering wet months such as January, October, November, and December. A comparison of the number of hotspots to monthly average rainfall shows an inverse relationship between the two. When rainfall increases, the number of hotspots detected will decrease, and vice versa. For further study, the validation needs to be done directly to the field, especially about the truth of the detection of hotspots by MODIS satellites.

IOP Conf. Series: Materials Science and Engineering **982** (2020) 012013 doi:10.1088/1757-899X/982/1/012013

References

[1] Joosten, T. A. Sirin, D. Charman, H. Minayeva, M. Silvius, Paris F, and L. Stringer (Eds.). 2007. Assessment on Peatlands, Biodiversity and Climate Change: Main Report. Global Environment Centre, Kuala Lumpur and Wetlands International. Wageningen.

[2] Istomo. 2006. Improvement of Peat Mining Material Resources: Research on Exploitation of Peat Mining Materials in Humbang Hasundutan Regency, North Sumatra Province. North Sumatra: Cooperation between the Department of Mining and Forestry of Humbang Hasundutan Regency of North Sumatra Province and the Faculty of Forestry of IPB.

[3] Fire Fight South East Asia. 2002. Pengadilan Pelaku Kebakaran Hutan dan Lahan : Sebuah Studi Kasus Mengenai Proses Hukum di Riau,Indonesia. Fire Fight South East Asia. WWF. IUCN. European Union.

[4] Justice, C., Giglio, L., Boschetti L., Roy, D., Csiszar, I., Morisette, J & Kaufman, Y. 2006. MODIS Fire Products, Algorithm Technical Background Document, Version 2.3.

[5] Liu, C-Y., Aryastana, P., Liu, G-R., Huang. W-R,. 2020. Assessment of satellite precipitation product estimates over Bali Island, Atmospheric Research, Volume 244, ISSN 0169-8095, https://doi.org/10.1016/j.atmosres.2020.105032.

[6] Juaeni, I., Hardjana, T., Nurzaman., Suryantoro, A., Martono. & Noersomadi. 2010. Pengembangan Pemanfaatan Data TRMM Untuk Menunjang Ketahanan Pangan. Program Insentif Riset Direktorat Jenderal Pendidikan Tinggi (DIKTI). Pusat Pemanfaatan Sains Atmosfer dan Iklim, Lembaga Antariksa dan Penerbangan Nasional (LAPAN).

[7] Sitanggang G.T., Rahmawaty, Rauf, A. 2013. Pemetaan Potensi Karbon di Lahan Gambut Topogen Pada Berbagai Kecamatan di Kabupaten Humbang Hasundutan, Provinsi Sumatera Utara. jurnal.usu.ac.id/index.php/PFSJ/article/view/4522

[8] Heryalianto, S.C. 2006. Studi Tentang Sebaran Titik Panas (Hotspot) Sebagai Penduga Kebakaran Hutan dan Lahan di Propinsi Kalimantan Barat Tahun 2003 dan Tahun 2004. Skripsi Sarjana, Institut Pertanian Bogor. Bogor.

IOP Conf. Series: Materials Science and Engineering **982** (2020) 012014 doi:10.1088/1757-899X/982/1/012014

Mapping of Tornado Wind Vulnerability using Satellite Data (Study case of Humbang Hasudutan Regency, North Sumatera)

Yahya Darmawan[1,4], Lamtupa Nainggolan[2], Thahir Daniel Hutapea[1], Erwin Eka Syahputra Makmur[1], Ivan Mambaul Munir [3]

[1] Research and Development Center, The Agency for Meteorology, Climatology, and Geophysics of the Republic of Indonesia, Jl. Angkasa I No.2, Kemayoran, Jakarta Pusat 10720, Indonesia.
[2] Education and Culture Board, Medan City Government, Sidorame Barat II, Medan Perjuangan, Medan City, North Sumatra 20233, Indonesia.
[3] Banten Assessment Institute for Agricultural Technology (AIAT), Indonesian Agency for Agricultural Research and Development, Ministry of Agriculture Republic of Indonesia, JL. Cipta Yasa, KM. 01, Singamerta, Ciruas, Serang, Banten, 42182, Indonesia.

[4]yahya.darmawan@bmkg.go.id

Abstract. Tornado wind often occurs in Humbang Hasudutan (Humbahas), North Sumatra. Even though the map of tornado vulnerability has constructed by The National Agency for Disaster Countermeasure (BNPB), but several regions did not yet have this map due to limited observation data. This study's objective is to map the vulnerability of tornado in Humbahas using satellite data where the satellite data has more significant quality in area coverage and time-series data. The Composite Mapping Analysis (CMA) has been used to model the vulnerability map. There were three factors has used in this model, such as rainfall, land cover, and slope. Based on the tornado vulnerability map, there was an 802,406.24 ha area (74.63%) in Humbahas, classified as "high" risk to Tornado disaster. There was 271,945.46 ha (25.29%), which assigned as "moderate" risk. About 786.51 ha (0.07%) of the Humbahas area has a "low" risk of tornado disaster. As a result, the study area has a "moderate" to "high" risk of tornado wind disaster, then the adaption and mitigation plans need to be done in handling tornado hazard in this region.

1. Introduction

In recent years, several regions worldwide have experienced a variety of natural disasters such as tsunamis, earthquakes, forest fires, cyclones, drought, and tornado. Besides, a disaster will have an impact on the economy because of the damage to infrastructure. Also, there is a tendency for natural disasters to spread more and more intensely according to predictions of climate change. These are the two things that encourage the emergence of initiatives from governments from various countries to reduce the risk of disasters as set out in the HFA (Hyogo Framework for Action). One of the priority programs is identifying, assessing, and monitoring disaster risks and improving early warning capabilities. Indonesia is one of 160 countries that have ratified the HFA and put it in the form of Law number 24 of 2007 concerning disaster management (BNPB Head Regulation no. 3 of 2012) [1].

The mandate of the Act was followed up with the preparation of the 2010-2014 National Disaster Management Plan, which included 4 (four) issues, namely: threat, vulnerabilities, and capacity as well as disaster risk. Furthermore, BNPB issued PERKA (Head Regulation) BNPB number 02 of 2012, which is more specific about dealing with 13 disasters such as earthquakes and extreme weather (tornado) to social conflicts. The disaster risk index and classification (low, moderate, and high) for each disaster have been calculated based on specific indicators. For example, a tornado risk index is determined through a combination of 3 indicators and their weighting using the AHP (Analytical Hierarchy Procedure) technique [2]. The three indicators which be used by BNPB are slope, land cover, and annual rainfall for a land. The risk index is then mapped using the GIS technique. In a disaster management scenario, the vulnerability map and capacity have a vital role because a mitigation and adaption plan will be determined based on those maps [1].

In response to the release of the risk index map produced by the BNPB, Paharuddin, et al. (2015) has verified the tornado disaster map from the BNPB [2]. Based on their verification results, they concluded that the accuracy of the tornado disaster index by BNPB was still low. The low accuracy of the BNPB tornado risk map because this map cannot identify un-observed areas due to limited of observation data. Based on the tornado disaster index map produced by BNPB in 2010, there are still many areas that do not have a tornado disaster index because of limited observation data in the region, including Humbang Hasudutan (Humbahas) Regency (Figure 1). Compared to the observation data, satellite data has a more robust profile in spatial coverage and time series.

In this study, the rainfall data of observation data will be replaced by the rainfall from satellite data of Climate Vulnerabilities Group InfraRed Precipitation with Station data (CHIRPS). CHIRPS is a new 30+ year quasi-global groundwater level dataset. It has area scanning from 50°S to 50°N in all longitudes, which is the daily data that has duration from 1981 to present. For spatial resolution, CHIRPS incorporates 0.05° resolution satellite imagery, which validates by station data to create gridded groundwater level time series for many applications. Since February 12th, 2015, version 2.0 of CHIRPS is finished and opened to the public. Compared to other satellite data (GSMaP, IMERG, and CHIRPS,) CHIRPS has a better correlation to the observation data in Indonesia, especially for monthly periods [3]. Moreover, the slope data also provided based on the Shuttle Radar Topography Mission (SRTM). This study aims to map the tornado wind vulnerability by using satellite data in Humbang Hasudutan (Humbahas) Regency. This study will fill the tornado map gap due to limited data in some regions in Indonesia (Figure 1).

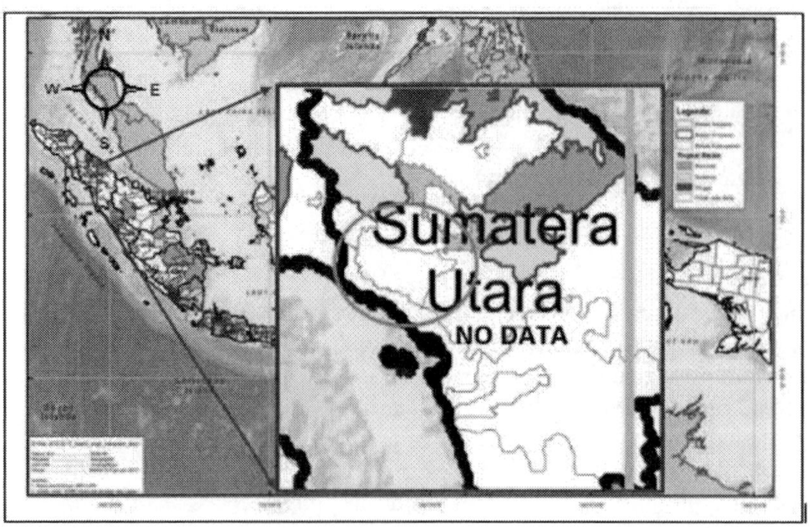

Figure 1. Tornado Disaster Risk Map from BNPB (2010). No data (brown circle)

IOP Conf. Series: Materials Science and Engineering **982** (2020) 012014 doi:10.1088/1757-899X/982/1/012014

2. Materials and Methods

2.1. Study area

The study area is located in the central part of North Sumatra Province. Humbang Hasundutan Regency is located on the lines 20 1' – 20 28 'LU and 98^0 10' – 98^0 58 'BT. The area of Humbang Hasundutan Regency is 251,765.93 Ha and consists of 10 sub-districts (Pakkat sub-district, Onanganjang sub-district, Sijamapolang sub-district, Lintongnihuta sub-district, Paranginan sub-district, Doloksanggul sub-district, Pollung sub-district, Parlilitan sub-district, Tarabintang sub-district, and Baktiraja sub-district).

2.2. Data used

We used several data, including land cover data, precipitation, Humbahas Regency administration map data, and elevation. land cover data was obtained from the Director-General of Forestry Planning in 2011. The precipitation data relied on satellite processing, which was available Climate Vulnerabilities Group InfraRed Precipitation with Station data (CHIRPS) in 2006-2019 with a spatial resolution of 5 km. Elevation data is based on the Shuttle Radar Topography Mission (SRTM) with a spatial resolution of 30 meters.

2.3. Methods

The method used in this research is the multi-criteria analysis by looking at the main factors which causing tornado disasters in an area. Tornado analysis will also be conducted to see tornado vulnerability areas. Digital Elevation Model-Shuttle Radar Topographic Mapping (DEM-SRTM) extracted the resulting slope information. At the same time, the CHIRPS was extracted by rainfall information. In comparison, information on land openness was obtained from the Director-General of Forestry Planning in 2011. From the analysis of the causes of tornado coupled with the three tornado variables mentioned above, then a multi-criteria analysis was carried out with a spatial analysis of tornado events so that tornado vulnerability models will be generated.

Determination of tornado vulnerability models in the Humbahas Regency uses several variables, including rainfall, land cover, and slopes. The tornado vulnerability modeling process requires each tornado variable's weight, and each tornado variable has a criterion class. Ranking and score of each criterion & each factor refer to BNPB (2010) scoring. For more details, the stages or steps in the Composite Mapping Analysis (CMA) process. Weight calculation for tornado vulnerability modeling using Composite Mapping Analysis (CMA) which follows these steps [4]:

1. Determination of tornado event maps based on location and tornado frequency that ever occurred in the study area.
2. Assigning several factors which support the potential zone of Tornado based on BNPB (2010).
3. Ranking and scoring of each criterion and each factor refer to previous research.
4. Furthermore, the tornado distribution map is overlaid with each tornado variable, where tornado variables include: rainfall, land openness, and slopes. The overlapping process carried out with each of these tornado variables results in a tabulation of the overlay results for each tornado variable.
5. A composite of all variables will be performed, so that the total weight of each variable causing tornado can be obtained.
6. Based on the class division, each total polygon value is categorized into five tornado disaster risk level classes, namely: "Low," "Moderate," and "High."

3. Results

3.1. Weight Calculation in the Tornado Vulnerability Model

Composite Mapping Analysis (CMA) has been used to calculate the tornado vulnerability weight. It has been done based on the frequency of tornado events occurring in the field, average rainfall, land coverage, and slope. Tornado histories over Humbahas can be seen in Table 1. The historical data obtained from the news on the internet that was successfully collected. The limitations of historical tornado data in the disaster database of BPNB is one of the challenges for collecting the tornado events in the past. From the 11 of tornado events, ten tornado events occurred in the Doloksanggul sub-district. This data will be a guideline that becomes the standard and verification of tornado models. For the Land cover, it was used in the land openness based on land cover from the Director-General of Forestry Planning in 2011.

Table 1. History of the Tornado Event

No	Date	Village	Sub-district	Impact
1	09-May-16	Saitnihuta	Doloksanggul	1 elementary school and 11 houses damaged
2	26-Apr-16	Janji	Doloksanggul	
3	26-Apr-16	Matiti 1	Doloksanggul	
8	15-Mar-15	Purba Dolok	Doloksanggul	
4	03-Sep-14	Pasaribu	Doloksanggul	14 houses damaged
5	03-Sep-14	Matiti 1	Doloksanggul	
6	03-Sep-14	Matiti 2	Doloksanggul	
7	07-Dec-12	Doloksanggul	Doloksanggul	12 houses damaged
9	02-Nov-12	Matiti II	Doloksanggul	
10	2012	Sitio	Lintongnihuta	11 houses damaged
11	15-Oct-11	Saitnihuta	Doloksanggul	1 elementary school badly damaged

For the Humbahas, the area is classified into five classes of land coverage, which consists of dryland agriculture, including open land, rice fields, settlements, dryland forests, and industrial forests. From the results of the classification of land coverage that the Humbahas regency is dominated by rice fields, settlements, mixed gardens, and ponds, as shown in Figure 2.

The slope condition in Humbahas Regency is produced from the extraction of Digital Elevation Model - Shuttle Radar Topographic Mapping (DEM-SRTM) spatial resolution of 30 meters, where the Humbahas Regency area consists of 5 slope classes, consisting of 0-8% class, 8-15%, 15-25%, and 25-45%, and> 45% where the Humbahas Regency is dominated by slopes with grades 0-8% and classes 8-15%, while slopes with classes 25-45% and > 45% are rarely found in the Humbahas district. The slope map in Humbahas can be seen in Figure 3.

For the average annual rainfall in the Humbahas Regency, it is produced by the Climate Vulnerabilities Group InfraRed Precipitation with Station data (CHIRPS) extraction with a data period of 13 years (1 January 2006 - 31 December 2019), where the Humbahas Regency area is divided into 5 bulk classes of rainfall data, consisting of classes <1000 mm / year (very dry), 1000 - 1500 mm / year (dry), 1501 - 2000 mm / year (quite wet), 2001 - 2500 (wet),> 2500 (very wet). For Humbahas, this is dominated by rainfall in the "wet" class (2001 - 2500) and "very wet". The slope map in Humbahas can be seen in Figure 4.

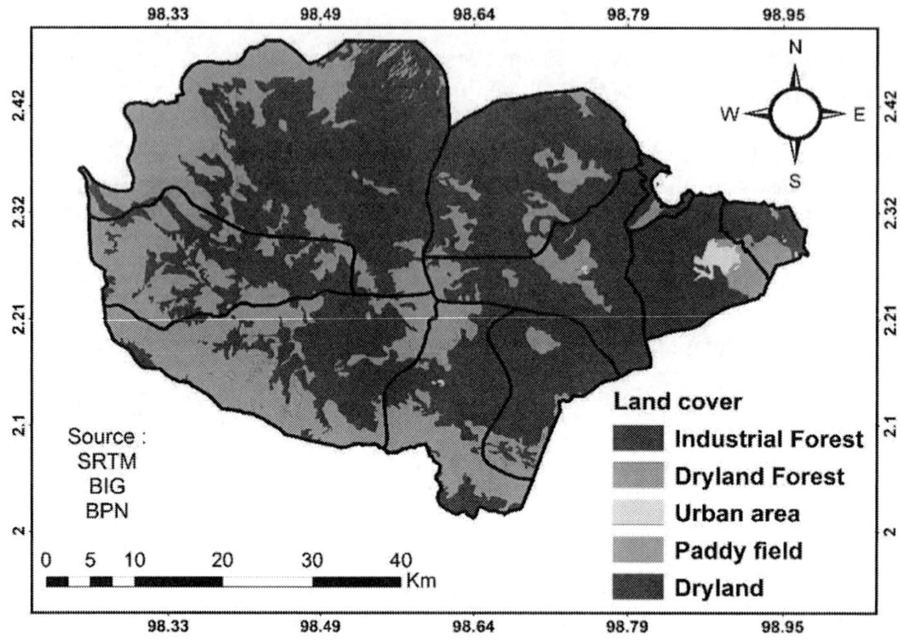

Figure 2. Land cover in the Humbang Hasudutan Regency

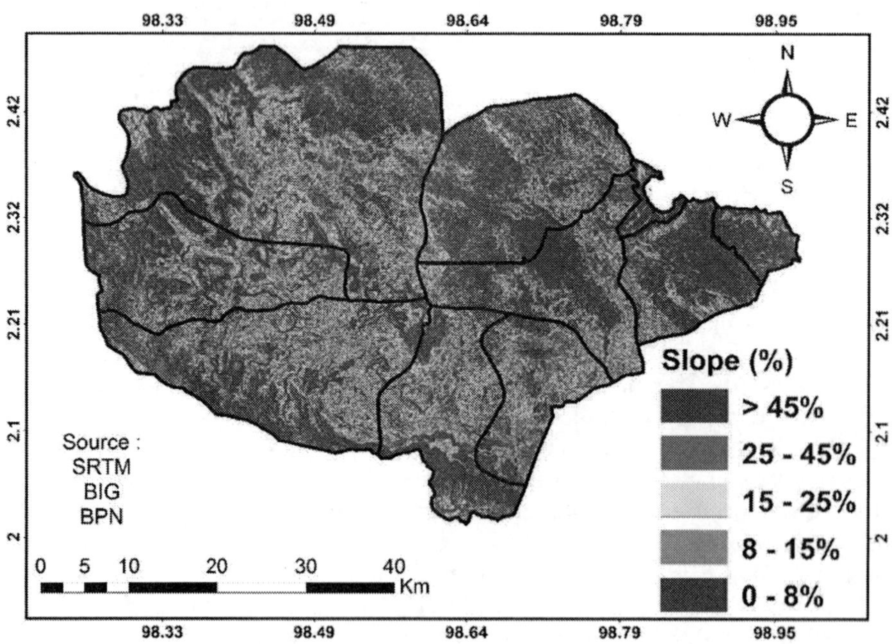

Figure 3. Slope in the Humbang Hasudutan Regency

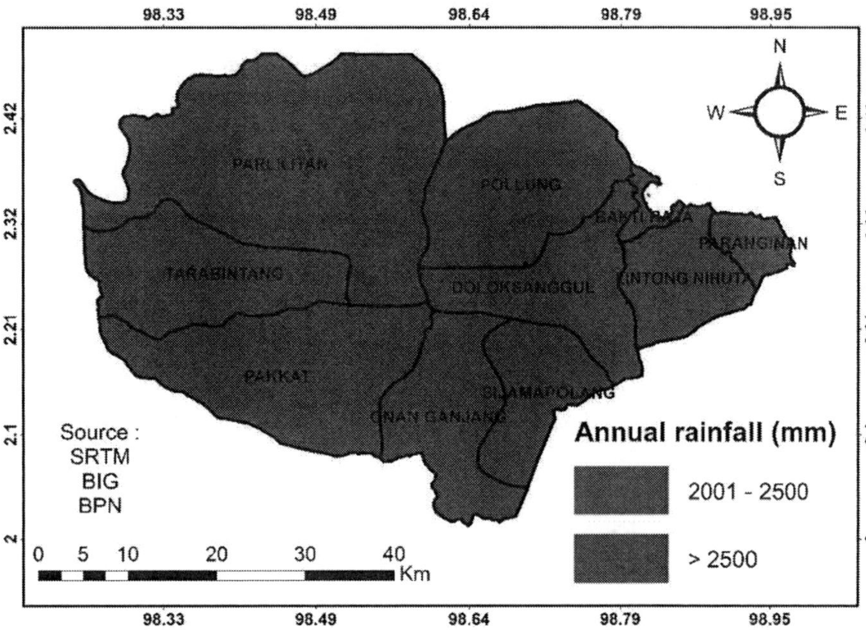

Figure 4. The annual average in the Humbang Hasudutan Regency

In this study, the weighting of the three variables using the weighting given by BNPB (2010) is as follows:

Table 2. Weighting for Tornado vulnerability map

No.	Variable	Weight
1	Land cover	33.3
2	Slope	33.3
3	Rainfall	33.3

Based on the recommendations of BNPB (2010), it is known that the weights used for the three variables are the same that is equal to 33.3. Mathematically the tornado vulnerability model in Humbahas can be formulated as follows:

$$S_{CMA} = (33.3 \times SKL) + (33.3 \times SL) + (33.3 \times SCH) \quad (1)$$

Where S_{CMA} is tornado CMA model vulnerability scoring, SCH is rainfall score value, SKL is a score of Land Cover and SL is slope score. For the weighting and scoring of tornado vulnerability maps in Humbahas Regency based on each variable from BNPB (2010) where the results of weighting and scoring calculations on each variable and tornado criteria can be seen as in Table 3.

According to the total scoring and weighting, tornado vulnerability can be explained into three (3) of tornado vulnerability classes, namely: "low", "moderate" and "High" tornado risks. The tornado class interval was calculated by the maximum score (355) which subtracted by the minimum score (211), then the output is divided by the number of classes the minimum score for each tornado variable divided by the number of classes (3 classes). The tornado vulnerability classes are obtained as in Table 4.

Table 3. Weighting and Scoring on all three research variables

No.	Variable	Criteria	Score	Weight
1.	Land Cover (LC)	Dry land / open land / water	5	33.3
		Paddy fields (cropland)	4	
		Urban area	3	
		Settlement	2	
		Dryland Forest	1	
		Industrial Forests		
2	Slope (S)	Flat - Ramps 0 - 8%	5	33.3
		Choppy 8 - 15%	4	
		A bit steep, hilly 15 - 25%	3	
		Steep 25 - 45%	2	
		Very Steep > 45%	1	
3	Precipitation (P)	> 2500 (very wet)	5	33.3
		2001 - 2500 (wet)	4	
		1501 - 2000 mm / year (quite wet)	3	
		1000 - 1500 mm / year (dry)	2	
		<1000 mm / year (very dry)	1	

Table 4. Tornado Vulnerability Class Intervals in Humbahas Regency

No.	Class Interval	Level of Tornado Risk
1.	< 211	Low
2.	≥ 211 - < 355	Moderate
3.	≥ 355	High

3.2. Tornado Disaster Risk Map in Humbahas Regency

In principle, tornado vulnerability maps are determined based on the static and dynamic variables. The static variables used include slope, which refers to the physiographic conditions. Furthermore, the dynamic variables in this modeling are land cover which refers to the land openness and rainfall as a common preliminary phenomenon a tornado wind. The final map of the tornado vulnerability map is shown in Figure 5. Based on the tornado vulnerability map, the areas which have the "high" and "low" of tornado vulnerability can be found mostly in the eastern part of Humbahas Regency, especially in the Districts of Doloksanggul, Pollung, Bakti Raja, Paranginan and some part of Parlilitan. For the west part of Humbahas regency, it mostly has a moderate vulnerability of tornado disaster.

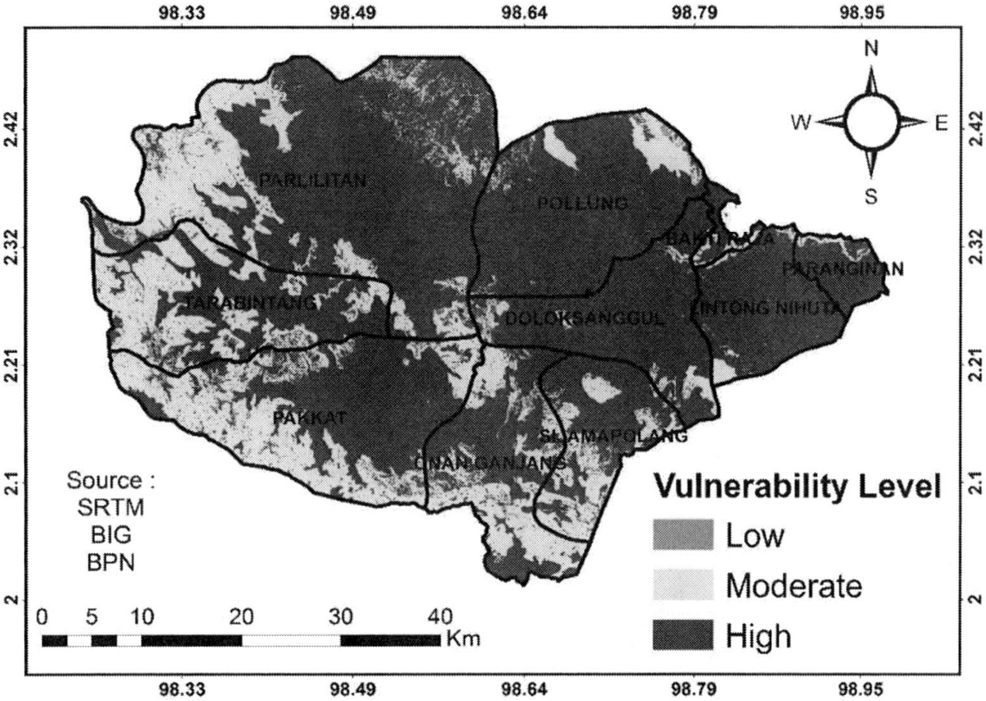

Figure 5. Tornado Disaster Risk Map in Humbang Hasudutan Regency

The details of tornado vulnerability maps in Humbahas Regency are as in Table 5.

Table 5. Percentage by Area of Tornado Vulnerability in Humbang Hasudutan Regency

Level of Risk	Area (ha)	Percentage (%)
Low	786,51	0,07
Moderate	271.945,46	25,29
High	802.406,24	74,63
Total	1.075.138,21	100

Based on the tornado vulnerability map in Humbahas, the "High" tornado risk class area is 802,406.24 hectares or 74.63%, the "Moderate" tornado risk class is 271,945.46 hectares or 25.29%, and the tornado risk class "Low" by 786.51 hectares or by 0.07%. From the results of the tornado vulnerability areas mentioned above, it can be said that in Humbahas Regency is included in the category of regions with high tornado wind hazard levels, so there is a need for anticipation in handling tornado vulnerability in the region. The area and percentage of tornado disaster risk areas based on sub-districts in Humbahas can be seen in Table 6.

Table 6. Tornado Vulnerability Map in Humbahas Regency per sub-district

No	Sub-district	Low		Moderate		High		Total	
		Area (ha)	(%)	Area (ha)	(%)	Area (ha)	(%)	Area (ha)	(%)
1	Bakti Raja	19,73	0,21	5372,15	56,63	4095,15	43,17	9487,03	100
2	Doloksanggul	16,474	0,01	5805,88	3,56	157438,5	96,43	163260,87	100
3	Lintongnihuta	58,674	0,03	4675,64	2,66	170807,2	97,30	175541,51	100
4	Onan Ganjang	41,70	0,05	42998,27	56,53	33021,04	43,41	76061,01	100
5	Pakkat	83,444	0,07	97466,67	76,30	30193,79	23,64	127743,91	100
6	Paranginan	30,307	0,04	2085,98	2,43	83795,97	97,54	85912,25	100
7	Parlilitan	165,81	0,08	63370,15	30,14	146709,1	69,78	210245,08	100
8	Pollung	309,627	0,24	4574,71	3,57	123260,2	96,19	128144,56	100
9	Sijamapolang	35,881	0,07	16735,28	33,80	32736,01	66,12	49507,17	100
10	Tarabintang	24,87	0,05	28860,72	58,62	20349,24	41,33	49234,82	100

4. Discussions

From the research results, the tornado wind hazard map for the Humbahas district has shown that the Humbahas Regency has a high risk of tornado disasters. Based on tables 6, as much as 74.63% of the Humbahas Regency has a high risk of tornado winds. The tornado wind risk spatial pattern in Humbahas District has a dominant pattern towards the east of the Humbahas Regency. Based on table 1, which shows the historical record of tornado events in the Humbahas Regency. Doloksanggul sub-district is one of the sub-districts that used to verify this tornado vulnerability map. This model's result is acceptable since the Doloksanggul sub-district has several history disasters, for example, in three villages that have recorded historical tornado causes The villages of Matiti I, Matiti II, and Saitnihuta.

The high risk of tornado wind disasters in Humbahas is mainly due to the slope, land openness, and high rainfall. Ilham & Kadir (2009) stated that when the wind blows, it can turn into a tornado when the landscape in its path is open and climatic conditions such as air pressure and extreme temperatures [5]. Usually, this incident is accompanied by lightning and thunderstorm. The indicative trigger of a tornado is due to the layered forms of cumulonimbus clouds. In this case, the land openness factor (land cover) positively influences the formation of wind into tornado winds. By the presence of vegetation on open and flat-land surfaces, it will reduce the local microclimate and the possibility of tornado winds forming in the area.

Based on the results of the overlaid of land cover, most of the land cover in the Humbahas Regency is dryland agriculture, open land, and bush/shrub. This type of land cover is one of the best factors which contribute to the tornado disasters in the Humbahas district. Besides the land openness factor (land cover), the next factor is the slope. When an area has a high slope, it means that there are variations in the topography and relief of the earth's surface in the region. When the wind blows through an area which has various surface and different tilt, it will directly reduce the wind speed and reduce some of the available energy of the wind. This mechanism influences the chance of tornado wind formation [4]. In the Humbahas district, the eastern area of the Humbahas district has a slope (0 - 8%), while the southern area of the Humbahas district has a higher slope than the eastern part. These factors also triggered many tornado disasters in the eastern region of Humbahas.

For dynamic factors such as rainfall, the Humbahas have rainfall in the wet category, which is very wet (> 2001 mm / year). With such rainfall conditions, the cumulonimbus cloud formation process likely has the opportunity to form during the rain. In general, the rainfall in the eastern area of Humbahas is lower than the rainfall in the western area of the Humbahas district. From this discussion, it can be concluded that the influence of rainfall factors has a smaller effect than two other factors, namely the neck slope and land openness. This means that further research needs to be evaluated on the weighting of BNPB (2010) of 33.3% for all three variables. For the Humbahas, the area is classified into five classes of land coverage, which consists of dryland agriculture, including open land, rice fields, settlements, dryland forests, and industrial forests. From the results of the classification of land coverage that the Humbahas regency is dominated by rice fields, settlements, mixed gardens, and ponds (Figure 2).

5. Conclusions

Three variables can be used for making a vulnerability map for a tornado disaster in Humbahas, such as land cover, slope, and rainfall. Due to limited observation data, an unmapped area of the Tornado vulnerability map can be filled by using the satellite data. Most of the Humbahas Regency has a high-risk level for the tornado. Based on the tornado vulnerability map, the "High" tornado risk class area is 802,406.24 hectares or 74.63%, the "Moderate" tornado risk class is 271,945.46 hectares or 25.29%, and the tornado risk class "Low" by 786.51 hectares or by 0.07%.

6. References

[1] Badan Nasional Penanggulangan Bencana (BNPB), 2009. Data Kebencanaan di Indonesia Tahun 2000 – 2009. Pusdatin-BNPB, Jakarta.

[2] Paharuddin, Hassanudin, Nur Hasanah. 2015. Verifikasi Peta Indeks Bencana Puting Beliung. Prosiding Seminar Nasional Geofisika 2014.

[3] Liu, C-Y., Aryastana, P., Liu, G-R., Huang. W-R. 2020. Assessment of satellite precipitation product estimates over Bali Island, Atmospheric Research, Volume 244, ISSN 0169-8095, https://doi.org/10.1016/j.atmosres.2020.105032.

[4] Elena B. C., 2002. Flood Vulnerability, Vulnerability, and Risk Assessment in the city of Turrialba, Costa Rica, Thesis S2 – ITC Netherland.

[5] Ilham, W & Kadir, S. Jurnal Hutan Tropis Borneo. Volume 10 No 28 Edisi Desember 2009. Investigasi Dan Model Terpadu Untuk Menduga Dampak Angin Puting Beliung Di Kawasan Hutan Dan Pedesaan Di Provinsi Kalimantan Selatan.

IOP Conf. Series: Materials Science and Engineering **982** (2020) 012015 doi:10.1088/1757-899X/982/1/012015

Simple Correlation Between Weather and COVID-19 Pandemic Using Data Mining Algorithms

Ari Fadli[1,3], Azis Wisnu Widhi Nugraha[1], Muhammad Syaiful Aliim[1], Acep Taryana[1], Yogiek Indra Kurniawan[2], Widhiatmoko Herry Purnomo[1]

[1]Electrical Engineering Department, Universitas Jenderal Soedirman, Indonesia
[2]Informatics Department, Universitas Jenderal Soedirman, Indonesia

[3]arifadli@unsoed.ac.id

Abstract. Currently, Indonesia is still struggling against the Coronavirus, the number of cases of the Coronavirus is increasing every day. The virus that began to attack around December 2019 has become a pandemic and is causing serious problems because there is still no vaccine to prevent transmission. The Government of the Republic of Indonesia as the party responsible for this pandemic has made various efforts including the lockdown method in the form of a Large-Scale Social Restriction (PSBB) policy. Besides that, another effort is to present a digital information portal that can be accessed and able to present the development status of coronavirus patients. With the development of information and communication technology, this research is a find correlation between weather and Covid-19 Using Data Mining Algorithm in processing data sources from the government to produce knowledge that can be used to make decisions. The research data used came from Covid-19 surveillance data of the Ministry of Health of the Republic of Indonesia and the climatic conditions in Surabaya Indonesia in the form of the average temperature (°C), the average humidity (%), the average duration of sunlight (hour) taken from the Department of Meteorology of the Republic of Indonesia. Based on the results of the study it was found that the weather affected the number of Covid-19 patients. The highest order of correlation is the average duration of sunlight (hour), the average temperature (°C), the average humidity (%).

1. Introduction

Coronavirus (COVID-19) is a new type of virus that was first identified in Wuhan, China in December 2019. The virus which then spread rapidly throughout China and all countries in the world has been declared a pandemic by the World Health Organization (WHO) [1]. Based on a study entitled "The Novel Corona Virus Originating in Wuhan, China: Challenges for Global Health Governance.", Said that on December 31, 2019, China reported a case of pneumonia, in Wuhan, Hubei Province, China, to the World Health Organization (WHO) is caused by the 2019-nCoV virus [2].

In Indonesia, the first COVID-19 case was identified on March 2, 2020. Until March 17, 2020, there were 134 confirmed cases from eight provinces, namely Bali, Banten, DKI Jakarta, West Java, Central Java, West Kalimantan, North Sulawesi, and Yogyakarta [3]. Figure 1 shows the distribution area of COVID-19 in Indonesia.

Some research in the field of Artificial Intelligence to overcome the Covid-19 epidemic is to use a model approach to identify the administration of "drug-repurposing" drugs to infected patients [4]. Other

IOP Conf. Series: Materials Science and Engineering **982** (2020) 012015 doi:10.1088/1757-899X/982/1/012015

studies use generative models to obtain compound entities to fight coronavirus COVID-19 [5]. Other research in the field of Deep Learning to make computed tomography (CT) image processing models, the results of the study show that the model is proposed when using 499 training data and 131 test data obtained accuracy reaches 0.901 accuracies with positive predictive value 0.840 and negative predictive value 0.982 [6]. In line with this research, Deep Learning shows high performance because they can We achieved an Area Under Curve (AUC) result of 0.994 with 94% sensitivity and 98% specificity (at threshold 0.5). [7].

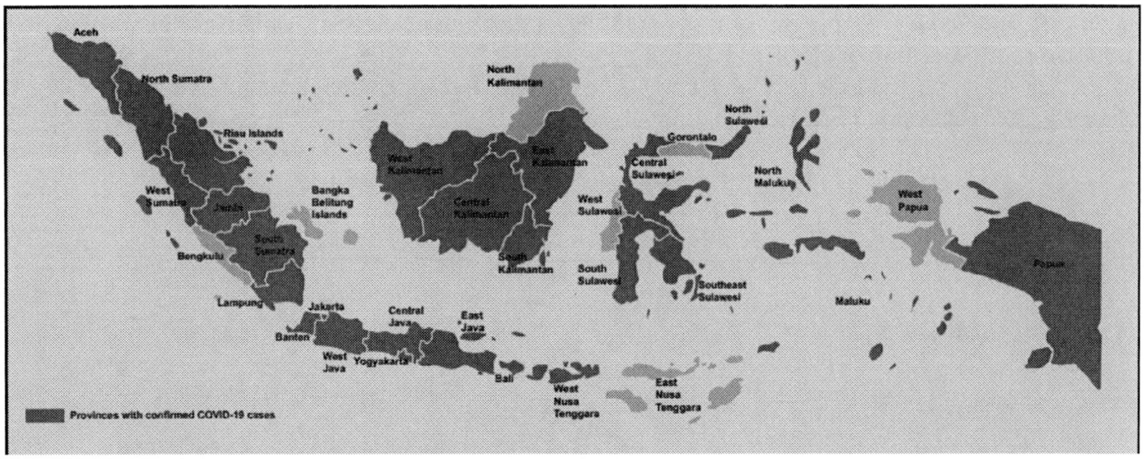

Figure 1 Geographic distribution of confirmed COVID-19 cases in Indonesia, as of 26 March 2020.
Source: https://www.covid19.go.id/2020/03/26/2112/

In other studies using the Naive Method algorithm, the prediction results obtained from the classification of COVID-19 infected status obtained positive and negative data with each data that is positive 55.48% of 249 data and negative 44.52% of 249 data. In other words, if classified as a number, positive classification is 139 people and negative classification is 110 people from a total of 249 data. [8]. Besides, research using the C4.5 Algorithm produced a test with a confusion matrix of 3 (three) classes from the COVID-19 supervision category, the calculation yielded an accuracy of 0.9286 (92.86%) [9].

Currently, the topic of research on the correlation between weather conditions and the spread of viruses is very interesting, some of which are researchers [10-15]. In the case of the West Nile virus in the United States and Europe [16-17] studied the relationship between climatic conditions and SARS-CoV and obtained results that climate variables can be a cause of biological interactions between SARS-CoV and humans. Other research results also state that weather correlates very significantly with changes in death rates due to pneumonia [18]. Viral transmission is caused by several factors such as temperature, humidity, and population density affected [19]. Tosepu studied the correlation between weather conditions and COVID-19 Jakarta, Indonesia, and found significant results [20].

In this study, the ID3 algorithm was used to find a correlation between the number of confirmed COVID-19 patients per day with changes in weather conditions. The number of COVID-19 patients used came from COVID-19 surveillance data of the Ministry of Health of the Republic of Indonesia and the weather change variables used were the average temperature (° C), the average humidity (%), and the average duration of sunlight (hour) in Surabaya taken from Department of Meteorology, Republic of Indonesia.

IOP Conf. Series: Materials Science and Engineering **982** (2020) 012015 doi:10.1088/1757-899X/982/1/012015

2. Material and Methods

2.1. Research Design

Data mining (knowledge discovery in databases (KDD)) is an activity of collecting historical data to find sequences, patterns, or variable relationships in large data. This data mining proceeds can be used to help make decisions in the future [21][22].

One of the advantages of the decision tree algorithm is its ability to produce easy-to-understand knowledge structures with low computational costs for classifying new cases, the ability to handle symbolic and numeric input variables, providing a clear indication of which attributes are most important for prediction or classification [23].

In this study, we used the Decision Tree ID3 algorithm to classify the data and extract the rules for the data set, as shown in Figure 2.

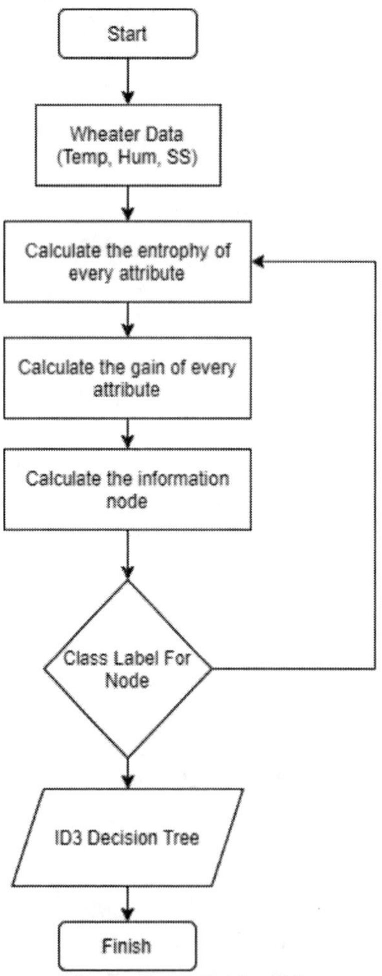

Figure 2. ID3 Algorithm

The decision tree resembles a tree structure. The tree is a hierarchical organization of node and link collection, where each node, except the root node, has one incoming link. Each node is a predictive feature and the link represents its respective value conditional variable. The decision rules of the decision tree are a way of representing knowledge. Based on the decision tree and based on testing can build decision rules.

The ID3 algorithm does not guarantee an optimal solution. It selects the best attributes from the given set. It then divides the dataset across each iteration. ID3 can adjust training data. As a solution for this is, rather than a larger tree, a smaller tree is preferred. Although this algorithm determines a solution, it does not always guarantee an optimal solution.

2.2. Data Collection

Figure 3 shows the distribution map of Covid-19 Patients in Surabaya.

Figure 3. Distribution map of Covid-19 in Surabaya City

The number of cases confirmed every day is collected from the official website of the city of Surabaya on page https://lawancovid-19.surabaya.go.id/visualisasi. The computerized data set at weather data was obtained from the Department of Meteorology, Ministry of the Republic of Indonesia. Data consists of a temperature average (°C), humidity average (%), duration of sunlight average (ss). The data used are from 1 June 2020 to 1 July 2020.

2.3. Study Area

This study was conducted in Surabaya. Surabaya is one of the major cities in Indonesia. The city of Surabaya as the capital of East Java Province is located on the north coast of East Java Province or precisely between 7 ° 9′-7 ° 21 ′ South Latitude and 112 ° 36 ′ - 112 ° 54 ′ East Longitude.

2.4. ID3 Algorithm

The ID3 (Iterative DiChaudomiser 3) algorithm was first developed by J. Ross Quinlan at the University of Sydney [24]. He first presented ID3 in 1975 in a book, Machine Learning [24], volume 1, number 1. ID3 is based on the Concept Learning System (CLS) algorithm. ID3 is a supervised learning algorithm, [25] constructing a decision tree. The resulting tree is used to classify future samples. The ID3 algorithm builds a tree based on information (information acquisition) obtained from training instances and then uses the same to classify test data [25]. ID3 algorithm generally uses nominal attributes for classification without missing values. Figure 4 shows the ID3 algorithm pseudocode [26].

```
Inputs:
R: a set of non- target attributes, C: the target attribute,
S: training data.

Output: returns a decision tree

Start
Initialize to empty tree;
   If S is empty then
       Return a single node failure value
   End If

If S is made only for the values of the same target then
   Return a single node of this value
End if

If R is empty then
   Return a single node with value as the most common value of the target
attribute values found in S
End if
   D ? the attribute that has the largest Gain (D, S) among all the
attributes of R
  {dj j = 1, 2, ..., m}  ? Attribute values of D
  {Sj with j = 1, 2, ..., m}  ?The subsets of S respectively constituted
of dj records attribute value D

Return a tree whose root is D and the arcs are labeled by d1, d2, ..., dm
and going to sub-trees ID3
             (R-{D}, C, S1), ID3 (R-{D} C, S2), .., ID3 (R-{D}, C, Sm)
End
```

Figure 4. ID3 Algorithm Pseudo Code

The selection of attributes in ID3 is used by the statistical property, which is called information gain. Gain is used to measuring how well an attribute separates training data into the target class. The attribute with the highest information will be selected. The gain value is calculated using the concept of information theory called entropy. Entropy is calculated using equation (1).

$$Entropy(S) = -P_a log_2 P_a - P_b log_2 P_b \text{ .... } (1)$$

Based on the formula above, P_a is the probability of sample S having a positive class. P_a is calculated by dividing the number of positive samples (S+) by the total number of samples (S) such that. $P_a = S+/S$. P_b is the probability of sample S having a negative class. P_b is calculated by dividing the number of negative samples (S-) by the total sample number (S) so that $P_b = S-/ S$. The leaf part of a decision tree. In the ID3 algorithm, the entropy reduction is called gain information. A comparison of sample S with attribute A can be calculated information gain using equation (2).

$$Gain(S, A) = Entropy(S) - \sum_{V \in Nilai(A)} \frac{Sv}{S} Entropy(Sv) \text{ .... } (2)$$

3. Results and Discussion

Figure 5 shows several patients obtained from the official website of the city of Surabaya on page https://lawancovid-19.surabaya.go.id/visualisasi.

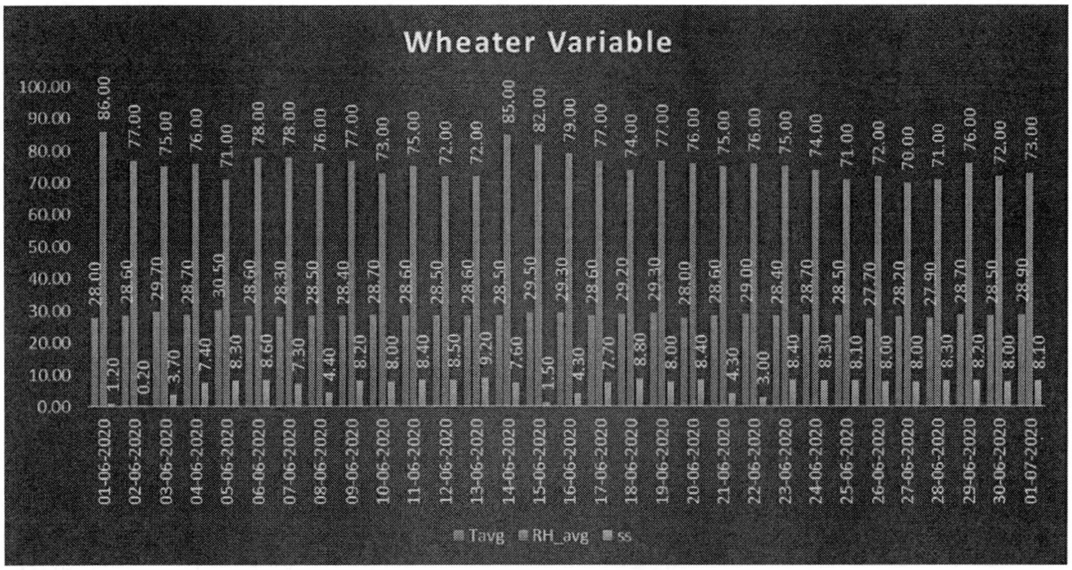

Figure 5. Wheater Variable

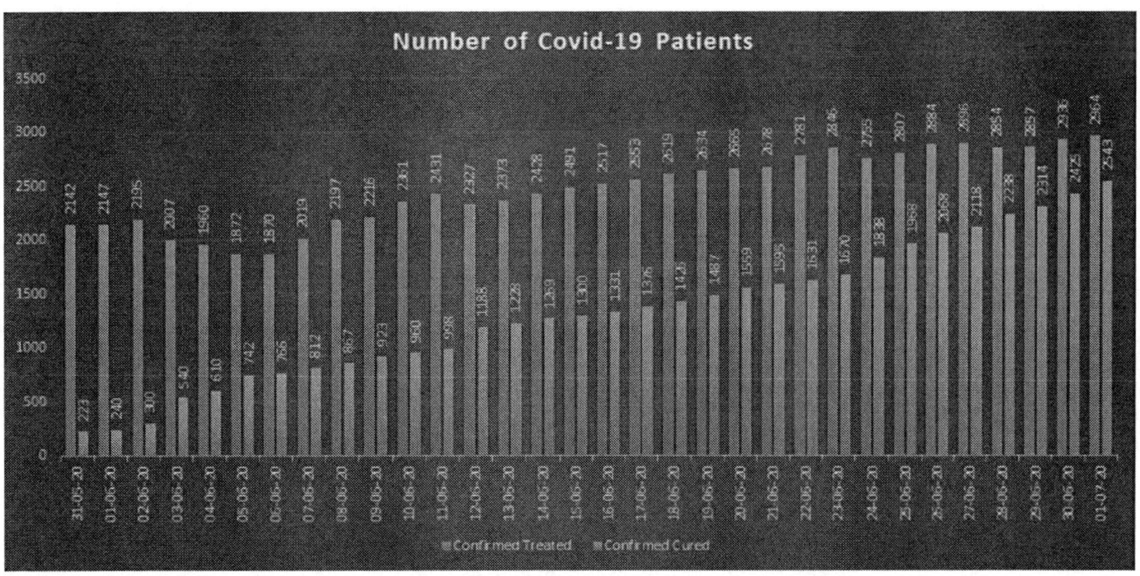

Figure 6. Number of Covid-19 Patients

Figure 6 shows the weather variable used was obtained from the Ministry of Health of the Republic of Indonesia and climatic conditions in the city of Surabaya in the form of temperature average (°C), humidity average (%), duration of sunlight average (ss) taken from the Department of Meteorology of the Republic of Indonesia. Data will be tested using the Decision Tree method with the ID3 algorithm using the Rapid Miner tool. The data used are from 1 June 2020 to 1 July 2020. Table 1 shows the dataset used in this study.

Table 1. Dataset

Tanggal	Tavg	RH_avg	ss	Kelas
01-06-2020	28 - 28.9	8.60	1.20	No
02-06-2020	28 - 28.9	7.70	0.20	Yes
03-06-2020	29 - 29.9	7.50	3.70	Yes
04-06-2020	28 - 28.9	7.60	7.40	No
05-06-2020	30 - 30.9	7.10	8.30	Yes
06-06-2020	28 - 28.9	7.80	8.60	No
07-06-2020	28 - 28.9	7.80	7.30	Yes
08-06-2020	28 - 28.9	7.60	4.40	Yes
09-06-2020	28 - 28.9	7.70	8.20	Yes
10-06-2020	28 - 28.9	7.30	8.00	No
11-06-2020	28 - 28.9	7.50	8.40	No
12-06-2020	28 - 28.9	7.20	8.50	Yes
13-06-2020	28 - 28.9	7.20	9.20	No
14-06-2020	28 - 28.9	8.50	7.60	No
15-06-2020	29 - 29.9	8.20	1.50	No
16-06-2020	29 - 29.9	7.90	4.30	No
17-06-2020	28 - 28.9	7.70	7.70	Yes
18-06-2020	29 - 29.9	7.40	8.80	Yes
19-06-2020	29 - 29.9	7.70	8.00	Yes
20-06-2020	28 - 28.9	7.60	8.40	Yes
21-06-2020	28 - 28.9	7.50	4.30	No
22-06-2020	29 - 29.9	7.60	3.00	No
23-06-2020	28 - 28.9	7.50	8.40	Yes
24-06-2020	28 - 28.9	7.40	8.30	Yes
25-06-2020	28 - 28.9	7.10	8.10	Yes
26-06-2020	27 - 27.9	7.20	8.00	No
27-06-2020	28 - 28.9	7.00	8.00	No
28-06-2020	27 - 27.9	7.10	8.30	Yes
29-06-2020	28 - 28.9	7.60	8.20	No
30-06-2020	28 - 28.9	7.20	8.00	Yes
01-07-2020	28 - 28.9	7.30	8.10	Yes

The strong motivation for this research is because it is interesting to conduct a study on COVID-19 in Indonesia. To achieve For this purpose, we are looking for a correlation between weather variables, namely the length of the sun's exposure, average temperature and average humidity for and confirmed cases of COVID-19 using the ID3 algorithm. This algorithm was built to understand the variables that most influence the COVID-19 growth curve.

Based on Figure 4, the ID3 Algorithm finds that in a row the weather variables that affect the Covid-19 growth curve are the average duration of sunlight (hours), average temperature (° C), average humidity (%), The resulting decision tree is shown in Figure 7.

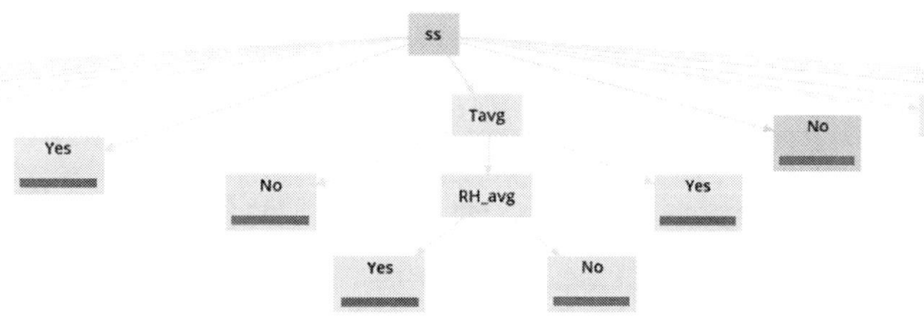

Figure 7. Decision Tree

Figure 8 is the confusion matrix used to evaluate the performance of the ID3 algorithm in producing a model to describe the correlation between weather variables and Covid-19. Based on the confusion matrix, the accuracy value is 96.67%.

accuracy: 96.77%

	true No	true Yes	class precision
pred. No	14	1	93.33%
pred. Yes	0	16	100.00%
class recall	100.00%	94.12%	

Figure 8. Confusion Matrix

4. Conclusion

Weather is an important factor in determining the number of confirmed Covid-19 cases in Surabaya. Weather variables such as the average duration of sunlight (hours), average temperature (° C), average humidity (%) was significantly correlated with the number of patients confirmed cured with successive gain values of 0.683, 0.450, 0.030, and with accuracy 96.77%.

5. References

[1]. LI, Q., GUAN, X., WU, P., WANG, X., ZHOU, L., TONG, Y., REN, R., LEUNG, K.S., LAU, E.H., WONG, J.Y. AND XING, X., 2020. Early transmission dynamics in Wuhan, China, of novel coronavirus–infected pneumonia. *New England Journal of Medicine.*

[2]. Phelan AL1,2, Katz R1, Gostin LO2. (2020), Novel Coronavirus Originating in Wuhan, China: Challenges for Global Health

[3]. KOMPAS, 2020. UPDATE: Tambah 17 Pasien, Total Ada 134 Kasus Positif Virus Korona. [online] Tersedia pada: KOMPAS.COM <https://gaya.tempo.co/read/877228/9-jutaorang-di-indonesia-mengalamidepresi/full&view=ok>. Last Accessed on 17th March 2020

[4]. B. R. Beck, B. Shin, Y. Choi, S. Park, and K. Kang, "Predicting commercially available antiviral drugs that may act on the novel coronavirus (2019-nCoV), Wuhan, China through a drug-target interaction deep learning model," BioRxiv, 2020

[5]. A. Zhavoronkov, V. Aladinskiy, A. Zhebrak, B. Zagribelnyy, V. Terentiev, D. S. Bezrukov, D. Polykovskiy, R. Shayakhmetov, A. Filimonov, P. Orekhov, Y. Yan, O. Popova, Q. Vanhaelen,

A. Aliper, and Y. Ivanenkov, "Potential COVID-2019 3C-like protease inhibitors designed using generative deep learning approaches," ChemRxi, 2 2020.

[6]. C. Zheng, X. Deng, Q. Fu, Q. Zhou, J. Feng, H. Ma, W. Liu, and X. Wang, "Deep learning-based detection for COVID-19 from chest CT using weak label," MedRxiv, 2020

[7]. O. Gozes, M. Frid-Adar, N. Sagie, H. Zhang, W. Ji, and H. Greenspan, "Coronavirus detection and analysis on chest CT with deep learning," arXiv preprint arXiv:2004.02640, 2020.

[8]. Santoso P, Fauziah F, Nurhayati N. Application of Data Mining Classification for Covid-19 Infected Status Using Algortima Naïve Method. mantik [Internet]. 1May2020 [cited 18Jul.2020];4(1, May):267-75. Available from: http://iocscience.org/ejournal/index.php/mantik/article/view/740

[9]. Wildan Wiguna1; Dwiza Riana, DIAGNOSIS OF CORONAVIRUS DISEASE 2019 (COVID-19) SURVEILLANCE USING C4.5 ALGORITHM, Jurnal PILAR Nusa Mandiri Vol. 16, No. 1 March 2020

[10]. Wang, Y., Wang, Y., Chen, Y., & Qin Q. (2020). Unique epidemiological and clinical features of the emerging 2019 novel coronavirus pneumonia (COVID-19) implicate special control measures. Journal of Medical Virology, 92(6), 568– 576. http://dx.doi.org/10.1002/jmv.25748.

[11]. Tomar, A., & Gupta, N. (2020). Prediction for the spread of COVID-19 in India and effectiveness of preventive measures. Science of the Total Environment, 725, 138762.

[12]. Gupta, S., Raghuwanshi, G. S., & Chanda, A. (2020). Effect of weather on COVID-19 spread in the US: A prediction model for India in 2020. Science of the Total Environment, 728(2020), 138860.

[13]. Ma, Y., Zhao, Y., Liu, J., He, X., Wang, B., Fu, S., et al. (2020). Effects of temperature variation and humidity on the death of COVID-19 in Wuhan, China. Science of The Total Environment, 724, 138226. http://dx.doi.org/10.1016/j.scitotenv.2020.138226.

[14]. Poole, L. (2020). Seasonal influences on the spread of SARS-CoV-2 (COVID19), causality, and forecastabililty (March 15, 2020). SSRN Electronic Journal, http://dx.doi.org/10.2139/ssrn.3554746.

[15]. Sajadi, M. M., Habibzadeh, P., Vintzileos, A., Shokouhi, S., Miralles-Wilhelm, F., & Amoroso, A. (2020). Temperature and latitude analysis to predict potential spread and seasonality for COVID-19. SSRN Electronic Journal, http://dx.doi.org/10.2139/ssrn.3550308.

[16]. Epstein, P. R. (2001). West Nile virus and the climate. Journal of Urban Health, 78, 367– 371.

[17]. Yuan, J., Yun, H., Lan, W., Wang, W., Sullivan ,S. G., Jia, S., & Bittles, A. H. (2006). A climatologic investigation of the SARS-CoV outbreak in Beijing, China. American Journal of Infection Control, 34(4), 234– 236. http://dx.doi.org/10.1016/j.ajic.2005.12.006.

[18]. Bull, G. (1980). The weather and deaths from pneumonia. Lancet, 315, 1405– 1408.

[19]. Dalziel, B. D., Kissler, S., Gog, J. R., Viboud, C., Bjørnstad, O. N., & Metcalf, C. J. E. (2018). Urbanization and humidity shape the intensity of influenza epidemics in U.S. cities. Science, 362, 75– 79.

[20]. Tosepu, R., Gunawan, J., Effendy, D. S., Imran Ahmad, L. O. A., Lestari, H., Bahar, H., & Asfian, P. (2020). Correlation between weather and Covid-19 pandemic in Jakarta, Indonesia. Science of the Total Environment, 725, 138436.

[21]. H. Siahaan, H. Mawengkang, S. Efendi, A. Wanto, and A. P. Windarto, "Application of Classification Method C4 . 5 on Selection of Exemplary Teachers," in IOP Conference Series, 2018, pp. 1–6.

[22]. M. I. Zulfa, A. Fadli, and Y. Ramadhani, "Classification model for graduation on time study using data mining techniques with SVM algorithm," AIP Conf. Proc., vol. 2094, 2019.

[23]. S. Sudirman, A. P. Windarto, and A. Wanto, "Data Mining Tools | RapidMiner : K-Means Method on Clustering of Rice Crops by Province as Efforts to Stabilize Food Crops In Indonesia," *IOP Conference Series: Materials Science and Engineering*, vol. 420, no. 12089, pp. 1–8, 2018.

[24]. Ankur Shrivastava and Vijay Choudhary ,Comparison between ID3 and C4.5 in Contrast to IDS Surbhi Hardikar, VSRD-IJCSIT, Vol. 2 (7), 2012, 659-667.
[25]. J.R. QUINLAN, Induction of Decision Trees, 1986, Machine Learning 1:81-106
[26]. Hssina Badr, Hanane Ezzikouri Mohammed Erritali "A comparative study of decision tree ID3 and C4.5, (IJACSA) International Journal of Advanced Computer Science and Applications, Special Issue on Advances in Vehicular Ad Hoc Networking and Applications, 2014

Acknowledgments
We would like to thank LPPM Unsoed for research funding in the scheme Riset Dosen Pemula.

IOP Conf. Series: Materials Science and Engineering **982** (2020) 012016 doi:10.1088/1757-899X/982/1/012016

Design of Open Loop Single Axis Solar Tracker System

Rizal Rinaldi [1], Bandiyah Sri Aprillia[1,2], Cahyantari Ekaputri [1], Muhamad Reza[1]

[1]School of Electrical Engineering, Telkom University, Bandung, Indonesia

[2]bandiyah@telkomuniversity.ac.id

Abstract. The amount of power output in a solar power system depends on the intensity of solar radiation. The earth's movements however cause changes in the intensity of solar radiation received by solar panels on the daily. Therefore, the implementation of a controller that can track the position of the sun (solar tracker) is believed to increase the power output of solar panels. This research focuses on the design of a photovoltaic panel drive system using an Android-based application as a monitoring media. The application is used to monitor the current temperature and provide a desired angle input. Based on the test results, the average power output of a fixed PV reached only 17.15 Watt, while on a single axis PV, average output reached 21.50 Watt. It can be concluded that in the single axis PV, average power output increased by more than 25% compared to the fixed PV.

1. Introduction

Indonesia is a tropical country that has solar radiation intensity exposure reaching 4 kWh / m^2 / day in each area [1]. Thus, to take advantage of this fact, technology to convert solar energy into electrical energy has been widely implemented in Indonesia. One of the means of converting solar radiation into electrical energy commonly used are solar panels. Solar panel is a device consisting of several semiconductor-based solar cells that are able to convert sunlight radiation into electrical energy based on the principle of the photovoltaic effect. In general, solar power integration can be connected on-grid or standalone [2-4]. In standalone solar power systems, solar panels are connected to batteries or other energy storage media to be connected to a converter to meet load requirements [5]. On-grid solar power systems on the other hand involve solar panels being connected to the State Power Plant (PLN) network. The implementation of electric energy source control methods in an on-grid solar power system can also improve the efficiency of energy use, because energy flow can be adjusted according to load requirements [2, 6].

The amount of power output in a solar power system depends on the intensity of solar radiation [7, 8]. The earth's movements however cause changes in the intensity of solar radiation received by solar panels on the daily. Therefore, the implementation of a controller that can track the position of the sun (solar tracker) is believed to increase the output power of solar panels. In theoretical analysis and experimental evidence, the solar tracker is able to increase the output power of solar panels by up to 60% [9, 10].

To produce optimal power from photovoltaics we can use a control system that functions as a driving and monitoring. This research focuses on the design of a photovoltaic panel drive system using an Android-based application as a monitoring media. The android application is used to monitor the current temperature and provide the desired angle input.

IOP Conf. Series: Materials Science and Engineering **982** (2020) 012016 doi:10.1088/1757-899X/982/1/012016

2. Method

A single axis solar tracker is controlled by an open loop control system. Mechanical parts are used by servo motors that are connected to bike wheels to make solar panels move to follow the sun's movements. Bike gear is used to reduce torque load on servo motors.

It is important to use gear because they are engine elements used to transmit motion and power from one shaft to another without slippage. The use of gear transmission is the appropriate choice because the distance between the drive and the driven components are not far apart, making it compact. For this study, straight gears are used. Using 15 gear drivers, for 18 gear driven drives, the gear ratio is 15/18 = 0.83. The gear driver rotates 1.2 times for every 1 gear driven rotation.

Connections between gears are maintained by using a belt (chain). The chain reduces slip between gears. The gear chain uses iron to reduce maintenance costs. To determine the length of the belt:

$$L = \pi(r_1 + r_2) + 2x + \left\{\frac{(r_1 - r_2)^2}{x}\right\} \quad (1)$$

where:

x = shaft axis distance (mm)

r_1 = small shaft radius (mm)

r_2 = large shaft radius (mm)

L = belt length (mm)

Therefore, the length of the belt (chain) is 133.97 cm

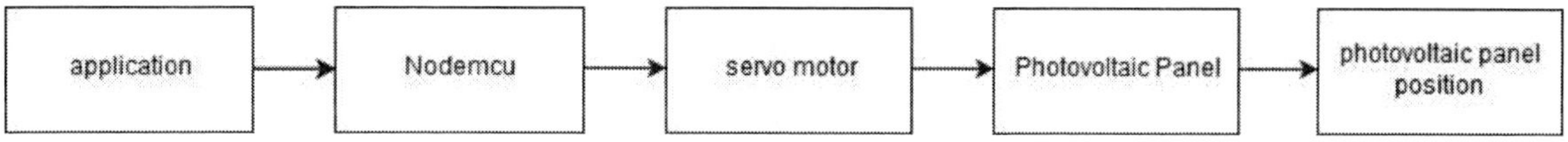

Figure 1. Block diagram of a single axis PV drive system.

The application used in this Android-based system can be controlled via mobile phones. In the app, there are features which can control the angle and monitor the temperature readings on the panel. Controls from the smartphone are very important because temperature and PV angle can greatly affect the output voltage generated by the panel.

The use of node MCU on the system is intended as communication media with the microcontroller. This communication uses the application media on mobile phones. Smartphones would send data through the application and then forward it to the controller. The data sent would be processed by node MCU and produce an angular output.

This system uses a motor as a drive which can later move the plan to a system in the form of a 100 Wp solar panel with a mass of 7 kg. Servo motor specifications on the system have a torque that can rotate that weighs 23 kg and can rotate up to 180 degrees of angle. The PV position would be adjusted to correspond to the input command given by the application and driven by a servo motor.

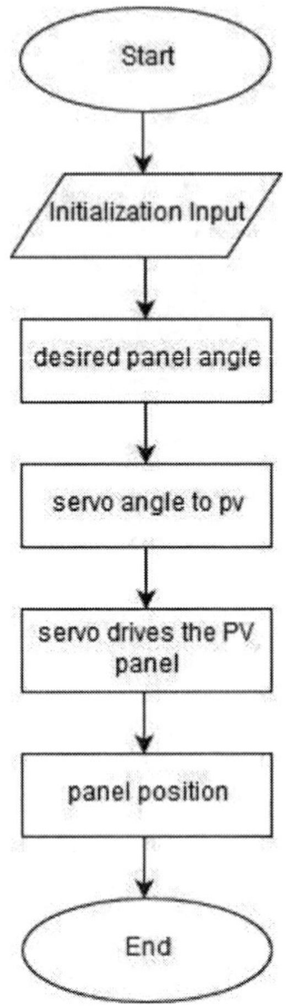

Figure 2. System Flowchart

Based on Figure 2, the system will identify the angle value. Furthermore, this angle value is in the form of data sent through the mobile application and then processed by the controller. If the data has reached the controller, the data in the form of an angle value will be a reference to the system. The servo will move if there is a deviation from the reference angle value and will adjust back. This angle will affect the position of the panel so that the resulting output voltage is always at maximum values.

Figure 3. Overall Drive System Mechanism Design

Figure 3 represents a mechanical merging of the whole system that has been connected with solar panels, modules and a 23 kg servo motor. In this mechanism, both sides of the triangle can be folded and then inserted into the iron beam at the bottom.

3. RESULT AND DISCUSSION

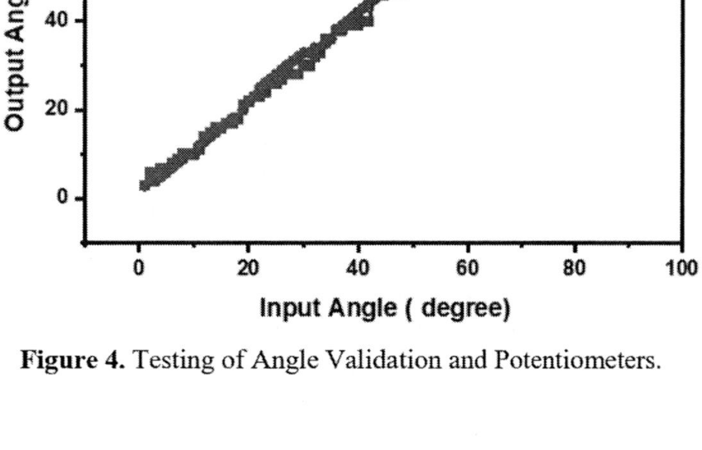

Figure 4. Testing of Angle Validation and Potentiometers.

Based on Figure 4, it can be seen that the error in an open loop drive system compared to the results entered from the application with an angle value has an average error value of 1.94% and compared with a potentiometer reaching an average error value of 1.4%. This error value is obtained from the average value of the whole test.

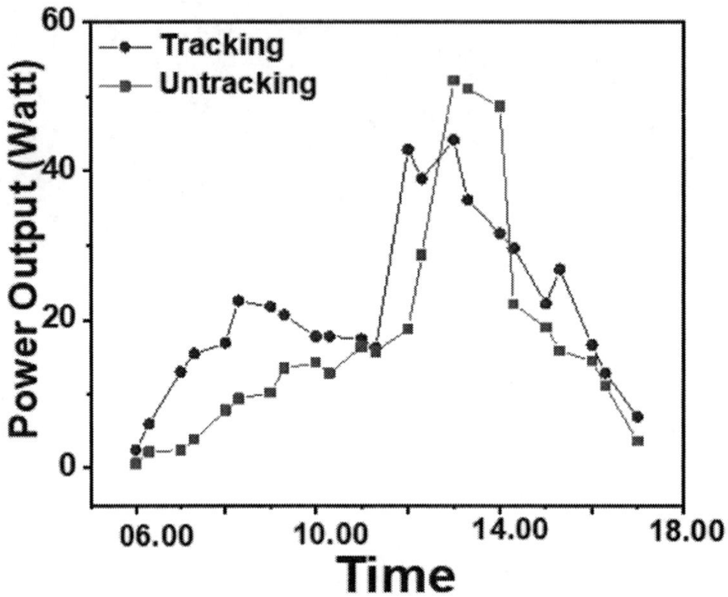

Figure 5. Solar Panel Power Output untracking and tracking systems.

To measure the power output between a fixed and single axis PV as in Figure 5, current and voltage measurements are taken. Figure 5 shows a slight increase in power output for the single axis PV.

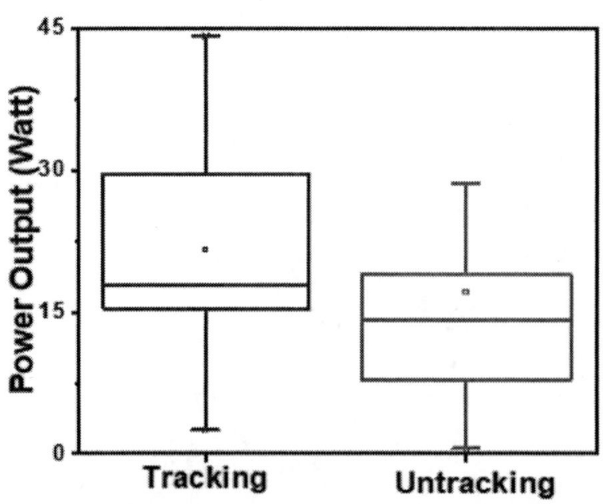

Figure 6. Statistical study of Solar Panel Power Output for untracking and tracking systems.

To investigate variations in solar panel output during the day, statistical studies have been provided in Figure 6. In general, the distance between the various parts of the box shows the degree of spread and slope in the data, and shows the outlier. In addition, from the bottom up, each line represents the minimum value, the bottom quartile, the median upper quartile, and the maximum values respectively [11]. An average power output of 17.15 Watt is obtained from the fixed system, while the single axis system shows an average power output of 21.50 Watt. An increase of more than 25% for the single axis systems

4. CONCLUSSION

Based on the results of testing a single axis PV drive system, the error in the open loop drive system compared to the results entered from the application with angle values has an average error value of 1.94% and compared with a potentiometer reaching an average error value of 1.4%. This error value is obtained from the average value of the whole test. The average power output in the fixed system reached 17.15 Watt, while in the single axis system the average power output reached 21.50 Watt. It can be concluded that in the single axis system the average power output has increased by more than 25% compared to the fixed system.

References

[1] N. Handayani, and D. Ariyanti 2012 Potency of solar energy applications in Indonesia. *International Journal of Renewable Energy Development* **1**(2) 33-38.
[2] Rahmawati, D., et al. 2020 Design of Automatic Switch System of Solar Panel and Power Plant
for Residential Load using Artificial Neural Network *MS&E* **771**(1): p. 012008
[3] Aprillia, B.S. and M.A.F. Rigoursyah 2020 Design On-Grid Solar Power System for 450 VA Conventional Housing using HOMER Software *MS&E.*
[4] Al Nabulsi, A. and R. Dhaouadi, 2012 Efficiency optimization of a DSP-based standalone PV system using fuzzy logic and dual-MPPT control *IEEE Transactions on Industrial informatics* **8(3)** 573-584
[5] Afif, A., B. Aprillia, and W. Priharti, 2020 Design and Implementation of Battery Management
System for Portable Solar Panel with Coulomb Counting Method *MS&E.*
[6] Silalahi, D., et al. 2020 Design of Automatic Switch System of Residential Load From Solar Cell and Power Plant Resources using Neural Network *MS&E* **771**(1) p. 012006.
[7] Aprillia, B.S., M.R. Zulfahmi, and A. Rizal, 2019 Investigasi Efek Partial Shading Terhadap Daya Keluaran Sel Surya *Jurnal Elektro dan Mesin Terapan* **5**(2) 9-17.
[8] Tan, D.W.W.W 2006. A simplified type-2 fuzzy logic controller for real-time control *ISA transactions* **45**(4) 503-516.
[9] Fuentes, M., et al. 2014 Design of an accurate, low-cost autonomous data logger for PV system
monitoring using Arduino™ that complies with IEC standards *Solar Energy Materials and Solar Cells* **130** 529-543.
[10] Purwadi, A., et al. 2011 Prototype development of a Low Cost data logger for PV based LED Street Lighting System *Proceedings of the 2011 International Conference on Electrical Engineering and Informatics.*
[11] Pangaribuan, P., B. Aprillia, and A. Wibowo 2020 Automatic Gate System Based on Water Flow Within the Intake of a Micro-Hydro Power Plant. *MS&E* **771**(1) p. 012004,

IOP Conf. Series: Materials Science and Engineering **982** (2020) 012017 doi:10.1088/1757-899X/982/1/012017

Control System for Smart Home Energy Management

Alfian Teguh Suryanto [1], Bandiyah Sri Aprillia[1,2], Cahyantari Eka Putri[1]

[1]School of Electrical Engineering, Telkom University, Bandung, Indonesia

[2]bandiyah@telkomuniversity.ac.id

Abstract. The Increasing energy demand for residential load is often associated with a lot of energy waste due to lack of a good energy management system. In this work, we propose an intelligent energy management system (EMS) which is designed to control the electrical energy source from solar energy and the power plant from PLN resources both automatically and manually based on the Internet of Things. This control system can be controlled remotely during the process of connecting and disconnecting backup power sources that can be accessed through smartphone devices. Based on the test results of the control system of displacement of energy sources, the DC voltage sensor accuracy value is 99.44% and the PLN voltage measurement has an accuracy of 98.31%. In the control system transfer to a backup power source with manual control has an average duration of time of 0.74 seconds while the automatic control has an average duration of time of 0.98 seconds to 1.33 seconds and for automatic control with priority reading the battery capacity of 12.75 Volts can do transfer from the PLN power source to the backup power source and transfer to the PLN source if the voltage is less than 10.5 Volts with a 100% success rate.

1. Introduction

In a smart home, productivity and quality of community life depends mainly on the level of comfort inside in particular from visual and thermal comfortableness [1-3]. For example, regulations related to lighting and air conditioning systems have a major role to determine optimum design of comfort level of the building [4].

Electricity distributed by the State Electricity Enterprise (PLN) is mostly produced from coal, diesel, oil, and MFO. Meanwhile, alternative energy sources, which are environmentally friendly, are just started to be promoted through Ministerial Regulation (Permen) Number 49 of 2018 regarding the Use of Roof Solar Power Systems by consumers of PT PLN. However, the amount of power output in a solar power system depends on the intensity of solar radiation [5]. Therefore, the integration of solar power as an additional energy source can be combined with the main electricity source, namely PLN, known as On-Grid System [6].

In this study the energy system produced by solar panels will be directly connected to the battery and a converter is distributed to the load. The implementation of electric energy source control methods for on-grid solar power generation systems can also increase the efficiency of energy use, because energy flow can be adjusted according to load requirements [7].

In this study, a control system is designed to be able to transfer electricity from the main power source to a backup power source that has been stored on the battery. This control system can be operated manually or automatically and accessed remotely with Internet of Things technology. The use of Internet-based transfer control systems would control all electrical equipment in the home by

IOP Conf. Series: Materials Science and Engineering **982** (2020) 012017 doi:10.1088/1757-899X/982/1/012017

disconnecting and connecting backup power sources through a smartphone application. With this system, the user can choose the control features found on the smartphone and is able to reduce costs and the excessive use of electrical energy from PLN.

2. Design of the system

2.1 Hardware Design

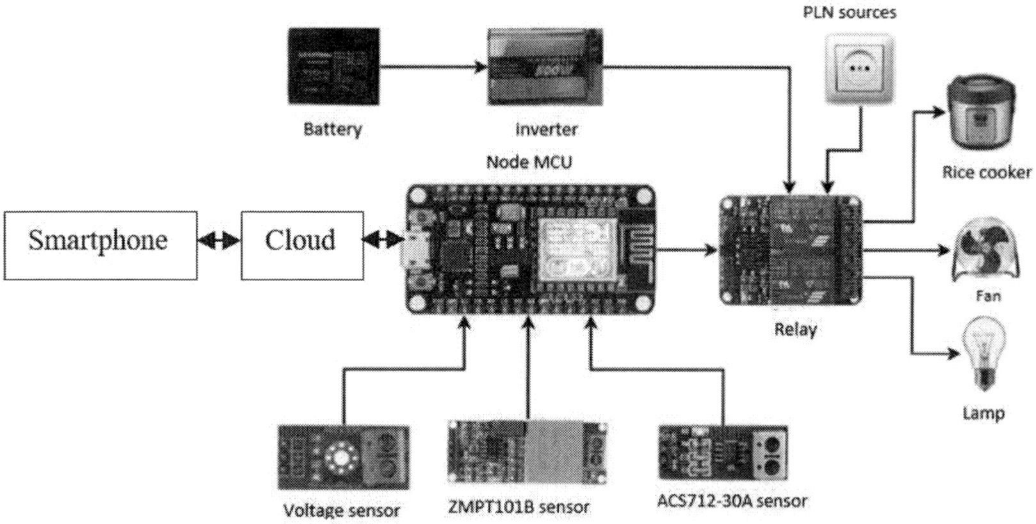

Figure 1. Hardware design of electric power transfer control system.

The Smartphone application in Figure 1 functions as a medium that sends input selected by the user to perform manual or automatic switching and receive the results of sensor measurements through the cloud server. The cloud server functions as a place to exchange data that is received or sent to the Node MCU smartphone and microcontroller application. The Node MCU microcontroller functions as an input data processor that has been received from the cloud server and provides output to the relay to transfer electrical power and will send sensor measurement data and displacement indicators to the cloud server.

The ZMPT101B sensor functions to measure AC voltage at PLN to be able to carry out the displacement automatically, the ACS712-30A sensor functions to measure AC current in electrical equipment supplied by PLN, and the voltage sensor functions to determine the available battery voltage. The relay functions as an output in connecting and disconnecting backup power sources by receiving input from Node MCU.

2.2 Software Design

This smartphone application is designed for users to carry out control systems for the transfer of electrical resources for homes based on the internet of things.

IOP Conf. Series: Materials Science and Engineering **982** (2020) 012017 doi:10.1088/1757-899X/982/1/012017

Figure 2. Software Design

Figure 2 shows the software design that will be used to make transfers manually or automatically by selecting a control system to transfer electricity to the home.

IOP Conf. Series: Materials Science and Engineering **982** (2020) 012017 doi:10.1088/1757-899X/982/1/012017

3. Analysis

3.1. Sensor Calibration

The purpose of this test is to determine the accuracy of the sensor readings that will be compared with multimeter and power meter.

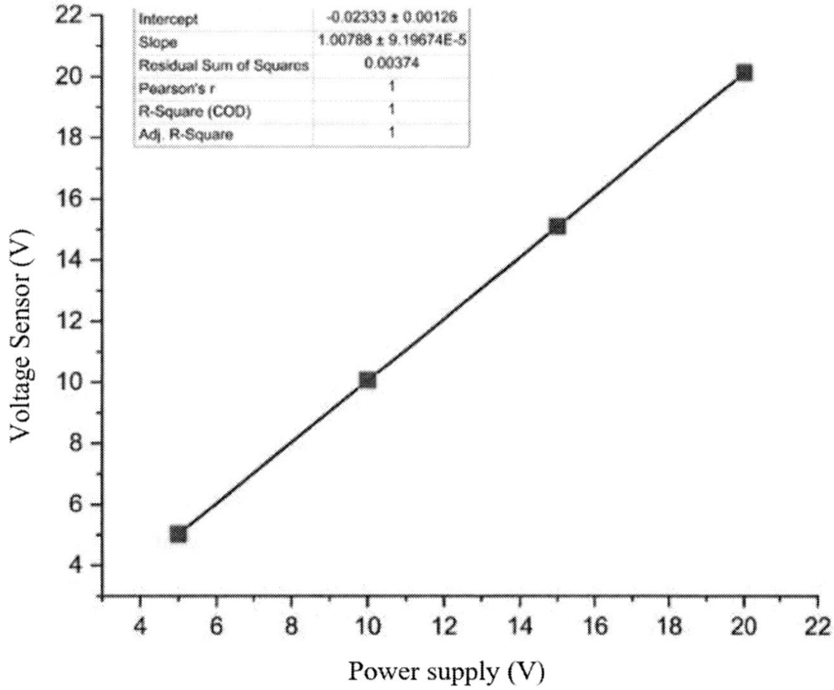

Figure 3. Voltage Sensor Calibration

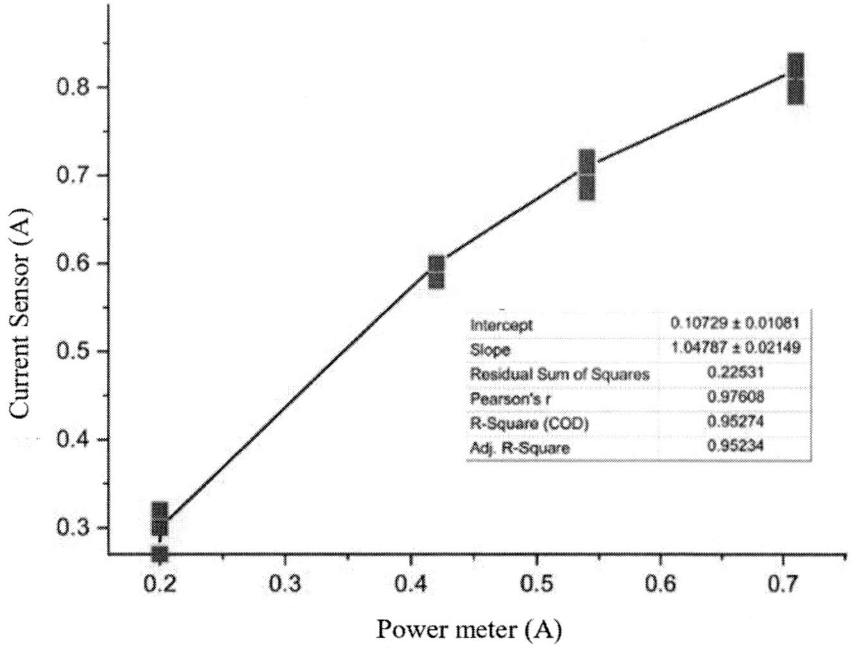

Figure 4. Current Sensor Calibration

In testing the voltage sensor to the power supply, the average error value percentage is 0.56%. The biggest error value is 0.14 Volt with a voltage sensor accuracy of 99.44%. Testing the ACS712-30A sensor to the power meter obtained an average error value percentage of 31.37%. The largest error value on the ACS712-30A sensor is 0.18 A with the accuracy of the AC voltage sensor of 68.63%.

ZMPT101B sensor testing of the power meter obtained an average value of percent error of 1.59% in 3500 (VA) homes. The biggest error value at 3500 (VA) is 4.25 Volts with an AC voltage sensor accuracy of 98.422% and the ZMPT101B sensor testing of the power meter shows an average percent error value of 1.88% at 2200 (VA). The biggest error value in a house of 2200 (VA) is 5.18 Volts with an AC voltage sensor accuracy of 98.12% while the power meter shows the density value in the measurement of the PLN voltage.

3.2. Automatic Control System with Battery Capacity Calibration
This test aims to determine the duration required by the system to transfer electricity automatically by calculating the capacity of available batteries.

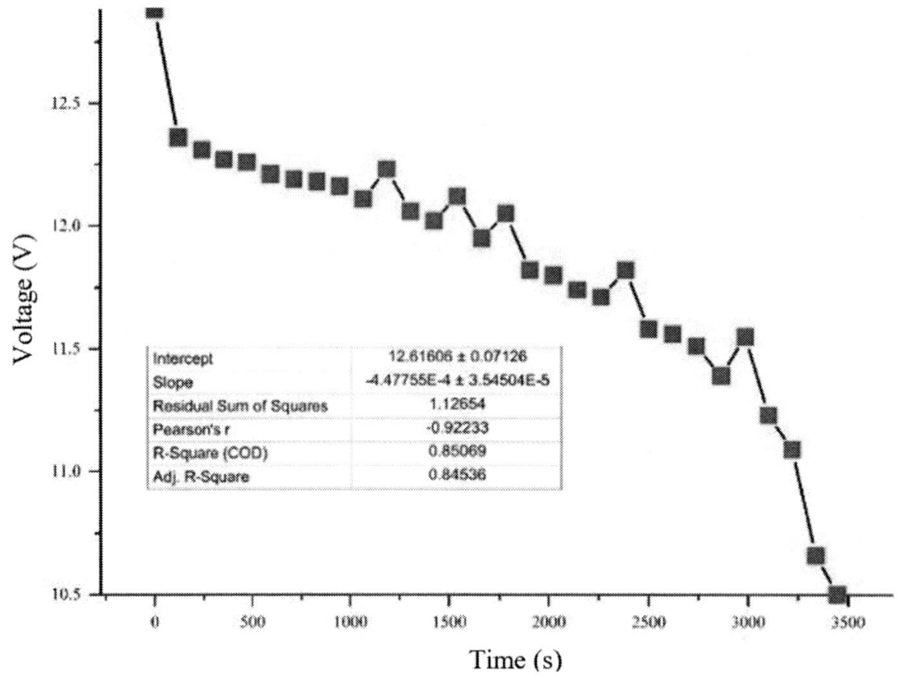

Figure 5. Calibration of Automatic Control System with Battery Capacity

The use of batteries in supplying electrical equipment such as 32 watt lamps, fans, and rice cookers has a voltage drop from 12.75 Volts to 10.5 Volts with a battery load supply duration of 57 minutes 43 seconds.

4. Conclusion
The system has worked properly where the system can transfer electricity resources manually and automatically. The DC voltage measurements can read the voltage value on the battery with good accuracy. The accuracy value of the sensor when compared to the power supply is 99.44%. Measurement of AC voltage in the form of a ZMPT101B sensor can read PLN voltage values with good accuracy. The sensor accuracy value when compared to the power meter has an accuracy level of 98.41%. While the electric power transfer control system for home on manual control has an average time of 0.74 seconds, the automatic control has an average time of 0.98 seconds to 1.33 seconds.

Finally, our system can transfer from the PLN power source to the backup power source showed by dropping of the battery capacity of 12.75 to 10.5 Volts with a 100% success rate.

References

[1] Wang, L., Z. Wang, and R. Yang, Intelligent multiagent control system for energy and comfort management in smart and sustainable buildings. IEEE transactions on smart grid, 2012. 3(2): p. 605-617.
[2] Harper, R. (Ed.). (2006). Inside the smart home. Springer Science & Business Media.
[3] Davidoff, S., Lee, M. K., Yiu, C., Zimmerman, J., & Dey, A. K. (2006, September). Principles of smart home control. In International conference on ubiquitous computing (pp. 19-34). Springer, Berlin, Heidelberg.
[4] Wang, Z., R. Yang, and L. Wang. Multi-agent control system with intelligent optimization for smart and energy-efficient *buildings*. in *IECON 2010-36th annual conference on IEEE industrial electronics society*. 2010. IEEE.
[5] Sri Aprillia, B., C. Ekaputri, and M. Rafiqy Zul Fahmi, *Solar Cell Output Optimization using Light Convergence Method.* MS&E, 2020. **771**(1): p. 012010.
[6] Aprillia, B.S. and M.A.F. Rigoursyah, *Design On-Grid Solar Power System for 450 VA Conventional Housing using HOMER Software.* 2020.
[7] Silalahi, D., et al., *Design of Automatic Switch System of Residential Load From Solar Cell and Power Plant Resources using Neural Network.* MS&E, 2020. **771**(1): p. 012006.

IOP Conf. Series: Materials Science and Engineering **982** (2020) 012018 doi:10.1088/1757-899X/982/1/012018

Electricity Power Needs Evaluation in Kulonprogo Post Developing the Yogyakarta International Airport

Iyus Rusmana[1,2] **and Dulhadi**[1,3]

[1] Elecrical Department of Institut Teknologi Nasional Yogyakarta

[2]iyusrusmana@itny.ac.id, [3]dulhadi@itny.ac.id

Abstract. The planned construction of the Kulonprogo district's strategic zone, which began in 2013, this Regency will certainly experience a large development, so for that, it needs to be accompanied by adequate electricity preparation. Based on the real conditions, the speed of construction is experiencing some obstacles, so there is only phase 1 of NYIA airport has been completed by 2020, while for the Steel Industry Center construction, Iron Sand Processing, Adikarto Port, there are also have many obstacles. This also affected the domestic and non-domestic residents' activities, especially in terms of economic progress in the district. This study intended to evaluate the preparation of electricity power so its preparation is expected to be more efficient. From the results of this study at the end of 2032, at least 626 MVA of electricity power needs to be prepared to support an estimated peak load of 438 MVA. This research was conducted quantitatively and aimed for giving some references to be used by PT PLN (Persero) in reviewing Electricity Supply Business Plan (RUPTL).

1. Introduction

1.1 Background

Kulonprogo Regency is an underprivileged area in the Special Region of Yogyakarta (DIY) compared to Sleman, Bantul, and Yogyakarta Regencies. The condition of the district's infrastructure is very minimalized, while the potential of the region is quite large and has a great opportunity to be developed. (1)

There is some awareness to be concerned. The Regional Government of Kulon Progo Regency through its policies mentioned in Kulonprogo Regional Regulation No: 1 of 2012 regarding the Spatial Planning of the Kulon Progo Region for 2012 - 2032, has established a strategic area. In its Regional Regulation (PERDA) it is defined that: "Regency Strategic Area is defined as an area which has spatial planning should be prioritized because it has a very important influence within the district scope in terms of economy, social, culture, and/or environment".(1)

There are Strategic plans planned by the government as below:

a. Construction of Southern Cross Roads (Jalur Jalan Lingkar Selatan/JJLS),

b. International Airport Development,

c. Ocean and Fisheries Port Development
d. Special Economic Zone Plan,
e. Development of the Glagah-Congot Tourism Area,
f. Iron sand mining with its processing industry

From the above data that really needs attention is the construction of the New Yogyakarta International Airport, Special Economic Zone consisting of the Development Plan for the Processing of the Iron Sand Industry, Steel Industry, and Large Industries in the Lendah and Sentolo areas. Because of this development, it will greatly impact economic improvement, besides that, in terms of electricity, its capacity is quite large. To support these developmental needs, which is very important to note is the availability of electric power and its distribution network. For this reason, an in-depth study is needed (1)

To support these developmental needs, the availability of electric power and its network distribution are very important concerns. Based on those reasons, an in-depth study is needed to be conducted.

1.2 Literature Review

The role of electrical energy is very important for increasing economic activity and national resilience, so that energy management which including the supply, utilization, and exploitation must be carried out fairly, sustainably, rationally, optimally, and in an integrated manner as it stated in the Law of the Republic of Indonesia Number 30 Year 2007, concerning Energy.(2)

Refer to the appendix B.5 on 2015-2024 PLN RUPTL, it stated that the electricity system development of PT PLN (Persero) in the Special Region of Yogyakarta is intended to serve the load growth, a new substation will be built at 660 MW.(3)

Table 1. Development of 150/20 KV substation in DI Yogyakarta

No	Sub Station	Voltage (v)	Information	Capacity (MV/LB)	COD (year)
1	Wirobrajan	150/20 kV	Ext	60	2015
2	Kentungan	150/20 kV	Ext	60	2015
3	Wates	150/20 kV	Uprate	60	2015
4	Godean	150/20 kV	Ext	60	2016
5	Kentungan	150/20 kV	Ext	60	2018
6	Wates	150/20 kV	Ext	2 LB	2019
7	Kentungan Baru/ Kalasan	150/20 kV	New	60	2019
8	Wates	150/20 kV	Ext	60	2019
9	Bantul Baru	150/20 kV	New	60	2021
10	Kentungan Baru/ Kalasan	150/20 kV	Ext	60	2024
	Sum			660	

Source: 2015-2024 PLN RUPTL[11]

From the data above, it is clear that the electric power that will be prepared by PLN until 2019 is 240 MVA. This seems to have not been realized, and also many large projects such as the Processing of Iron Sand, Steel Mills have not yet been realized either.

The development of the Kulonprogo Airport Area is included in the PPP Project Book BAPPENAS Program, 2010-2014. It means that it provided an opportunity for private and foreign investors interested in managing the airport under the Technical Implementation Unit (UPT) of the Directorate General of Civil Aviation. The Kulonprogo Airport Masterplan has located about 30 km on the west of Yogyakarta city and has 3,600 meters' runway. The runway is very adequate for large size aircraft and international standards. On the east-west side runway, plus two perimeters of 900 meters each for a total of 4,400 meters. The runway is the first thing to be done, then following with the perimeter and lighting. The area of the airport covered around 350 hectares, equipped with 7 taxiways with 4 interconnecting taxiways. There are required airport facilities that should be provided such as apron, terminal building, commercial building, technical building, and flight operations support with Air Traffic Control (ATC) and parking facilities for visitors. The airport is capable of serving 61 million passengers every year. (4)

Tanjung Adikarto Pelabuan area is located in Karangwuni Village, Wates District. The total area for infrastructure development is 83 Ha. Its location has good accessibility and easy to reach, connected to cities in southern Java by the south Java regional transportation routes. It is located 2 km south of the South Java National road and is lined by the Southern Cross Road Development Plan (JJLS). Planned to be able to land ships with weights up to 150 GT. It has 6 hectares parking area and can be expanded up to 15 hectares

Mining Allocation Area according to Kulon Progo Regency Regulation Number 1 Year 2012 (abbreviated as KPP) is an area that has the potential for solid, liquid, or gas mining material resources based on maps / geological data and is the place where all stages of mining activities include research, general investigations, exploration, production/exploitation, and post-mining operations, both inland and water areas, and not limited by land use, both cultivation and protected areas

The area around the iron sand mining is planned to be developed into 2,000 Ha National Steel Industrial Zone (Sentolo, Lendah, and Galur Districts) initiated by the Ministry of Industry of the Republic of Indonesia.

1.3 Basic Theory

1.3.1 Arithmatic Method

Arithmetic method is a method used to calculate the projected population in Kulon Progo Regency from 2019 to 2032 with a tendency to follow the data set by BPS Kulonprogo Regency, while from 2010 to 2018 using data taken from BPS reports

This method is used assuming that the population in the future will increase by the same amount every year. The following is the formulation of the arithmetic method. (5)

$$P_n = P_0(1 + rn) \qquad (2.1)$$

with,

P_n : total population in year n
P_0 : total population in the initial year (base)
r : average population growth in 2020 - 2032
n : time period between the base year and year n (in years)

To determine the amount of electricity needed in Kulon Progo Regency, it is needed an estimate of the population (projected population) until 2032 and basic assumptions for domestic and non-domestic electricity demand.

1.3.2 Domestic, Non-Domestic and Industrial Estate Electricity

The electricity consumption in 2010 to 2019 is based on the data from the Central Bureau of Statistics (6-10) and from 2020 to 2032 which is analyzed based on assumptions by the Regional /

City Spatial Detailed Plan Guidelines and also based on the National Technical Shutter Reference that is in force today, namely:

a. Domestic Categories
   Domestic (initial/standard) needs will be divided according to the type of house, as follows:(5)
   i. 10% luxury home section type (assumed 2200 VA);
   ii. Type of medium housing 30% part (assumed 1300 VA);
   iii. Type of simple house 60% of the part and settlement that already exists 1 part (assumed 900 VA and 450 VA).
   iv. For each type of house above will be multiplied by the Electrification Ratio (RE) of 85%, for 2018 and 100% for 2032
b. Non-Domestic Categories
   The needs are defined as government/office needs, commercial, social and public facilities, and street lighting with the assumption of need that is 30% of domestic needs.(5)
c. Industrial Estate Category
   Referring to the Decree of the Minister of Industry No: 35 / M-IND / PER / 2010, Table 4. Regarding General Service Technical Standards, Page 31, stated that the electricity demand for industrial estates is 0.15 to 0.2 MVA per Ha. (11-14)
d. International Airport
   The international Airport Passenger Capacity about 61 million passengers per year, Electricity Power is around 75 MPA. (15)
e. The estimated reduction in transmission of electric power is assumed to be 4% from domestic and non-domestic demand. The 4% power loss consists of 2.5% technical losses and 1.5% non-technical losses (electricity theft). (16)

Domestic electricity needs were calculated using arithmetic method, while projected electricity needs for industrial areas and airport electricity needs were calculated using a comparison method, which is to compare with similar objects that already exist.

1.4 Problem Identification

Based on PERDA no 1, regarding strategic areas. What has been realized is Airport Development phase 1, while others have not been realized as desired, as an example of the construction of the Iron Sand factory and the existence of the Tanjung Adikerto Port development. So this affects the effectiveness of electricity supply in Kulonprogo Regency

1.5 Research Objectives and Benefits

The purpose of this study is intended to obtain macro data on the need for electricity infrastructure in Kulonprogo Regency to support the development of the Kulonprogo Regency strategic area. The benefits of this research can be used as one of the basic considerations for executives in determining their policies.

2. Research methods

2.1. Research Methods

This study used qualitative and quantitative methods for collecting the data. The qualitative methods applied a questionnaire for interviewing the targeted respondents among the parties and also site survey to complete the qualitative data. The calculation from survey data considered as quantitative data.

2.2. Research Steps

IOP Conf. Series: Materials Science and Engineering **982** (2020) 012018 doi:10.1088/1757-899X/982/1/012018

The study was conducted in the following stages:

a. Phase 1, reviewing secondary data
b. Phase 2, interviewing with Kulonprogo Regency BAPPEDA office, PT PLN UPJ WATES, Kulonprogo Regency's Department of Industry and ESDM to obtain more details information on the development plan profile which can be derived from the secondary data
c. Phase 3, conducts a study and identification of primary data obtained from:
    i. Regional Planning and Development Agency, Kulonprogo Regency
    ii. PT PLN UPJ Wates
    iii. Department of Industry and Trade and ESDM Kulonprogo
d. Phase 4, carrying out surveys and measurements
e. Phase 5, carries out analysis and calculation of measurement data

The research stages as described below:

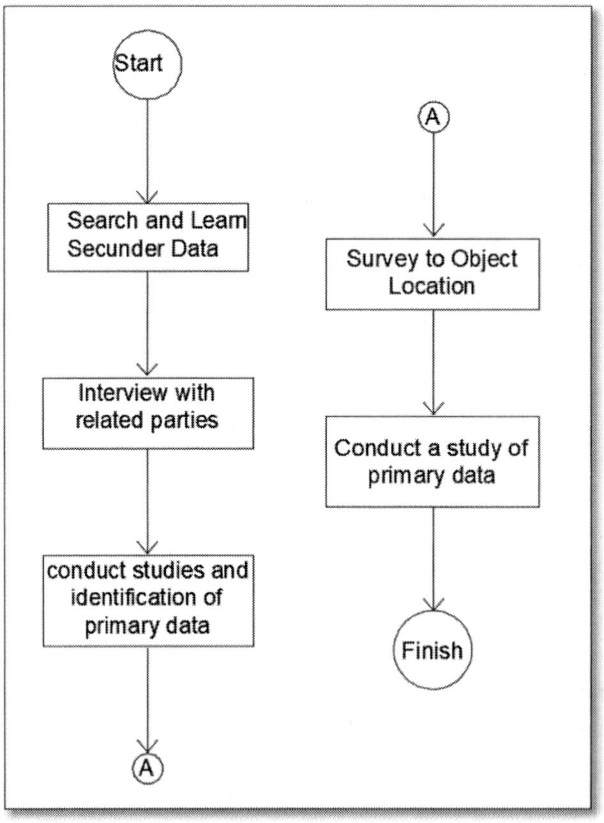

Figure 1. Research Stages

3. Analysis

The following are research data including analysis of domestic, non-domestic electricity power needs, industrial electric demand, air sport electricity needs, and substation requirements as well as transmission and distribution networks.

To determine the amount of electricity needed in Kulon Progo Regency, it is needed an estimate of the population (projected population) until 2032 and basic assumptions for domestic and non-domestic electric demand and energy losses.

3.1. Projected population

To calculate domestic and non-domestic electricity power requirements, up to 2032 data on the number of households is needed. Data from 2010 to 2018 can be obtained from BPS data, while 2019 to 2032 is calculated based on projections. As for the data on the addition of the population is calculated based on the population increase in previous years. The results of the projections are shown in table 02.

Table 02: Total Population Projection Result by 2032

Sub Diatrict	Average Growth Rate for 2020-2032	Basic Year		Total Population Projection Result				
		2010	2018	2022	2026	2030	2031	2032
1 Temon	1,34%	24.471	27.310	30.646	32.322	34.089	34.546	35.009
2 Wates	1,34%	44.123	49.090	51.287	54.091	57.049	57.814	58.589
3 Panjatan	1,24%	33.492	36.946	40.691	42.916	45.263	45.870	46.484
4 Galur	1,04%	29.196	31.715	34.420	36.303	38.288	38.801	39.321
5 Lendah	1,20%	36.547	40.212	43.224	45.588	48.081	48.725	49.378
6 Sentolo	1,31%	44.653	49.353	52.552	55.426	58.457	59.241	60.034
7 Pengasih	1,34%	45.308	50.412	54.538	57.520	60.666	61.479	62.303
8 Kokap	0,54%	31.193	32.553	38.185	40.273	42.475	43.045	43.621
9 Girimulyo	0,61%	21.944	23.018	26.131	27.560	29.067	29.457	29.851
10 Nanggulan	1,21%	27.316	30.076	32.253	34.016	35.877	36.357	36.845
11 Kalibawang	0,63%	26.865	28.242	31.873	33.616	35.454	35.929	36.411
12 Samigaluh	0,94%	24.744	26.649	30.036	31.679	33.411	33.859	34.313
Kulon Progo	**0,48**	**389.852**	**425.576**	**465.835**	**491.310**	**518.179**	**525.122**	**532.159**

Source: Analysis 2020

3.2. Projected population, domestic and non-domestic power requirements

From the above data we can calculate the domestic and non-domestic electricity power requirements as shown in table 03 and table 04, as below:

Tabel 03 Domestic and non Domestic Electricity Needs by 2020

	Sub District	Projection Result 2020	Number of Family Head	Domestic				Non Domestic (30%)	Total Domestic & Non Domestic	Trans. + Distr. Shrinkage (4%)	Total Elecricity Power Need (MVA)
				900 VA (60%)	1300 VA (30%)	2200 VA (10%)	Number of ER (ER 79%)				
1	Temon	29.841	7.460	4.028.478	2.909.456	1.641.232	6.863.333	2.059.000	8.922.332	356.893	9
2	Wates	49.939	12.485	6.741.811	4.869.086	2.746.664	11.486.048	3.445.814	14.931.862	597.274	16
3	Panjatan	39.622	9.905	5.348.958	3.863.137	2.179.205	9.113.040	1.822.608	10.935.648	437.426	11
4	Galur	33.516	8.379	4.524.684	3.267.827	1.843.390	7.708.721	1.541.744	9.250.465	370.019	10
5	Lendah	42.089	10.522	5.681.951	4.103.632	2.314.869	9.680.362	1.936.072	11.616.434	464.657	12
6	Sentolo	51.172	12.793	6.908.170	4.989.234	2.814.440	11.769.476	3.530.843	15.300.318	612.013	16
7	Pengasih	53.105	13.276	7.169.202	5.177.757	2.920.786	12.214.196	2.442.839	14.657.035	586.281	15
8	Kokap	37.182	9.295	5.019.522	2.416.807	1.022.495	6.767.060	2.030.118	8.797.177	351.887	9
9	Girimulyo	25.444	6.361	3.435.000	1.653.889	699.722	4.630.889	1.389.267	6.020.156	240.806	6
10	Nanggulan	31.405	7.851	4.239.711	2.041.342	863.645	5.715.758	1.714.728	7.430.486	297.219	8
11	Kalibawang	31.035	7.759	4.189.776	2.017.299	853.473	5.648.438	1.694.531	7.342.970	293.719	8
12	Samigaluh	29.247	7.312	3.948.392	1.901.078	804.302	5.323.018	1.596.905	6.919.923	276.797	7
Strategic Region		**453.597**	**113.399**	**61.235.656**	**39.210.544**	**20.704.223**	**96.920.338**	**23.607.564**	**115.204.885**	**4.608.195**	**127**

Source: Analysis 2020

Tabel 04. Domestic and Domestic Electricity Needs in 2032

IOP Conf. Series: Materials Science and Engineering **982** (2020) 012018 doi:10.1088/1757-899X/982/1/012018

Sub District	Projection Result 2018	Number of Family Head	Domestic 900 VA (60%)	Domestic 1300 VA (30%)	Domestic 2200 VA (10%)	Domestic Number of ER (ER 100%)	Non Domestic (30%)	Total Domestic & Non Domestic	Trans. + Distr. Shrinkage (4%)	Total Elecricity Power Need (MVA)
1 Temon	35.009	8.752	4.726.197	3.413.364	1.925.487	10.065.048	3.019.514	13.084.563	523.383	14
2 Wates	58.589	14.647	7.909.469	5.712.395	3.222.376	16.844.241	5.053.272	21.897.513	875.901	23
3 Panjatan	46.484	11.621	6.275.380	4.532.219	2.556.636	13.364.235	4.009.270	17.373.505	694.940	18
4 Galur	39.321	9.830	5.308.344	3.833.804	2.162.659	11.304.807	3.391.442	14.696.249	587.850	15
5 Lendah	49.378	12.345	6.666.046	4.814.367	2.715.797	14.196.209	4.258.863	18.455.072	738.203	19
6 Sentolo	60.034	15.009	8.104.642	5.853.353	3.301.891	17.259.886	5.177.966	22.437.852	897.514	23
7 Pengasih	62.303	15.576	8.410.884	6.074.527	3.426.656	17.912.067	5.373.620	23.285.687	931.427	24
8 Kokap	43.621	10.905	5.888.886	4.253.085	2.399.176	12.541.147	3.762.344	16.303.491	652.140	17
9 Girimulyo	29.851	7.463	4.029.931	2.910.506	1.641.824	8.582.260	2.574.678	11.156.938	446.278	12
10 Nanggulan	36.845	9.211	4.974.014	3.592.344	2.026.450	10.592.809	3.177.843	13.770.651	550.826	14
11 Kalibawang	36.411	9.103	4.915.431	3.550.033	2.002.583	10.468.047	3.140.414	13.608.461	544.338	14
12 Samigaluh	34.313	8.578	4.632.241	3.345.507	1.887.209	9.864.957	2.959.487	12.824.444	512.978	13
Strategic Region	**532.159**	**133.040**	**71.841.464**	**51.885.502**	**29.268.745**	**152.995.711**	**45.898.713**	**198.894.424**	**7.955.777**	**207**

Source: Analysis 2020

3.3. Projected electricity power requirements at the end of the 2032 year

The following are the results from the calculation of electrical power requirements in Kulonprogo Regency according to the data from table 01 to table 04 and also power requirements for large and small industrial zones, maritime industry, and electric power needs at NYIA Airport.

Tabel 05 Electricity Power Needs of Domestic and non Domestic, Industrial Zone and Airports

Location	Alocation	Land Area (Ha)	Coefficient of Land Use (%)	Electricity per Ha (MVA)	Installed Power (MVA)	Demand Factor
Desa Tuksono, Sentolo	Industrial Zona	200	65%	0,14	18	70%
Galur, Lendah, Sentolo	Iron Sand Processing Industry and Steel Industry	2000	65%	0,14	182	60%
Sentolo, Lendah	Large Industry	4796	65%	0,14	436	60%
Naggulan	Industry	700	65%	0,14	64	60%
Temon	Maritime Industry	500	65%	0,14	46	50%
	Airport Electricity Power				75	90%
Kulonprogo	Domestic and Non-Domestic Electricity Power				207	50%

Source: Analysis 2020

From table 05 the results of this analysis can be evaluated more deeply. By calculating the demand load, demand factor, and simultaneous factor, it is obtained:

- Projected Instaled Electricity Load = 1.028 MVA
- Projected Demand Load = 626 MVA
- Projected Demand Factor = 61 %
- Projected Simultaneous Factor = 70 %
- Projected Peakload estimation = 438 MVA

4. Conclusion

This study concluded some points as below:

a. The electricity demand in Kulonprogo Regency at the end of 2032 can be predicted as follows:

i. Projected Instaled Electricity Load = 1.028 MVA
ii. Projected Demand Load = 626 MVA
iii. Projected Demand Factor = 61 %
iv. Projected Simultaneous Factor = 70 %
v. Projected Peakload estimation = 438 MVA

b. The power capacity of the 46 MVA substation, which has been upgraded on November 2014 to 90 MVA, will not be able to supply electricity needs in Kulonprogo Regency after the enactment of the Kulonprogo Regency Regulation No. 1 of 2012 concerning the Kulonprogo Regency Spatial Planning Year 2012 – 2032

c. Industrial area power supply and NYIA airports should be separated from other power supplies to maintain stability and continuity of power distribution.

References

1. Pemda. PERATURAN DAERAH KABUPATEN KULON PROGO NOMOR 1 TAHUN 2012. 2012.
2. Presiden. "Undang-undang No 30 tahun 2007 Tentang Energi".pdf. 2007.
3. PLN P. RUPTL PLN 2015-2024. 2015.
4. BAPPENAS. PPP Book 2010-2014 Infrastructure Project In Indonesia.pdf. 2010.
5. BPSI. "Pedoman-penghitungan-proyeksi-penduduk-dan-angkatan-kerja".pdf. 2010.
6. BPS. "*Kabupaten Kulon Progo Dalam Angka 2015*".pdf. 2015.
7. BPS. "Kabupaten Kulon Progo Dalam Angka 2017".pdf. 2017.
8. BPS. "Kabupaten Kulon Progo Dalam Angka 2019".pdf. 2019.
9. BPS. "Kabupaten Kulon Progo Dalam Angka 2020".pdf. 2020.
10. BPS. "Kabupaten Kulon Progo Dalam Angka 2016".pdf. 2016.
11. Menperindag. "Permen_deprin_35_2010".pdf. Keputusan Menteri. 2010.
12. Otsuka A. Demand for industrial and commercial electricity: evidence from Japan. Journal of Economic Structures. 2015;4(1).
13. Kwanda T. "Pengembangan Kawasan Industri Indonesia".pdf. DIMENSI TEKNIK ARSITEKTUR. 2000;28.
14. Suwandi. Pengelolaan Konsumsi Energi Listrik Untuk Kawasan Industri Dengan Metode Kombinasi Open System Dan Closed System. CITEE 2017. 2017.
15. Ortega Alba S, Manana M. Energy Research in Airports: A Review. Energies. 2016;9(5).
16. Hakimah Y. "Analisis Kebutuhan Energi Listrik di Sumatra Selatan".pdf. Desiminasi Teknologi. 2019;7.

IOP Conf. Series: Materials Science and Engineering **982** (2020) 012019 doi:10.1088/1757-899X/982/1/012019

Hotspot Detection Method in Large Capacity Photovoltaic (PV) Farm

P A A Pramana[1,2], R Dalimi[1]

[1]Universitas Indonesia, Depok 16424, Indonesia

[2]putu.agus91@ui.ac.id

Abstract. The obligation to use low carbon emissions power plants encourages the increased utilization of renewable energy generation. Among the whole renewable energy plants, photovoltaic (PV) is a modular plant that is easy to implement, which the utilization reaches 100GW in the year 2017. By the increasing use of PV globally, the health of PV modules needs to be a concern because, during the operation, PV modules can experience various faults. Almost 50% of the overall fault is the hotspot which is very hard to detect ~~in~~ on a wide area PV farm. For example, a 30 MW PV generation with an area of 60 hectares and composed of 126000 modules (consists of millions of cell), the existing hotspot detection methods takes up to 210 days. The long time and not continuous detection lets the hotspot to degrade and burn the modules. To prevent ~~the~~ catastrophic failure due to hotspot, a detection method ~~which~~ that can detect the fault quickly is needed. The proposed method, thermal imaging using a fish-eye lens could be used in this case as it has a very wide angle of view, which allows monitoring all of the PV modules in one detection period.

1. Introduction

The obligation to use power plants that produce low carbon emissions encourages the increased use of renewable energy generation such as geothermal, photovoltaic (PV), biomass, tidal, and wind. Among the other renewable energy plants, PV is a modular plant that is easy to implement for various power generation ranges, starting from tens of kilowatts of power to the utility-scale plants with power reaching tens of megawatts. Until the year of 2017, the global utilization of PV reaches 100GW[1].

With the increasing use of PV, the health of PV modules needs to be a concern because, during its operation, PV modules can experience various faults. Almost 50% of the overall fault is the hotspot that is very hard to detect in a wide area PV farm. For example, a 30 MW PV generation with an area of 60 hectares and composed of 126000 modules (consists of millions of cell), the existing hotspot detection methods takes up to 210 days. These methods are unable to detect the fault quickly and continuously so that the hotspot fault can degrade and burn the module.

One case that has been found in Indonesia, the degradation of PV modules can cause a decrease in peak PV power to 94% of its installed capacity after a five-year operating period. On the other hand, faults that cause the burning of the PV module is shown in [2] and [3]. Reference [2] describes an increase in the number of fires that occur in PV installations for various power capacities. This increase experienced a peak with the emergence of more than 700 fire cases. Meanwhile, reference [3] explained the large fires on the PV system in California that was started by the arcing failure.

Considering the importance of the health of the PV module especially in avoiding degradation and fires, this paper provides a review of the faults that can create a hotspot, review of its existing detection methods, and the proposed method to detect hotspot fault quickly.

2. Hotspot Fault

Classification of faults in PV modules that potentially create hotspots can be divided into two parts, namely faults that occur due to physical problems in the module as well as a fault caused by environmental factors where the PV module operates. According to [4], hotspots faults events have a percentage of 49% compared to all fault events in the PV modules. The percentage of hotspot fault due to physical problems in the module is 25% and hotspot fault due to diode damage by 24%.

The physical problem that occurs in the PV module can be in the form of encapsulation material damage, delamination, cracking, interconnection failure, corrosion, bypass diode failure, mismatch fault, and arc fault. Besides, physical faults in PV modules can trigger a short circuit, such as ground fault, phase fault, and open circuit fault. On the other hand, fault in PV modules caused by environmental factors is the shading and soiling. Shading and soiling are the PV module covering by the shadow of an object that is located around the PV. Shading can be temporary or permanent and both can cause hotspots on the PV module. Explanations of these faults are given as follows.

2.1. Encapsulation material damage

In general, the arrangement of PV modules from the top (facing the sun) to the bottom sequence is the glass layer, encapsulation material, PV cells, encapsulation material, and the back sheet. Damage to encapsulation material can be in the form encapsulant broken so that the glass material can be in contact with the active PV cells. This event can occur by various factors such as the defect due to excessive pressure during production, the ambient temperature and humidity that are too high (so that it can cause melting in the encapsulation material), as well as the entry of salt particles between the encapsulation material and PV cells. The presence of salt material coverings and the entry of water vapor in the encapsulation gap can trigger the hotspot and increase the reflection of light to the PV module so that it can reduce the output power generated by the PV. To detect this fault, a thermographic method can be used because the fault will produce a hotspot [5-7].

2.2. Delamination

Delamination is the loss of adhesion between the materials that construct the PV module, especially in the back sheet and encapsulation materials. The delamination process can also occur due to production defects and the use of PV operations in very hot weather. Loose of adhesion between layers in the PV module can trigger the penetration of water (in the form of water favor in high humidity area or in the form of liquid water when the rain happen), this potentially creates fault with the cell hotspot. Besides, gaps arising from the loss of bond between layers can cause bubbles due to chemical reactions and expansion of water favor which are trapped in the gap [8-13].

2.3. Cracking

Cracked PV material generally occurs in the protective glass layer but it also probably occurs in the active PV cell layer. This damage can occur in the installation process or before the installation process such as during the manufacturing process, packaging, or during the module transportation process. In the installation process, cracks can occur due to mismatch of the location of the bolt in the frame so that the module experiences excessive bending and results in cracking. After the installation process or when the PV is operating, cracks can occur due to the hit of other objects on the PV module such as birds crashing and the cat that runs on the module. Crack that occurs in the module can induce water penetration, create shading, and triggering hotspots [2] [9] [14-16].

2.4. Interconnection failure

PV modules are series and parallel circuits of many cells so that many conductor connections between cells are potentially damaged. Besides, this interconnection damage can also occur in the connecting conductors between modules. This interconnection problem can occur due to imperfections in the soldering process during manufacturing, errors in the transportation process, as well as repeated mechanical stress. Interconnection problems can cause an increase in resistance of a conductor junction so that it can potentially cause hotspot fault [8] [17-20].

2.5. Corrosion

The delamination, cracks, and encapsulation events allow water penetration into the module through the edge of the module frame. The influx of water causes an increase in humidity of the PV module, which potentially causes corrosion and affects the conductivity of the material in the module layer, especially in the metallic material which constructs the module. Corrosion can cause leakage currents that can reduce the power production of PV modules and can increase the resistance value of the cell conductor which potentially produce hotspots [18-19] [21-27].

2.6. Bypass diode failure

PV modules are composed of parallel circuits of PV cell series arrangement (string). At the same irradiation conditions, the output voltage between the strings will have the same value. However, if one string is covered by the shadow of an object, the output voltage of the string will be lower than the other one. Because each string is arranged in parallel, when the shading happened, there will be a current flow from the healthy string to the string that experiences shading (reverse bias). To prevent reverse bias, a bypass diode is installed. In a healthy condition, the diode will block the reverse bias current flow so that there is no heating in the PV module. Failure at the bypass diode will cause a hotspot on the PV module which will degrade the module [6] [28-33].

2.7. Mismatch fault

Mismatch fault can be divided into two conditions, namely mismatch in the electrical parameters of the module and mismatch that occurs in the conductor junction of the PV module. This electrical parameter mismatch occurs due to differences in voltage, current, or power generated by cells contained in the module. The difference in electrical parameters can be caused by several factors such as the degradation of the PV module due to delamination, encapsulation damage, and corrosion. Meanwhile, mismatch in the soldering process (conductor junction) can occur due to production defects in the module. This mismatch disturbance can produce hotspot because it potentially creates reverse bias current flow and increases the resistance of module conductor material [8] [34-37].

2.8. Arc fault

An arc fault is an electric arc event between two separate metal ends at a certain distance in the PV module. This arc fault can occur because the potential difference that occurs between the two metals has a value greater than the breakdown voltage of the insulating material between the two ends of the metal. In the PV module, an arc can occur between the conductor connections on the module as well as between the conductor to the body of the PV module. This fault can produce very high temperatures up to hundreds of degrees Celsius so that it can burn PV modules [38-44].

2.9. Shading and soiling

In addition to fault that occurs in the PV module physically (internal), hotspot fault can also occur due to environmental influences such as shading and soiling in the PV module. Shading in the PV module can be divided into two conditions, namely dark shading and transparent shading. Dark shading is shading that blocks 100% of the sunlight that going towards the module like the shadow of trees, while transparent shading is shading that transmits a small portion of sunlight to the PV module, such as smoke or fog covering around the PV module. Besides, transparent shading can occur due to the presence of liquid that dries up on the surface of the PV modules such as liquid due to bird droppings. On the other hand, soiling is the sticking of dust particles/soil which has a small size that can stick to and cover the PV module. Soiling will block the sunlight received by the PV module and in humid conditions can trigger the growth of moss on the PV surface.

Shading and soiling that occur in cells in the PV module will cause these cells to have increased resistance when compared to other cells that do not experience shading. Increasing the value of resistance in a small number of cells in a PV module will result in differences in current and voltage characteristics of the PV string. A string that experiences shading will produce a lower total voltage value compared to a healthy string so that there will be a reverse bias current flowing from the healthy string to the shading one [33] [45-47].

The reverse bias process will cause heating on the shaded cell so that it can produce hotspots on the cell. As described in the previous section, the reverse bias process is generally prevented by using a bypass diode on each string. However, the addition of a bypass diode still has the probability to fail if the voltage difference between the healthy string and the shading string exceeds the limit of the voltage capability of the diode (breakdown voltage), or if the bypass diode is damaged [48-49].

3. PV Hotspot Detection Method

3.1. Existing method

The technology used to detect the faults that cause hotspots on PV is the thermal imaging method. Thermal imaging is a non-destructive measurement technique, which provides the temperature distribution features of the PV module. This method can be used as a contactless method to diagnose some fault that generates heat on the PV module when the PV operates in normal condition.

Thermal imaging technology utilizes infrared cameras (with electromagnetic wavelengths about 8 to 14um) to detect the phenomenon of hotspots that occur in the PV module. PV cells that experience hotspots will have a higher temperature increase compared to the normal cells around them so that when it is observed with an infrared camera, color differences will appear between hot spotted cells and normal cells.

In Indonesia, hotspot fault detection methods for PV modules currently use manual thermal imaging techniques that utilize infrared cameras, which are carried by the operator that walking around the PV farm. The operator observes the PV module one by one with an infrared camera to see the occurrence of hotspots on the PV.

In other countries, hotspot detection methods are using thermal imaging through a drone that has been equipped with infrared cameras. Drones are flown at certain times (not operated continuously) to monitor the occurrence of hotspots on the PV module[26].

The method of detecting hotspots using infrared both manually and using drones can only work at certain times and requires a long time to detect a fault on a large PV area. At present, to detect the fault in a 30 MW PV generation with an area of 60 hectares and composed of 126000 modules, the manual detection method takes about 210 days and the detection method with drones will take about 30 days. These methods cannot detect the fault quickly and continuously because these methods need a movement from one location to the other. Hotspot fault can occur intermittently influenced by environmental conditions. Therefore, fault detection equipment is needed to be able to work quickly and continuously.

3.2. Proposed concept

The current hotspot fault detection technology takes more time to detect the hotspot fault. To complete the gap of the existing detection technology, a hotspot detection method using a fisheye lens equipped with infrared sensors is proposed. The fisheye lens has a wide area detection characteristic, which can have a field of view (FOV) of 180°. Therefore all objects in front of the lens will be observable. The illustration of the detection method is given in figure 1.

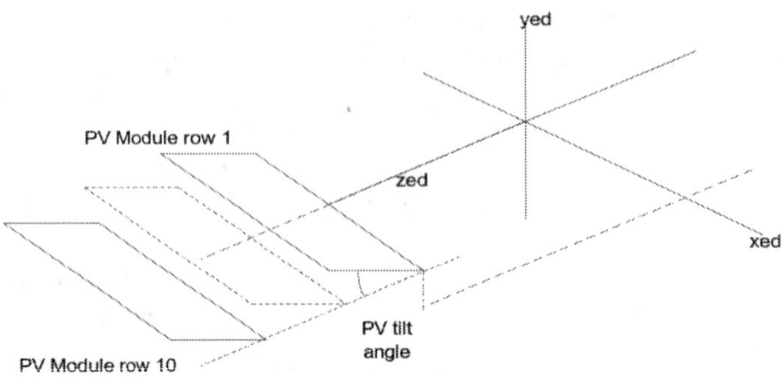

Figure 1. Hotspot detection using a fisheye lens.

Zed, Yed, and Xed is the z, y, x coordinate position of the fisheye lens relative to the PV module row. PV module row 1 is the nearest PV module row to the fisheye lens and the PV module row 10 is the farthest one. Base on the preliminary study, with the special configuration (the value of Zed=0, Yed=10m, xed=0, 100m length of PV module row, and 10mm fisheye lens focal length) the proposed method is possible to detect whole of PV module. The view of the whole PV module row as the fisheye lens perspective view is given in figure 2.

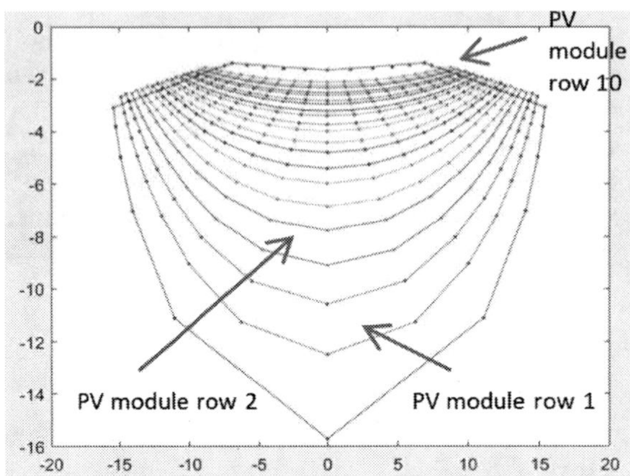

Figure 2. The view of PV module row in fisheye lens perspective

In figure 2, (both x and y-axis are in mm) all PV module rows can be monitored using a single fisheye lens. If this model is implemented to the infrared material lens, when a hotspot occurs in one module, the proposed method will utilize the image processing to localize the hotspot. After the hotspot location is definite, the algorithm will calculate the exact module position in the PV farm. Thus, the hotspot location that occurs in a single module could be detected. Finally, it is expected, the process of detecting hotspot faults will be faster.

4. Conclusion

An explanation of the faults types that produce hotspots on the PV module has been carried out, from these results it can be seen that hotspots can occur due to the module physical fault as well as environmental influences such as shading and soiling. Hotspot fault occurs of almost 50% of all fault types in the PV system. The current hotspot fault detection methods utilize manual thermal imaging or drones thermal imaging, where these methods cannot detect the fault quickly and continuously. A fisheye lens has been modeled to monitor all of the PV modules. The result shows that the fisheye lens could simultaneously monitor all of the PV modules rows which has a length of 100m each. Future research will try to find the accuracy of the monitoring system which is affected by the infrared lens material, hotspot temperature, and the lens position relative to the PV module rows.

Acknowledgment
The authors would like to express special thanks and acknowledgment to the Indonesia Endowment Fund for Education (Lembaga Pengelola Dana Pendidikan Indonesia) which gave the financial support for this research.

References
[1] IRENA 2019 *Future of Solar Photovoltaic: Deployment Investment Technology Grid Integration and Socio-Economic Aspects*

[2] Mellit A, Tina G M and Kalogirou S A 2018 *Fault detection and diagnosis methods for photovoltaic systems: A review* Renew. Sustain. Energy Rev. **91**

[3] Nair S S 2018 *A Survey report of the firefighters on fire hazards of PV fire* IEEE Int. Conf. Syst. Comput. Autom. Networking ICSCA.

[4] Deng S et al 2017 *Research on hotspot risk for high-efficiency solar module* SNEC Shanghai China.

[5] Tanahashi T Fukumoto Y, Tamai F, Masuda A and Kondo M 2017 *Causes of Degradation Identified by the Extended Thermal Cycling Test on Commercially Available Crystalline Silicon Photovoltaic Modules.*

[6] Triki-lahiani A, Abdelghani A B and Slama-belkhodja I 2017 *Fault detection and monitoring systems for photovoltaic installations : A review* Renew. Sustain. Energy Rev. no. July pp. 0–1.

[7] Beinert A J, Romer P, Heinrich M, Mittag M, Aktaa J and Neuhaus D H 2020 *The Effect of Cell and Module Dimensions on Thermomechanical Stress in PV Modules* IEEE J. Photovoltaics **10** 1 pp. 70–77.

[8] Abdulmawjood K, Refaat S S, and Morsi W G 2018 *Detection and prediction of faults in photovoltaic arrays: A review* Proc. - 2018 IEEE 12th Int. Conf. Compat. Power Electron. Power Eng. CPE-POWERENG 2018 pp. 1–8.

[9] Chattopadhyay S et al. 2018 *Correlating Infrared Thermography With Electrical Degradation of PV Modules Inspected in All-India Survey of Photovoltaic Module Reliability 2016* IEEE J. Photovoltaics vol. PP pp. 1–9.

[10] Quarter P B et al. 2014 *Light Unmanned Aerial Vehicles (UAVs) for Cooperative Inspection of PV Plants* **4** no. 4 pp. 1107–1113.

[11] Dhoke A and Mengede A 2018 *Degradation analysis of PV modules operating in Australian environment* 2017 Australas. Univ. Power Eng. Conf. AUPEC 2017 vol. 2017-Novem pp. 1–5.

[12] Dhoke A, Sharma R, and Saha T K, 2018 *PV module degradation analysis and impact on settings of overcurrent protection devices* Sol. Energy **160** no. December 2017 pp. 360–367.

[13] Madeti S R, and Singh S N 2017 *A comprehensive study on different types of faults and detection techniques for solar photovoltaic system* Sol. Energy **158** no. August pp. 161–185.

[14] Brooks W S M. Lamb D A, Irvine S J C, and Abstract A 2015 *IR Reflectance Imaging for Crystalline Si Solar Cell Crack Detection* pp. 1–5.

[15] Types P V A, Alam M K, Khan F, Member S, Johnson J, and Flicker J 2015 *A Comprehensive Review of Catastrophic Faults in Mitigation Techniques* pp. 1–16.

[16] Wang W, Member S and Liu A C 2015 *Fault Diagnosis of Photovoltaic Panels Using Dynamic Current-Voltage Characteristics* **8993** no. APEC pp. 1–39.

[17] Quater P B, et al. 2014 *Light Unmanned Aerial Vehicles (UAVs) for Cooperative Inspection of PV Plants* **4** 4 pp. 1107–1113.

[18] Alsafasfeh M, Abdel-qader I and Bazuin B 2017 *Fault Detection in Photovoltaic System Using SLIC and Thermal Images* pp. 672–676.

[19] Han H et al. 2018 Analysis of the Degradation of Monocrystalline Silicon Photovoltaic Modules After Long-Term Exposure for 18 Years in a Hot – Humid Climate in China **8** 3 pp. 806–812.

[20] Schuss Cc et al. 2017 *Defect localisation in photovoltaic panels with the help of synchronized thermography* I2MTC 2017 - 2017 IEEE Int. Instrum. Meas. Technol. Conf. Proc.

[21] Yi Z and Etemadi A H 2017 *Line-to-line fault detection for photovoltaic arrays based on multi-resolution signal decomposition and two-stage support vector machine* IEEE Trans. Ind. Electron. **64** 11.

[22] Wang W, Member S and Liu A C 2015 *Fault Diagnosis of Photovoltaic Panels Using Dynamic Current-Voltage Characteristics* **8993** no. APEC pp. 1–39.

[23] Roy S, Alam M K, Khan F, Johnson J and Flicker J 2018 *An Irradiance-Independent Robust Ground-Fault Detection Scheme for PV Arrays Based on Spread Spectrum Time-Domain Reflectometry (SSTDR)* IEEE Trans. Power Electron. **33** 8 pp. 7046–7057.

[24] Lin X, Wang Y and Pedram M 2014 *Designing Fault-Tolerant Photovoltaic Systems* no. July pp. 76–84.

[25] Solórzano J and Egido M A 2013 *Automatic fault diagnosis in PV systems with distributed* MPPT **76** pp. 925–934.

[26] Lešetick J, Poulek V and Sedl J 2019 *Monitoring of Defects of a Photovoltaic Power Plant Using a Drone* vol. 0.

[27] Harrou F, Sun Y, Taghezouit B, Saidi A and Hamlati M 2017 *Reliable fault detection and diagnosis of photovoltaic systems based on statistical monitoring approaches* Renew. Energy.

[28] Fadhel S et al. 2018 *Data-driven approach for isolated PV shading fault diagnosis based on experimental I-V curves analysis* Proc. IEEE Int. Conf. Ind. Technol. vol. 2018-Febru pp. 927–932.

[29] Hosseinzadeh M and Salmasi F R 2016 *Simulation Modelling Practice and Theory Determination of maximum solar power under shading and converter faults — A prerequisite for failure-tolerant power management systems* **62** pp. 14–30.

[30] Kumar B P, Ilango G S, Reddy M J B and Chilakapati N 2018 *Online fault detection and diagnosis in photovoltaic systems using wavelet packets* IEEE J. Photovoltaics **8** 1 pp. 257–265.

[31] Hu Y, Cao W, Member S, Wu J, Ji B and Holliday D 2014 *Thermography-Based Virtual MPPT Scheme for Improving PV Energy Efficiency Under Partial Shading Conditions.* **29** 11 pp. 5667–5672.

[32] Heidari N, Gwamuri J, Townsend T and Pearce J M 2015 *Impact of Snow and Ground Interference on Photovoltaic Electric System Performance*l. **5** 6 pp. 1680–1685.

[33] Modules B P, Zhang Y, Yu Y, Meng F and Liu Z 2019 *Experimental Investigation of the Shading and Mismatch Effects on the Performance of IEEE.* J. Photovoltaics. pp. 1–10.

[34] Escribano M M, Solano M G, Laita I D L P, Alvarez J M, Marroyo L and Pigueiras E L 2018 *Module temperature dispersion within a large PV array: Observations at the amareleja PV plant IEEE* J. Photovoltaics **8** 6 pp. 1725–1731.

[35] Dhimish M, Holmes V, Mehrdadi B, Dales M and Mather P 2017 *Output Power Enhancement for Hotspotted Polycrystalline Photovoltaic Solar Cells* **4388** no. c.

[36] Belhadj C A, Banat I H and Deriche M 2017 *Spot Phenomena based on the Bishop Model* pp. 222–227.

[37] Appiah A Y, Zhang X, Ayawli B B K and Kyeremeh F 2019 *Review and Performance Evaluation of Photovoltaic Array Fault Detection and Diagnosis Techniques* Int. J. Photoenergy vol. **2019** pp. 1–19.

[38] Garoudja E, Harrou F, Sun Y, Kara K, Chouder A and Silvestre S 2017 *Statistical fault detection in photovoltaic systems* Sol. Energy **150** pp. 485–499.

[39] Akram M N and Lotfifard N 2015 *Modeling and Health Monitoring of DC Side of Photovoltaic Array* pp. 1–9.

[40] R. Hariharan M. Chakkarapani G. S. Ilango C. Nagamani and S. Member A Method to Detect Photovoltaic Array Faults and Partial Shading in PV Systems pp. 1–8 2016.

[41] Pillai D S, Blaabjerg F and Rajasekar N 2019 *A Comparative Evaluation of Advanced Fault Detection Approaches for PV Systems* IEEE J. Photovoltaics **9** 2 pp. 513–527.

[42] Murtaza A F, Bilal M, Ahmad R and Sher H A 2019 *A Circuit Analysis based Fault Finding Algorithm for Photovoltaic Array under L-L/L-G Faults* IEEE J. Emerg. Sel. Top. Power Electron. vol. PP no. c pp. 1–1.

[43] Chen S, Member S, Li X, Member S and Xiong J 2017 *Series Arc Fault Identification for Photovoltaic System Based on Time-Domain and Time-Frequency-Domain Analysis* pp. 1–10.

[44] Lu S, Sirojan T, Phung B T, Zhang D and Ambikairajah E 2019 *DA-DCGAN: An Effective Methodology for DC Series Arc Fault Diagnosis in Photovoltaic Systems* IEEE Access. **7** no. c pp. 45831–45840.

[45] Platon R, Martel J, Woodruff N and Chau T Y 2015 *Online Fault Detection in PV Systems* no. October pp. 1200–1207.

[46] Bonsignore L, Davarifar M, Rabhi A, Tina G M and Elhajjaji A 2014 *Neuro-Fuzzy fault detection method for photovoltaic systems* Energy Procedia **62** pp. 431–441.

[47] Salazar A M 2016 *Hotspots Detection in Photovoltaic Modules Using Infrared Thermography* vol. **10015**.

[48] Hu Y et al. 2015 *Online Two-Section PV Array Fault Diagnosis with Optimized Voltage Sensor Locations* vol. **0046** no. c.

[49] Meyer E L and Dyk E E V 2004 *Assessing the reliability and degrada- tion of photovoltaic module performance parameters* IEEE Trans. Rel. **53** 1 pp. 83–92 Mar.

IOP Conf. Series: Materials Science and Engineering **982** (2020) 012020 doi:10.1088/1757-899X/982/1/012020

CUK Converter for Power Factor Correction Using *Moth Flame Optimization-PI Controller*

Adinda Aisyah Pratiwi[1], Moh. Zaenal Efendi[2], Farid Dwi Murdianto[3]

[1,2,3]*Department of Electrical Engineering, Politeknik Elektronika Negeri Surabaya, Indonesia*

[1]adindaaisyah98@gmail.com, [2]zen@pens.ac.id, [3]farid@pens.ac.id

Abstract. Electronic equipment in household loads in the form of non-linear loads that require a circuit that requires a power supply in the form of a direct voltage (DC) that requires an AC to DC rectifier in the form of a full-wave rectifier. However, the installation of capacitor values in the circuit can cause the waveforms of the input current to become uneven or distorted so that the emergence of harmonics which results in a low power factor with the improvement of the power factor. It affects the output voltage to the load so we need control for the voltage regulator. The solutions to overcome the problem of input current defects can be overcome using a power converter used as a PFC. In this paper, *CUK Converter* used as PFC is expected to get close to 1 (one) results in increasing power factor and the need for a voltage regulator to regulate the voltage requirements for loads that are utilized by the *Moth Flame Optimization* algorithm method - PI method. This *CUK Converter* functions in DCM (Discontinuous Conduction Mode) which makes the system resistive. In conclusion from the results of simulations that have been done, the results obtained from the use of *CUK Converter* as PFC can improve the power factor from pf value 0.731 to 0.994 and with MFO-PI control it can maintain a constant output voltage of 12 volts with a current of 5 amperes.

1. Introduction

In electric power systems, there are two types of electrical loads called linear load and non-linear load. The load is called linear if the current is directly proportional to the load voltage and causes the current waveform to be the same as the waveform of the load. However, in non-linear loads, the current waveform is not the same as the voltage waveform or the distorted current waveform. The non-linear load that is commonly used in electrical equipment is one of them in the form of a diode rectifier circuit that will result in a non-pure sine input current. Non-sine currents indicate harmonics. Whereas at the household load it is often needed a series of diode rectifiers, namely electronic equipment such as television, computers, and others. With the use of a diode that causes the current wave to be distorted, this results in a low power factor value.

To overcome the high harmonic problem and increase the power factor of the rectifier circuit, one can be done by adding a converter that is used as Active *Power Factor Correction* (PFC) [1]. Adding a converter to the circuit can improve the input current so that it is sinusoidal in shape, a power factor closes to one, and low harmonics (according to international standards, IEC1000-3-2) [2][3]. The DC-DC converter circuit to improve the power factor that is often used is the *Buck Converter*, *Boost Converter* and *Buck-Boost Converter* [4-7]. The most popular converter among PFC converters is the boost converter, but the output voltage is higher than the peak supply voltage [8].

In this paper, a research was made on improving power factors using two converters in the form of *CUK Converter* and *Buck Converter*. *CUK Converter* is used as a PFC and *Buck Converter* is used as a voltage regulator. The *CUK Converter* can overcome the weakness of the *Boost Converter* in terms of noise that EMI usually does [9]. *CUK Converter* which is used as a PFC must have a resistive or near resistive impedance. For the use of *CUK Converter* as PFC, it is operated in a non-continuous condition or Discontinuous Conduction Mode (DCM) [10]. When the rectifier circuit is connected and supplies a resistive load, the load current that flows back to the source has a sine waveform and resembles the input voltage waveform so that its power factor value approaches 1 (unity). *Buck Converter* is operated as a voltage regulator that serves to maintain the stability of the output voltage which is controlled using the *Moth Flame Optimization-PI Controller* method. The first *Moth Flame Optimization* method was developed by Mirjalili. This method is based on the moth navigation method in nature. Moths fly at night while maintaining a fixed angle to the moon, which is an effective mechanism for reaching straight lines at great distances [11]. Intended this method is used to find the fitness value used on the *PI* controller which later with the combination of the algorithm with the controller will be able to maintain the stability of the output voltage connected to the load.

2. Modelling and Design of DC-DC *CUK Converter*

2.1. CUK Converter as Power Factor Correction Converter

CUK Converter is a type of DC-DC converter which is a perfect complement of *Buck-Boost Converter*. The converter can increase and decrease the voltage and have negative polarity [12]. The *CUK Converter* circuit can be seen in Figure 1. The inductor on the input side works as a filter on the DC supply to prevent harmonics.

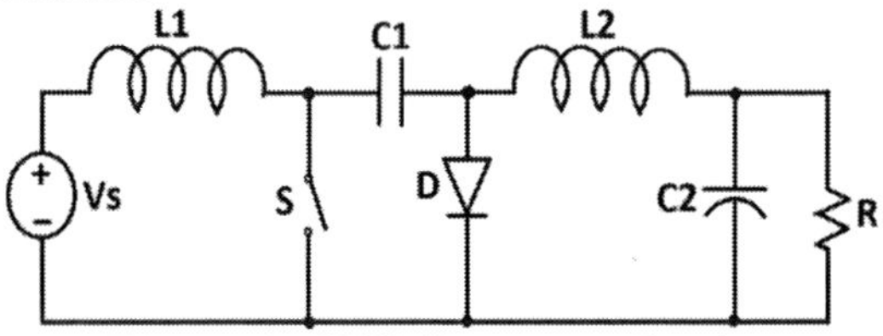

Figure 1. *CUK Converter* Circuit

The switching topology of the DC-DC *CUK Converter* circuit is shown in Figure 2 (a) and (b) [13].

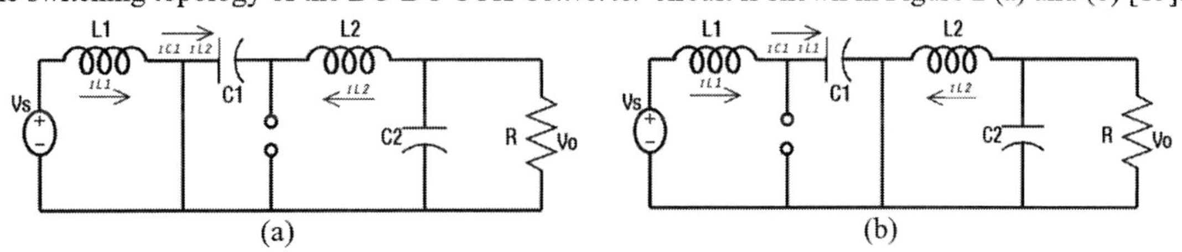

Figure 2. (a) *CUK Converter* is ON condition (b) *CUK Converter* is OFF condition

In the on the condition, the switch is closed so that the current is passed, the diode does not work, and the Ci capacitor is discharged by the current from the Lo inductor. When the condition is off, the switch is open so that the diode flows currently from the Li and Lo inductors. In a state of shutdown, capacitors are also needed by the current from the Lo inductor.

To be operated as a PFC, a *CUK Converter* must have a resistive or near resistive impedance. *CUK Converter* is operated in DCM conditions, having an input current (I_{in}) expressed in Equation 1

$$I_{in(t)} = \frac{D^2T}{2L_1}\left(\frac{e}{1-\left(\frac{e}{vci}\right)}\right) \tag{1}$$

The input impedance value is expressed in Equation 2:

$$Z_{in(t)} = R_{in} + jX \tag{2}$$

From a reactance value of less than one, the reactance value can be ignored in Equation 3 and Equation 4, thus:

$$Z_{in(t)} = R_{in} \tag{3}$$

$$r_{i(t)} \approx \frac{2L_1}{D^2T} \tag{4}$$

Then the input resistance value can be stated in Equation 5:

$$r_{i(t)} = \frac{2L_1}{D^2T}\left(1 - \frac{e}{vc_1}\right) \tag{5}$$

If value $\left|\frac{e}{vc_1}\right| << 1$ then the input resistance value becomes :

$$r_{i(t)} \approx \frac{2L_1}{D^2T} \tag{6}$$

From Equation 6 it can be seen that the value of input resistance depends on the value of the inductor, duty cycle, and switching period. If the duty cycle value is constant then the input resistance value will also be constant. The input current will follow the shape of the input voltage. Then, there will be a reduction in harmonics on the input side. There will also be an improvement in the power factor.

2.2. Design PFC DC-DC CUK Converter

The following designs for PFC DC-DC *CUK Converter* are shown in Table 1.

Table 1. Design PFC DC-DC *CUK Converter*

Variable	Value
Input Voltage (V_s)	99.03 V
Output Voltage (V_o)	50 V
Output Current (I_o)	0.9615 A → 1.5 A
Switching Frekuensi (f_s)	100 kHz
Ripple Input Current	10%
Ripple Output Current	11%
Power	66 W
Inductor Input (L_1)	760μH
Capacitor Input (C_1)	≥0.59μF
Inductor Output (L_2)	750μH
Capacitor Output (C_2)	≥5.1μF
k	0.9393
L_m	1.5424 mH

The rating of inductors and capacitors can be found by following equation[14].

$$L_2 = \frac{V_o.(1-D)}{f.(\lambda_1 i_1 + \lambda_2 i_2)} \tag{7}$$

$$L_m = L_2 \tag{8}$$

$$L_1 = \left(V_s - \frac{L_m \lambda_2 i_2 f}{D}\right) \times \left(\frac{D}{\lambda_1 i_1 f}\right) \tag{9}$$

$$k = \sqrt{\frac{L_2}{L_1}} \tag{10}$$

IOP Conf. Series: Materials Science and Engineering **982** (2020) 012020 doi:10.1088/1757-899X/982/1/012020

$$C_2 \geq \frac{1-D}{(\Delta V_o / V_o) 8 L_2 f^2} \tag{11}$$

$$C_1 \geq \frac{V_o D}{R f \Delta V_{C_1}} \tag{12}$$

2.3. *Operation and Design of Voltage Regulator Buck Converter*

Buck Converter is a type of DC-DC converter that works to reduce voltage or step down by setting the duty cycle [15]. Shown in Figure 3.

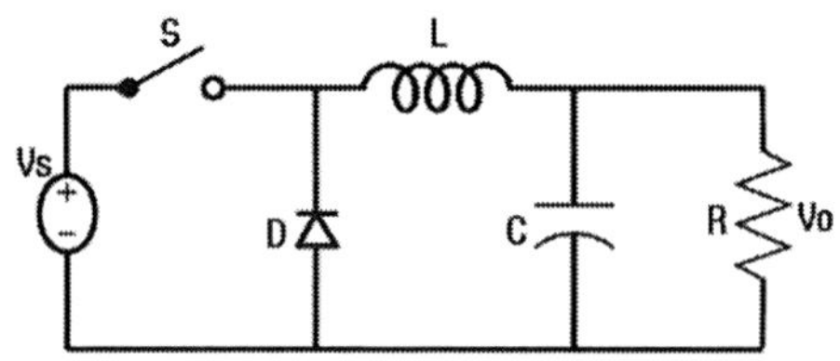

Figure 3. *Buck Converter* Circuit

Buck Converter can produce a maximum voltage value equal to the input voltage. In this paper, the *Buck Converter* is used for voltage regulators connected to the load, following the design of the *Buck Converter* in Table 2.

Table 2. Design *Buck Converter* for voltage regulator

Variable	Value
Input Voltage (V_s)	65 V
Output Voltage (V_o)	12 V
Output Current (I_o)	4.167 A → 5 A
Switching Frekuensi (f_s)	100 kHz
Ripple Current	12.5%
Ripple Voltage	0.1%
Daya	66 W
Inductor (L)	139.669µH
Capacitor (C)	65.104µF

The rating of inductors and capacitors can be found by following equation [16].

$$L = \left(\frac{1}{f_s}\right) \times (V_s - V_o) \times \left(\frac{V_o + V_f}{V_s + V_f}\right) \times \left(\frac{1}{\Delta i_L}\right) \tag{13}$$

$$C_o = \frac{\Delta i_L \times T}{8 \Delta V_o} = \frac{\Delta i_L}{8 \times f \times \Delta V_o} \tag{14}$$

3. Closed Loop Control Scheme and Design of Controller

3.1. Closed Loop Scheme of Proposed Method

This section explains about the closed control system of PFC *CUK Converter* by using the *Moth Flame Optimization-PI*. Shown in Figure 4.

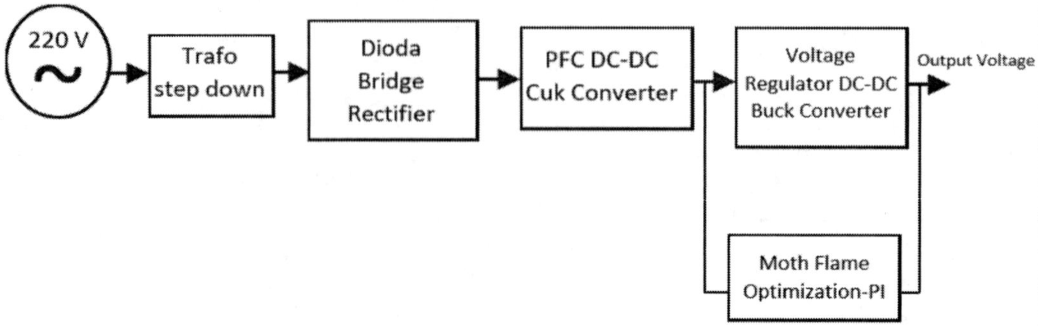

Figure 4. PFC *CUK Converter* circuit configuration with *Moth Flame Optimization-PI Controller*

The input source from PLN is derived using a step-down transformer. Then, the AC source is converted to a DC source bypassing the bridge rectifier diode. The output of the rectifier circuit is a DC voltage which will be the input voltage of the *CUK Converter* which functions as a PFC. To make a converter as a PFC, the converter must be operated as if it is resistive. The converter must be operated in DCM (*Discontinuous Conduction Mode*), then connected to the *Buck Converter* voltage regulator. The voltage regulator will be kept constant at 12 volts. To maintain a constant voltage MFO-PI (*Moth Flame Optimization-PI*) is used. MFO here is used to determine the optimal value of the PI which will be used to control the output voltage.

3.2. Design Moth Flame Optimization-PI

The *Moth Flame Optimization* (MFO) algorithm is an algorithm adapted from the lateral movement of moths around bright objects such as candles, which are found in nature [20]. This MFO was first proposed in 2016 by Mirjalili [11]. In this MFO algorithm moth and flame are the main roles and both are considered as a result of this algorithm. However, there is a difference between the way they are updated through each iteration. Moths are search agents (e.g. they are parameter values of problems) and the flame is the best position of moths throughout the search space (it can be assumed that moths embed the best solution they obtained using flame) [17]. Because the MFO algorithm is a population-based algorithm, here is a matrix representing the population of Moth:

$$M = \begin{bmatrix} m_{1,1}, & m_{1,2} & ,\ldots, & m_{1,d} \\ m_{2,1}, & m_{2,2} & ,\ldots, & m_{2,d} \\ \vdots & & \vdots & \\ \vdots & & \vdots & \\ m_{n,1}, & m_{n,2} & ,\ldots, & m_{n,d} \end{bmatrix} \tag{15}$$

n = Number of *Moth*
d = Number of variabel (dimension)

For all Moth, it is assumed that an array to store the values match, as follows:

$$OM = \begin{bmatrix} OM_1 \\ OM_2 \\ . \\ . \\ OM_n \end{bmatrix} \tag{16}$$

Another component in the proposed algorithm is Flame. This matrix is considered similar to the Moth matrix which is as follows:

$$F = \begin{bmatrix} FL_{1,1}, FL_{1,2} ,\dots, FL_{1,d} \\ FL_{2,1}, FL_{2,2} ,\dots, FL_{2,d} \\ \vdots \quad\quad \vdots \\ \vdots \quad\quad \vdots \\ FL_{n,1}, FL_{n,2} ,\dots, FL_{n,d} \end{bmatrix} \tag{17}$$

n = Number of *Moth*
d = Number of variabel (dimension)

It is known that the dimensions of the arrays M and F are the same. For the Flame , it is assumed that it is in the form of arrays to store matching values, as follows:

$$OF = \begin{bmatrix} OFL_1 \\ OFL_2 \\ . \\ . \\ OFL_n \end{bmatrix} \tag{18}$$

The flowchart of the MFO algorithm is shown in Figure 5.

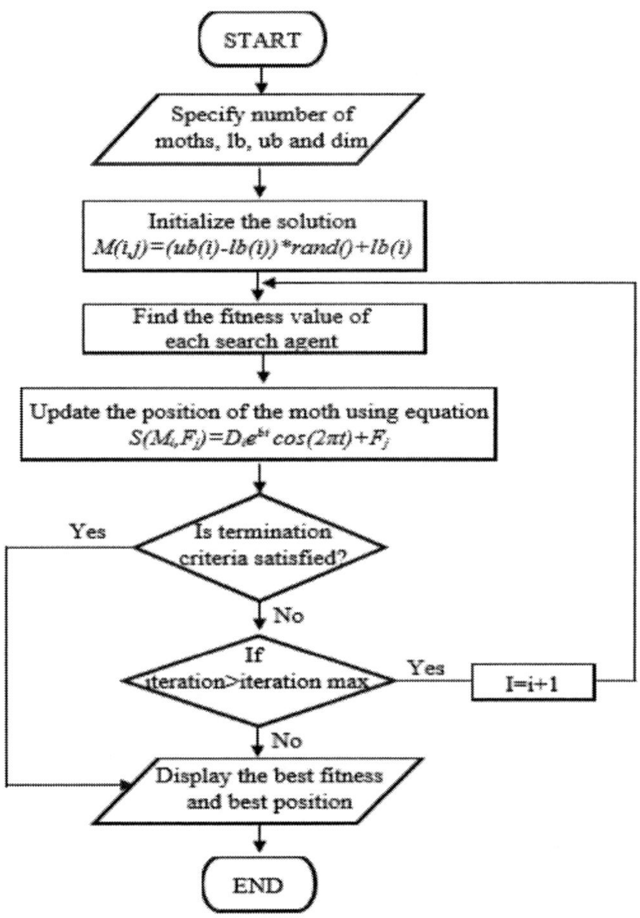

Figure 5. Flowchart *Moth Flame Optimization* [17]

To get the optimal value of PI control using *Moth Flame Optimization*, needed a limit of Kp and Ki values [18].

$$\left.\begin{aligned} K_P^{MIN} \le K_P \le K_P^{MAX} \\ K_I^{MIN} \le K_I \le K_I^{MAX} \end{aligned}\right\} \quad (19)$$

MIN = for the minimum value of the gains
MAX = for the maximum value of the gains.

In this process, the MFO algorithm is used to enhance the PI controller to control the working of the *Buck Converter* voltage regulator. This simulation is done with many Moth populations of 50 and many iterations of 500.

In Figure 6. is a toolbox from *Moth Flame Optimization* to find fitness values from Kp and Ki with predetermined parameters

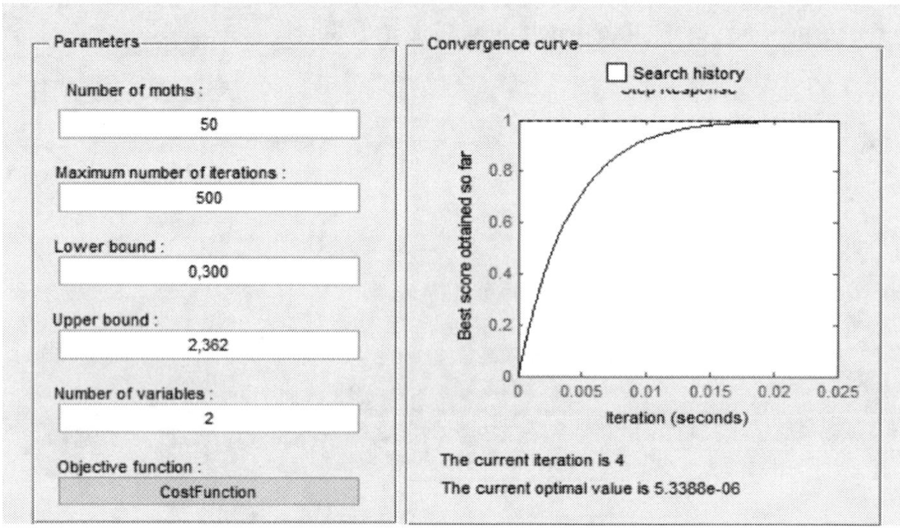

Figure 6. *Moth Flame Optimization* Toolbox

4. Simulation Result and Discussion

The proposed PFC *CUK Converter* is designed with DC-DC *CUK Converter* operated in DCM (*Discontinuous Current Mode*) mode which is directly connected to the rectifier output with a *CUK Converter* input of 99V and by connecting directly to the voltage regulator in the form of *Buck Converter* which is used to stabilize output voltage using the MFO-PI Controller (*Moth Flame Optimization - PI Controller*) control. By using a 100kHz switching frequency. The proposed method is expected to get the optimal value of the Kp and Ki parameters that will be used to control the voltage regulator system. And for series images can be seen in Figure 7.

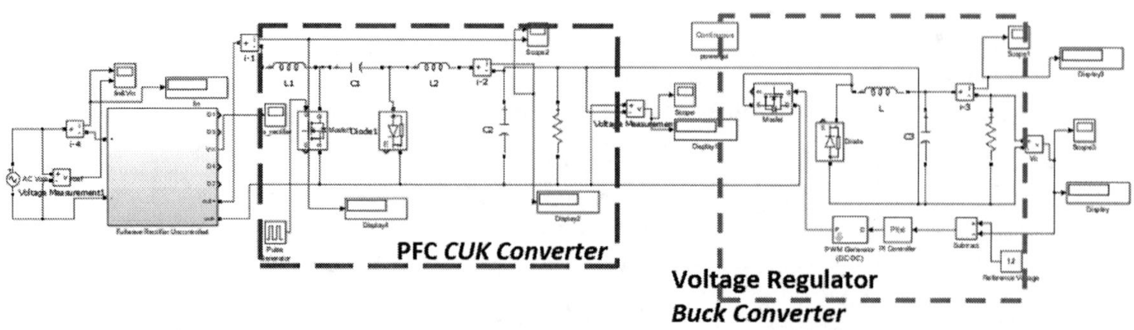

Figure 7. Model of Proposed PFC *CUK Converter*

Figure 7. Describes the simulation image of the whole system consisting of a *CUK Converter* as a *Power Factor Correction* (PFC) and a *Buck Converter* as a voltage regulator controlled using *Moth Flame Optimization - PI Controller* which in the MFO optimization algorithm uses a toolbox and is used to adjust the output voltage to be stable at 12 Volts with Kp and Ki settings on the PI controller toolbox.

Figure 8. (a) shows a 110V AC source voltage waveform and (b) shows an input current waveform distorted due to being connected to a 1 phase full-wave rectifier. Non-sine input waveforms due to the harmonic current content of the input current [19]. The distortion in the input current in Figure 8.(b) below is caused by the installation of a filter capacitor on the rectifier output side which causes the input current to flow when the capacitor is charging. So that the waveform becomes pulsating as a result the distortion factor does not approach 1(unity) but approaches 0. Which causes the pf to below.

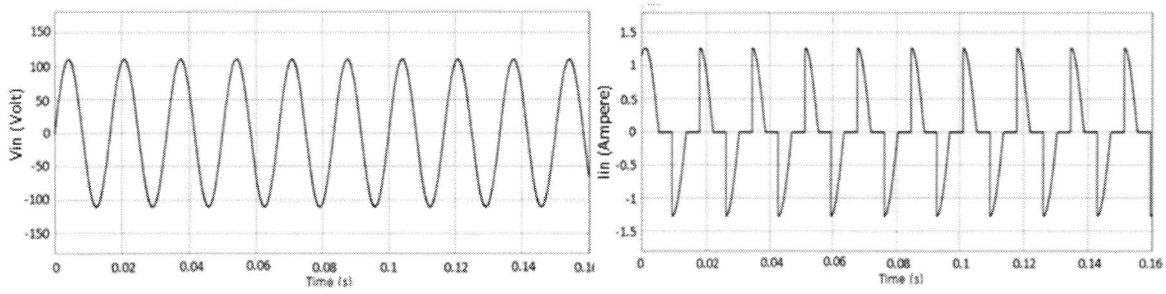

(a) *(b)*

Figure 8. (a) Source Voltage Waveform (b) Input Current Waveform without PFC *CUK Converter* at source voltage 110V

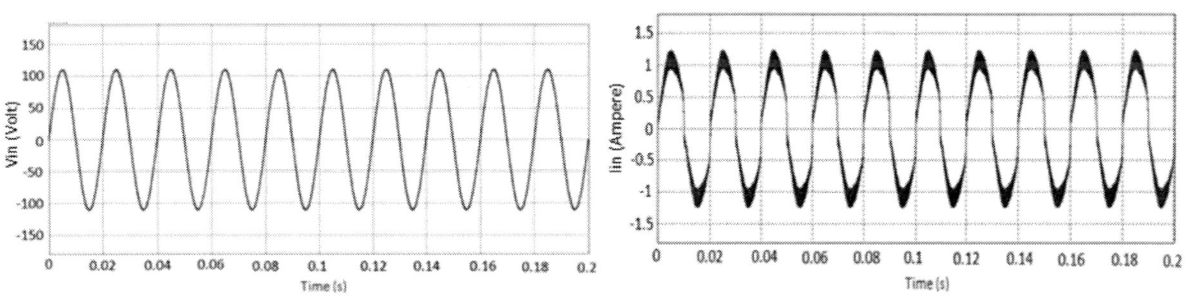

(a) *(b)*

Figure 9. (a) Source Voltage Waveform (b) Input Current Waveform with PFC *CUK Converter* at source voltage 110V

Figure 9. shows the source voltage and input current wave connected to the PFC *CUK Converter*. Based on data obtained with 110V input voltage that the THD value of 1.58% is shown in Figure 10. In Figure 9 it can be observed the difference in the input current wave current before PFC installation and after PFC installation. When after the existence of the PFC circuit, the input current wave approaches sinusoidal so that the distortion factor approaches 1 and the displacement factor approaches 1 so that the pf value approaches 1. So that by installing the PFC circuit, the harmonic spectrum contained in the input current waveform becomes low or close to 0, the harmonic content is small.

IOP Conf. Series: Materials Science and Engineering **982** (2020) 012020 doi:10.1088/1757-899X/982/1/012020

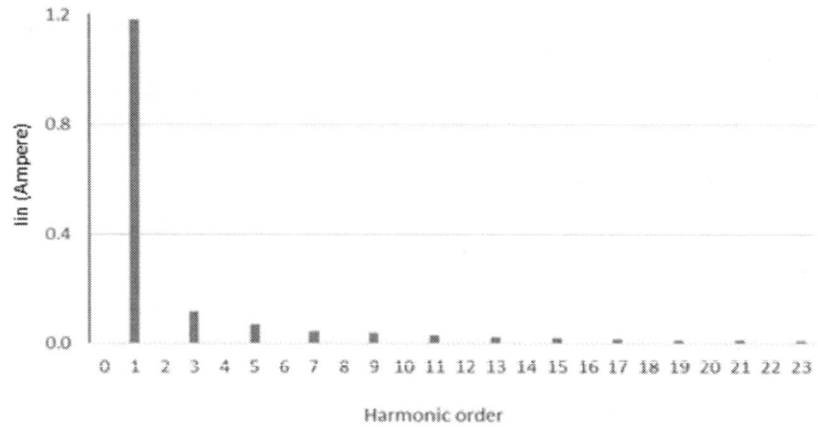

Figure 10. Input Current Waveform and Harmonic Spectrum at a Source Voltage of 110V

Figure 11. shows the output voltage and output voltage waves with Vo values of 12 Volts and Io of 5 Ampere according to load requirements. The results of the simulation on the *Buck Converter* voltage regulator using the *Moth Flame Optimization-PI Controller* method can successfully control the output voltage by the set point. With the *Moth Flame Optimization-PI Controller* with various load changes, the output voltage and bias current remain constant according to the set point of 12Volt and 5 Ampere.

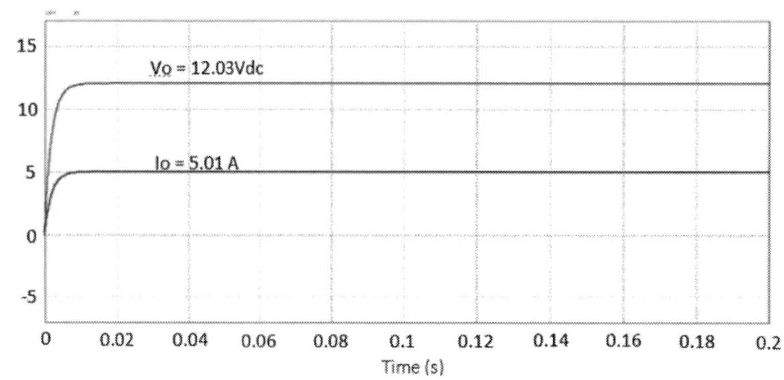

Figure 11. Output Voltage and Output Current Waveform at a source voltage of 110V

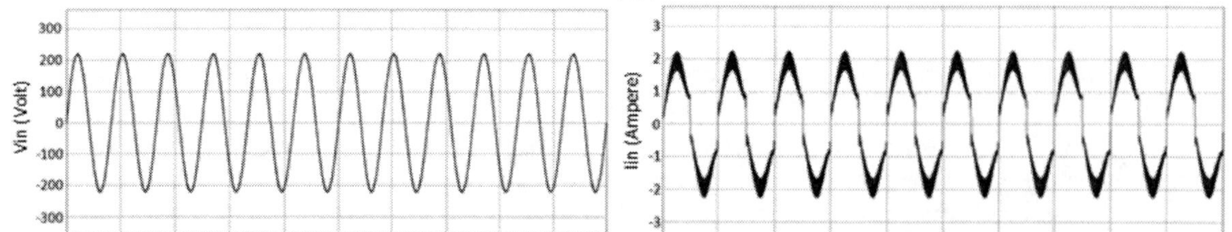

Figure 12. (a) Source Voltage Waveform (b) Input Current Waveform with PFC *CUK Converter* at Source Voltage 220V

Figure 12. represent the waveform when the source voltage changes with a value of 220Vrms. Figure 14. displays the current harmonics spectrum and obtained THD of 1.61%. Figure 13 is the output voltage waveform and input current when the source is 220V.

IOP Conf. Series: Materials Science and Engineering **982** (2020) 012020 doi:10.1088/1757-899X/982/1/012020

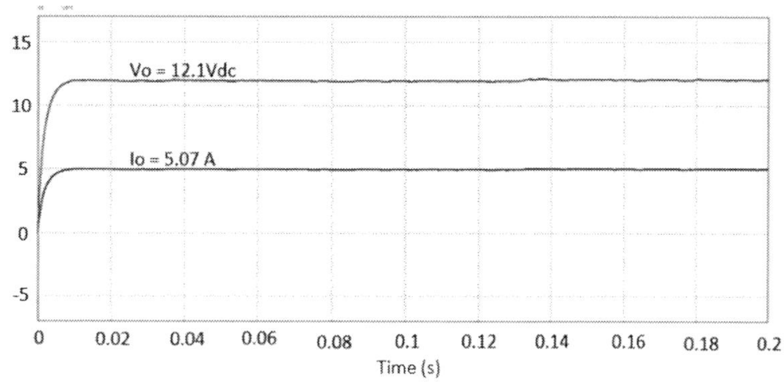

Figure 13. Output Voltage and Output Current Waveform at a Source Voltage of 220V

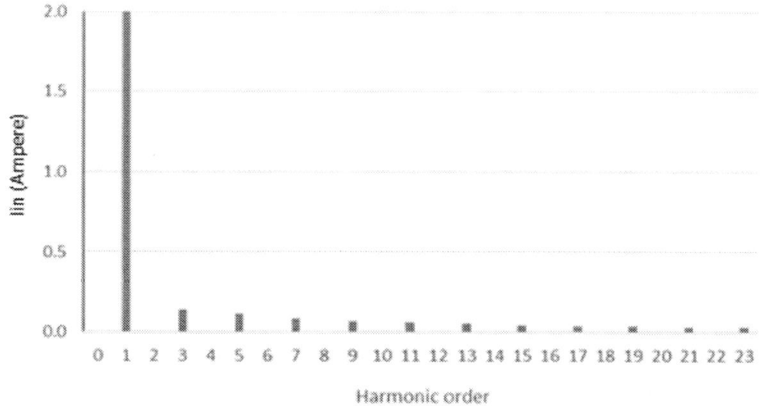

Figure 14. Input Current Waveform and Harmonic Spectrum at a Source Voltage of 220V

Figure 15 is a comparison of class D harmonics with the comparison when using the *CUK Converter*. It can be seen that this converter repairs or reduces harmonics components so as to increase the value of the power factor.

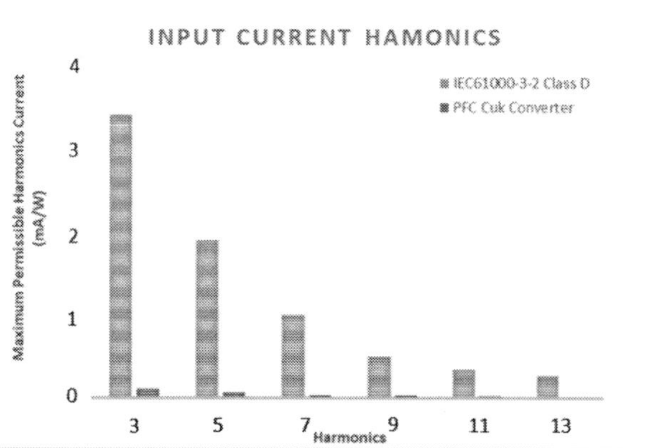

Figure 15. Comparison of class D harmonics with converter current harmonic testing

IOP Conf. Series: Materials Science and Engineering **982** (2020) 012020 doi:10.1088/1757-899X/982/1/012020

Table 3. shows the comparison after and before the installation of the *CUK Converter* with the total harmonic current distortion (THDi) value and the power factor (pf) value. It can be observed that the value of pf generated by the installation of a *CUK Converter* is equal to 0.98 and 0.99 which is affected by the current wave approaching or already the same as the voltage waveform in the form of sinusoidal. And THD values are 1.58% and 1.61% which are relatively small harmonic components.

Table 3. THD and Power Factor

Performance	Without PFC *CUK Converter*	With PFC *CUK Converter*	
		110V	220V
THDi	5.7 %	1.58%	1.61%
Power Factor	0.731	0.994	0.982

5. Conclusion

The simulation results that have been submitted in this paper that discuss *CUK Converter* as PFC and with *Moth Flame Optimization-PI* for voltage regulators. This simulation experiment was conducted with MATLAB / Simulink and PSIM software by existing simulation results. From this design, the value of the power factor obtained using the *CUK Converter* can increase to 0.99 with a THD value of 1.58% and meet class D. IEC 61000-3-2 standards and for the utilization of the *Moth Flame Optimization* method that can determine the Kp and Ki is optimal, can maintain the output voltage according to the design.

References

[1] N A Bachok, M R Sahid and J A Aziz 2015 *Bridgeless SEPIC Power Factor Correction with Coupled Inductors* (Johor Bahru: IEEE Conference on Energy Conversion (CENCON)) pp 160-164

[2] Electromagnetic Compatibility (EMC) 2005 Part 3-2: *Limits—Limits for Harmonic Current Emissions*, Int. Std. IEC 61000-3-2

[3] M Z Efendi, F D Murdianto, F A Fitri, L Badriyah 2020 *Power factor improvement on LED lamp driver using BIFRED converter* (Yogyakarta : TELKOMNIKA) pp.571-578.

[4] U Kamnarn and V Chunkag 2009 *Analysis and Design of a Modular ThreePhase AC-to-DC Converter Using CUK Rectifier Module With Nearly Unity Power Factor and Fast Dynamic Response* (IEEE Transactions On Power Electronics)vol. 24 No. 8 pp. 2000 – 2012

[5] M Gotfryd January 2000 *Output Voltage and Power Limits in Boost Power Factor Corrector Operating in Discontinuous Inductor Current Mode* (IEEE Transaction on Power Electronics) Vol.15 No.1

[6] H Endo, T Yamashita and T Sugiura 1992 *A High Power Factor Buck Converter* (Toledo : IEEE Applied Power Electronics Conference) pp.1071-1076

[7] Y M Jiang and F C Lee 1993 *A new control scheme for Buck+Boost power factor correction circuit* (Blacksburg : Proceedings of the Virginia Power Electronics Seminar) pp. 189-193.

[8] D Wijeratne and G Moschopoulos 2011 *A Novel Three-Phase Buck-Boost AC-DC Converter* (Fort Worth : Applied Power Electronics Conference and Exposition (APEC), Twenty-Sixth Annual IEEE) pp 513-520

[9] F S Dos Reis, J Sebastian, J Uceda, 1993 *Characterization of Conducted Noise Generation for Sepic, Cuk and Boost Converters Working as Power Factor Preregulators* (Maui : IECON) pp. 965-970.

[10] N Jayaram and D Maksimovic 1998 *Power Factor Correction based on Coupled-Inductor Sepic and Cuk Converter with Nonlinear-Carrier Control* (Anaheim : Applied Power Electronics Conference and Exposition (APEC), Thirteenth Annual IEEE) pp 468-474

[11] S Mirjalili 2015 Know. Syst. *Moth-flame optimization algorithm: a novel nature-inspired heuristic paradigm* (Elsevier BV) vol. 89 pp. 228-249

IOP Conf. Series: Materials Science and Engineering **982** (2020) 012020 doi:10.1088/1757-899X/982/1/012020

[12] D W Hart 2015 McGRAW-HILL International *Power Electronics*

[13] M A Z A Rashid, A Ponniran, M K R Noor, J N Jumadril, M H Yatim and A N Kasiran 2019 *Optimization of PFC Cuk Converter parameters design for minimization of THD and voltage ripple* (International Journal of Power Electronics and Drive System (IJPEDS)) vol. 10 No. 1 pp. 514-521

[14] A Shrivastava and B Singh 2010 5th *PFC Cuk converter based electronic ballast for an 18 W compact fluorescent lamp* (Mangalore : International Conference on Industrial and Information Systems) pp. 393-397

[15] I Jamil, Z Jinquan and R Jamil 2013 *Analysis,Design and Implementation of Zero-Current-Switching Resonant Converter DC-DC Buck Converter* (International Journal of Electrical and Electronics Engineering (IJEEE)) ISSN 2278-9944 Vol.2 Issue 2 pp.1-12, 2013.

[16] Y Gu and D Zhang 2014 *Voltage Regulator Buck Converter with a Tapped Inductor for Fast Transient Response Application* (IEEE Transactions on Power Electronics) Vol 29 Issue 12 pp 6249 - 6254

[17] M Sharma, R K Bansal, S Prakash, S Dhundhara 2018 *Frequency Regulation in PV integrated Power System using MFO tuned PIDF controller* (Kurukshetra : IEEE 8th Power India International Conference (PIICON))

[18] S Chatterjee and M A Dalel, 2019 *MFO Based Traditional PID Controller for Automatic Voltage Regulator* (Journal of The Gujarat Research Society) ISSN 0374-8588 Vol. 21 Issue 8s pp. 736-745

[19] A ELSebaay, M Ramadan, M A A Adma 2017 *Studying the Effect of Non-Linear Loads Harmonics on Electric Generator Power Rating Selection* (European Scientific Journal) ISSN 1857-7881 vol.13 No.18

[20] K Kamalapathi, N Priyadarshi, S Padmanaban, J B Holm-Nielsen, F Azam, C Umayal, V K Ramachandaramurthy 2018 *A Hybrid Moth-Flame Fuzzy Logic Controller Based Integrated Cuk Converter Fed Brushless DC Motor for Power Factor Correction* (Article Electronics MDPI) pp 1-19

IOP Conf. Series: Materials Science and Engineering **982** (2020) 012021 doi:10.1088/1757-899X/982/1/012021

Investigation of Wind Power Potential in Al-Aqiq, Saudi Arabia

Saeed Ahmed Al-Ghamdi[1,2]

[1]Electrical Engineering Department, Faculty of Engineering, Albaha University, Albaha, Saudi Arabia

[2]sasg2000@gmail.com

Abstract. Saudi Arabia authorities recognize the importance of renewable energy, especially wind, and they intend to invest in this promising power sector. This paper presents a wind data analysis in terms of annual, seasonal, and diurnal variations at Al-Aqiq, which is located in the highlands of the southwestern region of the Kingdom of Saudi Arabia. In the analysis, the diurnal variation of wind speed at 3 m above the roof surface shows that wind speed remains above 2.5 m/s during the period 10:00 AM to 10:00 PM. The overall wind speed is found to be 2.68 m/s, 4.2 m/s, and 5.48 m/s at heights of 3 m, 10 m, and 50 m, respectively, with the main wind direction ranging approximately from south-southeast to south-southwest. The highest average monthly wind speed is recorded in March, with values of 3.09 m/s, 5.11 m/s, and 6.43 m/s at heights of 3 m, 10 m, and 50 m, respectively, and with southwest being the main wind direction. The wind data analysis is then used to estimate wind energy production. Wind energy generation is considered for 21 wind machines of different sizes. At a height of 10 m above ground level, the highest energy estimate is generated using an Enercon E141/4200 wind machine, with a rate of 4898.96 MWh/Year. The lowest energy estimate is generated using a Soyut Wind 100 wind machine, with a rate of 165.54 MWh/Year. The highest capacity factor (26.0%) is recorded for an Aeronautica Windpower 33-225 wind machine and the lowest capacity factor (5.9%) is recorded for an Enercon E126/7580. The analysis shows that the power factor is low for wind turbines located at 3 m above roof surface level, but higher heights at Albaha University, Al-Aqiq, may be suitable for wind farm development.

1. Introduction

Wind energy applications require open areas or available shores for wind energy plants. Saudi Arabia is a vast country with wide-open areas and long shores. The wind speed in most of these areas is high enough to make the application of wind energy economical. Saudi Arabia authorities recognize the importance of renewable energy, especially wind, and they intend to invest billions of SAR in this promising power sector. Even though Saudi Arabia has huge resources of oil, it is interested in playing an active part in the development of new technologies for exploiting and utilizing renewable sources of energy.

The electricity produced by wind could save oil that could then be exported to increase national income. Also, the production of electric power from wind energy would reduce the environmental pollution generated by conventional power plants. Recently, a significant amount of studies have focused on evaluating the applications of wind energy systems in Saudi Arabia, most of which have recommended wind as a promising and economical source of energy in the country [1 - 10].

Alamdari et al. [11] investigated the most important characteristics of wind energy at sixty-eight different locations in Iran by recording wind speed data for heights of 10 m, 30 m, and 40 m at ten-minute time intervals. Similarly, Lashin and Shata [12] analyzed wind speed data measured at a height of 19 m in coastal Port Said in Egypt and found that the wind speed was greater during spring months than in winter months, which is the opposite of the prevailing wind speed parameters in most European countries. Aman et al. [13] estimated the wind energy potential in Karachi, Pakistan, using wind speed data obtained from the Pakistan Metrological Department (PMD) for various heights (i.e., 10 m, 30 m, 50 m, 75 m, and 100 m) over four years. Statistical calculations conducting using SPSS (Statistical Package for the Social Sciences) software indicated that the city has an enormous wind potential. Also, Đurišić and Mikulovic [14] analyzed wind energy resources in the South Banat region using wind parameter measurements taken at the village of Bavaniste. Data were collected at the heights of 10 m, 40 m, 50 m, and 60 m during 2009 and 2010. The results indicated that the region of South Banat has a high wind energy potential and represents a promising region for the development of wind farms. Finally, Yaniktepe et al. [15] investigated wind characteristics and the wind energy potential in Osmaniye, Turkey.

The total energy generated by a turbine over some time can be calculated by adding up the energy corresponding to all possible wind speeds in the region in which the system is operational. Hence, along with the power characteristics of the turbine, the probability density corresponding to different wind speeds should also be included in energy calculations. Knowing the wind speeds of a certain city is important in determining the characteristic speeds of the turbine, which are its cut-in velocity, rated velocity, and cut-out velocity. Also, the effective utilization of wind energy entails a detailed knowledge of the wind characteristics at a particular location. Although the wind resource potential in Saudi Arabia is significant, many issues are surrounding its development [16-20]. These include the intermittency of resources, its seasonal and diurnal characteristics, its geographically remote locations, and the electrical grid infrastructure that must be used to transmit the wind energy to load areas. All of these issues pose significant technical barriers to the full development of Saudi Arabia's wind potential.

This work aims to assess wind energy potential in the Al-Aqiq region, KSA, in terms of annual, seasonal and diurnal variations and to analyze wind availability, and then use the results of this assessment to estimate wind energy in the Al-Aqiq area.

2. Site Description and Weather Station Program

Al-Aqiq is an important city in the south-western region of the Kingdom of Saudi Arabia (see figure 1) in which Albaha University and King Abdul-Aziz airport are located. The latitude and longitude of the city are 20° 10′ 45.588″ North and 41° 38′ 8.196″ East, and it is 1680 m above sea level. For the current study, data were collected regarding meteorological activity recorded in the period 2015 to 2019. The measured parameters included wind speed, wind direction, temperature, relative humidity, pressure, rain, and many others at 10 m above the ground surface. A total of 8,760 hourly records of wind speed and direction were used.

The data were collected from the Albaha University site (Al-Aqiq area) (see figure 2). The measuring device was fixed to the roof at a height of 3 m above the roof surface. The device (Tier 1 – Research Station, Albaha University — Solar Resource Monitoring Station, https://rratlas.energy.gov.sa/) was used to measure temperature, humidity, rain rate, wind speed, wind direction, barometer, solar radiation, and ultraviolet radiation. This station is the most complete and complex of the monitoring systems in the area and provides the highest quality data with reasonable measurements. The measuring device was set to record a reading every hour and then extract that reading.

Al-Aqiq's climate is fairly hot in summer and fairly cold in winter. Regarding general weather conditions, the recorded temperature varied from a minimum of 7.65°C to a maximum of 35.90°C, with an average of 23.53°C. The highest temperature was recorded in June, July, and August, and the lowest was recorded in January. The surface pressure changed from 829.09 to 840.09 mBar with a mean value of 834.51 mBar. The lowest pressure values were recorded in July and August, and the highest pressure was in December. The relative humidity varied between 6.56% and 89.32%, with an average value of 30.25%. The lowest humidity percentage was recorded in June and July and the highest value was recorded in December (see figure 3).

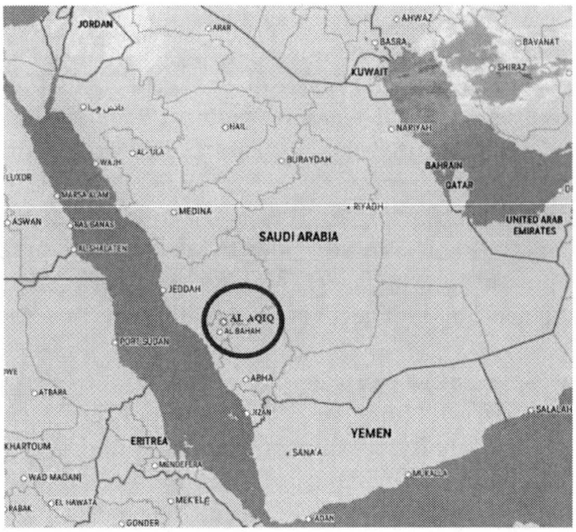

Figure 1. Geographical locations of meteorological stations.

Figure 2. Location of the wind-solar monitor at Albaha University (Al-Aqiq).

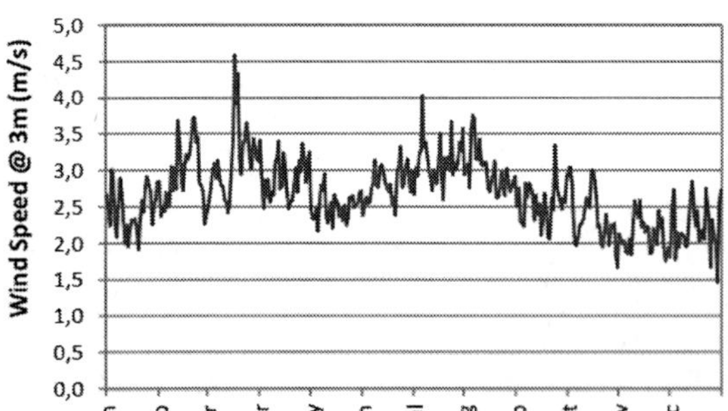

Figure 4. Average daily wind speed during a year (average from 2015 to 2019).

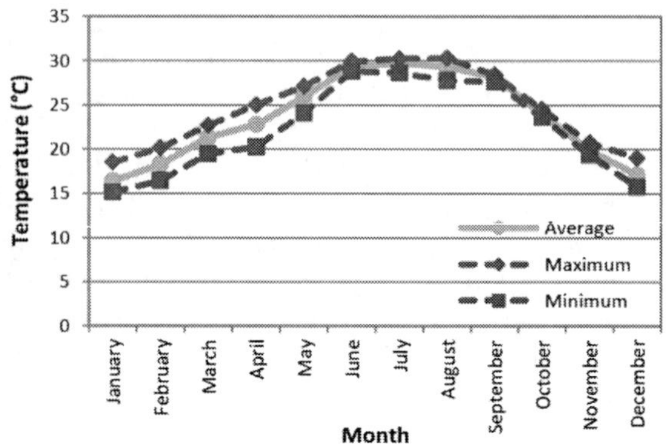

(a) Average air temperature.

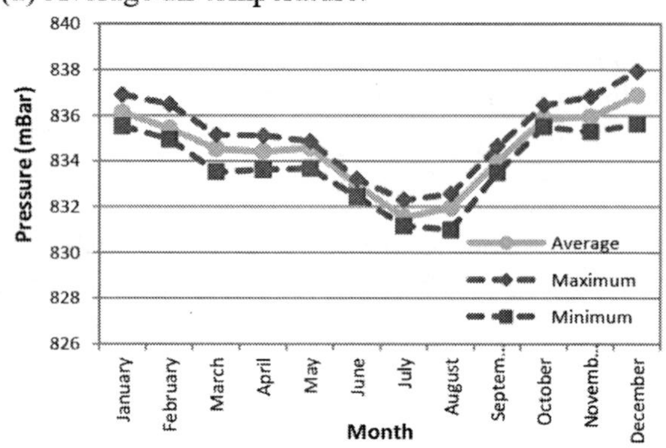

(b) Average station pressure.

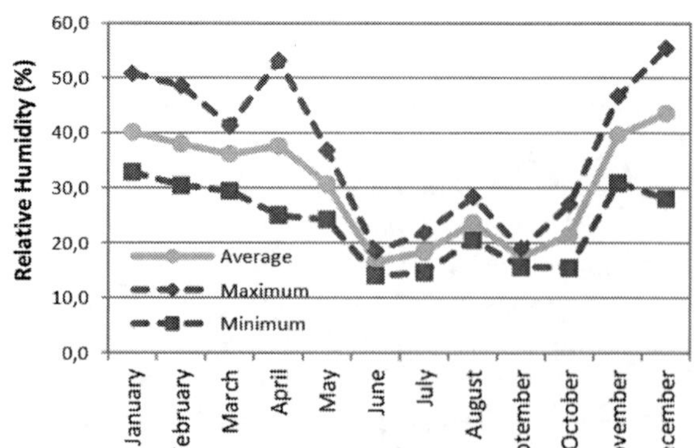

(c) Average relative humidity.

Figure 3. Average monthly temperature, station pressure, and relative humidity during a year (average from 2015 to 2019).

The wind speed was found to reach a maximum of 6.78 m/s, with an average speed of 2.67 m/s at 3 m above the roof ground surface (see figure 4). The main wind direction was between south-southwest and south-southeast. The extracted data were loaded and analyzed using a graphical user interface program called Weather Station Program (WSProg), which was built in-house using Matlab, to calculate the average of each factor by hours, then by day, and finally by month [21].

The main interface of WSProg is shown in figure 5. The WSProg program contains the following modules: (i) Wind Rose Diagram, (ii) Bar & Line Charts, (iii) Calculating Wind Speed Averages & Direction, and (iv) Calculating Power Generated from Turbines. In the Wind Rose Diagram module, the user can generate a wind rose diagram for one day, one week, one month, any season, and one year. In the Bar & Line Charts module, the user can generate a bar or line chart for one day, one week, one month, any season, and one year. In the Calculating Wind Speed Averages & Direction module, the user can extract average wind speed for one day, one week, one month, any season, and one year. Using the Calculating Power Generated from the Turbines module, it is possible to predict the power generated by some common machines.

Figure 5. Main Interface of the Weather Station Program.

3. Estimating the Power Curve of a Wind Turbine

For energy calculation purposes, several wind turbines were selected. These were: Aeronautica Windpower (2), Dewind (2), Enercon (6), Leitwind (1), Soyut Wind (3), Wobben (2), Nordex (3), and SouthWest (2) — making up a total of 21 wind machines. The technical data and power curves of these machines are available on the Internet. Wind turbines of different sizes and made by various manufacturers were selected to estimate the potential energy produced if they were installed at Al-Aqiq, KSA. The technical data and specifications for some of the selected wind turbines are summarized in table 1. Figure 6 shows the power curve for the selected wind turbines. For energy analysis, the wind speed bins were centered at 0.0, 0.5, 1.0, 1.5, 2, 2.5, … etc.

4. Results and Discussion

This section covers the annual, seasonal, and diurnal variation of mean wind speed; the wind availability in terms of a frequency distribution; and the energy calculations conducted for the selected wind machines. Wind machines were selected to cover a wide range of rated powers, from 100 kW to 7580 kW. The rotor diameters ranged from 1.17 m to 141 m, and cut-in wind speeds ranged from 1.5 m/s to 3.6 m/s.

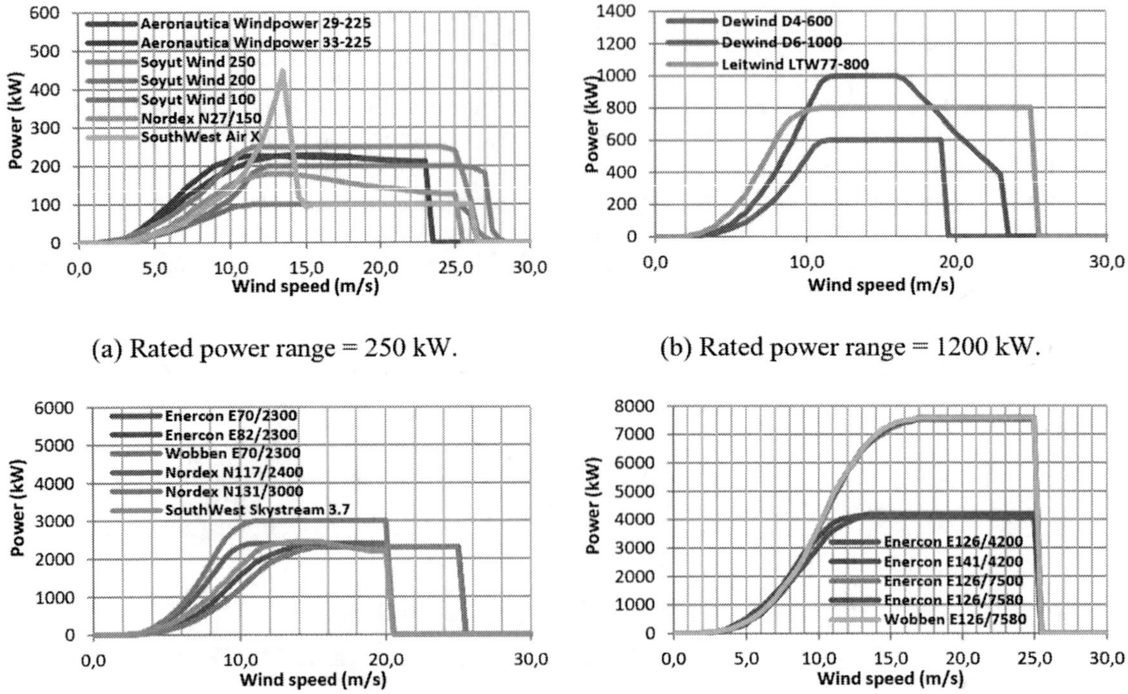

(a) Rated power range = 250 kW. (b) Rated power range = 1200 kW.

(c) Rated power range = 3500 kW. (d) Rated power range = 7850 kW.

Figure 6. Sample of wind turbine power curves, kW.

Table 1. Sample of wind turbine specifications

No	Type/Model	Main data		Rotor					Wind speeds (m/s)		
		Rated power (kW)	Rotor diameter (m)	Number of blades	Type	Swept area (m²)	Power density (m²/kW)	Max speed (rad/min)	Cut-in	Rated	Cut-off
1	Aeronautica Windpower 29-225	225	29	3	Stall	660.5	2.94	37.9	2	14	23
2	Aeronautica Windpower 33-225	225	33	3	Stall	855.3	3.8	33	2	12	23
3	Dewind D4-600	600	48	3	Pitch	1809.6	3.02	29.2	2.5	12	19
4	Dewind D6-1000	1000	62	3	Pitch	3019.1	3.02	25.2	2.5	12	23
5	Enercon E70/2300	2300	71	3	Pitch	3959.2	1.72	21.5	2	15.5	25
6	Enercon E82/2300	2300	82	3	Pitch	8281	2.3	18	2	14	25
7	Enercon E126/4200	4200	127	3	Pitch	12667.7	3.02	11.6	2.5	14	25
8	Enercon E141/4200	4200	141	3	Pitch	15614.5	3.72	10.6	2	14	25
9	Enercon E126/7500	7500	127	3	Pitch	1266.7	1.69	NA	2.5	17	25
10	Enercon E126/7580	7580	127	3	Pitch	12667.7	1.67	12.1	2.5	17	25
11	Leitwind LTW77-800	800	77	3	Pitch	4656.6	5.82	16.8	2.5	11	25
12	Soyut Wind 250	250	41.5	3	Stall	1333.2	5.33	NA	1.5	12	28
13	Soyut Wind 200	200	36.8	3	Stall	1063.6	5.32	NA	1.5	12.5	28
14	Soyut Wind 100	100	26	3	Stall	530.9	5.31	NA	1.5	11.5	26.5
15	Wobben E126/7580	7580	127	3	Pitch	12667.7	1.67	12.1	2.5	17	25
16	Wobben E70/2300	2300	71	3	Pitch	3959.2	1.72	21.5	2	15.5	25
17	Nordex N27/150	150	27	3	Stall	572.6	3.82	36	3	13	25
18	Nordex N117/2400	2400	117	3	Pitch	10751.3	4.48	13.2	3	11	20
19	Nordex N131/3000	3000	131	3	Pitch	13478.2	4.48	11.6	3	11.5	20
20	SouthWest Air X	280	1.17	3	NA	1.07	NA	NA	3.6	12.5	26
21	SouthWest Skystream 3.7	2400	3.72	3	Stall	10.87	NA	50 - 330	3.5	13	20

4.1. Analysis of wind speed variation

Table 2 shows the average wind speed and direction for the station located at Albaha University in Al-Aqiq (average of each month during the period 2015 to 2019). The seasonal changes in mean wind speed provide insight into the availability of suitable wind throughout the year. The monthly mean wind speed values for 10 m and 50 m were obtained from NASA Prediction of Worldwide Energy Resources (https://power.larc.nasa.gov/).

The monthly mean wind speed values for 3 m, 10 m, and 50 m above the ground are plotted in figure 7. Higher wind speed of 3.13 m/s, 5.1 m/s, and 6.43 m/s were observed at heights of 3 m, 10 m, and 50 m, respectively. These values were recorded during January, February, and March. Lower wind speeds of 2.12 m/s, 3.38 m/s, and 4.33 m/s were observed at heights of 3 m, 10 m, and 50 m, respectively. These speeds occurred during May and November. The main wind direction ranged approximately from south-southeast to south-southwest: the wind blew from a southwest direction in the period January to June, from a northwest direction in the period July to September, and from a south-southeast direction in the period October to December.

Table 2. Average wind speed and direction (from 2015 to 2019).

Month	Wind Speed (m/s)			Wind Direction
	At height 3 m	At height 10 m	At height 50 m	
January	2.42	4.62	5.99	SW
February	3.09	4.92	6.31	SW
March	3.13	5.11	6.43	SW
April	2.90	4.38	5.46	SW
May	2.50	3.38	4.33	SW
June	2.81	3.72	4.93	SW
July	3.12	4.61	5.85	NW
August	2.99	3.84	4.92	NW
September	2.55	3.63	5.03	NW
October	2.35	4.17	5.89	E
November	2.12	3.85	5.21	E
December	2.22	4.11	5.44	SE
Average	2.68	4.20	5.48	SW

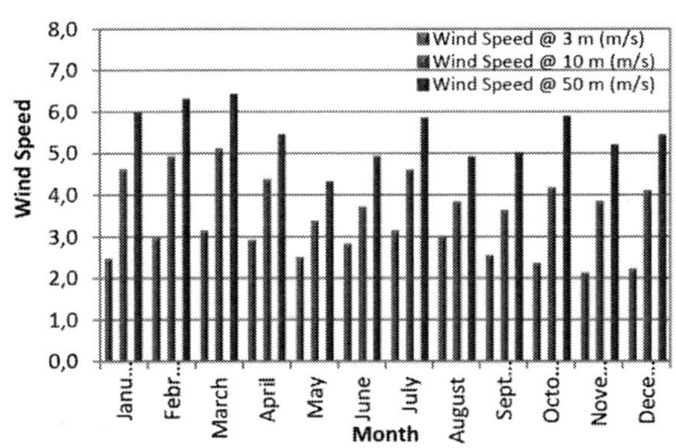

Figure 7. Monthly variation of mean wind speed at Albaha University, Al-Aqiq (from 2015 to 2019).

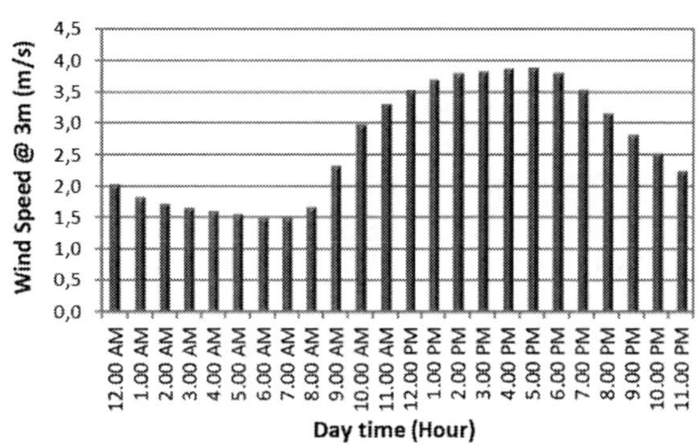

Figure 8. Diurnal variation of hourly mean wind speed at Albaha University, Al-Aqiq (from 2015 to 2019).

The diurnal variation of wind speed provides information about the availability of suitable winds during the entire 24 hours of the day. To study this pattern, overall hourly mean values of wind speed are shown in figure 8. The figure shows that the wind speed at a height of 3 m above the roof surface remained above 2.5 m/s from 10:00 AM to 10:00 PM and below it during the rest of the hours of the day. During the entire day, the mean wind speed varied between a minimum of approximately 1.5 m/s at 06:00 AM and a maximum of approximately 3.87 m/s at 05:00 PM.

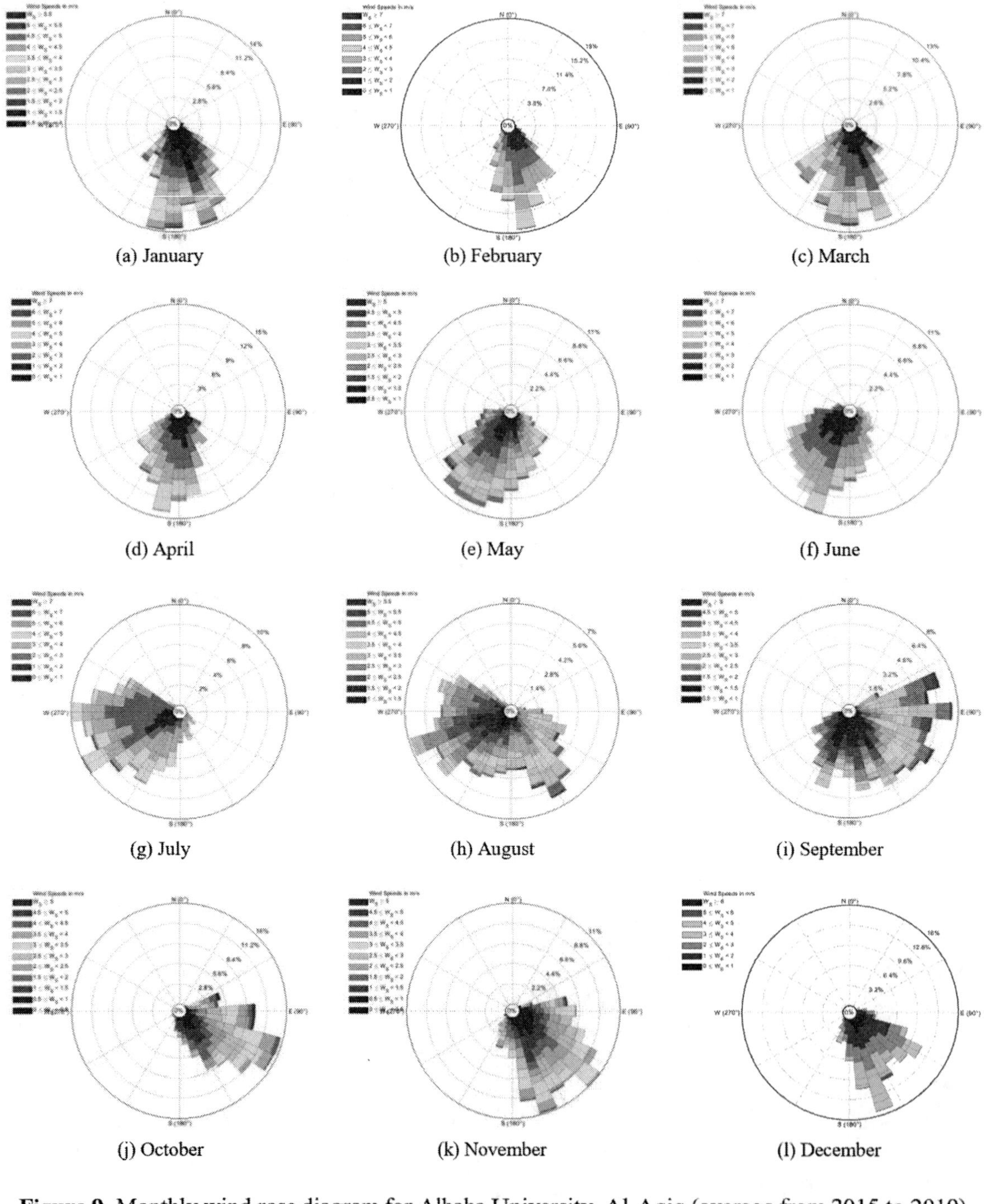

(a) January (b) February (c) March

(d) April (e) May (f) June

(g) July (h) August (i) September

(j) October (k) November (l) December

Figure 9. Monthly wind rose diagram for Albaha University, Al-Aqiq (average from 2015 to 2019).

A wind rose diagram provides information about the occurrence of the number of hours or percentage of time during which wind remains in a certain wind speed bin in a particular wind direction. Wind roses are constructed using hourly mean wind speed and corresponding wind direction values. Like wind speed, wind roses also vary from one location to another and are known as a form of a meteorological fingerprint. Hence, an in-depth examination and understanding of the wind rose are extremely important for siting wind turbines effectively. For instance, if the wind rose diagram shows that a large share of wind or wind energy comes from a particular direction, then the wind turbines should be placed or installed against that direction.

Figure 9 shows the wind rose diagrams generated for Albaha University in Al-Aqiq (an average from 2015 to 2019). It was observed that the main wind direction ranged approximately between south-southeast and south-southwest. From January to April, the wind direction was southerly. The highest monthly wind speeds were recorded in March, with values of 3.13 m/s, 5.11 m/s, and 6.43 m/s at heights of 3 m, 10 m, and 50 m, respectively, and with a predominantly south-westerly direction. The lowest monthly wind speeds were recorded in May and November, with values of 2.12 m/s, 3.38 m/s, and 4.33 m/s at heights of 3 m, 10 m, and 50 m, respectively, and with predominantly south-westerly and easterly wind directions. The highest seasonal wind speed was 6.778 m/s with a mainly south-southeasterly direction and recorded in winter (see figure 10).

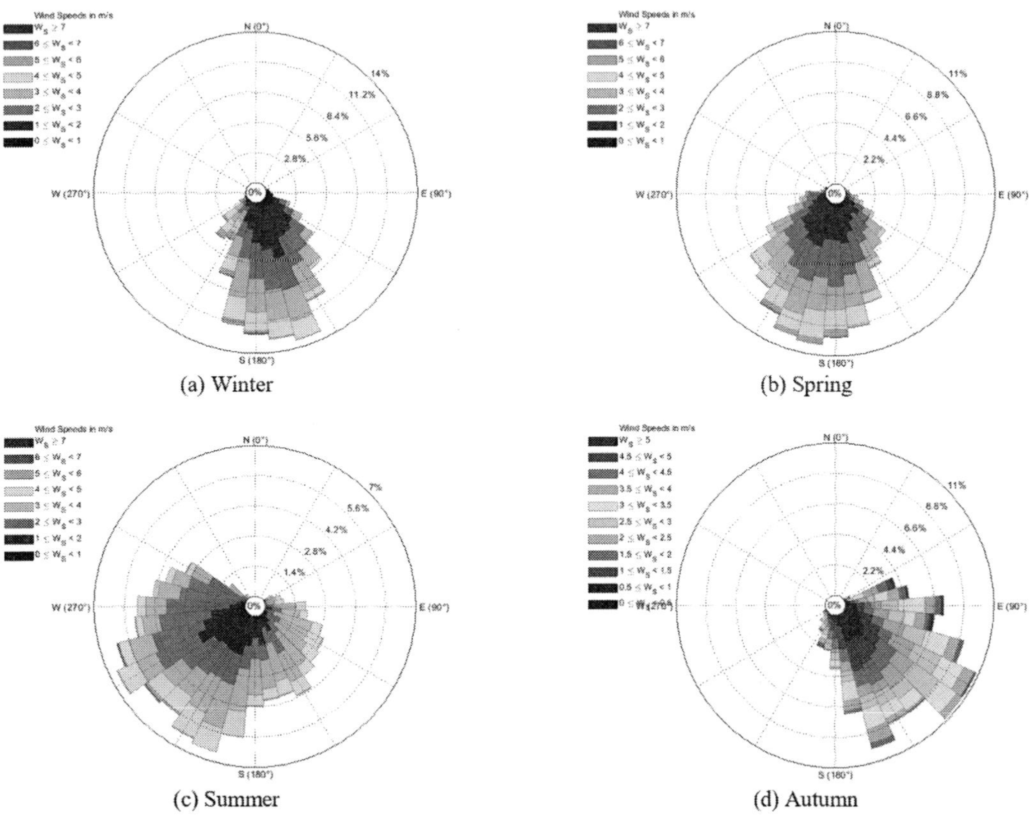

(a) Winter (b) Spring

(c) Summer (d) Autumn

Figure 10. Seasonal wind rose diagram for Albaha University, Al-Aqiq (average from 2015 to 2019).

Table 3. Average seasonal wind speeds at heights of 3 m, 10 m, and 50 m and expected electric power generated during different seasons.

Season	Wind direction	@ 3 m			@ 10 m			@ 50 m		
		Ave wind speed	%	kW	Ave wind speed	%	kW	Ave wind speed	%	kW
Winter	SSE	2.72	10.17	80403	4.73	18.05	74831	6.08	20.72	77942
Spring	SSW	2.79	10.47	85142	4.81	16.37	88098	6.17	19.87	94081
Summer	SSW	2.77	12.29	106742	3.99	18.31	109423	5.09	22.25	112949
Autumn	SE	2.91	7.31	70566	4.03	15.53	77814	5.25	18.30	73195
Year	SSE-SSW	2.67	40.24	342853	4.20	68.25	350166	5.48	81.14	358167

The lowest seasonal wind speed was 4.968 m/s with a mainly southeasterly direction and recorded in Autumn. It is evident from this figure that the wind in the area blows predominantly from the south-southwest and south-southeast. Hence, wind turbines should be installed against these directions — provided there are no or minimal high rise buildings and the terrain is smooth — for optimal energy output. It was found that the percentage shares for wind were 25%, 25%, 26%, and 24% for directions of south-southeast, south-southwest, and southeast during winter, spring, summer, and autumn, respectively. Table 3 lists the average seasonal wind speed at heights of 3 m, 10 m, and 50 m and the expected electric power generated during the different seasons.

In order to construct the wind rose diagram and analyze the frequency distribution, all hourly average wind speeds and wind directions were used, and the resulting wind rose diagram is shown in figure 11. It is evident from this figure that the wind blew predominantly from south-southeast to south-southwest direction for approximately 60% of the time, and from different directions the rest of the time. Hence, wind turbines could be installed against these directions — provided there are no or minimal high rise buildings or very large trees and the terrain is smooth — for optimal energy output.

4.2. Wind energy calculations

Wind turbines can safely produce energy at particular wind speed, usually between 10.5 m/s and 16 m/s. The following formula illustrates factors that are important to the performance of a wind turbine. The rotor swept area, A, is important because the rotor is the part of the turbine that captures the wind energy. So, the larger the rotor, the more energy it can capture.

$$P_w = \frac{1}{2}\rho\, A\, V^3 C_p \tag{1}$$

where:

P_w	=	power output, watts;
C_p	=	maximum power coefficient, ranging from 0.25 to 0.45, dimension less;
ρ	=	air density, kg/m^3;
	=	1.226 kg/m^3 at sea level;
	=	1.055 kg/m^3 at 1680 m height;
A	=	rotor swept area, m^2;
V	=	wind speed, m/s.

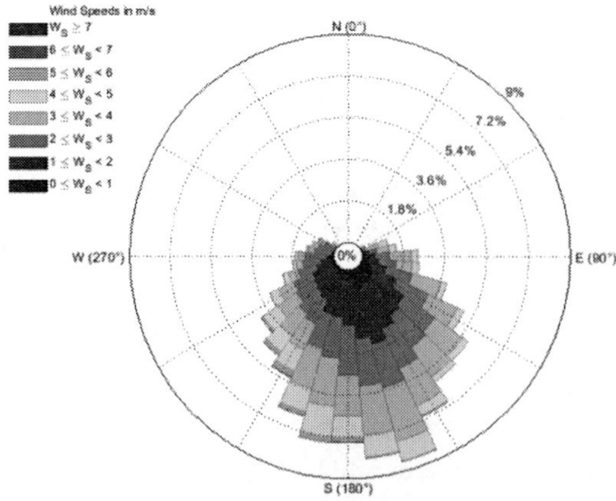

Figure 11. The wind rose diagram for Albaha University, Al-Aqiq (average from 2015 to 2019).

The best measure of wind turbine performance is the annual energy output. An estimate of the annual energy output from a wind turbine (kWh/year) is the best way to determine whether a particular wind turbine will produce enough electricity to meet consumer requirements. Capacity factor (CF) estimation was calculated for each machine and listed in table 4.

A wind turbine power curve can help estimate expected energy production. The estimation is based on the average annual wind speed at a site, the frequency distribution of the wind, and the number of hours that the wind will blow at each speed during an average year [22].

Table 4. The energy is produced from the wind turbines and estimation of capacity factor.

No	Type/Model	At height 3 m		At height 10 m		At height 50 m	
		Energy (MW/Year)	CF (%)	Energy (MW/Year)	CF (%)	Energy (MW/Year)	CF (%)
1	Aeronautica Windpower 29-225	119.494	6.1	419.943	21.3	699.857	35.5
2	Aeronautica Windpower 33-225	**155.547**	**7.9**	**512.292**	**26.0**	**832.159**	**42.2**
3	Dewind D4-600	107.626	2.0	531.894	10.1	1175.400	22.4
4	Dewind D6-1000	148.986	1.7	862.619	9.8	1922.111	21.9
5	Enercon E70/2300	326.569	1.6	1409.515	7.0	3082.936	15.3
6	Enercon E82/2300	346.597	1.7	1759.422	8.7	3878.652	19.3
7	Enercon E126/4200	865.137	2.4	3959.403	10.8	8211.143	22.3
8	Enercon E141/4200	1156.672	3.1	4898.962	13.3	9648.598	26.2
9	Enercon E126/7500	886.162	1.3	4013.793	6.1	9020.766	13.7
10	Enercon E126/7580	**762.019**	**1.1**	**3946.612**	**5.9**	**9015.680**	**13.6**
11	Leitwind LTW77-800	340.784	4.9	1435.567	20.5	2631.089	37.5
12	Soyut Wind 250	148.191	6.8	388.681	17.7	661.007	30.2
13	Soyut Wind 200	86.782	5.0	204.470	11.7	358.089	20.4
14	Soyut Wind 100	58.097	6.6	165.537	18.9	286.697	32.7
15	Wobben E126/7580	930.010	1.4	4232.793	6.4	9329.714	14.1
16	Wobben E70/2300	260.377	1.3	1324.173	6.6	2990.259	14.8
17	Nordex N27/150	48.847	0.6	556.426	6.4	1383.521	15.8
18	Nordex N117/2400	28.381	2.2	214.538	16.3	434.186	33.0
19	Nordex N131/3000	677.207	3.2	3419.442	16.3	6769.479	32.2
20	SouthWest Air X	805.250	3.1	4193.251	16.0	8367.633	31.8
21	SouthWest Skystream 3.7	53.203	2.2	197.991	8.1	402.991	16.4

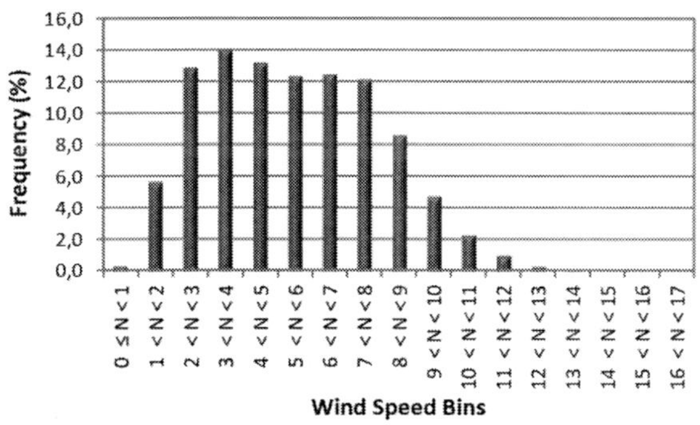

(a) At 3 m.

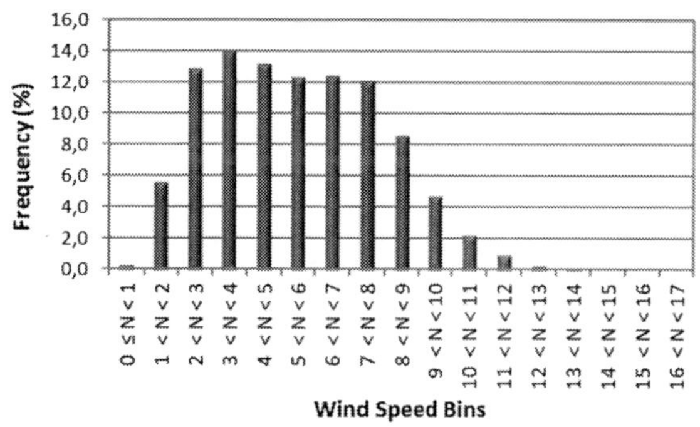

(b) At 10 m.

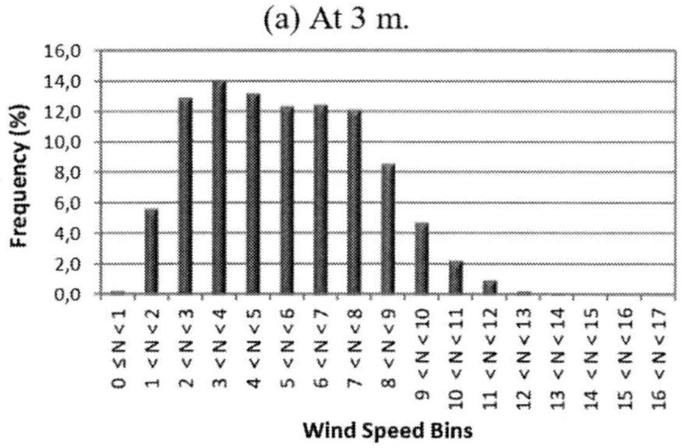

(c) At 50 m.

Figure 12. Percent frequency distribution of wind speed at 3 m, 10 m, and 50 m above ground level in different bins.

The Weibull distribution of wind at Albaha University in Al-Aqiq was derived from wind rose analysis presented in the previous sections. The percent frequency distribution of mean wind speed in the different bins is shown in figure 12. Figure 12(a) shows that at 3 m, the wind speed remained between 0.0 m/s and 2.0 m/s for 31.9% of the time and above 2.0 m/s for almost 68.1% of the time during the entire period of data collection. Figure 12(b) shows that at 10 m, the wind speed remained between 0.0 m/s and 2.0 m/s for 14.5% of the time and above 2.0 m/s for almost 85.5% of the time during the entire period of data collection. Figure 12(c) shows that at 50 m, the wind speed remained between 0.0 m/s and 2.0 m/s for 6.0% of the time and above 2.0 m/s for almost 94.0% of the time during the entire period of data collection. It is also clear from this figure that the wind remained above 2.0 m/s for almost 40.2%, 68.3%, and 94.0% of the time at 3 m, 10 m, and 50 m, respectively.

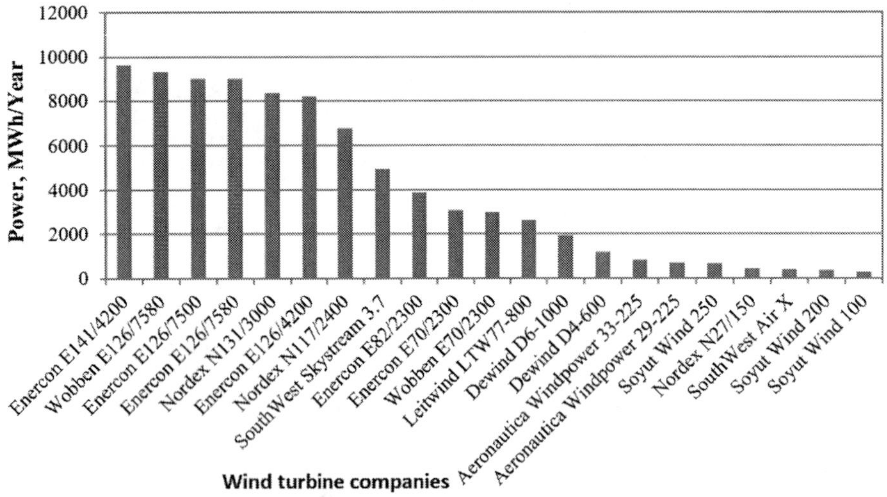

(a) Power generated, MWh/Year.

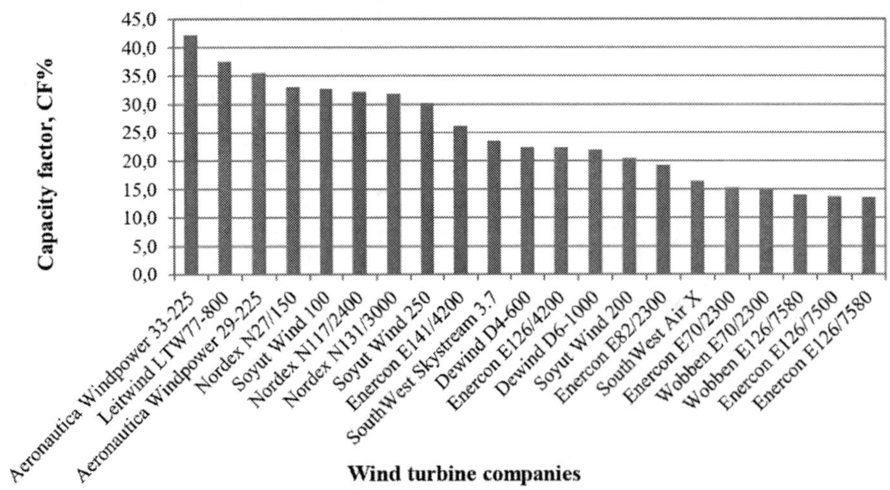

(b) Capacity factors, %.

Figure 13. The power generated and capacity factor for some selected wind turbines.

The wind speed remained above 3.0 m/s for almost 40.2%, 68.3%, and 81.0% of the time at 3 m, 10 m, and 50 m, respectively. Finally, about 0.03%, 12.6%, and 35.6% of the time, wind speed remained above 6.0 m/s at 3 m, 10 m, and 50 m, respectively. Since many modern wind turbines usually start producing energy above 1.5-2.0 m/s, the 51.6% availability of wind speed above the cut-in-speed of wind turbines is a fair indication of Albaha University, Al-Aqiq, being a suitable site for wind farm development.

The total energy output per year of the different-sized wind turbines was calculated. Then, a capacity factor analysis was conducted, as presented in table 4. The capacity of a wind energy conversion system is obtained by dividing the actual energy produced by the rated power and number of hours in a year, as in Eq. (2). Capacity factors are different from the average power of a station: the calculation includes the entire duration time, including the whole day, not just the time the turbine is in full working condition and not just the period in which it actively produces power.

$$CF = \frac{Pw}{P_r \times 24 \times 365} \times 100\% \quad (2)$$

where,

CF	=	capacity factor, %;
P_r	=	wind turbine rated power, kW;
P_w	=	power output, kWh/Year.

Power was calculated for the 21 turbine models made by Aeronautica Windpower, Dewind, Enercon, Leitwind, Soyut Wind, Wobben, Nordex, and SouthWest. The estimated power of the different wind turbines is listed in table 4. The highest energy estimation was generated for the Enercon E141/4200 wind machine, with rates of 9648.60 MWh/Year, 4898.96 MWh/Year, and 1156.67 MWh/Year at 50 m, 10 m, and 3 m, respectively. The capacity factors were calculated as 26.2%, 13.3%, and 3.1% at 50 m, 10 m, and 3 m, respectively. The second highest energy estimation was generated for the Wobben E126/7580 wind machine, with rates of 9329.71 MWh/Year, 4232.79 MWh/Year, and 930.01 MWh/Year at 50 m, 10 m, and 3 m, respectively. The capacity factors were calculated as 14.1%, 6.4%, and 1.4% at 50 m, 10 m, and 3 m, respectively.

The lowest energy estimation for 10 m and 50 m was generated for the Soyut Wind 100 wind machine, with rates of 165.54 MWh/Year and 286.697 MWh/Year with capacity factors of 18.9% and 32.7% at 10 m and 50 m, respectively. The lowest energy estimation for 3 m was generated for the Nordex N27/150 wind machine, with a rate of 28.38 MWh/Year with a capacity factor of 2.2% at 3 m, as shown in figure 13.

The highest capacity factor was recorded for the Aeronautica Windpower 33-225 wind machine, with rates of 7.9%, 26.0%, and 42.2% at 3 m, 10 m, and 50 m, respectively. Also, the estimated energy was recorded as 155.547 MWh/Year, 512.29 MWh/Year, and 832.16 MWh/Year at 50 m, 10 m, and 3 m, respectively. The lowest capacity factor was recorded for the Enercon E126/7580 wind machine, with rates of 1.1%, 5.9%, and 13.6% at 3 m, 10 m, and 50 m, respectively. Besides, estimated energy was recorded as 762.02 MWh/Year, 3946.61 MWh/Year, and 9015.68 MWh/Year at 50 m, 10 m, and 3 m, respectively.

5. Conclusion

This study investigated the wind characteristics and wind potential for Albaha University in Al-Aqiq, KSA, using data recorded during the years 2015–2019. The average wind speed and wind direction for the station located at Albaha University, Al-Aqiq, from 2015 to 2019 were calculated. The overall wind speed was found to be 2.68 m/s, 4.2 m/s, and 5.48 m/s at heights of 3 m, 10 m, and 50 m, respectively. These values were recorded during January, February, and March. The main wind direction in these months ranged approximately from south-southeast to south-southwest. The highest average monthly wind speed was recorded in March, with values of 3.09 m/s, 5.11 m/s, and 6.43 m/s recorded at 3 m, 10 m, and 50 m, respectively.

The main wind direction during this month was southwesterly. The lowest average monthly wind speed was recorded in November, with values of 3.09 m/s, 5.11 m/s, and 6.43 m/s recorded at 3 m, 10 m, and 50 m, respectively. The main wind direction during this month was easterly. The diurnal variation of the wind speed at 3 m revealed that the wind speed remained above 2.5 m/s from 10:00 AM to 10:00 PM. During an entire day, the mean wind speed varied between a minimum of approximately 1.5 m/s at 06:00 AM and a maximum of approximately 3.87 m/s at 05:00 PM. Wind energy generation was considered for 21 wind machines of different sizes. The capacity factor analysis showed, in general, that small wind machines have a higher capacity factor. The capacity factor was low for wind turbines located at 10 m above ground level, which is a fairly good indication of Al-

Aqiq, KSA, being a suitable site for wind farm development. Generally, wind turbines located at higher levels generate higher power and have a better capacity factor.

6. References

[1] Said S, El-Amin I and Al-Shehri A, editors 2004 *Renewable energy potentials in Saudi Arabia.* (Beirut regional Collaboration Workshop on energy efficiency and renewable energy technology, American University of Beirut)
[2] Rehman S and Al-Abbadi NM 2008 *Renew. Energy* **33(12)** 2653-2660
[3] Alawaji SH 1996 *Renew. Energy* **7(4)** 319-328
[4] Rehman S, Halawani T and Mohandes M 2003 *Renew. Energy* **28(4)** 573-583
[5] Rehman S and Ahmad A 2004 *Energy* **29(8)** 1105-1115
[6] Rehman S, El-Amin I, Ahmad F, et al. 2007 *Renew Sust Energ Rev* **11(5)** 937-950
[7] Al-Abbadi NM 2005 *Renew. Energy* **30(10)** 1489-1499
[8] Şahin AZ and Aksakal A 1998 *Renew. Energy* **14(1-4)** 435-440
[9] Rehman S and Al-Abbadi N 2009 *Energy & Environment* **21(8)** 1257
[10] Rafique M, Rehman S, Alam M, et al. 2018 *Energies* **11(8)** 2147
[11] Alamdari P, Nematollahi O and Mirhosseini M 2012 *Renew Sust Energ Rev* **16(1)** 836-860
[12] Lashin A and Shata A 2012 *Renew Sust Energ Rev* **16(9)** 6660-6667
[13] Aman M, Jasmon G, Ghufran A, et al. 2013 *Renew Sust Energ Rev* 18 528-542
[14] Đurišić Ž and Mikulović J 2012 *Renew Sust Energ Rev* **16(5)** 3014-3023
[15] Yaniktepe B, Koroglu T and Savrun M 2013 *Renew Sust Energ Rev* **21** 703-711
[16] Ahmed A and Khalid M 2019 *Renew Sust Energ Rev* **100** 9-21
[17] Alfawzan F, Alleman JE, and Rehmann CR 2019 *Energy Sci Eng*
[18] Allhibi H, Chowdhury H, Zaid M, et al. 2019 *Energy Procedia* **160** 746-751
[19] Chen W, Castruccio S, Genton MG, et al. 2018 *J Geophys Res-Atmos* **123(12)** 6443-6459
[20] Mosly I and Makki A 2018 *Sustainability* **10(11)** 4269
[21] Al-Ghamdi SA, Abdel-Latif AM, Hazza GAW, et al. 2017 *J. Basic & Appl. Sci.* **1(2)** 1-8.
[22] Rehman S 2004 *Energ Convers Manage* **45(13-14)** 2019-2032

Acknowledgment
The author is thankful to the Renewable Resource Atlas of Saudi Arabia, King Abdullah City for Atomic and Renewable Energy (K.A.CARE), Al-Riyadh, Saudi Arabia, for providing the requisite data.

IOP Conf. Series: Materials Science and Engineering **982** (2020) 012022 doi:10.1088/1757-899X/982/1/012022

Design of 1 kW Buck-Boost Chopper with PI Control for Photovoltaic Power Conversion

Suroso[1,2], Winasis[1], Priswanto[1], and Ignatius Daniel Purnama[1]

[1]Electrical Engineering Department, Jenderal Soedirman University, Indonesia

[2]suroso.te@unsoed.ac.id

Abstract. The design of a 1 kW buck-boost chopper with proportional-integral (PI) control is presented and discussed in this paper. The buck-boost chopper was proposed as a photovoltaic power converter to achieve a stable dc output voltage. In the design, the circuits employed IGBT power switch to function the PWM control signals to adjust the chopper's output voltage. The circuits were tested to investigate the output waveform set at 24 V from various dc input voltage values. The chopper circuits were connected to ten solar PV modules with a total capacity of 1 kW. Simulation test results confirmed that the system was able to output 24 dc voltage with output current about 41 A. Furthermore, a prototype of the buck-boost chopper was set up and tested. The circuits worked well to output a stable 24 V DC voltage from PV output voltage varied 15 V-32 V.

1. INTRODUCTION

The development of new renewable energy technologies has been able to produce controllable DC power that can be stored in batteries or work as DC power supplies. One of the most popular is the photovoltaic (PV) or solar panel systems [1-3]. The working principle of the solar panel is that if sunlight hits the panel, electrons in the solar cell will flow from negative to positive region. As a result, the output terminal of the solar panel will generate electrical potential. The generated electrical energy of solar panels changes depending on the count of solar cells combined in the solar panel. Inherently, they work producing DC power with unstable voltage because of the fluctuated solar radiation. To utilize this energy source, the PV voltage should be kept stable. So, the generated energy can be delivered to the load or to be stored in the battery systems well. In some applications, the photovoltaic system is connected via power inverter to the power grid to supply AC power [4-6].

Along with this, the development of power electronics, control algorithms, and circuit technologies has been able to produce a high performance of DC power supply, which is generated through the conversion of DC input voltage to a higher or lower DC voltage. This DC voltage conversion is commonly referred to as DC-DC converter or specifically called as buck-boost Chopper. In its development, the application of DC-DC converter has enabled an electronic device to function in a small size using a small battery energy source where the output voltage can be changed according to application requirements [7, 8].

The buck-boost chopper system is a non-isolated switching type DC regulator that can answer the need for a voltage source in the form of a variable output voltage. With the buck-boost chopper system, the output voltage value can be set to be greater or smaller than the input voltage value by controlling the pulse width or duty cycle of the PWM (pulse width modulation) control signal [9-12]. The design of a 1 kW buck-boost chopper prototype for photovoltaic power conversion is discussed in this paper.

2. CIRCUIT DESIGN

Fig. 1 shows the developed buck-boost chopper circuits in this project. The main components of these circuits are dc voltage source, power inductor, capacitor, power IGBT, power diodes, power resistor as load, and dc voltage sensor. Two dc voltage sources are connected to the circuits to investigate the work of chopper circuits for multi-sources with different dc input voltage values. This condition will be found in a real situation such as in photovoltaic power systems.

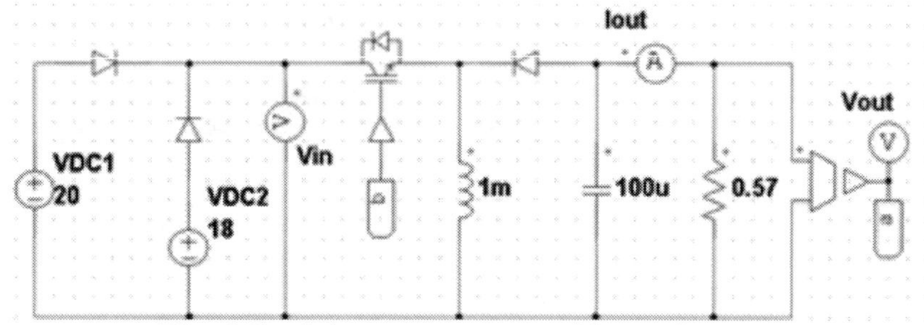

Fig. 1 Buck-boost converter with two inputs

Table 1 presents the parameters of chopper circuits that will be developed. The dc output voltage value is predetermined at 24 V with a ripple value of about 1%. While the output current value is set at a maximum of 42 A to achieve 1 kW output power. To minimize the size of the inductor and capacitor, the IGBT switch is operated at 31 kHz switching frequency.

To obtain the desired circuits, the circuit parameters are calculated as follows. The inductance value of power inductor (L_{min}) of circuits can be determined as:

$$L_{min} = \frac{(1-D)^2 R}{2f} \quad (1)$$

The capacitance of the capacitor (C) is:

$$Capacitance > \frac{Iout}{V_{ripple} \times f} \quad (2)$$

The duty cycle (D) of IGBT gating signal is:

$$D = \frac{Vo}{Vo + i} \quad (3)$$

Where f is the switching frequency, V_o is the output voltage, and I_{out} is the output current.

To regulate the output voltage of chopper circuits, a proportional-integral (PI) controller was applied as in Fig. 2. The output voltage is sensed by the voltage sensor, and its value is fed back and compared with the reference value. The errors will be modulated by a triangular carrier signal to create a control signal of IGBT switch.

Table 1. Parameters design of chopper

Input voltage, V_{in}	20 V
Output voltage, V_{out}	24 V
Voltage ripple	1%
Output current, I_{out}	42 A

Switching frequency	31 kHz

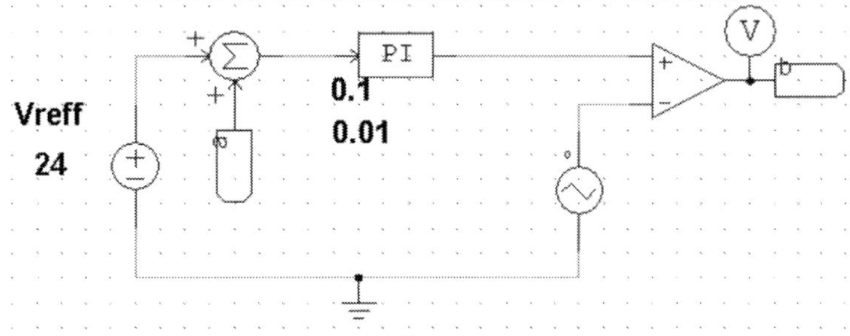

Fig. 2 PI controller

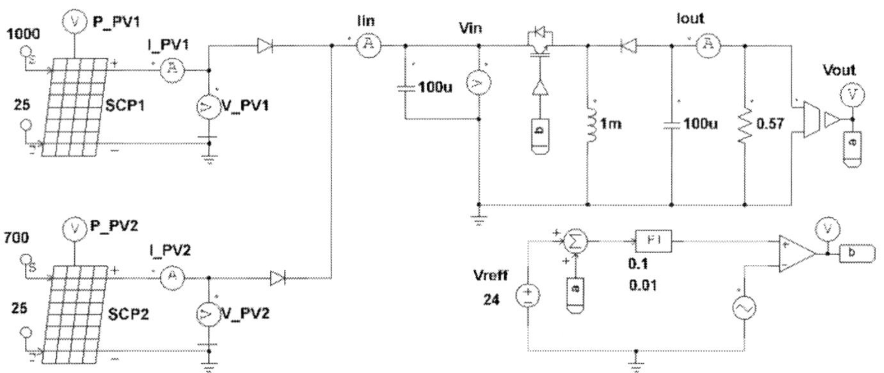

Fig. 3 Chopper circuits connected to PV systems

Fig. 3 depicts the buck-boost converter connected to two photovoltaic system PV1 and PV2 each constructed by ten PV modules. PV modules used in this research are Kyocera KD 140 series.

Moreover, a prototype of the buck-boost chopper was built in the laboratory using power IGBT as a controlled power switch. The circuit parameters are the same as in Table 1. The voltage sensor was implemented using the resistive voltage divider. The PI controller was implemented by using ARDUINO UNO as shown in Fig. 4. The gate drive circuits of IGBT is realized using optocoupler TLP 250.

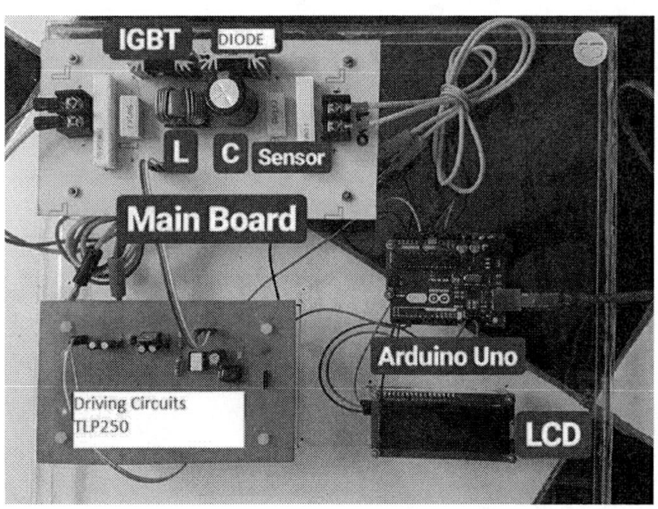

Fig. 4 Prototype of the buck-boost chopper

3. TEST RESULTS

To examine the proposed system, some computer simulation and experimental tests were carried out for both "boost mode" and "buck mode" of chopper circuits as follow:

3.1. Boost Mode Test

Fig. 5 presents the computer simulation test results of the chopper circuits connected to photovoltaic systems, i.e. PV1 and PV2. The figure presents the output power of photovoltaic P_{PV1} and P_{PV2}, output voltage of photovoltaics V_{PV1} and V_{PV2}, input and output voltage of chopper V_{in} and V_{out}, and current of photovoltaic systems I_{PV1} and I_{PV2}. The chopper circuits changed the output voltage of PV system from 20 V to become 24 V which is the boost operation mode of the chopper circuits.

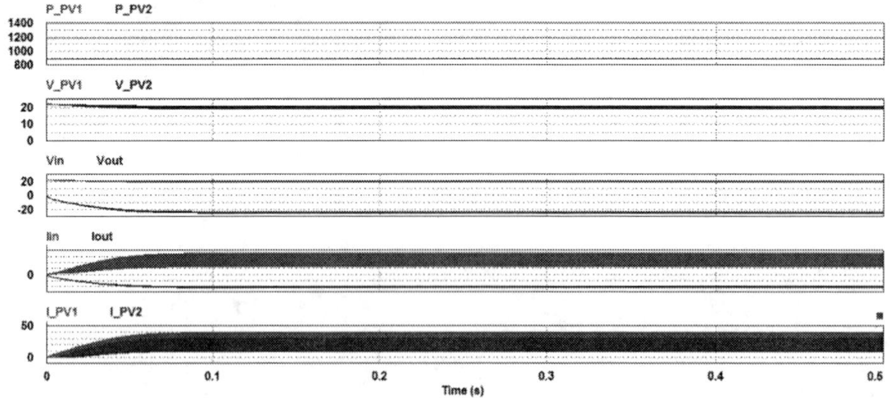

Fig. 5 Simulation test waveforms of boost operation mode

3.2. Buck Mode Test

Test results of the system during buck mode operation is shown in Fig. 6. Similarly, the output power of photovoltaic P_{PV1} and P_{PV2}, output voltage of photovoltaics V_{PV1} and V_{PV2}, input and output voltage of chopper V_{in} and V_{out}, and current of photovoltaic systems I_{PV1} and I_{PV2} can be observed in this figure. However, in this operation the voltage of photovoltaics was reduced from 40 V to be 24 V. It is the buck operation mode of chopper circuits. Hence, the two operation modes of chopper circuits have been confirmed by computer simulations.

Moreover, the chopper prototype was tested experimentally. Fig. 7 is the gating signal of chopper circuits at duty cycle 0.5. Fig. 8 presents the measured waveform of the dc input voltage during boost operation. The voltage was boosted up from 10 V to be 23.2 V. Fig. 9 depicts the input and output voltage of chopper circuits at buck operation mode. The voltage was decreased from 32.6 V to become 23.8 V. The error value of the output voltage during open circuits condition is presented in table 2.

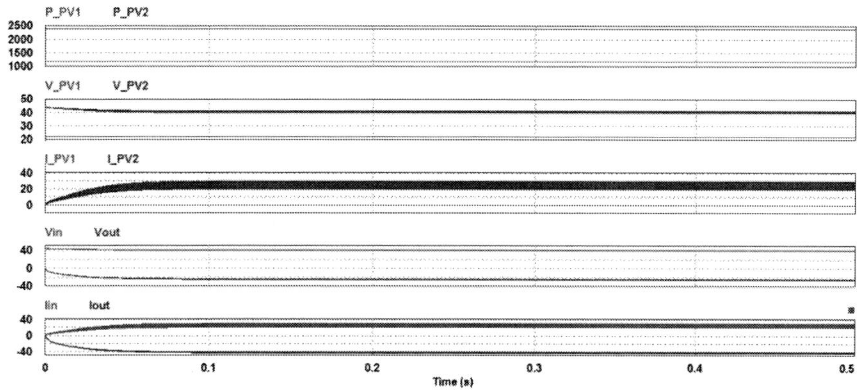

Fig. 6 Simulation test waveforms of buck operation mode

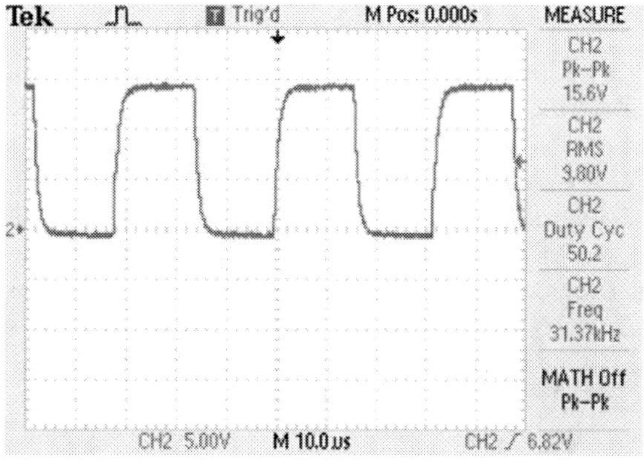

Fig. 7 Gating signal of IGBT with duty cycle 50%

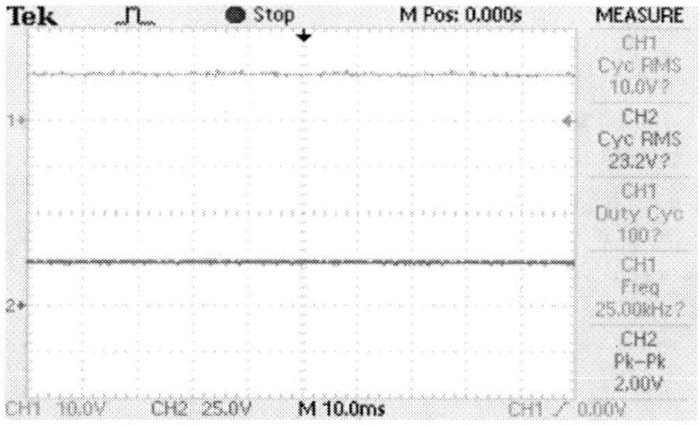

Fig. 8 Test results of boost operation: (1) input, and (2) output voltage

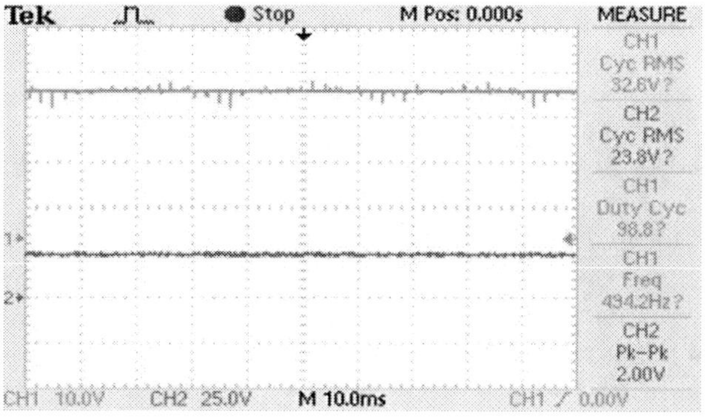

Fig. 9 Test results of buck operation: input and output voltage

Table 2. The error of the output voltage at open circuit

V_{in} (V)	Setpoint	V_{out} (V)	Error (%)	Mode
9	24 V	23.2	3.3%	Boost
15	24 V	23.6	1.6%	Boost
18	24 V	23.7	1.25%	Boost
28	24 V	23.5	2.1%	Buck
32	24 V	23.8	0.83%	Buck

4. CONCLUSION

The design of Buck-Boost Chopper is influenced by the value of the duty cycle and the value of components that complete the system. Buck-Boost Chopper circuit with PI-based Arduino Uno microcontroller control can balance the output voltage with a 24-volt setpoint value. In the Buck-Boost Chopper test with PI control and loaded with 1 resistor (10 Ω), 2 resistors are installed in series (20 Ω), and 2 resistors are installed parallel (5 Ω) with an input voltage of 15V - 32V already working properly. Because it can produce an output voltage with a 24 Volt setpoint value, with a maximum error percentage of 3.3%.

References

[1] K. Ogura, T. Nishida, E. Hiraki, M. Nakaoka, and S. Nagai, "Timesharing boost chopper cascaded dual mode single-phase sine wave inverter for solar photovoltaic power generation system", In Proc. Of PESC'04, Aachen, Germany, pp: 4763 – 4767, 2004.

[2] Suroso, D. T. Nugroho, Amran, and T. Noguchi, "Parallel operation of current-source inverter for low-voltage high-current grid-connected photovoltaic system", International Journal of Electrical and Computer Engineering (IJECE) Vol. 9, No. 4, pp. 2220-2229, 2019.

[3] N. Mutoh, M. Ohno, and T. Inoue, "A method for MPPT control while searching for parameters corresponding to weather conditions for PV generation systems", *IEEE Transactions on Industrial Electronics*, vol. 53, no. 4, pp. 1055-1065, 2006.

[4] S. R. Mohamed, P. A. Jeyanthy, and D. Devaraj, "Hysteresis-based voltage and current control techniques for grid connected solar photovoltaic systems: comparative study", *International Journal of Electrical and Computer Engineering (IJECE)*, vol. 8, no. 5, pp. 2671-2681, 2018.

[5] Suroso and T. Noguchi, "A new three-level current-source PWM inverter and its application for grid connected power conditioner", Energy Conversion and Management, vol. 51, issue 7, pp. 1491-1499, 2010.

[6] R.T.H. Li, H. S. Chung, and T.K.M. ,Chan, "An Active Modulation Technique for Single-Phase Grid-Connected CSI", IEEE Trans. on Power Electron., vol. 22, no. 4, pp: 1373 - 1382, 2007.

[7] W. Wu, J. Ji, and F. Blaabjerg, "Aalborg Inverter - A New Type of "Buck in Buck, Boost in Boost" Grid-Tied Inverter", *IEEE Transactions on Power Electronics*, vol. 30, Issue: 9, p.p. 460-467, 2015.

[8] Sairatun N. S., Golam S., Md. Ashraful H., and Md Saidul Hasan, "Design and Analysis of a DC -DC Buck Boost Converter to Achieve High Efficiency and Low Voltage Gain by using Buck Boost Topology into Buck Topology", 2018 International Conference on Advancement in Electrical and Electronic Engineering (ICAEEE), p.p. 1-4, 2018.

[9] C. Restrepo, et al., "A non-inverting buck-boost dc-dc switching converter with high efficiency and wide bandwidth," IEEE transactions on Power Electronics, vol. 26, p. 2490, 2011.
[10] G. Sizikov, et al., "Efficiency optimization of integrated DC-DC buck converters," 17th IEEE International Conference on Electronics, Circuits, and Systems (ICECS), p.p. 1208-1211, 2010.
[11] K. I. Hwu, and T. J. Peng, "A Novel Buck–Boost Converter Combining KY and Buck Converters", IEEE Transactions on Power Electronics, 2012, Vol. 27, Issue: 5, p.p. 2236-2241.
[12] M. Uno, and A. Kukita , "Single-Switch Voltage Equalizer Using Multistacked Buck–Boost Converters for Partially Shaded Photovoltaic Modules", IEEE Transactions on Power Electronics, vol. 30, issue 6, p.p. 3091 - 3105, 2015.

IOP Conf. Series: Materials Science and Engineering **982** (2020) 012023 doi:10.1088/1757-899X/982/1/012023

Standalone Photovoltaic System Cost Optimization for Matantimali Village Central Sulawesi

Bandiyah Sri Aprillia [1,2], I Made Hendry keswara[1], Jangkung Raharjo[1], Mohamad Ramdhani[1], Kharisma Bani Adam[1] and Efri Suhartono[1]

[1]School of Electrical Engineering, Telkom University, Bandung, Indonesia

[2]bandiyah@telkomuniversity.ac.id

Abstract. Electricity is very limited in Matantimali village where it is located on the slopes of Gawalise, Marawola District, Sigi Regency, Central Sulawesi Province. In this village, the state electricity company (PLN) can only provide electricity during the day from 06:00 to 18:00. During night, the diesel generator becomes the main energy source for the village providing electrical energy from 18:00 until 06:00. However, the use of diesel generators contradicts with the government commitment to generate environmental eco-friendly power generation. This study aims to analyse and to optimize the price of off-grid solar systems which will be implemented in the village of Matantimali as a potential replacement of the existing diesel based electric energy. We find in this study that 11 PV units with a total power of 3,686 Wp, two 2400 W powered inverter, and a 16-unit battery cover are the required specification for optimal power of 1300 VA load. The Off-grid solar panel installation system creates annual cost savings of IDR 3,543,397 compared to the use of conventional electricity using diesel generators and PLN.

1. Introduction

Increasing global energy demand is one of the biggest challenges experienced by modern society. To meet this demand, nearly 80% of our energy consumption comes from fossil fuels (gasoline, coal and natural gas). However, these energy sources are non-renewable. The most dangerous aspects related to its use are air pollution and rising global temperatures [1].

Matantimali Village, located in Marawola Sub district, Sigi Regency, Central Sulawesi Province, is on the Gawalise Mountains at an altitude of 1,500 meters above sea level. The location of the village is very far from the city resulting in very limited availability of electricity. PLN can only provide electricity during the day, from 06:00 to 18:00. During the off hours, the community's solution is to use diesel generators to obtain electricity.

Renewable energy is an alternative method to generate electricity. Among renewable energy choices, solar energy can directly convert solar radiation energy into electrical energy [2]. Therefore, the electrical energy produced by solar power station connected to the network is very dependent on solar radiation [3]. In solar power systems, solar panels can be connected on-grid or off-grid [4,5]. In the Off-grid system, solar panels are connected to batteries as storage media or energy banks [5,6]. Whereas on the On-grid system require solar panels to be connected to the PLN grid and can be regulated according to their energy requirements load.

The integration of hybrid renewable energy in remote and rural areas can supply the needed power needs and reduce emissions. Hence solar, wind, hydro and biomass energy, as well as hybrids, can effectively electrify rural areas [7]. To maintain the level of hybrid energy storage in meeting peak loads during the use of biomass, wind, and solar energy sources during periods of low or no solar radiation or during periods of low wind it is strongly recommended to use storage media [8]. The integration of hybrid power plants sourced from renewable energy (solar panels) and conventional power (diesel generators) sourced from fuel is also one of the beneficial solutions to meet the needs of daily electricity loads in remote areas such as Sirilogui Village, located in the District of North Siberut, Mentawai Islands Regency [9]. However, the use of generators is very contradictory to the government's commitment to generate environmentally friendly power because the generators used are based on fossil fuel (biodiesel) and if the generator is damaged, then the village has no electricity at night.

Based on BPPT data, BMG of Matantimali Village has solar radiation intensity reaching 5,512 Wh/m^2. Therefore, the appropriate method to generate power for the village would be a photovoltaic solar power station. This research aims to design an off-grid photovoltaic system with the lowest possible cost planning. The results of this study are expected to be considered appropriate electrification designs for the village of Matantimali, Central Sulawesi, Indonesia

2. Research Methodolohy

Research methodology during the study is shown in Figure 1 as follows:

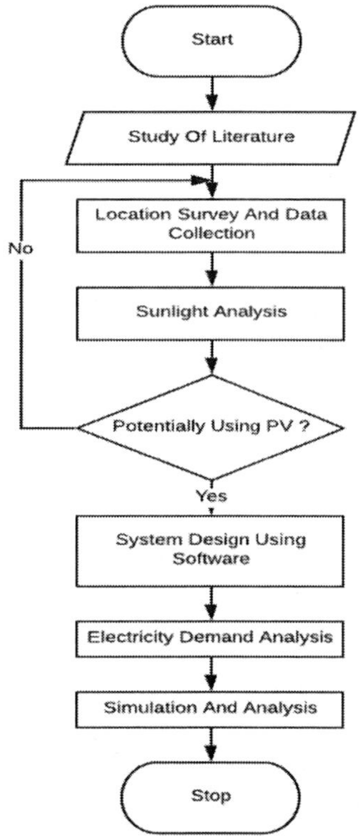

Figure 1. Research Methodology

Figure 1 describes the research method carried out by the author. Starting from studying literature and looking for references from books, journals, etc. The survey was conducted at the location to

IOP Conf. Series: Materials Science and Engineering **982** (2020) 012023 doi:10.1088/1757-899X/982/1/012023

collect data and to analyse the intensity of solar radiation at the location of the house that will be reviewed in this study. Before simulation and analysis, we must first analyse the electricity needs of one house in the village of Matantimali. After analysing the electricity requirements, it is continued by designing the optimum off-grid PV system.

2.1. System Design

Broadly speaking, the system design contains data of daily electricity load (kWh), location in the form of a map, and also the cost and specifications of components.

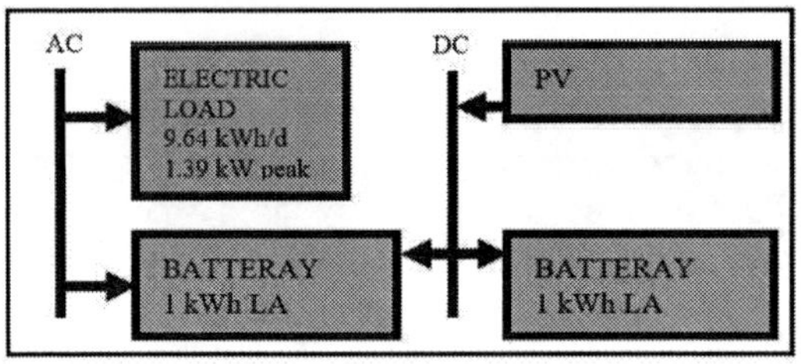

Figure 2. System Design

Figure 2 shows the design of a solar electric energy installation system consisting of solar panels, converters and batteries to meet load requirements. The solar panel functions as a DC input connected to the battery and the voltage on the battery is converted to AC voltage by the inverter to meet load requirements.

2.2. Solar Panel Installation Design

Figure 3. Photovoltaics Installation Design

Figure 3 shows the solar panel installation design in a house located in the village of Matantimali. Solar panels would be installed on the roof of the house with 2 inverters. Green dots are used to represent trees around the solar panels.

2.3. Residential load profile in Matantimali Village

Electricity load study is a method used to calculate daily energy consumption. Electrical energy consumption in table 1 is obtained by listing electrical loads (load groups, type of load, load power rating) and making a load profile as shown in Figure 4 which contains daily load period.

Table 1 shows the daily use of household electricity in Matantimali Village from 00:00 to 23:00.

Table 1. Matantimali Village Daily Energy Consumption

No	Time	Loads	Total (W)
1.	00.00	Lights 10 x 5 = 50 Refrigerators 300 x 1 = 300	50+300 = 350
2	01.00	Lights 10 x 5 = 50 Refrigerators 300 x 1 = 300	50+300 = 350
3	02.00	Lights 10 x 5 = 50 Refrigerators 300 x 1 = 300	50+300 = 350
4	03.00	Lights 10 x 5 = 50 Refrigerators 300 x 1 = 300	50+300 = 350
5	04.00	Lights 10 x 5 = 50 Refrigerators 300 x 1 = 300	50+300 = 350
6	05.00	Lights 10 x 5 = 50 Refrigerators 300 x 1 = 300	50+300 = 350
7	06.00	Lights 10 x 5 = 50 Refrigerators 300 x 1 = 300	50+300 = 350
8.	07.00	Lights 10 x 5 = 50 Refrigerators 300 x 1 = 300 Rice Cooker 400 x 1 =400	50 + 300 + 400 = 750
9.	08.00	Lights 10 x 5 50 Refrigerators 300 x 1 = 300 Rice Cooker 400 x 1 = 400 Hand phone Chargers 5 x 1 = 5	50 + 300 + 400 + 5 = 755
10.	09.00	Lights 10 x 5 50 Refrigerators 300 x 1 = 300 Rice Cooker 400 x 1 = 400 Handphone Chargers 5 x 1 = 5 TV 60 x 1 = 60	50 + 300 +400 +5 +60 = 815
11.	10.00	Lights 10 x 5 = 50 Handphone Chargers 5 x 1 = 5 TV 60 x 1 = 60	50 + 5 +60 = 215
12.	11.00	Lights 10 x 5 = 50 Fans 45 x 1 = 45 Handphone Chargers 5 x 1 = 5 TV 60 x 1 = 60	50 + 45 + 5 + 60 = 160
13.	12.00	Lights 10 x 5 = 50 Fans 45 x 1 = 45 Handphone Chargers 5 x 1 = 5	50 + 45 + 5 + 60 = 160
14.	13.00	Lights 10 x 5 = 50 Kipas 45 x 1 = 45 Handphone Chargers 5 x 1 = 5 TV 60 x 1 = 60 Clothes Iron 300 x 1 = 300	160 + 600 = 760
15.	14.00	Lights 10 x 5 = 50 Kipas 45 x 1 = 45 Refrigerators 300 x 1 =300 Handphone Chargers 5 x 1 =5 TV 60 x 1 =60 Clothes Iron 300 x 1 =300	400 + 360 = 760
16.	15.00	Lights 10 x 5 =50 Refrigerators 300 x 1 =300 Handphone Chargers 5 x 1 = 5 TV 60 x 1 =60	350 + 5 + 60 = 415

17.	16.00	Lights 10 x 5 = 50 Refrigerators 300 x 1 = 300	50+300 = 350
18.	17.00	Lights 10 x 5 = 50 Refrigerators 300 x 1 = 300	350
19.	18.00	Lights 10 x 5 = 50 Refrigerators 300 x 1 = 300	350
20.	19.00	Lights 10 x 5 = 50 Refrigerators 300 x 1 = 300	350
21.	20.00	Lights 10 x 5 = 50 Refrigerators 300 x 1 = 300	350
22.	21.00	Lights 10 x 5 = 50 Refrigerators 300 x 1 = 300	350
23.	22.00	Lights 10 x 5 = 50 Refrigerators 300 x 1 = 300	350
24.	23.00	Lights 10 x 5 = 50 Refrigerators 300 x 1 = 300	350
TOTAL			9640

Data of a day's worth of electricity consumption was obtained to calculate the daily energy consumption which was then arranged in table 1.

2.4. System Specification

Table 2. System Specification

Components	**Specification**
Inverter	DC 2400 W
PV	340 Wp
Battery	12V/100Ah

Table 2 shows PV system specifications used in this research. Components used consist of inverter, PV, and batteries.

2.5. Solar Panel Requirements Calculation

The solar panels are designed with 100% of the load requirements. The solar panel calculations needed are as follows:

Total solar panel energy = 100% x 1,3 x Total power in one day
= 100% x 1.3 x 9,640
= 12,532 Wh/day (1)

Note: Percent supply of solar panels = 100%
Note: 1,3 is obtained from a one – power factor.

Wp is required = Total solar panel energy / Solar panel generator factor = 12,532/3.4
= 3,686 Wp (2)

The number of solar panels = 3,686/340 = 10.841 $\approx$ 11 pcs (3)

2.6. Components Prices

Component prices are adjusted to the specifications needed to make costs within reason and in accordance with system requirements.

Table 3. Price Table and Component Specification

Components	Specifications	Prices Per-Components (IDR)	Amount	Total Price (IDR)
Inverter	2400 W	6,675,000	2	13,350,000
Solar Panels	340 Wp	2,280,000	6	25,080,000
Batteries	12 V ; 100 Ah	3,685,194.45	4	14,740,777.80
Total				53,170,777.8

Component prices consisting of: inverter, solar panels, and batteries are searched in accordance to the system requirements. The total cost required is IDR 37,615,777.8.

Total electric load in one year = 9,640 x 1 year = 9,640 x 366 days = 3,528.24 kWh/year (4)

Total power needed = 3,528.24 kWh/year x 100% x 1.3 = 4,586.712 kWh/y (5)

Total power = Total electricity load in one year – Total power needed in PV

= 1,058.472 kWh/year (6)

Power pricing of PLN = IDR 1600/kWh

Cost of Energy (each unit) = (Total Power x Price of PLN power)/Total power needed in PV) = (1,058.472 x 1600)/4,586.712 = IDR 369.230/kWh.

2.7. PLN and Diesel Generator Costs

Matantimali Village is a village that uses PLN and diesel generators as a daily electricity supply with a total load of 1300 VA. PLN supplies from 06.00 until 18.00. During those 12 hours according to the data that has been obtained, the power load is 5.83 kWh using the PLN electricity network. So, it can be calculated:

The cost of PLN electricity in one day = price of 1 kWh x load power consumption = IDR 1600 x 5.83 = IDR 9,328.

1 year = 366 days.

So, in one year the cost that must be spent for electrical energy from PLN supplies is IDR 9,328 x 366 = IDR 3,414,048.

Whereas from 18:00 to 6:00 the village of Matantimali uses a generator to generate electricity. Generators used have 80% efficiency of the maximum power load installed in each house. Then the generator needed for a load of 1300 VA or 1300 Watt is a generator that has 2500 Watt power.

The generator uses biodiesel fuel. So the price that can be issued by heating 12 hours on the generator can be calculated:

2500 Watt generator power = 2.5 kWatt.

Fuel used for 12 hours = diesel consumption per kilowatt hour x generator power in kilowatt x heating time = 0.21 x 2.5 x 12 = 6.3 liters.

Price of Bio Solar = IDR 9,400

So, the price of generator fuel issued in 1 day = fuel used for 12 hours' x the price of generator fuel (biodiesel) = 6.3 x 9,400 = IDR 59,220 / day.

IOP Conf. Series: Materials Science and Engineering **982** (2020) 012023 doi:10.1088/1757-899X/982/1/012023

1 year = 366 days
Therefore, the cost of generator fuel in one year is IDR 59,220 x 366 = IDR 21,674,520.

Table 4. Conventional Electricity Cost Calculation

Supply	Period	Load or Fuel	Yearly Cost
PLN	12 hour	5.83 kWh	IDR 3,414,048
Diesel Generators	12 hour	6.3 Liter	IDR 21,674,520
Total			IDR 25,088,568

3. RESULT AND DISCUSSION

HOMER application was used to simulate the installation of a circuit that has been made and to calculate the Net Present Cost (NPC) and the cost of power per kWh (COE). The results of the HOMER simulation are arranged to have a possibility on the NPC and COE values. NPV and COE are affected by the intensity of sunlight and the fraction of renewable energy used.

Table 5. HOMER Application Standalone PV System Simulation Results

Components	HelioScope (PV) (unit)	Battery (unit) 1 kWh LA	Converter (Inverter) (KW)	NPC (IDR)	COE (IDR)	Operational Cost (IDR/year)	Installation Cost (IDR)	Ren fraction (%)	Initial Cost (IDR)
PV + Battery +Inverter	11	16	4.8	325 M	4,775	11.9 M	97.4 M	100	25,080,000

Table 5 shows the Off-Grid system configuration in the house located in Matantimali village using the HOMER application. The configuration results obtained by the system can meet 100% of the electrical needs with daily electricity load consumption of 9,640 Wh/day with the installation of 11 units of PV configuration with 3,686 Wp, inverter powered 2400 W, and batteries 16 units in total and an expense procurement cost of up to IDR 90,700,000. Off-grid installation systems will cover the expensive costs of the electrical energy gains in PV. Based on calculation, the total cost to be paid for 25 years for the installation of the Off-Grid system is IDR 325,000,000. The batteries are used to store electricity as much as 16 units. The greater the profit (greater Rent Friction), the lower the operational costs of the system, but the higher the present total net cost (NPC), therefore the higher the operational costs.

Table 6 shows the Off-grid system installation by calculating the NPC and COE on the solar panel installation in yearly usage and the total annual cost amounts to IDR 21,465,170.

Table 6. Pricing Comparison with Conventional Method

System	Total cost in IDR/Year	Total Cost	Savings in IDR/Year
Solar Panel Installations	21,465,170	21,465,170	3,543,397
PLN	3,414,048	25,088,568	
Diesel Generator	21,674,520		

Table 6 shows the comparison of Off-grid system installations with conventional electrical energy in annual usage. The total profit can be obtained reaching IDR 3,543,397 per year with the configuration of the standalone PV system.

4. CONCLUSSION

The power requirements of 1300VA in Matantimali Village can be fulfilled with the Off-grid solar power system. Based on the analysis results of the configuration of 11 PV units with 3,686 Wp power, two 2400 W power inverter, and 16 units of batteries would produce optimum results. Our proposed the Off-grid solar panel installation system will save the amount of IDR 3,543,397 per year compared to conventional electricity in the form of diesel generators and PLN grid.

References

[1] Nicoletti G 2015 A technical and environmental comparison between hydrogen and some fossil fuels Energy Conversion and Management. 89 p 205-213.

[2] Ozoegwu C, C Mgbemene and P A Ozor 2017 The status of solar energy integration and policy in Nigeria Renewable and sustainable energy reviews. 70 p 457-471.

[3] Aprillia B S, C Ekaputri and M R Z Fahmi Solar Cell Output Optimization using Light Convergence Method.

[4] Aprillia B S and M A F Rigoursyah 2020 Design On-Grid Solar Power System for 450 VA Conventional Housing using HOMER Software.

[5] Al Nabulsi A and R Dhaouadi *Efficiency optimization of a DSP-based standalone PV system using fuzzy logic and dual-MPPT control.* IEEE Transactions on Industrial informatics 2012 8(3) p 573-584.

[6] Zubi, G., Dufo-López, R., Pasaoglu, G., & Pardo, N. (2016). Techno-economic assessment of an off-grid PV system for developing regions to provide electricity for basic domestic needs: A 2020–2040 scenario. *Applied Energy*, *176*, 309-319.

[7] Hossain F M 2015 *Impact of renewable energy on rural electrification in Malaysia: a review.* Clean Technologies and Environmental Policy. 17(4) p 859-871.

[8] Balamurugan P, S Ashok and T Jose 2009 *Optimal operation of biomass/wind/PV hybrid energy system for rural areas* International Journal of Green Energy 6(1) p 104-116.

[9] Sari D P and R Nazir 2015 *Optimalisasi desain sistem pembangkit listrik tenaga hybrid diesel generator-photovoltaic array menggunakan HOMER (studi kasus: Desa Sirilogui, Kabupaten Kepulauan Mentawai)* Jurnal Nasional Teknik Elektro. 4(1).

The Electrical Property of SiO_2/Graphene/PBA-NHS/anti O and K *E. coli* Antibodies as Sensing Layer for *Escherichia coli* Bacteria Sensor

K. M. Wibowo[1], N. M. Rosni[2,,3,6], A. Muslihati[4], G. Romadhona[1], Royan[1], Fatiatun[5], M. Z. Sahdan[2]

[1]Electromedical Engineering, Universitas Muhammadiyah Purwokerto (UMP), Purwokerto, Indonesia
[2]Department of Electronic, Faculty of Electrical and Electronic Engineering, UTHM, Parit Raja, Batu Pahat, Johor, Malaysia
[3]Microelectronics and Nanotechnology Shamsuddin Research Center (MiNT-SRC), UTHM, Parit Raja, Batu Pahat, Johor, Malaysia
[4]Faculty of Applied Science and Technology, UTHM, Pagoh, Johor, Malaysia
[5]Department of Physics Education, Faculty of Education and Teaching, Universitas Sains Al-Qur'an, Wonosobo, Indonesia

[6]E-mail: ge160103@uthm.edu.my

Abstract. This paper explains the effects of BSA/PBA-NHS/Graphene/SiO_2 used as a sensing layer for *E. coli* bacteria sensor based on the electrical properties. In this research, graphene is used as a sensing layer due to its superior properties especially its biocompatibility. The biocompatibility of graphene was examined by topological properties by AFM. Moreover, it was also confirmed by Raman spectroscopy. The topological structure and Raman spectroscopy showed good results which is indicate the graphene layer suitable for biosensing applications. Additionally, it was also confirmed that the antibody was successfully immobilized on the modified SiO_2/graphene/PBANHS. PBANHS is used as a linker due to its cost-effective and time-saving. The electrical properties of the sensing layer before and after the immobilized antibody shows a significant difference. When the number of *E. coli* bacteria increases, the current increase also became higher. The change in resistance gradually increases from 0.04 to 2.62 Ω when the number of bacteria increases from 55 to 1200 CFU/ml.-+.

1. Introduction

Recently, research on graphene such as for applications nanoelectronics [1], sensors [2], composite materials [3], biosensing [4], and energy harvesting technology [5] still increasing. The graphene-based sensor is one of the issues that attract the attention of researchers to develop into many applications such as gas sensors, electrochemical sensors, biosensors, and many more. Graphene-based biosensor like bacteria sensor was still the most interest area to explore more. This is due to the statistic that there are still many fatalities both sick and died caused by bacteria diseases or infection. In 2007, the reports from the Joint Monitoring Programme (JMP), every year there were about 289,000 children <5 years die from diarrheal diseases (caused by *E. coli* bacteria). Its equal to 800 children per day, or 1 child every 2 minutes. The unsafe water supply, inadequate sanitation, and poor hygiene are contributing 88% of these deaths [6]. Over the last decades, a significant amount of research has been undertaken to establish reliable and effective techniques for detecting water pollutants [7].

The sensing layer plays an important role in the biosensor itself. This is due to it will directly in contact with the bio component and gives a signal for the senor. There were many sensing materials used in the research such as polyvinyl alcohol (PVA) hydrogel nanofibers [8], metal/oxide [9], and carbon [4]. PVA has outstanding biocompatibility; however, it still needs a complex setup. Moreover, metal/ metal oxide offers high endurance toward extreme environments. However, not all metals are suitable for bacterial detection. Moreover, some metals could kill bacteria instead of detectings, such as Ag (silver), brass, and copper (Cu) [10]. The most commonly used carbon as a sensing layer in graphene due to its superior properties like high specific face area and high mobility, and low electrical noise.

Some linkers have been developed in the biosensor formation depend on the purpose. 1-Pyrenebutyric acid N-hydroxysuccinimide ester (PBANHS) is one of the developed linkers as an activator reagent for the carboxylic acid which consists of hydrophobic headgroup π-π stacking system. Succinimidyl ester group is possessing a binding to nucleophilic substitution strongly by amine groups on the antibody. However, the use of PBANHS can save both sensing layer preparation time reaction and cost without using EDC (1-ethyl-3-(3-dimethyl aminopropyl carbodiimide hydrochloride) coupled to (NHS) N-hydroxysuccinimide reaction to activate carboxylic acid group [11].

Furthermore, the antibody was also explored by Oh *et.al* on a gold sensing layer through 11-mercaptoundecanoic acid and protein G linker. The *E. coli* antibody was immobilized over the optical system based on the surface plasmon resonance, a reflection of light from the thin metal film that will shift the reflectance index when any binding occurred on the sensor. The sensor can sense *E. coli* with a limit of detection of 10^4 cell/ml [12]. The antibody of *E. coli* can also be functionalized as a bioreceptor in bacteria detection on graphene film using 1-pyrenebutanoic acid succinimidyl ester (PASE) linker which provide amine linkage for antibody immobilization by π-π interaction with graphene film [13].

Theoretically, the material measured has a low in resistivity if the material is a conductive material, where the charge freely to move. On the other hand, if the material is non-conductive, the resistivity will be measured high. Bacteria, a living things organism, supposedly will act as a good conductor and has a low resistivity. In this paper we describe the functionalized of BSA/PBA-NHS/Graphene/SiO_2 as a sensing layer for *E. coli* bacteria detection based on electrical properties. Moreover, the biocompatibility of the sensing layer was also briefly discussed through Raman spectroscopy and topological properties.

2. Experimental Method

The experiment was begun by preparing the sensing layer of graphene and a sample of *E. coli* bacteria. The graphene used is a multilayer (p-type) commercial SiO_2/graphene P3106 from Graphenea, USA with a size of 0.5 × 0.5 cm^2. Bacteria *E. coli* ATCC 25922 (American Type Culture Collection, USA) was used in this research. The suspension and culture of bacteria were conducted by Chromocult Coliform Agar (CCA) 2.65 gr (*w/v*) (Merck, USA) and nutrient broth (Sigma Aldrich, USA) respectively. The number of *E. coli* bacteria was stated in the colony-forming unit (CFU).

The graphene sensing layer development was firstly initiated by mix 5% Pyrene Butyric Acid (PBA) with 5% N–hydroxysuccinimide ester (NHS) as a cross-linker on graphene. After that, 5.15 ml of 97% PBA was added into 94.85 ml of dimethylformamide (DMF) 99.8% solvent and 5.10 ml of 98% NHS was added into 94.89 ml dimethylformamide (DMF) 99.8% solvent which then both PBA and NHS solution were blended. PBA (257354) 97% and NHS (130672) 98%. and DMF was purchased from Sigma Aldrich, USA. Then, 0.5 × 0.5 cm^2 SiO_2/graphene film was incubated in PBANHS solution for 2 h at room temperature. It was then washed once with DMF, twice with ethanol 96%, and once more with isopropanol 99%, then blow-dried using nitrogen gas. After the

IOP Conf. Series: Materials Science and Engineering **982** (2020) 012024 doi:10.1088/1757-899X/982/1/012024

cross-linking process, the modified SiO_2/graphene/PBANHS was incubated in a mixture solution of antibody and Phosphate Buffer Saline (PBS) (pH 7.4; 1.0 M) (Sigma Aldrich, USA) solvent. The antibody was anti–O and K *E. coli* polyclonal antibodies ab31499 (liquid phase) (Abcam, USA). Anti–O and K *E. coli* polyclonal antibodies were diluted in PBS (1:2000 v/v). 1 μl of antibody was added into 2000 μl PBS solvent. The modified SiO_2/graphene/PBANHS was put into the antibody dilution overnight at 4° C. Then, the layer was rinsed with DI water and PBS. Next step, 80 mg of Bovine Serum Albumin (BSA) was mixed into 2 mL of DI water 96% (A2153) (Sigma Aldrich, USA) in lyophilized powder to block nonspecific agents.

The characterizations that were carried out in this research are topologically imaged by AFM, Raman spectroscopy, and the electrical properties. Topological characterization is important to indicate the topographic information of the sample. Based on local properties such as friction, height, and magnetism that the AFM can provide, the presence and quality of the anti-*E.coli* antibodies on the graphene layer can also be confirmed. The topological image of the sensing layer was observed using Park System XE-100 (Park system Ltd, Korea non-contact mode since the bacteria was the presence on the sample, it acts as the precaution to avoid the contamination to the cantilever tip. The surface roughness analysis before and after immobilized was done by 3D topographic image and by using height parameters such as average roughness (R_a), root mean square roughness (R_q) and ten-point average roughness (R_z). The height parameter was chosen instead of the case of roughness indicator because it is the most significant parameter among the local properties.

Raman spectroscopy analysis was performed on the sensing layer to confirm the functionalization process of the linker molecules and antibodies that lying on the top of single graphene. To obtain more accurate data from Raman spectroscopy, calibration was performed before the usage. The standard sample used for calibration, was a graphene/SiO_2/Si thin film because graphene on silicon has a unique Raman peak located at three locations which are D, G, and 2D peak and also strong Raman peak in around 520 cm-1 area that corresponds to single crystal silicon. The Raman spectrometer used in this experiment was the Horiba XploRA PLUS.

The electrical characterization of the functionalized sensing layer with anti-*E. coli* antibodies on the graphene layer were carried out using a two-point probe (Newport Corp., USA). The current and voltage electrodes on the sensing layer were measured and act as a reference. Using the two-probe method, the fixed connector platform was set up to measure the electrode of the sensing material before and after immobilization. Besides, a two-point probe also was utilized to test the resistance of the sensing layer when the sensing layer was treated with different concentrations of *E. coli* bacteria. The water sample with 1.0 ml was dropped on the sensing chamber. The resistance of the source and drain electrode was measured by scanning the potential difference between -3 V to 3 V. The I-V curve that appeared on the screen monitor using Oriel instruments software was recorded to know the resistance value which is the inverse of the slope in the I-V curve.

3. Results and Discussion

3.1 Topological structure of sensing layer

The topological structure of the graphene/SiO_2/Si layer and after linked with the anti-*E.coli* antibodies with the BSA was observed qualitatively 3D image using atomic force microscope (AFM). The average roughness (Ra), maximum peak-to-valley height (Rz), and root-mean-square roughness (Rq) of bare graphene and with immobilization of the anti-*E.coli* antibodies and the BSA were listed in Table 1 as the quantitative results. It indicated that the coverage of denatured BSA film and anti-E.*coli* antibodies on graphene/SiO2/Si surface is in nanoscale in size.

The height profile result corresponding by horizontal line scan based on the image in Figure 1 is illustrated in Figure 2. Bare graphene in Figure 1 (a) has a thin and flat appearance with a height of approximately 2.5 nm which may correspond to the multilayer state of graphene. The height of the surface was increased due to the conjugated with the anti-*E. coli* antibodies by around 6.0 nm which was in line with the previous research [14][15]. Based on experimentally result reported by Wu *et al.*

IOP Conf. Series: Materials Science and Engineering **982** (2020) 012024 doi:10.1088/1757-899X/982/1/012024

the thickness of PBA-NHS linker was around 0.5 nm while the thickness of anti-E. *coli* antibodies were around 4.2 nm [14].

Also, the following roughness parameters are presented in Table 1. The topography images gave rough estimation for both graphene and after immobilized with anti-*E. coli* antibodies and BSA. The results of Ra, Rz, and Rq demonstrated the increase in the surface roughness of the sensing layer after sensing immobilization. However, the quantitative data roughness in Table 1 shows that the value of Ra and Rq was not changed significantly compared to the value of Rz. It can be concluded that for the sensing layer that had high porosity, the value of Rz showed more significant change when compared to Ra and Rq value. AFM result confirmed that anti-*E. coli* antibodies were successfully attached to graphene thin film based on the distributions of the particles.

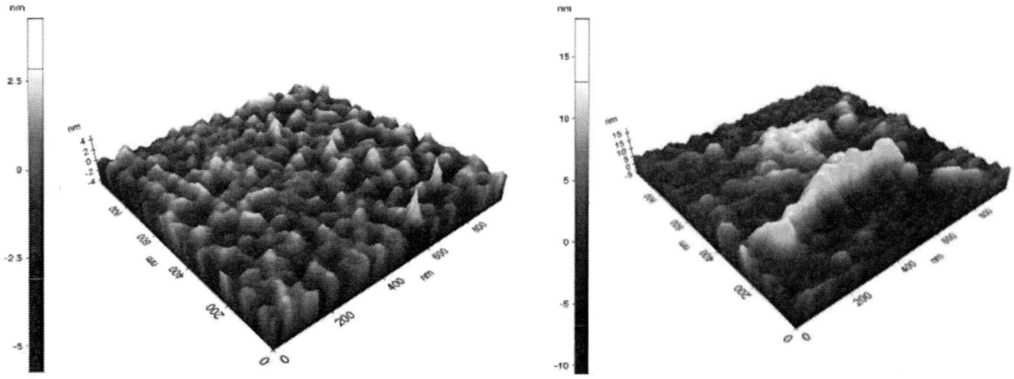

(a) (b)

Figure 1. Topology of (a) graphene/SiO2/Si, and (b) anti-*E. coli* antibodies with BSA/PBA-NHS/graphene/SiO_2/Si

Table 1. The variables of Ra, Rq, and Rz based on surface treated

Treated of sensing layer	R_a (nm)	R_q (nm)	$\mathbf{R_z}$ **(nm)**
Graphene/SiO_2/Si	0.669	0.986	2.327
Anti-*E. coli* antibodies with BSA/PBA-NHS/graphene/SiO_2 /Si	1.955	2.876	11.169

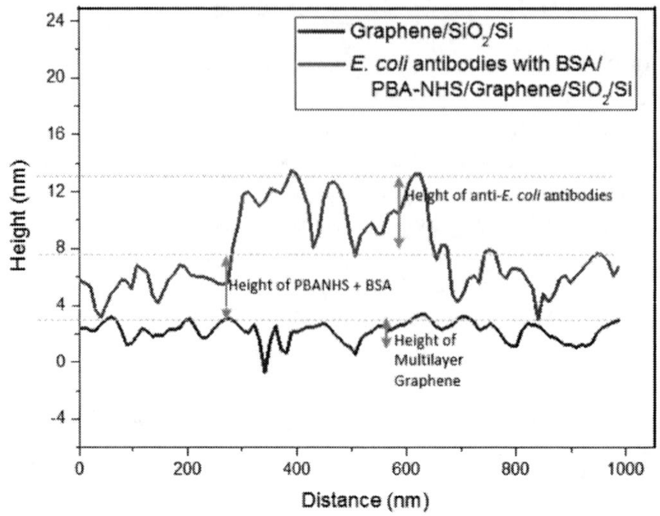

Figure 2. Height profiles of the line scans (arrows) in Figure 1

3.2. Raman spectra of sensing layer

Figure 4 shows the Raman spectrum of the graphene/SiO_2/Si layer and after immobilized with the anti-*E. coli* antibodies and the BSA with the scanning image in Figure 3. The peaks of D, G, and 2D peaks located at 1348.46, 1580.48, and 2718.20 cm^{-1} respectively show the common Raman peak of graphene. The intensity ratio of G peak was slightly higher than 2D peak indicated that the graphene sample was multilayer graphene [16]. Graphene quality is indicated by the low-intensity ratio of D and G peaks (ID/IG) of graphene which is lower than 0.1 a.u in Raman spectrum [17]. The intensity ratio of D and G peaks (ID/IG) of this graphene used was 0.08 a.u indicated that the high quality of graphene was applied. Smaller ID/IG peak intensity ratio can be assigned to lower defects or disorders in the graphitized structure [18].

When the anti-*E. coli* antibodies were linked with the graphene layer by PBA-NHS linker, the change of peak intensity can be observed. From the result, it can be observed that the D, G, and 2D peaks were increased and slight right shifted. This attribute to the aromatic pyrene group presented on PBA-NHS resonance stacked with π-π interaction of graphene as reported by Wu *et al.* [19].

The Raman markers of the side chain groups of antibodies as a protein derived from the localized vibrational mode of specific groups. Raman marker bands of aromatics in protein were well recognized in visible Raman spectra [20][21][22]. The Raman spectra of the anti- *E. coli* antibodies and the BSA in both compositions were shown in Table 2 with its band assignment. It can be seen that the conformation of amide I and amide III region with the backbone skeletal stretch acted as a fingerprint that represented anti-*E. coli* antibodies and BSA as a protein were presented on the sensing layer [23][24][25].

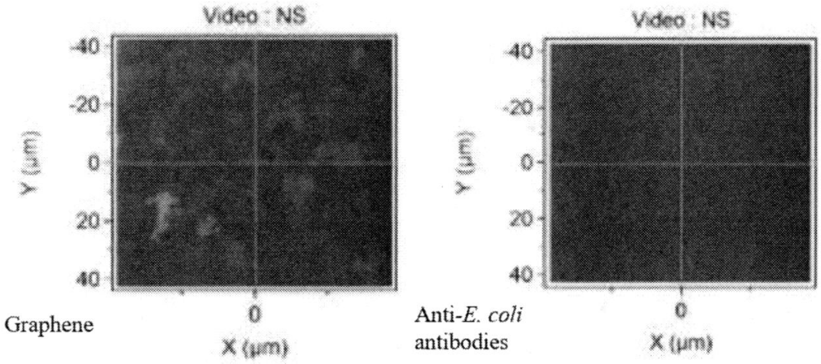

Figure 3. Raman image of graphene before and after undergoing immobilization

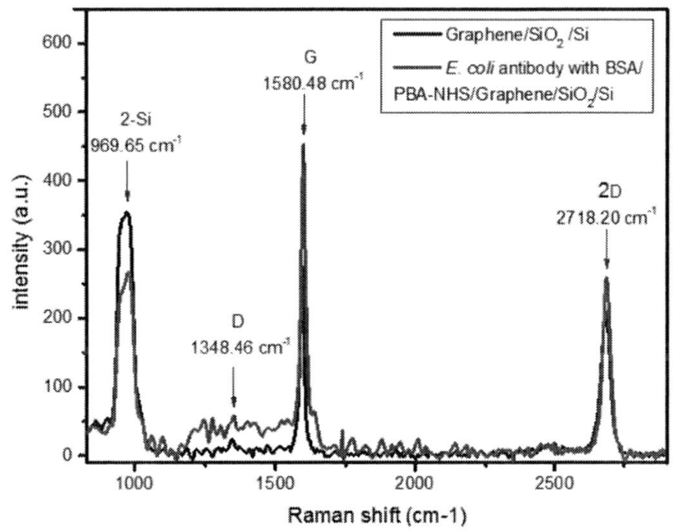

Figure 4. Raman shift of graphene and after added anti-*E. coli* antibodies and BSA as the protein

Table 2. Raman spectra of anti-*E. coli* antibodies and BSA as the protein

Raman peak (cm^{-1})	Tentative assignment
1062 1101	Backbone skeletal stretch. Assigned to C_α-C_β, C_α-N stretches
1241 1277 1309 1309 1397 1426 1514 1547	Amide III region. Assigned to NH bending and C_α-N stretching mode
1641	Amide I region. Assigned to C=O stretching of carbonyl group

3.3 Electrical properties

To monitor the magnitude of each subsequent functionalization the I-V measurement was performed before and after anti-*E. coli* antibodies. The resistances were calculated from the I-V slope and transferred in Table 3 up to ten samples to know the value of the differences of all initial immuno-sensor samples. The driving current at the given voltage shows the decreasing value after the immobilization process. Hence, it's worth noting that the resistivity of the sample increased after the immobilization process. This observation was attributed to the additional substance that contains charge either via the π-π stacking interaction between linker and pyrene or via anti-*E. coli* antibodies and BSA stacking. The decrease of current after the anti-*E. coli* antibody immobilization agreed with a previous study reported by Huang *et al.* [26].

From the results can be observed that the initial resistivity from each immobilized sample was different from each other. The device for measuring the electrical change was utilized with the calibration function using the Wheatstone bridge that read the initial value by null value so that the resistivity measurement is valid.

Table 3: Resistivity of sensing layer by I-V characterizations before and after the immobilization process

Number of samples	Resistivity of bare graphene (Ω)	Resistivity of graphene with anti-*E. coli* antibodies (Ω)	Resistivity change of the samples (ΔΩ)
1	7.428	8.150	0.722
2	6.954	7.947	0.993
3	6.631	8.634	2.003
4	7.122	7.804	0.682
5	6.616	7.670	1.054
6	6.348	7.362	1.014
7	6.735	7.698	0.963
8	6.981	8.490	1.509
9	7.149	7.885	0.736
10	7.416	8.362	0.946

The I-V characteristics of the sensing layer before and after the exposure to the different concentrations of *E. coli* bacteria are shown in Table 3. When in contact with the *E. coli* bacteria, the I-V graph shifts to the y-axis which shows the increasing current. When the number of *E. coli* bacteria increases, the increase of current becomes higher as well. This corresponds to research undertaken by Akbari et al. which shows that the I-V curve of graphene shifts upwards after increasing the concentration of *E. coli* bacteria [27]. In their research, the charge transfer between the graphene immunosensor and the low concentration of *E. coli* bacteria is between 100 to 105 CFU/ml. A similar result was also shown by Wu *et al.*, who measured the current with the *E. coli* bacteria concentration of 103 to 105 CFU/ml. The result of their study indicates that increasing the number of *E. coli* bacteria leads to an increase in the current [19][28]. Hence, the results show a decreasing graphene resistivity. This is attributed to the negatively charged *E. coli* bacteria inducing holes in the graphene. The linear pattern indicates the metallic nature of the sheets and the formation of ohmic contact between the sheets and the electrode [18]. From the I-V graph, the slope can be used to calculate the resistance of the graphene sensor. The change in resistance gradually increases from 0.04 to 2.62 Ω when the number of bacteria increases from 55 to 1200 CFU/ml as shown in Table 4.

Table 4. The change of resistance caused by differences concentrations of *E. coli* bacteria

Number of Bacteria (CFU/ml)	Change of resistance (ΔR)
55	0.04
160	0.13

290	0.51
540	1.26
1200	2.62

4. Conclusion

From this experiment, it was observed that graphene material has good biocompatibility toward the *E. coli* antibody which is proved by topological imaging by AFM and the fingerprint by Raman Spectroscopy. Furthermore, the electrical property of graphene was also increased due to the immobilization of the *E. coli* antibody. Increasing the number of *E. coli* results in decreasing the resistivity of the sensing layer (SiO_2/Graphene/ PBA-NHS/anti O and K *E. coli* antibodies). It was verified that the graphene sensing layer very promising to detect *E. coli* bacteria.

References

[1] P. Sharma, "Graphene for Nanoelectronic Applications," vol. 6886, 2015.
[2] H. J. Yoon, D. H. Jun, J. H. Yang, Z. Zhou, S. S. Yang, and M. M. C. Cheng, "Carbon dioxide gas sensor using a graphene sheet," *Sensors Actuators, B Chem.*, vol. 157, no. 1, pp. 310–313, 2011.
[3] C. M. Santos, J. Mangadlao, F. Ahmed, A. Leon, R. C. Advincula, and D. F. Rodrigues, "Graphene nanocomposite for biomedical applications: fabrication, antimicrobial and cytotoxic investigations," *Nanotechnology*, vol. 23, p. 395101, 2012.
[4] P. K. Basu, D. Indukuri, S. Keshavan, V. Navratna, S. R. K. Vanjari, S. Raghavan, and N. Bhat, "Graphene based *E. coli* sensor on flexible acetate sheet," *Sensors Actuators, B Chem.*, vol. 190, pp. 342–347, 2014.
[5] G. S. Selopal, R. Milan, L. Ortolani, V. Morandi, R. Rizzoli, G. Sberveglieri, G. P. Veronese, A. Vomiero, and I. Concina, "Graphene as transparent front contact for dye sensitized solar cells," *Sol. Energy Mater. Sol. Cells*, vol. 135, pp. 99–105, 2015.
[6] WHO and UNICEF, "WHO/UNICEF Joint Monitoring Programme (JMP)," 2017. [Online]. Available: https://washdata.org.
[7] M. R. Nurliyana, M. Z. Sahdan, K. M. Wibowo, A. Muslihati, H. Saim, S. . Ahmad, Y. Sari, and S. Mansor, "The Detection Method of Escherichia coli in Water Resources : A Review The Detection Method of Escherichia coli in Water Resources : A Review," *J. Phys. Conf. Ser.*, vol. 995, pp. 1–11, 2018.
[8] P. M. Shaibani, K. Jiang, G. Haghighat, M. Hassanpourfard, H. Etayash, S. Naicker, and T. Thundat, "The detection of Escherichia coli (*E. coli*) with the pH sensitive hydrogel nanofiber-light addressable potentiometric sensor (NF-LAPS)," *Sensors Actuators, B Chem.*, vol. 226, pp. 176–183, 2016.
[9] N. Uria, N. Abramova, A. Bratov, F.-X. Muñoz-Pascual, and E. Baldrich, "Miniaturized metal oxide pH sensors for bacteria detection," *Talanta*, vol. 147, pp. 364–369, 2016.
[10] M. Yasuyuki, K. Kunihiro, S. Kurissery, N. Kanavillil, Y. Sato, and Y. Kikuchi, "Antibacterial properties of nine pure metals: A laboratory study using Staphylococcus aureus and Escherichia coli," *Biofouling*, vol. 26, no. 7, pp. 851–858, 2010.
[11] S. Jahanbani and A. Benvidi, "A novel electrochemical DNA biosensor based on a modi fi ed magnetic bar carbon paste electrode with Fe 3 O 4 NPs-reduced graphene oxide / PANHS nanocomposite," vol. 68, pp. 1–8, 2016.
[12] Y. Kim, Y. M. I. N. Bae, and W. O. N. H. Lee, "Detection of Escherichia coli O157 : H7 Using Immunosensor Based on Surface Plasmon Resonance," vol. 12, pp. 780–786, 2002.
[13] W. Song, D. W. Li, Y. T. Li, Y. Li, and Y. T. Long, "Disposable biosensor based on graphene oxide conjugated with tyrosinase assembled gold nanoparticles," *Biosens. Bioelectron.*, vol. 26, no. 7, pp. 3181–3186, 2011.

[14] G. Wu, X. Tang, Z. Lin, M. Meyyappan, and K. W. C. Lai, "The effect of ionic strength on the sensing performance of liquid-gated biosensors," in *2017 IEEE 17th International Conference on Nanotechnology (IEEE-NANO)*, 2017, pp. 242–245.
[15] A. Huang, W. Li, S. Shi, and T. Yao, "Quantitative fluorescence quenching on antibody-conjugated graphene oxide as a platform for protein sensing," *Sci. Rep.*, vol. 7, p. 40772, 2017.
[16] E. W. Hill, A. Vijayaragahvan, and K. Novoselov, "Graphene sensors," *IEEE Sens. J.*, vol. 11, no. 12, pp. 3161–3170, 2011.
[17] S. Afsahi, M. B. Lerner, J. M. Goldstein, and J. Lee, "Novel graphene-based biosensor for early detection of Zika virus infection," *Biosens. Bioelectron.*, vol. 100, pp. 85–88, 2018.
[18] O. Akhavan and E. Ghaderi, "Escherichia coli bacteria reduce graphene oxide to bactericidal graphene in a self-limiting manner," *Carbon N. Y.*, vol. 50, no. 5, pp. 1853–1860, 2012.
[19] G. Wu, M. Meyyappan, and K. W. C. Lai, "Graphene field-effect transistors-based biosensors for Escherichia coli detection," in *2016 IEEE 16th International Conference on Nanotechnology (IEEE-NANO)*, 2016, pp. 22–25.
[20] P. Carey, *Biochemical applications of Raman and resonance Raman spectroscopes*. Elsevier, 2012.
[21] A. T. Tu and A. T. Tu, "Raman spectroscopy in biology: principles and applications," 1982.
[22] Z. Wen, "Raman spectroscopy of protein pharmaceuticals," *J. Pharm. Sci.*, vol. 96, no. 11, pp. 2861–2878, 2007.
[23] T. Miyazawa, T. Shimanouchi, and S. Mizushima, "Normal vibrations of N-methylacetamide," *J. Chem. Phys.*, vol. 29, no. 3, pp. 611–616, 1958.
[24] R. C. Lord and N.-T. Yu, "Laser-excited Raman spectroscopy of biomolecules: I. Native lysozyme and its constituent amino acids," *J. Mol. Biol.*, vol. 50, no. 2, pp. 509–524, 1970.
[25] T. G. Spiro and B. P. Gaber, "Laser Raman scattering as a probe of protein structure," *Annu. Rev. Biochem.*, vol. 46, no. 1, pp. 553–570, 1977.
[26] Y. Huang, X. Dong, Y. Liu, L.-J. Li, and P. Chen, "Graphene-based biosensors for detection of bacteria and their metabolic activities," *J. Mater. Chem.*, vol. 21, no. 33, pp. 12358–12362, 2011.
[27] E. Akbari, A. Nikoukar, Z. Buntat, A. Afroozeh, and A. Zeinalinezhad, "Escherichia coli bacteria detection by using graphene-based biosensor," *IET Nanobiotechnology*, vol. 9, no. 5, pp. 273–279, 2015.
[28] K. M. Wibowo, M. Z. Sahdan, N. I. Ramli, A. Muslihati, N. Rosni, V. H. Tsen, H. Saim, S. A. Ahmad, Y. Sari, and Z. Mansor, "Detection of Escherichia Coli Bacteria in Wastewater by using Graphene as a Sensing Material," in *Journal of Physics: Conference Series*, 2018, vol. 995, no. 1, p. 12063.

Sensitivity of Optical Fiber Sensors to Deflection of Reinforced Concrete Beam

Farida Asriani[1,3]. Winasis[1] and Gandjar Pamudji[2]

[1] Electrical Engineering Departement. Universitas Jenderal Soedirman
[2] Civil Engineering Departement. Universitas Jenderal Soedirman

[3]farida.asriani@unsoed.ac.id

Abstract. The capacity of concrete beams will decrease by many external factors. To find simple and reliable method to monitor the quality of concrete beams is a challenging task. An optical-based fiber sensors is very interesting to develop for such task because of its many advantages. In this study, the optical fiber sensors were embedded in reinforced concrete beams to detect and to monitor deflection of the beam where a straight-line configuration of optical fiber was used. We perform experimental work to test performance of the use of optical fiber sensor by collecting data from flexural testing the concrete beam with the Universal Testing Machine (UTM). While the concrete deflection was measured by (linear variable differtial transducer (LVDT) as elongation unit, the fiber optic sensor output was observed in volts unit. We test the sensitivity of the optical fiber sensor by analyzing the relationship graph between the changes in the deflection of the concrete beam and output voltage of optical fiber sensors. The results show that optical fiber sensors have good sensitivity to detect and monitor concrete beam deflection.

1. Introduction

Sensors are a major component in the development of a monitoring system. At present, research on the use of fiber optic sensors (FOS) is being developed. The advantage of this FOS is better resistance to electromagnetic interference. including storms. and the potential ability to live in a harsh environment. FOS is also resistant to corrosion when used. in open structures (such as bridges and dams) that can be used throughout the lifetime of concrete.

Fiber optic sensors that are widely applied include micro bending sensors. wave sensor. FBGs. and fiber optics interferometer [1]. Fiber optic sensors provide a good response to loads applied to concrete with embedded fiber optic sensors. Changes in load indicate changes in the intensity of light that is passed on the optical fiber sensor[2].Optical fiber sensors embedded in lightweight concrete have been applied to detect damage or cracks in concrete. Measurements were made using an optical power meter. The experimental results show that when cracks occur in concrete there is a significant decrease in the output of fiber optic sensors [3].

The use of optical fiber sensors in monitoring concrete structures has been developed previously. Detection concrete collapse by monitoring the concrete strain has been carried out. In this study, fiber optic sensors are coated with chemical material[4][5]. Lightweight concrete damage detection with

IOP Conf. Series: Materials Science and Engineering **982** (2020) 012025 doi:10.1088/1757-899X/982/1/012025

optical fiber sensors with other methods was develop. When the quality of concrete decreases, the optical fiber sensor output power also decreases [6].

In a reinforced concrete beam monitoring system, the compressive strength, strain, and deflection parameters are the main parameters monitored. Single mode type optical fiber has been applied to detect strain in lightweight concrete. The value of power loss generated by optical fiber due to loading is directly proportional to the value of the load given. The test results show that optical fiber with A wave configuration has good sensitivity to changes in load given to concrete.[7]. Distributed optical fiber sensors have also been applied to measure strain on reinforced concrete wall reinforcement [8].

Measurement of reinforced concrete beam deflection has been carried out in the laboratory, so there are many problems if direct measurements will be applied for the purposes of continuous monitoring of reinforced concrete structures. For this reason, it is necessary to develop a sensor system embedded in concrete. In this paper, the sensitivity of fiber optic sensor sensitivity to the deflection parameters of reinforced concrete beams will be examined.

2. Experimental Method

The sensitivity of the fiber optic sensor to reinforced concrete beam deflection is determined by taking data about the concrete deflection and the optical fiber sensor output power. To achieve this, the outline is divided into 3 steps, namely the design of fiber optic sensors. making reinforced concrete beam samples with embedded sensors and data acquisition and analysis.

2.1. *Design of Fiber Optic Sensor Configuration*

The optical fiber sensors used in this study is single mode with a length of 4 meters. The configuration applied is a straight line as shown in Figure 1. optical fiber sensors are implanted straight in the beam.

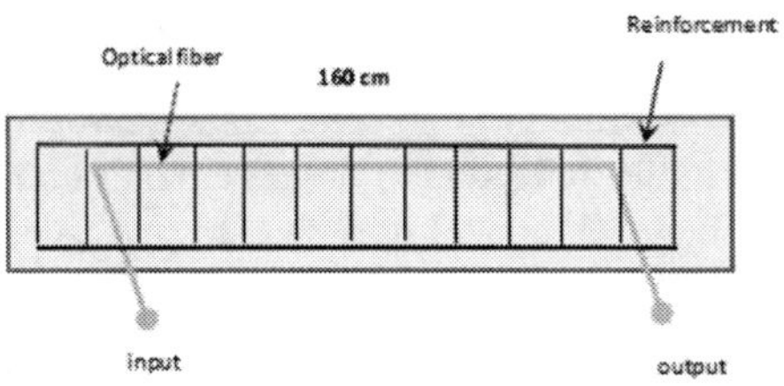

Figure 1. Stright line configuration of optical fiber

The optical fiber sensor consists of two terminals. the input terminal that conected to laser source 1310nm. and the output terminal connected to the photodetector to convert the signal from the laser wave to the voltage level. The middle part of the fiber optic sensor is peeled off the jacket and the peeled part is installed embedded in a concrete beam.

2.2. *Design of Reinforced Concrete Beams with an Embedded Sensor*

Concrete beams used in this study are lightweight reinforced concrete beams with 20MPa compressive strength. In order to reach a lightweight, plastic gravel is used [9] The size of the concrete blocks made is 8x16x160 cm. Fiber optic sensors are implanted in concrete with the straight configuration in Figure 2. The mix design of concrete is shown in Table 1.

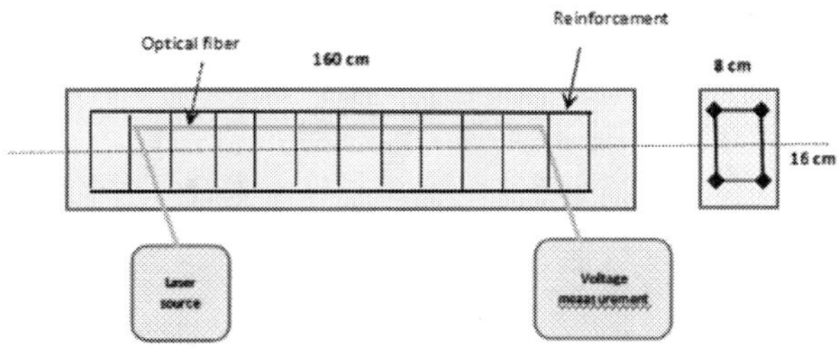

Figure 2. Model of concrete beam samples with embedded sensors

Table 1. The mix design of concrete.

materials	Concrete Volume (Kg) - Plastic	
	1 M3	2*Sil 10/20 & * 1Sil 15/30
Cement	500	17,14
sand	845	28,96
Naturak gravel	0	0
Plastic gravel	432	14,8
SP Sika LN	12	0,4
water	150	5,14

2.3. *Data Acquisition and Analysis*

Data acquisition is done by flexural testing using UTM. Concrete deflection is measured using LVDT. The observed data are changes in concrete deflection and changes in the optical fiber sensor output voltage due to loading on the concrete beam. The data acquisition scheme is shown in Figure 3.

IOP Conf. Series: Materials Science and Engineering **982** (2020) 012025 doi:10.1088/1757-899X/982/1/012025

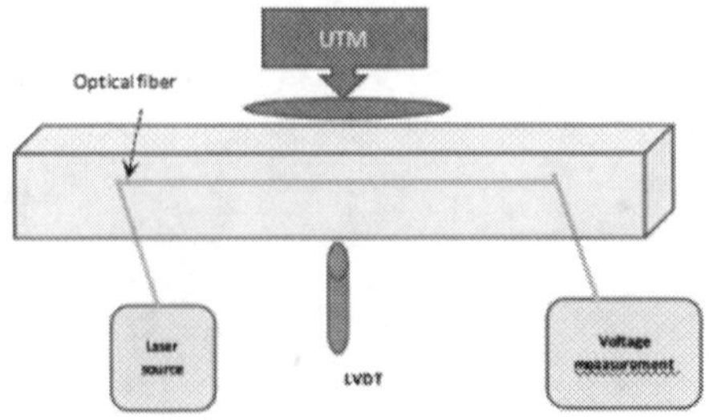

Figure 3. Data retrieval scheme

3. Result and Discussion

3.1. Data

The fiber optic configuration in the concrete is shown in Figure 4, and the reinforced concrete block with embedded sensors is shown in Figure 5.

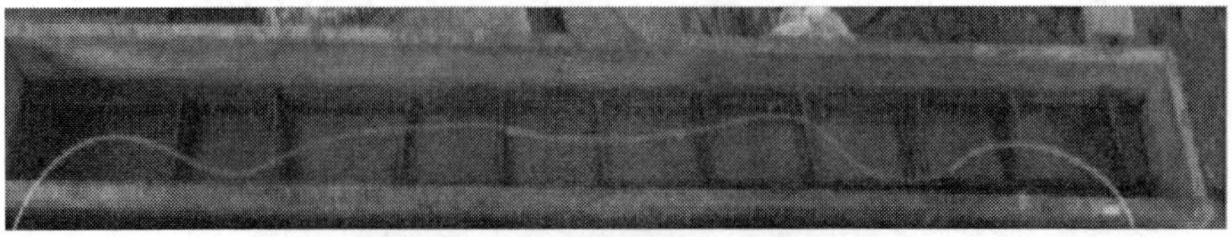

Figure 4. fiber optic configuration inside the concrete mold

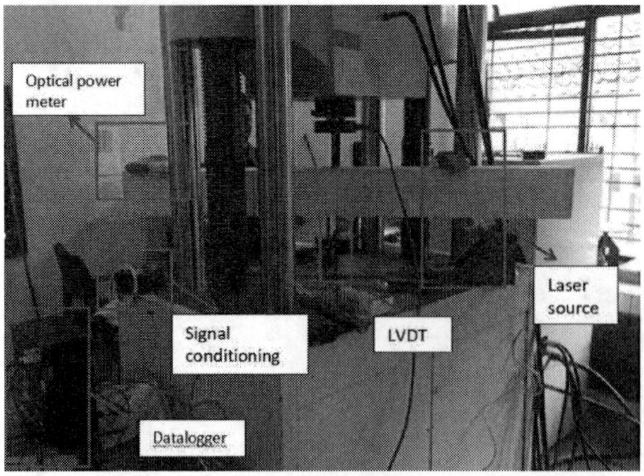

Figure 5. Flexural testing concrate beam sample with embedded fiber optic

In this study data were taken from 3 concrete beam samples with embedded sensors. The results of the data from the first concrete beam are shown in Table 2. The second concrete beam in Table 3 and the third concrete beam in Table 4.

Tabel 2. The first concrete beam sample data

IOP Conf. Series: Materials Science and Engineering **982** (2020) 012025 doi:10.1088/1757-899X/982/1/012025

no	load (kN)	deflection (mm)	Output voltage optical fiber sensors (volt)
1	0.004941041	0.00000487	0.37
2	0.11957637	0.22377928	0.56
3	0.57090282	0.32311198	1.12
4	1.7325027	0.83081281	1.49
5	3.65113	1.196736	1.68
6	6.0351171	1.8259418	2.43
7	8.5939245	2.5912743	2.61
8	10.107606	3.0649996	3.17
9	14.518381	4.418685	3.55
10	16.03191	5.8213181	3.73
11	16.907057	8.036375	3.92
12	17.398458	9.9108582	4.11
13	17.840471	13.267413	4.48

Tabel 3. Second concrete beam sample data

no	load (kN)	deflection (mm)	Output voltage optical fiber sensors (volt)
1	0.068897821	0.036128148	0.4
2	0.51017541	0.52297401	0.6
3	1.1495951	0.72594935	0.81
4	2.6247454	1.1949829	1.01
5	3.5582852	1.4596689	1.41
6	5.3155766	2.051162	1.81
7	7.4279442	2.8705878	2.01
8	9.6917734	3.6616268	2.21
9	10.854418	4.0441976	2.62
10	13.480095	4.8764434	2.82
11	15.15642	5.4440699	3.02
12	15.846836	6.1979642	3.22

13	16.391153	6.8814449	3.62
14	16.709873	7.5666599	3.82
15	17.527981	12.060499	4.23
16	17.753162	22.424404	4.43
17	18.186796	38.390656	4.83

Tabel 4. Third concrete beam sample data

no	load (kN)	deflection (mm)	Output voltage optical fiber sensors (volt)
1	0.82641274	0.036763895	0.35
2	2.0133584	0.21275349	0.53
3	3.6153882	0.54891306	1.06
4	5.2496629	0.95169693	1.24
5	7.0092497	1.5461302	1.42
6	8.5411425	2.0465524	1.77
7	9.6772785	2.417012	1.95
8	11.71697	3.1238008	2.13
9	13.75316	3.8872695	2.3
10	14.499065	4.8545823	2.66
11	14.689847	6.7961712	3.19
12	15.386856	8.257062	3.54
13	16.247229	12.41567	4.08
14	16.62388	27.62908	4.25
15	17.472326	55.628048	4.61

Based on the test results with UTM, we obtain that the reinforced concrete beam has a maximum average load capacity of 17,833kN. While the mean deflection of reinforced concrete beams when loading is 35.62 mm. The average sensor output voltage when the concrete beam reaches its maximum capacity reaches the average of 4.64volt. From table 1, table 2 and table 3 it can be seen that both deflection and sensor output voltage increases with increasing loading test

3.2. Sensitivity Analysis of Optical Fiber Sensors

Sensitivity analysis is done by graphing the relationship between the deflection of the concrete beam and the optical fiber sensors output voltage. To determine the sensitivity, a linear region determined from the response of the fiber optic sensor, then a linear approach analysis is performed.

3.2.1. Embeded Sensor 1. Embedded sensor 1, is an optical fiber sensors that is embedded in the second concrete beam. The graph results for Embedded sensor 1 are shown in Figure 6.

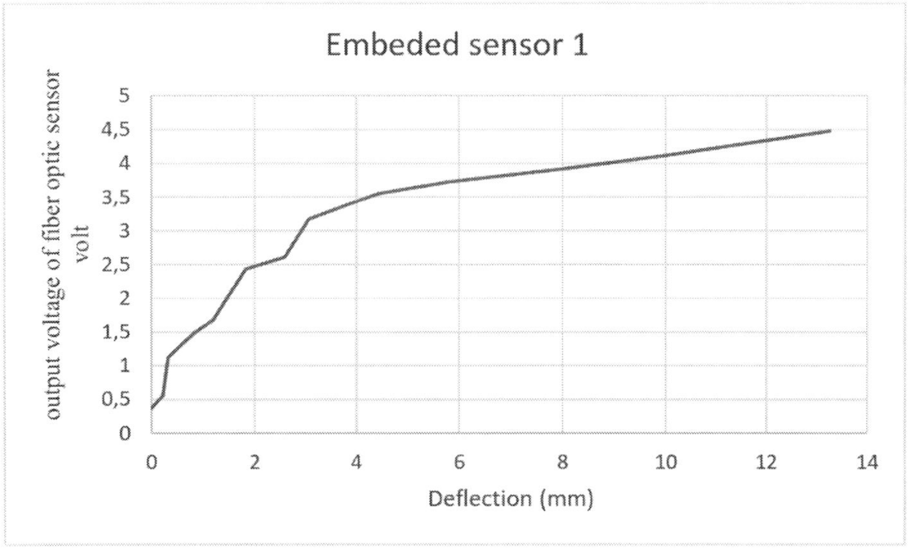

Figure 6. Graph of the relationship between the deflection of a concrete beam and the optical fiber sensor output voltage from first sensor.

In Figure 6 can be seen that there are two different linear regions. This shows that the senor has different sensitivity values for two different linear regions. Graph linear 1a where the deflection value is less than 5 mm. The linear region 1b where the deflection value is more than the same as 5 mm. The results of the linear approach are shown in Figure 7 and Figure 8

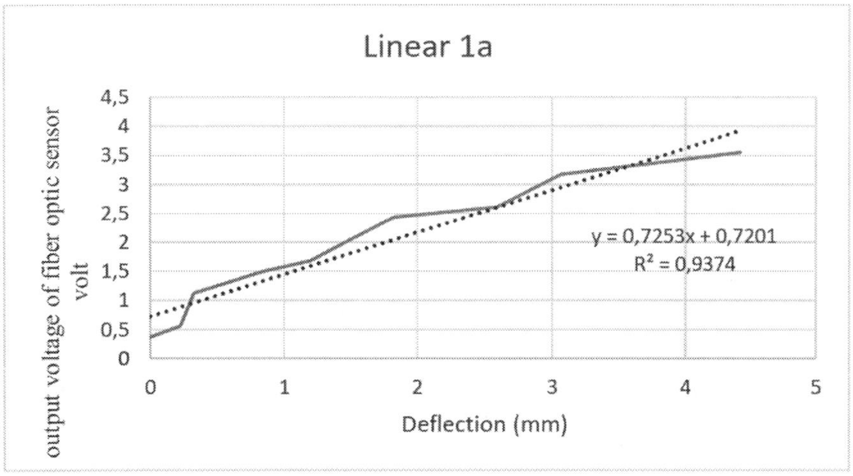

Figure 7. Linear approach 1a

In the linear region 1a in the deflection range below 5mm, the optical fiber sensor has a sensitivity of 0.7253volt / mm.

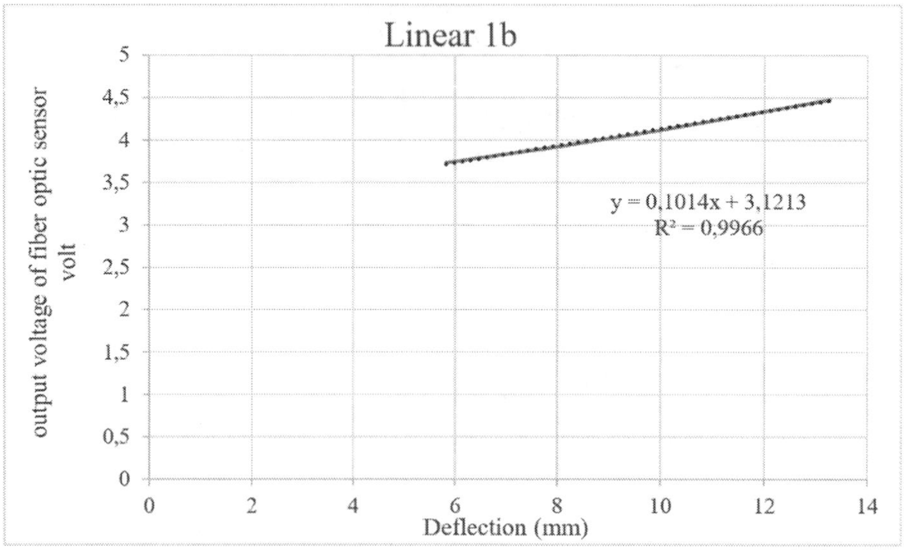

Figure 8. Linear approach 1b

In the linear region 1b, where the deflection range is more than or equal to 5mm, the fiber optic sensor has a sensitivity of 0.1014volt / mm. For embedded sensor 1, the sensitivity ratio of linear region 1a to linear region 1b reaches 7.15: 1

3.2.2. .Embedded Sensor 2. Embedded sensor 2, is an optical fiber sensor that is embedded in the second concrete beam. The graph results for Embedded sensor 2 are shown in Figure 9.

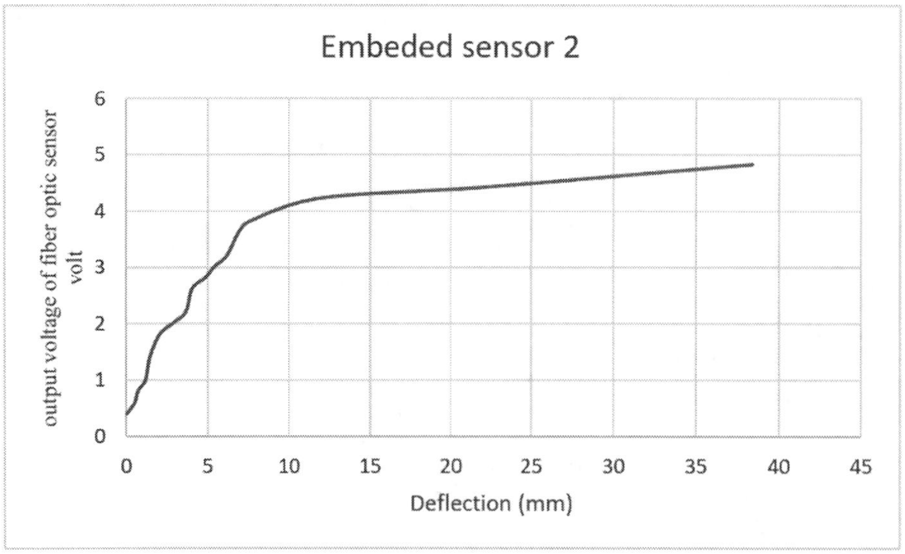

Figure 9. Graph of the relationship between the deflection of a concrete beam and the optical fiber sensor output voltage from second sensor.

In Figure 9, there are two linear region. Liear 2a show in Figure 10 and linear 2 b show in Figure 11.

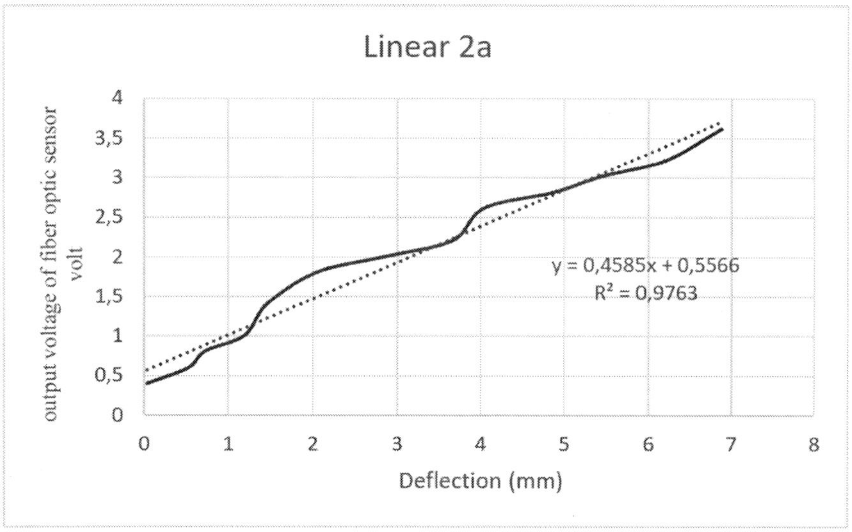

Figure 10. Linear approach 2a

In the deflection range of concrete beams less than 8mm, optical fiber sensors embedded in concrete have a sensitivity of 0.4585volt / mm.

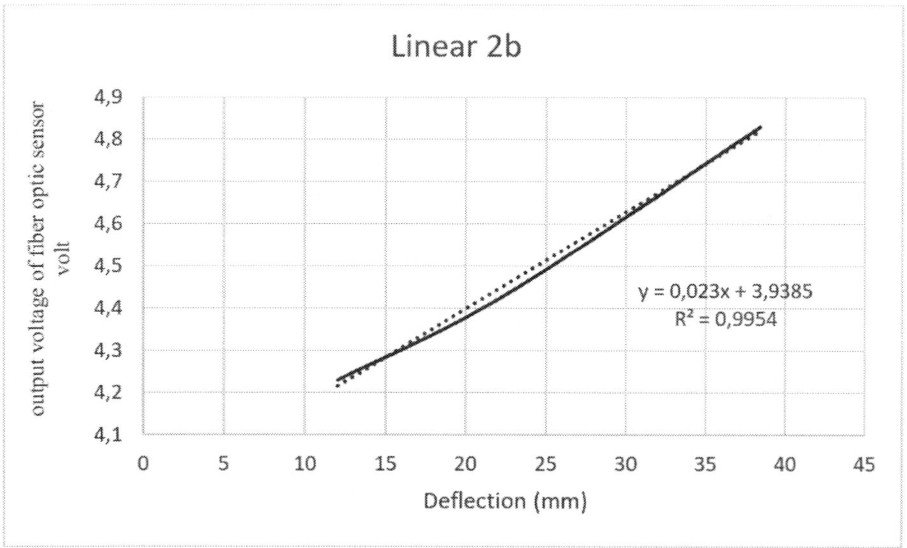

Figure 11. Linear approach 2b

In the deflection range of concrete beam more than the same as 8mm, optical fiber sensors embedded in concrete have a sensitivity of 0.023volt / mm. The ratio of linear sensitivity 2a to linear 2b is 19.93: 1

3.2.3. Embeded Sensor 3. The Embedded sensor 3 is an optical fiber sensors that embedded in the third concrete beam. The graph results for Embedded sensor 3 are shown in Figure 12.

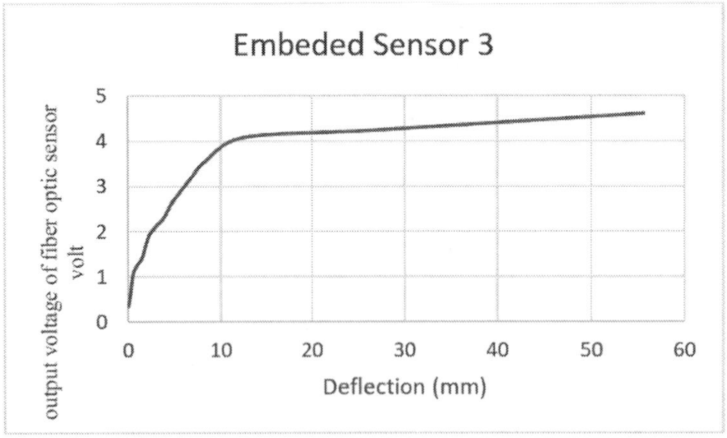

Figure 12. Graph of the relationship between the deflection of a concrete beam and the optical fiber sensor output voltage from third sensor.

Figure 12 shows the existence of two linear regions. The linear region in the deflection range is less than 9 mm, referred as the linear region 3a shown in Figure 13, and the linear region in the deflection range is greater than 9mm called the linear region 3b shown in Figure 14.

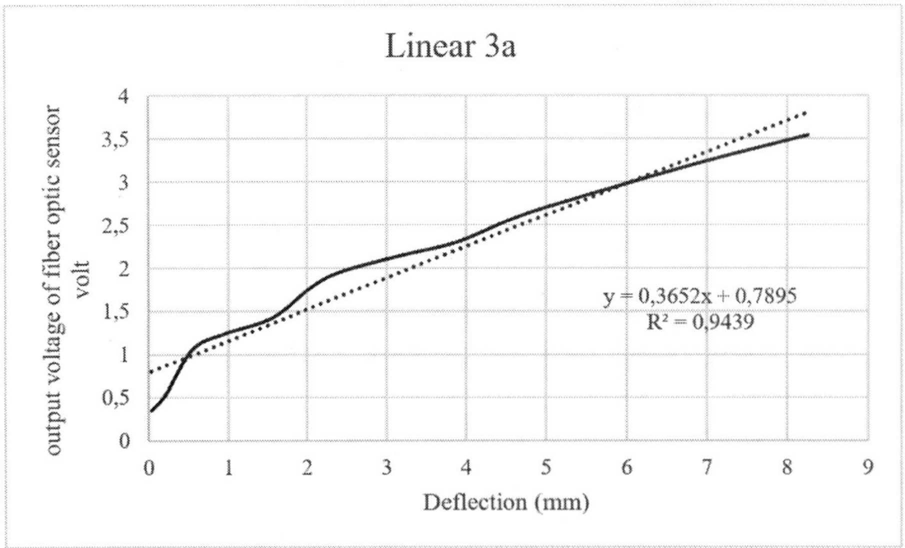

Figure 13. Linear approach 3a

In the deflection range of less than 9mm, the fiber optic sensor has a sensitivity of 0.3652 volts / mm.

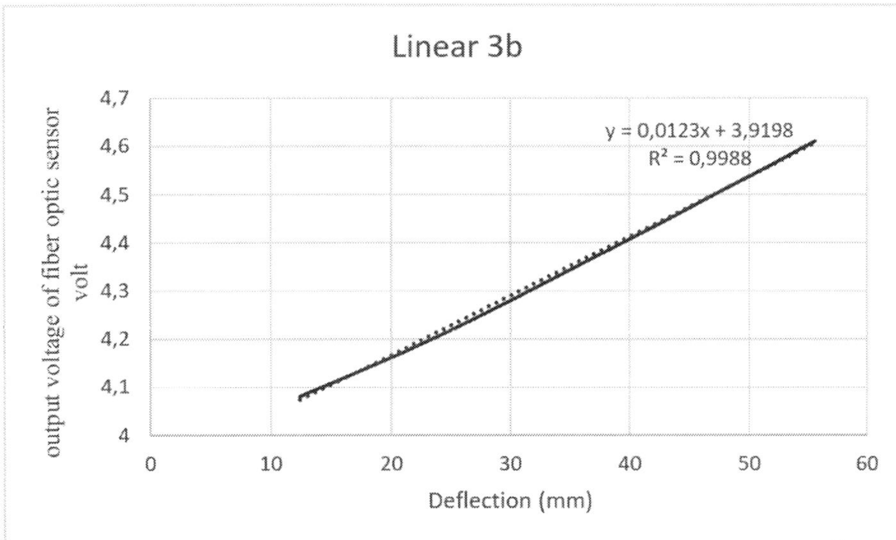

Figure 14. linear approch 3b

In the deflection range above 9 mm, the optical fiber sensors have a sensitivity of 0.0123volt / mm. The sensitivity ratio of 3a to 3b is 29.69: 1.

Based on data retrieval from the three sensors, it can be seen that the sensor has a good sensitivity in the linear a region with an average sensitivity of (0.7253volt / mm + 0.4585volt / mm + 0.3652 volts / mm) / 3 = 0.5613volt / mm. The average of effective working area is for deflection value less than 7.33mm

4. Conclusion

Optical fiber sensor embedded with a straight configuration in concrete has a sensitivity related to the changes of reinforced concrete beam deflection in the number of 0.5613volt / mm and the average of effective working area is for deflection value less than 7.33mm

References

[1] K. Fidanboylu and Efendioglu, H.S., "Fiber Optic Sensors and Their Applications," 5th International Advanced Symposium (IATS'09). May, 2009.

[2] S. Ahmad, M. U. Rizvi, S. Mohd, A. Husain, and A. Anwar, "Study of Fiber Optic Sensor Using Concrete Beams," pp. 2319–2322, 2014.

[3] Asriani Farida, Pamudji Gandjar, and Susilawati, H. "Pengembangan Deteksi Kerusakan Beton yang Menggunakan Agregat Sampah Plastik Dengan Sensor Fiber Optik Multi Mode Tertanam," in *Seminar Nasional Ketekniksipilan Bidang Vikasional III*, 2016.

[4] C. G. Berrocal, I. Fernandez, and R. Rempling, "Crack monitoring in reinforced concrete

beams by distributed optical fiber sensors," *Struct. Infrastruct. Eng.*, vol. 0, no. 0, pp. 1–16, 2020.

[5] Barrias A, Casas Joan R, andVillalba S, "Embedded Distributed Optical Fiber Sensors in Reinforced Concrete Structure," Sensors pp. 1–22, 2018.

[6] F. Asriani, G. Pamudji, H. Susilawati, and S. W. Ismani, "Damage Detection Tool Design of Lightweight Concrete Using Optical Fiber Sensor and Phototransistor," *Ijitee*, vol. 1, no. 1, pp. 8–12, 2017.

[7] Asriani Farida, Pamudji Gandjar, and Susilawati Hesti, "Fiber Optik Singlemode Sebagai Sensor Regangan yang Tertanam di dalam Beton," in *Seminar Nasional Teknologi Terapan 2016*, 2016.

[8] J. E. Woods, D. T. Lau, X. Bao, and W. Li, "Measuring strain fields in FRP strengthened RC shear walls using a distributed fiber optic sensor," vol. 152, pp. 359–369, 2017.

[9] H. Purnomo, G. Pamudji, and M. Satim, "Influence of uncoated and coated plastic waste coarse aggregates to concrete compressive strength," *MATEC Web Conf.*, vol. 101, 2017.

[10] P. Gandjar, H. Bimasena, P. A. Yuta, and P. Heru, "Bond-Slip Behavior of Steel Bar Embedded in Lightweight Concrete Using Sand Coated Polypropylene Coarse Aggregate," *Mater. Sci. Forum*, vol. 929, pp. 103–108, 2018.

IOP Conf. Series: Materials Science and Engineering **982** (2020) 012026 doi:10.1088/1757-899X/982/1/012026

Performance Comparison of File Transfer Protocol Service Between Link State and Distance Vector Routing Protocol in Software Defined Network

Rohmat Tulloh[1,3], Jafaruddin Gusti Amri Ginting[2], Asep Mulyana[1], Muhammad Lutfi[1]

[1] School of Applied Science, Telkom University, Bandung, Indonesia
[2] Faculty of Telecommunication and Electrical Engineering, Institut Teknologi Telkom Purwokerto, Central Java 35147, Indonesia

[3]rohmatth@telkomuniversity.ac.id

Abstract. Software-Defined Network (SDN) is a new concept in computer networks where the network control function is separated from the data forwarding function (data plane). The control plane and data plane which are separated from each other is the answer for faster, more flexible, and secure internet service. File transfer protocol (FTP) is an internet protocol that runs at the application layer that is used to exchange data between server and client. Open Shortest Path First (OSPF) is a routing protocol for internet networks that belongs to the Interior Gateway Protocol (IGP) group and uses the Link State Routing algorithm (LSR). Routing Information Protocol (RIP) is a type of Distance Vector Routing protocol that is still used today. This study aims to compare the performance of FTP services using OSPF and RIP on SDN networks and conventional networks. The research was conducted with direct implementation on the device with a planned topology. The measurement results show that FTP using OSPF routing has better results than using RIP. FTP that runs on SDN also has a better quality of service value compared to conventional networks.

1. Introduction

Software-Defined Network (SDN) is a new technology on computer networks, where this technology separates the function of the data Forwarding (data plane) with the network control (control plane) which can be programmed. With the SDN, a network infrastructure could have centralized control and more efficiency in the configuration of a network. The separate Control Plane and Data Plane make the network architecture more flexible, cost-effective, and innovative [1, 2]. As shown in Figure 1, it appears that one of the most important things in SDN is the centralized controller, which runs the routing protocol algorithm. In conventional networks, routing protocols are run on each router, so the routing process is determined by the router itself. Research on routing protocol or types of service performance in SDN has been carried out such as using the RIP [3-5] or OSPF routing protocol [6-11]. The research of FTP service on conventional networks is also carried out using various simulations such as bandwidth analysis using [12], FTP over p2p [13], or compare the performance of HTTP services with FTP on simulations using Opnet [14]. Most research is usually done by using a simulator or emulator.

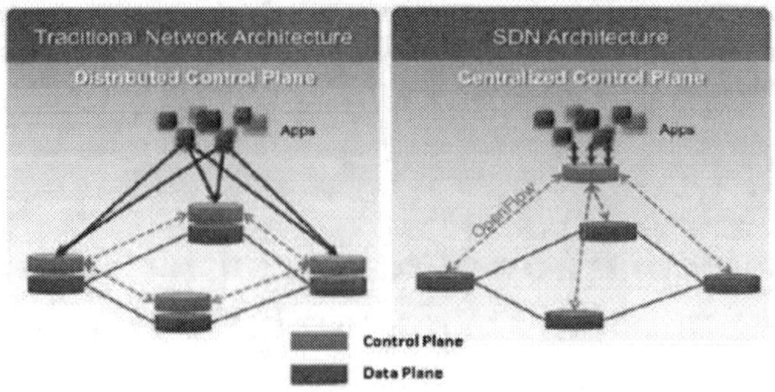

Figure 1. Comparison of conventional and SDN Network Architecture

The study aims to compare the performance of FTP services using OSPF and RIP on SDN, as well as on conventional or traditional networks. The research was carried out with direct implementation on the device with a planned topology. Upload and download scenarios on FTP servers, to see whether OSPF and RIP routing protocols could run on FTP services with good results on SDN and conventional networks.

2. Research Method

2.1 Software Defined Network (SDN)

In conventional network technology, where the control plane and the data plane are made in one device, but on the SDN network has two advantages. First, the control plane and the data plane are separated. Second, each device's control plane is put together and put together in a centralized controller. So the controller could control many devices at one command.

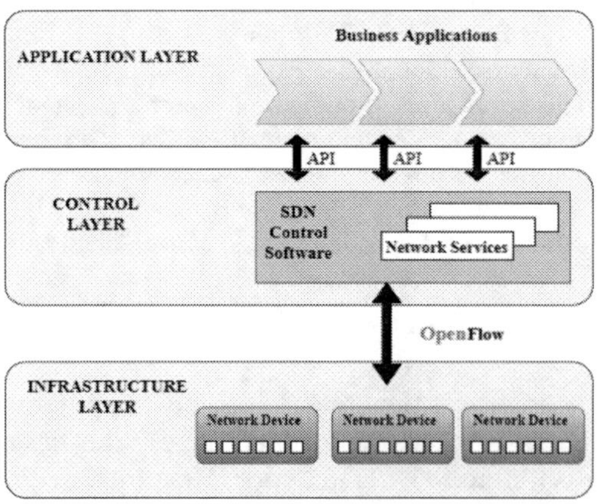

Figure 2. SDN Architecture [2]

As shown in Figure 2, the control plane is centered in the control layer. The centralized controller manages all network devices at the infrastructure layer. Various services in the application layer could run on SDN, through various APIs provided.

The separation between the control plane and the forwarding plane, where the control plane contains RF Client, Quagga, RF Server, and RF proxy, while the forwarding plane uses OpenFlow which is responsible for connecting the Control Plane and the Data Plane. Openflow has become a standard in SDN-based research [15-18].

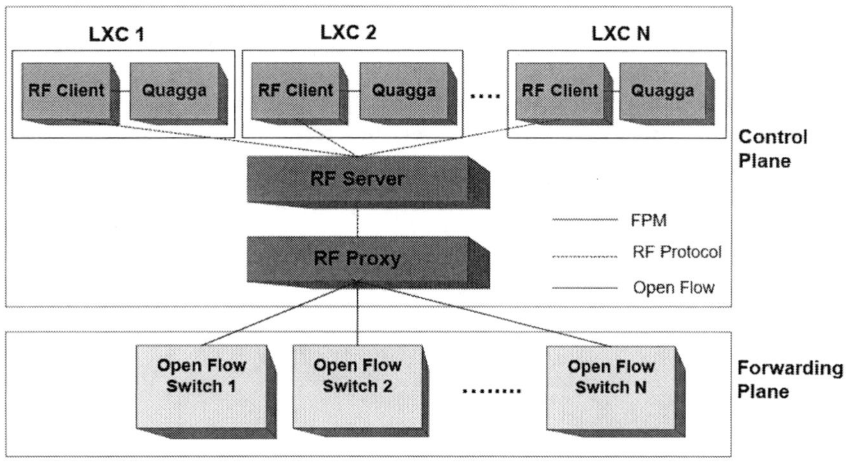

Figure 3. SDN controller workflow

The function of each block in figure 3 are as follows:

a) RF client functions as a daemon in the VM, it's a function to detect the changes in Linux ARP and routing tables. When there is a routing update, the routing information is sent to the RF Server.
b) RF Server is a standalone application that holds central control connected to the RF Client. Where RF server makes mapping between RF clients with interfaces, the appropriate switches, and ports.
c) RF Proxy is an Application Controller (POX controller) that forward protocol policies from the RF Server to the data plane.

2.2 Routing Protocol

Open Short Path First (OSPF) is a routing that uses the link-state algorithm, which works effectively on large scale networks. OSPF requires information on the weighting of each link on the network obtained from neighboring nodes. Routing Information Protocol (RIP) is a type of routing protocol that is effective in small scale networks and uses distance vector routing protocols as a reference in determining its path. The OSPF protocol is one of routing Interior Gateway Protocol (IGP) for the internet that is used to distribute routing information across interconnected networks. OSPF collects information for all paths to be passed from an existing router. To determine the shortest path, OSPF requires weighting each link on the network. RIP is one type of routing protocol that is effective in small scale networks and uses distance vector routing protocols as a reference in determining its path. Currently, RIPv2 also supports Ipv6 [19].

2.3 File Transfer Protocol (FTP)

File Transfer Protocol (FTP) is an internet protocol that runs at one application layer, it's a function to exchange data between servers and clients in a network. FTP before sending data, communication session will occur first in the transport layer using TCP. Usually, FTP uses two ports, port 20 and 21. Port 21 is used to "listen" and port 20 for the data line. FTP provides users access to a large number of concurrent accesses by computer. In this research, the FileZilla application is used to implement the FTP server and client.

IOP Conf. Series: Materials Science and Engineering **982** (2020) 012026 doi:10.1088/1757-899X/982/1/012026

3. System Model

3.1 Network Topology Design

The hardware and software are used in this study consisted of 2 laptops, each of which was used as a controller and FTP client, 5 switches as data planes, and some software with the specifications are listed in table 1.

Table 1. Device specification

Component	Specification
Controller device	Intel(R) Core(TM) i5-6200HQ CPU @ 2.30GHz, RAM 12 GB, NVIDIA GeForce 920M
Forwarding plane	6 Switch TP-Link WR1043ND v1
Router	5 Mikrotik RB951Ui-2^{nD}, RAM 64 MB
Operating system	Ubuntu 12.04 LTS 32 bit & Ubuntu 12.10 Quantal 64 bit (Mininet)
Virtual machine	Vmware
Emulator	Mininet
Openflow Controller	POX controller
FTP software	FileZilla
Routing engine	Quagga

The first network topology design is implemented on a conventional network using five Mikrotik RB951Ui routers that function as data planes as well as control planes. Figure 4 shows a FileZilla-based FTP server and FTP client on the other side. Topology is designed to provide several routing alternatives, so the OSPF and RIP routing protocols could determine the best route. On conventional networks, it does not have a centralized controller like the SDN network, because the device itself regulates where the data flow must be sent.

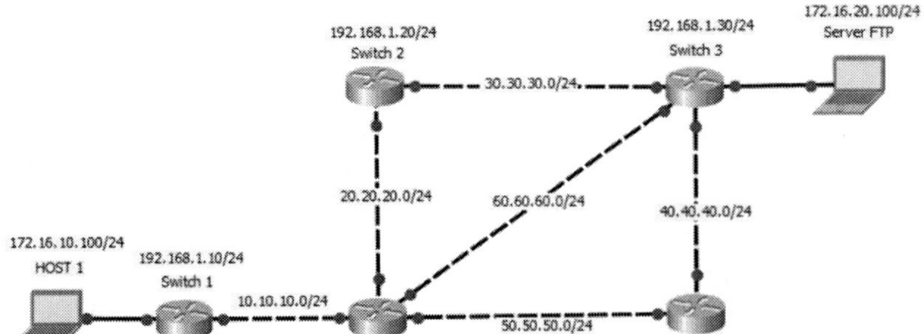

Figure 4. Conventional Network Topology

In the second scenario is designed SDN-based network topology. In figure 5, a computer is functioning as a control plane with route flow software installed in it. The data plane function uses 5 conventional switches that are adjusted to support SDN. Openflow is used to connect between the control plane and the data plane. Unlike the conventional network, SDN simply uses a switch as a data plane for a different network.

IOP Conf. Series: Materials Science and Engineering **982** (2020) 012026 doi:10.1088/1757-899X/982/1/012026

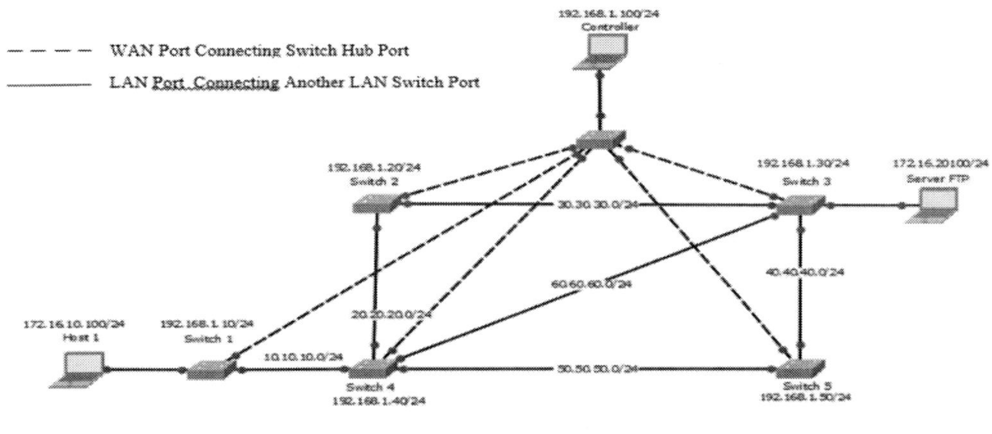

Figure 5. SDN Network Topology

3.2 Measurement Parameters

The examination in this study uses the ITU G.1010 standard as in table 2. The measured test parameters are throughput, delay, and packet loss. Using FTP services from server to client. The test scenario is comparing OSPF and RIP routing protocols in conventional and SDN networks. Network conditions are adjusted to real conditions by adding background traffic from 0 Mbps to 100 Mbps. The test is conducted 30 times for each measurement point and the average measurement results are taken.

Table 2. Measurement parameters [20]

No	Parameter	ITU-T G.1010
1	Throughput	NA
2	Delay	Preferred <15s, acceptable < 60 s
3	Jitter	NA
4	Packet Loss	0%

Throughput is the rate of data that is successfully delivered to the receiver over a communication link during a certain observation time. Throughput is measured in bits per second (bps). The higher throughput means better network performance. The formula to measure the throughput is written as below.

$$Throughput = \frac{Rs}{Lo} \quad (1)$$

Where:
Rs = The data that are successfully received by the receiver.
Lo = length of observation time.

Delay is the time taken by data to travel across the network from sender to receiver. Delay is measured in the second unit. The higher delay cause quality of service in the network more unacceptable The formula to measure the delay can be written as below.

$$Delay = Tr - Ts \quad (2)$$

Where:
Tr = The time when data is received
Ts = The time when data is sent

IOP Conf. Series: Materials Science and Engineering **982** (2020) 012026 doi:10.1088/1757-899X/982/1/012026

Packet Loss is defined as the percentage of data fail to reach it's intended destination during transmission. Several reasons cause packet loss in a communication network, such as network congestion, software bugs, the problem with network hardware, and security threat. Packet loss and throughput are opposite on each-other, high packet loss causes low throughput at receiver. The formula to measure packet loss can be written as below.

$$Packet\ loss = \frac{Pt - Pr}{Pt}\ x\ 100\% \quad (3)$$

Where:
Pt = Transmitted data from sender
Pr = Received data at the receiver

3.3 Port Configuration

The port on the TP-Link WR1043ND v1 wireless router device is configured by separating the ports that are connected to the controller, port for configuration, and port for the Local Area Network. The configuration is conducted on WAN port which has been set before. The configurations in open vSwitch are dpid, IP controller, OpenFlow version. In this research using Open vswitch version 2.3.90, while the Open Flow (OF) version supports OF 1.0, OF1.1, OF1.2, or OF1.3. For the OpenvSwitch configuration as shown in Figure 6.

```
root@OpenWrt:~# ovs-vsctl add-br s1
root@OpenWrt:~# ovs-vsctl set bridge s1 other-config:datapath-id=0000000000000005
root@OpenWrt:~# ovs-vsctl set-controller s1 tcp:192.168.1.100:6633
root@OpenWrt:~# ovs-vsctl set bridge s1 protocols=OpenFlow10
root@OpenWrt:~# ovs-vsctl add-port s1 eth1.1
root@OpenWrt:~# ovs-vsctl add-port s1 eth.1.3
root@OpenWrt:~# ovs-vsctl add-port s1 eth1.4
root@OpenWrt:~# ovs-vsctl add-port s1 eth1.5
```

Figure 6. OpenvSwitch Configuration

To check the OpenvSwitch configuration results, can use the "ovs vsctl show" command, then the results are like in figure 7. From figure 7, we can see some information that we can check, such as the IP controller, interface of each ethernet port that has been set up to ensure that each port runs properly.

```
root@OpenWrt:~# ovs-vsctl show
f510a9c5-0088-4023-8769-6d9d63b15a88
        bridge "s1"
                Controller "tcp:192.168.1.100:6633"
                Port "eth1.4"
                        Interface "eth1.4"
                Port "eth1.5"
                        Interface "eth1.5"
                Port "eth1.3"
                        Interface "eth1.3"
                Port "s1"
                        Interface "s1"
                                type: internal
                Port "eth1.1"
                        Interface "eth1.1"
```

Figure 7. Open Switch Configuration Results

4. Results and Discussion

This section discusses the measurement results in a predetermined test scenario. Shown in figure 8 results of SDN implementation. Network conditions are adjusted to real conditions by adding background traffic from 0 Mbps to 100 Mbps with an addition every 20 Mbps each test. Adding background traffic is using the Iperf application. The bandwidth topology is set at 100Mbps according to gigabit ethernet device specifications.

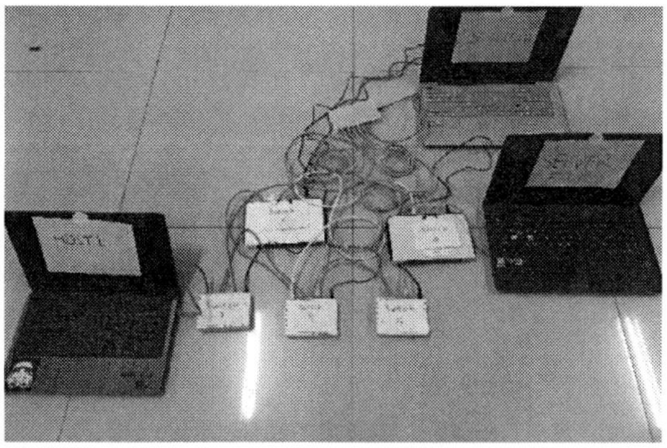

Figure 8. Results of SDN network implementation

4.1 Delay

Delay is the time of sending packets from one source point to the destination point. The delay measured in this study is a one-way delay from FTP server to the client. The measurement results of delay in figure 9 show that conditions without background traffic, the smallest delay is owned by the OSPF protocol on the SDN network that is equal to 95ms, then followed by the RIP protocol on SDN of 97 ms. Delay in OSPF and RIP in conventional networks yields the same value, which is 105 ms. Adding background traffic to all protocols causes an increase in delay, which is near linear. Overall, the smallest delay is generated by the OSPF protocol on SDN, followed by RIP on SDN, OSPF on conventional networks, and finally, RIP on conventional networks. Overall, the delay value still meets the ITU-T G.1010 standard for data services.

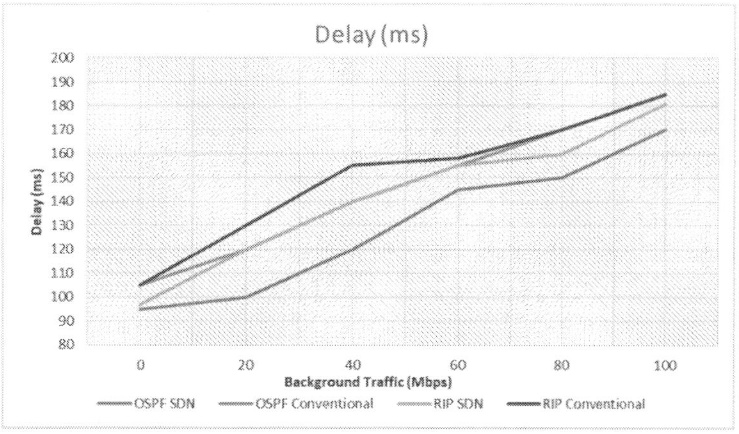

Figure 9. Delay measurement results

4.2 Throughput

Throughput is a comparison between the number of packets received per observation time. By another definition, throughput is the average speed of data received by a node in a certain observation interval. Throughput measurements are conducted by providing background traffic from 0 Mbps (without background traffic) to 100 Mbps with an increase every 20 Mbps. In the initial condition without background traffic, OSPF protocol in SDN provides the greatest throughput value, which is 93.4 Mbps. followed by RIP in SDN that is 84.2 Mbps, OSPF in conventional networks that are 82.5 Mbps, and RIP in conventional networks that are 80.5 Mbps. Adding background traffic causes a reduction value of throughput for all conditions. However, on 100Mbps background traffic, OSPF in SDN still has the best throughput value followed by RIP in SDN. OSPF and RIP in conventional networks have almost the same value. For overall results can be seen in Figure 10.

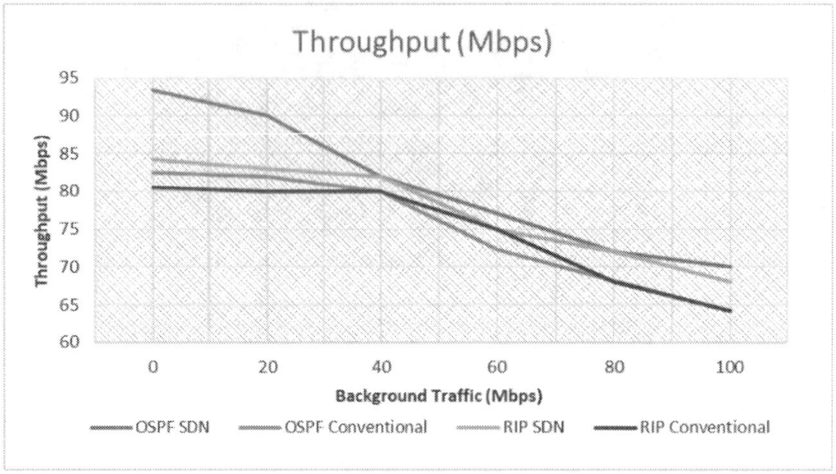

Figure 10. Throughput measurement results

4.3 Packet Loss

Packet loss is the percentage comparison between the number of packets that have failed to send and the number of packets that have been sent successfully. The greater the value of packet loss, of course, further reduce the value of network QoS. From the examination results that have been conducted for all scenarios, it is found that the value of packet loss is 0%. This means that almost no packet data is lost or failed. This reminds us that the FTP service uses TCP protocol instead of UDP. TCP has the characteristic to repeat sending failed packets until all packets arrive at the destination node.

5. Conclusions

In this study, we have succeeded in implementing the FTP service at SDN on conventional devices that exist today. The separation between the control plane and the data plane was successful and the FTP service was able to run well on SDN. QoS measurement results such as delay and throughput, show that FTP runs on SDN have a better QoS value compared to conventional networks. FTP services that use OSPF (link-state routing) routing on SDN have better results than using RIP (distance vector) on SDN. From four test scenarios, the routing protocol that has the best QoS in the sequence is the OSPF protocol in SDN, RIP protocol in SDN, OSPF protocol in conventional networks and finally RIP protocol in conventional networks.

References

[1] S. Singh and R. K. Jha 2017 A survey on software defined networking: Architecture for next generation network Journal of Network and Systems Management, vol. 25, no. 2, pp. 321-374

[2] V. Monita, I. D. Irawati, R. Tulloh 2020 Comparison of routing protocol performance on multimedia services on software defined network," Bulletin of Electrical Engineering and Informatics, vol. 9, no. 4, pp. 1612-1619

[3] L. Han, S. Sun, B. Joo, X. Jin, and S. Han 2016 QoS-aware routing mechanism in OpenFlow-enabled wireless multimedia sensor networks," International Journal of Distributed Sensor Networks, vol. 12, no. 7, pp. 1-11

[4] D. Gopi, S. Cheng, and R. Huck 2017 Comparative analysis of SDN and conventional networks using routing protocols," International Conference on Computer, Information and Telecommunication Systems (CITS), pp. 108-112

[5] E. Amiri, M. R. Hashemi, and K. R. Lejjy 2018 Policy-based routing in RIP-hybrid network with SDN controller," 4th National Conference on Applied Research Electrical Mechanical Computer and IT Engineering, pp. 1-8

[6] O. Marzuqi, A. Virgono, and R. M. Negara 2019 Implementation model architecture software defined network using raspberry Pi: A review paper," Telkomnika (Telecommunication Comput. Electron. Control., vol. 17, no. 3, pp. 1136–1141

[7] A. A. Khan, M. Zafrullah, M. Hussain, and A. Ahmad 2018 Performance analysis of OSPF and hybrid networks," International Symposium on Wireless Systems and Networks (ISWSN), vol. 2018, pp. 1-4

[8] R. Adrian, A. Dahlan, and K. Anam 2018 OSPF cost impact analysis on SDN network," 2nd Int. Conf. on Information Technology, Information Systems and Electrical Engineering (ICITISEE), vol. 2018, pp. 198-201.

[9] S. Abdallah, A. Kayssi, I. H. Elhajj, and A. Chehab 2018 Network convergence in SDN versus OSPF networks," 5th International Conference on Software Defined Systems, pp. 130-137

[10] A. Rego, S. Sendra, J. M. Jimenez, and J. Lloret 2017 OSPF Routing Protocol Performance in Software Defined Networks," *2017 Fourth International Conference on Software Defined Systems (SDS).*, pp. 131–136

[11] R. Tulloh, H. Tussyadiah, R. W. Hutabri and R. M. Negara 2018 Load distribution analysis on bipartite topology using floodlight controller," Journal of Theoretical and Applied Information Technology, vol. 96, no. 5, pp. 1238-1252

[12] N. Udayakumar, A. Khera, L. Suri, C. Gupta and T. Subbulakshmi 2018 Bandwidth Analysis of File Transfer Protocol, International Conference on Communication and Signal Processing (ICCSP), Chennai, 2018, pp. 0791-0795, doi: 10.1109/ICCSP.2018.8524583.

[13] T. Wilson, E. Abdelfattah and S. Hamada 2018 Evaluation of routing protocols with FTP and P2P," 2018 IEEE Long Island Systems, Applications and Technology Conference (LISAT), Farmingdale, NY, 2018, pp. 1-5, doi: 10.1109/LISAT.8378009.

[14] P. Wei, Z. Hong and M. Shi 2016 Performance analysis of HTTP and FTP based on OPNET IEEE/ACIS 15th International Conference on Computer and Information Science (ICIS), Okayama, 2016, pp. 1-4, doi: 10.1109/ICIS.2016.7550952.

[15] M. Karakus and A. Durresi 2017 Quality of service (QoS) in software defined networking (SDN): A survey," Journal of Network and Computer Applications, vol. 80, pp. 20-218.

[16] N. McKeown et al.2008 OpenFlow: Enabling innovation in campus networks," ACM SIGCOMM Computer Communication Review, vol. 38, no. 2, pp. 69-74

[17] M. Alsaeedi, M. M. Mohamad, and A. A. Al-Roubaiey 2019 Toward adaptive and scalable OpenFlow-SDN flow control: A survey," IEEE Access, vol. 7, pp. 107346-107379

[18] M A H. Al-Junaid et al 2019 Software Defined Networks Security: Link Failure Analysis in SDN," IOP Conf. Ser.: Mater. Sci. Eng. 557 012039

[19] G. Malkin and R. Minnear 1997 RFC 2080: RIPng for IPv6," Network Working Group, IETF, pp. 119

[20] ITU-T Recommendation G.1010 2001 Series G: Transmission systems and media, digital systems and networs: Quality of service and performance: End-user multimedia QoS categories," International Telecommunication Union, pp. 1-18

IOP Conf. Series: Materials Science and Engineering **982** (2020) 012027 doi:10.1088/1757-899X/982/1/012027

Performance analysis of erbium-doped fiber amplifier (EDFA) and hybrid optical amplifier in NG-PON2 based on TWDM-PON system

Rafiendra Bagas Aryandhika[1], Anggun Fitrian Isnawati[1], Dodi Zulherman[1,2]

[1]Fakultas Teknik Telekomunikasi dan Elektro, Institut Teknologi Telkom Purwokerto, Jl. D.I. Panjaitan No. 128 Purwokerto, Central Java-Indonesia 35147

[2]zulherman.dodi@ittelkom-pwt.ac.id

Abstract. One of the milestones in the optical access network evolution is the Next-Generation Passive Optical Network Stage 2 (NG-PON2) that has 40 Gbps as minimum upstream/downstream capacity. The optical amplifiers have to implement to combat the power degradation in the transmission line. This research aims to analyze the performance of EDFA boosters and hybrid optical amplifiers, which are implemented on TWDM-PON systems based on NG-PON2. Our work used 80 Gbps data speed, which each channel had 10 Gbps on eight channels with a channel space of 100 GHz. The optical link used in this simulation was 60 km with the Erbium-Doped Fiber Amplifier (EDFA) and Hybrid Optical Amplifier (HOA). The performance analysis is based on the variation of ONU and amplifier. Based on the results, the best value in the downstream is the transmission line using EDFA with the received power of -19.184 dBm, Q Factor of 7.975, and BER of 5.852×10^{-16}. In upstream transmission, the best value obtained was a hybrid optical amplifier with a received power value of 9.025 dBm, Q Factor of 69.64, and BER of 0. The optical amplifier increased the system performance when implemented on TWDM-PON systems based on NG-PON2.

1. Introduction

Recently, information and communication technology has grown exponentially [1] forced by the customer needs increasing for the high-speed and huge capacity internet. The passive optical network (PON), for instance, EPON and GPON, is the most widely deployed all over the globe [1] to bring high-speed internet. The first generation PON is based on Time Division Multiple Access (TDMA) by providing a 1 Gbps upstream rate and a 2.4 Gbps downstream rate. NG-PON1 can increase data speeds up to 10 Gbps for both streams. There are two scenarios for NG-PON1 technology, namely EPON that is deployed to XG-EPON and GPON used to XG-PON. For enhanced GPON can be deployed to XG-EPON, which is another potential path that can be considered. However, NG-PON1 will not meet the demand for bandwidth and future Quality of Service (QoS) requirements. Therefore, research is investigating the options for NG-PON2 with some of the technologies that NG-PON2 uses with standardization institutions, i.e., International Telecommunications Union (ITU) and Institute of Electrical and Electronics Engineers (IEEE) working together with Full-Service-Access-Network [2]. The Time and Wavelength Division Multiplexing Passive Optical Network (TWDM-PON) can be recommended for the design of multiplexing NG-PON2. This technology develops with the latest developments of PON technology that can transmit information with a higher speed of more than 40 Gbps for the downstream and 10 Gbps for the upstream side [3].

In recent years, several researchers that analyze the TWDM-PON system in access networks have been published [3,4,5,6]. Hidayat [3] in 2018 examined the impact of the Erbium-Doped Fiber Amplifier (EDFA) and Raman Optical Amplifier (ROA) in the NG-PON2 system using 4-channels with the 40 Gbps downstream and 10 Gbps upstream. They employ a splitting ration 1:256 in TWDM-PON with a booster amplifier for downstream and a pre-amplifier for upstream. In this work, the EDFA and the ROA were observed by Q-factor and BER parameters. Based on the discussion, the system with EDFA had a better performance than the system that used ROA as an optical amplifier. Pratiwi [4] in the 2018 simulated TWDM-PON system to evaluate the performance of the NG-PON2-based system with two variations of bitrate, several variations of the length of the optical link, three scenarios of network capacity and the scenario of the optical amplifier. Based on the discussion, the system performance was enhanced by using EDFA as a pre-amplifier. This system can increase the capacity without the lost performance that was required in the ITU-T standard. Pertiwi [5] in 2018 study the effect of EDFA as a booster amplifier and as a pre-amplifier in the TWDM-PON based in NG-PON2. In this work, Pertiwi observed the effect amplifier for several variations of the length of an optical link and ONT capacity. The length of the optical link with the best performance was 20 km, and both access networks with 64 ONT and 128 ONT can be met the standard of network performance. In 2017, Yasyir [6] simulated the TWDM-PON based NG-PON2 system to obtain the performance of the system that uses EDFA as an optical amplifier. Based on the discussion, EDFA employment can increase system performance.

This work analyzes the performance of the TWDM-PON system using EDFA on the OLT as a booster amplifier and HOA as a pre-amplifier in the Optisystem simulator. To analyze the performance, we used a variety of the length of optical link to obtain received power, BER, and Q-factor values. For a better understanding, the rest of this paper is organized as follows. In section 2, we propose a method that uses in this work, and the results of the simulation are discussed in section 3. Finally, the conclusions are presented in section 4.

2. Proposed Method

This study used a model to analyze the performance of TWDM-PON system that uses NG-PON2. In this section, we elaborate on how simulation was conducted, the NG-PON 2-based network system design model which refers to the figure 1. The TWDM-PON system, in accordance with NG-PON 2, is composed of three main blocks, the transmitter block of Optical Line Terminal (OLT), the transmission block of Optical Distribution Network (ODN) and the receiver block in the form of Optical Network Unit (ONU). The OLT used eight channels with different wavelengths and a bit rate on each channel of 10 Gbps on the downstream and upstream sides. The ODN section used three splitting points, with a total of 64 and 128 users with the most remote transmission distance of 60 km. In this design, it was using an additional optical amplifier that aims to increase the signal receiving power. Types of optical amplifiers were EDFA and Hybrid Optical Amplifier. The OLT block consists of CW Laser, Mach-Zehnder Modulator, NRZ Pulse Generator, and Pseudo-Random Binary Sequence (PRBS). An Optical Amplifier block used EDFA and HOA, which consist of Erbium-Doped Fiber Amplifier (EDFA) and Raman Optical Amplifier (ROA), ideal mux dan pump laser array. EDFA used a length of five meters with Pump Power 100 mW. The Raman Optical Amplifier used a length of 10 km. The transmission media, which is a network connection between the central to the customer, was also called the Optical Distribution Network (ODN). This block consists of Optical Attenuator, Bidirectional Optical Fiber, and Splitter. On the bidirectional optical fiber had a reference wavelength of 1550 nm with a variation of distance from 10 km to 60 km. ONT block was called Optical Network Terminal or Optical Network Unit (ONU). On the main device at ONT is a photodetector that has the function of detecting light emitted from the transmitter as well as converting it into an electric current.

IOP Conf. Series: Materials Science and Engineering **982** (2020) 012027 doi:10.1088/1757-899X/982/1/012027

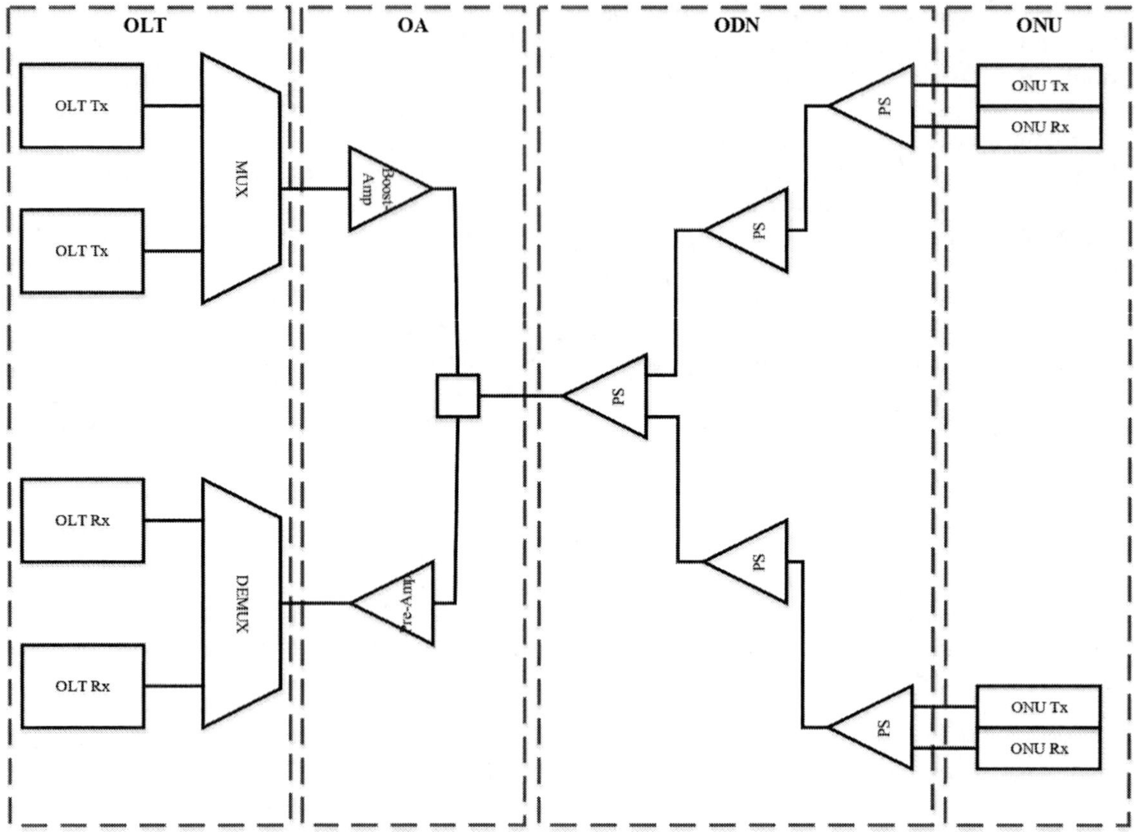

Figure 1. TWDM-PON architecture

ONT consists of Bessel Optical Filter, Photodetector APD, Low Pass Bessel Filter, 3R (Reamplification, Reshaping and Retiming), and BER Analyzer. The system configuration in Optisystem shows in table 1.

Table 1. System Design Parameter

	Parameter	Unit	Value
OLT	Bitrate in each channel	Gbps	10
	Optical launch power	dBm	6
	Channel spacing	GHz	100
	Bandwidth	GHz	20
	Number of channels	-	8
	Wavelength Downstream	nm	1596.34 – 1598.89
	Wavelength Upstream	nm	1532.68 – 1535.04
EDFA	Fiber length	m	5
	Pump power	mW	100
	Pump wavelength	nm	1480
ROA	Fiber length	Km	10
	Pump power	mW	100

Table 1. System Design Parameter

	Parameter	Unit	Value
Transmission link	Pump wavelength	nm	1480
	Optical link	Km	10 – 60
	Attenuation	dB/km	0.3
	Dispersion	ps/(nm.km)	17.46
Splitter	Instertion loss	dB	3
	Output port	-	2, 4, 8
ONT	Responsivity	A/W	0.059
	Ionization ratio	-	0.45
	Resistance	Ohm	50

3. Result and Discussion

The system performance analysis used in this research was the results of the parameters of Q-factor, Bit Error Rate (BER), and received power. The power used in this test was six dBm with optical links variation of 10, 20, 30, 40, 50, and 60 km. For a minimum value on the Q Factor optical link has the value must be < 6 to meet the eligibility standards of the transmitted signal quality. Q Factor itself has a relation with the value of BER that has a maximum value of 1×10^{-9} to be able to meet the standard.

3.1. Performance of the system design without an amplifier

In the first scenario was the simulated system 80 Gbps TWDM-PON without using amplifier on the downstream or upstream. The following are the test parameters used on the system without using an amplifier Received Power, Q Factor, Bit Error Rate (BER).

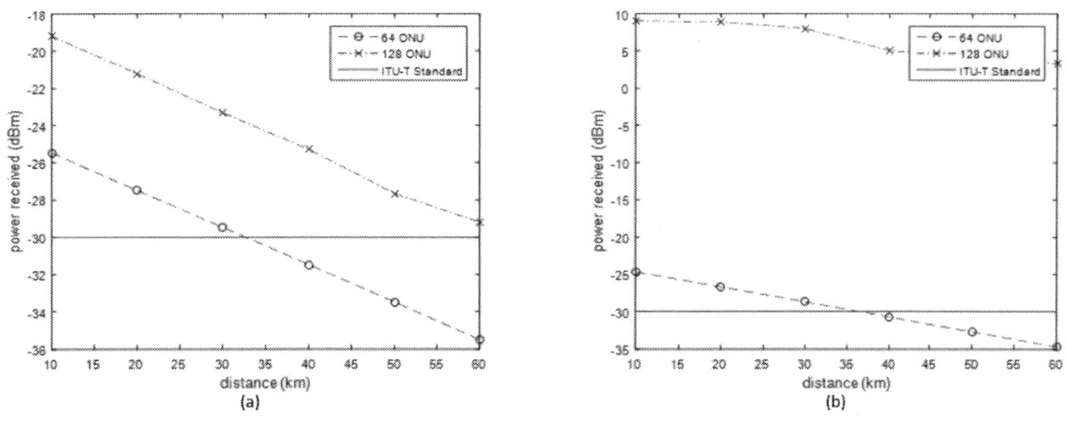

Figure 2. Received power in the system without amplifier (a) downstream received power, (b) upstream received power

Based on the results of the simulation in Figure 2 (a), the value received power for the system, without using an amplifier on the downstream side, was obtained the value of received power at 64 ONU with the highest value received power of -25.459 dBm at a distance of 10 km while for the smallest value of -29.458 dBm at a distance of 30 km. Whereas at 128, ONU obtained the highest received power value from the distance of 10 km, amounting to -19.184 dBm to a distance of 60 km, amounting to -29.188 dBm. On both, the ONU has already qualified the value of received power from 64 ONU only up to 30 km while the 128 ONU qualifies from a distance of 10 km to 60 km. In this case, the enormous received power value is influenced by a long-range Link. Connection received power with distance compared to the upside, which is the length of a link, then the value received power obtained will be smaller and vice versa.

Based on the simulation results in Figure 2 (b), for the received power value without using amplifier on the upstream side, where obtained the value of received power at 64 ONU without using amplifier obtained the highest value received power of-24.686 dBm on the distance of 10 km while for the smallest value of-28.690 dBm at a distance of 30 km. While at 128 ONU obtained, the highest received power value from the distance of 10 km is 9.025 dBm to a distance of 60 km is at 3.314 dBm. On both, the ONU has already qualified the value of received power from 64 ONU only up to 30 km while the 128 ONU qualifies from a distance of 10 km to 60 km. In this case, the large received power value is influenced by a long-range Link. Connection received power with distance compared to the upside, which is the length of a link, then the value received power obtained will be smaller and vice versa.

Based on the simulation that has been done with the first scenario, the system without using an amplifier on the downstream side and upstream side obtained the Q Factor result as follows in figure 3.

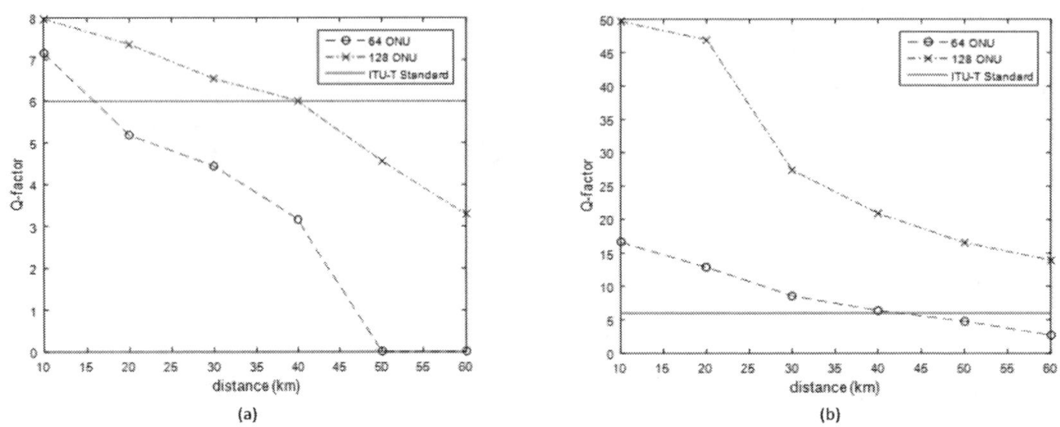

Figure 3. Q-factor in the system without amplifier (a) downstream Q-factor, (b) upstream Q-factor

The Q Factor value gained in simulated results with a bit rate of 80 Gbps on the downstream side, indicating that the scenario I without using the amplifier at 64 ONU obtained a value of 7.159 at a distance of 10 km where the distance is fulfilled Terms of Q factor while the other distance does not meet its standard value. At the same time, the 128 ONU obtained the highest score of 7.975 at a distance of 10 km and for the smallest value of 6.011 at a distance of 40 km. Based on the results of the Q factor shown on the chart, for users, 64 ONU only meet the standard at a distance of 10 km while for users 128 The ONU meets the standard ranging from 10 km to 40 km. This is because the increasing number of users used then for the higher Q factor value will be further the resulting distance. The Q Factor value obtained in the simulated results with a bit rate of 80 Gbps on the upstream side, indicating that scenario I without using the amplifier at 64 ONU obtained the highest value of 16.635 at a distance of 10 km and the smallest value obtained is of 6.368 at a distance of 40 km. Where at a distance of 10 to 40 km already qualified Q factor while the other distance does not meet its standard value. In comparison, the 128 ONU obtained the highest score of 49.624 at a distance of 10 km and for the smallest value 13.950 at a distance of 50 km. Based on the results of the

Q factor shown on the chart, for users 64 ONU that meet Q factor standards at a distance of 10 km to 40 km Whereas for users 128 ONU who meet Q factor standard from ranging from 10 km to 60 km. This is because the more number of users used then for the higher the Q factor value will be further the resulting distance.

Bit Error Rate (BER) is a comparison of the bit rate loss that will occur in a digital transmission system that has a standard parameter value of 10-9. If the value system generated on BER Analyzer is smaller to generate the value of the BER, then otherwise the system is not good because it is less qualified value is generated. Based on the simulation of the scenario, I obtained a result as follows in figure 4.

The value of the simulated results in the scenario I in the user 64 ONU obtained a value of 3.423 x 10^{-13} at a distance of 10 km while the other distance does not meet the standard value of BER. In user 128, ONU obtained the highest value of 7.764 x 10^{-10} at a distance of 40 km while for the smallest value of 5.852 x 10^{-16} at a distance of 10 km. Based on the results shown in the graph form, the user 64 ONU that meets the standard values only At a distance of 10 km while the user 128 ONU meets the standard value from 10 km to 40 km. The relationship between Q Factor and BER is inversely proportional, where the greater the Q Factor value, then the resulting BER value will be smaller. Where the smaller the value of the BER will be better with a small likelihood of an error occurring on a network. The value of the simulated results in the scenario I in user 64 ONU obtained the highest value of 9.472 x 10^{-11} at a distance of 40 km while for the smallest value is obtained at 1.675 x 10^{-62} at a distance of 10 km. At the user, 128 ONU obtained the highest value is of 1.373 x 10^{-44} at a distance of 60 km while for the smallest value of 0 at a distance of 10 km.

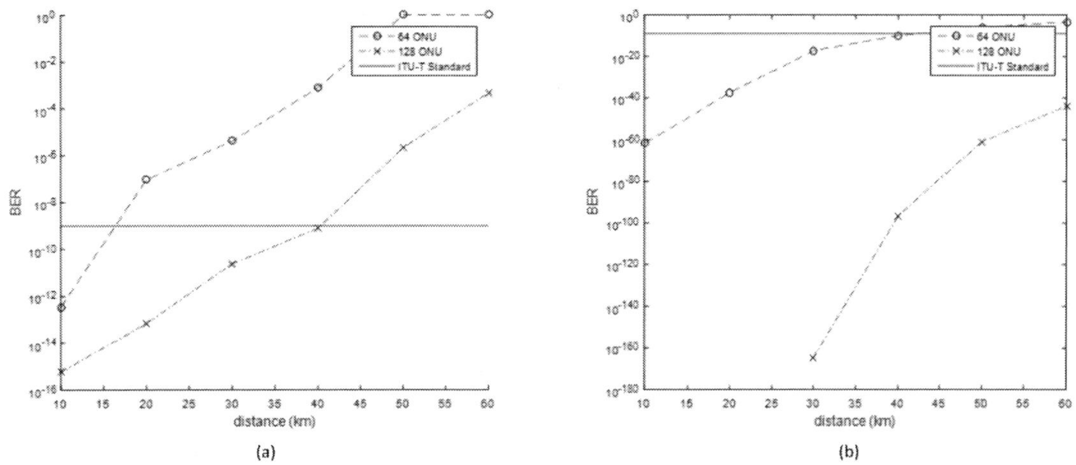

Figure 4. BER value for the system without amplifier (a) downstream BER, (b) upstream BER

Based on the results shown in the form of graphs, the user 64 ONU that meets the standard values from the range of 10 km to the distance 40 km the rest is not meeting BER standard value. In comparison, the user 128 ONU meets the standard values from a distance of 10 km to 60 km.

3.2. Performance of the system design using amplifier

The second scenario conducted simulated system 80 Gbps TWDM-PON using the EDFA amplifier on the downstream side and HOA on the upstream side. This simulation uses three divider points with the number of users, i.e., 64 and 128 ONU with a variation distance of 10, 20, 30, 40, 50, and 60 km. Where the system is simulated by using pre-defined parameters. The test parameters used in the system use the amplifiers Received Power, Q Factor, Bit Error Rate (BER).

The second scenario was using the EDFA amplifier on the downstream side and HOA on the upstream side. The Send power is the maximum left power value and the appropriate receiving power to be accepted by the sensitivity of the photodetectors to the device that is worth-30 dBm. To view the value received, power can be used component Optical Power Meter (OPM).

Figure 5 (a) is simulated results on the downstream side in the form of graphs where the value of received power in the second scenario at 64 ONU with the highest received power value of-16.174 dBm at a distance of 10 km. The lowest value is amounting to-26.295 dBm at a distance of 60 km. In user 128, ONU obtained the highest value of-19.184 dBm at a distance of 10 km while the smallest value-29.382 dBm at a distance of 60 km. In the results of this simulation seen from both users, 64 and 128 ONU meet the standard value Up to 60 km. Where the relationship is received power with the distance inversely due to the more extended a link, then the resulting value received power will be smaller as well as vice versa. Therefore, the received power measurement is crucial to identify how the sensitivity of the photodetectors still receives the receiving power.

In figure 5 (b) is simulated results on the upstream side in the form of graphs where the value of received power in the second scenario at 64 ONU with the highest received power value of 8.677 dBm at a distance of 10 km. In contrast, the lowest value is 2.201 dBm at a distance of 60 km on user 128 ONU obtained the highest value of 9.025 dBm at a distance of 10 km while the smallest value is 3.314 dBm at a distance of 60 km. In the results of this simulation seen from both users 64 and 128, ONU meets the standard value of sensitivity Minimum range of up to 60 km. where the connection received power with distance is reversed because the length of a link then the value received power will be smaller as well as vice versa.

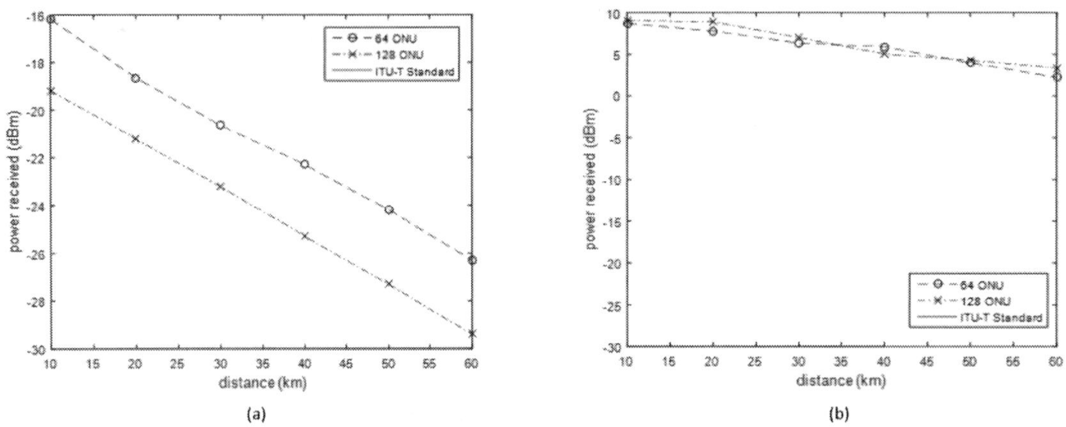

Figure 5. Received power in the system with the amplifier (a) downstream received power, (b) upstream received power

Based on the simulation that has been done with a scenario using an additional amplifier that is EDFA on the downstream side and HOA on the upstream side obtained the result of Q Factor as follows

IOP Conf. Series: Materials Science and Engineering **982** (2020) 012027 doi:10.1088/1757-899X/982/1/012027

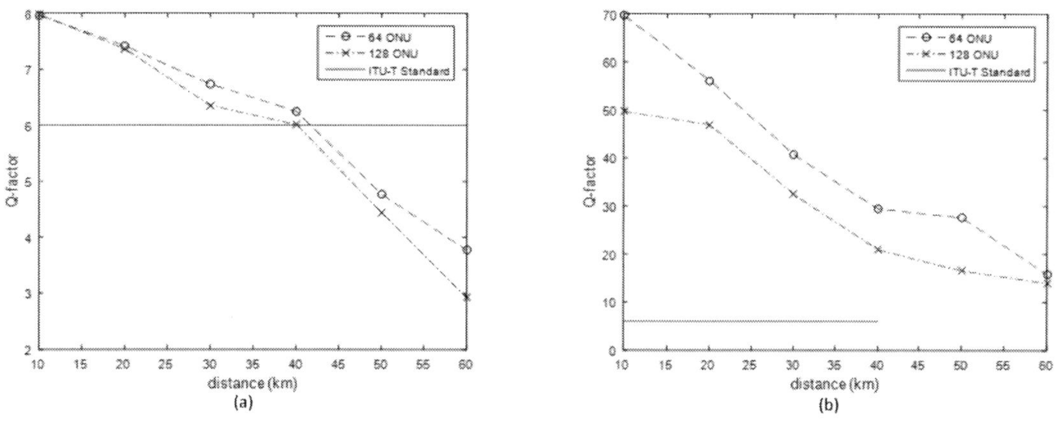

Figure 6. Q-factor in the system without amplifier (a) downstream Q-factor, (b) upstream Q-factor

Figure 6 (a) is a simulated result of Q Factor value obtained at 80 Gbps bit rate on the downstream side, indicating that the second scenario with an additional amplifier on the downstream side and upstream side. In the user, 64 ONU obtained the highest value of 7.961 at a distance of 10 km while for the lowest value of 6.248 at a distance of 40 km. While at 128 ONU obtained the highest value of 7.975 at a distance of 10 km and for the smallest value is 6.011 at a distance of 40 km. Based on the results of the Q factor shown on the chart, the user 64 ONU meets the standard at a distance of 10 km to 40 km while for users 128 ONU that meet the standard range from 10 km to 40 km more at a distance of 50 and 60 km dike Two users do not meet the minimum Q factor value. This is because the more number of users used then for the value of the larger Q factor will be further the resulting distance. Based on the simulation results of the Q Factor value obtained at 80 Gbps bit rate on the upstream side, indicating that the second scenario with the additional amplifier on the downstream side and upstream side. At 64 ONU users obtained the highest value of 69.642 at a distance of 10 km while for the lowest value of 15.783 at a distance of 60 km. While at 128 ONU obtained the highest value of 49.624 at a distance of 10 km and for the smallest value That is 13.950 at a distance of 60 km. Based on the results of the Q factor shown on the chart, user 64 and 128 ONU are compliant with the minimum Q factor standard value. This is because the more number of users used then for the value of the larger Q factor will be further the resulting distance.

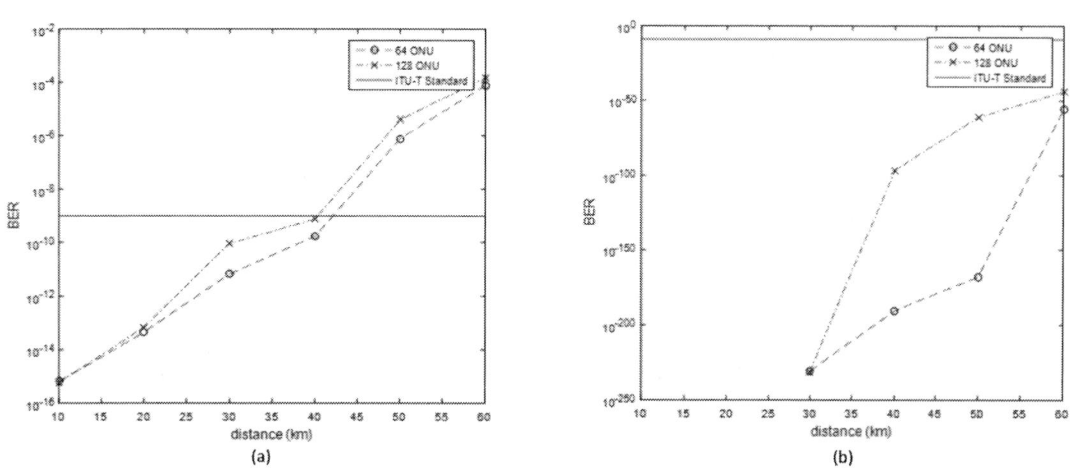

Figure 7. BER value for the system with the amplifier (a) downstream BER, (b) upstream BER

Based on the results of the simulation BER parameter on the user 64 ONU obtained the highest value of 1.699 x 10^{-10} at a distance of 40 km while for the lowest value of 6.460 x 10^{-16} at a distance of 10 km. for user 128 ONU obtained the highest value of 7.764 X 10^{-10} at a distance of 40 km while for the lowest value of 5.852 x 10^{-16} at a distance of 10 km. Judging from the graph results above can be concluded that from both users 64 and 128 ONU value is generated only up to a distance of 40 km. where the relationship between Q factor With BER is inversely proportional, which for the value of the AIR obtained the smaller then, the better signal is transmitted because of the possibility of a very small error. Based on the results of the simulation BER parameter on the user 64 ONU obtained the highest value of 1.769 x 10^{-56} at a distance of 60 km while for the lowest value of 0 at a distance of 10 km. for user 128 ONU obtained the highest value of 1.373 x 10^{-44} at a distance 60 km whereas for the lowest value is 0 at a distance of 10 km. Judging from the above graph, results can be concluded that from both users, 64 and 128 ONU values are produced from a distance of 10 km to a distance of 60 km. Where for the value of the obtained, the smaller, The better the signal is transmitted due to the possibility of a tiny error? This is because the value of the Q-factor in both users has a significant value where the value of the value will affect the value of the given BER will be smaller, which allows the lack of errors.

4. Conclusion

Based on the result of simulation and discussion in the previous section, the employment of optical amplifiers in TWDM-PON based on NG-PON2 can increase the system performance in all parameters. The proposed system meets the ITU-T standard for BER and Q-factor both system 64 ONU and system 128 ONU. Based on the standard, the value of the Received Power parameter obtained the best result in the downstream transmission at 64 ONU with a value of -16.174 dBm and 128 ONU with a value of -19.184 dBm. For upstream transmission at 64 ONU obtained the best value of 8.677 dBm, and at 128 ONU generated a value of 9.025 dBm. Based on the standard value of the parameter Q Factor obtained the best result is downstream transmission at 64 ONU with a value of 7.961, and at 128 ONU obtained a value of 7.975. For upstream transmission at 64 ONU obtained the best value of 69.642, and at 128 ONU produced a value of 49.624. Based on the standard value of the Bit Error Rate (BER) parameter obtained the best results downstream transmission at 64 ONU with a value of 6.46 x 10^{-16} and at 128 ONU obtained a value of 5.85 x 10^{-16}. For upstream transmission at 64 ONU obtained the best value of 0 and at 128 ONU generated value of 0.

5. Acknowledgment

The authors thank goes to Lembaga Penelitian dan Pengabdian Masyarakat (LPPM), Institut Teknologi Telkom Purwokerto for funding this project.

References

[1] A.O. Maria, M.Y. Pinto, G. V. Arevalo, “Experimental evaluation of a NG-PON2 Network” in *2018 IEEE Third Ecuador Technical Chapters Meeting (ETCM)*, Cuenca, 2018

[2] M. A. G. Huda Saleh Abbas, “RMIT University Melbourne,” The Next Generation of Passive Network: A Review.

[3] A. Hidayat, Perbandingan Performansi EDFA - ROA pada Sistem TWDM-PON Berbasis Next Generation Passive Optical Network Stage 2, 2018.

[4] E. Y. D. Pratiwi, Simulasi Sistem TWDM-PON Menggunakan Hybrid Optical Amplifier pada Next Generation Passive Optical Network Stage 2 (NG-PON2), 2018.

[5] L. H. Pertiwi, Pengaruh EDFA sebagai booster amplifier dan pre-amplifier pada sistem 40 Gbps TWDM-PON berbasis NG-PON2, 2018

[6] M. Yasyir, Simulasi dan analisis pengaruh EDFA pada sistem 80 G TWDM-PON berbasis Next Generation Passive Optical Network Stage 2, 2017

[7] G. Keiser, OPTICAL FIBER COMMUNICATIONS, Singapore: Mc Graw Hill Higher

Education.
[8] R. Kjaer, P. Jeppesen, L. K. Oxenlowe and B. Palsdottir, Raman Amplification in Optical Communication Systems, 2008.

IOP Conf. Series: Materials Science and Engineering **982** (2020) 012028 doi:10.1088/1757-899X/982/1/012028

Performance Analysis of DDM Method to Reduce The Doppler Effect on Vehicular

Muhammad Ardi Rahmadiansyah[1,2], **Titiek Suryani**[1,3]

[1]Department of Electrical Engineering, Faculty of Intelligent Electrical and Information Technology,
Institut Teknologi Sepuluh Nopember, Surabaya, Indonesia

[2]ardi.rahmadiansyah16@mhs.ee.its.ac.id, [3]titiks@ee.its.ac.id

Abstract. VANET is an innovation that supports vehicle to vehicle and vehicle to infrastructure communication. The main application of VANET technology is to realize better and safer traffic management. However, due to the high mobility of the vehicle, this technology causes the Doppler effect, when the spectral signal enlarges in the OFDM system. The Doppler effect occurs due to the influence of the use of Rayleigh fading vehicle-to-vehicle channels on the VANET environment and the effect of changing channel movement conditions causing time-varying channels. It also triggers the spread of Doppler to be larger than the symbol duration. This certainly can disrupt the orthogonality of the OFDM system which is the basis of VANET communication. The effect is the occurrence of ICI (inter-carrier-interference) causing the performance of OFDM to decrease. This research proposes the Direct Development Method (DDM) scheme, a technique to compensate for the Doppler effect which is different from the general scheme. In this study, the scheme used is to focus the corrector on the data frame before sending it. The simulation is carried out by testing the modulation which refers to the IEEE 802.11p standard with 4 different vehicle speed conditions namely 24, 64, 144, and 294 km/h. Analysis of the performance of the application of the DDM scheme on modulation based on the effect of the speed provided gives better results compared to the modulation performance without the compensation of the Doppler (Non-DDM) or conventional effects on the SNR requirement side to achieve the BER 10^{-3} target.

1. Introduction

Vehicular ad-hoc network (VANET) is part of the development of ITS technology (Intelligent Transport System) which is presented to improve safety and comfort in driving. Until now, the development of communication quality between vehicles continues to be considered and formed a standard of wireless communication in a vehicle or standard environment Wireless Access for Vehicular Environment (WAVE) to support vehicle to vehicle (V-to-V) and vehicle to infrastructure (V-to-I) communications that work at a frequency of 5.885 GHz as the basis of an Intelligent Transport System (ITS) [1]. The WAVE standard consists of two standards, IEEE 1609 and IEEE 802.11p. IEEE 1609 relates to standardization at the top layer and IEEE 802.11p explains all operations at the Medium Access Control (MAC) and physical (PHY) layers. The machinery used at the base layer for VANET communication is Orthogonal Frequency Division Multiplexing (OFDM) [2].

Until now, VANET has constraints on the performance side, namely when two or more vehicles communicate with each other with high mobility, with the condition of the vehicles moving away from or near each other resulting in a Doppler or Doppler Shift (DS) frequency shift. It has an impact on the transmitted signal in the receiver side that will shift or cause the orthogonality of the OFDM system to decrease. This causes many errors so that the system performance decreases [3]. The Doppler frequency shift can affect carrier frequency, subcarrier frequency, and symbol duration on OFDM systems, resulting in decreased data performance received [4].

Based on some researches on techniques or models of Doppler effect compensation as a challenge to reduce problems in VANET, the Doppler effect compensation can refer to schemes in three domains, namely the time, frequency, and space domain [5]. One of the Doppler frequency shift coverage in the frequency domain is the bandpass modulation scheme and the coding of the baseband used.

Albarazi et al., (2011) analyzed the compensation of the Doppler effect on VANET by proposing adaptive modulation schemes with variations in channel codes in various baseband modulations by allocating power and selecting appropriate rainy modulation to overcome the fading of signals that have deteriorated when arriving at the receiver's side. So if a system needs a particular BER, the power sent has to be estimated based on channel conditions or designated to each subcarrier considering the channel circumstances. For instance, BPSK and QPSK, which are categorized as low-order modulation, are more resistant to noise but smaller in sending the amount of data. In contrast to 16-QAM high order modulation, the data transmission rate is large but susceptible to noise [6].

Novfitri, (2014) [9] mentioned that the adaptive modulation used power allocation counterfeited on the OFDM system. As a result, adaptive modulation methods provide excellent performance for several small symbols. It performs an adaptive transmission scheme on V-to-V with a calculation of three-channel circumstances [10]. In this study, variations speed is used in those three conditions. Thus, variations in V-to-V conditions with higher speeds are required to examine the adaptive modulation performance. V-to-V communication reproduction with a correlated double-ring channel model has been conducted in the presence of Rayleigh and Rician fading at various speeds and a different amount of scatterer. The channel response results show the correlated double-ring channel model that can be used in the high-speed environment in-vehicle communication [11].

The study describes the Direct Derivation Method (DDM) method to solve Doppler shifts on VANET. Here, the Doppler shift was exposed up to 1400 Hz by modulation scheme choice. Doppler shifts can be overcome by choosing the best modulation scheme outcome. Each of them then correlated to a compensation arrangement without a Doppler shift.

This research will analyze the implementation of the Direct Development Method (DDM) with the OFDM system in the VANET environment with V-to-V Rayleigh fading channel conditions. This method is used as compensation for Doppler shift conditions before the data is transmitted. Direct Development Method (DDM) is a Doppler shift compensation technique with SymMax values established based on the amount of coherence time generated in each channel and given variable corrector to avoid the Doppler effect of the transmitted data, so the value of the Doppler shift can be corrected. Usage testing of this method is carried out by applying modulation in accordance with the rules of IEEE 802.11p as seen from BER on the SNR of several conditions that indicate the performance of the proposed system.

This paper is divided into several sections. Section II explains the planned model system and the derivation of the mathematical representation of channel models. Elaboration of DDM simulation outcome arrangement in OFDM system over V-to-V Rayleigh fading channel is presented in Section III. Ultimately, Section IV concludes this research.

2. Model System

2.1. OFDM for VANET Model System

IOP Conf. Series: Materials Science and Engineering **982** (2020) 012028 doi:10.1088/1757-899X/982/1/012028

VANET technology has been used in this research, OFDM which supports IEEE 1609 and 802.11p which is made specifically for DSRC message types. Figure 1 shows the DSRC channel design consisting of SSH and CCH channels [7]. In this figure, the SSH allocation channel is divided into two requesting inter-vehicle communication applications (V-to-V) set on channel 172. Meanwhile, communication between vehicles and telecommunications infrastructure (V-to-I) is established on channel 184 so that it can be moved between V-to-I and V-to-V communication. The V-to-V service places two distribution blocks on channels 174 and 176. Furthermore, channels 180 and 182 are for the placement of the V-to-I service channel. Each service channel is provided by the control. Channel 178 is trying to improve safety and for parameters of the 802.11p standard as shown in table 1.

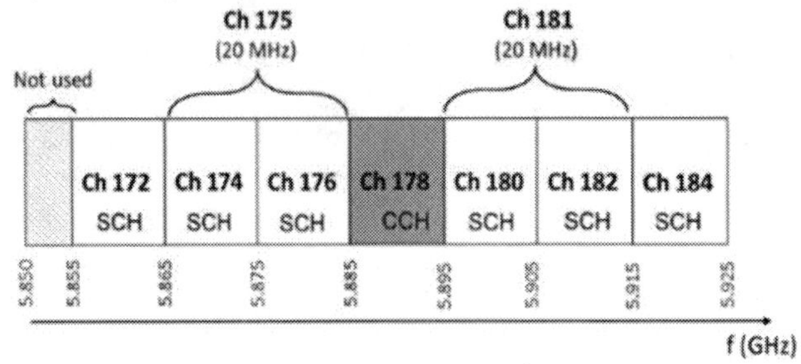

Figure 1. DSRC allocation channel

The carrier frequency used is 5.885 GHz with a bandwidth of 10 MB. The CCH and SCH channels have periods with intervals of less than 100ms. Where the division of time in the CCH and SCH intervals is divided based on Guard Interval (GI) of 4ms, so 27 Mbps is the maximum data speed that can be relayed through this innovation. The concept of time division in the CCH and SCH interval can be seen in Figure 2

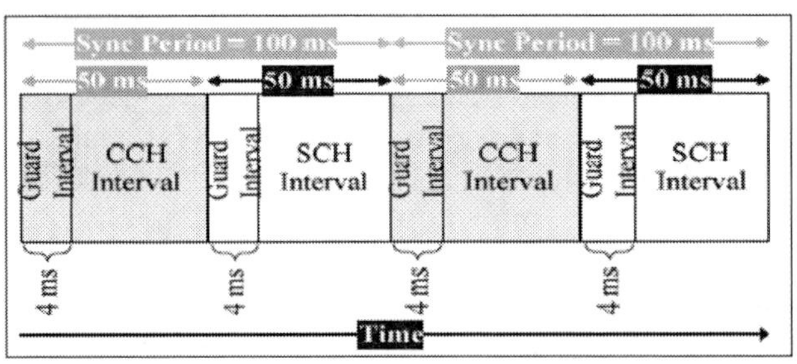

Figure 2. Division of time into CCH and SCH interval

Table 1. Physical layer parameter 802.11p

Parameters	DSRC 802.11p
Information Data Rate	3 – 27 Mbps
Bit rate (Mb/s)	3, 4.5,6,9,12,18,24,27
Modulation mode	BPSK, QPSK, 16QAM, 64QAM
Code rate	1/2, 2/3,3/4
Number of subcarriers	52
OFDM Symbol duration	8μs
Guard time	1.6μs

FFT Period	6.4μs
Preamble duration	32μs
Bandwidth	10 MHz
Subcarrier spacing	0.15625 MHz

Based on the outline of this research, the system modeling shown in Figure 3, about the OFDM system for suppliers, is implemented with the application of DDM. The transceiver block is based on IEEE 802.11p which uses OFDM waves at a carrier frequency of 5.885 GHz. The wireless channel adapts the condition of the cellular channel to cellular using a correlated double ring from changes in the speed of both automobiles adjusted by Doppler.

The use of the Rayleigh fading V-to-V channel is added to the channel to see the performance resistance of the proposed system. Additionally, AWGN noise jamming is added after passing through the channel because at that time the signal condition is weak. Channel Estimator block containing channel state information (CSI) which functions as a switch modulator as a form of automation response to the effects of the Doppler shift experienced. Channel estimation is considered perfectly calculated at the receiver before going through the system block. On the receiver OFDM side, the process occurred in contrast to the previous process, the data signal that has passed through the channel will be converted again into parallel data according to the specified subcarrier value. Next, the CP data is separated by the original data and through the demodulator, the FFT process. Then bandpass demodulation works according to the input of the channel estimator. The results obtained by demapping then become serial data in the form of bits which are then compared with the input data on the transmitter to see the value of the bit error rate (BER).

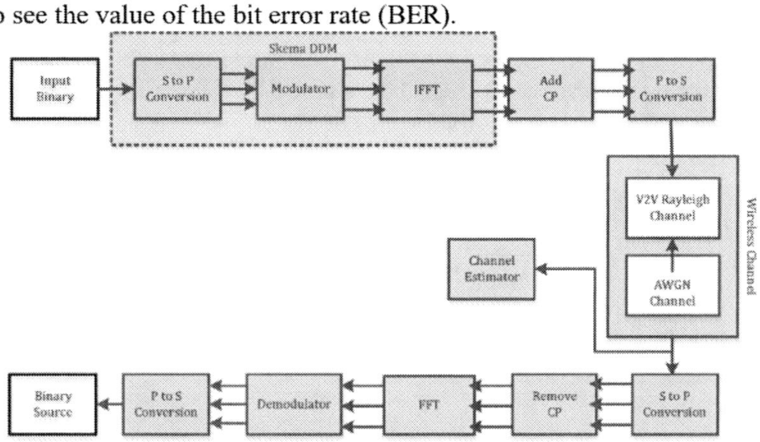

Figure 3. Transceiver OFDM using DDM scheme.

2.2. Doppler Frequency Shift

Channel condition in vehicular communications is dynamic or time-varying [7]. In the planned system can experience the Doppler shift to conduct the modulation degradation in OFDM. The amount of the Doppler shift is related to relative motion between the transmitter, receiver, and vehicle speed [8]. The Doppler shift (f_d) can be computed by this formula :

$$f_d = f_c \frac{v}{c} \cos\alpha \tag{1}$$

where f_c is source frequency of DSRC at 5.885 GHz, v is the difference of acceleration between vehicles, c indicates a speed of light $3\times10^8 m/s$, and $\alpha \in [0, \pi]$ denotes the angle of the velocity vector. To ensure the maximal Doppler shift achievement, it should happen when $\alpha=0$. Doppler shift is expressed by normalized maximum Doppler shift (f_d) In the MATLAB simulation process. It can be calculated by comparing the division of the Doppler shift value with the subcarrier spacing (F_s)

$$V(t)=\sum_{i=1}^{N_s-1} A_i \cos(\omega_i t+\phi_t) \quad (2)$$

A_i is the value of the amplitude of the signal V, $\omega_i t$ represent the angular frequency ϕ_t is the phase of the i - subcarrier, N_s shows the number of subcarriers used in OFDM, then each subcarrier must be orthogonal to one another. This can be obtained in the condition of $f_1=\omega_i 2\pi$, which is the integer multiplication of $1/2T$, with T (the period of the data) and f_1 (the frequency range with a value of $R_S=1/T$).

If we proceed to equation (2) with processing at up converter side to produce 5,8 GHz frequency, then the formula should be

$$V_i=A_i \cos(\omega_i t+\phi_t)e^{-(j2\pi f_c)\tau_i} \quad (3)$$

if the value of $A_i \cos(\omega_i(t)+\phi(t))=S_i$ then

$$V_i=S_i e^{-(j2\pi f_c)\tau_i} \quad (4)$$

with the value of $e^{-(j2\pi f_c)\tau_i}$ and τ_i is frequency carrier and τ_I represent the carrier frequency and delay to the i_{th} component of the received signal. If the vehicle moves at a constant velocity v, and each vehicle moves straight with $\alpha = 0$ then the value of the transmitted signal is affected by the value of $\frac{v}{c}\cos\alpha$ that the equation (4) becomes:

$$V_i=S_1 e^{-(j2\pi f_c)\left(\tau_i-\frac{v}{c}\cos\alpha\right)} \quad (5)$$

$$V_i=S_1 e^{-(j2\pi f_c)} e^{(j2\pi f_c)\frac{v}{c}\cos\alpha} \quad (6)$$

By replacing equation (1) to (6), it becomes,

$$V_i=S_1 e^{-(j2\pi f_c)\tau_i} e^{(j2\pi\Delta f)} \quad (7)$$

with $j2\pi\Delta ft$ is a value from phase changing at the transmitted signal. if the value of $t=\frac{1}{4}\Delta f$, then the equation becomes

$$V_i=S_1 e^{-(j2\pi f_c)\tau_i} e^{(j2\pi\Delta f/4\Delta f)} \quad (8a)$$

$$V_i=S_1 e^{-(j2\pi f_c)\tau_i} e^{(j\pi/2)} \quad (8b)$$

With $e^{(j\pi/2)}=j$ (Euler equation) the equation becomes

$$V_i=jS_1 e^{-(j2\pi f_c)\tau_i} \quad (9)$$

From the derivation, assumption the period is ¼ of the Doppler shift, consider of coherence time then the equation becomes :

$$T_c=\frac{1}{4\Delta f} \quad (10)$$

During the delivery process of OFDM symbol, a discrete time channel impulse is defined as,

$$h(n,t)=\sum_{i=1}^{L}h_i(n)\delta(\tau-\tau_1) \tag{11}$$

2.3. *DDM (Direct Development Method) Scheme*

Direct Development Method (DDM) is a technique to reduce Doppler impact which refers to [3]. DDM uses the concepts of equations (3) to (10). This technique uses a correction system (system corrector) which will later offset the data affected by the Doppler shift on the wireless channel when transmitted steps in generating a DDM system.

From these steps, can be determined based on DSRC, the frequency used is 5.855 GHz and the maximum speed of each vehicle is 250 km/hour. And *SymMax* value refers to equation (12).

$$SymMax=\frac{Transmission\,time\,duration-GI}{Interval\ Symbol} \tag{12}$$

The key to this method is the correction of DS in frames that have been divided according to the *SymMax* Sub Period before the data transmitted results from *SymMax* even given the variable *j*. after giving corrector j, the data is transmitted on the wireless channel determined later on the wireless media, *SymMax* will even meet the factor *j* derived from the Doppler effect. After the correlation *j* is multiplied by the *j* factor of the Doppler effect, the DS can be ignored and the data matches the transmitted signal.

2.4. *Channel Model*

The channel model used in this study is the Rayleigh fading V-to-V channel which was adapted from the application of the concept of mobile to mobile channel ring [12] as shown in Figure 4. The part of the sum of sinusoid (SoS), can be written mathematically in the following equation [13]:

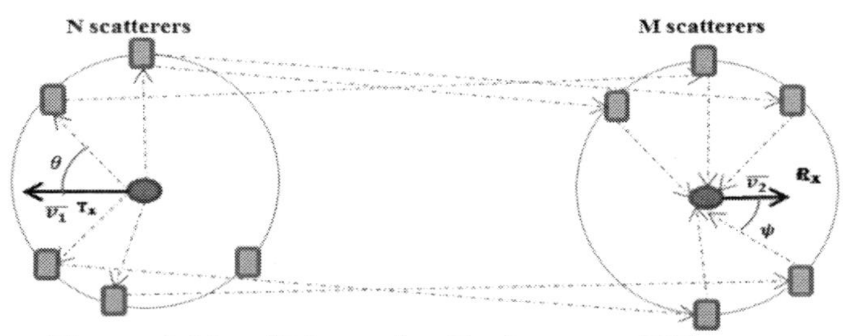

Figure 4. V-to-V Scattering Environment [13]

$$h(t)=\sqrt{\frac{2}{MN}}\sum_{m=1}^{M}\sum_{n=1}^{M}e^{\left[j\left(2\pi f_1 t\cos\alpha_n+2\pi f_2 t\cos\beta_m+\phi_{nm}\right)\right]} \tag{13}$$

$$\alpha_n=\frac{2n\pi-\pi+\theta_n}{4N} \tag{14}$$

$$\beta_m=\frac{2\left(2m\pi-\pi+\psi_m\right)}{4M} \tag{15}$$

M and N are the values of the specified scatterer M=N=8. α_n and β_m are the initial angles of departure of each scatterer part of θ_n and the angle of arrival of each scatterer part of ψ_m

$\theta_n, \psi_m, \theta_{nm}$ are random variables that are independent of each other in the range $-\pi$ to π In this simulation, because channel modeling is an SoS iteration, will be carried out 50 times to get the

IOP Conf. Series: Materials Science and Engineering **982** (2020) 012028 doi:10.1088/1757-899X/982/1/012028

appropriate results. f_1 and f_2 are normalized Doppler frequencies generated from each vehicle. For this reason, the relationship between Doppler frequency and the velocity of each vehicle can be presented in equation (13).

f_c is the carrier frequency (5.885 GHz), v is the relative speed of each vehicle, and c the speed of light $\left(3\times10^8\right)$. The value of the normalized Doppler frequency is obtained by applying the following equation

$$f_d\ norm = \frac{f_d}{\Delta f} \tag{14}$$

where Δf is the change in frequency that has been obtained at the receiver.

3. Simulation Result

In this section explains the influence of the Doppler shift in the V-to-V communication system can affect the phase of the signal being sent but with the application of the DDM scheme the effect of the shift can be suppressed, to show this, the scheme of the DDM is compared to the system without applying the DDM scheme as seen in Figure 5 below shows the graph of the comparison between DDM and Non-DDM scheme with testing on BPSK modulation with predetermined speed variations. Table 2 shows the need for the SNR value of each test which shows that the implementation of the DDM scheme still provides better performance when compared to Non-DDM modulation.

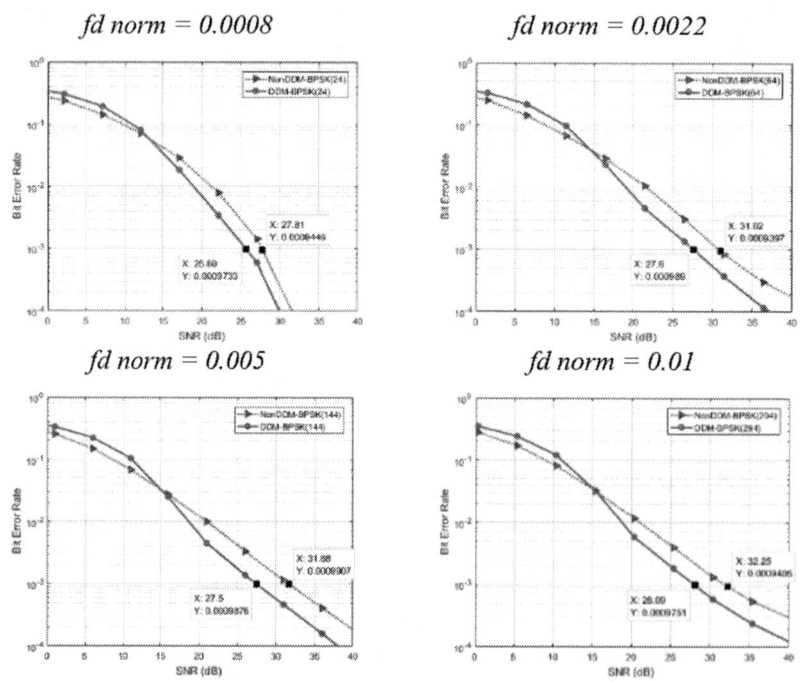

Figure 5. Comparison of BER and SNR functions in DDM and Non-DDM schemes with BPSK Modulation.

Table 2. DDM vs Non-DDM Response With BPSK Modulation Using V-to-V Channel

Mode Transmisi BPSK	SNR [dB]

f_d norm	speed	DDM	Non DDM
0.0008	24 km/h	25.69	27.81
0.0022	64 km /h	27.6	31.02
0.0051	144 km/h	27.5	31.68
0.01	294 km/h	28.09	32.25

Furthermore, in Figure 6 below shows the graph of the comparison between the DDM and Non-DDM schemes with testing on QPSK modulation with a predetermined speed variation. Table 3 shows the SNR value of each test which shows that the implementation of the DDM scheme still provides better performance when compared to Non-DDM modulation.

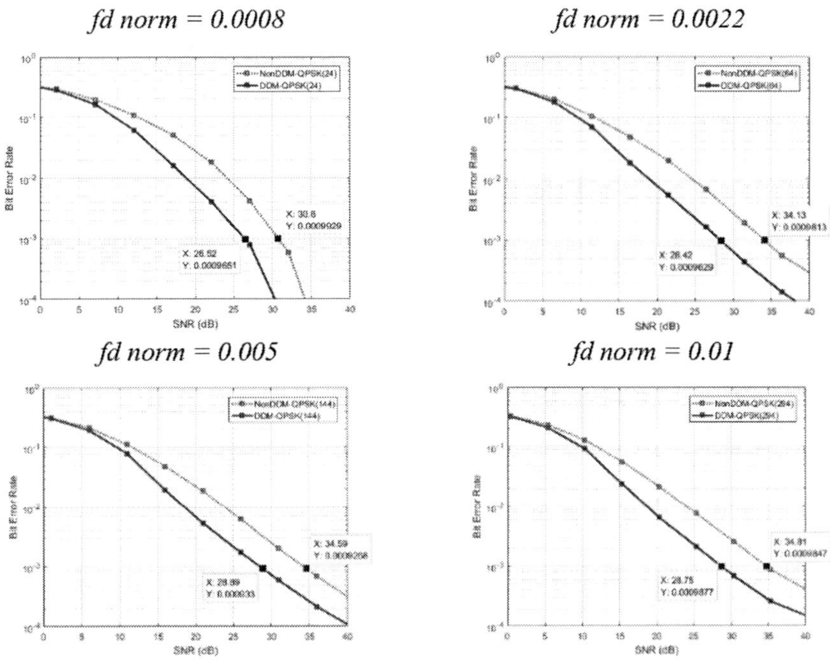

Figure 6. Comparison of BER and SNR functions in DDM and Non-DDM schemes with QPSK Modulation.

Table 3. DDM vs Non-DDM Response With QPSK Modulation Using V-to-V Channel

Mode Transmisi QPSK		SNR [dB]	
f_d norm	speed	DDM	Non DDM
0.0008	24 km/h	25.52	30.8
0.0022	64 km/h	27.6	34.13
0.0051	144 km/h	28.89	34.59
0.01	294 km/h	28.75	34.81

And the final test results of the comparison between the DDM and Non-DDM schemes can be seen in Figure 7 below, the results graph with 16QAM modulation still provides better performance when compared with Non-DDM modulation can be seen in table 4 for the SNR value of each test.

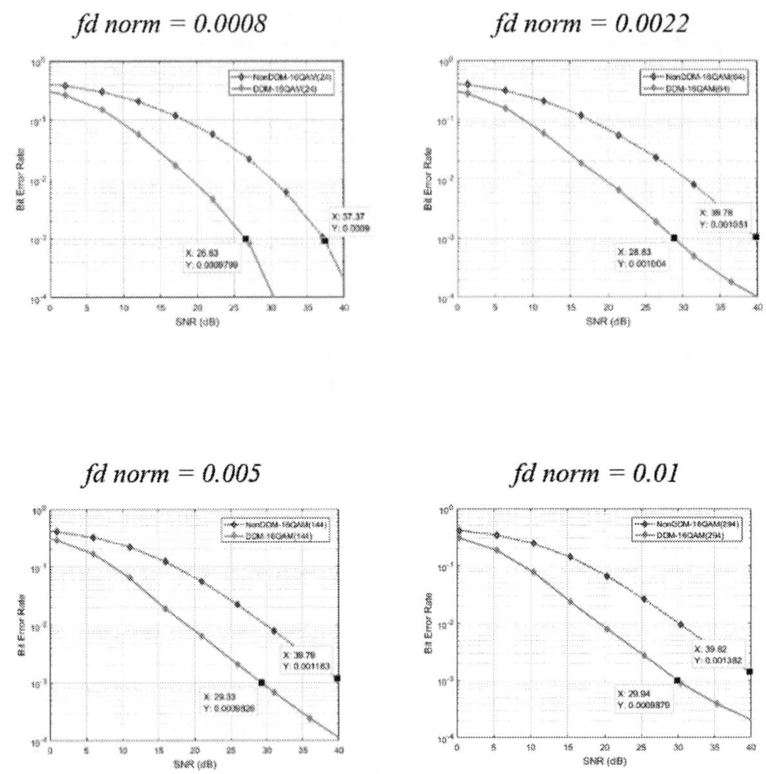

Figure 7. Comparison of BER and SNR functions in DDM and Non-DDM schemes with 16QAM Modulation.

Table 4. DDM vs Non-DDM Response With 16QAM Modulation Using V-to-V Channel

Transmision Mode 16-QAM		SNR [dB]	
f_d norm	speed	DDM	Non DDM
0.0008	24 km/h	26.63	37.37
0.0022	64 km/h	28.83	39.78
0.0051	144 km/h	29.33	39.79

4. Conclusion

The application of the DDM scheme is based on channel achievement allocation for V-to-V communication which is affected by Doppler frequency. The final results of the performance analysis of the three modulations with the application of the DDM scheme compared to Non-DDM on the fourth value of the speed conditions tested on the DDM scheme provide better performance when compared to the Non-DDM modulation. It gives more efficient results in terms of SNR for this communication system.

References

[1] Campolo C, Molinaro A, Scopigno R.“Vehicular ad hoc Network: Standarts, Solutions, research”. New York. Springer. 2015

[2] Feukeu EA, Djouani K, Kurien A. "Doppler Effect Analysis and Modulation Code Derivation". 7th IST-AWSN. Procedia Computer Science 10 (2012) 1032-1038

[3] Feukeu EA, Djouani K, Kurien A ." Compensating the effect of Doppler shift in a vehicular network", AFRICON,9-12 Sept 2013

[4] Fuqin. X and Monty. A, "The Effect of Doppler Frequency Shift, Frequency Offset of the Local Oscillators, and Phase Noise on the Performance of Coherent OFDM Receivers," National Aeronautics and Space Administration Washington, DC 20546–0001

[5] F. Nyongesa, K. Djouani, T. Olwal, and Y. Hamam."Doppler Shift Compensation Schemes in VANETs" -Article Review.Tshwane University of Technology, Private Bag X680, Pretoria 0001, South Africa .2015

[6] K. Al-barazi, U. Mohammad, and N. Al-holou, PhD. " Doppler Shift Impact On Vehicular Ad-hoc Networks", Canadian Journal on Multimedia and Wireless Networks, Vol.2, No.3, August. 2011.

[7] A. M. S. Abdelgader and W. Lenan, "The Physical Layer of the IEEE 802 . 11p WAVE Communication Standard : The Specifications and Challenges," Proceedings of the World Congress on Engineering and Computer Science., vol. II, pp. 22–24, 2014.

[8] C. College and R. Korea, "Doppler Effect on V2I Path Loss and V2V Channel Models," International Conference on Information and Communication Convergence, no. 1, pp. 898–902, 2016.

[9] Novfitri, A., Suryani, T., & Suwadi. (2018). Performance Analysis of Vehicle-to-Vehicle Communication with Adaptive Modulation. 2018 Electrical Power, Electronics, Communications, Controls and Informatics Seminar, EECCIS 2018, 187–191. https://doi.org/10.1109/EECCIS.2018.8692895

[10] Scott Carr Ken Lye, Shee Eng Tan. Yit Kwong Chin, Bih Lii Chua, and Kenneth Tze Kin Teo. "Performance Analysis of intelegent transport system (ITS) with Adaptive Transmition shceme," 2012 4th International conference on Computational Intellegence. Communication system and network.

[11] Pamungkas, W., & Suryani, T. (2018). Doppler effect in VANET technology on high user's mobility. 2018 International Conference on Information and Communications Technology, ICOIACT 2018, 2018-Janua, 899–904. https://doi.org/10.1109/ICOIACT.2018.8350663

[12] Alenka Zajic. "Mobile-to-Mobile Wireless Channels." 2013. Artech House. Boston. London.

[13] Chirag S. Patel. Gordon L.Stuber. and Thomas G.Pratt. "Simulation of Rayleigh-Faded Mobile-to-Mobile Communication Channels." IEEE Transactions on communications. vol. 55. no. 11. November 2005.

IOP Conf. Series: Materials Science and Engineering **982** (2020) 012029 doi:10.1088/1757-899X/982/1/012029

Calibrating the Standard Propagation Model (SPM) for Suburban Environments Using 4G LTE Field Measurement Study Case in Indonesia

Muntaqo Alfin Amanaf[1,2], Alfin Hikmaturokhman[1], Achmad Fadhlan Septian[1]

[1]Faculty of Telecommunication and Electrical Engineering, Institut Teknologi Telkom Purwokerto, Central Java 35147, Indonesia

[2]muntaqo@ittelkom-pwt.ac.id

Abstract. Effective and efficient network planning for appropriate area coverage can use the standard propagation model (SPM) calibrating or tuning method. SPM functions to increase the accuracy of predictions from the appropriate coverage planning results in field implementation. This study uses a calibration method resulting from field measurements with 4G LTE 1800 MHz technology. The results of the simulation found that the percentage of SPM propagation model for Reference Signal Received Power (RSRP) is above -80 dBm, the difference is 4.30%, RSRP values range from -80 dBm to -95 dBm at 21.52%, -95 to -110 dBm at 25.45%, and RSRP is smaller than -110 dBm at 0.37% by measuring field signal distribution. Coverage prediction of Calibrated SPM propagation model has the higher accuracy than COST 231, it represents the accuracy of SPM about 87% and COST 231 about 79%. The improvement of RSRP in the simulation is caused by changes in the correction values K1 to K7 which are in accordance with the criteria of the Purwokerto City area after calibrating.

1. Introduction

Telecommunications technology changes the human life trends that make the telecommunication services to be the primary needs. Long Term Evolution (LTE) is the fourth generation mobile technology designed to improve information access services to the internet network (mobile broadband) which can fulfill the human needs for telecommunication services. LTE technology has criteria like latency less than 10 ms, the data rate reached 100 Mbps when users were moving and 1 Gbps when users were static. LTE Bandwidth is flexible and has a better efficiency spectrum than 3.5 G technology from release 6. LTE can work in a variety of frequency spectrum and this technology can collaborate with non-3GPP systems that already exist [1][2].

LTE Network Service criteria in Indonesia now is still not optimal. It needs optimal network coverage planning to make this technology can cover all area network. Coverage planning is a method of network planning for predicting and simulating the coverage of each eNodeB in close to real conditions. eNodeB is the transceiver station to connect the user equipment with the LTE network. The first step in the coverage planning process is calculating the link budget. This Link budget calculation estimates the value of the Maximum Allowed Path Loss (MAPL) between a transmitter and a receiver on the downlink and uplink communication. The output of the link budget calculation is the number of eNodeB to cover all area by we know the cell radius. The cell radius can be estimated based on a review of the area and the propagation model used. This process needs the accurate propagation model as the real condition to get the accurate cell radius and coverage prediction each eNodeB. In this study, we calibrating the standard propagation model (SPM) in 1800 MHz frequency by tuning the parameter K1-K7 of SPM formula using radio planning software based on the measurement data drive test method in real condition [3].

SPM is a model of development of propagation modeling formula path loss Hata. It is the Empirical model of the propagation path loss at frequency 1500-2000 MHz. it defines large-scale fading in a signal received with the distance 1-20 km. SPM Equation assumes that the proper settings in some parameters for intended for factor K and the effect on the propagation of the signal. SPM can enhance the accuracy of the predictions in a planning network with the process of calibrating or tuning the propagation model. Calibrating is an approach of calculation that affects a formula from the propagation model, this calculation can conduct an approach quite close to the real measurement results. Previous research[4] already calibrated the SPM Model for Built-Up and Urban Environments using Field measurement and geospatial data.

That research shows that after proper calibration, the SPM provides a much better fitness, achieving average Mean Absolute Error (MAE), Root Mean Square Error (RMSE) and Standard Deviation (SDE) values and significantly better than the reference values without calibration. Other research from [5] Standard Propagation Model (SPM) is proposed to simulate 4.5G coverage available for the entire Singapore airspace up to 400 feet altitude. This calibrated SPM considers path loss correction due to angular-depression between Unmanned Aerial Systems (UAS) altitude and cell tower height to improve coverage prediction accuracy. The accuracy of less than 8dB error invariance between simulation and experimental results have been achieved. The calibrated model will be used to support the uTM-UAS to indicate suitable airspace in terms of communication coverage and performance to ensure safe UAS operations.

2. Literature Review

The propagation model is an empirical mathematical formula to simulate the characteristics of radio wave propagation based on altitude, distance, frequency, and other conditions. There are several types of propagation models used in cellular network planning such as Okumura-Hata, COST 231, and Standard Propagation Models [4] [6].

In this study, the propagation model for network planning uses the SPM propagation model with equation (1) below.

$$L = K1 + K2\log d + K3\log h_{te} + K4.Diffraction\ Loss + K5\log d.\log h_{te} + K6.h_{re} + K7.\log h_{re} + K_{clutter}.f_{clutter} + K_{hill,LOS} \quad (1)$$

The model parameters are defined as follows :

K1	: Constant offset (dB)
K2	: Multiplying factor for log d
d	: Distance between receiver and transmitter (m)
K3	: Multiplying factor log h_{te}
h_{te}	: Effective height of the transmitter antenna (m)
K4	: Multiplying factor for diffraction calculation
K5	: Multiplying factor for log d . log h_{te}
K6	: Multiplying factor for h_{re}
K7	: Multiplying factor for log h_{re}
h_{re}	: Effective height of the receiver antenna (m)
$K_{clutter}$	: Multiplying factor for $f_{clutter}$
$f_{clutter}$	: Average of weighted losses due to clutter
$K_{hill,LOS}$	: Corrective factor for hilly regions

IOP Conf. Series: Materials Science and Engineering **982** (2020) 012029 doi:10.1088/1757-899X/982/1/012029

Refers to equation (1), the correction factor of the UE antenna height and clutter is ignored so the equation of the path loss model is given as follows

$$L = A_1 + A_2 \log(f) + A_3 \log h_{te} + [B_1 + B_2 \log h_{te}][\log d] \quad (2)$$

Equation (2) has the values:
A1 = 46.30
A2 = 33.90
B1 = 44.90
B2 = -6.55

So, the equation between COST 231 and SPM was formulated as the equation (3)

$$L = A_1 + A_2\, log(f) - 3B_1 + [A_3 - 3B_2][log\, h_{te}] + B_1\, log\, d + B_2\, log\, h_{te} \,.\, log\, d \quad (3)$$

$K_1 = A_1 + A_2 \log(f) - 3B_1$
$K_2 = B_1$
$K_3 = A_3 - 3B_2$
$K_5 = B_2$

The formula for implementing LTE 1800 given as equation (4) belows

$$L = 22 + 44.9\, log(d) + 5.83\, log\, h_{te} - 6.55\, log(d) \,.\, log\, h_{te} \quad (4)$$

The Approaching method used the standard propagation model to calibrate K1 until K7 parameter like the measurement data drive test in certain areas. it made the prediction coverage more accurate to plan the coverage of the eNodeB in certain areas. Table 1 and Table 2 are the range of calibrated SPM parameter.

Table 1. SPM Range Reference [7]

Parameter	Minimum	Typical	Maximum
K2	20	44.9	70
K3	-20	5.83	20
K4	0	0.5	0.8
K5	-10	-6.55	0
K6	-1	0	0
K7	-10	0	1

Table 2. K1 Parameter of Standard Propagation Model [7]

Frequency	900 MHz	1800 MHz	1900 MHz	2100 MHz
K1	12.5	22	23	23.8

3. Methodology

3.1. LTE Networks Planning and Link Budget

The network planning process uses existing data which is simulated with the Atoll radio network planning software to get coverage predictions. the simulation of radio network planning was configured by model propagation at the 1800 MHz LTE frequency. The existing data use to give information about the real condition like in the measurement data. The software Atoll can simulate the existing network like the real condition on surface and altitude in certain areas. In this study, we simulate the coverage prediction of the existing network in Purwokerto by calibrating the correction factor value of SPM like the measurement data drive test. The value calibrating data Correction factor can represent the condition in the typical area, especially at Purwokerto.

In coverage planning, the first process to do is calculating the link budget. The purpose of calculating the link budget is to get the real Most Allowable Path loss (MAPL), this is the maximum loss value between the transmitter and receiver for uplink and downlink directions. From the calculation of MAPL it can predict cell radius based on the real area and the propagation model used. If we get the cell radius, we can get the number of sites to plan how many sites or eNodeB we needed to cover all areas. the calculation of the MAPL is calculated on the downlink side only[2].

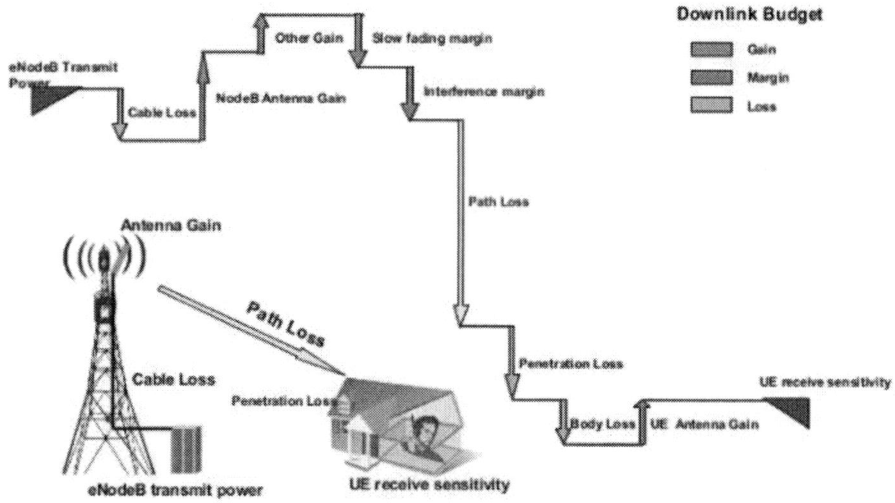

Figure 1. Link Budget Downlink [8]

The link budget downlink equation is given as follow:

EIRPD	= PUE – GeNodeB – Lb	(5)
SRMINDL	= RXSen – GUE– Lc+ IM	(6)
RXsen	= SINR + RxNF – NTH + 10 log (15000)	(7)
MAPLDL	= EIRPDL – SRMINDL – L – MSF	(8)

Equation (5), (6), (7) and (8) are defined as:

PUE	: UE transmitter power each resource block
GUE	: UE antenna gain
GeNodeB	: eNodeB antenna gain
Lc	: feeder cable loss
Lb	: body loss
EIRPDL	: effective isotropic radiated power downlink
SINR	: Signal to Interference ratio
RxNF	: receiver noise figure
NTH	: thermal noise
RXSEN	: receiver sensitivity
SRMINDL	: minimum signal reception downlink
IM	: interference margin
L	: penetration loss
MSF	: shadow fading margin
MAPLDL	: maximum allowable path loss downlink

3.2. Propagation Model Tuning

The Propagation Model Tuning method is an approach to calibrate the propagation model calculations and formula as the real propagation in a certain area. This calculation can take an approach that is quite close to the measurement results in the field, the calibrating parameter that affects the propagation model is considered. This method uses a field measurement data (drive test) which contains information on signal level distribution from measurement data [9].

4. Results & Analysis

4.1. Measurement data using Drive Test

Drive test is a measurement method to find out the real performance of the network based on the user experience of handphones in a cellular network on the side of the air interface (MS to BTS or vice versa) using a tool called drive test tools. The results of the driving test will be used as an evaluation of the existing network to improve the performance and maintenance of the network.

In this study, There were three measurement areas. measurement areas were in Sokaraja Tengah, Sampoerna Berkoh, and Teluk site. the signal quality measurement used Nemo Outdoor drive test tools. The drive test tools would record all the activities carried out by the handset (mobile phone) because the drive test tools could be directly integrated with the notebook. The parameter measured to calibrated the SPM Model as a reference signal received power (RSRP).

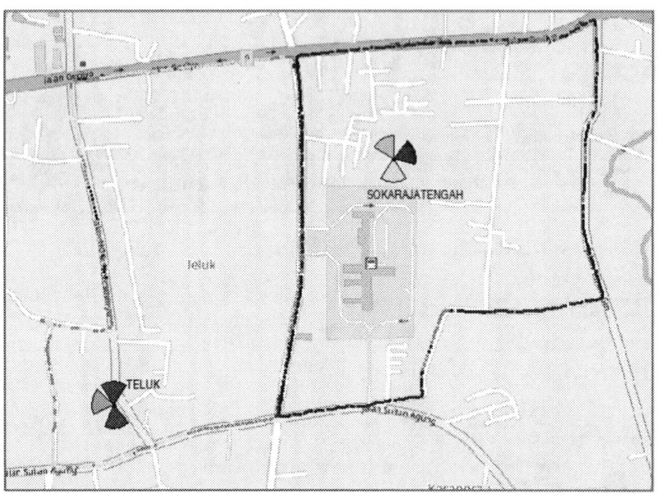

Figure 2. RSRP's measurement data drive test plotting at a digital map.

The Range of RSRP key performance indicator (KPI) was from -45 to -110 dBm. The Classification of RSRP KPI to obtain good or poor criteria on the RSRP LTE network can be seen in table 3.

Table 3. The RSRP Category for every level Signal [10]

Category	RSRP Range Value (dBm)
Excellent	-80 =< x
Good	-95 =< x < -80
Fair	-110 =< x < -95
Poor	X < -110

It was the power of the signal received from the eNodeB towards the UE. The closer the serving cell the better the signal strength would be received by the UE, and vice versa if the serving cell was getting farther away, the worse signal would be received. The RSRP calculation values could be formulated in equation (9)[8] :

$$RSRP = RSSI - 10.\log(12.N) \tag{9}$$

The correction factor constant value is determined by approaching the constant value function. A function of this constant value becomes a reason to increase or decrease the modified parameter, we called it the calibrating standard propagation model. The standard propagation model is calibrated by the radio network simulation software Atoll in two scenarios. The first scenario is a simulation of coverage prediction using COST 231 propagation model before using the standard propagation tuning model. The second scenario is a simulation of the calibrating parameter from standard propagation model tuning according to the range of parameters permitted by the standard propagation model so the values of the correction parameters are matched with the measurement area. The simulation at scenario 1 uses the K Parameter value from COST 231 Parameter model like table 4 below.

Table 4. The K Parameter COST 231 Propagation Model

Parameter	Value
K1	22
K2	44.9
K3	5.83
K4	0
K5	-6.55
K6	0
K7	0

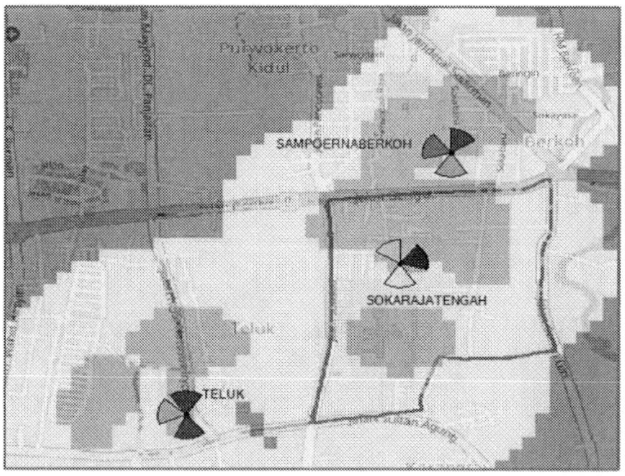

Figure 3. Scenario 1 coverage prediction and measurement data plotting before calibrating.

Signal levels prediction at some sites have the RSRP value above -80 dBm is 0%. The RSRP Signal level from -80 dBm until -95 dBm is 29.77%, The RSRP Signal level from -90 dBm until -110 dBm is 65.30% and The RSRP Signal level below -110 dBm are 4.93%. The RSRP prediction produced by Figure 3 is the propagation COST 231 with separation distance resulting in 231 propagation COST that has been calculated is 1.9 km with the transmitter antenna height on average is 52 meters, and the receiver antenna height is 2 meters. Approach to the measurement signal with the drive test method and the simulation results have results that are not as good as the results of the simulation signal distribution in figure 4.

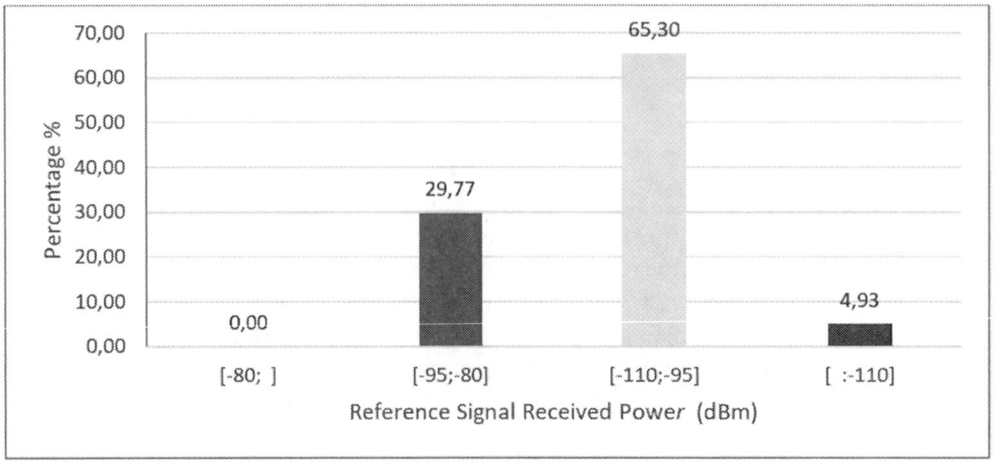

Figure 4. Scenario 1 RSRP histogram before calibrating.

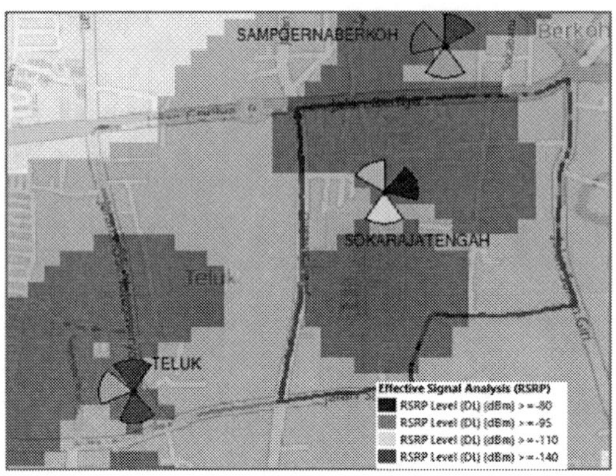

Figure 5. Scenario 2 Coverage Prediction after calibrating.

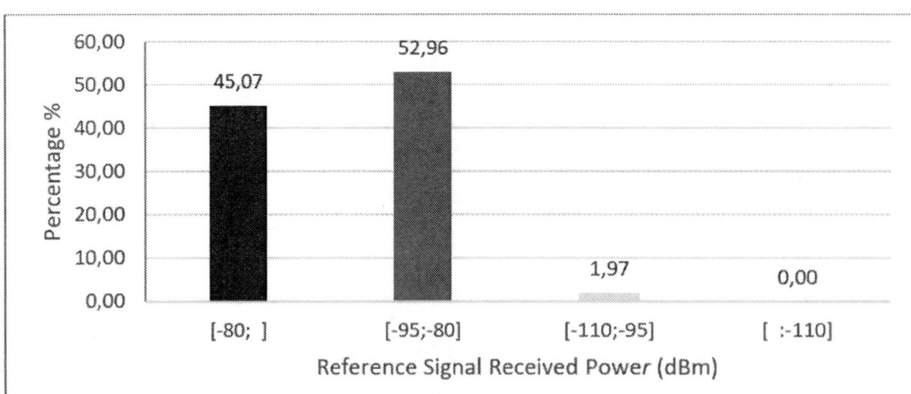

Figure 6. Scenario 2 RSRP histogram after calibrating.

Signal spread on the Sokaraja Tengah site, Sampoerna Berkoh, and Teluk this second scenario has an approach between measured signals and simulation signals that approach each other's value. RSRP with values above -80 dBm in the area reviewed is 45.07%. RSRP is the range of -80 dBm to -95 dBm at 52.96%, RSRP -95 dBm to -110 dBm at 1.97%, and below RSRP -110 dBm has a percentage of 0%. This value is higher than the results in scenario 2 which includes 3 sites reviewed. This is caused by the propagation of the model used calibrated drive test data with the correction value used that has been tuned using a range of minimum to maximum values according to the standard propagation model rules. From the results of the calibrating the new correction values are obtained according to the characteristics for the city of Purwokerto as table 5 follows:

$$L = 22 + 44.9\log(d) + 5.83\log h_{te} - 6.55\log(d).\log h_{te} \quad (10)$$

$$L = 22 + 50\log(d) - 4.72\log h_{te} - 10\log(d).\log h_{te} \quad (11)$$

Table 5 is The result of SPM Parameter Calibration for Suburban typical area especially at Purwokerto, that value is the new correction parameter obtained for constant offset (K1) is 22 dB due to the 1800 MHz frequency used in this study. The multiplier of log d (K2) is 51, the multiplier factor of the newly generated log hte (K3) is -4.98, the multiplier factor of diffraction calculation (K4) is 0, the multiplier of log d multiplied by log hte (K5) has a value by -10. The parameter K6 or the multiplier of the receiver antenna and K7 which is the multiplier of the hre log is 0 because the value has no impact on path loss. The Separation distance generated by SPM tuning propagation that has been calculated by equation (10) and (11) is 6.6 km with the transmitter antenna height about 52 meters, and the receiver antenna height about 2 meters.

Table 5. The Calibration of SPM Parameter for Suburban typical area especially at Purwokerto.

Parameter	Value
K1	22
K2	51
K3	-4.98
K4	0
K5	-10
K6	0
K7	0

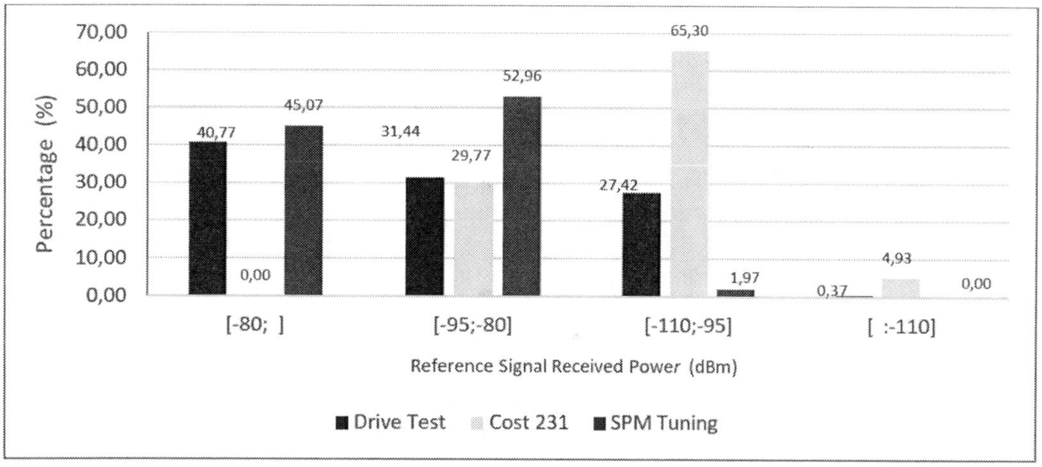

Figure 7. The comparison of drive test measurement RSRP Histogram between COST 231 and SPM Tuning RSRP Prediction.

The comparison of coverage prediction in figure 7 using the propagation model COST 231 with a range of RSRP values greater than -80 dBm has not approached the actual measurement results (drive test) compared to the use of SPM propagation models that have been calibrated. The results of the SPM for RSRP are above -80 dBm, the difference is 4.30% with the drive test data. The RSRP value in the range of -80 dBm to -95 dBm in the 231 propagation COST model obtained a better difference of 1.67% and the SPM propagation model of 21.52%. RSRP is -95 to -110 dBm in the SPM propagation model is much better than the use of the propagation model COST 231. SPM propagation has a difference of 25.45% with the results of actual measurements or drive tests. Poor coverage with RSRP values above -110 dBm in the SPM propagation model getting a difference of 0.37% and the propagation model COST 231 amounting to 4.56%. it can analyze that SPM has a prediction accuracy of about 87% and COST 231 about 79%. It concludes the Calibrated SPM makes a higher accuracy to measurement data than COST 231 model.

5. Conclusion

The results of calibrated SPM Parameter values for Suburban Purwokerto typical area are K1= 22, K2= 51, K3= -4.98, K5= -10, and zero value for the K4, K6, and K7 parameter. Every area has a different calibrated parameter. In this study, The Results of coverage prediction and measurement data can be seen that SPM has been calibrated and tuned with prediction accuracy about 87% and COST 231 about 79%. The SPM correction Parameter that has been calibrated with the results of the driving test in this simulation can increase the accuracy of the results of network planning.

6. References

[1] N. Shabbir, M. T. Sadiq, H. Kashif, and Rizwan Ullah. 2011. "Comparison of Radio Propagation Models for Long Term Evolution (LTE) Network," *Int. J. Next-Generation Networks*, vol. 3, no. 3, pp. 27–41.

[2] A. Hikmaturokhman, L. Wardhana, B. Aginsa, and A. Dewantoro. 2014. *4G Handbook Edisi Bahasa Indonesia*, 1st ed. www.nulisbuku.com.

[3] M. M. Elbagir and K. H. Bilal. 2014. "LTE Radio Planning Using Atoll Radio Planning and Optimization Software," *Int. J. Sci. Res.*, vol. 3, no. 10, pp. 1460–1464.

[4] S. I. Popoola, A. A. Atayero, N. Faruk, C. T. Calafate, E. Adetiba, and V. O. Matthews. 2017. "Calibrating the Standard Path Loss Model for Urban Environments using Field Measurements and Geospatial Data," *World Congr. Eng. 2017 Vol I WCE 2017, July 5-7, 2017, London, U.K. Calibrating*, vol. I, pp. 5–10.

[5] M. D. Villaluz, L. Gan, J. Sia, S. K. Tan, S. Foo, and K. H. Low. 2018. "Preliminary 4.5G Cellular Network Assessment with Calibrated Standard Propagation Model (SPM) for uTM-UAS Operations in Singapore Airspace," *2018 Int. Conf. Unmanned Aircr. Syst. ICUAS 2018*, pp. 796–805.

[6] M. A. Amanaf, E. S. Nugraha, and D. Kurnianto. 2018. "Analisis Simulasi Model COST-231 Multiwall Pathloss Indoor Berbasis Wireless Sensor Network pada Aplikasi Absensi Mahasiswa dengan Tag RFID Menggunakan RPS (Radiowave Propagation Simulator). Analysis of Wireless Sensor Network-based Indoor COST-231 Mult," vol. 16, no. 1, pp. 17–26.

[7] Forsk. 2016. "Atoll 3.3.2 User Manual for Radio Networks," France.

[8] O. Arifianto. 2017. "Analisa Perancangan Jaringan LTE-Advanced Carrier Aggregation Pada Frekuensi 850 MHz dan InterBand 850 MHZ Dengan 2300 MHz," IT Telkom Purwokerto.

[9] Z. Liang. 2018. "RNP Model Tuning Guide," *Huawei Technol. Co., Ltd*, p. 85.

[10] A. PURNAMA, E. S. NUGRAHA, and M. A. AMANAF. 2020. "Penerapan Metode ACP untuk Optimasi Physical Tuning Antena Sektoral pada Jaringan 4G LTE di Kota Purwokerto," *ELKOMIKA J. Tek. Energi Elektr. Tek. Telekomun. Tek. Elektron.*, vol. 8, no. 1, p. 138.

IOP Conf. Series: Materials Science and Engineering **982** (2020) 012030 doi:10.1088/1757-899X/982/1/012030

Energy Consumption Performance on IEEE 802.11ah Networks with Station Grouping Scheme

A S Aminah[1,2], D Perdana[1], I Wahidah[1]

[1]Departement of Electrical Engineering, Telkom University, Bandung, Indonesia

[2]ayayusa@student.telkomuniversity.ac.id

Abstract. Sending packages from the sender to the recipient at the station requires adequate resources. To ensure messages in the form of text, data, images, and videos are received in real-time, reliably, and flexibly, power performance with low energy consumption is required at the station. IEEE 802.11ah can be one of the wireless technologies to overcome this problem by saving low power consumption, the transmission range of 1 km and access points can reach 8191 stations. The purpose of this study was to determine the performance of energy-saving consumption by station grouping which is regulated in the regulation of the number of RAW slots. From the results of this study, energy savings were obtained by dividing the two groups when the station numbered 900 stations with an energy consumption of 8.105 joules while in one group with a smaller station scale the total energy consumption of 8.124 joules was obtained. The second scheme is by dividing slots (NRawSlot) to reduce energy consumption in large station capacities. The NRawSlot division is divided into seven-slot with energy consumption of 8,105 joules. It is proven that energy consumption is more efficient by grouping with more station capacity.

1. Introduction

The increase in internet users is arguably very much. One person can use more than one laptop and handphone. All of these devices must be connected to the internet with various services used such as sending packages in the form of data, voice, video, and others. the pandemic in 2020 that had not been predicted before, had resulted in changes inhabits. all on-site activities become online. From workers, students, employers, companies, etc. quickly adjusting the situation by switching to a digital platform so that activities continue to run normally. According to sources, Cisco IBSG predicts that there will be 25 billion devices connected to the Internet by 2015 and 50 billion by 2020. It is important to note that these estimates do not take into account rapid advances in the internet or device technology; the numbers presented are based on what is known to be true today.

To support devices in cellular networks, a large fee is required. Therefore, people will look for alternative bands at cheaper prices. IEEE 802.11ah is a new WLAN network protocol and is the result of the development of the IEEE 802.11-2007 WLAN standard. IEEE 802.11ah is designed to work on sub gigahertz frequency spectrums, where the lower frequency spectrum is juxtaposed with pre-existing WLAN standards. this causes the energy consumption used to also be lower [2].

One of the new standards in wireless is IEEE 802.11ah, which is to provide low energy consumption, handling many users, and a higher coverage area up to 1000 meters[1]. To address the collision problem caused by the increasing number of contenders, IEEE 802.11ah introduces a Restricted

Access Window (RAW) mechanism to realize the grouping based MAC protocol[3]. In particular, sensors are partitioned into groups, each assigned to a RAW slot. The clients in a group can only contend for access in the assigned RAW slot, as a result alleviating the cost (i.e., overhead and energy consumption) of contention[4]. In the RAW mechanism, there is a RAW group. In the RAW group, there are slots called RAW slots. RAW slots are small groups that are in one RAW group that is used to limit the channel access of each group in one RAW group [5]. Each RAW slot has a duration or time interval and turns to access the channel. If the RAW slot is not in a condition allowed to access the channel, then the RAW slot is not allowed to access the channel and all stations in the slot are in sleep [6]. With this RAW slot mechanism, it can limit channel access aimed at reducing collisions and saving energy [7]. The purpose of this study was to determine the performance of energy-saving consumption by station grouping which is regulated in the regulation of the number of RAW slots. Several previous studies have discussed the grouping of IoT sensor types with a grouping scheme and the division of stations based on traffic [4]. In research [8], research focuses on improving throughput and delay on the 802.11e standard. In this study, the focus is more on station grouping to save energy.

2. Method

2.1. 802.11ah

IEEE 802.1ah is designed to operate in the sub gigahertz frequency band, with communication coverage up to 1 kilometer, has a data rate above 100 Kbps, low power consumption by adopting a power-saving strategy, capable of serving up to 8192 stations through a hierarchical identifier structure, and solutions economical for manufacturing network devices. IEEE 802.11ah has a one-hop network topology and data transmission is short and not too frequent (data packet size approaches 100 bytes and interarrival time is greater than 30 seconds) [7].

2.2. Restricted Access Window (RAW)

To support several devices associated with the AP, TGah has developed a new mechanism to reduce congestion in channel access. In this mechanism during the time window (RAW), a group of stations is allocated with a specific time slot that is used to access the channel. Stations are not permitted to access the channel when outside the RAW slot of the station. Besides, in the beacon interval, several RAWs are allowed to access the channel [7].

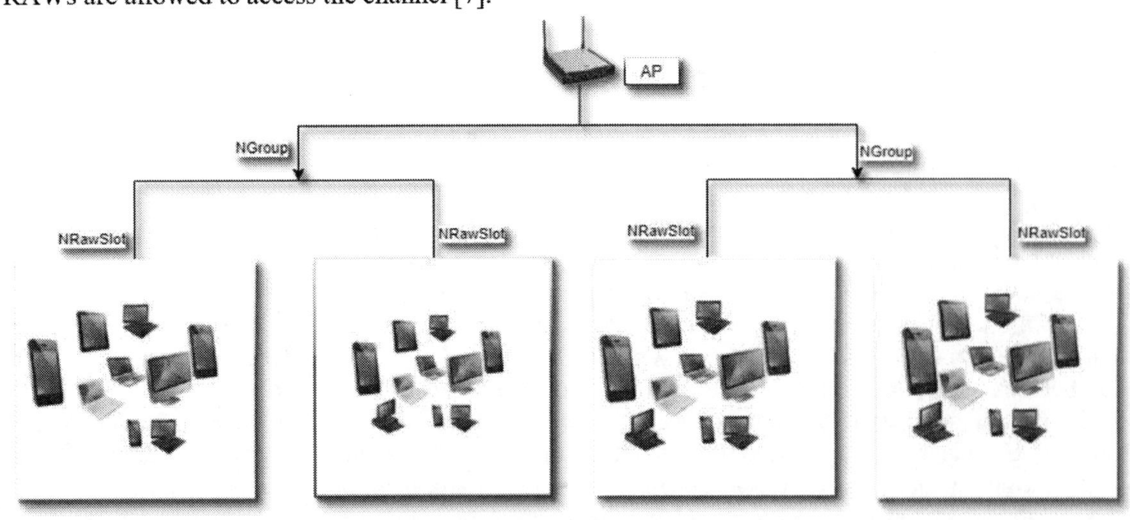

Figure 1. Grouping Station by 2 NRawSlot

When an STA gets an uplink data from the upper layer, the STA can request channel access at the beginning of the RAW slot allocated for the STA. But STAs must be allowed to access channels in certain RAW slots. The STA will stop immediately to transmit the packet after the specified time for the slot has ended. It is also possible that STA might not be allowed to use RAW slots [9].

Station grouping can be done by the AP that assigns groups to the station by using the RAW Parameter Set (RPS) element that is broadcasted in the beacon frame. In addition to group assignment information, the RPS element contains essential control information about RAW operations including slot duration and several RAW slots (NRAW) [10]. This information is used by stations to calculate the total duration of RAW (TRAW) [7].

2.3. Simulation Scenarios

In this research, energy consumption savings will be made by using a grouping scheme. This grouping scheme is carried out with several kinds of scenarios. The first scenario is done by adjusting the change in the number of Groups as Ngroup. In dividing the number of groups into two groups, after that, there will be changes in the number of RAW Slots that differ between two to seven RAW Slot as NRawSlot. In addition to determining the two-division changes, group changes will be made to the number of stations. Because in this study the aim is to make energy savings in a large amount of capacity.

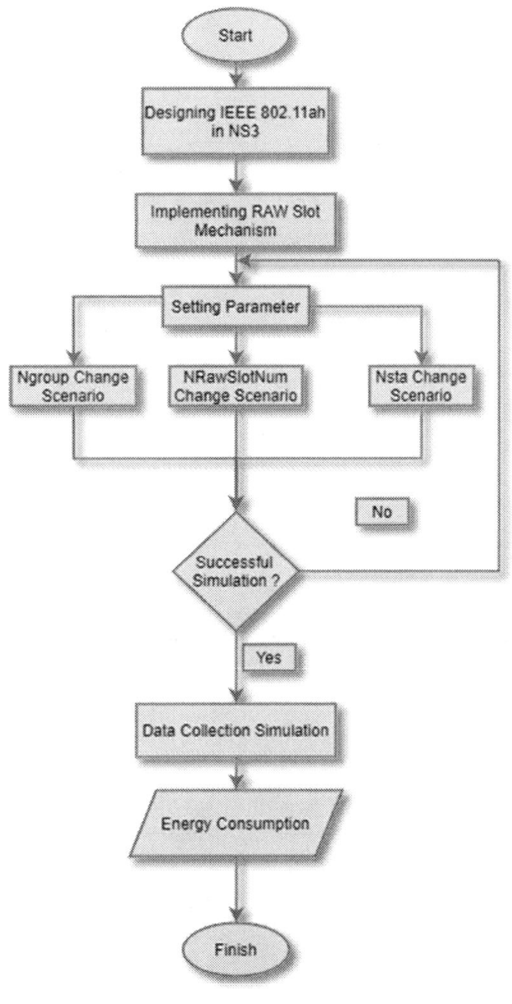

Figure 2. Energy Consumption Simulation Scenario

Table 1. Common Scenario Parameters

Parameters[1]	**Value**

Ngroup	1,2
NrawSlotNum	6 and 7
Nsta	700, 800 and 900
Distance Access Point with Stasiun	~ 100 meters

[1]Ngroup = Number of RAW groups; NrawSlotNum = Number of slots per RAW group; Nsta = Number of total stations.

3. Result And Discussion

3.1. Analysis of the First Scenario: Changes in the Amount of Ngroup and Nsta

In this scenario, a change in the number of Ngroup and Nsta is implemented, where the value of Ngroup = 1 and Ngroup = 2. Changes are also made to the number of Nsta. changes in the number of NSTAS are 700, 800, and 900. Scenarios are performed on a network with 802.11ah standard.

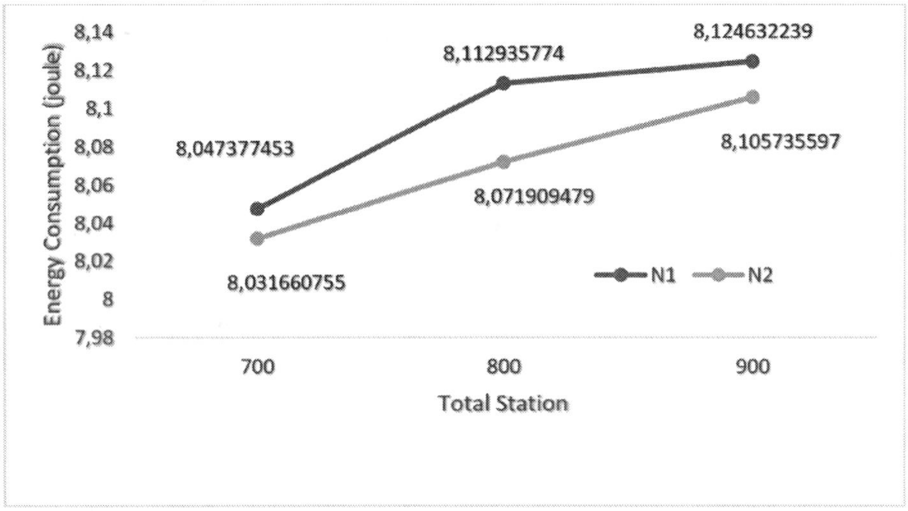

Figure 3. Energy Consumption by Station Grouping

In Figure 3, two-line graphs represent the two types of the number of Ngroup divisions. With the increase in the number of group divisions, the energy consumption will be lower along with the increasing number of stations. The division of several STAs into groups can help network performance to be better in packet delivery. This is because increasing the number of Ngroups reduces contention, collision, and supports retransmission of failed packets. Without the division of Ngroups, increasing the number of STAs will make STAs wait even longer to be able to deliver their packages. However, the development of re-transmission activities can make the throughput decrease and increase the delay time. Then, the use of RAW also allows the STA to be idle or sleep or inactive if their RAW slot is not in a state of having to send packets, where this can affect the value of energy consumption to be smaller as the number of sta increases [9].

3.2. Second Scenario Analysis: Change in the Amount of Share Nsta

In this scenario, it is implemented by changing the number of Nsta. This scenario consists of 6 and 7 RAW slots and uses a change in the number of NSTA with a total of 700, 800, and 900 stations.

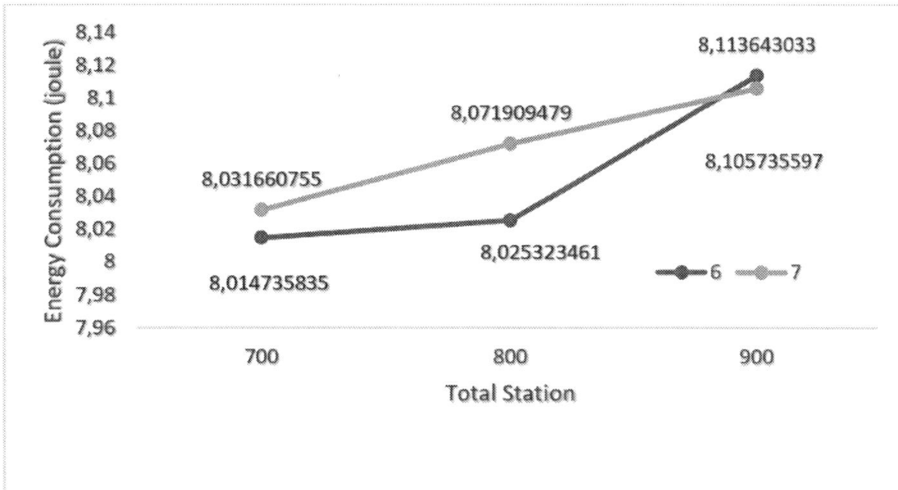

Figure 4. Energy Consumption by Nsta Grouping

3.3. *Third Scenario Analysis: Change in the Number of NRawSlot Shares*

In this scenario, it is implemented by changing the number of NRawSlot and Nsta. This scenario consists of 6 and 7 RAW slots and uses a change in the number of NSTA with a total of 700, 800, and 900 stations.

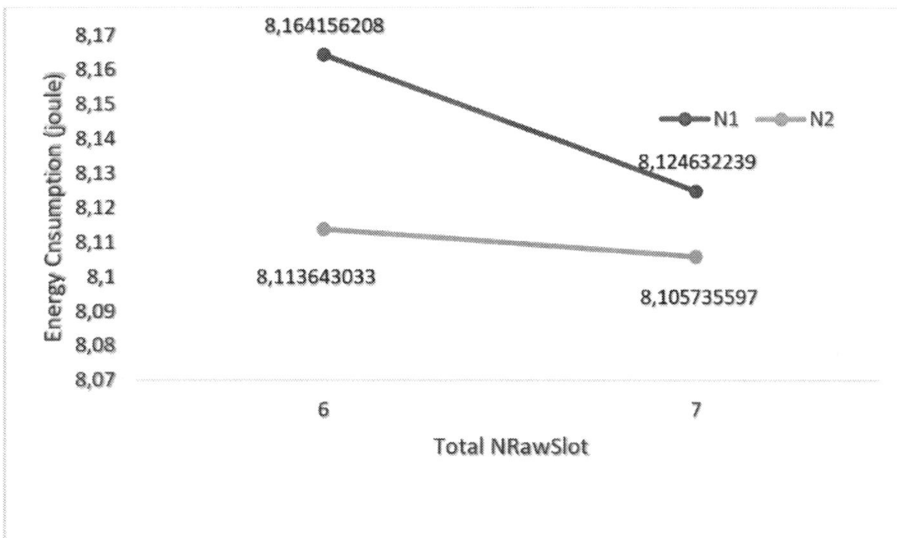

Figure 5. Energy Consumption by NRawSlot Grouping

3.4. *Energy Consumption Performance Analysis*

By the initial objective to see the performance of energy consumption in 802.11ah with the station grouping method, simulations were carried out with several scenarios of changes in the number of Ngroups, NRawSlot, and NSTA. This test is carried out in various steps. initially with a smaller number of stations. Starting from 50,100,150, 200, 250, and 300 then with the division of NGroup 1 and 2, and NRawSlot 2 to 7 slots. The data presented in the research above is the second test with an increase in the number of stations. 802.11ah is recommended to be used in supporting IoT and a large number of stations (up to 8191 stations). The increase in stations from 700, 800, and 900 stations, of course, was accompanied by changes to the NRawSlot and Ngroup. the more NGroups and NrawSlots, the lower the energy consumption is consumed. On the IEEE 802.11ah network, dividing stations into the same group to access channels simultaneously can reduce collisions. thereby achieving better performance. proven by the simulation results above.

4. Conclusion

In this paper, we introduce the savings in energy consumption at the station. By using a grouping scheme it is proven that consumption is lower when the station is getting bigger in number. The NRawSlot, Ngroup grouping scheme depends on the number of jugs stations. For further research, it can also be seen from a load of each station. Because it will also affect the energy consumption that will be spent at each station.

References

[1] Putra M A P, Perdana D, Negara R M 2017 *Performance Analysis of Data Traffic Offload Scheme on Long Term Evolution (LTE) and IEEE 802.11AH*

[2] Weiping Sun M C 2013 *IEEE 802.11ah: A Long Range 802.11 WLAN at Sub 1 GHz.*

[3] IEEE P802.11ah/D2.0 2014 *IEEE Draft Standard for Information Technology - Telecommunications and Information Exchange Between Systems - Local and Metropolitan Area Networks - Specific Requirements - Part 11: Wireless LAN Medium Access Control (MAC) and Physical Layer (PHY) Specifications: Amendment- Sub 1 GHz License-Exempt Operation*. pp. 1–582.

[4] Tung C C, Chi H L, Kate Ching J L, and Wen T C *Load-Balanced Sensor Grouping for IEEE 802.11ah Networks*

[5] Tian L, Latre S, Deronne S, and Famaey J 2017 *Research Gate.Retrieved from Implementation and Validation of an IEEE 802.11ah Module for NS-3*

[6] Tian L, Famaey J, Latré S 2016 *Evaluation of the IEEE 802.11ah Restricted Access Window Mechanism for dense IoT networks*. Seventeeth International Symposium on a World of Wireless, Mobile and Multimedia Networks (WoWMoM), Ciombra, Portugal

[7] Ud-Din M Q 2015 *Enhancements and Challenges in IEEE 802.11ah-a Sub Gigahertz Wi-Fi for IoT Applications*. Faculty of Computing and Eletrical Engineering.

[8] Riyadh Qashi M B, and Hanssgen K 2011 *Case Study : The Effect of Variable Priority Parameters on the QoS of WLANs IEEE 802.11e EDCF.* IEEE, 104 -108

[9] Daneshfar N 2015 *Performance Enhancement Mechanism of IEEE 802.11ah Machine Communication System*

[10] Park M 2014 *IEEE 802.11ah : Energy Efficient MAC Protocols for Long Range Wireless LAN.* IEEE

IOP Conf. Series: Materials Science and Engineering **982** (2020) 012031 doi:10.1088/1757-899X/982/1/012031

FEMA 310 Tier 1 seismic evaluation of existing building: A case study of a 7-story academic RC building of Jenderal Soedirman University, Indonesia

Yanuar Haryanto[1,2,4], Hsuan-Teh Hu[1,3], Arnie Widyaningrum[2], Esya Alfidiah[2] and Banu Ardi Hidayat[1]

[1]Department of Civil Engineering, National Cheng Kung University, Tainan, Taiwan
[2]Department of Civil Engineering, Jenderal Soedirman University, Purwokerto, Indonesia
[3]Department of Civil and Disaster Prevention Engineering, National United University, Tainan, Taiwan

[4]yanuar.haryanto@unsoed.ac.id

Abstract. A need to predict the seismic vulnerability of current buildings has led to a heightened interest in study that deals with seismic vulnerability assessment methods. The easiest and fastest method, known as a walk-down survey or a street survey, involves just the superficial information accumulated during a short inspection of a building. This paper picks a fairly new seven-story academic building made of reinforced concrete (RC) located at the Jenderal Soedriman University in Purwokerto, Indonesia, which was designed in 2015 using construction details illustrative of that exact period, for FEMA 310 Tier 1 assessment in the context of Performance-Based Earthquake Engineering (PBEE). Mandatory checklists as a role of the region of seismicity and the building's level of performance are done. The outcomes after a procedure projected the building in the case-study demonstrated a range of possible flaws and seismic vulnerability. Therefore, detailed comprehensive evaluation is intensely endorsed.

1. Introduction

Earthquakes threaten both buildings and people [1-6]. The amount of deaths as well as injuries caused by earthquakes is directly linked to the building's vulnerability as well as the readiness to deliver first aid [7]. Hence, the seismic vulnerability assessment of the current buildings and particularly of the strategic buildings, e.g., the academic buildings, is of the utmost urgency. Academic buildings ought to be deemed as the top vital buildings in a society [8], as they play a vital part in the educational procedures and could be used as emergency shelters if an earthquake occurs. In that case, academic buildings require a comprehensive strategy for assessing their ability to confront possible earthquakes. Academic buildings are often particularly complicated structures and they have extremely varied roles, which range from functions common to class rooms, as well as meeting rooms, science labs, and offices to warehouses in which dangerous and degradable substances get stored. Because of their unique construction, academic buildings are sometimes quite vulnerable to seismic occurrences.

The requirement of foreseeing the seismic vulnerability of current buildings has brought heightened interest in studies which deal with developing seismic vulnerability assessment methods. Several methods have been suggested in the literature, and these can normally be put into 3 types. The easiest and fastest method, known as a walk-down survey or a street survey, involves only some superficial information accumulated from a short examination of the buildings. The quantity of floors, as well as vertical and plan irregularities, where the building is located, how old the building is, the building's fundamental structure, as well as the obvious material and the quality of its workmanship are the standard parameters utilized. FEMA P-154 [9], FEMA 310 Tier 1 [10] assessment and the Japanese assessment [11] falls into that classification. The objective of rapid evaluation methods is used to classify or rank extremely susceptible buildings which deserves more investigation [12].

In the past few years, because of the significance of educational facilities, a lot of scientists have concentrated on the seismic vulnerability of academic buildings. O'Reilly et al. [13] analyzed the seismic evaluation of 3 current school structures which are representative of Italian school construction stock. They assessed the expected annual loss (EAL) as well as the school's collapse safety. Michel et al. [14] established reasonable probabilistic procedures centered scenarios for mid-size construction stock and did several chosen damage scenarios utilizing school structures in the Basel city. A methodology was submitted by Jaimes and Nino [15] particularly for Mexican public school buildings to complete a cost-benefit assessment to evaluate the options for either retrofitting or reconstructing school buildings concentrated on the justification of direct physical losses due to future seismic occurrences. De Angelis and Pecce [16] established a cataloging of the seismic vulnerability of nonstructural elements in school buildings dependent on survey forms taken to assemble a nonstructural index along with a priority ranking which demonstrates the top vulnerable classification of nonstructural elements.

Mazumder et al. [17] submitted a seismic vulnerability evaluation method on a tiny scale for Chittagong University of Engineering and Technology (CUET) campus region in Bangladesh. At the close of the survey they established that the majority of the structure had excellent performance scores ranging from a level 1 evaluation. Amongst the structures surveyed, merely two percent of these structures can be placed in this classification where they are extremely vulnerable to earthquakes. After the earlier researches and as an extension which enlarges the extent of the work done by Haryanto et al. [18], this paper chooses a comparatively new seven-story academic structure made of reinforced concrete (RC) at the Jenderal Soedirman University in Indonesia. It was designed back in 2015 with the construction specifics being typical of that time period, for FEMA 310 Tier 1 [10] assessment in the context of Performance Based Earthquake Engineering (PBEE). Necessary checklists as a role of the region of seismicity and building's performance level were done and the process results were reviewed.

2. Methodology

2.1. Description of the case-study building

The current academic structure studied, which was the Integrated Laboratory (Faculty of Economics and Business), is a seven-story RC building in Purwokerto, Indonesia; construction started in 2105. This structure is around 5810 m^2 in area along with being thirty meters high. Average slab elevation is 120 mm along with a 175 mm spacing out of longitudinal and transverse reinforcements. The structure is built using momnet resisting frames as a lateral force resistance system in 2 horizontal directions. Figure 1 illustrates an isometric view as well as a plane view of a typical floor level of the building. Member dimensions, structural detail illustrations, foundation specifics as well as additional data was assembled from the structure's owner, which is Jenderal Soedirman University in Indonesia.

2.2. Rapid Visual Screening

Rapid Visual Screening (RVS) in agreement with FEMA P-154 [9] is an example of the extremely recommended means of doing a seismic vulnerability evaluation and it can be done without doing structural computations, and instead utilizing a sidewalk visual survey of the structure as well as filling out an information collection form from the surveyor [19]. That provides a notion of where an in-depth research must be completed grounded on the prioritization of the weakness and suggesting actions for the equivalent. RVS survey sheets may be personalized centered on their context [20]. For instance, in India screening for masonry structures along with buildings made of reinforced concrete are individually treated for 5 different seismic regions [21,22]. In Japan, this form is centered on seismic index (intensity, ductility, and consistency) of structures [23] and in Canada both of the structural parameters (stiffness and regularity) along with nonstructural parameters (occupancy and falling hazards) of the structure are considered [24].

(a) Isometric view of the building

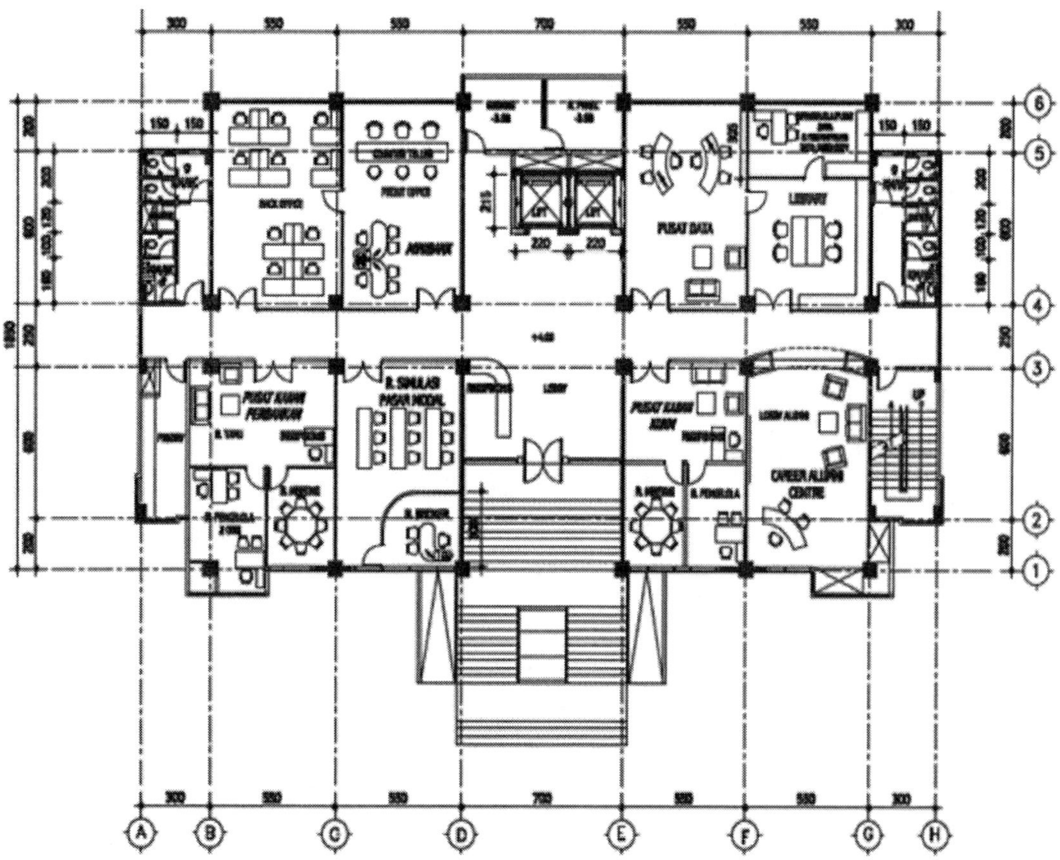

(b) A typical floor plane

Figure 1. Illustration of the case-study building

IOP Conf. Series: Materials Science and Engineering **982** (2020) 012031 doi:10.1088/1757-899X/982/1/012031

The total order for executing the RVS process can be seen in FEMA P-154 [9]. After the project's scope and budget are defined, enactment of the RVS program persists with further pre-field actions, as follows:

1) Pre-field development, to include selecting and developing a record-keeping procedure, development of any desired electronic scoring devices, as well as the assemblage and development of maps detail the local seismic risk data;
2) Choice of a Data Collection Form centered on the seismic risk and review and revision of the Data Collection Form for the specific requirements of the RVS system;
3) Choice and training of the screening personnel;
4) Procurement and the review of pre-field information, to include a review of accessible structure files and databases to gather existing data on the structures that is to be screened (e.g., the address, the lot number, amount of levels, and the date it was designed) and identifying the types of soil for the area being surveyed; and
5) Evaluation of current structure plans, if they are available.

2.3. Tier 1 Evaluation

Tier 1 evaluation (screening phase) comprises several groups of checklists which let a quick assessment of the building's structural, nonstructural, as well as the foundation and the geologic risk aspects. As seen in table 1, the Basic Structural Checklist will be done for structures in areas of low seismicity getting assessed on the Immediate Occupancy Performance Level and structures in areas of both moderate as well as high seismicity. The Supplemental Structural Checklist will be done as well as a Basic Structural Checklist for structures in areas of moderate seismicity getting assessed for the Immediate Occupancy Performance Level and structures in areas of high seismicity.

The intention of the screening phase of FEMA 310 [10] is to detect those structures which seem to conform to the authorized seismic performance. During the screening phase, a building's structural type is selected for each of the buildings utilizing one of twenty-four types of building. A FEMA change for 310 [10] in comparison to FEMA 178 [25] adds brand-new building types. The layout of the subject structure is first assessed in contrast to the design requirements of the model's building codes dependent on the building's structural type. If the structure is recognized as having been conceived and built in agreement with the selected model building code for that particular building style, called the benchmark structures, and after that the structure is considered to have passed the structural design constraints [26].

Table 1. Checklists required for a Tier 1 evaluation [10]

Region of seismicity	Level of performance[2]	Required Checklist[1]					
		Region of low seismicity	Basic structural	Supplemental structural	Geologic site hazard and foundation	Basic nonstructural	Supplemental nonstructural
Low	LS	✓					
	IO		✓		✓	✓	
Moderate	LS		✓		✓	✓	
	IO		✓	✓	✓	✓	✓
High	LS		✓	✓	✓	✓	
	IO		✓	✓	✓	✓	✓

[1]A checkmark designates that the checklist must be completed for a Tier 1 Evaluation as a function of the region of seismicity and level of performance.
[2]LS = Life-Safety; IO = Immediate Occupancy

Limited structural computations, known as Quick Checks [10], are utilized within this phase of the assessment. Rough computations are done to approximate the seismic forces opposed by the main lateral force resisting aspects. These forces are then calculated dependent on an unlimited earthquake base shear yet subsequent forces are diminished at the element level and then contrasted with an acceptable force. The intention is these Quick Checks can be done with a small amount of computational energy and it does not necessitate a comprehensive structural assessment of the building.

3. Result and discussion

3.1. Final level 1 score

Figure 2 indicates the chosen Data Collection Form. Centered on the structure type, the case-study structure accepted as C1 (moment-resisting frame) and C3 (concrete frame structures with unreinforced masonry infill walls), so the basic scores are 1.7 and 1.4, respectively. Moreover, the assessment results reveal that with the RVS technique, the structure has a final level 1 score, S_{L1}, +3.6, and +1.4 for structure type C1 and C3 respectively. As the final level 1 score, S_{L1}, for structure type C3 less than the cut-off score, 2.0, a detailed structural evaluation is necessary. Essentially, the score is an assessment of the likelihood (or chance) that the structure will collapse if ground motions happen that equal or surpass the maximum considered earthquake targeted risk (MCE_R) ground motions. For instance, a final score of +1.4 indicates a probability of $10^{1.4}$, or 1 in roughly 4.00, that the structure will collapse if these ground motions happen. These score estimates are centered on limited noted and analytical information, hence the likelihood of collapse is consequently an estimate. It should be stated that the prior study done by Haryanto et al. supplied more details of the RVS findings for the case-study structure along with a further 8 academic structures at the Jenderal Soedirman University in Indonesia [18].

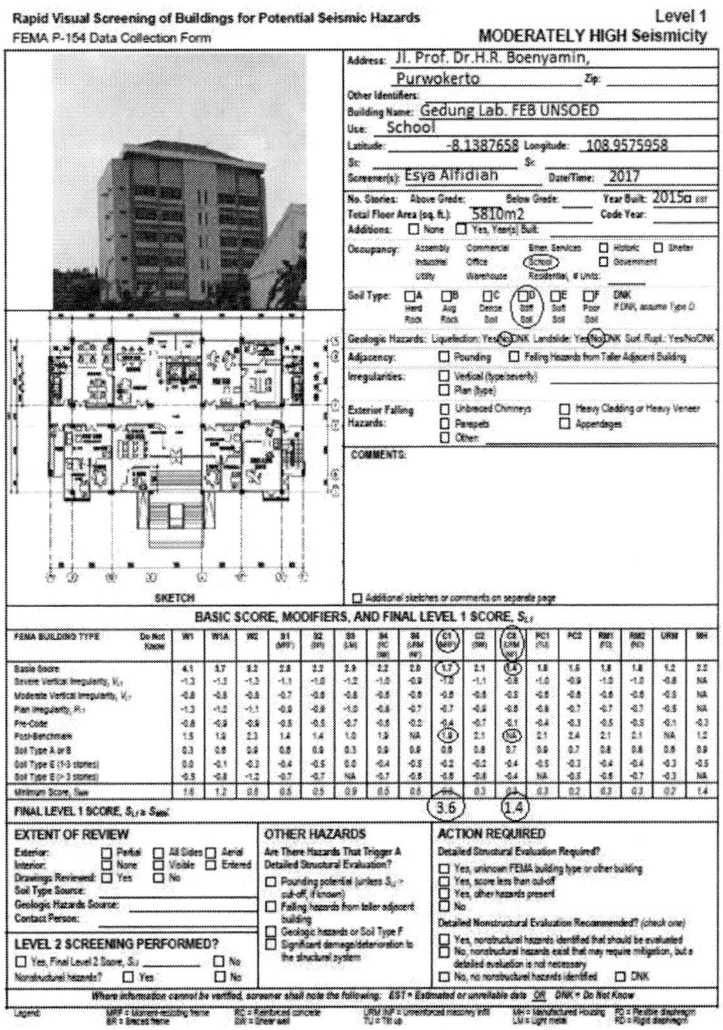

Figure 2. Data collection form for the case-study building

IOP Conf. Series: Materials Science and Engineering **982** (2020) 012031 doi:10.1088/1757-899X/982/1/012031

3.2. Screening phase

It is crucial to collect pertinent data of a structure as much as feasible via illustrations, analysis, design computations, a soil report (if offered), inspection and prior inspections report, and prior repair works [27]. At minimum one location official visit will be made to assess exposed conditions of the configurations, construction features, site and its base, and the adjoining structures [28]. The most crucial step for suitable condition assessment of a structure is identifying any current damages and the probable causes of the damage. Table 2 shows the checklists considered necessary in this research for the moment resisting frame in in high seismicity. With a structure in high seismicity, it is required to finalize the components for low along with moderate seismicity, along with the elements of high seismicity like a strong column and a weak beam, or column and beam bar splices and the spacing of stirrups. Each checklist evaluation statement will be marked as being Compliant (C), being Noncompliant (NC), being Unknown (U), or being Not Applicable (N/A).

Tabel 2. Tier 1 checklist

No	Parameters	Statements
Basic Structural Checklist		
1.	Load path	C
2.	Adjacent buildings	C
3.	Soft story	NC
4.	Geometry	C
5.	Vertical discontinuities	C
6.	Mass	C
7.	Torsion	C
8.	Deterioration of concrete	C
9.	Redundancy	C
10.	Interfering wall	NC
11.	Shear stress check	NC
12.	Axial stress check	C
13.	Connections of concrete coloums	C
Supplemental Structural Checklist		
1.	Flat slab frame	C
2.	Shortcaptive coloums	C
3.	Coloumns-bar splices	C
4.	Coloumns-tie spacing	NC
5.	Stirrup spacing	C
6.	Joint reinforcing	NC
7.	Joint eccentricity	C
8.	Strirrup and tie hooks	C
9.	Deflection compatibility	NA
10.	Flat slabs	C
11.	Diaphgrahm reinforcement at openings	NA
12.	Lateral load at pile caps	NA

Basic Non- Structural Checklist		
1.	Unreinforced mansory	NC
2.	Integrated ceilings	C
3.	Support	C
4.	Canopies	C
5.	Heavy equipment	C
Geologic and Foundation Chekclist		
1.	Liquefaction	C
2.	Slope failure	C
3.	Surface faultrupture	C
4.	Foundation performance	C
5.	Deterioration	C
6.	Pole foundation	NA
7.	Overturning	NA
8.	Ties between foundation elements	C
9.	Deep foundations	NA
10.	Sloping sites	C

The case study applies Tier 1 screening and the checks of the axial and shear stresses of the columns are completed. The Integrated Laboratory, Faculty of Economics and Business, Jenderal Soedirman University, Indonesia doesn't comply with a few elements of life safety performance goal. Following the procedure, it was discovered that a few deficiencies existed in the evaluated building. For example, the deficiencies formed the Basic Structural Checklist as follows: the stiffness of the lateral-force-resisting system in the first level is less than 70 percent of the stiffness in an adjoining level above. The next deficiency is that not all the infill walls were put in moment frames and separated from the other structural components. The 3rd deficiency was the shear stress of the concrete columns, computed utilizing the Quick Check process was discovered to be over 100 psi or $2\sqrt{f_c^{'}}$ for several levels. The design expert might decide to (1) report the deficiencies and advise mitigation or (2) conduct to bury in the 2nd phase (Tier 2). For Tier 2, the evaluation of the deficiencies will be done centered on the obligations of evaluation determined in Tier 1. In this occurs, the building assessment necessitates the as-built dimensions, the reinforcement particulars of all the structural components, a soil report and basic tests to ascertain the concrete's strength.

4. Conclusions

This paper examined the seismic performance of a representative 7-story moment-resisting frame or concrete frame building with unreinforced masonry infill walls in Purwokerto, Indonesia that designed for an academic building in Jenderal Soedirman University during 2015. Rapid Visual Screening (RVS) and Tier 1 evaluation (screening phase) were carried out following procedures contained in two modern codes or guidelines: FEMA P-154 and FEMA 310. The results indicated:

- The building have a final level 1 score, S_{L1}, +3.6, and +1.4 for building type C1 (moment-resisting frame) and C3 concrete frame building with unreinforced masonry infill walls) respectively.
- It has been found several deficiencies in the assessed building form the screening phase.
- The design professional may choose to report deficiencies and recommend mitigation or conduct to inter in the second phase (Tier 2).

Acknowledgments

The authors expressed appreciation and gratefulness to the Institute for Research and Community Service (LPPM) of Jenderal Soedirman University for providing the funding for the research. The first author also acknowledge the support of the Bank SinoPac Taiwan in the form of the Establishment of Bank SinoPac Outstanding Talents Scholarship in National Cheng Kung University.

References

[1] Haryanto Y, Sudibyo G H and Effendi F C A 2017 *Procedia Eng.* **171** 1025
[2] Haryanto Y, Gan B S, Wariyatno N G and Indriyari E W 2017 *ARPN J. Eng. Appl. Sci.* **12** 4858
[3] Haryanto Y, Gan B S, Wariyatno N G and Indriyari E W 2018 *MATEC Web Conf.* **192**
[4] Widyaningrum A, Haryanto Y and Hermanto N I S 2018 *MATEC Web Conf.* **195**
[5] Haryanto Y, Hu H-T, Han A L, Hidayat B A and Indriyari E W 2019 Aceh Int. J. Sci. Technol. **8** 1
[6] Widyaningrum A, Haryanto Y, Sudibyo G H and Hermanto N I S 2019 J. Phys. Conf. Ser. **1367**
[7] Perrone D, Aiello M A, Pecce M and Rossi F 2015 *Structures* **3** 57
[8] Samadian D, Ghafory-Ashtianty M Naderpour H and Eghbali M 2019 *Soil Dyn. Earthq. Eng.* **127**
[9] Federal Emergency Management Agency 2015 *FEMA P-154 Rapid Visual Scre ening of Buildings for Potential Seismic Hazards: A Handbook* (Washington, DC: Federal Emergency Management Agency)
[10] Federal Emergency Management Agency 1998 *FEMA 310 Handbook for the Seismic Evaluation of Buildings: A Prestandard* (Washington, DC: Federal Emergency Management Agency)
[11] Ohkubo M 1991 *Report No. SSRP-91/02 Current Japanese System on Seismic Capacity and Retrofit Techniques for Existing Reinforced Concrete Buildings and Post-Earthquake Damage Inspection and Restoration Techniques* (San Diego: Department of Applied Mechanics and Engineering Sciences, University of California)
[12] Yakut M 2004 *Eng. Struct.* **26** 1447
[13] O'Reilly G J, Perrone D, Fox M, Monteiro R and Filiatrault A 2018 *Eng. Struct.* **168** 142
[14] Michel C, Hannewald P, Lestuzzi P, Fäh D and Husen S 2017 *Bull. Earthq. Eng.* **15** 1471
[15] Jaimes M A and Niño M 2017 *Bull. Earthq. Eng.* **15** 3919
[16] De Angelis A and Pecce M 2015 *Nat. Hazards* **79** 333
[17] Mazumder R K, Khair A, Sakib N, Bhuiyan M A R and Alam M J 2014 *J. of South Asian Dis. Stud.* **6** 9
[18] Haryanto Y, Hu H-T, Han A L, Hidayat B A, Widyaningrum A and Yulianita P E 2020 *Civ. Eng. Dim.* **22** 13
[19] Sarmah T and Das S 2018 *Proc. Eng.* **212** 214
[20] Lang K 2002 *Seismic Vulnerability of Existing Buildings* (Zurich: Swiss Federal Institute of Technology)
[21] Bureau of Indian Standards 2009 *IS 13935: 2009 Indian Standard Guidelines for Repair and Seismic Strengthening of Buildings* (New Delhi: Bureau of Indian Standards)
[22] Jain S K, Mitra K, Kumar M and Shah M 2010 *Earthq. Spectra* **26** 709
[23] Japan Building Disaster Prevention Association 2001 *Seismic Evaluation and Retrofit* (Tokyo: Japan Building Disaster Prevention Association)
[24] Institute for Research in Construction 1993 *Manual for Screening of Buildings for Seismic Investigation* (Ottawa: National Research Council of Canada)
[25] Kehoe B 2004 *Proc. 13th World Conf. on Earthquake Eng.* (Vancouver: International Association for Earthquake Engineering)
[26] Federal Emergency Management Agency 1994 *FEMA 178 NEHRP Handbook for the Seismic Evaluation of Existing Buildings* (Washington, DC: Federal Emergency Management Agency)
[27] American Society of Civil Engineers 2014 *ASCE/SEI 41-13 Seismic Evaluation and Retrofit of Existing Buildings* (USA: American Society of Civil Engineers)
[28] Chakrabarti A 2007 *Handbook on Seismic Retrofit of Buildings* (India: Central Public Works Department & Indian Building Congress)

IOP Conf. Series: Materials Science and Engineering **982** (2020) 012032 doi:10.1088/1757-899X/982/1/012032

Load-carrying capacity and failure mode of composite steel-concrete truss element under monotonic loading

Nanang Gunawan Wariyatno[1,3], Yanuar Haryanto[2,3,5], Ay Lie Han[1], Buntara Sthenly Gan[4] and Gathot Heri Sudibyo[1,3]

[1]Department of Civil Engineering, Diponegoro University, Semarang, Indonesia
[2]Department of Civil Engineering, National Cheng Kung University, Tainan, Taiwan
[3]Department of Civil Engineering, Jenderal Soedirman University, Purwokerto, Indonesia
[4]Department of Architecture, Nihon University, Koriyama,, Indonesia

[5]yanuar.haryanto@unsoed.ac.id

Abstract. Composite steel-concrete trusses are among the most economical systems for building, particularly for greater spans, allowing proper internal space utilization without limiting columns. This paper examines 12 lab tests on two types of composite steel-concrete truss elements compared with the conventional steel truss element as a control specimen. The experiments investigate the load-carrying capacity of the truss element with different configurations under monotonic loading. The study on the specimen's failure modes was based on the welded joints. The obtained results were compared with analytical evaluations of typical resisting mechanisms of steel-concrete composite, reinforced concrete, and steel structures with updated Indonesian design codes. The basic features of the steel-concrete truss elements discussed quantitatively and qualitatively.

1. Introduction

Composite materials are widely used to design truss structures since they have suitable properties [1]. Composite steel-concrete trusses are highly economical for designing structures with spans of up to 20 m [2]. These structural elements are used for longer spans of about 30 m, ensuring effective utilization of internal space with no impact to the columns [2]. Furthermore, the trusses not only facilitate the achievement of the exact limits in building height, but are also useful in managing complex systems involving electricity, heat, ventilation and communication. The filigree steel truss and slim piers are used to support composite steel bridges with carriageway deck to increase their fitness, especially the ordinary concrete bridges. It accounts for the technical and architectural aspects with concession between landscape protection and hard transport basics. Therefore, a composite truss bridge, with a quick engineering assembly is economically and aesthetically appropriate.

Several studies focused on the strategies for designing composite truss structures. For instance, Shütze [3] developed and implemented equilateral triangle composite truss beams with cross-sectional bases by joining individual parts to bonded joints, and supporting them with gusset plates. The structured beams were used in the rigid airship frame. Weaver and Jensen [4] used a braiding machine to create a truss beam by employing IsoTruss®, a concept that is currently under commercial development. This is advantageous because it does not require the designing of separate individual members to be connected. In other studies, internal mandrels have been used to design a similar basic structure [5–7], through winding-based techniques. The mandrel geometries are, therefore, created by strengthening the nodes of the non-continuous cross-section. As a result, the complex multi-component solutions are easily dissembled once the composite is cured.

Another studies have examined types of braided truss structures that facilitate geometrical simplification. This is achieved by using a circular cross-sectioned mandrel [8]. The circular cross-sections are flexible and weaker because they function as beams under combined axial and bending loading. Although many studies have examined composite truss elements' structural behavior, accurate knowledge of these elements' behavior is vital. It is also critical to study how different parameters affect the composite truss elements' efficiency. This paper mainly deals with the prime significance for the behavior of composite steel-concrete trusses subjected to monotonic loads. Specifically, the paper primarily addresses the investigation of the load-carrying capacity and failure mode mechanisms in composite steel-concrete trusses in experimental three-point bending tests on specimen representative of the elements stated.

2. Material and methods

Tests were conducted on a series of composite steel-concrete truss elements in Howe and Waren configuration. The top chord was integrated with a concrete material (figure 1 and figure 2). The total length of the element and span between support axes was 1000 mm, while the height and width were 250 mm. The concrete material was designed to have 50 mm wide and 70 mm thick. The average compressive strength of used concrete determined on the cubic samples was 14.19 MPa and 17.15 MPa for Howe and Waren configuration specimens. The bottom chords and webs were made of steel bars with 12 mm diameter having a yield strength of 300 MPa, while a 10 mm diameter steel bars with a yield strength of 353 MPa were used as the top chords. Finally, all the members were connected using a welded joint.

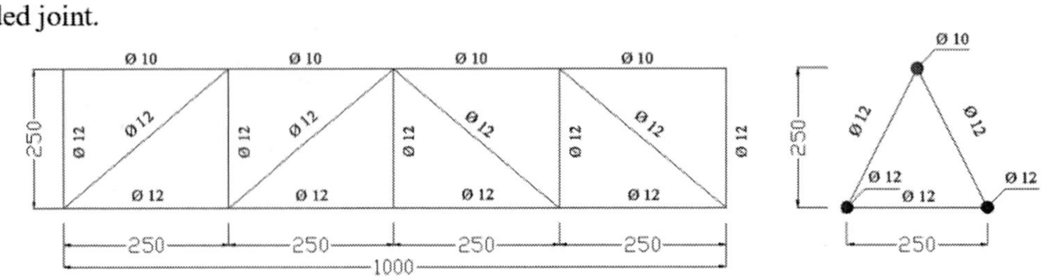

(a) TH-S specimen

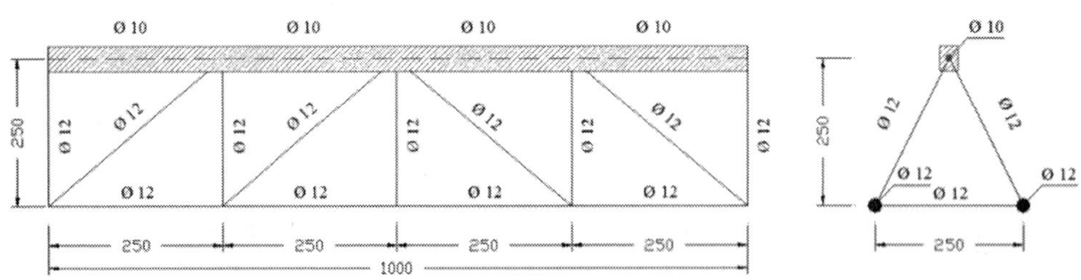

(b) TH-C specimen

Figure 1. Geometric characteristics of the Howe configuration specimens

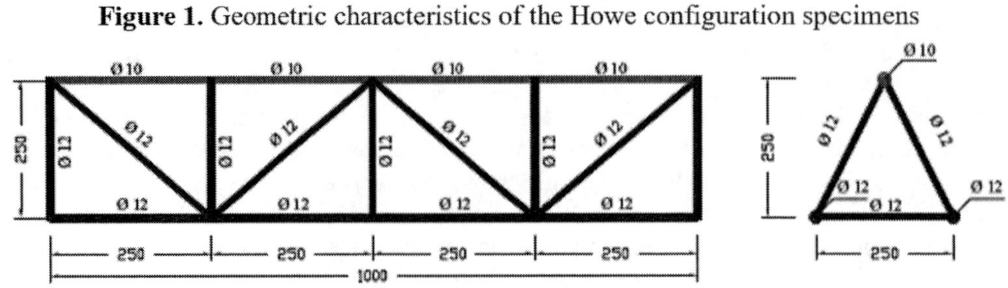

(a) TW-S specimen

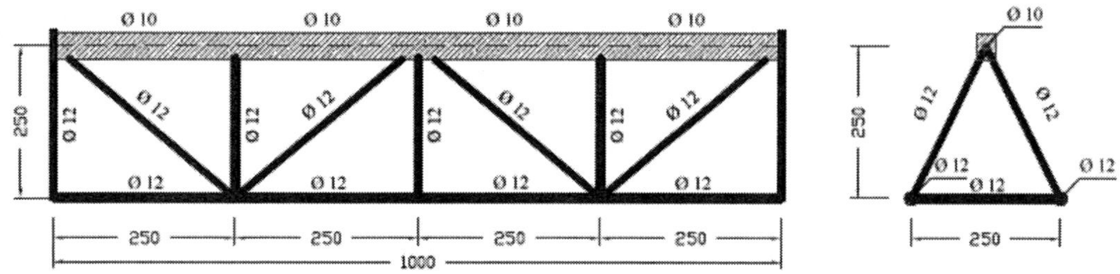

(b) TW-C specimen

Figure 2. Geometric characteristics of the Waren configuration specimens

The first letter of the specimen code indicates it is a truss element while the type of configuration is shown by the second letter, followed by whether it is a conventional steel truss element or composite steel-concrete truss element shown by the last letter of the specimen code. During the tests, the specimens were under a monotonic load with a scheme analogical to three-point bending a freely supported beam, as shown in figure 3. The load from Universal Testing Machine (UTM) was applied to the upper concrete part over the central node of the steel-concrete truss element. The experiments we performed at the Jenderal Soedirman University, Indonesia, as a research program to validate methods and criteria for element typology's structural design.

Figure 3. The scheme of test

3. Result and discussion

3.1. Load-carrying capacity

At the time of failure of the specimen, the load level was the final loading level, taken as the ultimate load-carrying capacity. The ultimate load-carrying capacity of each specimen is presented in table 1, where the average load-carrying capacity of the conventional steel truss element with Howe configuration (TH-S) could reach 43.43 kN, while the average value for the composite steel-concrete truss element with Howe configuration (TH-C) was 58.43 kN. The measured value for the TW-C specimen was 34.54% higher than the TW-S specimen. The average load-carrying capacity of the composite steel-concrete truss element with Waren configuration (TW-C) was improved by about 56.66% compared to the conventional steel truss element with Waren configuration (TW-S). This was mainly caused by the placement of concrete material in the top chords of the truss elements that enhanced the members' compression resistance.

IOP Conf. Series: Materials Science and Engineering **982** (2020) 012032 doi:10.1088/1757-899X/982/1/012032

Table 1. The ultimate load-carrying capacity

No.	Specimen code	Load-carrying capacity (kN)	Average value (kN)
1.	TH-S-01	39.50	
2.	TH-S-02	45.50	43.43
3.	TH-S-03	45.30	
4.	TH-C-01	65.50	
5.	TH-C-02	54.60	58.43
6.	TH-C-03	55.20	
7.	TW-S-01	34.40	
8.	TW-S-02	36.30	35.33
9.	TW-S-03	35.30	
10.	TW-C-01	59.50	
11.	TW-C-02	59.10	55.33
12.	TW-C-03	47.40	

3.2. Mode of failure

The specimens were designed for compression failure. For this reason, the welded joint was strengthened in the best way to avoid tension failure. The welding process was conducted two times because, during the preliminary test, there was a tension failure on the welded joint. The welding process had to be strengthened again (re-welding). When the final tests were performed, all the truss elements failed on the compression chords. The failure mode was the compression chords experiencing buckling, as shown in figure 4 and figure 5.

(a) TH-S specimen (b) TH-C specimen

Figure 4. Failure mode of Howe configuration specimen

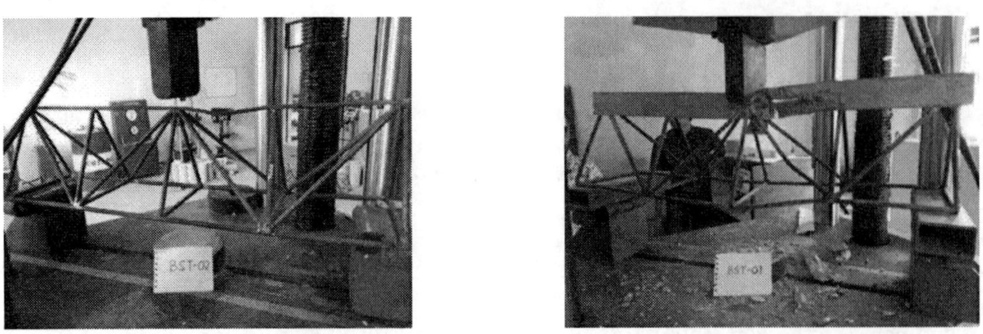

(a) TW-S specimen (b) TW-C specimen

Figure 5. Failure mode of Waren configuration specimen

3.3. Comparison with analytical modeling

The experimental results were compared with the analytical modeling of typical resisting mechanisms of steel-concrete composite, reinforced concrete, and steel structures. They were then updated based on Indonesian design codes, developed using ETABS, as shown in figure 6.

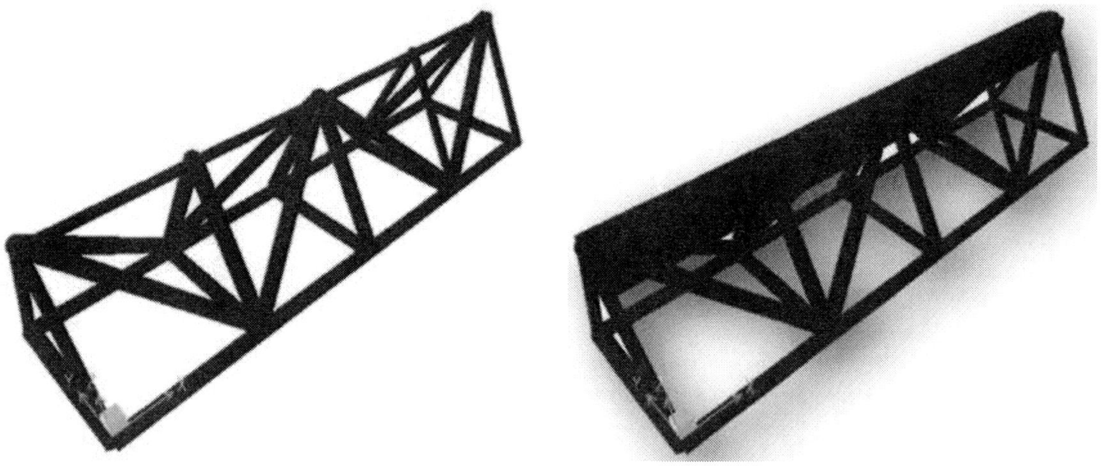

(a) Convensional steel truss model (b) Composite steel-concrete truss model

Figure 6. Analytical modeling

The analytical evaluation showed that the TH-S and TW-S model's load-carrying capacity was 27.13% and 8.36% lower than the average experimental results. However, there was an increase in the TH-C and TW-C model's load-carrying capacity by about 10.95% and 6.42% compared to the average value of the experimental results. This is because the material properties were homogeneous in all chords in the developed analytical models. The graphical comparison of the load-carrying capacity for all models is shown in figure 7.

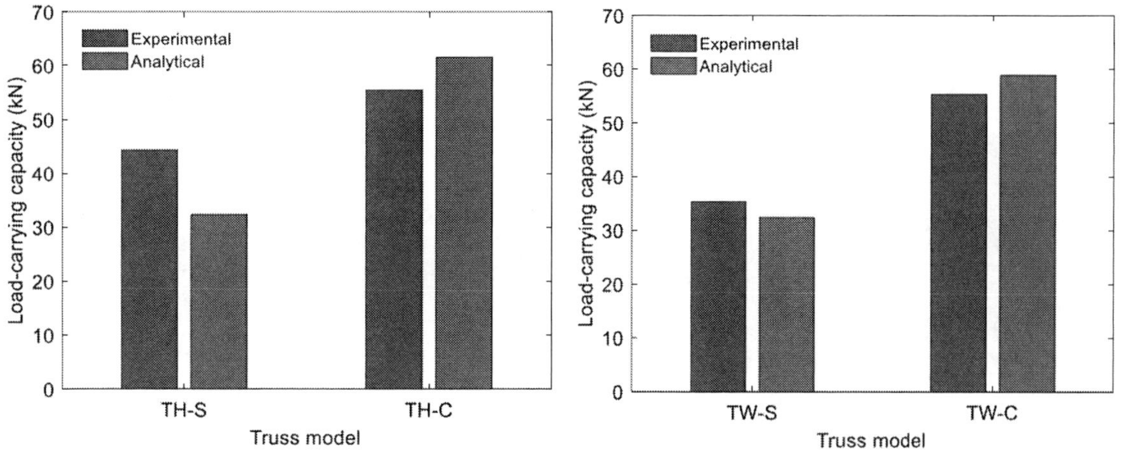

(a) Howe configuration (b) Waren configuration

Figure 7. Comparison of load-carrying capacity

4. Conclusions

This study tested 12 specimens on two type configurations of composite steel-concrete truss structures under monotonic loading. Furthermore, it examined the load capacity of the failure modes. The results were compared with the analytical modeling of the steel structures. The study made the following conclusions:

- The average load-carrying capacity was 34.54% and 56.66% higher due to the placement of concrete material in the truss elements' top chords for Howe and Waren configurations, respectively.
- The compression chords experienced buckling, leading to compression failure in all specimens.
- There was a reasonable difference between the analytical evaluation and experimental test results due to the homogeneous material properties for all chords in the models.

Acknowledgments
The research team appreciated the funding from the Institute for Research and Community Service (LPPM) of Jenderal Soedirman University.

References

[1] Sombabu Y 2018 *Int. J. Eng. Tech.* **4** 484
[2] Bouchair A, Bujnak J and Duratna P 2012 *Procedia Eng.* **40** 96
[3] Schütze R 1997 *Mater. Des.* **18** 231
[4] Weaver T and Jensen D 2000 *J. Aerosp. Eng.* **13** 23
[5] McCune D. 2005 *Manufacturing Quality of Carbon/Epoxy IsoTruss® Reinforced Concrete Structures* (Birmingham: Brighman Young University)
[6] Lai C, Wang J, Liu C, Fan H, Xu B and Wu K 2015 *Compos. Sci. Technol.* **113** 63
[7] Sui Q, Fan H and Lai C 2015 *Compos. Sci. Technol.* **118** 207
[8] Gurley A, Beale D, Broughton R and Branscomb D. 2015 *J. Mech. Des.* **137** 101401

IOP Conf. Series: Materials Science and Engineering **982** (2020) 012033 doi:10.1088/1757-899X/982/1/012033

Capacity-based seismic design of a middle-rise residential building in an area of moderately-high seismicity

Banu Ardi Hidayat[1,2], Yanuar Haryanto[1,3,6], Nuroji[2], Sukamta[2], Wisnu Kurnia Praja[2,4] and Hsuan-Teh Hu[1,5]

[1]Department of Civil Engineering, National Cheng Kung University, Tainan, Taiwan
[2]Department of Civil Engineering, Diponegoro University, Semarang, Indonesia
[3]Department of Civil Engineering, Jenderal Soedirman University, Purbalingga, Indonesia
[4]PT Adhi Karya (Persero) Tbk, Jakarta, Indonesia
[5]Department of Civil and Disaster Prevention Engineering, National United University, Miaoli, Taiwan

[6]yanuar.haryanto@unsoed.ac.id

Abstract. As Indonesia is located across three active tectonic plates, it is prone to a high potential earthquake threat. Strict regulations have been introduced to prevent damage to buildings caused by seismic activity. This study aims to examine the behavior of a middle-rise residential building in a moderately-high seismicity area (i.e. Semarang City) and the design of the Reinforced Concrete (RC) members under seismic load. The design was based on the capacity method and the strong column-weak beam concept was applied to the plastic hinge effect that occurred in the beam. Therefore, the analytical computation was performed, taking the structural seismic response into account. Also, the code-based design calculation of RC members was achieved. The measured structural response showed that the building fulfilled the requirements stated in the Indonesia Code of Earthquake-based Design. The first two modes of the building translated in the Y and X directions, which is the preferred response under the earthquake. The story-drifts that occurred were within the acceptable limits and the structural stiffness in the X and Y directions were found to be similar. Finally, the exact design of RC members was found to satisfy the code.

1. Introduction

Indonesia is vulnerable to high potential earthquake hazards because it is situated across three active tectonic plates [1–5]. Recent seismic activities in Indonesia have lead to excessive damage to structures and caused loss of life. In order to prevent this damage, strict regulations are contantly being reviewed and altered to meet the current conditions [6]. Large earthquakes have hit Indonesia many times in recent decades [7–9]. The seismic forces exerted on high buildings will have more complex effects than those seen in low-rise buildings, as many structural elements may have been damaged and weakened [10]. Consequently, the strength and capacity of the higher buildings under seismic loading need to be more carefully observed and calculated.

Many structures are designed on stiffness and strength criteria, with the stiffness being attributed to the serviceability limit state, which maintains the displacements under the loading impact and keeps them within the acceptable limit. However, strength is related to the ultimate limit state and assumes that the forces in the structure, due to loading, remain in the elastic range [11]. By using the capacity design method, the flexural capacity of critical member sections, based on hypothetical behavior under seismic load, can be achieved. This is expressed in the assumption that the static equivalent of seismic load increases slowly until the near-collapse state of the building is reached and the object still remains elastic. Also, the plastic hinges effects are expected to occur simultaneously at the predetermined locations (assumed to be located at the end of the beam), to process the strong column-weak beam mechanism. Finally, the ductile behavior mechanism can be formed and the structure can possess an acceptable seismic resistance.

Numerous researchers have studied the capacity-design method on buildings to obtain the resistant design of the structures [12–16]. However, studies on the more complex conditions of the structural model should be considered to enrich the understanding of the seismic loading effects on Reinforced Concrete (RC) buildings. Hence, this study addresses the structural behavior of middle-rise residential buildings located in moderately-high seismicity (i.e. Semarang City) and soft soil conditions. Also, the structural design of the RC members under earthquake loading can be calculated based on the capacity design method. Numerical simulation, using ETABS software, was performed to develop a model of the RC building. The code-based design calculation was also achieved and depends on the structural model, which takes the suitable structural behavior under seismic conditions into account.

2. Numerical simulation

The numerical model in this research is taken from a residential RC building with 17 stories and the special moment frame system. The total height of the building is 51 metres, with an area of 28 x 60 m in the X and Y direction, respectively. From the car parking spaces on the basement floor up to the third floor, the floor composition becomes half-floors to account for the movement of the cars into the parking spaces. The base of the actual building is not at the same level, some structures are lower than others. This may cause the center of mass to not be the same as the center of stiffness, which may result in rotation behavior. The soil condition is assumed to be soft, which is likely to create significant site amplification effects during the earthquake events [17] and can increase the inter-story drifts and lateral deflections in mid-rise buildings [18].

The analytical solution was carried out with the appropriate finite elements model created in the ETABS simulation program. The building was designed with the beam-slab option, which accounts for the RC columns, beams, slabs, and shear walls in the three-dimensional analysis. Linear models considering elastic behavior are used for structural development. The models are comprised of frame and shell elements to represent structural components. The lateral stability of the building can be achieved by the dual system, combining the special moment frame system and the RC shear walls. The concrete's compressive strength and the yield strength of the reinforcement are recorded as 25 MPa and 400 MPa, respectively. Thus, all ordinary RC beam members are modeled as frame elements, however, the shear walls are modeled as linear, thin shell elements. The seismic loads absorbed in the building were expected to be accommodated by the combined system.

The additional dead load and live load applied in the model is based on the Indonesian Standard Code for Loading in the Building, with the values for apartment buildings being 125 kg/m^2 and 250 kg/m^2, respectively. However, the input for the earthquake load analysis used the dynamic spectrum response method, applied to the moderately-high seismicity area in Semarang City with a damping value of 5%. To simulate the expected arbitrary earthquake loading, the seismic loading influence on the major effective direction of 100% is assumed to be occurring at the same time as the effect of only 30% effective loading, perpendicular to the main direction [19]. The plastic hinge effect, expected to occur at the end of the beam to account for the method of the strong column with the weak beam, and the structural components can display ductile behavior. Figure 1(a) to 1(c) depict the structural model for the mid-rise residential building, in terms of the structural plan layout, side view, and three-dimensional isometry, respectively.

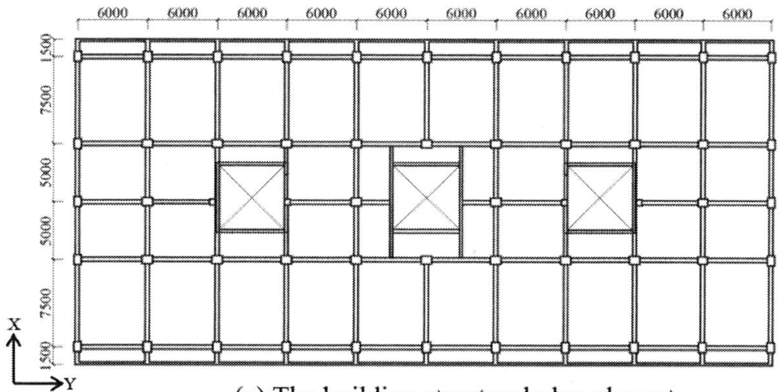

(a) The building structural plan playout

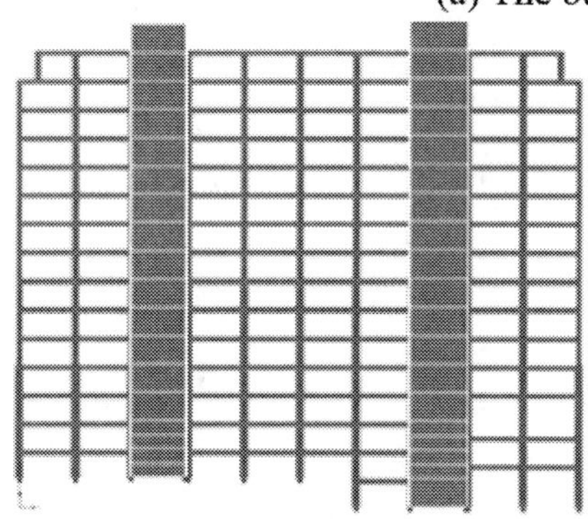

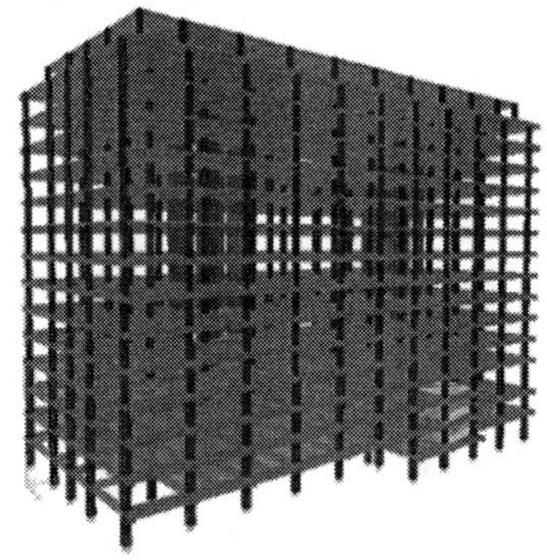

(b) The side view of the building (c) The 3D isometry of the building

Figure 1. The structural model for the residential building.

3. Overall response evaluation

3.1. Modal analysis

The mode shapes of the building are shown in Figure 2. Thus, the natural periods and modal participating mass ratios of the first ten modes are shown in Table 1.

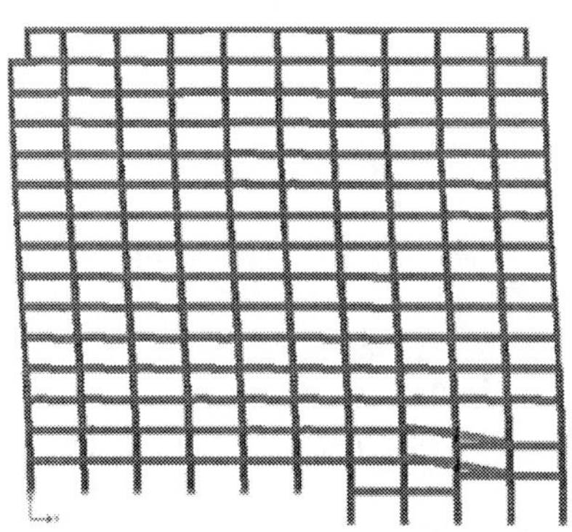

(a) Translation in Y direction, with $T_{1(cracked)} = 1.225$ s

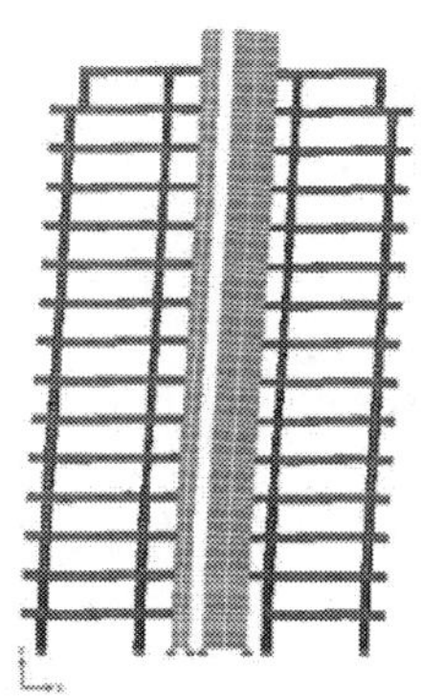

(b) Translation in X direction, with $T_{2(cracked)} = 1.2$ s

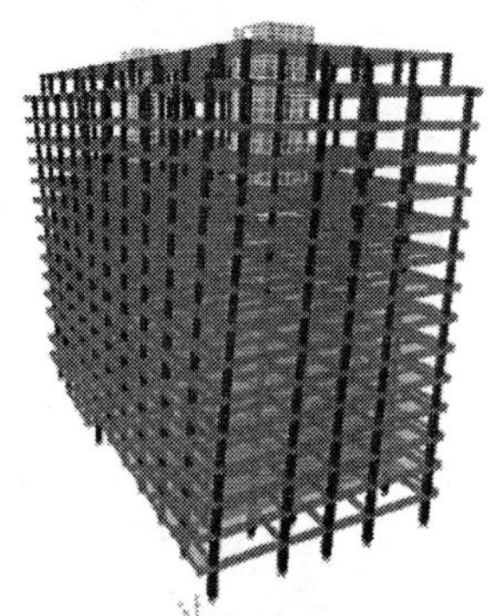

(c) Torsion, with $T_{3(cracked)} = 1.12$ s

Figure 2. Mode shapes and natural period of the residential building.

Modal analysis was performed to determine the vibration modes of a building. A combination of mass, from 100% of dead load and superimposed dead load plus 30% of the live load, is considered in modal analysis. The first two modes of the building translated in the Y and X directions, respectively, gave the preferred response under earthquake loading. Furthermore, the rotation behavior appears in the third mode, as presented from the tiny value of the participating mass with no dominant value for the X and Y direction. Approximately 66.91% of the total mass was determined to be participating in the first translation modes, showing that the higher modes were significant in the seismic response. Sufficient numbers of modes are considered so that more than 90% of the mass is participating in a total participating mass ratio in each orthogonal direction.

Table 1. Mass participation ratio for the structural building model.

Mode	Period (s)	Modal mass participation ratio			
		Y (%)	X (%)	Cumulative Y (%)	Cumulative X (%)
1	1.225	0.017	0.669	0.017	0.669
2	1.200	0.648	0.024	0.665	0.693
3	1.120	0.058	0.006	0.724	0.699
4	0.352	0.136	0.002	0.860	0.701
5	0.339	0.013	0.018	0.873	0.720
6	0.330	0.000	0.140	0.873	0.861
7	0.178	0.040	0.000	0.914	0.861
8	0.175	0.007	0.000	0.921	0.862
9	0.157	0.001	0.056	0.922	0.918
10	0.116	0.002	0.001	0.924	0.919

3.2. Base shear

The elastic seismic base shear above ground level, resulting from static lateral force procedure, and response spectrum analysis, with an R-value as the modification coefficient, are summarized in Table 2. It was mentioned in the Indonesian Standard Code of Earthquake-resistant Design [20] that the comparison value between the dynamic and static procedures for base shear force is over 85%.

Table 2. The ratio of base shear force above ground level.

Base shear	Equivalent static (V)	Dynamic response (V_t)	Ratio (V_t/V)
F_x (kN)	18351.55	19116.11	1.04
F_y (kN)	18720.55	19500.48	1.04

The calculated results show that the value of the comparison between the static and dynamic base shear meets the requirements in the code, which was found to have been exceeded by 85%. This means that the magnification of earthquake forces is not required. It can also be seen from Table 2 that the distribution of base shear in a residential building shows results that are not much different for the X and Y direction. This means that the seismic effects on the building are almost the same in both directions, regarding the base shear force.

3.3. Story drifts

Story drifts, due to earthquake loading along the height, are plotted in Figure 3 and it was checked whether the drifts were within the acceptable limits or not. The requirements stated in the Indonesian Code [20] can be stated as equation (1). The symbol of ρ stands for the redundancy factor which should be obtained from several preconditions mentioned in the standard code and the value of ρ in this work can be taken as 1.3. Thus, the height of the floor below the corresponding story was signified by the h_{sx} notation.

$$\Delta_i < (0.02h_{sx})/\rho \tag{1}$$

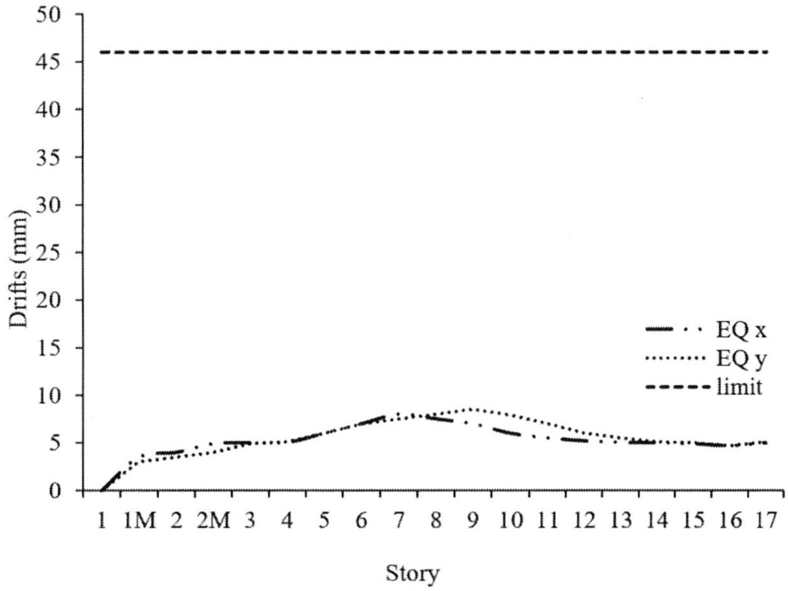

Figure 3. The story drifts in the X and Y direction due to seismic loading.

Figure 3 shows that the story drifts are within the acceptable limit, which is 2% under seismic loading. The maximum calculated story drifts in the X and Y directions are 8.34 mm and 8.67 mm, respectively. This is still below the acceptable limit of 46.15 mm, as calculated with equation (1).

4. Structural design

4.1. RC columns

Theoretically, the minimum value of the axial force acting on the columns should be more than 10% of the column's sectional area multiplied by the compressive strength of the concrete. The columns will transfer the combination of the axial force from the gravity load and the seismic force and also the flexural force from the beams to the foundations. The total area of longitudinal reinforcement obtained in the column is limited to between 1% and 4% by comparing the area of the column itself. The shear reinforcement calculation is based on equation (2), along with the total shear force, the moment acting on the top and bottom of the column, the clear span, the strength reduction factor, and the shear force contributed by the steel and the concrete are symbolized by V_u, M, l_n, $Ø$, V_s, and V_c, respectively.

$$V_u = \frac{M_{top} + M_{bot}}{l_n} \text{ and } V_s = \frac{V_u}{\phi} - V_c \tag{2}$$

Furthermore, in order to model the behavior of the columns, some control calculations need to be conducted, including the slenderness ratio, the bi-axial flexural check and a comparison between the flexural strength of the column and the beams. The calculation of the axial-flexural interaction diagram can then be evaluated and checked against their force demands. Table 3 outlines the reinforcements detail of the columns used in the RC building; three types of columns were considered in the design.

IOP Conf. Series: Materials Science and Engineering **982** (2020) 012033 doi:10.1088/1757-899X/982/1/012033

Table 3. The cross-section and reinforcement details of the RC columns (unit: mm).

Beam type	On the end side	On the middle side
C1 950x950	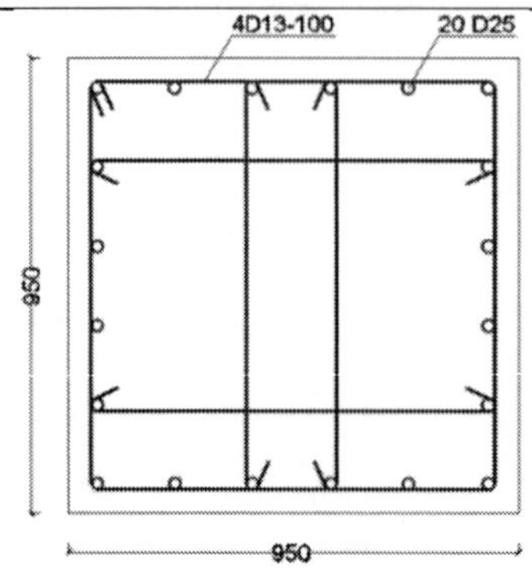	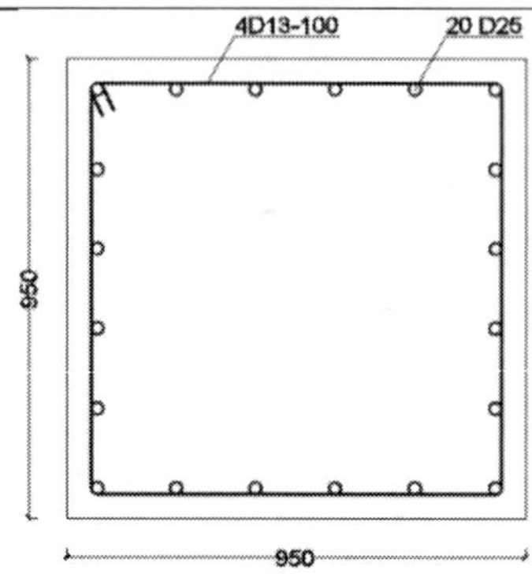
C2 700x900	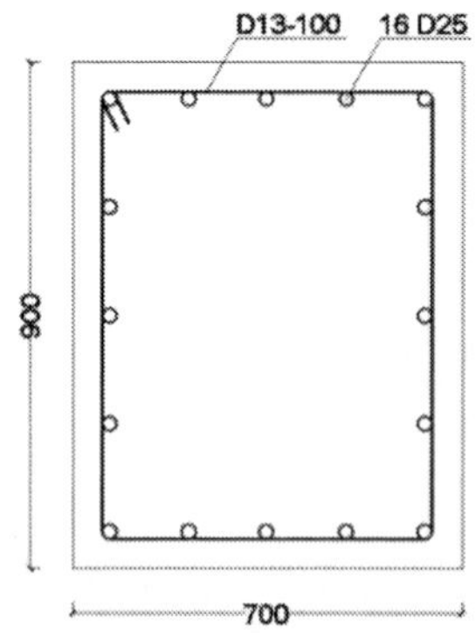	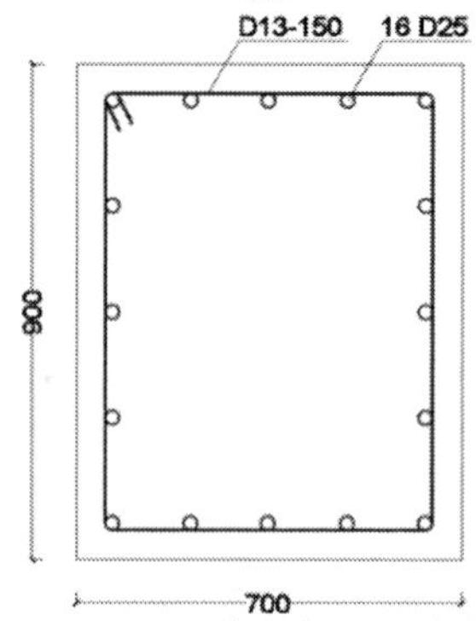
C3 600x800	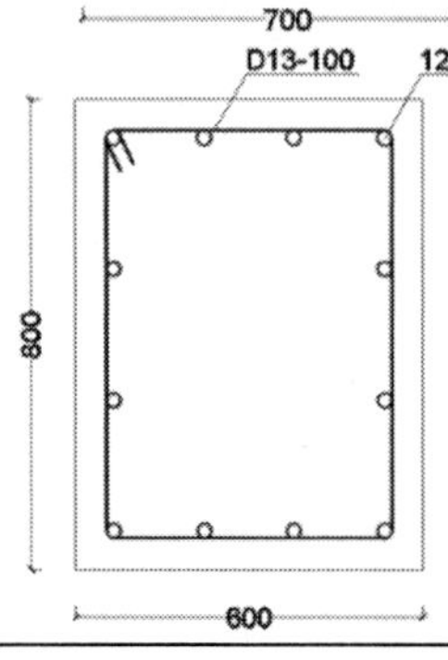	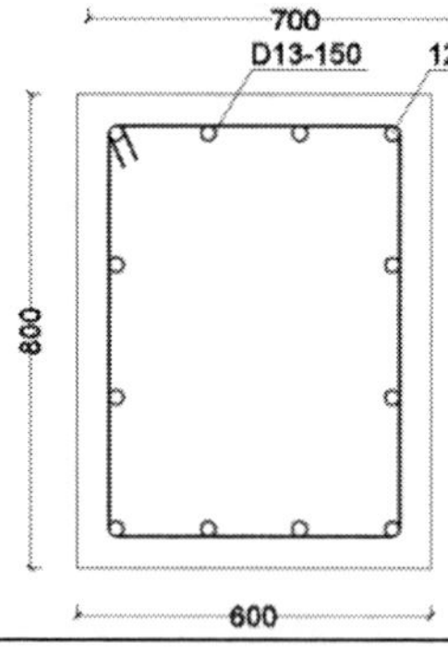

4.2. RC beams

The reinforcement embedded in the RC beams is needed to resist stresses under the combination of gravity loading and earthquake forces. As the assumed placement of the plastic hinges is at the end of the beams, the beam ends should counter the positive and negative moments. Consequently, the use of the double reinforced section on the top and bottom of the beam is considered.

The calculation of the flexural reinforcement needed is based on the stress-strain diagram shown in Figure 4. The method adopted requires the steel in compression to be calculated, along with the compression zone of the concrete and the tensile reinforcement, see equation (3) and (4). The letter code symbolized the concrete's compressive strength, the yield strength of the reinforcement, the elastic modulus of the concrete, the area of compressive steel and the tensile steel, and the steel strain for f'_c, f_y, E, A_s', A_s, and ε_s, respectively. The acquired area of the reinforcements should be checked to the minimum and maximum boundary area of the steel for the section. Furthermore, the control of strength and strain (and also the moment capacity) is needed to prove whether the reinforcement is in accordance with the assumptions or not.

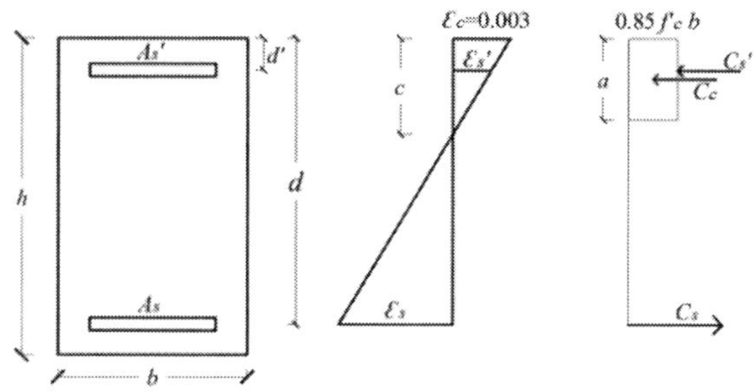

Figure 4. Stress-strain diagram for doubly reinforced beam section.

$$C_s' + C_c = C_s \tag{3}$$

$$A_s'\varepsilon_s'E + 0.85f'_c ab = A_s f_y \tag{4}$$

Shear reinforcement is also needed, to prevent failure in shear and increase the ductility of the beams. The shear force of the beams can be calculated by using equation (5), where M_{pr} represents the probable moment caused by the combination of the 120% of dead load and the 100% of live load, l_n denotes the clear span of the beam, V_u for the ultimate shear force of the system, V_c stands for the concrete's shear force, and $\emptyset$ denotes the strength reduction factor.

$$V_{sway} = \frac{M_{pr1} + M_{pr2}}{l_n} \pm \frac{V_u + l_n}{s} \quad \text{and} \quad V_s = \frac{V_{sway}}{\phi} - V_c \tag{5}$$

Several types of primary and secondary beams were taken into account in this research. Following the design methods mentioned above, the results are summarized in Table 4, including the dimensions, flexural reinforcement and shear reinforcement of the RC beams in the building. The end-side length of the reinforcement was calculated from 25% of the clear span, located on each side of the beams.

Table 4. The cross-section and reinforcement details of the RC beams (unit: mm).

<table>
<tr><th>Beam type</th><th>On the end side</th><th>On the middle side</th></tr>
<tr><td>B1 350x700</td><td>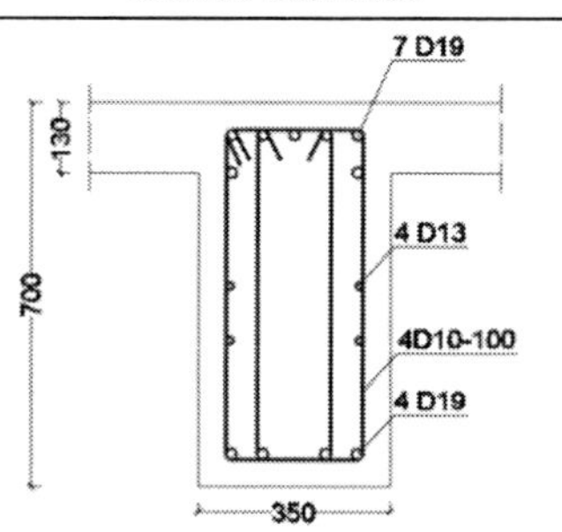</td><td>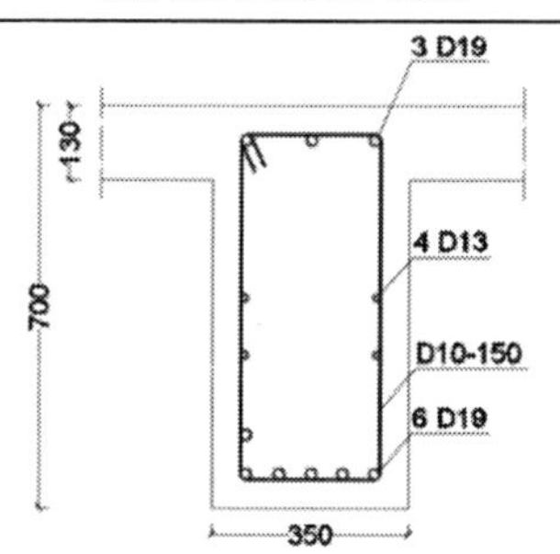</td></tr>
<tr><td>B2 350x600</td><td>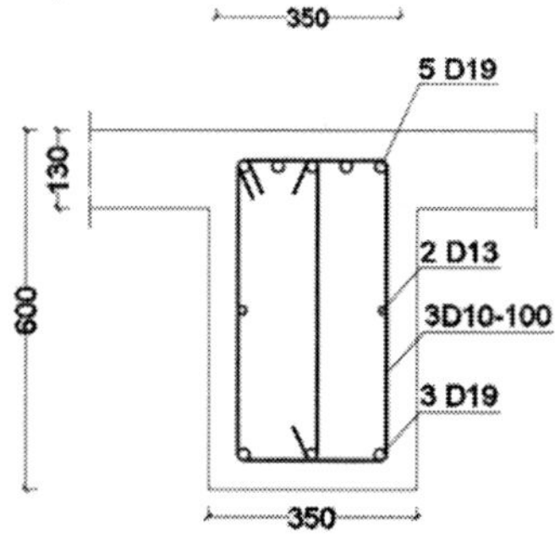</td><td>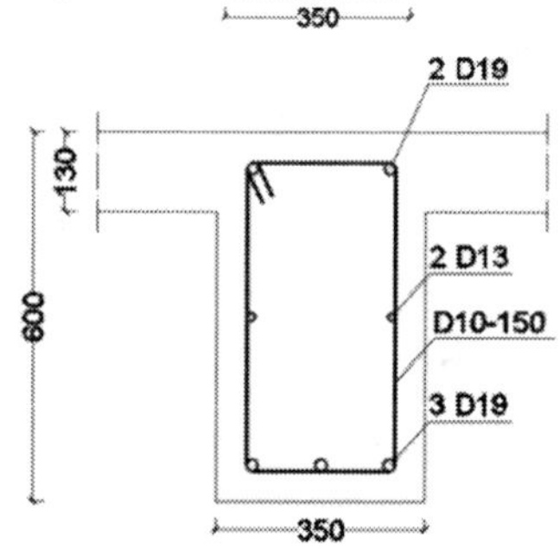</td></tr>
</table>

Beam type	On the end side	On the middle side
B3 250x450	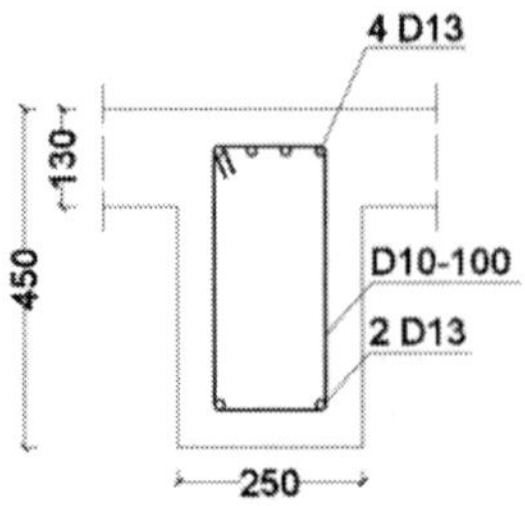	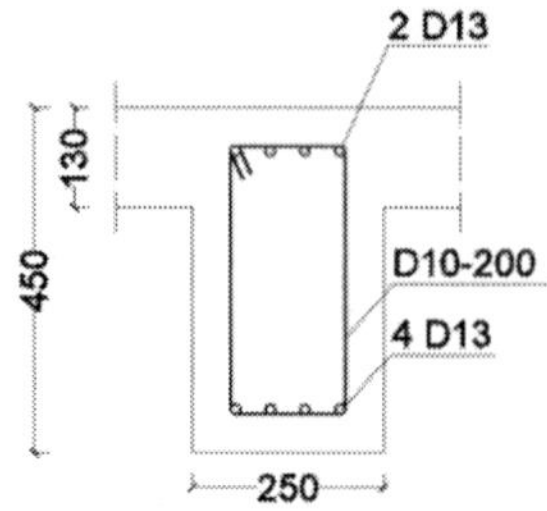
B4 200x400	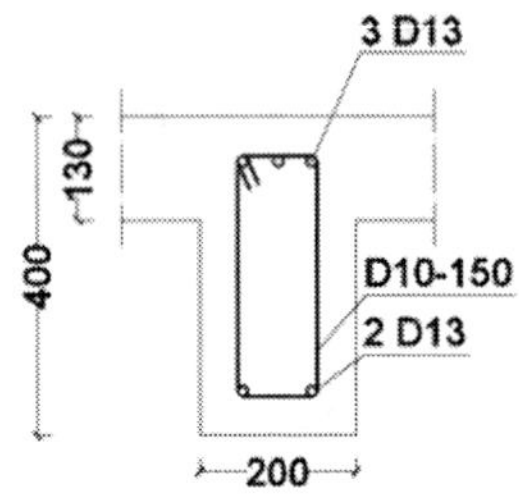	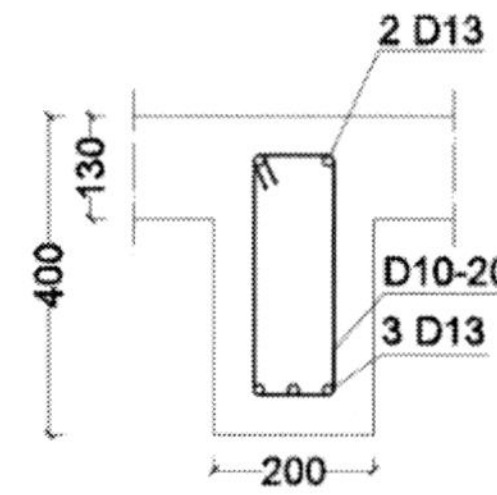

5. Conclusions

This paper focused on the structural behavior and design of RC members of a middle-rise residential building situated in a moderately-high seismicity area. The building was designed using the capacity-design method and by considering the placement of the plastic hinges at the end of the beams. The dual system combined the special frame system and the shear walls placement to withstand the earthquake loading impact in the building. Therefore, the numerical simulation, using ETABS software, was performed considering the seismic response and the soft soil condition. The structural design was calculated based on the new standard code for earthquake and RC design in Indonesia. The simulation results indicated that the RC building performed well under the designed earthquake load. The structural behavior was found to fulfill the requirements written in the code and performed within the acceptable limits. The structural design of the RC beams and columns was also found to satisfy the code.

Acknowledgments

The first author acknowledges the partial financial support of the Bank SinoPac Taiwan in the form of the Establishment of Bank SinoPac Encouragement Student Aid in National Cheng Kung University.

References

[1] Haryanto Y, Sudibyo G H and Effendi F C A 2017 *Procedia Eng* **171** 1025.
[2] Haryanto Y, Gan B S, Wariyatno N G and Indriyari E W 2017 *ARPN J Eng Appl Sci* **12** 4858.
[3] Haryanto Y, Gan B S, Wariyatno N G and Indriyari E W 2018 *MATEC Web Conf* **192**.
[4] Widyaningrum A, Haryanto Y and Hermanto N I S 2018 *MATEC Web Conf* **195**.
[5] Widyaningrum A, Haryanto Y, Sudibyo G H and Hermanto N I S 2019 *J Phys Conf Ser* **1367**.
[6] Haryanto Y, Hu H-T, Han A L, Wariyatno N G, Sudibyo G H, Hidayat B A and Naqiyah K 2019 *IOP Conf Ser Mater Sci Eng* **615** 012070.
[7] Tudjono S, Han A L, and Hidayat B A 2015 *Proceedia Eng* **125** 1070-1075.
[8] Hidayat B A, Hu H-T, Han A L, Haryanto Y, Widyaningrum A and Pamudji G 2019 *IOP Conf Ser Mater Sci Eng* **615** 012073.
[9] Haryanto Y, Hu H-T, Han A L, Hidayat B A, Widyaningrum A, and Yulianita P E 2020 *Civ Eng Dimens* **22**(1) 13-21.
[10] Haryanto Y, Hu H-T, Han A L, Atmajayanti A T, Galuh D L C and Hidayat B A 2019 *J Teknol* **81**(4) 143-154.
[11] Priyanka B and Mayur B 2016 *Int J Eng Res Appl* **6**(4) 78-84.

[12] Mirzaei A, Sangree R H, Velchev K, Comeau G, Balh N, Rogers C A and Schafer B W 2015 *J Constr Steel Res* **115** 81-91.
[13] Bakhaswala H, Shah V, Chandiwala A, and Desai D 2017 *Int J Adv Eng Res Dev* **4**(11) 327-333.
[14] George S S and Varghese V 2012 *IJITEE* **1**(2) 211-215.
[15] Rajawat U S and Tiwari S 2016 *Int J Res* **3**(8) 515-518.
[16] Singanapalli S K, Devadaas C S C, and Mahesh S 2020 *J Eng Sci* **11**(2) 509-511.
[17] Al-Hussaini T M and Khan K 2010 *Proc. 5th Int. Conf. on Recent Advances in Geotechnical Earthquake Engineering and Soil Dynamics* (San Diego: USA).
[18] Fatahi B, Tabatabaiefar S H R and Samali B 2014 *Geomech Eng* **6**(3) 293-320.
[19] Haryanto Y, Hu H-T, Han A L, Hidayat B A and Indriyari E W 2019 *Aceh Int J Sci Technol* **8**(1) 1-11.
[20] National Standardization Agency of Indonesia 2012 *Indonesian National Standard 1726:2012, the Procedures for Earthquake Resistance Planning for the Building and Non-building Structures.* (Jakarta: National Standardization Agency of Indonesia). (In Indonesian).

IOP Conf. Series: Materials Science and Engineering **982** (2020) 012034 doi:10.1088/1757-899X/982/1/012034

Potential tsunami hazard in Ujung Kulon National Park

Wahyu Widiyanto[1,2,3] **and Shih-Chun Hsiao**[1]

[1] Department of Hydraulic and Ocean Engineering, National Cheng Kung University, Tainan, Taiwan

[2] Department of Civil Engineering, Universitas Jenderal Soedirman, Purwokerto, Indonesia

[3]wahyu.widiyanto@unsoed.ac.id

Abstract. Ujung Kulon National Park (UKNP) is a natural world heritage site located at the western tip of Java Island and on the edge of the Indian Ocean. This area of 122,956 ha has the potential for tsunami hazard originating from Mount Anak Krakatau and the Sunda Arc subduction zone. Almost no residents live in the UKNP region, however, it is the only place on earth that remains a habitat for the Javan rhino (*Rhinoceros sundaicus*) whose population is less than 100. This study aims to discuss the potential for tsunami hazards in UKNP originating from the earthquakes in Indian Ocean. The shallow water equation model was used to simulate the generation and propagation of tsunami waves. A total of 50 numerical gauges with intervals of 4 km were used as assessment points placed along the 190 km coastline of UKNP. From the simulation it can be seen that the height of the tsunami reaches 12.9 m around the coast of UKNP causes this area is vulnerable to tsunami hazards. This information can be used as consideration for the management of the UKNP area so that it can continue to preserve flora and fauna, especially to avoid the extinction of the Javan rhino.

1. Introduction

A peninsula located at the western end of the island of Java has become the home of one-horned rhinoceros that remains on earth. Actually this area is also a habitat for other animals such as Java bulls, monkeys, deer, and mouse deer [1]. However, the one-horned rhino or Javan rhino (*Rhinoceros sundaicus*) is the animal that gets the most attention because its population is very vulnerable to extinction. Therefore UNESCO established this area as a natural world heritage site in 1992 [2]. Previously this area had long been designated as a national park by the Indonesian government with the official name Ujung Kulon National Park.

Historically, this region was not a safe area from natural hazards. Earthquakes often occur but did not directly affect the life of plants and wildlife. While the more obvious influence was the eruption of volcanoes and tsunamis that have struck several times. During the Krakatoa eruption in 1883 which was one of the most severe eruptions on earth, this area was reported to have suffered severe environmental damage [3]. The traces of the tsunami in Ujung Kulon due to the 1883 eruption were proven by Paris et al through the study of tsunami deposits [4]. Likewise in the latest tsunami event in 2018 [5][6], this region also experienced the highest tsunami wave runup compared to other coastal areas in the Sunda Strait as reported by [6]. [7] and [8] also stated that this area was at risk of experiencing a tsunami that could affect the lives of protected animals in the national park. The scenario that fits the 2018 tsunami in the Sunda Strait was simulated in 2012 by [9]. Based on these simulations, the tsunami caused by the flank collapse of Mount Anak Krakatau directed radially until it reached the coast at the southern tip of the island of Sumatra (Lampung Province) and the western end of the island of Java (Banten Province).

Javan rhinos are often near the beach to meet the needs of salt minerals. Plants as Javan rhino food exist in the distance of 0-600 m from the beach. In addition, 61.9% distribution of Javan rhinos is 0-1000 m from the coast [10]. Figure 1 is a photo report of the death of a rhino in 2018 [2]. Whereas Figure 2 illustrates the destruction of forests in the UKNP due to the 2018 tsunami [6]. The tsunami can directly hit an individual Javan rhino to cause its death, besides that an important thing is the possibility of forest destruction which is a food source for wildlife, especially Javan rhino. In the effort to conserve the Javan rhino, tsunami risk analysis is one of the factors that need to be considered in addition to other factors that have been analyzed, for example regarding the quality of food [11] and habitat suitability [12].

Figure 1 Death of a Javan rhino on Ujung Kulon beach (courtesy: UKNP 2018)

Figure 2 Forest destruction as a habitat for Javan rhinos in the UKNP due to the 2018 tsunami (photo is the result of a survey by Widiyanto et al 2018)

This article discusses the potential for tsunami hazard in the UKNP due to earthquakes obtained from simulations. In this study only tsunamis originating from earthquakes were simulated. The results of modeling and analysis can be used as consideration for the management of the national park, especially in realizing a safer second habitat outside the existing national park that is prone to tsunamis.

2. Method

In this study, tsunami waves propagation were calculated by the long wave equation or shallow water equation. Considering that the simulation area is large enough, a spherical coordinate system was applied. The nonlinear shallow water equation using spherical coordinates can be stated:

$$\frac{\partial \eta}{\partial t} + \frac{1}{R\cos\varphi}\left\{\frac{\partial P}{\partial \psi} + \frac{\partial}{\partial \varphi}(\cos\varphi\, Q)\right\} = -\frac{\partial h}{\partial t} \quad (1)$$

$$\frac{\partial P}{\partial t} + \frac{1}{R\cos\varphi}\frac{\partial}{\partial \psi}\left\{\frac{P^2}{H}\right\} + \frac{1}{R}\frac{\partial}{\partial \psi}\left\{\frac{PQ}{H}\right\} + \frac{gH}{R\cos\varphi}\frac{\partial \eta}{\partial \psi} - fQ + F_x = 0 \quad (2)$$

$$\frac{\partial Q}{\partial t} + \frac{1}{R\cos\varphi}\frac{\partial}{\partial \psi}\left\{\frac{PQ}{H}\right\} + \frac{1}{R}\frac{\partial}{\partial \psi}\left\{\frac{Q^2}{H}\right\} + \frac{gH}{R\cos\varphi}\frac{\partial \eta}{\partial \psi} - fP + F_y = 0 \quad (3)$$

Where η is the water surface elevation; (P,Q) denote the volume fluxes in X (West-East) direction and Y (South-North) direction, respectively; (φ,ψ) denote the latitude and longitude of the Earth; R is the radius of the Earth; g is the gravitational acceleration and h is the water depth. And the term $-\partial h/\partial t$ reflects the effect of transient seafloor motion, can be used to model landslide-generated tsunamis. F represents the Coriolis force coefficient due to the rotation of the Earth

Equations (1) - (3) are then solved by the COMCOT (Cornell Multi-grid Coupled Tsunami model) [13]. This model is one of the most widely used tsunami models and has successfully modeled many tsunami cases. Tsunami generation originating from seabed deformation was modeled with elastic finite fault plane theory proposed originally by Mansinha and Smylie (1971) and then improved by Okada (1985) [13]. This approach requires input in the form of epicenter, focal depth, fault length and width, strike angle, dip angle, slip angle, rake angle, and slip.

The tsunami propagation model from source to points assessed requires bathymetric data. This study uses Batnas (6 arcsec resolution) as bathymetry data. These data are sourced from the Geospatial Information Agency (Indonesian: BIG). Bathymetry data were used to form the simulation layer. In this study, three layers of simulation are used which form a nested grid system. COMCOT uses explicit staggered leap-frog finite difference schemes to solve numerical equations (1) - (3). From this simulation a tsunami wave height will be obtained in the assessed area, in this study the Ujung Kulon National Park.

The simulation domain is shown in Figure 3 where the observed coastal area is part of the UKNP area. At Ujung Kulon Peninsula and Panaitan Island, 50 numerical gauges were placed at intervals of 4 km along the coastline to review tsunami wave heights that occurred due to an earthquake with magnitudes of Mw 8 and Mw 8.7. Selection of earthquake magnitude based on historical and potential earthquake that can occur due to ruptures in the Sunda Strait segment.

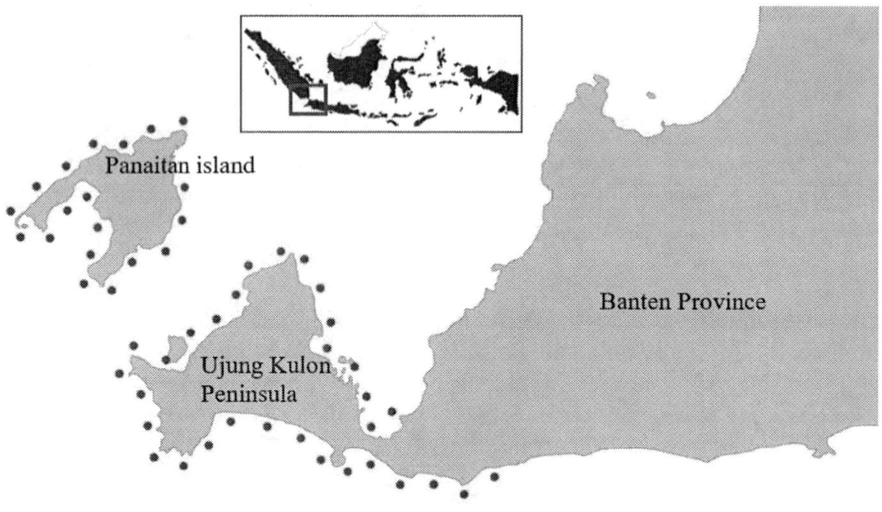

Figure 3. Study area covering a small part of Indonesia, east side of Sunda Strait and west tip of Java Island. Red-dots are numerical gauges for detecting water surface elevation.

3. Result

Figure 4 is a visualization of the results of the COMCOT model simulation. For the first simulation, an Mw 8 magnitude earthquake with epicenter at coordinates 104.547612 E and -7.277303 S was simulated occurring in the middle of the Sunda Strait tectonic segment. This magnitude selection is based on the history of tsunamis that occurred along the subduction zone of southern Java. Two tsunamis in modern era occurred in 1994 in the south of Banyuwangi with a magnitude of Mw 7.8 and in 2006 in the south of Pangandaran with a magnitude of Mw 7.7. Although the magnitude is not so large to generate a tsunami, in reality the wave heights and runups that occur were significant. In addition, the epicenter was determined randomly but according to historical earthquake data in the tectonic segment. Rupture due to earthquake with magnitude was estimated by scaling relation [14] as shown in Equation 4-5:

$$log\ L = a + b\ M_w \tag{4}$$

$$log\ W = a + b\ M_w \tag{5}$$

IOP Conf. Series: Materials Science and Engineering **982** (2020) 012034 doi:10.1088/1757-899X/982/1/012034

where L is rupture length, W is rupture width, a and b are regression coefficient and M_w is moment magnitude.

Calculation using the equations obtained length and width of 155 km and 66 km respectively. Deformation of the seafloor with this area is significant enough to change the sea level above it so that it can be the beginning of a tsunami generation (Figure 4). Changes in sea level then spread towards the coast as a tsunami wave, and arrived around Ujung Kulon Peninsula and Panaitan Island earlier before the tsunami entered the Sunda Strait (Figure 5). The direction of propagation is influenced by the angle of direction of the rupture (strike angle) and bathymetry.

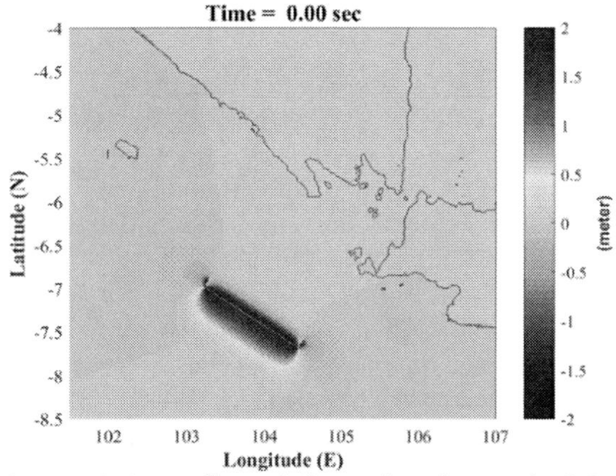

Figure 4 A sea floor deformation due to the Mw 8 earthquake produced changes in the surface of the water which would propagate as a tsunami wave.

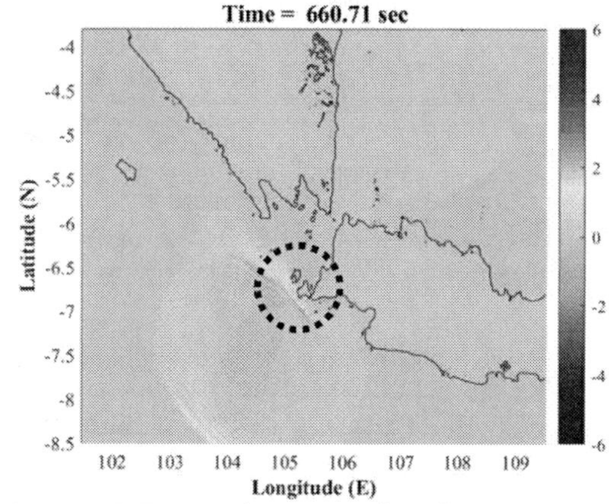

Figure 5 Tsunami propagation from the Indian Ocean entering the Sunda Strait. Ujung Kulon Peninsula and Panaitan Island were seen to have been hit by the tsunami even though in this case the waves were not so high.

The second simulation used a greater earthquake magnitude of M_w 8.7 with the same epicenter as the first simulation. The choice of magnitude corresponds to the maximum potential for earthquake magnitude in the Sunda Strait segment. The length and width of the rupture is greater than the magnitude of M_w 8, which is 299 km and 112 km, respectively. Dislocation and deformation are also greater, giving rise to higher tsunami waves (Figure 6 and Figure 7). Water level profiles due to tsunami waves on two numerical gauges are shown in Figure 7. One of the numerical gauges is located on the southern coast of Ujung Kulon peninsula, the other on southern tip of Panaitan Island. From this profile, it can be seen the maximum wave height and the arrival time of the tsunami. In this case, the tsunami came sooner on Panaitan Island than in Ujung Kulon peninsula, each 14 minutes and 17 minutes from the time the earthquake occurred. The amplitude and wave height in Ujung Kulon Peninsula is significant, 7.4 m and 12.9 m respectively. It is greater than in Panaitan Island, reaching 5.6 m and 8.4 m for amplitude and height of tsunami wave respectively. The tsunami wave heights for Mw 8.0 were lower than Mw 8.7. They were 4.5 m and 5.2 m for Ujung Kulon Peninsula and Panaitan Island respectively, which were not significant different in both location. These wave amplitudes, wave heights and arrival times can be different if we simulate with different parameters. It is possible to obtain a higher wave height. Therefore, it is necessary to simulate various scenarios. A probability tsunami hazard assessment (PTHA) may be done for this region. Since PTHA includes intensive simulation work, for efficiency, the study area can be expanded, for example by including the entire Sunda Strait area.

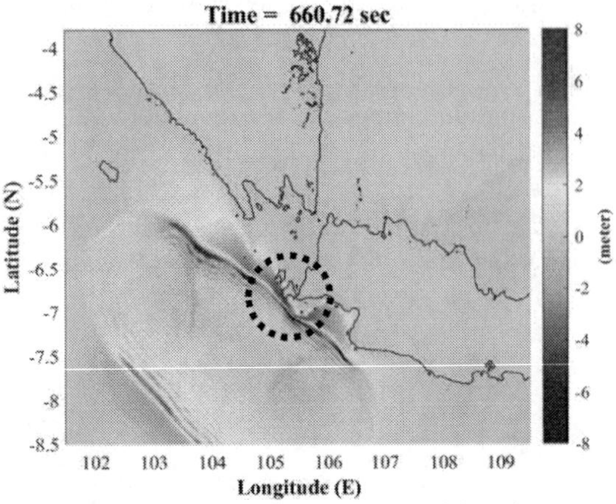

Figure 6 Tsunami waves generated by the Mw 8.7 earthquake seen higher than the magnitude of Mw 8. Black-dots circle is UKNP

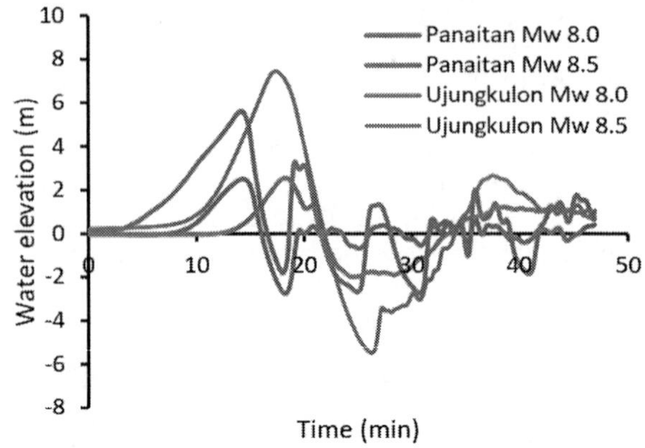

Figure 7 The height of the tsunami on Panaitan Island and Ujung Kulon due to the Mw 8.0 and Mw 8.7 earthquakes

From the simulation results it can be seen that the UKNP area has a risk of tsunami hazard. Likewise, the results of the field survey by Widiyanto et al. illustrate that part of the forest in Ujung Kulon National Park was destroyed by the tsunami in December 2018 [6]. Fortunately, the last population of Javan rhinos escaped extinction. This suggests that providing a second habitat outside the Ujung Kulon peninsula for the Javan rhino is very urgent. Research on the suitability of the release location has been investigated, for example by Ramono et al. who conducted a field survey and proposed relocation sites on Java, namely on Mount Honje, Mount Halimun, Masigit Kareumbi and Leuweung Sancang [15]. Therefore, it is time for the authorities to adopt policies to accelerate the realization of the second habitat of Javan rhinos.

4. Conclusion

The simulation results show that the UKNP area has a potential tsunami event with a significant height reaching nearly 13 m with a relatively short tsunami arrival time around 15 minutes. Mitigation needs to be done to save the Javan rhino habitat. Relocation of populations to second habitats located in areas farther from the coast is urgent.

This simulation only uses earthquakes with magnitudes of M_w 8 and M_w 8.7. The simulation needs to be extended for earthquakes with the same magnitude but different parameters causing the seabed deformation. Likewise, it can be continued for other magnitudes or worse scenarios need to be simulated also to find out the worst impact on UKNP. If deemed necessary, the simulation can be reproduced with the magnitude of other earthquakes to form a tsunami hazard probability for UKNP that extends to the Sunda Strait or the West Coast of Lampung which also has a UNESCO heritage national park namely the Bukit Barisan Selatan National Park. Another future work needs to be conducted will be tsunami simulation and analysis related to Mount Anak Krakatau, both due to the landslide cliffs and eruptions that directly change the sea surface.

References

[1] Hariyadi, ARS, Priambudi, A, Setiawan, R, Daryan, D, Yayus, A, and Purnama, H, *Estimating the population structure of Javan rhinos (Rhinoceros sondaicus) in Ujung Kulon National Park using the markrecapture method based on video and camera trap identification* Pachyderm No. 49 January–June 2011

[2] Balai Taman Nasional Ujung Kulon 2020 *Sejarah dan Status Kawasan* https://www.ujungkulon.org/ last access: 9 July 2020

[3] Verbeek, R. D. M. 1884 *The Krakatoa eruption*, Nature, 10–15

[4] Paris, R, Wassmer, P, Lavigne, F, Belousov, A, Belousova, M, Iskandarsyah, Y, Benbakkar, M, Ontowirjo, B, and Mazzoni, N. *Coupling eruption and tsunami records: the Krakatau 1883 case study, Indonesia* Bull Volcanol (2014) 76:814 DOI 10.1007/s00445-014-0814-x

[5] Widiyanto W, Santoso P B, Hsiao S C and Imananta R T 2019 *Post-event field survey of 28 September 2018 Sulawesi earthquake and tsunami* Nat. Hazards Earth Syst. Sci. 19 2781–2794 https://doi.org/10.5194/nhess-19-2781-2019

[6] Widiyanto W, Hsiao S C, Chen W B, Santoso P B, Imananta R T and Lian W C 2020 *Run-up, inundation, and sediment characteristics of the 22 December 2018 Sunda Strait tsunami, Indonesia* Nat. Hazards Earth Syst. Sci. 20 933–946 https://doi.org/10.5194/nhess-20-933-2020

[7] Setiawan R 2018 *Preventing Global Extinction of the Javan Rhino: Tsunami Risk and Future Conservation Direction* Conservation Letters, 11(1)

[8] Gross M 2018 *Last call to save the rhinos* Current Biology 28, R1–R16

[9] Giachetti, T., Paris, R., Kelfoun, K., and Ontowirjo, B. 2012 *Tsunami hazard related to a flank collapse of Anak Krakatau Volcano, Sunda Strait, Indonesia*, Geol. Soc. London, Spec. Publ., 361(1), 79–90, doi:10.1144/sp361.7

[10] Sumunar, DRS, Salsabila, S Fitriana, H A Hamid 2019 *Landsat-8 Multispectral Satellite Imagery for Rhinoceros Sondaicus Habitat Spatial Distribution Modelling through Biophysical Parameters in Ujung Kulon National Park*, Indonesia IOP Conf. Series: Earth and Environmental Science 286 012044

[11] Hariyadi, ARS, Sajuthi, D, Astuti, DA, Alikodra, HS, and Maheshwari, H 2016 *Analysis of nutritional quality and food digestibility in male Javan rhinoceros (Rhinoceros sondaicus) in Ujung Kulon National Park*, Pachyderm No. 57 July 2015–June 2016

[12] Rahmat, UM, Santosa, Y, Prasetyo, LB, and Kartono, A P, 2012 *Habitat Suitability Modeling of Javan Rhino (Rhinoceros sondaicus Desmarest 1822) Ujung Kulon National Park* JMHT Vol. EISSN: 2089-2063 DOI: 10.7226/jtfm.18.2.129 XVIII, (2):129-137

[13] Wang, X., 2009. *COMCOT User Manual Ver. 1.7.* Cornell University, 6, pp.1–59

[14] Blaser, L. et al., 2010 *Scaling relations of earthquake source parameter estimates with special focus on subduction environment,* Bulletin of the Seismological Society of America.

[15] Ramono, W., Isnan, M. W., M., Sadjudin, H.R, Gunawan, H, Dahlan, E.N., Sectionov, Pairah, Hariyadi, A.R., Syamsudin, M., Talukdar, B.K., And Gillison, A.N. 2009. *Report on A Second Habitat Assessment for the Javan Rhinoceros (Rhinoceros sondaicus sondaicus) within the Island of Java.* International Rhino Foundation, Yulee, Fl, Usa.

IOP Conf. Series: Materials Science and Engineering **982** (2020) 012035 doi:10.1088/1757-899X/982/1/012035

Estimation of Groundwater Recharge in Semarang City, Indonesia

Sanidhya Nika Purnomo[1,2,3], Wei Cheng Lo[1]

[1] Department of Hydraulic and Ocean Engineering, National Cheng Kung University, Taiwan
[2] Civil Engineering Department, Engineering Faculty, Universitas Jenderal Soedirman, Indonesia

[3]sanidhyanikapurnomo@gmail.com

Abstract. In regional groundwater management, it is necessary to find out the water balance between the supply and demand of groundwater. One of the most important in analyzing the water balance of groundwater is groundwater recharge. Semarang city that has a groundwater management problem needs to know the groundwater recharge to analyze the groundwater balance. On the other hand, it is difficult to measure groundwater recharge accurately, consequently, the groundwater recharge should be estimated. There are several methods to estimate groundwater recharge. To verify the groundwater recharge estimation, more than one method of groundwater recharge estimation should be used. In this paper, the determination of groundwater recharge is using the Chaturvedi formula and Krishna Rao formula based on the amount of rainfall, and well level data method that based on the difference in water level. Based on the final results, the well level data method seems the best fit in estimating the groundwater recharge.

1. Introduction

Groundwater as part of the hydrological cycle is often overlooked because of its location, appearance, and invisible movement. However, groundwater is an important component in meeting raw water needs, especially in areas where surface water quality is poor.

As a renewable water source, knowing the amount of groundwater recharge is very important. But on the other hand, calculating groundwater recharge is not easy. For this reason, groundwater recharge analysis can be done by estimating and measuring using isotope, especially in carrying out groundwater management.

Groundwater recharge can be defined as water that is put into an aquifer when it rains, which occurs after the infiltration and percolation process, into the unsaturated zone [1]. Groundwater recharge depends on several factors, such as infiltration capacity, stochastic characteristics of rain, and climate factors [2]. There are several methods for estimating groundwater recharge, namely numerical approach, tracer method, hydrologic budget method, and multiple methods. Groundwater recharge can be estimated with several conventional methods, but unfortunately, these methods require a lot of analysis and hydrological data [3]. In the numerical approach method, the calibration process can be used to get the recharge flux needed to adjust the actual head [4]. In the tracer method, the most widespread use is in surface hydrology, where the water velocity is the greatest [5]. However, measurement using the tracer method can give results above or below estimation, because it does not measure recharge directly [6]. Groundwater recharge estimation can also use the hydrologic budget method, which is recommended by the Groundwater Estimation Committee, India. However, the use of this method is limited to an area, where water table surveillance is carried out frequently [7].

Unfortunately, it is very difficult to judge the accuracy of each method of estimating groundwater recharge. For this reason, it is strongly recommended to estimate groundwater recharge using several methods by expecting consistency in the results, although consistency itself should not be taken as an indication of the accuracy of groundwater recharge estimates [8]. In this paper, the determination of groundwater recharge is using the Chaturvedi formula based on water level fluctuation methods and the amount of rainfall, and well level data methods that are based on the difference in water levels.

2. Study Area

Semarang City, the capital of Central Java, has an area of 373 km^2. The northern part of Semarang City is bordered by the Java Sea, the western side is bordered by Kendal Regency, the eastern side is bordered by Demak Regency, and the southern side is bordered by Semarang Regency, as shown in Fig. 1. The height of Semarang City is between 0.75 m to 348.00 m above the coastline.

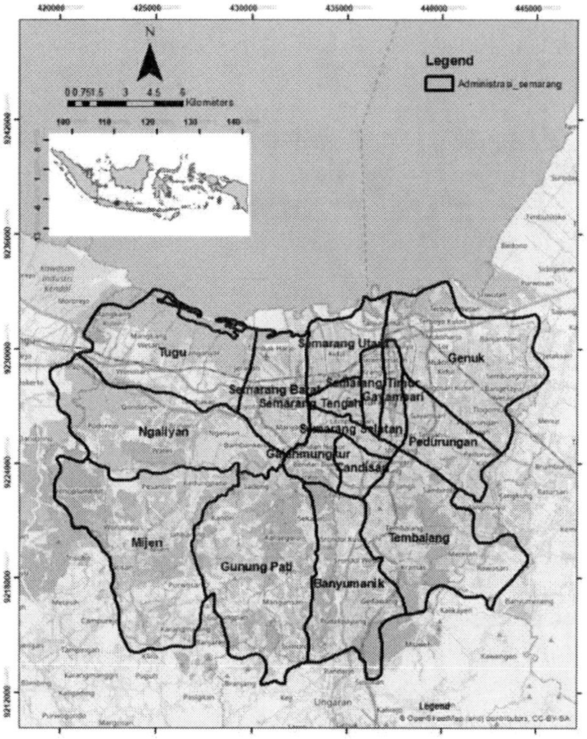

Figure 1. Semarang City.

3. Methodology

In Indonesia, rain is the most important component in recharging groundwater. In estimating groundwater recharge, it is easier to use empirical methods based on rain. This is because in Indonesia, rainfall data is easier to obtain when compared to other estimation methods. For this reason, in this paper, groundwater recharge estimation in Semarang City uses an empirical method based on the rain that occurred in Semarang City and its surroundings. However, to compare the empirical method with the actual conditions in the field, this paper also analyzes groundwater recharge estimation using the well level data method. However, to compare the empirical method with actual conditions in the field, this paper also analyzes groundwater recharge estimation using the well level data method.

Because of the limitations in estimating groundwater recharge, this study uses 3 empirical methods to estimate groundwater recharge. These methods are Chaturvedi formula, modified Chaturvedi formula, and well level data method.

3.1. Chaturvedi formula

In 1973, Chaturvedi derived the empirical relationship of recharge groundwater which is a function of annual precipitation based on the initial fluctuations in water and the amount of precipitation in Ganga-Yamuna as shown in [9]. The equation of the Chaturvedi Formula is shown in Eq. 1.

$$R = 2.0(P - 15)^{0.4} \quad (1)$$

where,

R = recharge due to annual precipitation, inches

P = annual precipitation, inches

Chaturvedi formula is ultimately used for preliminary estimations of groundwater recharge due to precipitation.

3.2. Krishna Rao

The Krishna Rao method (Eq. 2) provides groundwater recharge estimates in areas with limited and homogeneous climatology [9].

$$R = K(P - X) \quad (2)$$

Where:

$R = 0.20(P - 400)$ for areas with annual rainfall from 400 until 600 mm

$R = 0.25(P - 400)$ for areas with annual rainfall from 600 until 1000 mm

$R = 0.35(P - 600)$ for areas with annual rainfall above 2000 mm

3.3. Well level data method

Well the level data method is a method of estimating groundwater recharge based on differences in water level in the observation wells at the beginning and end of the same water year by considering soil porosity [10]. The equation of the well level data method is shown in Eq. 3.

$$R = (WL_2 - WL_1)p \quad (3)$$

where:

R = estimated recharge, inches.

WL_2= water level at the beginning of the water year, inches.

WL_1= water level at the end of the same water year, inches.

p = adjusting for porosity 0.44 [11].

Well data method is considered more accurate because it estimates groundwater recharge based on measurements of water level in the observation well and considers the soil porosity.

4. Result and Discussion

Rainfall data used in the analysis of groundwater recharge estimation using the Chaturvedi formula and Krishna Rao in Semarang City is annual rainfall data from 1970 to 2016. While the water level data used for the well level data method is the water level data at the observation well. In groundwater analysis, the more points of observation wells are used, the results will be better and more in line with real conditions. However, due to the limited data length and the number of observation wells, this paper only uses two observation wells located in areas with high water fluctuation. at SMKN 10 from 1986 until 2010 and Wot Gandul, from 1977 until 2010. The location of the observation wells SMK 10 and Wot Gandul can be seen in Figure 2.

Based on the Chaturvedi and Krishna Rao methods, it appears that the estimated groundwater recharge has a very large range, ranging from 263.72 mm to 1106.21 mm. The results of the analysis of groundwater recharge estimates based on the Chaturvedi, Krishna Rao method, and the deviation from the mean precipitation is shown in Figure 3. The Krishna Rao method estimates a greater

IOP Conf. Series: Materials Science and Engineering **982** (2020) 012035 doi:10.1088/1757-899X/982/1/012035

groundwater recharge than the Chaturvedi method. The estimated amount of groundwater recharge decreases and increases according to the deviations from the average precipitation because in the Chaturvedi and Krishna Rao methods the estimation of groundwater recharge is only based on precipitation.

Based on the well level data method, the estimation of groundwater recharge in the observation well of SMKN 10 and Wot Gandul (Figure 4) does not show the same trend with the Chaturvedi and Krishna Rao methods, where groundwater recharge follows the deviation of average precipitation. This happens because the recorded data taken from the observation well is not only influenced by precipitation but is also affected by various things, including hydrogeological conditions and evaporation in the area.

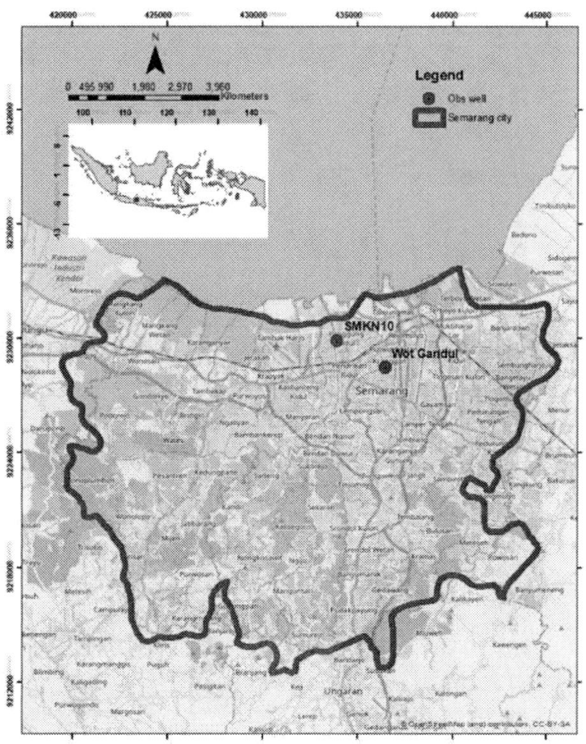

Figure 2. Observation wells of Semarang City.

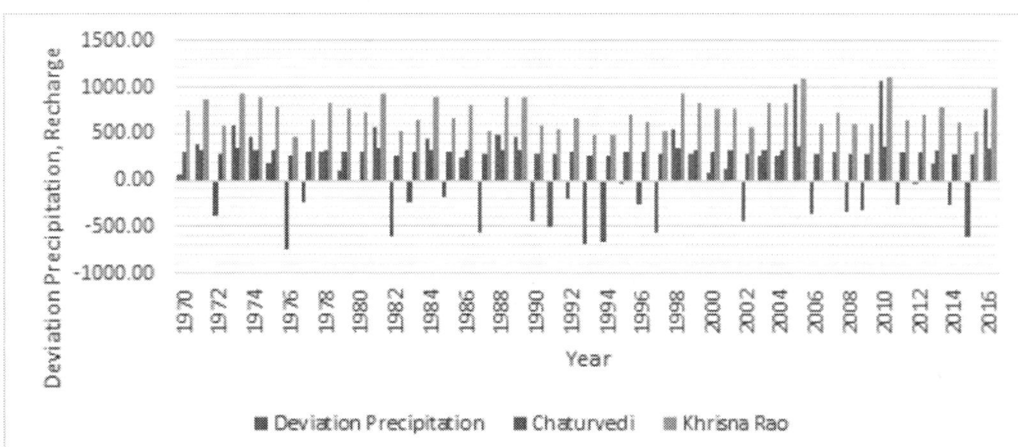

Figure 3. Groundwater recharge estimation based on Chaturvedi and Krishna Rao method, and deviations from the average of precipitation

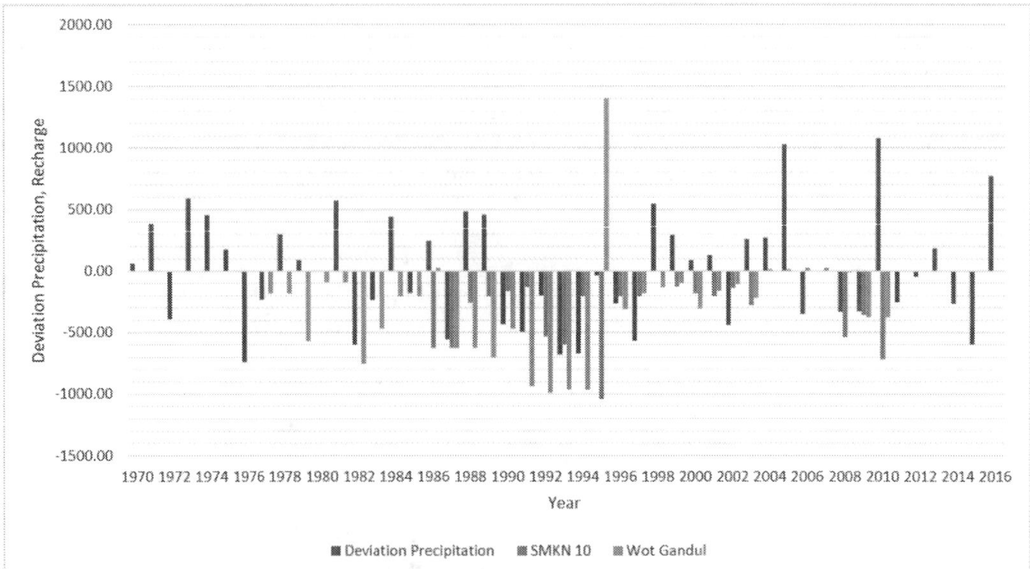

Figure 4. Groundwater recharge estimation based on water level data method and deviations from the average of precipitation

Meanwhile, the relationship between groundwater recharge estimation and annual precipitation is shown in Figure 5. The relationship shows that the increase in precipitation affects the groundwater recharge increase by the equation $y = -0.0002x^2 + 0.2283x + 2780$.

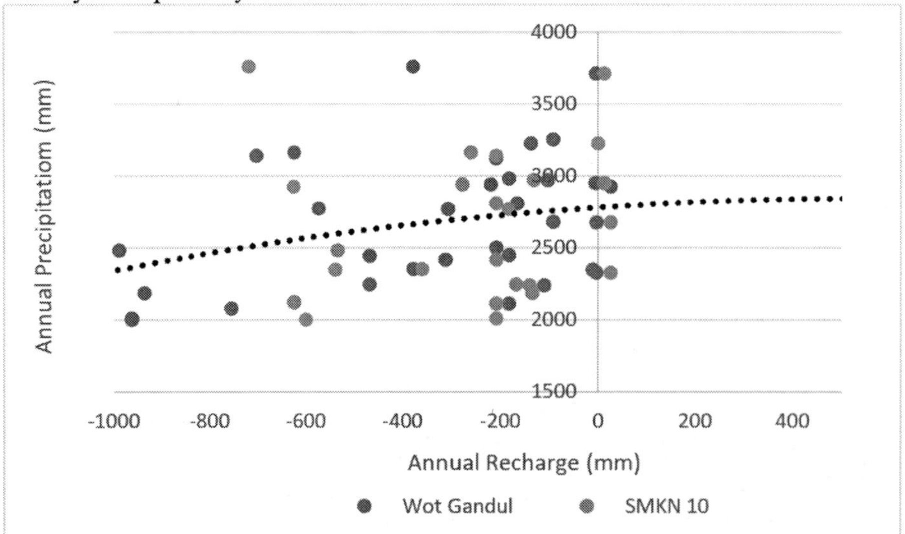

Figure 5. Relationship of groundwater recharge estimation and annual precipitation

5. Conclusions

Based on the results and discussion, it shows that the Krishna Rao method gives the largest groundwater recharge estimate when compared to the other two methods. The Chaturvedi method is a groundwater recharge estimation method that is quite easy to use but is more suitable for areas that have small evaporation, or the evaporation value can be ignored.

For areas that have high evaporation values such as Semarang City, the most appropriate method for estimating groundwater recharge is the well level data method. The well level data method can also be used in other areas. After all, it is considered more by local conditions, because it is based on groundwater level measurements in the observation well.

6. References

[1] ŞenZ. Unconfined Aquifers. Practical and Applied Hydrogeology. 2015. 209–278 p.
[2] ŞenZ. Climate Change, Droughts, and Water Resources. Appl Drought Model Predict Mitig. 2015;321–91.
[3] ChandR, ChandraS, RaoVA, SinghVS, JainSC. Estimation of natural recharge and its dependency on sub-surface geoelectric parameters. J Hydrol. 2004;299(1–2):67–83.
[4] SinghA, PandaSN, UzokweVNE, KrauseP. An assessment of groundwater recharge estimation techniques for sustainable resource management. Groundw Sustain Dev [Internet]. 2019;9(April):100218. Available from: https://doi.org/10.1016/j.gsd.2019.100218
[5] CookPG, SolomonDK. Recent advances in dating young groundwater. J Hydrol. 1997;191(1–4):245–65.
[6] LernerDN, IssarAS, SimmersI. Groundwater recharge: a guide to understanding and estimating natural recharge. 1990. 345 p.
[7] ShardaVN, KurotheRS, SenaDR, PandeVC, TiwariSP. Estimation of groundwater recharge from water storage structures in a semi-arid climate of India. J Hydrol. 2006;329(1–2):224–43.
[8] HealyRW, CookPG. Using groundwater levels to estimate recharge. Hydrogeol J. 2002;10(1):91–109.
[9] KumarCP. Estimation of natural ground water recharge. ISH J Hydraul Eng. 1997;3(1):61–74.
[10] QablawiB. A Comparison of Four Methods to Estimate Groundwater Recharge for Northeastern South Dakota. South Dakota State University; 2016.
[11] WardAD, TrimbleSW. Environmental hydrology. 2nd ed. Lewis Publishers; 2003.

IOP Conf. Series: Materials Science and Engineering **982** (2020) 012036 doi:10.1088/1757-899X/982/1/012036

Analyzing Tsunami Hazard using Numerical Modelling: Study Case Palu, Sulawesi Tengah, Indonesia

A Kurniawan[1,3], R A Satria[1] and M B Pratama[2]

[1]Ocean Engineering Program, Faculty of Civil and Environmental Engineering, Institut Teknologi Bandung, Bandung, Indonesia
[2]Institute for Energy Systems, School of Engineering, The University of Edinburgh, Edinburgh, United Kingdom

[3]alamsyah@ocean.itb.ac.id

Abstract. The present study aims to assess the Tsunami hazard in Palu due to the future Tsunamigenic event in terms of near-shore and far-field Tsunami runup, inundation height, and Tsunami arrival time along the coast of Palu. A numerical model is developed to simulate propagation and inundation in the study area. Four scenarios are conducted, which consists of the Tsunamigenic earthquake and aerial subsidence as an approach for mimicking hypothetical potential landslide. The outcome from simulation such as inundation height each grid points then calculated with building footprint stored data – derivation data from the terrain model to develop a Tsunami hazard map. Preliminary Tsunami Hazard Map is made to classify the Tsunami hazard level by existing classification added with estimated building impacted and arrival time ashore on each village in Palu City administrative region.

1. Introduction

The Tsunami in Palu, located at Sulawesi Tengah Indonesia, (Figure 1) on September 28, 2018, portrays a complex Tsunamigenic scenario where the earthquake might not be the primary cause of the Tsunami. A sinistral strike-slip plate boundary fault system like Palu-Koro rarely produces a devastating Tsunami. The motion of strike-slip faults is caused by shearing forces and have dominant horizontal deformation. It could not provide large vertical uplift on the seabed to make a sudden move for the water column above it and produce a large Tsunami. The earthquake magnitude itself is relatively considered as small in the context of Tsunami generation. Nonetheless, the earthquake could be a contributor for a multi-following peril – multiple submarine landslide and liquefaction in deltaic soils failure, which in this case could have a higher contribution to generating a Tsunami. One similar incident to this case is the 2010 Haiti earthquake, where research shows that even strike-slip earthquakes, can generate Tsunamis triggering aftershocks [1]. In Palu Tsunami case, some researchers conduct a post-disaster field survey and identify sites of potential subsidence along the coast [2].

Even though disaster is inevitable, the risk could be reduced. Tsunami is considered as a rare event and unique from its generation and impact towards the affected area. From one case to another, a lesson learns could be taken to improve either disaster awareness on society or mitigation action that ought to be performed [3,4]. Tailoring strategies to set a countermeasure the disaster is one key to reducing its risk. Nevertheless, examining the mechanisms and phenomena of disasters that occur could be an effective step – even though it is a longer project. Implementation of the Tsunami Hazard Map could be one of the key preliminary stages to raise awareness on Tsunami disaster risk reduction and spatial planning for coastal cities.

A Tsunami hazard map shows Tsunami affected area both in the context of its inundation depth and coverage. It calls for reference to evacuation planning strategies and set up a warning system. This product also an essence for spatial planning to set mitigation from Tsunami impacts. Based on the 2018 Palu Tsunami experiences, deprivation of Tsunami hazard maps, evacuation systems and miscarriage of warning systems leads to the high loss of life. The current advanced warning system is still hardly preventing huge casualties and catastrophic losses from the Palu case since Tsunami came faster than the warning itself. One of the preliminary stages for avoiding the damage is to identify uncertainties in Tsunami arrival times and potentially inundated areas. As most of the Tsunami hazard maps generally only visualize the impact of the worst possible scenario, this recent event proves that other sources' mechanisms could also cause fatal damage. A tsunami hazard map could be an effective medium to deliver far-reaching information for fulfilling disaster mitigation strategies in spatial planning policy. To help such efforts, this study is posed to address Tsunami parameters for various sources shown in the Tsunami hazard map.

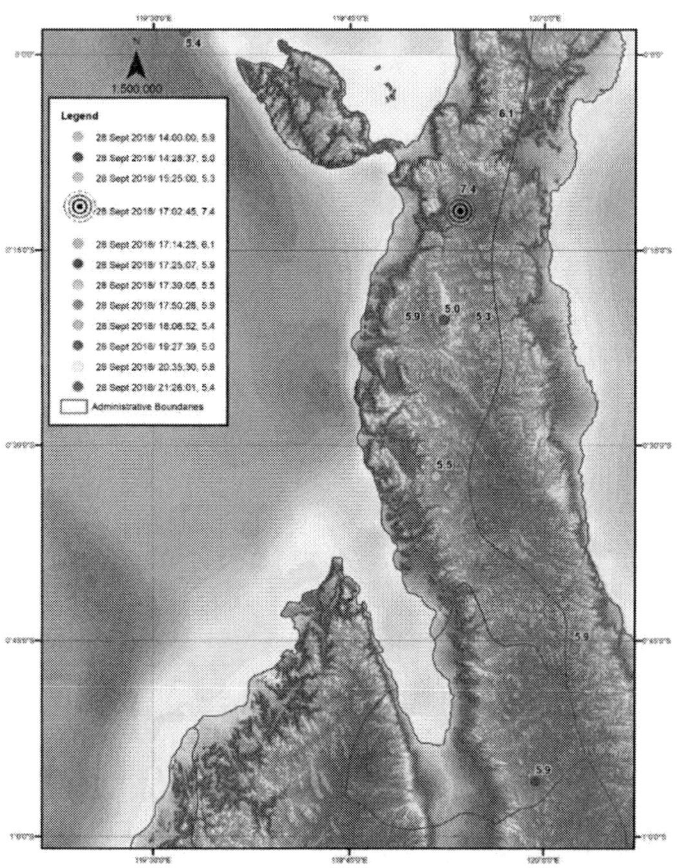

Figure 1. Map of Earthquake Epicenter in Sulawesi September 28, 2018
(Data Source: Badan Nasional Penanggulangan Bencana with modification)

2. Materials and Methods

The general objective of this research is to determine Tsunami prone areas based on inundation depth and its implication towards mitigation disaster action in Palu City. To do this, a numerical model to simulate the propagation of the Tsunami wave is performed. The submarine landslides were assimilated by using an empirical approach to executed the initial water wave. The hydrodynamic models using Delft3D-FLOW are set to compute Tsunami propagation and inundation [5]. Delft3D-FLOW is one of the most established models for conducting Tsunami simulations which has been validated for several Tsunami events such as the 2004 Aceh Tsunami [6]. This modeling tool can show maximum runup and flood depth for the specified domain area.

2.1. Numerical Model

The Global Domain model grid is developed to represent the various source scenarios that may potentially impact Palu City from outside the bay. It covers from 3° 33' 45.3433" N to 3° 56' 15.7605" S latitude and 116° 15' 31.5033" E to 124° 18' 52.2784" E longitude. With 221,125 rectangular grids counted, each grid has 1.11355 km-length and covers 1.24 km^2 area. The domain also rotates 10° to embody North Sulawesi Megathrust. This domain is not able to compute inundation on dry land. Global Domain is illustrated as in Figure 2 below.

The Regional Domain grid is used to investigate Tsunami behavior during propagation into the land area and generate possible landslide submarine-induced Tsunami generation scenarios within the bay. It spans from 119° 40' 19.4641" E to 119° 53' 56.7096" E longitude and 0° 34' 11.0964" S to 0° 54' 21.3421" S latitude. Each grid covers 0.01534 km^2 with 122 m length of a rectangular grid.

To connect the two domains, Delft3D-FLOW has a Domain-Decomposition (DD) boundaries feature [5]. DD boundaries are a feature as internal boundaries to communicate between separate models. DD boundaries connect the wide-area model and regional area model which is separated in the mouth of Palu Bay. This regional area model is developed by subdivision from the Global area model using domain decomposition technique. Thus, the model can be divided into numbers of smaller model domains.

The local Area grid is the refinement of the southern part of the Regional Domain model. Higher grid size resulting in less accurate in the calculation of flooded area. The grid is simplified into a rectangular of 40 m by 120 m. This horizontal refinement is deliberately to identify Tsunami propagation in shallow water areas onshore and provide better inundation parameter information. The grid in a shallow water area will be smaller compare to the deep ocean zone to give a satisfactory result. The farthest inland area only covers up to 2.3 km to the North East area. Figure 2 illustrated the Global and Regional Domain as well as the Local Area.

Palu Bay area has a complex and unique contour. The geomorphology of it beneath the sea surface is not flat. It shapes a ravine filled by an alluvially deltaic system from each mountainous side. Palu Bay has a 30 km-length with 7 km-width. Its thalweg is reaching up to 700 m along the bay. Bathymetry data in this study are compiled from several data sets. The GEBCO_2014 bathymetry dataset is available at 30-arc seconds resolutions (0.926 km) [7] and applied as preliminary bathymetry information for the deep-water area. Some areas are covered by other finer bathymetry data replacing them whenever it is available. The bathymetry of the entire bay of Palu was surveyed in 2014 and shared publicly by the BIG [8] (tides.big.go.id/DEMNAS). Some primary survey had been conducted by Indonesian agencies as well as researchers [9]. Coastal Research Group from the Ministry of Public Works and Housing had been surveying bathymetry along the southern part of the Palu waterfront area. *Badan Pengkajian dan Penerapan Teknologi* (BPPT) had also made a primary observation to obtain a higher resolution of bathymetry data in Palu Bay. All bathymetry data are representatively shown in Figure 3.

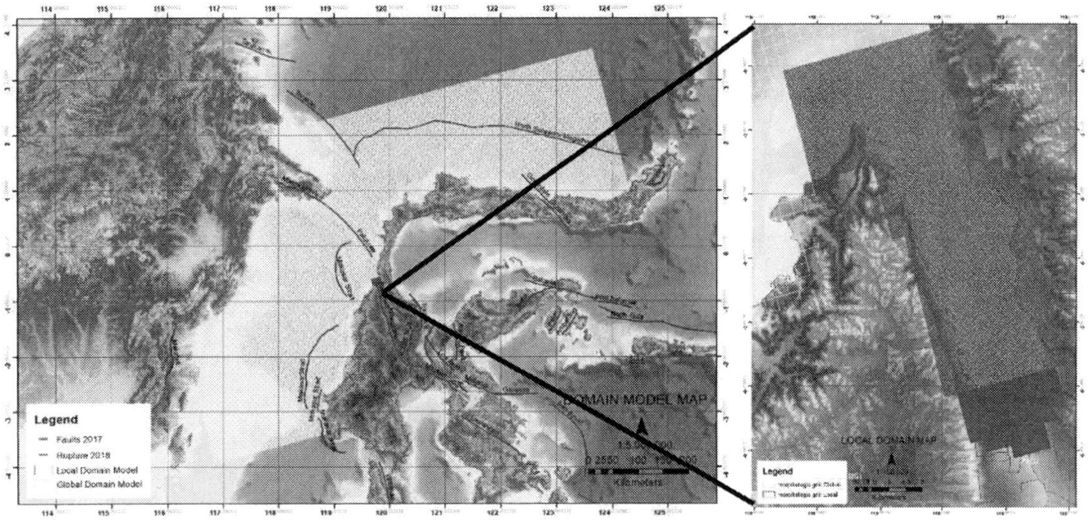

Figure 2. Model domains, Global (right) and Regional with Local Area (left)

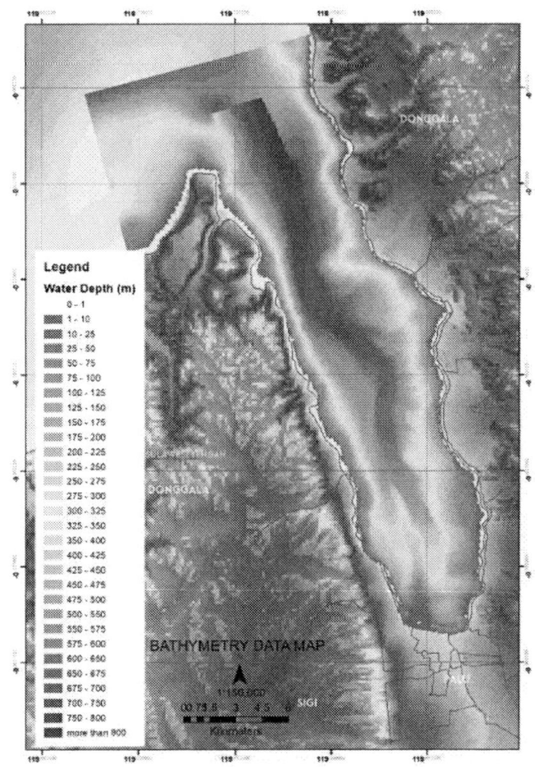

Figure 3. Bathymetry data map showing for Regional Model and Local Model Domain

2.2. Tsunami Sources

Any sudden disturbance factor that could deform a large water mass vertically in the sea from its equilibrium position can generate a Tsunami. Such sudden disturbance caused by geological process: earthquake, landslides, volcanic eruptions, or astronomical events like meteoritic impacts. A Tsunamigenic earthquake happens due to seafloor plates moving against each other. There are only three types of faults that can generate a Tsunami: a strike-slip earthquake on a vertical fault, a dip-slip earthquake on a vertical fault, and a thrusting earthquake on a dipping plane. Landslide generated Tsunami is mainly caused by an earthquake-triggered submarine-landslide in steep bathymetric. Submarine landslide is generally proclaimed as a cause of surprise Tsunami. The surprise Tsunami can be generated by even smaller expected earthquake magnitude and far away from the epicenter.

Tsunami modeling has long-established been performed in scenario-based. Modeling outcomes rely on available information and data. Even though reality condition contains a complex parameter, scenarios can be created to generated various alternative output and be able as comparison tools of options. In other words, scenarios here are functioned to simplify the complexity to produce an understanding.

A zero-water level is prescribed at the beginning of the simulations in all cases as only water level propagation is the only parameter considered in the modeling. The initial condition is the water surface perturbation at the initial time of the simulations which can be caused by Earthquake along the Faultline or landslide, e.g. using Okada (1975) [10].

Figure 4 shows the map of Faultline surrounding Sulawesi Island. Based on this, three simulation scenarios based on Faultline are carried out. North Sulawesi Subduction Zone (NSSZ), North Makassar Thrust (NMTH) and Palu-Koro Faultline Supersheer (PKFS) are based on previous studies [11-15]. The detail Faultline parameter for simulation scenarios can be seen in Table 1.

In comparison to earthquakes, there are many parameters such as slump density, total distance moved, grain size, material density, which are particularly difficult to observe, especially for the submarine landslide. Murty (2003) has identified a simple relationship between the volume of a landslide (V) and maximum resulting Tsunami wave height (H) based on observational data in published literature and numerical model [16]. He hypothesized that simply landslide volume is a key parameter that should exist as the source of Tsunami by mathematical expression as in (1),

$$H = 0.3945\ V \tag{1}$$

By conducting calculation using spatial analysis tools, there is an estimated 3,976,358 m^3 of coastal land subsidence. Therefore, the initial water level height is obtained with 1.57 m. Therefore, 1 (one) simulation scenario based on a possible landslide location in Pantoloan is added. The location can be seen in Figure 4.

Table 1. Faultline Parameter for Simulation Scenario

Scenario ID	Mw	Longitude (deg.)	Latitude (deg.)	Segment (km) Length	Segment (km) Width	Depth (km)	Strike (deg.)	Dip (deg.)	Slip (deg.)
NSSZ	9	120.166	1.577	200.53	110	30	72	20	90
NMTH	8	119.1724	-0.3731	100.53	50	20	11	20	90
PKFS	7.5	119.1724	-0.3731	100.53	50	20	11	20	90

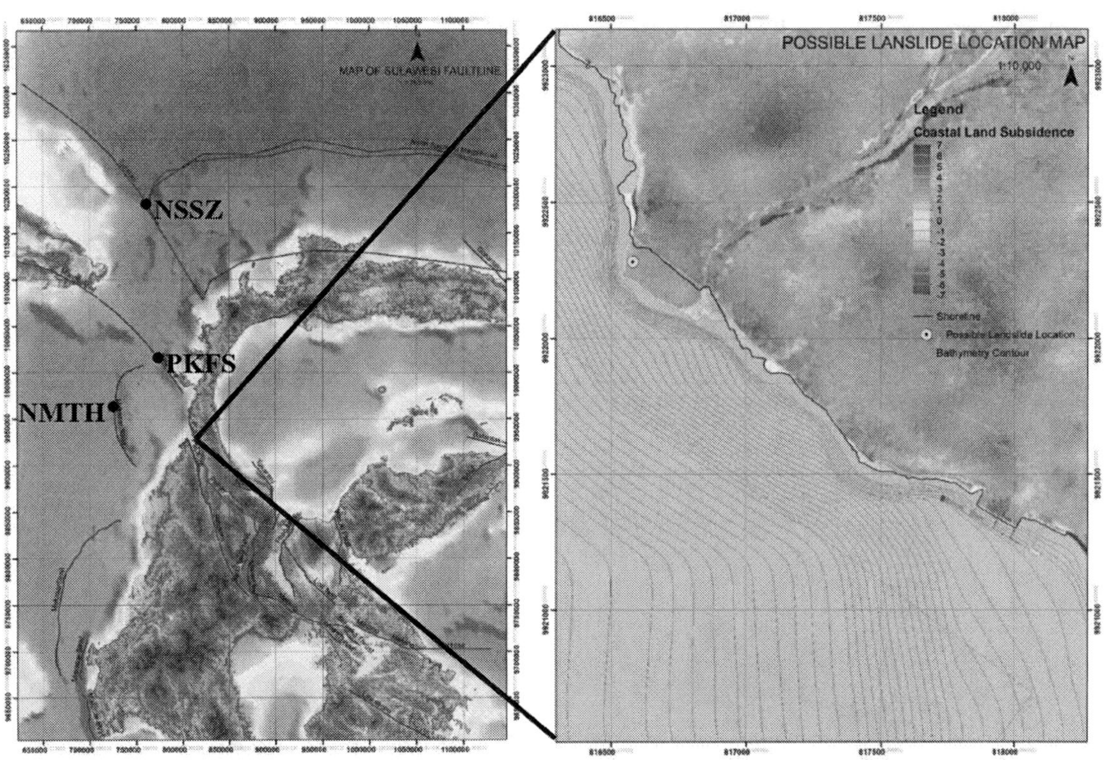

Figure 4. Map of Faultline surrounding Sulawesi Island, showing Tsunami based earthquake scenarios (right) and possible landslide locations in Pantoloan (left).

3. Results and Discussion

3.1. Initial Water Level Condition and Tsunami Propagation

The result of the formulation of the Okada (1975) based on the Faultline parameter in Table 1 creates an initial water level. For example, North Sulawesi Subduction Zone (NSSZ) creates as high as 4.97 m along 200 km continuously in the Celebes Sea in semi-spherical shape. Whereas, North Makassar Thrust (NMTH) creates an initial water level as high as 1.4 m along 100 km continuously in the Makassar Strait in semi-spherical shape, 33 km far from Palu Bay mouth. The Palu-Koro Faultline Supersheer (PKFS) case only generates water level deformation to 1.17 m in two-bubble shaped.
Different from the previous scenario, Scenario 4 does not use Global Domain and uses a different approach for generating water level in Local Domain. The initial water level is input manually and creates with the same shape as its possible landslide dimension. The initial water level for all scenarios is shown in Figure 5.

3.2. Model Verification

Any research comprehensive scientific journal which explained in detail regarding Tsunami numerical modeling in Palu area is considerably limited and consists of assumptions (e.g. [17]). The present model numerical scheme and setting parameters are based on Van Veen et al., (2006) [6]. Besides, the present model results can be accepted, qualitatively, based on hydrodynamic consideration. Hence, giving the Tsunamigenic source, the Tsunami propagation and inundation can be deemed valid.

3.3. Tsunami Propagation and Inundation

As the scenario NMTH gives the worst result, Tsunami propagation and inundation are illustrated based on this scenario. For this scenario, the first wave starts entering the Palu Bay from minutes 7 after the earthquake occurred. As happened in the previous scenarios, the water level takes around 8

IOP Conf. Series: Materials Science and Engineering **982** (2020) 012036 doi:10.1088/1757-899X/982/1/012036

minutes to reach the tip of Palu Bay in the Silebeta area. Nevertheless, the water level begins to increase as the passes Pantoloan and reach the Silebeta area with water level twice higher than its former condition. The snapshot of water level propagation in the Global Domain and within Palu Bay area is illustrated in Figure 6.

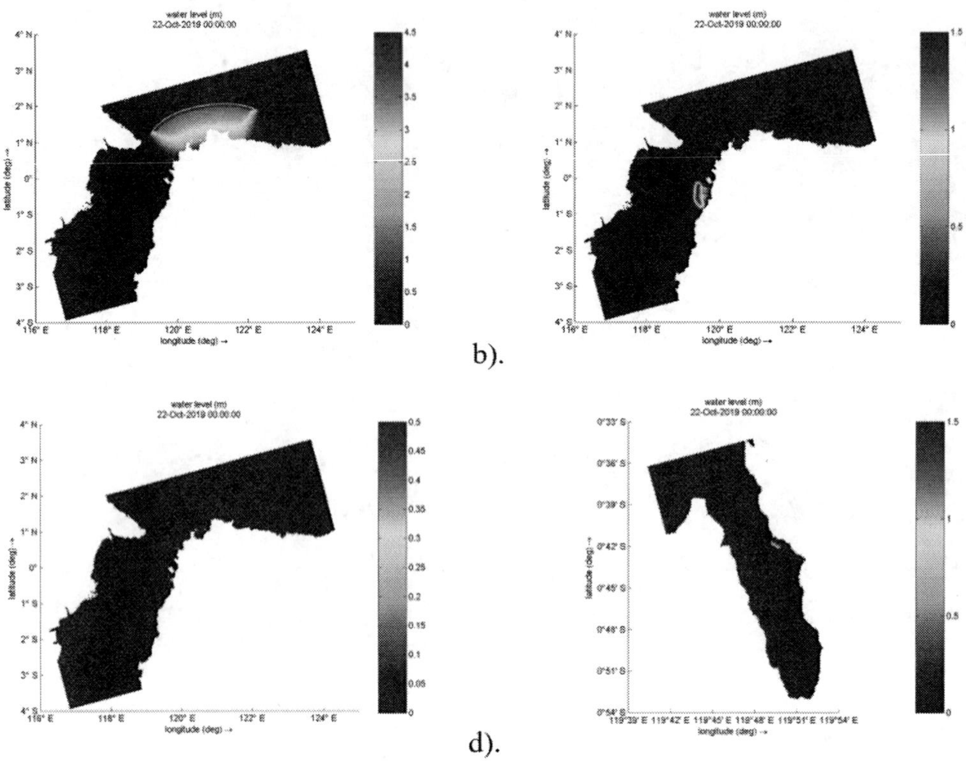

Figure 5. Initial water level condition showing for scenario a). NSSZ; b). NMTH; c). PKFS and d). Landslide at Pantoloan

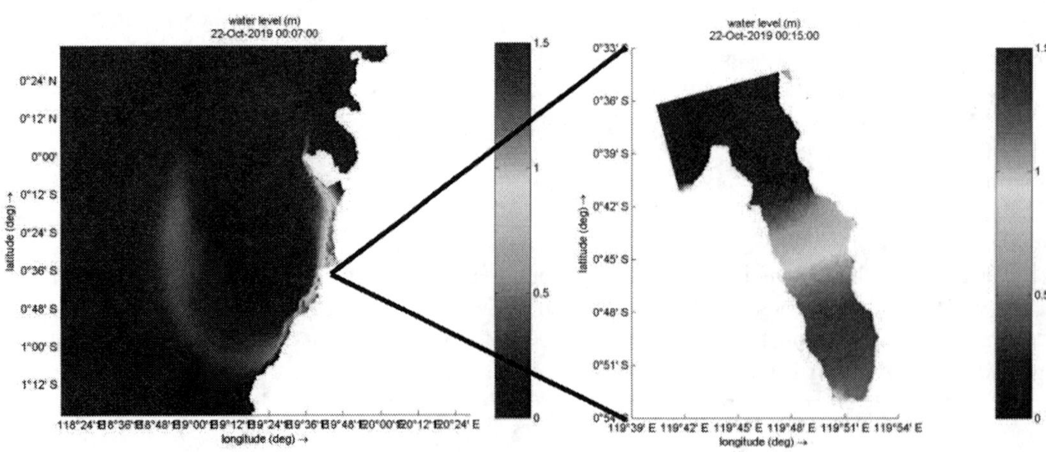

Figure 6. Tsunami propagation after 7 minutes in Global Domain (left) and 15 minutes in Regional Domain (right) for the scenario NMTH

The inundated height is then extracted and presented in the form of Inundation Map. Highest inundation point is found in Buluri Village with 3.6 m and maximum runup reach for 45 m from the coastal line. As contrast with Buluri village, maximum runup is found in Lere village whereas 3.5 m is identified as maximum inundation height. Furthermore, 2.2 km along the Palu River is inundated. However, two villages adjacent to Palu River in upper Stream – North Lolu and Ujuna, do not get any significant impact. The topographical condition reveals that both villages have a steep slope and higher elevation on the riverbank area makes them do not suffer severe damage in this scenario. The eastern part has no significant impact as well as the western part. Many 0.2 m inundation height points have been found in Wani. The inundation map for Silabeta area for Scenario NMTH is illustrated in Figure 7.

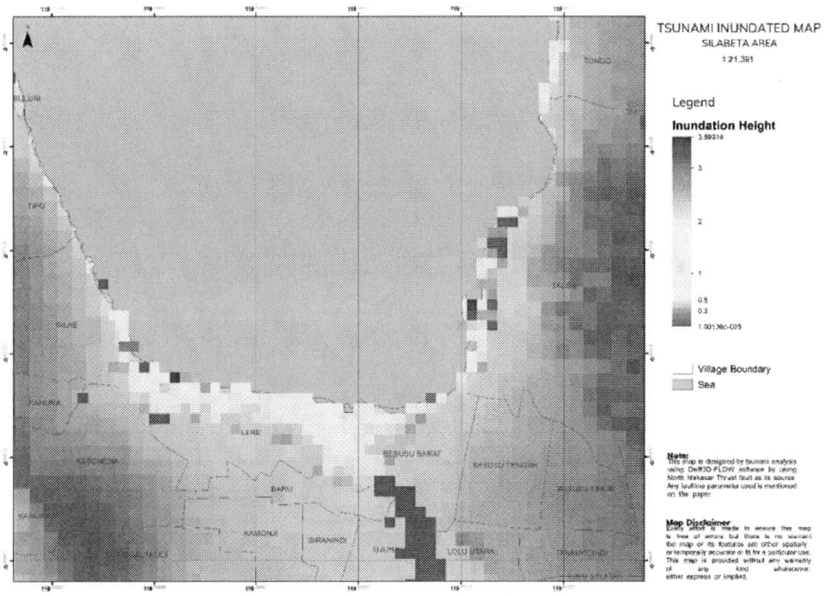

Figure 7. Inundation Map in Local Area Model for Scenario NMTH

3.4. Preliminary Tsunami Hazard Map

Tsunami hazard is one of the major concerning issues in Indonesia. In the past century alone, there have been 24 Tsunami events that occurred in Indonesia. One of the purposes of hazard maps is to provide residents with appropriate and sufficient information on disaster prevention to help them act appropriately at the occurrence of the disaster. To this end, Geological Agency has released a Tsunami hazard map that classified the hazard into three-level: High Tsunami hazard zone, Moderate Tsunami Hazard Zone and Low Tsunami Hazard Zone [18]. The classification is based on its inundation depth, which less than 1 m inundation is considered as a low Tsunami hazard zone, 1 to 3 m categorized as moderate Tsunami hazard zone and more than 3 m is considered as high Tsunami hazard zone. Based on research conducted by previous researchers [19,20], the inundation area is calculated in a grid size of the Local Domain (40 m x 120 m) to measure the Tsunami hazard level. The tsunami hazard level used is based on the definition made by the Geological Agency. Based on that, the number of buildings that have tendencies to be impacted by a certain level of Tsunami hazard is counted. This calculation was conducted for each sub-district and Tsunami arrival time was also provided. Preliminary of Tsunami Hazard Map in Lere Village is shown in Figure 8.

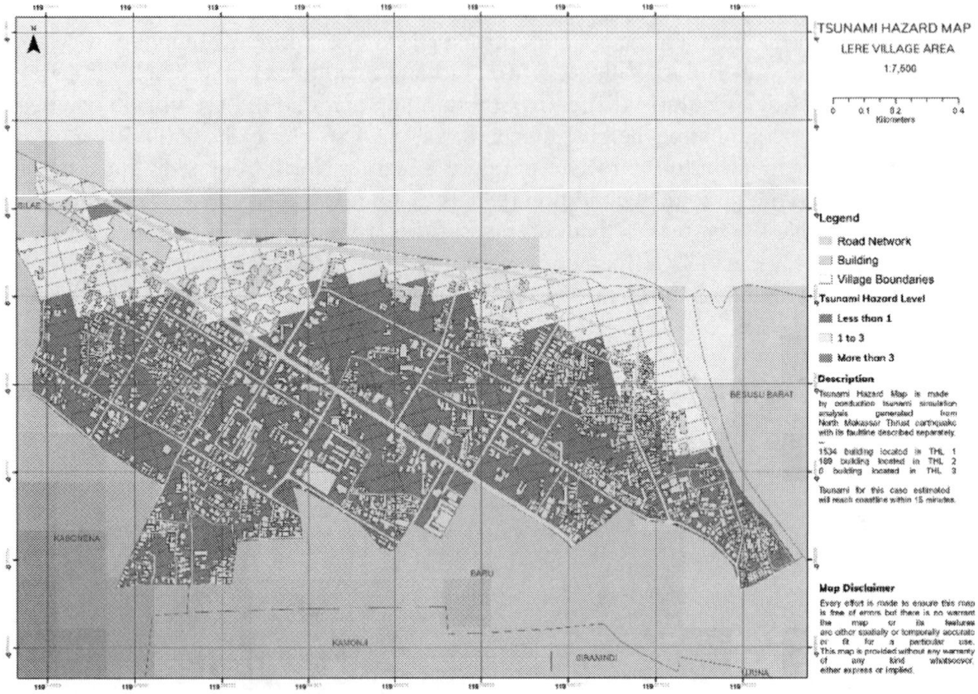

Figure 8. Preliminary Tsunami Hazard Map

4. Conclusions

The general objective of this research is to determine Tsunami prone areas based on the inundation depth and its implication towards mitigation disaster action in Palu City. To do this, a numerical model has been developed. Tsunami inundation modeling using four source scenarios is carried out to determine the Tsunami area for the coastal region of Palu.

Despite a parallel Faultline with Palu Bay, the scenario of North Makassar Thrust (NMTH) turned out to be inducing the most destructive result compare to other scenarios designed. Gentle slope bathymetry profile in front of Lere and Silae coastal region indeed caused the Tsunami wave flows rushed onshore rapidly and drawback in a long time. Palu River itself contributes to inundating adjacent areas. A more comprehensive check-in topographic dataset is required to identify whether it is natural terrain or misinterpretation from spatial analysis erroneous.

Numerous parameters should be considered to generate submarine or aerial landslide-induced Tsunami to give a more satisfactory result. One large landslide will in fact not produce significant deformation of water level as it propagates when it does quite far within the bay. The impact certainly felt on the land opposite.

Nonetheless, high-resolution topographic data is crucial to give more detail and accurate inundation area. Building attributes such as building height information extracted from high-resolution topographic datasets might be useful in the future to develop a Tsunami hazard map. This information can derive analysis such as Tsunami impact towards building on a macro scale or to determine which Tsunami-prone building corresponding to the level mentioned.

5. Acknowledgments

The authors gratefully acknowledge the support and contributions of the Ocean Engineering Program, Faculty of Civil and Environmental Engineering, Institut Teknologi Bandung.

6. References

[1] Brink U T, Wei Y, Fan W, Granja-Bruna J and Miller N 2020 Mysterious Tsunami in the Caribbean Sea following the 2010 Haiti earthquake possibly generated by dynamically triggered early aftershocks *Ear. & Pla. Sci. Letters* **540** 1-11.

[2] Arikawa T, Muhari A, Okumura Y, Dohi Y, Afriyanto B, Sujatmiko K A and Imamura F 2018 Coastal subsidence induced several Tsunamis during the 2018 Sulawesi Earthquake *J. of Dis. Res.* **13** 1-3.

[3] Esteban M, Tsimopoulou V, Mikami T, Yun N Y, Suppasri A and Shibayama T 2013 Recent Tsunamis events and preparedness: Development of Tsunami awareness in Indonesia, Chile and Japan *Int. J. of Dis. Risk Red.* **5** 84-97.

[4] Syamsidik, Rasyif T M, Suppasri A, Fahmi M, Al'ala M, Akmal W, Hafli T M and Fauzia A 2020 Challenges in increasing community preparedness against Tsunami hazards in Tsunami-prone small islands around Sumatra, Indonesia *Int. J. of Dis. Risk Red.* **47** 101572.

[5] Deltares 2014 *Delft3D-FLOW User Manual, Hydro-morphodynamics*, **3.15.34158.**

[6] Veen B V, Vatvani D and Zijl F 2014 Tsunami flood modelling for Aceh & West Sumatra and its application for an early warning system *Cont. Shelf Res.* **79** 46-53.

[7] Weatherall P, Marks K M, Jakobsson M, Schmitt T, Tani S, Arndt J E, Rovere M, Chayes D, Ferrini V and Wigley R 2015 A new digital bathymetric model of world's oceans *Ear. & Space Sci.* **2** 331-345.

[8] Badan Informasi Geospasial 2018, Indonesia, accessed 1 June 2020, <http://tides.big.go.id/DEMNAS>

[9] Takagi H, Pratama M B, Kurobe S, Esteban M, Aranguiz R and Ke B 2019 Analysis of generation and arrival time of landslide tsunami to Palu City due to the 2018 Sulawesi Earthquake *Landslides* **16** 983-991.

[10] Okada Y 1985 Surface deformation due to shear and tensile faults in a half-space *Bull. of the Seis. Soc. of Ame.* **75**(4) 1135-54.

[11] Harris R and Major J 2017 Waves of destruction in the East Indies: The Wichmann catalogue of earthquakes and Tsunami in the Indonesian Region from 1538 to 1877 *Geo. Soc. Lon. Spe. Pub.* **441**(1) 9.

[12] Latief H, Puspito N and Imamura F 2000 Tsunami catalog and zones in Indonesia *J. of Nat. Dis. Sci.* **22**(1) 25-43.

[13] Baeda A Y 2011 Seismic and Tsunami hazard potential in Sulawesi Island *J. of Int. Dev. And Coop.* **17**(1) 17-30.

[14] Amlani F, Bhat H S, Simons W J F, Schubnel A, Vigny C, Rosakis A J, Efendi J, Elbanna A and Abidin H Z 2020 Supershear Tsunamis and insights from the Mw 7.5 Palu Earthquake *arXiv – Geop* 1-13 arxiv.org/pdf/1910.14547.

[15] Fang J, Xu C, Wen Y, Wang S, Xu G, Zhao Y and Yi L 2019 The 2018 Mw 7.5 Palu Earthquake: A supershear rupture event constrained by InSAR and broadband regional seismograms *Rem. Sens.* **11**(11) 1330.

[16] Murty, T.S. 2003. Tsunami Wave Height Dependence on Landslide Volume. Pure and Applied Geophysics. 160. pp.2147 – 2153.

[17] Heidarzadeh M, Muhari A, Wujanarto A B 2019 Insights on the Source of the 28 September 2018 Sulawesi Tsunami, Indonesia Based on Spectral Analyses and Numerical Simulations *Pure Appl. Geophys.* **176** 25-43.

[18] Horspool N, Pranantyo I, Griffin J, Latief H, Natawidjaja D H, Kongko W, Cipta A, Bustaman A, Anugrah S D and Thio H K 2014 A probabilistic Tsunami hazard assessment for Indonesia *Nat. Haz. and Ear. Sys. Sci.* **14** 3105-22.

[19] Shuto N 1993 Tsunami intensity and disasters *Adv. in Nat. and Tech. Haz. Res.* **1** 197–216
[20] Papadopoulos G A and Imamura F 2001 A proposal for a new Tsunami intensity scale *ITS 2001 Proc. Session 5.*

IOP Conf. Series: Materials Science and Engineering **982** (2020) 012037 doi:10.1088/1757-899X/982/1/012037

Impact of Land Use Land Cover Changes on River Discharge at Brantas Catchment Area using SHETRAN Model

Suroso[1], Kharisma Panji Krishna[1,2], Purwanto Bekti Santoso[1]

[1] Civil Engineering Department, Faculty of Engineering, Universitas Jenderal Soedirman, Indonesia

[2] kharisma.krishna@mhs.unsoed.ac.id

Abstract. Changes in land use need to be assessed for future needs. One of them is in efforts to mitigate natural disasters. This research goal is analyzed the correlation between changes in land use land cover on the discharge in the river, in the range between 2001–2017. This research located at Brantas watershed, which is the largest watershed in East Java. It has an area of approximately 11,988 km^2. There are 24 million people who occupy this area and this is one of the national strategic watersheds. SHETRAN is modeling that is based on physical distribution. Reviewing spatial aspects, hydrological and climate data makes SHETRAN is comprehensive model. The method used is to combine input data of the digital elevation model, evaporation rate, rainfall data, land use land cover data, and soil properties classified using the British system. The results obtained for land use, the biggest change is for the grass area which increased by 80.49%, the forest area increased by 22.37%. As for river discharge, modeling results indicate that river flow rates upstream range 6–30 compare to downstream is between 1200–2200 meters^3/second.

1. Introduction

In the tropics, the impact of land use and climate change on watersheds can change the flow volume to be greater [18] due to large energy input and faster anthropogenic changes [12]. The most significant factors affecting surface runoff are changes in forests, dry agriculture, and urban areas. With increasing urbanization, intensification of industry and agriculture, increased runoff will increase the flow of nutrients and sediments to water bodies [33]. Uncontrolled logging makes forest vegetation from a hydrological point of view unstable. In a previous study, logging action could influence the impact of the 100-year return period rain. Land use evaluation is needed within a certain period in order to determine trends in natural change that can be used for mitigation needs in the future [32].

In addition to land use analysis, it is also necessary to develop a more up-to-date hydrological model to be implemented in Indonesia. There is a long history of research that has been carried out on automatic generation of river networks from digital elevation models (DEM) [10]. The basic method for extracting river channel networks then uses flow direction to calculate upstream contribution areas for each square grid [22]. Models are needed that can define channels to simulate hydrological mitigation needs.

SHETRAN is a physical-based distribution model (PBDM) for water flow, sediment and solute transport in river catchments [2]. Includes hydrological components for simulation: interception of rainfall by vegetation; evaporation and transpiration; snow formations and melted snow; land flow and channel; subsurface flow varies saturated; and river/aquifer. SHETRAN solves physics-based partial differential equations, for flow and transportation on rectangular finite elements.

The purpose of this study is to develop a modeling that was first carried out in Indonesia. The analysis carried out is an investigation of changes in land use that occurred on the results of SHETRAN for river discharge modeling. The research object chosen in the Brantas catchment area, East Java because in the recapitulation of existing data, the cities and regencies which include in Brantas catchment area account for approximately 50% of the total impact of buildings and humans due to flooding in East Java, and Brantas watershed is a watershed that is included in the national strategic area.

2. Data and Study Location

Brantas watershed chosen because it is one of major watersheds in Indonesia, which watershed area around of 11,988 km^2 or about 25% of East Java Province. Located at coordinates 110°30' East to 112°55' East and 7°01' East to 8°15' East. Covering 8 cities and 16 districts, upstream location is in Kediri Regency and the downstream is in Surabaya City.

Brantas watershed has great potential for various sectors, such as Surabaya city which is center of government in East Java, Kediri city which is the city with third highest Gross Regional Domestic Revenue, industrial and plantation cities in Batu and Malang, tourism and cultural cities in Ponorogo, and also other cities and districts. Brantas watershed also has a significant role in supporting East Java Province as a national food barn. In 2018, East Java Province will still be the backbone for rice, corn and cassava stockpiles by contributing ± 19.3% of the national food stock needs. In Brantas watershed there are also a total of 10 Irrigation Areas included in the Brantas watershed with an area around 113,368 Ha.

In terms of geography, Brantas watershed is crossed by volcanoes and mountains, which results in very varied topographic conditions ranging from flat, hilly, valley and mountainous. This causes the available land can not be fully cultivated to improve the welfare of the community, because there must be a protected area that must be protected and preserved to maintain environmental balance and prevent the recurrence of environmental damage, especially landslides and floods due to reduced land cover.

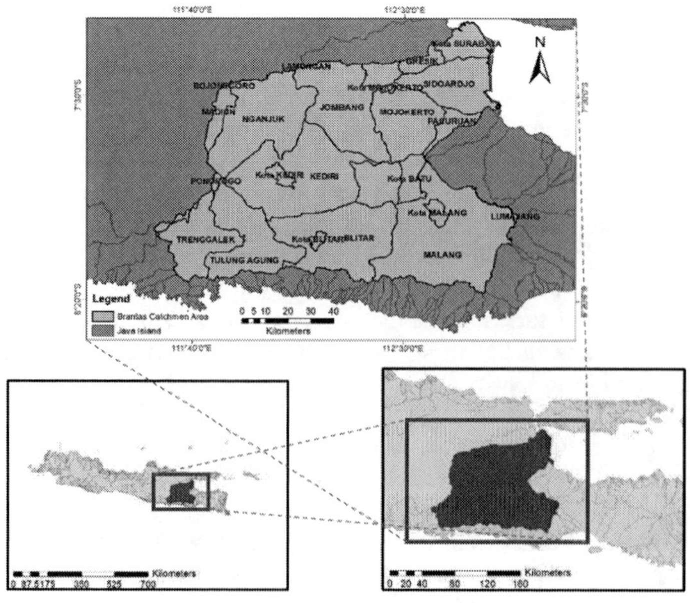

Figure 1. Research Location

3. Method

Hydrological model SHETRAN was utilized to perform runoff simulations for the Brantas Catchment Area. SHETRAN is a physically-based, distributed, deterministic, integrated surface and subsurface modelling system, designed to simulate water flow, sediment transport, and contaminant transport at the catchment scale ([2];[3]). SHETRAN has a modular design, in which each module or component is used to represent the different physical processes of the hydrological cycle in each part of the catchment. The methods used in SHETRAN to simulate the processes of its water flow component include Potential Evapotranspiration (PET), Actual Evapotranspiration (AET), Interception, Overland runoff, Infiltration/subsurface flow, River flow routing.

The catchment is conceptualized as an ensemble of columns and networks of stretches of channels, called river links. Each column is a stack of computational cells containing information of land cover/vegetation type at the top and sequential depth of soil/s horizons underneath it. Each river link runs along the edge of column tops. SHETRAN works out the water balance in each column. Spatially distributed data, including a digital elevation model, the soil and land use map were used in a raster format with a grid resolution maximum 200 × 200. The net outcome of hydrological processes at all the columns is the hydrological behaviour of the catchment.

In this model, the partial differential equations for flow and transport are solved by finite difference methods. The basin is discretized into a horizontal orthogonal river network and at each square grid in the vertical direction by a column of layers. A network of river links runs along the edges of grid squares representing the river network. The model explicitly incorporates spatial heterogeneity in topography, soil, land use and catchment properties with its grid-column structure. The model represents coupled surface/subsurface flow, allowing overland flow to be produced by rainfall excess over infiltration and by upward saturation of the soil column [4].

The interception of rainfall is represented by the modified Rutter model. The actual evapotranspiration is calculated from Potential Evapotranspiration (PET) as a function of soil water potential. The diffusive wave approximation of the St. Venant equation governs the overland and channel flow processes, while the Richards and Boussinesq equations give the one-dimensional flow in the unsaturated zone and two-dimensional flow in the saturated zone, respectively. The river aquifer interaction is calculated using the Darcy equation.

The procedure for the automatic generation of the river channel network in SHETRAN is different than typical approach. Therefore, the following procedure for the automatic generation of the river channel network in SHETRAN has been developed. Initially this follows a typical approach [8]: the pits in the DEM are removed, the flow directions are calculated and this gives an upstream contributing area for each grid square, and a procedure for channel extraction is defined.

In this study, the main steps undertaken from beginning to end are:

a) GIS processing to get ASCII files that will form coordinates as a basis for making maps on SHETRAN
b) Processing matrix as a record for input data that becomes a value for the output data formed
c) Configure XML to run SHETRAN model
d) Running SHETRAN to the standard version to get computational correction
e) Extract SHETRAN data results to be further processed for presentation of research data

4. Results and Discussion

4.1. Land Use Land Cover Change

Land use and cover data used is derived from MODIS which is a level 3 combination product from Terra and Aqua satellites (MCD12Q1). Provides information on annual land cover types with sinusoidal projections and spatial resolution of 500 meters, in HDF-EOS format, with an observation series from 2001 to 2017. The MCD12Q1 product is able to identify the geographical distribution of 17 land cover classes using a scheme defined by the International Geosphere Biosphere Program (IGBP). One MCD12Q1 product has several data layers, including the type of land cover, assessment layer and quality flag for each type of land cover. In this study, a similar approach was carried out by

IOP Conf. Series: Materials Science and Engineering **982** (2020) 012037 doi:10.1088/1757-899X/982/1/012037

classifying 17 types of land cover from MODIS MCD12Q1 data into 7 categories that have been classified for the SHETRAN program, shown in Table 1.

Table 1. Conversion MODIS Classification to SHETRAN LULC Classification

MODIS Classification	SHETRAN Classification
Croplands, Natural Vegetation Mosaics	Arable
Permanent Wetlands, Barren, Water Bodies	Bare Ground
Grasslands	Grass
Deciduous Needleleaf Forest, Deciduous Broadleaf Forest, Mixed Forest	Deciduous Forest
Evergreen Needleleaf Forest, Evergreen Broadleaf Forest	Evergreen Forest
Closed Shrublands, Open Shrublands, Woody Savannas, Savannas	Shrub
Urban and Built-Up Lands	Urban

Table 2 shows the results of an analysis for 17 years in the Brantas catchment area. Two types were degraded, it is an arable 9.17% and deciduous forest 11.26%. While those who experienced an increase there were five types, namely bareground 2.54%; grass 80.49%; evergreen forest 22.37%; shrub 10.77%; urban 8.74%. If calculated from the entire area of the Brantas catchment area, the changes in each type of LULC range from 0.06% to 5.38%. The arable area has become a changing area, be it an urban, shrub, or evergreen forest area. The addition of green forest areas that bring a lot of benefits, including groundwater storage that can affect river discharge.

Table 2. LULC change over the past 17 years (2001–2017) in Brantas Catchment Area

LULC	2001 (km^2)	2017 (km^2)	Change (km^2)	%Change of 2017	%Change of Total Area	% LULC of Total Area
Arable	8118	7373	-745	-9,17	-5,38	55,8
Bareground	316	324	8	2,54	0,06	2,3
Grass	51	93	41	80,49	0,30	0,4
Deciduous Forest	73	65	-8	-11,26	-0,06	0,7
Evergreen Forest	1334	1632	298	22,37	2,15	9,2
Shrub	2921	3235	315	10,77	2,27	23,8
Urban	1039	1129	91	8,74	0,66	7,8
Total	13850	13850				100

In Figure 2 below, shows the percentage of each type of LULC SHETRAN to the total area of the Brantas watershed. Visible changes in each type of LULC every year in 2001 to 2017.

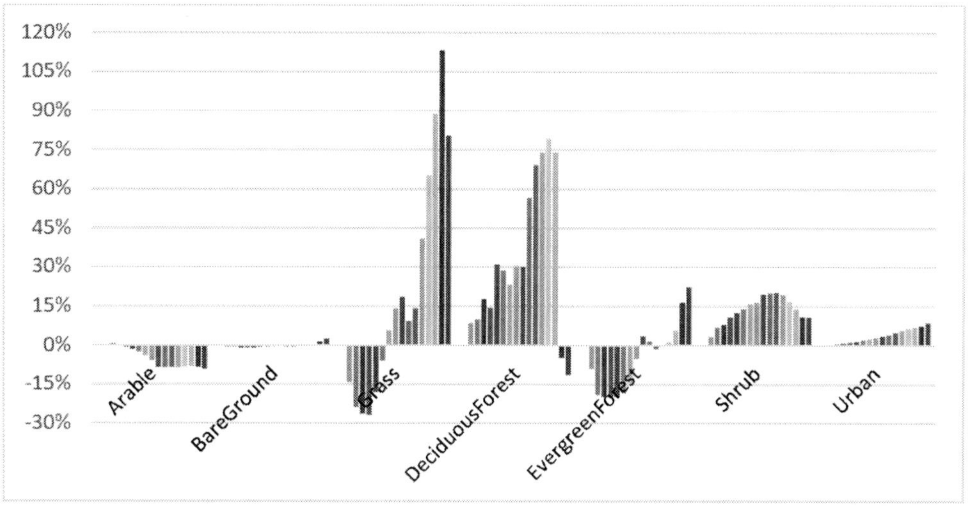

Figure 2. LULC change from 2001–2017 at Brantas Catchmen Area

4.2. River Discharge Analysis Based on SHETRAN Model

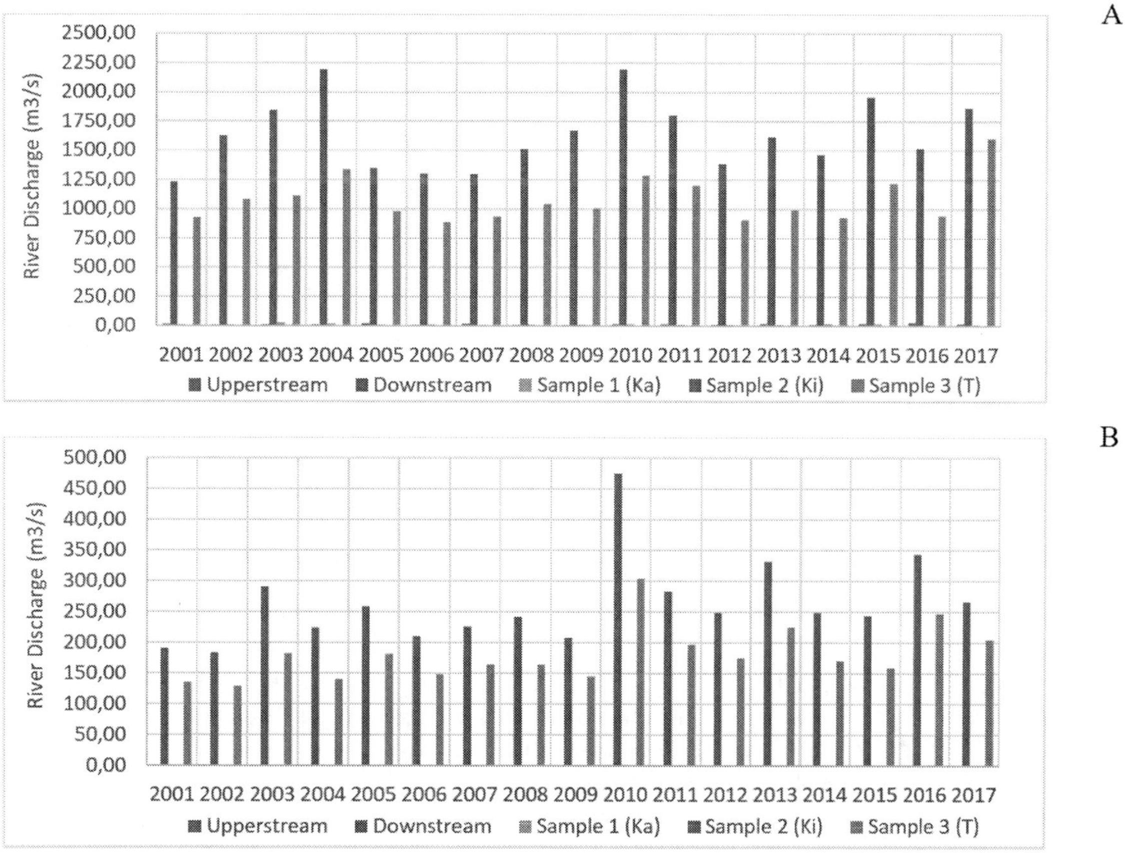

Figure 3. (A) Maximum river discharge, and (B) Average river discharge based on SHETRAN model (source : author)

In figure 3 (A), a very clear difference can be seen from the selected sample points. Discharge in the downstream area is different to more than 1000 meters^3/second when compared to upstream area. Sample points that are on the right and left side of the watershed also produce a discharge that is not too large. In line, figure 3 (B) show average river discharge produced by the SHETRAN model also shows a vast difference of more than 150 meters^3/second. In maximum flow analysis, discharges produced showed high results in 2004, 2010, 2015, and 2017. It is not much different from the data on rainfall and flood events.

The highest river discharge resulted by SHETRAN model in upper stream is 30,66 meters^3/second in 2016, which the lowest is 6,90 meters^3/second in 2008. In the outlet of Brantas catchment area, the highest river discharge resulted 2197,73 meters^3/second in 2010, and the lowest is 1230,87 meters^3/second in 2001. Still in the downstream area, the debit in the middle sample point obtained a large discharge, namely the lowest maximum discharge of 888.97 meters^3/second in 2006 and the highest maximum of 1600.15 meters^3/second in 2017. For the right sample point the maximum discharge results the highest is obtained 27.63 meters^3/second and the lowest maximum discharge is 8.77 meters^3/second. While the left sample, the results obtained 129.07 meters^3/second for the lowest and 303.96 meters^3/second for the highest maximum discharge, namely in 2002 and 2010.

For average discharge, from 2001-2017 shows a value that tends to stagnate, except in 2010 which found a significant increase in average discharge. Reviewing average daily discharge generated every year, the results obtained for outlets ranged from 200–350 meters^3/second, midpoint 150–250 meters^3/second, and for the right, left, and upstream samples ranging from 0.1–1.8 meters^3/second. The highest was in 2010, the average increase in debit at outlets and middle samples reached around 150 meters^3/second.

However, it should be noted that in actual cases, the sensitivity of a particular watershed to runoff generation is also impacted by other factors such as magnitudes of land use change, soil conditions, and different climatic factors [1].

5. Conclusion

Results of land use processing, obtained five types that have increased. The LULC types that have increased in sequence are: shrub, evergreen forest, urban, grass, and bareground. While those experiencing reductions along with the size of the area deficit: arable, and deciduous forest.

The most dominant area in the Brantas watershed is the arable type with an average of ± 55.8% of the total Brantas watershed area, then shrubs of ± 23.8%; evergreen forest ± 9.2%; urban ± 7.8%; bare-ground ± 2.3%; deciduous forest ± 0.7%; and grass ± 0.4%. Results of SHETRAN model on river discharge at each point show fluctuations because LULC every year are always changes. In 2010 or 2016, there was no significant forest growth in downstream location, while urban areas continued to increase. This result causes a lack of control over river discharge due to insufficient availability on absorption land. Meanwhile, the amount of discharge occurs when the condition of the largest deciduous forest like in 2014, or evergreen forest has not been degraded in a large amount like in 2001. It shows that good vegetation affects groundwater storage which will have an impact on surface water.

6. Acknowledgments

Financial support for completing this research is graciously provided by the Ministry of Research, Technology, and Higher Education of the Republic of Indonesia. Special thanks of gratitude is also expressed to our collaborator from School of Engineering Newcastle University namely Steven Birkinshaw, Chris Kilsby, and Andras Bardossy.

7. References

[1] Astuti I S 2019 Impact of land use land cover (LULC) change on surface runoff in an increasingly urbanized tropical watershed *Water Resources Management* 33(12) 4087-4103.
[2] Birkinshaw S J and Ewen J 2000 Nitrogen transformation component for SHETRAN catchment

nitrate transport modeling *Journal of hydrology* 230(1-2) 1-17.

[3] Birkinshaw S J, James P and Ewen J 2010 Graphical user interface for rapid set-up of SHETRAN physically-based river catchment model *Environmental Modelling & Software* 25(4) 609-610.

[4] Birkinshaw S J, Bathurst J C, Iroumé A and Palacios H 2011 The effect of forest cover on peak flow and sediment discharge—an integrated field and modelling study in central–southern Chile *Hydrological Processes* 25(8) 1284-1297.

[5] Bisri M 2017 Application of the Kineros model for predicting the effect of land use on the surface run-off Case study in Brantas sub-watershed, Klojen District, Malang City, East Java Province of Indonesia *Journal of Water and Land Development* 35(1) 3-9.

[6] Briassoulis H 2019 Analysis of land use change: theoretical and modeling approaches.

[7] Gumma 2019 Assessing potential locations for flood-based farming using satellite imagery: a case study of Afar region, Ethiopia *Renewable Agriculture and Food Systems* 1-15.

[8] Grimaldi S, Petroselli A, Alonso G and Nardi F 2010 Flow time estimation with spatially variable hillslope velocity in ungauged basins *Advances in Water Resources* 33(10) 1216-1223.

[9] Jati M I H, Suroso and Santoso P B 2019 Prediction of flood areas using the logistic regression method (case study of the provinces Banten, DKI Jakarta, and West Java) *In Journal of Physics: Conference Series* Vol. 1367 No. 1 p 012087 *IOP Publishing*

[10] Jenson S K and Domingue J O 1988 Extracting topographic structure from digital elevation data for geographic information system analysis *Photogrammetric engineering and remote sensing* 54(11) 1593-1600

[11] Karlsson I 2016 Combined effects of climate models, hydrological model structures and land use scenarios on hydrological impacts of climate change *Journal of Hydrology* 535, 301-317.

[12] Kondraju T T and Rajan K S 2019 Water Quality in Inland Water Bodies: Hostage to the Intensification of Anthropogenic Land Uses *Journal of the Indian Society of Remote Sensing* 47(11) 1865-1874.

[13] Kundu S 2017 Individual and combined impacts of future climate and land use changes on the water balance *Ecological Engineering* 105 42-57.

[14] Li Dan 2018 Adequacy of TRMM satellite rainfall data in driving the SWAT modeling of Tiaoxi catchment (Taihu lake basin, China) *Journal of Hydrology* 556: 1139-1152.

[15] Liu Xing 2017 Combining rainfall data from rain gauges and TRMM in hydrological modelling of Laotian data-sparse basins *Applied Water Science* 7.3: 1487-1496.

[16] Long H and Qu Y 2018 Land use transitions and land management: A mutual feedback perspective Land Use Policy 74 111-120.

[17] Marhaento H 2018 Hydrological response to future land-use change and climate change in a tropical catchment *Hydrological sciences journal* 63(9) 1368-1385.

[18] Marhaento H 2017 Attribution of changes in stream flow to land use change and climate change in a mesoscale tropical catchment in Java, Indonesia Hydrology research 48(4) 1143-1155.

[19] Marhaento H Attribution of changes in the water balance of a tropical catchment to land use change using the SWAT model *Hydrological processes* 31(11) 2029-2040.

[20] Martens Brecht 2017 GLEAM v3: Satellite-based land evaporation and root-zone soil moisture *Geoscientific Model Development* 10.5: 1903-1925.

[21] Michelson K and Chang H 2019 Spatial characteristics and frequency of citizen-observed pluvial flooding events in relation to storm size in Portland, Oregon *Urban Climate* 29 100487.

[22] Montgomery D R and Foufoula-Georgiou E 1993 Channel network source representation using digital elevation models *Water Resources Research* 29(12) 3925-3934.

[23] O'Loughlin F E 2016 A multi-sensor approach towards a global vegetation corrected SRTM DEM product *Remote Sensing of Environment* 182: 49-59.

[24] Op de Hipt F 2017 Applying SHETRAN in a tropical west African catchment (Dano, Burkina

Faso)—Calibration, validation, uncertainty assessment. *Water* 9(2) 101.
[25] Popp A 2017 Land-use futures in the shared socio-economic pathways *Global Environmental Change* 42 331-345.
[26] Pratidina G and Santoso P B 2019 Detection of satellite data-based flood-prone areas using logistic regression in the central part of Java Island *Journal of Physics: Conference Series* Vol. 1367 No. 1 p. 012086 *IOP Publishing.*
[27] Setyorini 2017 Simulating the impact of land use/land cover change and climate variability on watershed hydrology in the Upper Brantas basin, Indonesia *Applied Geomatics*, 9(3) 191-204.
[28] Siegel F R 2016 Land-Use Planning to Minimize Dangers to Citizens and Ecosystems". In Mitigation of Dangers from Natural and Anthropogenic Hazards *Springer Cham.*
[29] Talsma 2018 Partitioning of evapotranspiration in remote sensing-based models *Agricultural and Forest Meteorology* 260:131-143.
[30] Umiati T 2019 Spatial analysis and monitoring of drought using Standardized Precipitation Index in East Java *Journal of Physics: Conference Series* Vol. 1367 No. 1 p 012088 *IOP Publishing.*
[31] Usman 2015 Land use/land cover classification and its change detection using multi-temporal MODIS NDVI data *Journal of Geographical sciences* 25(12) 1479-1506.
[32] Yoshino K. 2017 Land use analysis using time series of vegetation index derived from satellite remote sensing in Brantas River watershed, East Java, Indonesia *Geoplanning: Journal of Geomatics and Planning* 4(2) 109-120.
[33] Zhang Y 2018 Simulation and assessment of urbanization impacts on runoff metrics: insights from landuse changes *Journal of hydrology* 560 247-258.

IOP Conf. Series: Materials Science and Engineering **982** (2020) 012038 doi:10.1088/1757-899X/982/1/012038

Detection of Flood Inundation in the Progo Watershed using the SHETRAN Model

Suroso[1], Refin Nadawaty[1,2], Purwanto Bekti Santoso[1]
[1]Department of Civil Engineering, Universitas Jenderal Soedirman, Purwokerto

[2]refinnadaw@gmail.com

Abstract. Floods are the most frequent disasters in Indonesia in the past 20 years. Based on the Indonesian Disaster Information Data (DIBI) compiled by BNPB, there are considered to be 9053 events (37.5%) of the total natural disaster events in Indonesia. The flood disaster has threatened and disrupted people's lives and even resulted in fatalities, environmental damage, property losses, and psychological impacts. Therefore several efforts of flood risk reduction are required. One of them is to develop a model to detect the depth of flood inundation. The purpose of this study is to apply the SHETRAN Model for flood inundation detection in the Progo Watershed. The SHETRAN model is a distributed hydrological model. The data required in this model are hydro-climatological data, including rainfall and evaporation, land use land cover, soil type, and topography data obtained from remote sensing. The results of this study indicate that in the range from 2001 to 2017, the maximum height of flood inundation between the data on flood events in the field and the phreatic depth of the results from the SHETRAN Model were 0.606 m which occurred on January 17, 2013 in Tirta Rahayu Village, Sub-district Galur, Kulon Progo Regency. While the maximum height of flood inundation results from the RStudio analysis on the Progo Watershed range from 2001 to 2017 were 3,856 m which occurred in 2010.

1. Introduction

Floods are the most frequent disasters in Indonesia in the past 20 years. Based on the Indonesian Disaster Information Data (DIBI) compiled by BNPB, there were 9053 events (37.5%) of the total natural disasters in Indonesia. Whereas in Central Java Province and Yogyakarta Special Region Province (DIY) flooding ranks third after landslides and tornadoes with 1449 events or around 23 percent of the total natural disaster events from early 2001 to the end of 2019 as shown in Figure 1. The frequency of floods in Central Java and DIY Provinces always changes every year. For that reason, flood disaster must always be a problem that needs attention at all times.

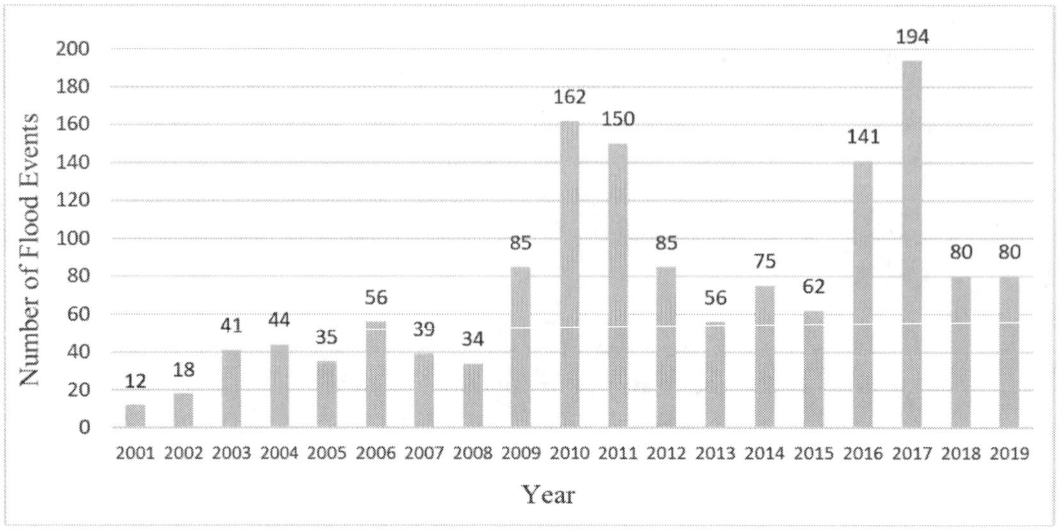

Figure 1. Frequency of Flood Events in Central Java and Yogyakarta in 2001-2019

Inundation is an event where water does not flow or water stops somewhere ([1];[12]). Thus, it is possible to have an inundation of water even though the water level in the river is still normal. Inundation is not a problem if it does not cause a problem, such as not causing casualties, losses, and not submerging community facilities that live in an inundation area [12]. Inundation becomes a problem and develops into a disaster or commonly called a flood if it interferes with human activities and even causes casualties and property.

In this research, flood modeling is developed using the SHETRAN Model. The SHETRAN Model is a spatial distributed hydrological model developed by Newcastle University ([2];[10];[14]). The SHETRAN physically-based distributed rainfall-runoff modelling system gives detailed simulations in time and space of water flow and sediment and solute transport in river catchments [2]. One of the advantages of this model is that only using input data from satellite data can produce multiple outputs in the hydrological cycle ([2];[3];[10]). The outputs of the SHETRAN Model include phreatic depth (m), overland flow (m^3/s), surface depth (m), soil moisture (m^3/m^3), and water table elevation (m). To detect flood inundation, the output of the SHETRAN Model used is phreatic depth (m) [3].

The data needed to use this model is hydroclimatological data. In this study, hydroclimatological data were obtained from satellite data obtained from related agencies such as the Meteorology, Climatology and Geophysics Agency (BMKG). Hydro-climatological data used include (1) topographic data obtained from Digital Elevation Model (DEM) data; (2) rainfall data and evaporation data obtained from Tropical Rainfall Measuring Mission (TRRM) data; (3) soil type data obtained from Hermonized World Soil Database (HWSD) data; and (4) land use data obtained from Land Use Land Cover (LULC) data.

River floods are common natural disasters that cause serious economic damage worldwide. In addition to direct economic damage, such as the destruction of physical assets, floods with long-lasting inundation cause direct and indirect economic losses within and outside the affected area. Direct economic losses include loss of opportunity, due to interruption of business activities, and the costs associated with emergency measures such as cleaning, while indirect economic losses affect sectors within the trade and supply network [12]. There are many factors that can cause flooding, but the causes of flooding cannot be completely eliminated. Therefore, efforts are needed to reduce the impact caused by flooding. One way to overcome the problem of flooding is to make flood modeling that can detect inundation. This study aims to apply the SHETRAN Model to detect flood inundation in the Progo Watershed.

2. Data and Study Location

2.1. Study Location
The research location was carried out in the Progo Watershed, located in Central Java Province and Yogyakarta Special Region. The Progo Watershed area covers 11 districts/cities, there are Bantul Regency, Yogyakarta City, Kulon Progo Regency, Sleman Regency, Purworejo Regency, Boyolali Regency, Magelang Regency, Magelang City, Semarang Regency, Temanggung Regency, and Wonosobo Regency as shown in Figure 2. The area of Progo Watershed is ± 2659 km^2 with the Progo River as the main river which has a length of ±138 km.
Geographically, the Progo Watershed is located at 109^0 59 'BT - 110^0 291' BT and 07^0 12' LS - 08^0 04' LS. The Progo Watershed has a tropical climate, with a rainy season between October and March, and a dry season between April and September. The amount of rain per year in the Progo Watershed varies from 1700 mm to 4000 mm per year, with monthly variations between 33 to 385 mm. Flooding is a routine problem that occurs in several areas within the Progo Watershed. This can happen because in the upstream part of the Progo Watershed is an agricultural area and many of them are turning into non-agricultural functions. Whereas in the downstream areas there was a reduction in irrigation areas due to the conversion of irrigated land into settlements due to population pressure.

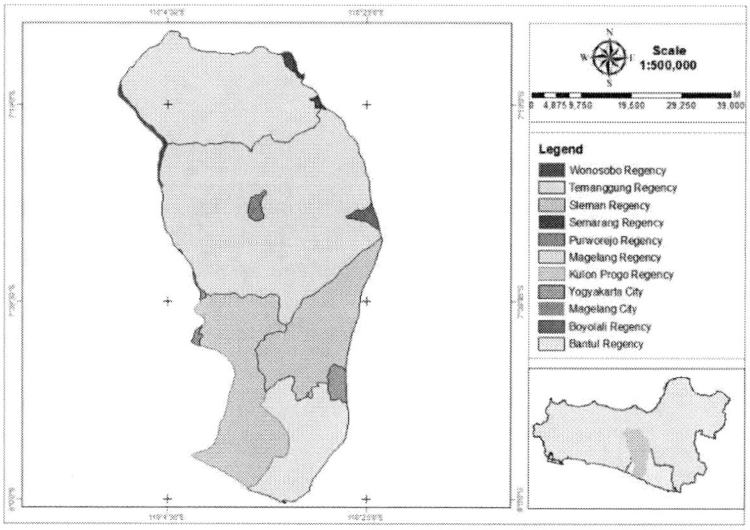

Figure 2. Progo Watershed Map

2.2. Data Collection
The data needed in this study can be explained as follows:

2.2.1. Topographic Data.
Topographic data is obtained from Digital Elevation Model (DEM) data, which is digital data that is able to describe the shape and condition of the earth's surface needed in flood modeling using the SHETRAN Model ([9];[11]). Therefore, the DEM data in this study was processed using ArcGIS software to describe the Progo Watershed so that it can be run on the SHETRAN model. The following is DEM map of Java Island can be seen in Figure 3.

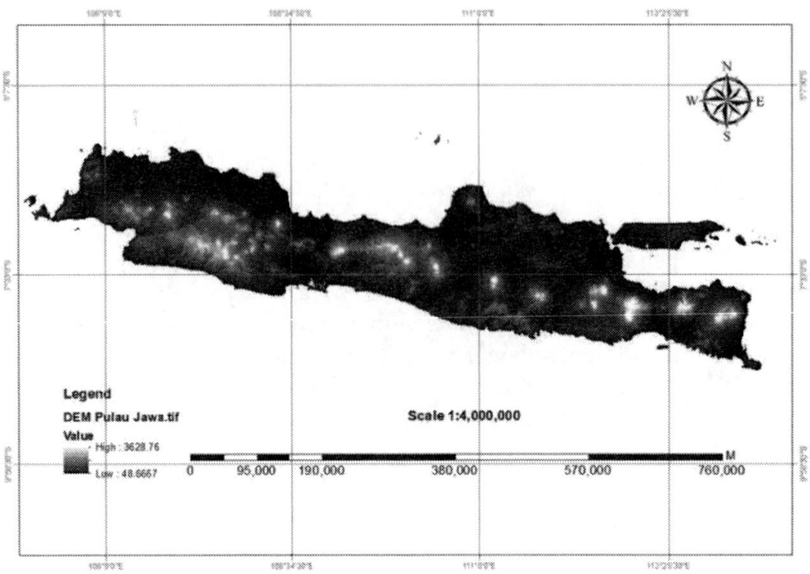

Figure 3. DEM Map of Java Island

2.2.2. Rainfall Data and Evaporation Data.
Rainfall data and evaporation data are obtained from Tropical Rainfall Measuring Mission (TRRM) data obtained using remote sensing technology ([7];[11]). In this study, the satellite data is processed using ArcGIS and Microsoft Excel to produce rainfall data and evaporation data on a daily scale in the Progo Watershed area so that it can be run on the SHETRAN Model. The following is TRMM map of Central Java Province and Yogyakarta Special Region can be seen in Figure 4.

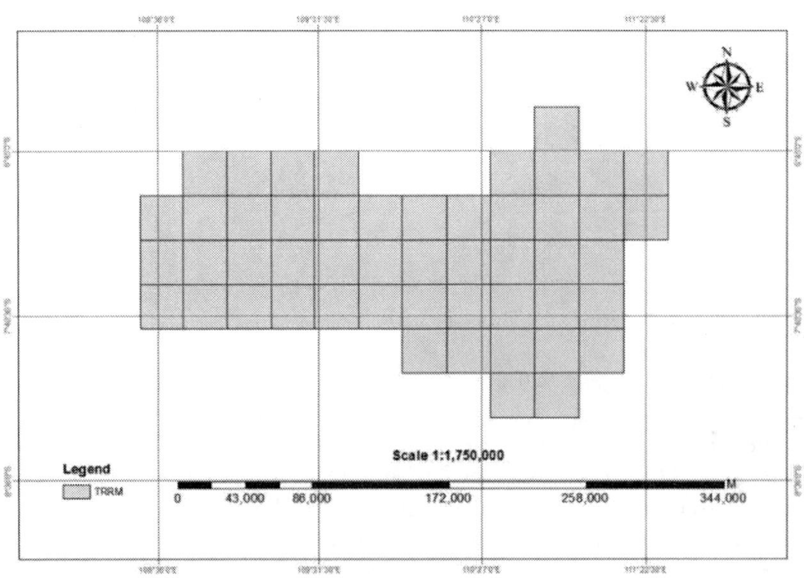

Figure 4. TRRM Map of Central Java Province and Special Region of Yogyakarta

2.2.3. Soil Type Data.
Soil type data is obtained from Hermonized World Soil Database (HWSD) data derived from satellite data ([6];[8];[14]). In this study, HWSD data was processed using ArcGIS and Microsoft Excel to determine raster data of soil types in the Progo Watershed area so that it can be run on the SHETRAN Model. The following is HWSD map of Java Island can be seen in Figure 5.

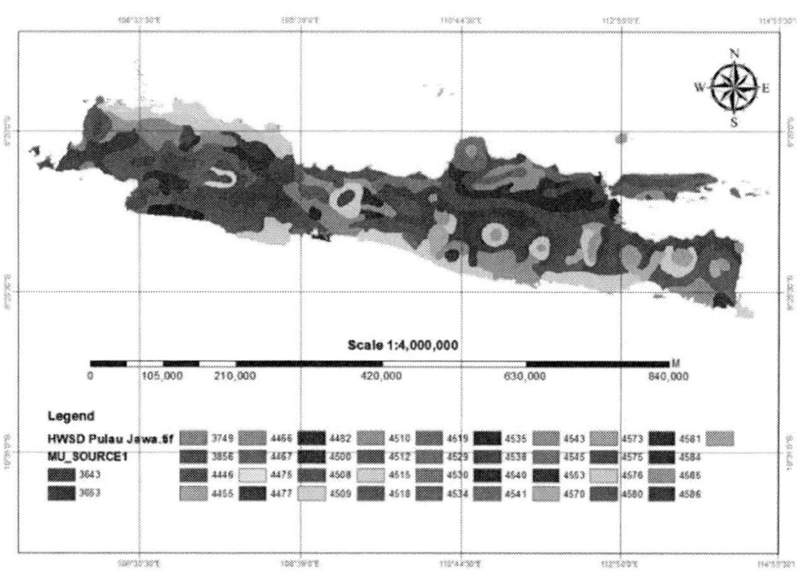

Figure 5. HWSD Map of Java Island

2.2.4. Land Use Data.
Land use data is obtained from TERRA MODIS satellite data in the form of Land Use Land Cover (LULC) data ([4];[5];[13]). In this study, the data used is LULC data on Java Island in the range of 2001 to 2017 that is processed using ArcGIS software to produce 7 classes of land use according to the SHETRAN Model requirements. The following is LULC Map of Java Island can be seen in Figure 6.

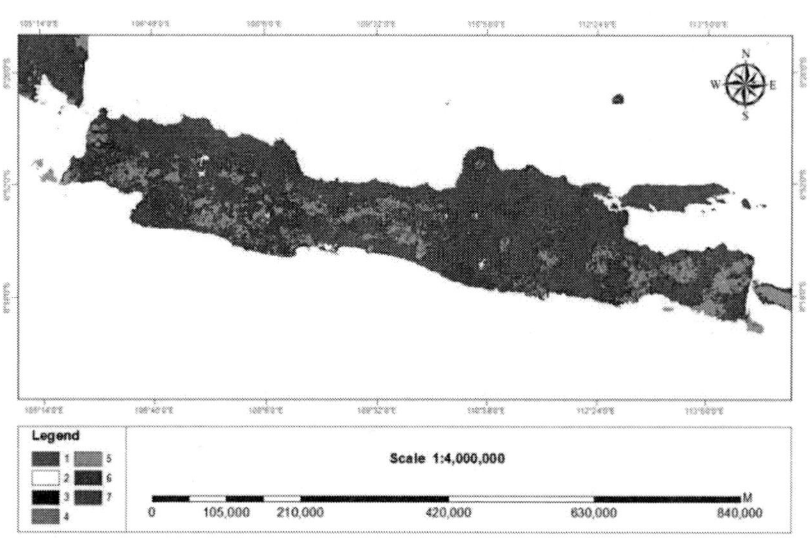

Figure 6. LULC Map of Java Island

2.2.5. Flood Event Data.
Flood event data is historical flood data that contains the time when a flood event occurred in an area at a certain time. Data on flood events was obtained from the National Disaster Management Agency (BNPB) and several well-known news portals such as Detik.com, Kompas.com, and others. In this study, the data used are data of flood events in the range of 2001 to 2017 in several cities/districts in the Progo Watershed area, which consists of data on flood events in Bantul Regency, Yogyakarta City, Kulon Progo Regency, Sleman Regency, Purworejo Regency, Boyolali Regency, Magelang Regency, Magelang City, Semarang Regency, Temanggung Regency, and Wonosobo Regency.

3. Method
As mentioned above, the purpose of this study is to detect flood inundation in the Progo Watershed using the SHETRAN Model. The method used is as follows:

3.1. Analysis of HWSD Data
HWSD data analysis was performed to determine the type of soil in the Progo Watershed area. Determination of soil type in this study uses the British Standards System. The stages of this analysis are HWSD data processed into raster data using ArcGIS software, then HWSD raster data is processed using Microsoft Excel to obtain the percentage of sand, clay, and silt used as a reference in using the soil texture triangle to determine soil types in the Progo Watershed.

3.2. Analysis of TRRM Data
TRRM data analysis was performed to determine rainfall and evaporation in the Progo Watershed. The stages of this analysis are that TRRM data is processed into raster data using ArcGIS software. Then the grid code from the TRRM raster data is used to determine daily rainfall and daily evaporation data located in the Progo Watershed area from 2001 to 2017 using Microsoft Excel.

3.3. Analysis of LULC Data
LULC data analysis was carried out to determine land use in the Progo Watershed. LULC data obtained from TERRA MODIS satellites are processed using ArcGIS software to obtain the type of land use according to the SHETRAN Model requirements.

3.4. Analysis of Flood Inundation
Flood inundation analysis is carried out to detect flood inundation in the Progo Watershed from 2001 to 2017. There are several steps that need to be done, the first stage is processing DEM, TRRM, HWSD, LULC data using ArcGIS software to produce output in ASCII format. The second stage is processing daily rainfall data and daily evaporation data of the Progo Watershed from 2001 to 2017 into output in CSV format. The third stage is creating an XML markup language, and the last stage is running the SHETRAN Model.

4. Results and Discussion

4.1. Analysis of HWSD Data
HWSD data that has been processed into soil raster data using ArcGIS software has a soil code (value) and output file in ASCII format. The Progo Watershed consists of nine soil codes that can be seen from the different colors on the soil raster map. Then, the Progo Watershed soil code is processed using Microsoft Excel to get the percentage of sand, clay, and silt in the top soil and sub soil parts used as a reference in using the soil texture triangle to determine the soil type in the Progo Watershed. The following is soil code map of the Progo Watershed can be seen in Figure 7.

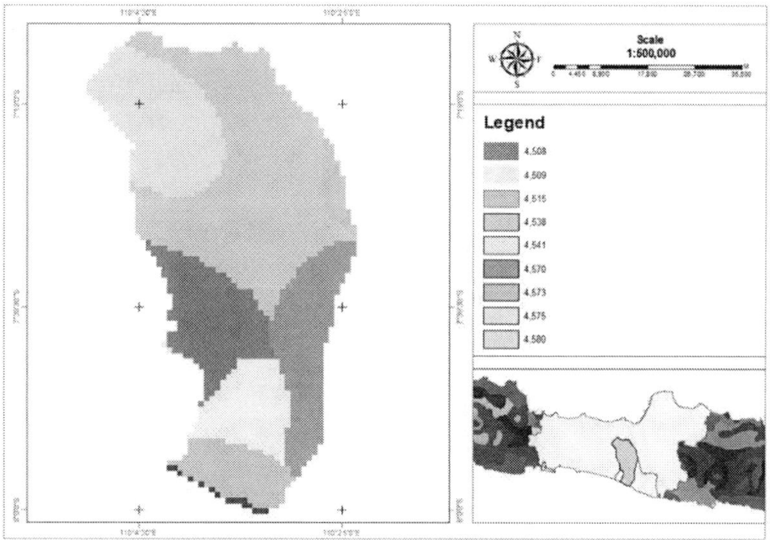

Figure 7. Soil Code Map of Progo Watershed

The Progo Watershed consists of 9 soil code which can be seen from the difference in color on the soil code map. The soil code is processed using Microsoft Excel so that it is known that the Progo Watershed consists of 5 types of soil, namely clay loam, clay, loamy sand, silt loam, and sandy clay loam.

4.2. Analysis of TRRM Data

TRRM data which is processed using ArcGIS software into rain raster data and evaporation raster data has grid code and output file in ASCII format. This grid code is used as a reference in determining daily rainfall and daily evaporation in the Progo Watershed using Microsoft Excel to be a file in CSV format. The following is map of TRRM grid code of the Progo Watershed can be seen in Figure 8.

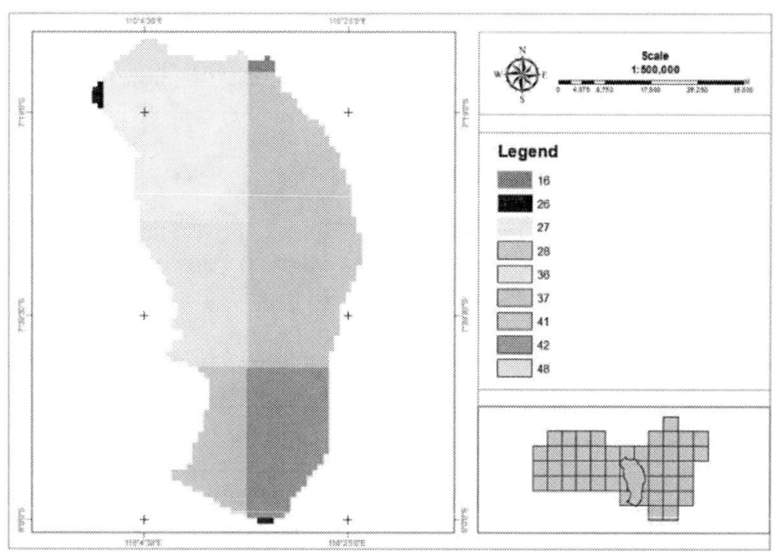

Figure 8. TRRM Grid Code Map of Progo Watershed

4.3. Analysis of LULC Data

LULC data that is processed using ArcGIS software will obtain area of the Progo Watershed data, land use maps, and output files in ASCII format from 2001 to 2017. The following is land use map of the Progo watershed in 2017 can be seen in Figure 9.

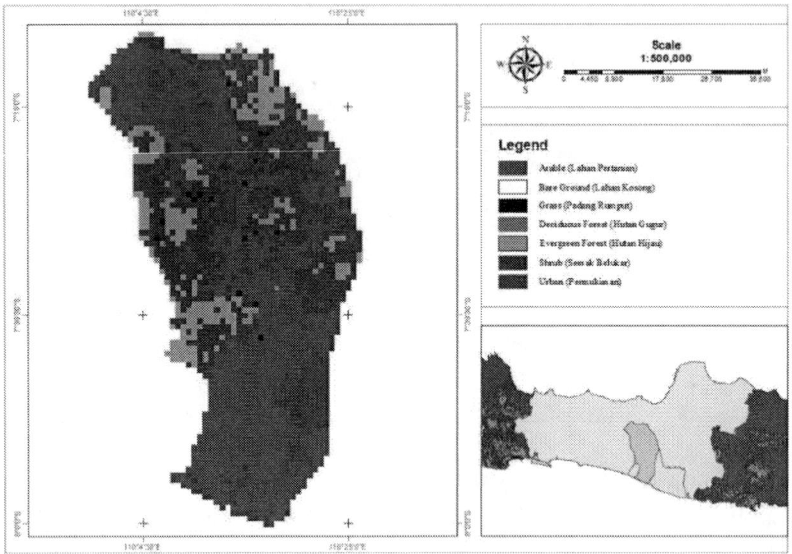

Figure 9. Progo Watershed Land Use Map in 2017

4.4. Analysis of Flood Inundation

Flood inundation in the Progo watershed is detected by the SHETRAN Model using a file in ASCII format containing hydroclimatological data (minimum DEM, average DEM, watershed map, soil, rainfall, evaporation, land use) and files in CSV format containing daily rainfall data and daily evaporation data to create files with the extension XML. Then, this XML file is run using the SHETRAN Model (SHETRAN Result Viewer 1.6.6) which can produce graphical outputs and phreatic depth grid maps so that it can be known the height of flood in a region. The height of flood inundation can be known at each point of the elements on the map displayed on the SHETRAN Model on a daily scale, so it is quite specific in detecting the height of flood inundation in an area.

To find out how accurate the SHETRAN Model is in detecting flood inundation, flood event data in the field is used as a reference in determining flood inundation using the SHETRAN Model. Flood event data taken is the flood event data for each region in the Progo Watershed in the range of 2001 to 2017, then adjusted to the element point above the map of an area in the SHETRAN Model. The following sample flood events with inundation heights above 0.1 m in the range from 2001 to 2017 have been correlated between flood event data in the field and the phreatic depth results from the SHETRAN Model can be seen in Table 4.

Table 1. Samples of Flood Event Data and Progo Watershed Inundation Height

No	Date	Village	Sub-district	Regency / City	Inundation Height (m)	Element
1	11/12/2005	Depok	Panjatan	Kulon Progo	0,105	578
2	11/12/2005	Tayuban	Panjatan	Kulon Progo	0.13	625
3	17/12/2005	Giri Peni	Wates	Kulon Progo	0.141	699
4	01/01/2011	Tirtonirmolo	Kasihan	Bantul	0.337	786

No	Date	Village	Sub-district	Regency / City	Inundation Height (m)	Element
5	25/02/2011	Panjatan	Panjatan	Kulon Progo	0.158	602
6	01/01/2012	Gadingsari	Sanden	Bantul	0.17	475
7	17/01/2013	Bumirejo	Lendah	Kulon Progo	0.141	606
8	17/01/2013	Banaran	Galur	Kulon Progo	0.314	468
9	17/01/2013	Wahyurejo	Lendah	Kulon Progo	0.589	533
10	17/01/2013	Panjatan	Panjatan	Kulon Progo	0.243	603
11	17/01/2013	Tirta Rahayu	Galur	Kulon Progo	0.606	507
12	20/01/2013	Bumirejo	Lendah	Kulon Progo	0.146	606
13	20/01/2013	Tirta Rahayu	Galur	Kulon Progo	0.32	468
14	20/01/2013	Ngentakrejo	Lendah	Kulon Progo	0.587	533
15	08/01/2014	Ngentakrejo	Lendah	Kulon Progo	0.237	533
16	25/04/2015	Banaran	Galur	Kulon Progo	0.341	452
17	25/04/2015	Panjatan	Panjatan	Kulon Progo	0.339	603
18	18/06/2016	Banaran	Galur	Kulon Progo	0.188	452
19	18/06/2016	Panjatan	Panjatan	Kulon Progo	0.157	603
20	02/12/2016	Wijirejo	Pandak	Bantul	0.143	616
21	28/11/2017	Bangunjiwo	Kasihan	Bantul	0.111	785
22	28/11/2017	Bumirejo	Lendah	Kulon Progo	0.208	606
23	28/11/2017	Banaran	Galur	Kulon Progo	0.331	452
24	28/11/2017	Panjatan	Panjatan	Kulon Progo	0.316	603
25	28/11/2017	Pendoworejo	Girimulyo	Kulon Progo	0.212	1108
26	28/11/2017	Wijirejo	Pandak	Bantul	0.119	616
27	09/12/2017	Bumirejo	Lendah	Kulon Progo	0.126	606

From 232 flood events in the Progo watershed recorded by online media, and after correlated with the SHETRAN Model it is known that 217 flood events in the field were detected there are inundation by the SHETRAN Model and the remaining 15 were not detected. The SHETRAN model detects flooding with phreatic depth. If the phreatic depth is positive (+), it means that the area does not have a flood, and the soil can still absorb water. Conversely, if the phreatic depth is negative (-) then there is a flood inundation.

In this study, a programming language was created and running using the RStudio program to determine the maximum height of flood inundation data in the Progo Watershed that occurred from 2001 to 2017. Data of the maximum height of flood inundation from the RStudio analysis results will be compared with the maximum height of flood inundation the result of correlation between flood event data in the field and the phreatic depth results from SHETRAN Model that can be seen in Table 4. To be able to do RStudio analysis, the map of Progo Watershed in ASCII format and files in H5 format are needed, which is one of the outputs of the standard version of the SHETRAN Model for the Rstudio programming language. After the programming language is created, then RStudio is running to produce an output file in CSV format that contains the maximum height of flood inundation of the Progo Watershed. The following graph shows the maximum flood inundation for Progo watershed 2001 - 2017 RStudio analysis results can be seen in Figure 10.

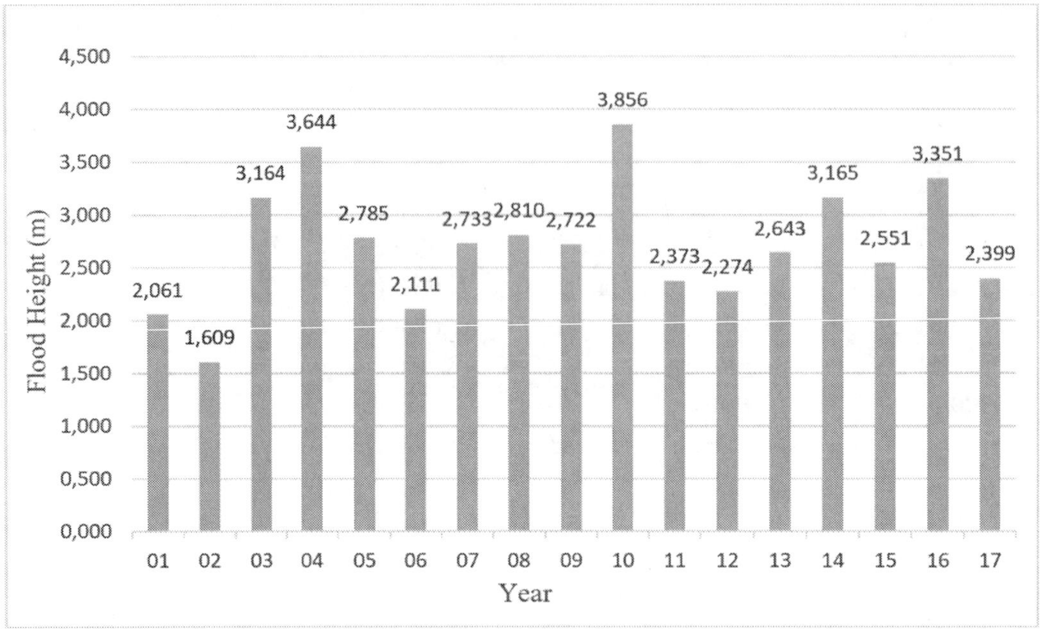

Figure 10. The Maximum Height of Flood Inundation in the Progo Watershed Chart for 2001 - 2017 Results of Rstudio Analysis

5. Conclusion

Based on the results of the research that has been done, it can be concluded that the results obtained are as follows:

- The maximum height of flood inundation results of the correlation between flood event data in the field with the phreatic depth results from the SHETRAN Model in the range from 2001 to 2017 is 0.606 m which occurred on January 17, 2013 in Tirta Rahayu Village, Sub-district Galur, Kulon Progo Regency.
- The maximum height of flood inundation for Progo Watershed in the range from 2001 to 2017 RStudio analysis results were 3.856 m which occurred in 2010.

6. Acknowledgments

Financial support for completing this research is graciously provided by the Ministry of Research, Technology, and Higher Education of the Republic of Indonesia and spesial thanks of gratitude is also expressed to our collaborator from School of Engineering Newcastle University namely Steven Birkinshaw, Chris Kilsby, Andreas Bardossy

7. References

[1] Bass B. and Bedient P. 2018 *Surrogate Modeling Of Joint Flood Risk Across Coastal Watersheds* Journal of Hydrology 558, 159-173.

[2] Birkinshaw S. J. James P. and Ewen J. 2010 *Graphical User Interface For Rapid Set-Up Of SHETRAN Physically-Based River Catchment Model* Environmental Modelling & Software 25(4) 609-610

[3] de Hipt F. O. Diekkrüger B. Steup G. Yira Y. Hoffmann T. Rode M. and Näschen K. 2019 *Modeling The Effect Of Land Use And Climate Change On Water Resources And Soil Erosion In A Tropical West African Catch-Ment (Dano, Burkina Faso) Using SHETRAN* Science of the Total Environment 653, 431-445

[4] Gumma M. K. Amede T. Getnet M. Pinjarla B. Panjala P. Legesse G. abd Siambi M. 2019 *Assessing Potential Locations For Flood-Based Farming Using Satellite Imagery: A Case Study Of Afar Region, Ethiopia* Renewable Agriculture and Food Systems 1-15

[5] Hentzej K. Thonfeld F. and Menz G. 2016 *Evaluating Crop Area Mapping From MODIS Time-Series As An Assessment Tool For Zimbabwe's "Fast Track Land Reform Programme"* PloS one 11(6)

[6] Jacobs L. Dewitte O. Poesen J. Delvaux D. Thiery W. and Kervyn M. 2016 *The Rwenzori Mountains, A Landslide-Prone Region* Landslides 13(3) 519-536

[7] Jati M. I. H. Suroso and Santoso P. B. 2019 *Prediction Of Flood Areas Using The Logistic Regression Method (Case Study Of The Provinces Banten, DKI Jakarta, And West Java)* Journal of Physics: Conference Series (Vol. 1367, No. 1, p. 012087) IOP Publishing

[8] Khanbilvardi, R. (2016). Guidance index for shallow landslide hazard analysis. Remote Sensing, 8(10), 866.

[9] O'Loughlin F. E. Paiva R. C. D. Durand M. Alsdorf D. E. and Bates P. D. 2016 *A Multi-Sensor Approach Towards A Global Vegetation Corrected SRTM DEM Product* Remote Sensing of Environment 182 49-59

[10] Op de Hipt F. Diekkrüger B. Steup G. Yira Y. Hoffmann T. and Rode M. 2017 *Applying SHETRAN In A Tropical West African Catchment (Dano, Burkina Faso)* Calibration validation uncertainty assessment Water 9(2) 101

[11] Pratidina G. and Santoso P. B. 2019 *Detection Of Satellite Data-Based Flood-Prone Areas Using Logistic Regression In The Central Part Of Java Island* Journal of Physics: Conference Series (Vol. 1367, No. 1, p. 012086) IOP Publishing

[12] Tanoue M. Taguchi R. Nakata S. Watanabe S. Fujimori S. and Hirabayashi Y. 2020 *Estimation Of Direct And Indirect Economic Losses Caused By A Flood With Long-Lasting Inundation: Application To The 2011 Thailand Flood* Water Resources Research 56(5) e2019WR026092

[13] Usman M. Liedl R. Shahid M. A. and Abbas A. 2015 *Land Use/Land Cover Classification And Its Change Detection Using Multi-Temporal MODIS NDVI Data* Journal of Geographical sciences 25(12) 1479-1506

[14] Zhang R. Santos C. A. Moreira M. Freire P. K. and Corte-Real J. 2013 *Automatic Calibration Of The SHETRAN Hydrological Modelling System Using MSCE* Water resources management 27(11) 4053-4068

DEGRADATION OF WASTEWATER CONTAINING AMOXICILLIN ANTIBIOTIC USING GREEN TECHNOLOGY "OZONATION"

[1,3]Ria Wulansarie, [2]Setijo Bismo, [1]Ismail Abdul Azis, [1]Devinda Rahmadhani

[1]Chemical Engineering Department, Engineering Faculty, Universitas Negeri Semarang, Indonesia
[2]Chemical Engineering, Faculty of Engineering, Universitas Indonesia

[3]ria.wulansarie@mail.unnes.ac.id

Abstract. In this research, wastewater containing amoxicillin antibiotic was treated using green technology "ozonation". The research was carried out with allowance for time variation (15, 30, 45, and 60 minutes), pH of wastewater (acidic, neutral, and alkaline), and ozone generator voltage (75, 150, and 225 Volt). The amoxicillin wastewater in this research was artificial. The concentration of amoxicillin was 50 ppm. Based on experimental result, the best condition of wastewater (containing amoxicillin antibiotic) ozonation was alkaline conditions with pH $\approx$ 10. The best percent degradation of amoxicillin antibiotic was 86.78 % with a final concentration of 6.61 ppm. The ozone generator voltage variations affect the amoxicillin removal in wastewater. The higher of ozone generator voltage, the percentage of amoxicillin removal was also increase.

1. Introduction

Hospital activities are closely related to waste production. Waste generated by hospitals can cause environmental pollution and can endanger living things including humans. Waste generated by hospitals can be grouped into three parts, namely solid waste, liquid waste, and clinical waste. The penicillin waste belongs to the liquid waste group. Penicillin was the first antibiotic discovered from Alexander Fleming in 1928 in London. Penicillin is included in the β lactam class of antibiotics because it has a built-in formula with a structure like the β lactam ring which is an absolute requirement to show its efficacy. The use of penicillin antibiotics can cause several dangers such as: symptoms of resistance, allergies and supra infection. Amoxicillin is one of penicillin antibiotics.

Penicillin antibiotic waste can cause environmental pollution. The antibiotic waste can cause bacteria that are resistant to antibiotics. This is certainly very dangerous for living things, especially humans if the resistant bacteria infect humans. Therefore, penicillin antibiotic waste needs to be set aside so as not to endanger living things and the surrounding environment. Research on penicillin waste removal has been done before. In 2004 Arslan-Alaton et.al eliminated penicillin by ozonation [1]. In that study the percentage of waste removal reached 56%. Then in 2004 Arslan-Alaton and Dogruel eliminated penicillin by ozonation, direct and indirect photolysis, phentone and phentone-like, photo-fenton and photo-fenton-like [1]. In that study the percentage of waste removal when using ozonation reached 49%. In 2005, kinetic of amoxicillin ozonation was also studied [2]. Ozone was also used for penicillin removal from wasterwater [3]. Then in 2006 Cokgor et. al. eliminating penicillin by ozonation [1]. In this study the percentage of waste removal reached 34%. Pharmaceuticals was also treated using ozone and advance oxidation processes [4]. Then in 2009

Yang excluded penicillin by microwave enhanced fenton-like [1]. The removal of amoxicillin was also studied using plasma treatment [5]. Amoxicillin removal in water also studied using medium-high frequency ultrasound and ozone based advanced oxidation [6]. Ozone technology was also used for tofu wastewater treatment [7]. In this study the percentage of waste removal reached 57.53%.

In this research, amoxicillin antibiotic wastewater was treated using ozonation with different ozone generator. The results of this study will be analyzed using a spectrophotometer to determine the concentration of amoxicillin.

2. Methodology/Experimental

Amoxicillin used in this research was artificial wastewater. In this research, ozonation was carried out with time variation of 15, 30, 45, and 30 minutes. The concentration of amoxicillin was 50 ppm. The pH variation used in this research were acid, neutral, and base (pH=3,4,7,9,10). Ozone generator used in this research was Resun type RSO-9805 made in Hong Kong. Voltage of ozone generator was varied into 75, 150, and 225 V. PH of wastewater was varied into acid, neutral, and base. The schematic design of ozonation process could be seen in Figure 1.

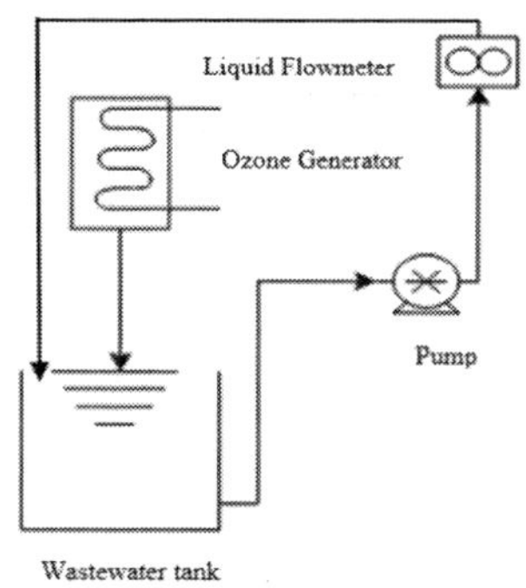

Figure 1. Schematic design of ozonation process

3. Results and Discussion

The ozonator used in this research was Resun type RSO-9805 made in Hong Kong. Measurement of ozonator products includes measurements of flow rate and ozonator productivity. The ozone productivity of the Resun RSO-9805 ozonator used in this study was 41,088 mg / hour. While the ozonator gas flow rate is 9 L / min. Ozone productivity of 41.088 mg / hour based on literature is sufficient to set aside antibiotic waste by 50 ppm [8].

The ozonation of wastewater containing amoxicillin was carried out under three conditions of pH: acidic, neutral and base (pH=3,4,7,9,10). The concentration of amoxicillin was 50 ppm. The results of ozonation wastewater containing amoxicillin could be seen in Figure 2.

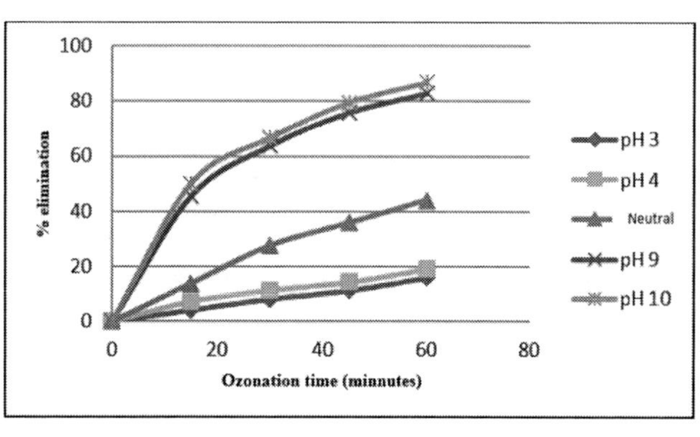

Figure 2. Percentage of

amoxicillin removal by ozonation

The removal of amoxicillin contained in the wastewater increases with increasing processing time, as shown at Figure 2. This indicates that the longer the removal time, the active species contact with amoxicillin compounds is also getting longer so that the amoxicillin that is set aside is also more. In this acidic condition, the dominant elimination reaction is direct ozonation [9]. In this direct ozonation, ozone will react exclusively with compounds that have specific groups through selective reactions such as electrophilic, nucleophilic, or dipolar additional reactions. At the allowance at pH = 3 the percentage of amoxicillin removal reached 16.11% and the concentration at the end of the allowance was 41.95 ppm. While the amoxicillin removal yield at pH = 4 reached 19.22% with amoxicillin concentration at the end of the allowance was 40.39 ppm. The results of ozone removal in acidic conditions tend to be small because the ozonation reaction that occurs is direct ozonation so that the dominant one to set aside the amoxicillin is ozone. Under acid conditions, the OH radical that are formed are lower than in neutral and basic conditions.

Next discussion was the treatment of wastewater containing 50 ppm amoxicillin in neutral. The intended neutral condition is the condition of liquid waste without treatment but the pH has been neutral, 7. In this neutral condition the allowance reaction that occurs is direct ozonation and indirect ozonation [9]. In direct and indirect ozonation, species that play a role in the removal process are not only ozone, but also OH radical which plays a role in the removal process. At this allowance at neutral pH the percentage of amoxicillin removal reached 44.08% and the concentration at the end of the allowance was 27.96 ppm. The yield of removal with ozone under neutral conditions is greater than the allowance under acidic conditions. That was because the ozonation reaction occurs in direct and indirect ozonation so that the species that play a role in removing amoxicillin are ozone and OH radical. In addition, in neutral conditions, the OH radical formed was greater than in acidic conditions.

Next, the treatment of liquid waste containing 50 ppm amoxicillin under alkaline conditions was also studied. The intended base condition is the condition of the waste which was regulated in such a way that the pH is 9 and 10 by adding NaOH. In these base conditions, the dominant allowance reaction is indirect ozonation [9]. In the indirect ozonation, the dominant species that plays a role in the removal process is OH radical, where the species is very strong in removing amoxicillin compounds. In the allowance at pH = 9 the percentage of amoxicillin removal reached 82.81% with a concentration at the end of the allowance of 8.60 ppm. Whereas at the allowance at pH = 10 the percentage of amoxicillin removal reached 86.78% with the concentration at the end of the allowance was 6.61 ppm. The yield of ozone removal under these basic conditions is greater than the allowance under acidic and neutral conditions. That is because the ozonation reaction that occurs is indirect ozonation so that the dominant species that play a role in removing the amoxicillin is OH radical which is a strong species, greater than ozone [10]. This is evidenced by the OH radical which formed under alkaline conditions tends to be greater than under acidic and neutral conditions.

The effect of pH on removal using ozone also studied in this research. Based on Figure 2., the percentage of amoxicillin removal was increase linearly with pH. The lowest percentage of amoxicillin removal occured at acidic condition while the highest percentage of amoxicillin removal occured under alkaline conditions, at pH = 10, which was capable of removing moxicillin from 50 ppm into 6.61 ppm. That was because the hydroxyl radical at base conditions formed more than the acidic and neutral conditions. The hydroxyl radical has a higher oxidation potential than ozone so that the removal of waste by ozonation is better in alkaline conditions. Based on literature, concentration of hydroxyl radical at 50 ppm can be used for remove amoxicillin and ampicillin [11].

The effect of variation in generator ozone voltage was also studied in this research. The result could be shown in Figure 3.

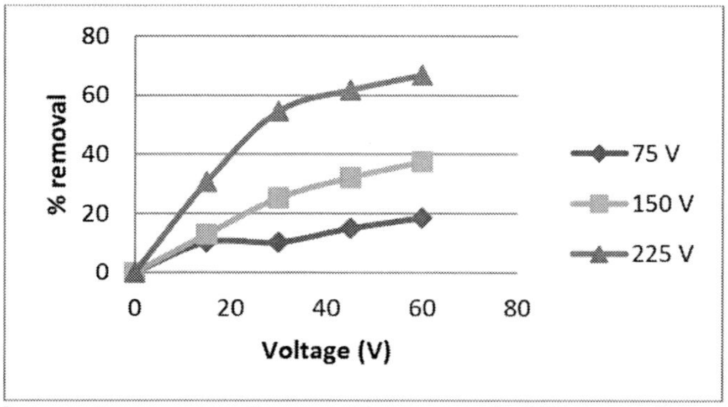

Figure 3. Effect of generator ozone voltage on the removal of 50 ppm of amoxicillin with ozone

Figure 3. shows that ozone generator voltage variations affect the amoxicillin removal in wastewater. On the graph it can be seen that the higher the voltage of the ozone generator, the better the yield. This is caused by the higher of ozone generator voltage, the more ozone is produced, so that the compounds that can be set aside are also large and the percentage of removal will increase.

4. Conclusion

According to this research, the percentage of 50 ppm amoxicillin removal using ozonation can achieve 86.78% at pH 10. The ozone generator voltage variations affects also the amoxicillin removal in wastewater. The higher of ozone generator voltage, the percentage of amoxicillin removal was also increase.

References

[1] Homem V and L Santos 2011 *Degradation and removal meth ods of antibiotics from aqueous matrices - A review Journal of Environmental Management* **92**: 2304-2347.

[2] Andreozzi R, Canterino M, Marotta R and Paxeus, N 2005 Antibiotic removal from wastewaters: The ozonation of amoxicillin *Elsevier*.

[3] Cokgor E U, Alaton I A, Karahan O, Dogruel S and Orhon D 2004 Biological treatability of raw and ozonated penicillin formulation effluent *Journal of Hazardous Materials* B116 (2004) 159–166.

[4] Keisuke Ikehata, Naeimeh Jodeiri Naghashkar and Mohamed Gamal El-Din 2006 Degradation of Aqueous Pharmaceuticals by Ozonation and Advanced Oxidation Processes A Review Ozone *Science and Engineering* 28:6 353-414 DOI: 10.1080/01919510600985937

[5] Magureanu M D P, N B Mandache, V David A, Medvedovici C, Bradu V I and Parvulescu 2011 Degradation of antibiotics in water by non-thermal plasma treatment *water research* **45**: 3407-3416.

[6] Kıdak R and Dogan S 2018 Medium-high frequency ultrasound and ozone based advanced oxidation for amoxicillin removal in water. *Ultrasonics Sonochemistry* 40 (2018) 131–139.

[7] Wulansarie R, Chafidz A, Hidayat P, Rengga W D P, Afrizal FJ, Dwipawarman A, Bismo S and Arifin R 2020 Effectiveness of Ozonation Process on Treating Tofu Industrial Liquid Waste: Effect of pH Conditions *Materials Science Forum* ISSN: 1662-9752 Vol. 981 pp 336-341.

[8] Dore M, Langlais B and Legube, B 1980 Mechanism of The Reaction of Ozone with Soluble Aromatic Pollutants *International Ozone Association*.

[9] Peratitus (ed.) 2003 *Ozone Reaction Kinetics for Water and Wastewater System* London: A CRC-Press.

[10] Manley T C and Niegowski S J 1967 *Ozone Encyclopedia of Chemical Technology* Vol. 14 2nd ed., pp 410–432 New York US: Wiley.

[11] Elmolla E S and Chaudhuri M 2010 Comparison of Different Advanced Oxidation Processes for Treatment of Antibiotic Aqueous Solution *Journal of Desalination* 256 (2010) 43 –47.

IOP Conf. Series: Materials Science and Engineering **982** (2020) 012040 doi:10.1088/1757-899X/982/1/012040

Utilization of *Mendong* Plant Activated Charcoal as SO_2 Gas Adsorbent: Preliminary Study

Ashadi Ashadi, Finly Khoirunnisa' Arabbani[a], Nurma Yunita Indriyanti, Sulistyo Saputro, Lina Mahardiani[b]
Chemistry Education Department, Universitas Sebelas Maret, Surakarta, Indonesia

[a]finlykhoirunnisa@student.uns.ac.id, [b]mahardiani.lina@staff.uns.ac.id

Abstract. This study aims to find alternative low-cost, easy and environmental-friendly adsorbent materials to eliminate SO2 gas, by utilizing *Mendong* plants as activated charcoal. Therefore, need to know the ability of this low-cost adsorbent in term of optimum concentration, adsorption efficiency, and the adsorption capacity of the charcoal to eliminate SO2. *Mendong* plants were prepared with a modified tool and macerated with $ZnCl_2$ 2.5% (w/v) for 24 hours. Then, the adsorption process with artificial SO2 waste was conducted for 1 hour. UV-Vis Spectrophotometer was used to determine the gas concentration. The results showed that the highest concentration of absorbed SO2 gas (C_t) is at the initial gas concentration 4.4006 µg/mL (mass $Na_2S_2O_3$ 0.015 g), with 3.008 µg/mL of gas is absorbed. The gas adsorption efficiency is 70.53% at the variation of initial gas concentration of 4.1400 µg/mL. While, the adsorption capacity of activated $ZnCl_2$-*Mendong* stem charcoal is 92.1123 (µg/g). The results of the characterization by FTIR showed that the *Mendong* plant $ZnCl_2$-activated charcoal is polar and aromatic, verified by the absorbance of hydroxyl group and aromatic skleleton of cellulose, hemicellulose and lignin.

1. Introduction

As human activity increases, high traffic activities cause air pollution problems. The key source of pollution which is road transport, is generated by freight transportation and logistic activities [1]. In addition, industrial sector, high fossil fuels consumption, and large urban populations also support the high production of gas pollutions, especially SO_2 gas [2][3]. SO_2 contribute a large proportion of worsening air quality in China [4][5]. SO_2 is formed when sulfur-containing components in fossil fuels such as coal and petroleum are oxidized through the combustion process [6][7]. SO_2 concentration in the air can be detected by taste at 0.35-1.05 ppm and odor threshold of 0.67-4.75 ppm [8]. According to WHO [9], daily concentration of SO_2 should not be exceeded over 20 µg/m^3.

The main effect of SO_2 gas pollutants on humans is irritation of the respiratory system such as coughing, mucus secretion, breathing difficult, reduce lungs function, chest tightness, alterations in the lungs' defenses, and even chronic lung disease for longer-term exposures of high concentrations of SO_2 [9][10][11]. In the atmosphere, SO_2 gas oxidized through homogeneous and heterogeneous pathways to form sulfate compounds. In this process, the intermediates (HSO_3) combine rapidly with water vapor in the atmosphere to form sulfuric acid aerosol [12]. This is process cause SO_2 gas easily becomes acidic, which can attack mucous membranes in the nose, throat and other respiratory tracts to the lungs [13]. To reduce the adverse effects caused by $_{SO2}$ gas, efforts should be made to reduce SO_2 gas pollution to the environment.

According to Xia [14], SO_2 with coconut shell activated carbon has the best reduction effect than other materials (coal, Shenxin coal, graphite, coke, and charcoal) with higher than 95% reduction. Other material such as hardened cement mortar (HCM), also be completely absorb SO_2 by the highly alkaline HCM slurry with the retained capacity 0.52–0.76 (g_{SO2}/g) [15]. Wang [16] also found that carbide slag-coal char pellet can give 100% of SO_2 removal efficiencies at 370 s.

Furthermore, other alternative material that simple, easy to use, and environmentally-friendly for absorbing SO_2 gas is to use *Mendong* plants (Fimbristylis globulosa). *Mendong* plants are grasses that grow in wet and muddy soils which usually grow 100 cm (Figure 1). *Mendong* is a type of grass, a family of Cyperaceans with short rhizomes, fibrous roots, and grooved. This plant is used to make traditional mats such as ropes, bags, hats and wallets because of its high fiber and its low cost [17]. Estimated production of *Mendong* in Java is about 14.000 tons per years [18]. With its economic potential, *Mendong* needs to be developed further to increase its use value.

Mendong plants contain 72.14% cellulose, 20.2% hemicellulose, 3.44% lignin, 4.2% extractive matter, and moisture content of 4.2%–5.2%. The *Mendong* fiber was arranged by the crystalline structure of cellulose. *Mendong* fibers have a higher crystallinity and crystalline size than straw fibers and wheat stalks [18]. Until now, *Mendong* also used as phytoremediation plant, microcrystalline cellulose, and polymer composite [17]. Thus, *Mendong* is considered to have potential as charcoal as it has relatively high content of cellulose. To the best of our knowledge there are no literatures reported related to utilization of *Mendong* plants adsorbent as gas elimination. Thus, in this study, *Mendong* plants will be used as activated charcoal to eliminate SO_2 as low-cost adsorbent in term of optimum concentration, adsorption efficiency, and the adsorption capacity of the charcoal.

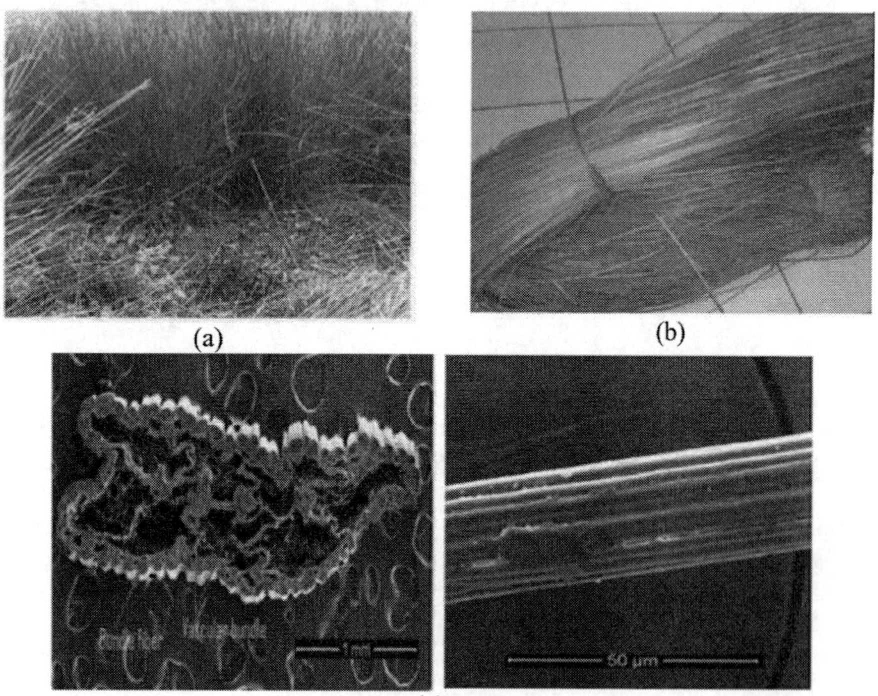

(c)

Figure 1. (a) Mendong plant, (b) Dried mendong straw and (c) Morphology of fiber in dried Mendong straw observed by SEM [18]

2. Method

2.1 Material preparation

Dried *Mendong* plants were burned with a modified tool. This tool is made of cans and closed to limit the air flow in the tool (Figure 2). The charcoal was mashed and sieved with 100-mesh sieve. Furthermore, it was activated using 500 mL of 2.5% (w/v) of $ZnCl_2$ solution by maceration method for 24 hours. The maceration method was chosen because it is a simple activation method only by soaking the sample with the selected solvent used in room temperature [19]. The result was neutralized, dried, and then formed into 3-mm pellets. FTIR analysis is carried out on inactivated *Mendong* plant charcoal (AMTA) and activated *Mendong* plant charcoal (AMA 2.5%).

Figure 2. Schematic figure of modified tool for *Mendong* plant combustion

2.2 Sulfur dioxide (SO_2) gas adsorption

The concentration of SO_2 in solution was determined by the Pararosaniline method with the analysis using UV-Vis spectrophotometer based on SNI 19-7119.7-2005 [20]. SO_2 gas source used is obtained from the reaction between $Na_2S_2O_3$ powders with HCl 1M solution. In this study, a preliminary study was conducted with variations in the initial concentration of SO_2 as adsorbate obtained from the variation of the mass of sodium thiosulfate used of 0.005; 0.015; and 0.025 g. Then, SO_2 gas is absorbed using TCM solution (tetrachloromerkurat ($HgCl_4^{2-}$). SO2 in the adsorbent solution is reacted with formaldehyde (HCHO) added with sulfuric acid solution (NH_2SO_3H) as a catalyst and pararosaniline hydrochloride ($C_{19}H_{17}N_3.HCl$) as a color indicator. This reaction will produce purple pararosaniline methyl sulfonate and its thickness will be measured using Visible Spectrophotometer at λ_{max} = 550 nm [19]. The schematic testing tool of adsorption is showed in Figure 3.

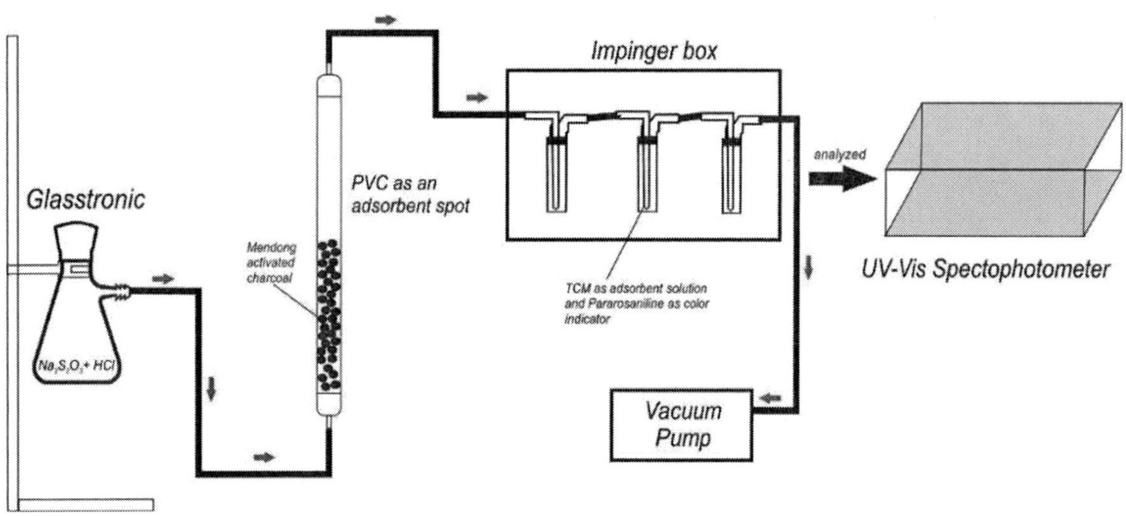

Figure 3. Schematic figure of testing tools for SO_2 adsorption using parrosaniline method [20]

2.3 Determination of the concentration of absorbed SO_2 gas (C_t)

The absorbency of the solution was measured with a UV-Vis spectrophotometer to determine the amount of SO_2 gas adsorbed. Determination of gas concentration was done by the calibration curve method. Calibration curves were made using standard $Na_2S_2O_3$ solution. The absorbed SO_2 gas concentration (C_t) is calculated from the difference of the initial gas concentration (C_o) and final gas concentration (C).

$$C_t = C_0 - C_1 \tag{1}$$

For the absorbed SO_2 level, it is calculated by

$$\%C_t = \frac{C_t}{C_0} x\ 100\% \tag{2}$$

Figure 4 showed the flow chart stage of the research.

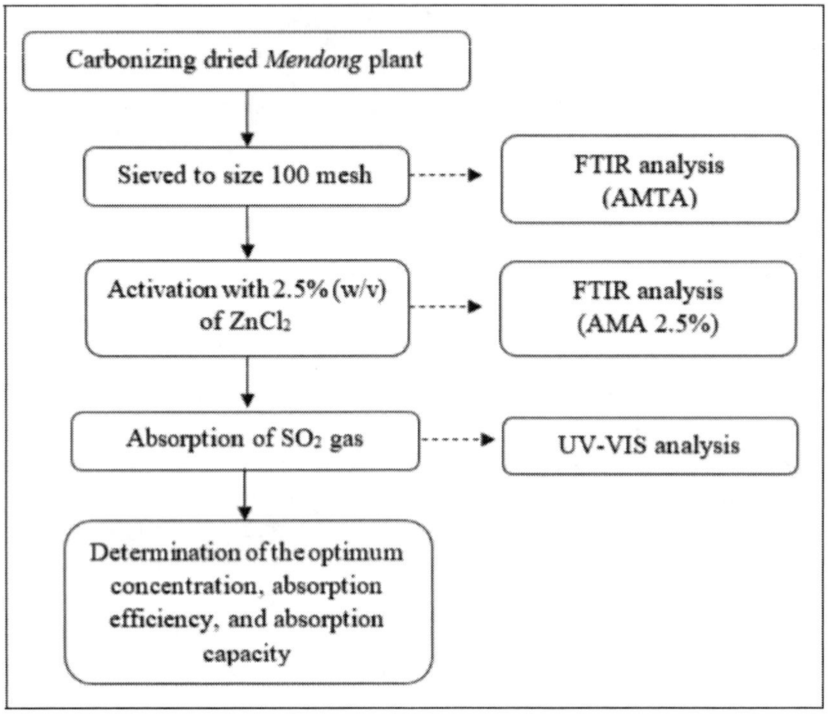

Figure 4. Flow chart of the research

3. Result and Discussion

The contacting process of SO_2 was carried out using a series of impinger set testing tools at a laboratory scale with artificial SO_2 gas. This series refers to SNI 19-7119.7-2005 by adding an adsorbent tube in the middle of the circuit. The adsorbent tube serves as a place for adsorbent, which is a 30-g active charcoal pellet of *Mendong* plants. Furthermore, the adsorbent tube was connected with 3 impinger bottles which already contain an adsorbent solution. This solution functions to capture sulfur dioxide gas which is not absorbed by the adsorbent.

After the contacting process had been completed, then the concentration of SO_2 gas was tested with a UV-Vis spectrophotometer. The absorbency measurement was performed with 3 variations of the initial SO_2 gas concentration, namely 4.1400 μg/mL (mass $Na_2S_2O_3$ 0.005 g), 4.4007 μg/mL (mass $Na_2S_2O_3$ 0.015 g), and 7.6734μg/mL (mass $Na_2S_2O_3$ 0.025 g). The result of adsorption of SO_2 based on concentration variation is shown in Table 1.

Based on Table 1, the sulfur dioxide gas concentrations obtained after the contact with *Mendong* plant activated charcoal are 1.2202 μg/mL (mass $Na_2S_2O_3$ 0.005 g), 1.3926 μg/mL (mass $Na_2S_2O_3$ 0.015 g), and 5.3110 μg/mL (mass $Na_2S_2O_3$ 0.025 g). Thus, the concentration of SO2 gas absorbed by the adsorbent in each variation of the initial gas concentrations of 2.9198 μg/mL, 3.0080 μg/mL, and 2.3623 μg/mL can be obtained. Based on the calculation results, it shows that the gas concentration mostly absorbed is a gas with a mass of sodium thiosulfate of 0.015 g.

Table 1. Concentration and adsorption efficiency of absorbed SO2 gas

C_0 (µg/mL)	C_1 (µg/mL)	C_1 mean (µg/mL)	C_t (µg/mL)	C_t mean (µg/mL)	$\%C_t$
	0.4324				
	0.4123	1.2529	2.8870		
4.1400	0.4083			2.9198	70.53
	0.4043				
	0.3922	1.1874	2.9525		
	0.3909				
	0.6075				
	0.3869	1.3746	3.0261		
4.4007	0.3802			3.0080	68.35
	0.6222				
	0.3976	1.4107	2.9900		
	0.3909				
	3.5781				
	1.4471	5.4227	2.2507		
7.6734	0.3976			2.3623	30.79
	3.4644				
	1.3134	5.1995	2.4739		
	0.4217				

Table 2. Adsorption capacity of *Mendong* plant activated charcoal against SO2 gas

Mass $Na_2S_2O_3$ (g)	C_0 (µg/mL)	C_1 (µg/mL)	C_t (µg/mL)	W (g)	q (µg/g)	Mean q (µg/g)
0.005	4.1400	1.2202	2.9198	30	97.3262	
0.015	4.4006	1.3926	3.0080	30	100.2674	92.1123
0.025	7.6733	5.3110	2.3623	30	78.7433	

In determine the adsorption efficiency of sulfur dioxide gas, at an initial concentration of 4.1400 µg/mL, 70.53% is absorbed. Meanwhile, the adsorption efficiency for the initial concentration of 4.4006 µg/mL is 68.38%, and the adsorption efficiency for the initial concentration of 7.6733 µg/mL is 30.79 %. Based on the efficiency results, it was obtained that the initial concentration of SO_2 gas of 4.1400 µg/mL, which is the lowest initial concentration, has the highest adsorption efficiency, while the adsorption efficiency for the gas with the highest initial concentration has the lowest efficiency. These results are less appropriate when related to the previous study by Basuki [21] who use local TiO_2-activated carbon insertion media with 15 cm length (with 15% of TiO_2 inserted), to absorb SO_2 by 714.46 µg with adsorption efficiency 76.75%. This is caused by the insertion of TiO_2 catalyst in the adsorbent as the gas adsorption medium, which was not used in this study.The other study of Elvitriana [22] which is able to absorb SO_2 gas by 2187 ppm with 100% adsorption efficiency after 5 minutes contact using activated charcoal adsorbents from banana peels. Elvitriana found that the best temperature of activation was 500°C for 45 minutes which produced adsorbent with adsorption capability 888mg/g.

Furthermore, according to Table 2 the adsorption of SO_2 gas using *Mendong* plant activated charcoal adsorbent was calculated based on the results of the concentration of the absorbed gas. The mass of the adsorbent used is 30 g. The adsorption capacity of *Mendong* plant activated charcoal at initial gas concentration 4.1400; 4.4006; and 7.6733 µg/mL are 97.3262; 100.2674; and 78.7433 µg/g. The highest adsorption capacity is at initial gas concentration 4.4006 µg/mL that directly proportional to the amount of SO_2 gas absorbed. Generally, at given temperature, high input concentrations will give higher adsorption capacity than at low input concentrations [23]. However, there is a limit mass of pollutant that can be adsorbed by an adsorbent. When the adsorbent receive more and more adsorbate and reach the end of bed, activated carbon filter will saturated, requiring either the bed to be replaced with new activated carbon or requiring the bed to be regenerated [23][24]. Based on this result, the calculation of average adsorption capacity of *Mendong* plant activated charcoal to SO_2 adsorption is 92.1123 µg/g.

To find out the characteristics of functional groups, FTIR test was conducted. The test was carried out on the *Mendong* plant charcoal with and without activation of $ZnCl_2$ 2.5% (w/v). The results of FTIR analysis are shown in Figure 5.

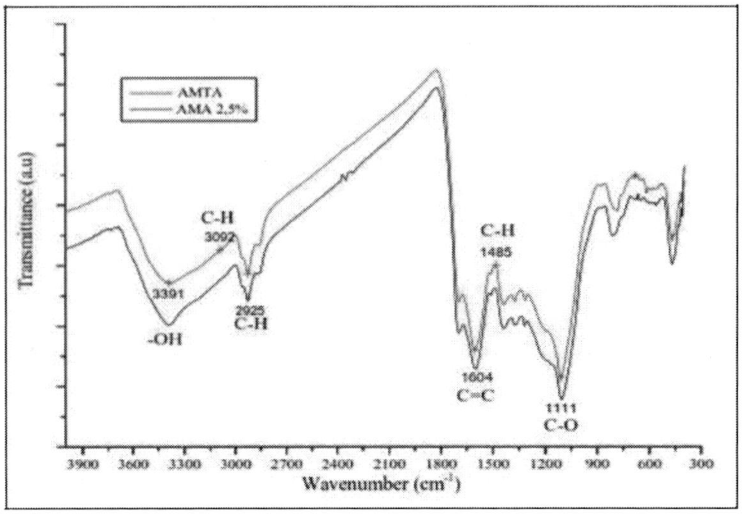

Figure 5. FTIR analysis of the *Mendong* plant charcoal (AMA and AMTA 2,5%)

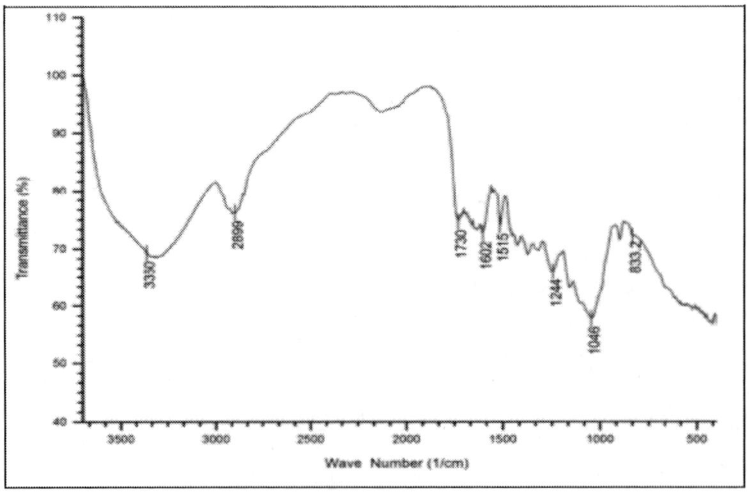

Figure 6. FTIR analysis of the *Mendong* plant fiber [25]

•

Figure 6 shows the chemical compoaition of *Mendong* plant fiber. According to Suryanto [25], the functional group of the fiber are shown in the Table 3.

Table 3. Functional group of *Mendong* plant fiber analyzed using FTIR [25]

Wavenumber (cm^{-1})	Functional group	Compound
3360	-OH	Cellulose, hemicellulose and lignin
2899	H-C-H	Aliphatic chain of cellulose, hemicellulose and lignin
1730	C=O	Aromatic ring of hemicellulose
1602	C=C	Aromatic ring of lignin [18]
1515	C=C	Aromatic ring of lignin [18]
1244	C-O-C	Lignin
1046	C-O	Cellulose
833	C-H	Lignin

Figure 5 shown that the activation treatment does not affect the functional groups on activated charcoal. This shows that there is no change in bonds or chemical structure of the two samples. There are only differences in bond vibration caused the absorbency of *Mendong* activated charcoal (AMA 2,5%) to be greater than the absorbency of *Mendong* inactivated charcoal (AMTA).

According to Figure 5 and Figure 6, there is no significant change of chemical composition between *Mendong* plant fiber and *Mendong* plant charcoal. Based on the appearance of the peak of the spectra in Figure 5, the adsorption band at the wave number of 3391 cm^{-1} is the peak of the hydroxyl group stretching vibration [25]. It is reffered to –OH stretching from cellulose, hemicellulose, and lignin [18]. The peak at 3092 cm^{-1} is attributed to C–H stretching vibration of aromatic hydrocarbons [18]. While, the peak at 2925 cm^{-1} most probably indicate the aliphatic C-H stretching [25][26]. The adsorption peak at 1604 cm^{-1} is assigned to the skeletal vibration of the aromatic rings of lignin [18][26]. In addition, C-H in plane deformation of cellulose, hemicellulose, and lignin is showed by the adsorption at 1485 cm^{-1} [18][27]. The bands ranging from 1163 to 1027 cm^{-1} are attributed to the asymmetric bridge oxygen and C-O stretching of cellulose and hemicellulose compounds [18][27]. Activation had an impact on the decreasing transmittance of the groups on the charcoal sample [28]. A slightly higer peak ratio for AMTA and AMA 2.5% was observed on 3391 cm^{-1} (hydroxyl group), and 2925 cm^{-1}/1604 cm^{-1} (aliphatic/aromatic) wavenumber pair. This indicated that AMA 2.5% was more polar and more aromatic than AMTA [29].

4. Conclusion and Future Perspective

Mendong plant is feasible to be applied as low-cost adsorbent material to reduce SO_2 gas contamination with 3.008 µg/mL concentration of SO_2 gas can be absorbed. While the highest efficiency of adsorption is 70.53%, and the adsorption capacity of SO_2 adsorption is 92.1123 µg/mL. Furthermore, the FTIR analysis found that *Mendong* plant charcoal have absorbance peak of –OH, –CH2. – CH3, C=C, and C–O groups refered to cellulose, hemicellulose and lignin compound. As *Mendong* plant showed potential adsorbent material for SO_2 gas, still *Mendong* plant charcoal needs further improvement to explore more about its utilization for eliminating other pollutants, such as gases, heavy metals, or organic substances. Optimization of carbonization temperature, methods and activation or modification may be suggested. In addition, it is also important to notice that low-cost adsorbent needs to be evaluated for its recovery so that applicable for large scale and commercialized. Ultimately, it is important for this material to be able to be implemented in real contaminated environment.

Reference

[1] Katarzyna G 2019 *Transp. Research. Proc* **39** 84-91,

https://doi.org/10.1016/j.trpro.2019.06.010

[2]. Zhao B, Wang S, Wang J, Fu J S, Liu T, Xu J, Fu X and Hao J 2013 *Atmos. Environ* **77** 453–463, https://doi.org/10.1016/j.atmosenv.2013.05.012

[3]. Yan Z, Can L, Nickolay A K and Joanna J 2017 *Atmos. Meas. Tech* **10** 445-458, 10.5194/amt-2016-226

[4]. Zhao S, Liu S, Hou X, Cheng F, Wu X, Dong S and Beazley 2018 *J. Environ. Pollut Nov 242 (Pt A)* 239-248, 10.1016/j.envpol.2018.06.085.

[5]. Li X, Wu X and Zhang F 2015 *Ener. Eco.* 49 451-459, 10.1016/j.eneco.2015.03.015

[6]. Likus-Cieślik J, Socha J, Gruba P and Pietrzykowski M 2020 *J. Environ. Pollut.* 258 113559, https://doi.org/10.1016/j.envpol.2019.113559

[7]. William B 2011 *Encyc. of Environ. Health* ed J O Nriagu (Amsterdam: Elsevier BV) p 635-45, https://doi.org/10.1016/B978-0-444-52272-6.00064-7

[8]. National Research Council (US) Committee on Acute Exposure Guideline Levels 2010 9, https://www.ncbi.nlm.nih.gov/books/NBK219999/

[9]. WHO 2018 https://www.who.int/news-room/fact-sheets/detail/ambient-(outdoor)-air-quality-and-health accessed 2 July, 2020

[10]. EPA 2018 https://www.epa.gov/SO2-pollution/sulfur-dioxide-basics, accessed 2 July, 2020

[11]. Bruce G M 2017 *Clean Coal Engineering Technology* (Oxford: Butterworth-Heinemann/Elsevier) pt 1 chapter 3 pp 105 https://doi.org/10.1016/B978-1-85617-710-8.00004-2

[12]. Daniel V 2014 *Fundamentals of Air Pollution (Fifth Edition)* (Cambridge: Academic Press/Elsevier) pt 2 chapter 6 6.1.5.3.3.5 https://doi.org/10.1016/B978-0-12-401733-7.00006-2

[13]. Martin C, Molénat J, Gascuel-Odoux C, Vouillamoz J-M, Robain H, Ruiz L, Faucheux M and Aquilina L 2004 *J. Hydro.* **326** 25-42, https://doi.org/10.1016/j.jhydrol.2005.10.040

[14]. Xia X, Zhao X, Zhou P, Feng T, Ma C and Song Z 2020 *J. Chem. Eng. and Proc. Intens.* **110** 107877, https://doi.org/10.1016/j.cep.2020.107877

[15]. Liu D, Zhu H, Wu K, Zhao X, Wang F and Liao Q 2020 *J. Hazard. Mater.* 392 122492 10.1016/j.jhazmat.2020.122492

[16]. Wang X, Li Y, Zhang W, Zhao J and Wang Z 2020 *J. Proc. Safe. and Env. Protec.* **134** 83-94 https://doi.org/10.1016/j.psep.2019.11.022

[17]. Suryanto H, Solichin S and Yanuhar U 2016 *Fiber Plants. Sustainable Development and Biodiversity* ed Ramawat K G and Ahuja M R (Switzerland: Springer, Cham) **13**(III) 35-52 10.1007/978-3-319-44570-0_3

[18]. Suryanto H, Marsyahyo E, Irawan Y S and Soenoko R 2014 *J. Nat. Fiber* **11**(4) 333-351 10.1080/15440478.2013.879087

[19]. Simone M, Rozangela C P and Sandra R S 2020 *Reference Module in Food Science* (Amsterdam: Elsevier) https://doi.org/10.1016/B978-0-08-100596-5.22816-3

[20]. Indonesian Nat. Standard 2005 *Udara Ambien-Bagian 2: Cara Uji Kadar Sulfur Dioksida (SO2)*
dengan Metoda Pararosanilin Menggunakan Spektrofotometer (Jakarta: Agency of Nat Standardization) SNI 19-7117.2-2005.

[21]. Basuki 2007 *Proc PPI - PDIPTN 0216 – 3128* (Yogyakarta: Pustek Akselerator dan Proses Bahan-BATAN) pp 105-114 http://digilib.batan.go.id/ppin/katalog/file/0216-3128-2007-2-105

[22]. Elvitriana, Vera V and Sari W 2017 *Proc. Sem. Nas. Pasca. (SNP) Unsyiah* (Banda Aceh: Unsyiah) pp A38-44 http://jurnal.unsyiah.ac.id/SNP-Unsyiah/article/view/6867

[23]. BAT-knowledge center 2020 *LUSS* (Belgium: VITO) https://emis.vito.be/en/bat/tools-overview/sheets/activated-carbon-adsorption#:~:text=As%20a%20reference%2C%20an%20adsorption,increase%2C%20the n%2 0capacity%20will%20fall.

[24]. EPA 2017 https://www.epa.gov/air-emissions-monitoring-knowledge-base/monitoring-control-technique-activated-carbon-adsorber accessed 3 August, 2020

[25]. Suryanto H, Yudy S I, Eko M and Rudy S 2013 *Proc. Nat. Conf. on Green Tech. 3 3,* https://www.researchgate.net/publication/282735207

[26]. Ananias F D J, Renata N O, Xavier D, Natália D S and José O B 2019 *Matéria (Rio de Janeiro)* **24**(3) e12387 https://doi.org/10.1590/s1517-707620190003.0700

[27]. Nicole L, David H, Timothy R and Thomas E 2006 *J. Agric. Food Chem* **54**(10) 3492–3497 https://doi.org/10.1021/jf053062n

[28]. Aziz Ş and Mehmet K 2015 *J. Chem.* 2015 651651 . http://dx.doi.org/10.1155/2015/651651.

[29]. Nadav I, Tarchitzky J, Lowengart-Aycicegi A and Chen Y 2013 *Irrig. Sci.* **31** 49–58, https://doi.org/10.1007/s00271-011-0291-3

Review of the seakeeping criteria for the study of a passenger ship criteria in Indonesian water

Jamal[1,2,3], A Sulisetyono[2], W D Aryawan[2]

[1] Naval Architecture Department. State Polytechnic of Bengkalis, Bengkalis 28711, Indonesia
[2]Naval Architecture Department, Faculty of Marine Technology, Institut Teknologi Sepuluh Nopember (ITS), Kampus ITS Sukolilo, Surabaya 60111, Indonesia

[3]jamal@polbeng.ac.id

Abstract. The human sickness of the ship's crews and passengers is considered as main requirements to be fulfilled in ship design stages which is related to the performance of ship motion when operating in seaway. This paper describe review some seakeeping criteria which are published by several institutions for the purpose of ship safety and comfort. Since these criteria were initially issued based on extreme sea conditions, it is necessary to study if they are applied to relatively moderate sea conditions in Indonesian. For this reason, a new approach is proposed to deal with a new criteria by measuring motions directly on the ship in operation. Briefly method is described as first, the ship motion in 6 degrees of freedom is measured by sensors which is installed in a particular place on the ship's board, for which the data generated include an angular velocity, a vertical and a tranversal accelerations of motions. Second, Fast Fourier Transform (FFT) algorithm is adopted to process the measured data which are presented in the form of time series and spectrum diagrams, and the Power Spectral Density (PSD) method is also used to eliminate nois data. Third, the ship motion criteria for the case of the passenger ship is determined and it compare to the existing criteria.

1. Introduction

Ship motion at sea can disturb the balance of crew members and passengers making Motion Induced Interruptions (MII). MII can cause decreased energy and increase fatigue, increase drowsiness, dizziness and can even interfere with gastric awareness which results in motion sickness incidences (MSI) and more seriously causes vomiting incidents.[1][2]. Sickness conditions that occur on the crew of the ship are things that must be considered because it involves the safety of the ship during voyage.

Anta˜o and Soares [3] explained that 60% of ship accidents occur in conventional ocean going ships and high speed craft were caused by human factors and were followed by equipment failure and environmental factors. According to Alert! [4], based on ship accident investigation reports, almost all ship accidents were caused by crew fatigue. Baker and Mc Cafferty's opinion [5] that fatigue must be closely related to the decrease in awareness of the situation and also the human error that caused the accident. According to Allen et. al. [6], crews ship fatigue has been investigated in the maritime domain, where the factors that cause fatique are as follows: lack of sleep, lack of rest due to limited time, poor sleep quality, vibration, room noise, ship motion, and high workload [4][7]. Haward [8] explained the main causes of sleep disorders were motion and noise. And motion can affect to decrease an operator performance and also safety [9].

Some suggestions have been given by related organizations for how to manage and reduce fatigue at sea during the ship's operation [4][7][10]. Proposed steps to be taken by passengers or crew include sleeping patterns, watching schedules, environmental manipulation, workload management, and food management. The ship design must consider to providing comfortable accommodation, minimizing vibration and noise, increase comfort in the room, and providing good work facilities to reduce workload.

Actually several regulations that support the human factor in ship design have been issued by the Classification Bureau. At least 93 documents have been identified as having relevance in dealing with human factors in relation to ship design [11] [12] (see Table 1). And 30 documents have briefly explored the design requirements of a ship which are intended include Habitability, Ergonomics, Maintainability, Controllability, Workability, Survivability, Maneuverability, System Safety, and Occupational Health and Safety [11].

Table 1 Summary related to human factor documents in ship design [11]

Classification Institions	No. of Document	Habitability	Ergonomics	Maintenability	Controllability	Workability	Survivability	Maneuverability	System Savety	Occupational Healt and Safety
DNV	18	1	0	1	7	6	3	3	11	0
ABS	17	6	2	1	4	7	3	1	9	3
LR	16	2	0	1	6	2	6	3	7	1
IMO	27	2	2	1	8	6	11	3	14	4
ISO (CT8)	13	7	0	0	2	3	4	0	2	0
ASTM	2	1	2	1	1	2	1	0	2	1
Total	93	17	6	5	28	25	29	10	45	9

Lloyd's Register (LR) [13] in 2008 had systematically issued references to human factors in ship design, such as: (i) Habitability, related to accommodation facilities that were of a standard and comfortable standard, including furniture and washing equipment, for example. Galley, Provisions under consideration regarding variations in room size, room shape, gender of seamen, and various causes of stress from the environment: heat, vibration and noise. (ii) Maintainability is a system for designing operational maintenance tasks to be safer, faster, and more effective, the goal being to improve the performance of equipment and systems. This system includes tool considerations, tool removal, access, operator expertise, and lifetime guarantee. (iii) Workability is a consideration that must be given to the operator, materials, usage procedures, tasks, equipment and including the software, and the environment in which the system is used. (iv) Controllability is designing a ship control center system which includes the design of an engine control room, design for a cargo control room, etc., a design can be in the form of integration between people or operators with equipment and systems such as communication equipment, control facilities, color or display in the form of text display. , image or video. (v) Maneuverability is a form of ship maneuver system design that has the most precise and consistent maneuverability, the design of the system includes the ship's manning and operational patterns, the main driving machine, the steering system, and also its propulsion system. And (vi) Survivability is the provision of safety equipment in the form of adequate fire fighting facilities, rescue systems, damage control, and safety arrangements on board.

In 2009 Lloyd's Register (LR) re-issued two additional factors related to human factors [14], such as: (i) occupational health and safety (OHS) is the protection of safety and welfare for workers, protection of work impacts, maintaining the work environment and protecting living conditions for health, (ii) Safety of the system is a risk consideration for people using the ship's system.

IOP Conf. Series: Materials Science and Engineering **982** (2020) 012041 doi:10.1088/1757-899X/982/1/012041

Guidelines and standards to improve the crew's habitability and comfort are ready applied on the ship by controlling environmental design and increasing the comfort rating of personnel, giving them time to sleep and rest better in order to improve the performance of these personnel. For crew comfort rating on board, the Det Norske Veritas (DNV) classification publishes additional comfort ratings divided into two groups: (i) noise rating and (ii) vibration rating with COMFV notation or COMF-C notation for room state notification [15]. For example DNV has divided into three comfort ratings: 1 (highest or very comfortable), 2 (moderate) and 3 (acceptable). For the measurement of noise rating regulated in the DNV classification must follow the ISO 2923 standard. The noise rating standards in DNV can be seen in table 2

Table 2 Noise level on accommodation in dB, referring to DNV 2009 [15]

Location	Comfort Rating nr (crn)		
	1	2	3
Wheel house	60	60	65
Crew cabins	50	55	60
Crew public spaces	55	60	65
Engine control room	70	70	75
Open deck recreation	70	70	75

In addition, the classification of the American Bureau of Shipping (ABS) has also provided several standard guidelines covering comfort and habitability [16] [17]. This guide has five aspects of ship criteria and layout namely noise, lighting, vibration throughout the body, accommodation, and climate in the room. For ships that have met the minimum threshold guidelines for the five aspects will be given the HAB notation by the ABS classification while the HAB + notation for stricter livability eligibility criteria. To measure comfort levels ABS recommends noise level standards provided by the International Maritime Organization (IMO). ABS and IMO have published a framework for considering ergonomics and work environments on ships in order to reduce accidents or workplace abuses [18]. IMO has also issued guidelines on noise level limits on ships [19] which can be seen in table 3.

Table 3 Noise level on board according to IMO 1981 [19]

Noise Level Limits	dB(A)
Machinery space (continuously manned)	90
Machinery space (note continuously manned)	110
Machinery control rooms	75
Workshops	85
Non-specified work space	90
Normally unoccupied space	90

2. Criteria of Ship motion performance

Research on the factors causing human sickness has been carried out by several researchers and institutions which then propose several criteria. The proposed criteria generally relate to restrictions on the behavior of ships when operating at sea in order to avoid the occurrence of human sickness on crew or passengers. A ship float in the high seas with 6 degrees of freedom [20] which are surge, sway, heave, roll, pitch, and yaw as shown in Figure 1. These motions can be grouped in rotational movements, i.e. the ship rotates with respect to the x, y and z axis (pitch, roll, yaw), and translational motion shifts on the ordinate axis.

Previous researches have discussion about a seakeeping or motions criteria of ship that is include Rumawas [21], Pattison and Sheridan [22], Ghaemi and Olszewski [23], Stevens and Parsons [24]. All their papers refer to the criteria of ship motion issued by the North Atlantic Treaty Organization

(NATO) STANAG 4154, the Nordic Co-operative Organization for Applied Research (NORDFORSK), the US Coast Guard Cutter Certification Plan, and Tasaki. The seakeeping criteria include a maximum roll amplitude, a pitch amplitude, a vertical and a lateral acceleration of motions.

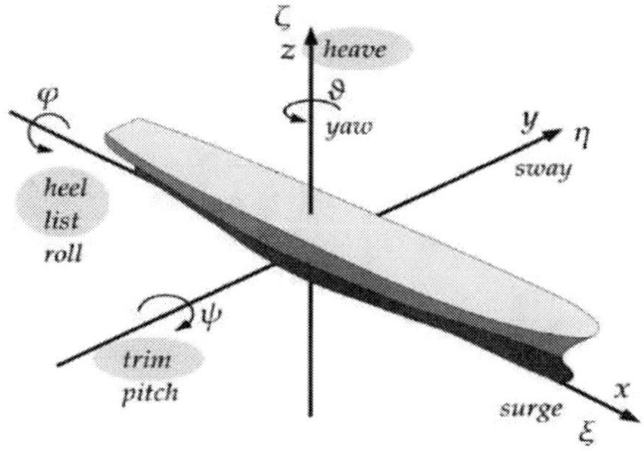

Figure 1 Ship motion with 6 degree of freedom [20]

Table 4 show the operability criteria set by NATO STANAG [25]. Ship design must refer to the seakeeping criteria outlined in the table adjusting the area of operation of the ship and the type of ship.

Table 4 Personnel criteria limits (NATO STANAG 4154) [25]

Rekommended Creteria	Limit	Location
Motion sickness incidence (MSI)	20% of crew @ 4 hrs	Task location
Motion induced Interruotion (MII)	1/m	Task location
Default Creteria	Root mean square (RMS)	
Roll	4^0	
Pitch	$1,5^0$	
Vertical acceleration	0,2 g	Bridge
Lateral acceleration	0,1 g	Bridge

The history related of NATO STANAG 4153 is written by Pattison and Sheridan [21]. Starting from the meeting of NATO subgroup 5 NG/6 in 1986, it was revealed that the navy found difficulties in the performance behavior of the crew during the study design. the next meeting made aware that some naval engineers were unable to effectively pursue the effects of the ship's movement on the crew. This organization understands the acceleration mechanism that causes a person to stumble incident, but cannot yet understand the implications of mental effects and physiological effects. To remedy this situation, protection experts who are now National Biotechnology Laboratories from the University of New Orleans, The New Orleans Bio Dynamics Laboratory (NBDL) at the American, British and Canadian Warship Operations Workshop (ABC-17) was invited to Halifax in late 1989. Fractional group in 1990, NATO made an experimental program planning on a ship motion simulator at NBDL. at the meeting the Dutch then joined the United States, Britain and Canada to form a naval group consisting of four countries, doctors, and behavioral scientists [21].

The results of the NATO meeting finally got an agreement on the criteria of the effect of ship motion on human performance. The criteria are defined into five factors as follows [21] such as (i) Motion induced interruptions (MII), (ii) Motion induced fatigue (MIF), (iii) Cognitive performance, (iv) Motion sickness incidence (MSI), and (v) Habituation.

In Table 5, the seakeeping criteria of NORDFORSK have been set for general operational restrictions on various types of vessels. Cretrian boundaries include the roll limit and the acceleration motion. For example in this criterion, the limitation of motion for fast small craft is to protect various types of personnel work for the movement of the roll limit of these criteria, such as Light manual work, heavy manual work, intellectual work, transit passengers, and Cruise liners. the purpose of this criteria is to expect the effectiveness of human comfort.

Table 5 Seakeeping criteria according to NORDFORSK 1987 [26]

General Operability Limiting Creteria for Ships	Marchant Ship	Navy Vessels	Fast Small Craft
RMS of vertical acceleration at FP	0.275 g ($L \leq 100$ m) 0.050 g ($L \geq 330$ m)	0.275 g	0.65 g
RMS of vertical acceleration at Bridge	0.15 g	0.20 g	0.275 g
RMS of lateral acceleration at Bridge	0.12 g	0.10 g	0.10 g
RMS of Roll	6.0 deg	4.0 deg	4.0 deg
Probability of slamming	0.03 ($L \leq 100$ m) 0.01 ($L \geq 330$ m)	0.03	0.03
Probability of deck wetness	0.05	0.05	0.05
Criteria with regard to acceleration and roll	Vert. Acc.	Lat. Acc.	Roll
Light manual work (RMS)	0,20 g	0.10 g	$6{,}0^0$
Heavy manual work (RMS)	0,15 g	0,07 g	$4{,}0^0$
Intellectual work (RMS)	0,10 g	0,05 g	$3{,}0^0$
Transit passengers (RMS)	0,05 g	0,04 g	$2{,}5^0$
Cruise liner (RMS)	0,02 g	0,03 g	$2{,}0^0$

Likewise for seakeeping criteria issued by the U.S. Coast guard cutter certification which contains roll / pitch movement criteria, at this time the seakeeping boundary is used by the US Navy and U.S. Coast Guard (USCG). Table 6 explains the criteria of MSI and MII with USN and USCG limits are equivalent, but in the table stated in different units, the NATO STANAG 4153 and NORDFORSK criteria use the root mean square (RMS) unit, while the US Coast Guart criteria uses the unit of amplitude single significant (SSA). Table 7 show the criteria of operability limitation for a ships introduced by Tasaki [23] that involve the probability occurrence of slamming, deck wetness and propeller emergence.

Table 6 Operability limiting criteria for ships according to USCG [23]

Creterion	Limit	Location
Vertical acceleration	0.4g SSA	forward perpendicular
Vertical acceleration	0.2g SSA	bridge
Lateral acceleration	0.2g SSA	bridge
Motion Sickness Incidence (MSI)	5% in a 30 minute exposure	task location
Motion Induced Interruption (MII)	2.1 tip per minute	task location
Roll	8.0° SSA	
Pitch	3.0° SSA	

Table 7 Operability limiting criteria for ships according to Tasaki [23]

Creterion	Limit	Location
Vertical acceleration	0.80g @ P=0.001	forward perpendicular
Lateral acceleration	0.60g @ P=0.001	bridge
Roll	25.0° @ P=0.001	
Slamming (probability)	0.01	
Deck wetness (probability)	0.01	
Propeller Emergence (probability)	0.1	

3. Ship motion measurement

The methodology used in evaluating ship motion is by direct measurement on the deck of the ship by using the ADIS16364 sensor with 6 axis motion. This sensor is equipped with an accelerometer sensor (3 axis) and gyroscope sensor (3 axis) (sensor specifications: see table 5 and figure 2). To measure the roll and pitch motion used gyroscope while the vertical movement and lateral acceleration use the accelerometer sensor

Placement of sensor measurement points according to NATO STANAG 4154 recommendations [25] Motion Sickness Incidence (MSI) and Motion Induced Interruotion (MII) measurements are at the crew assignment location, vertical acceleration and lateral acceleration measurements in the bridge space while roll and pitch measurements are not determined but at this research was carried out measurements in the bridge room.

Figure 2 ADIS16364 sensor [27]

Table 5 ADIS16364 6 axis spesification [27]

ADIS16364 spesification	
Triaxis digital gyroscope with digital range scalling	: +750/sec, +1500/sec, +3000/sec
Tight orthogonal alignment	: <0,050
Triaxis digital accelerometer	: +5.g
Calibration temperatur range	: -20°C to +70°C
Aplication	: - Medical Intrumentation, - Robotics, - Platform controls, - Navigation.

The data generated by ADIS16364 is in the form of angular velocity data generated by the gyroscope sensor to measure roll and pitch motion, in addition the data measured by the accelerometer sensor in the form of velocity to gravity is to measure vertical acceleration and tranversal acceleration movements. The results of field tests conducted by Rumawas [28] namely by direct testing above Offshore Supply Vessels (OSV), rotational speed is measured and recorded using the ADIS16364 sensor in degrees per second (deg / s), while vertical and lateral movements are accelerated in g-force (g). In his research the rotational speed is integrated into motion and presented in degrees (deg) which are recorded per 15 minutes and presented in the form of time series and spectrum diagrams, he uses the Fast Fourier Transform (FFT) algorithm and the result data is processed to eliminate irregular motion noise using Power Spectral Density (PSD) method with Matlab v 2012 [28] (see figure 3).

Furthermore, the statistical values for lateral and vertical acceleration are sought to obtain the RMS values according to table 6

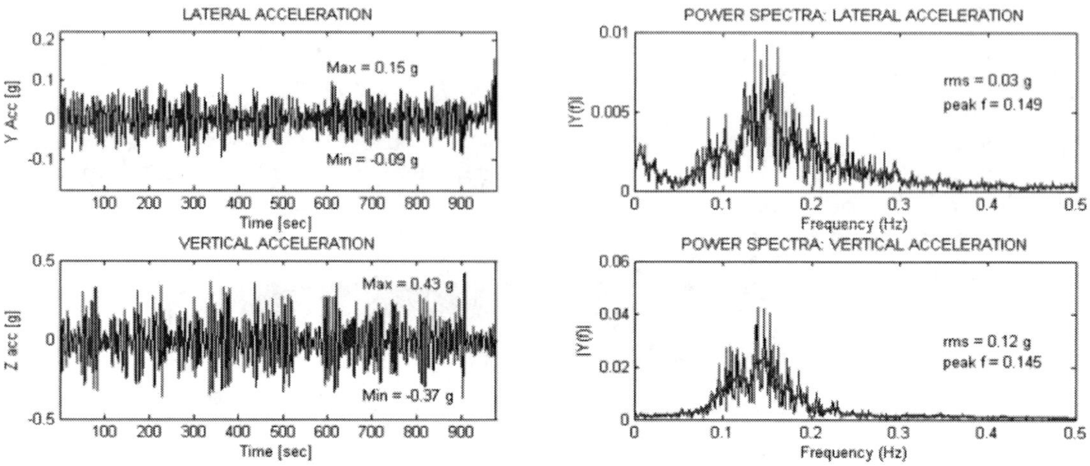

Figure 3 Lateral and vertical acceleration on OSV [28]

Table 6 Summary of lateral and vertical acceleration values using statistical analysis [28]

Ship	Season	Acceleration							
		Lateral acc.				Vertical acc. [g]			
		Min	Mean	Max	RMS	Min	Mean	Max	RMS
A	summer	-0.076	0.010	0.098	0.014	-0.149	-0.011	0.153	0.026
A	winter	-0.269	0.002	0.263	0.038	-0.720	-0.012	0.562	0.097

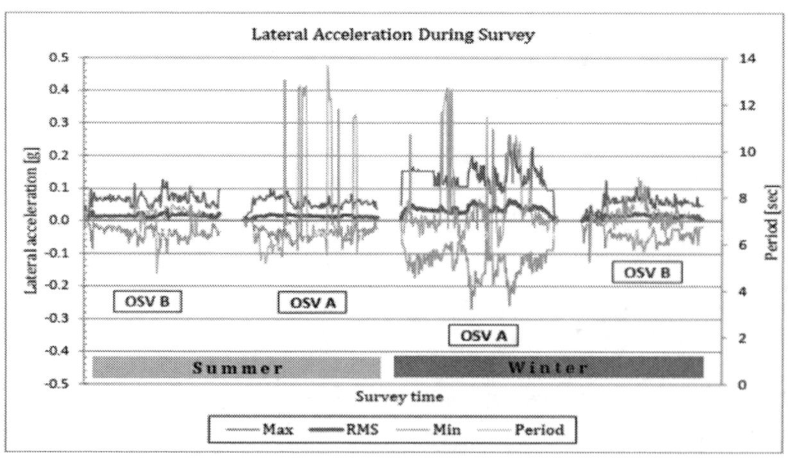

Figure 4 The survey results aboard the OSV to measure Lateral Acceleration [28]

McCauley and O'Hanlon have developed a formula to determine the value of motion sickness incidence (MSI), this formula is used to calculate the impact of ship movements on the percentage of personnel suffering from seasickness [14]. the following is the formula:

$$MSI = 100\left[0.5 + erf\left(2.5\, log_{10}\left(\frac{a_v}{g}\right) - \mu_{MSI}\right)\right] \quad (1)$$

where, erf is an error function, to get erf according to the following formula:

$$\mathrm{erf}(x) = \frac{1}{\sqrt{2\pi}} \int_0^x e^{(-x^2/2)} dx \tag{2}$$

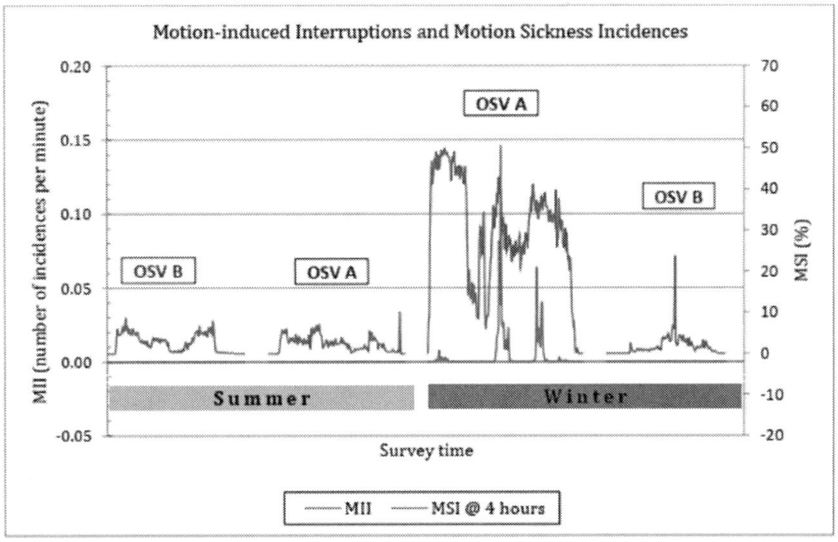

Figure 5 Survey results on OSVs in measuring MII and MSI values [28]

To achieve the representative criteria of ship motion for the Indonesia water, it is necessary to evaluate the motion of ships across three waters locations in Indonesia which have the characteristics of different waves, namely the Malacca Strait (the Indonesian sea at western), the Java sea (the central Indonesian sea), and the Banda sea (the Indonesian sea at eastern) (see figure 6).

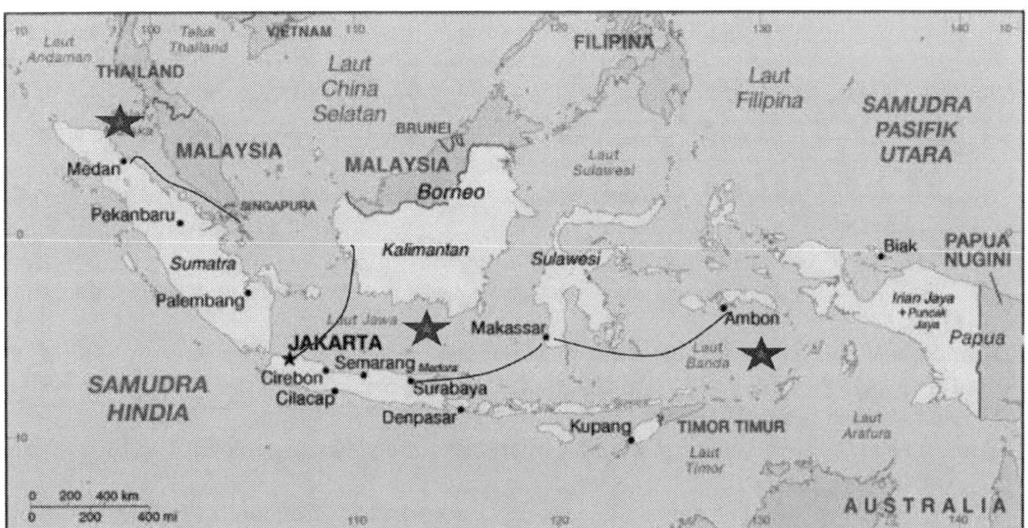

Notes: 1. ★ : Indonesia's sea area surveyed
2. —— : surveyed route

Figure 6 Survey area of the study

4. Conclusions

Seakeeping criteria which are widely used to evaluate a ship design performance in seaway is reviewed. Since the study limited on the passenger ship, the NORDFORSK 1978 criteria is suitable used in future study for cruise liner. The criteria was initially published based on North Sea conditions, so that necessary to study seakeeping criteria consider to the sea conditions in Indonesian. The direct measurement method is proposed to measure ship motion response in seaway. Fast Fourier Transform (FFT) algorithm is developed to identify the measured data, and the Power Spectral Density (PSD) method is adopted to eliminate noise data. This developed approach is expected to be suitable method in developing the seakeeping criteria of passenger ship in Indonesian water.

References

[1] Stevens S C and Parsons M G 2002 Effects of Motion at Sea on Crew Performance: A Survey. Marine Technology 39: 29-47.
[2] Haward B M, Lewis C H, et al. 2009 Motions and Crew Responses on an Offshore Oil Production and Storage Vessel. Applied Ergonomics 40(5): 904-914.
[3] Anta˜o P and Soares C G 2007 Causal Factors in Accidents of High-Speed Craft and Conventional Ocean-going Vessels. Reliability Engineering and System Safety 93 (2008) 1292–1304.
[4] Alert! 2007 Time to Wake Up to The Consequences of Fatigue. Alert! The International Maritime Human Element Bulletin, The Nautical Institute.
[5] Baker C C and McCafferty D B 2005 Accident Database Review of Human-Element Concerns: What Do The Results Mean for Classification? RINA, Royal Institution of Naval Architects International Conference - Human Factors in Ship Design, Safety and Operation, Febrary 23, 2005 - Febrary 24, 2005, London, United kingdom, Royal Institution of Naval Architects.
[6] Allen P, Wadsworthand E, Smith A 2008 Seafarers fatigue: a review of the recent literature."International Maritime Health 59 (1-4):11.
[7] Smith A, Allen P and Wadsworth E 2006 Seafarer Fatigue: The Cardiff Research Programme. Cardiff, UK, Centre for Occupational and Health Psychology, Cardiff University.
[8] Haward B M, Lewis C H, et al 2009 Motions and Crew Responses on an Offshore Oil Production and Storage Vessel. Applied Ergonomics 40(5): 904-914.
[9] Ross J M 2009 Human Factors Considerations for Marine Vehicle Design. International Conference on Human Factors in Ship Design, Safety and Operation. London, Royal Institution of Naval Architects: 29-37.
[10] ITF Seafarers 2012 Fatigue. Retrieved 29.01.2015, from http://www.itfseafarers.org/ITI-fatigue.cfm.
[11] Rumawas V and Asbjørnslett B E 2014 a Content Analysis Of Human Factors In Ships Design. Trans RINA, Vol 156, Part A3, Intl J Maritime Eng, Jul-Sep 2014. DOI No: 10.3940/rina.ijme.2014.a3.299.
[12] Kristiansen S 2005 Maritime transportation: Safety Management and Risk Analysis. Amsterdam, Elsevier.
[13] Lloyd's Register 2008 The Human Element: An Introduction. London, Lloyd's Register.
[14] Lloyd's Register 2009 The Human Element Best Practice for Ship Operators. London, Lloyd's Register.
[15] DNV 2009 Rules for Classification of Ships Comfort Class. Special Service and Type Additional Class Høvik, Norway, Det Norske Veritas. Part 5 Chapter 12.
[16] ABS 2001 Guide for Crew Habitability on Ships. Houston, TX, American Bureau of Shipping.

Hab 102.
[17] ABS 2001 Guide for Passenger Comfort on Ships. Houston, TX, American Bureau of Shipping. Comf 103
[18] IMO 2006 Framework for Consideration of Ergonomics and Work Environment. MSCMEPC.7/Circ.3.I.M. Organization.London. MSC-MEPC.7/Circ.3.
[19] IMO 1981 Code on Noise Levels On Board Ships. Resolution A.468(XII). IMO. Res A.468(XII)
[20] Habil N K 2012 Ship dynamics in waves (Ship Theory II). Rostock : Universitat Rostock Faculty of Mechanical Engineering and Marine Technology
[21] Rumawas V, Asbjørnslett B E and Klöckner C A 2016 Human Factors Evaluation in Ship Design: A Case Study on Offshore Supply Vessels in the Norwegian Sea, Part II: Multivariate Analyses and Structural Modelling. Naval Engineers Journal xxx(xx):pp.
[22] Pattison J H and Sheridan D J 2004 Human Performance Factors and Measures in Hull Form Selection. NATO unclassified/unlimited, RTO-MP-AVT-110, Meeting Proceedings-Paper 34 date 10/04/2004
[23] Ghaemi M H and Olszewski H 2017 Total Ship Operability –Review, Concept And Criteria. Polish Maritime Research. Special Issue 2017 S1 (93) 2017 Vol. 24; pp. 74-81 10.1515/pomr-2017-0024
[24] Stevens S C and Parsons M G 2002 Effects of Motion at Sea on Crew Performance: A Survey. Marine Technology, Vol. 39, No. 1, January 2002, pp. 29–47
[25] NATO 2000 Standardization Agreement (STANAG): Subject: Common Procedures for Seakeeping in the Ship Design Process, NATO, Military Agency for Standardization.
[26] NORDFORSK 1987 Assessment of Ship Performance in a Seaway. Copenhagen, Nordforsk.
[27] Analog Devices 2011 Six Degrees of Freedom Inertial Sensor.
[28] Rumawas V, Asbjørnslett B E, C A K 2009 Human Factors Evaluation In Ship Design : A Case Study On Offshore Supply Vessels In The Norwegian Sea , Part I : Theoretical Background. 1–20

IOP Conf. Series: Materials Science and Engineering **982** (2020) 012042 doi:10.1088/1757-899X/982/1/012042

Rifting and Lifting Neogene Age in Biak-Yapen Basin Based on Structural Trajectory Analogy in Biak and Supiori Island

R Y Saragih[1,4], Alviyanda[2], H L Sunan[3], E Slameto[1], and I Nurdiana[1]

[1] Geological Survey Centre, Bandung, 40122, Indonesia
[2] Geological Engineering Study Program, Institut Teknologi Sumatera, Lampung, 35365, Indonesia
[3] Geological Engineering Study Program, Jenderal Soedirman University, Jawa Tengah, 53122, Indonesia

[4] rahmatyantono2009@gmail.com

Abstract. The aim of this study is to calculate the rifting and lifting phase of Neogene age in the Biak - Yapen Basin region, related to the prediction of an oil and gas trap zone based on the analogy of the surface measurement trajectory. The methodology used is the measurement of geological structures on the surface and in the data analysis using Delvaux, 1997 and Zoback, 2007 concept. Based on the structure measurement in outcrop, there are 78 stations along Biak Island and Supiori Island, which will be used and projected as an analogy to the offshore basin. Statistically in the archipelago shows that there are three times rifting phase domination candidates categorized as maximum intensity and a lifting phase candidates are also categorized as maximum intensity. There are three times rifting phase with movement direction, σ-3 is relatively horizontal, consisting of NE-SW direction with an intensity of 15.60% rifting aged Middle to Late Miocene, ENE-WSW direction with intensity 13.30% rifting aged Late Miocene to Pliocene and finally from the W-E direction with an intensity of 17.80% rifting aged Pleistocene. The three rifting phases are followed by sedimentation of carbonate rock from the Middle Miocene to the Pleistocene aged which consists of Napisendi Fm, Korem Fm, Wardo Fm, and Mokmer Fm. The lifting phase once occurred with the compressional direction, σ-1 is relatively horizontal, directed at W-E with an intensity of 21.20% lifting aged Pleistocene to Resen. This phase has elevated formations from the age of the Late Oligocene to the Pleistocene in the archipelago.

1. Introduction

The Biak-Yapen Basin is located in the northern part of the mainland of the island of Papua. Based on Indonesian Sedimentary Basin Map published by Badan Geologi, 2009 (Figure 1). The Biak-Yapen basin covers the southern part of Biak-Supori Island and the North Western part of Yapen Island. This basin is classified as a Tertiary age Fore Arc Basin which is formed subduction activities in the northern part of mainland Papua [1] [2] [3] [4]. In addition, some researcher states that collision activity also occurred which causes the bedrock to the surface [3] [5] [6].

The aim of this research is to calculate the rifting and lifting phases at Neogene age as an analogy to the measurement of surface geological structures in the study area. And also as a prediction of the presence of hydrocarbon trap zones as an indicator from rifting and lifting activities in the Biak - Yapen Basin.

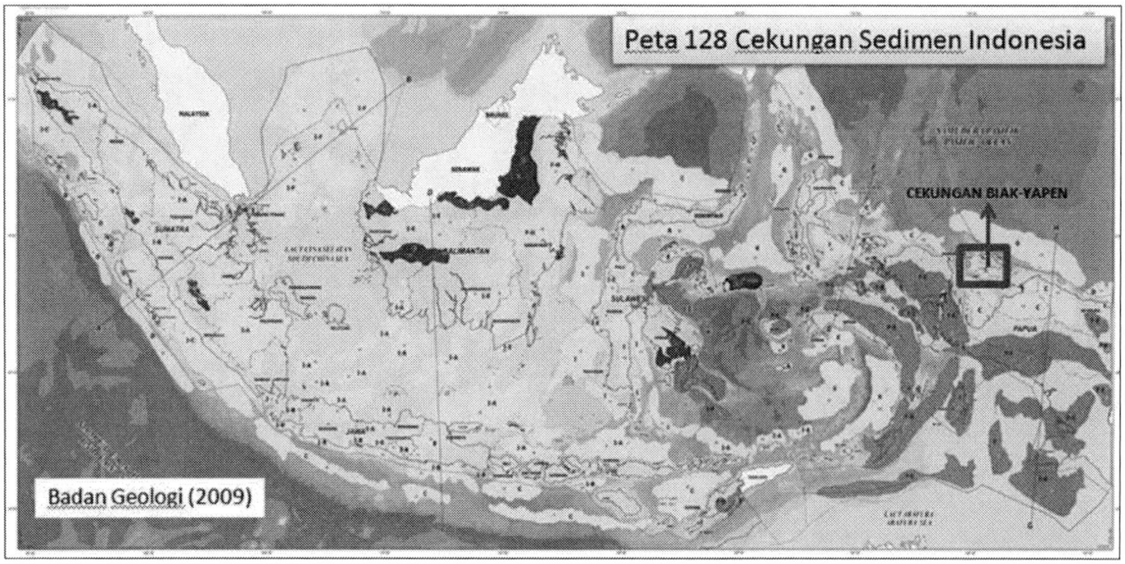

Figure 1. The map of Indonesian Sedimentary Basin was published by Badan Geologi [7].

The basin zoning is largely offshore and generally filled by groups of Cretaceous metamorphic rock, Eocene volcanic rocks, carbonaceous sedimentary rocks and non-carbonaceous clastic sedimentary rocks of Tertiary to Quaternary. This condition causes the basin to have a significant sediment thickness due to the intensive rifting from Neogene to the Pleistocene age. Early Tertiary sedimentary rocks have the characteristics of good source rock for petroleum systems and the maturation process is estimated to have occurred due to a very thick sediment loading process and also influenced by a quite well temperature at certain depth.

Administratively, the Biak-Yapen Basin is included in three regencies, namely Yapen Islands, Biak-Numfor and the Supiori, Papua Province (Figure 2). Because the basin covers two islands, namely Yapen Islands, and Biak-Supiori Islands, the research activity was focused on Biak-Supiori Islands, Papua Province located in Biak-Numfor Regency and Supiori Regency.

2. Regional Geology

2.1. Stratigraphy

The geology of the Biak Island and Supiori Islands area, including the stratigraphy and structures, has been described in the Regional Geological Map, Biak Sheet, Scale 1: 250,000 (Figure 3). Before the geological mapping of Biak and the surrounding areas, it is connected to the oil and gas resources. The regional stratigraphy of Biak Island and Supiori Island in [8] and [9] can be explained as follows:

- The Korido Metamorphic Rock (s) consists of phyllite, quartzite, chert, metamorphic tuff, and metamorphic greywacke and metamorphic sandstone. This formation is intruded by a small basalt dyke. This rock unit is unconformably overlain by the Auwewa Formation

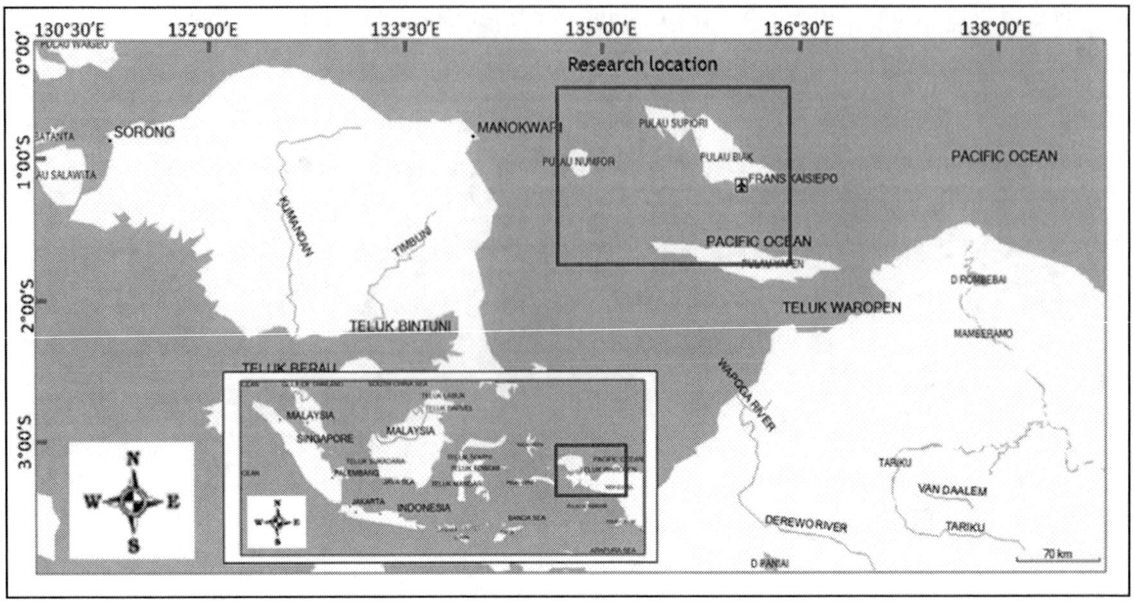

Figure 2. The research area map, administratively research area located in Biak-Numfor Regency and Supiori Regency.

- The Auwewa (Teoa) formation is a volcanic rock composed of lava and tuff, localized at the bottom of intercalation the breccia and conglomerate. This formation is estimated to be deposited in the terrestrial environment during the Eocene to the Early Oligocene, unconformity with the Korido metamorphic rock and unconformably overlies by the Supiori Group.
- The Wainukendi (Tomw) Formation consists of crystalline limestone, medium to roughly grained, local conglomerate lens and intercalation marl, fossiliferous limestone; and roughly grained greywacke. This formation is estimated to be the Late Oligocene to Early Miocene. This formation is deposited in a neritic environment, located unconformity on top of the Auwewa formation and conformably overlies or interfingering with the Wafordori Formation and subsequently conformably overlies by the Napisendi Formation.
- The Wafordori (Tmw) Formation consists of marl, part of tuff, slightly intercalation with sandstone and crystalline limestone. Marl is grey and brown colored, containing planktonic foraminifers' shows early Miocene age. This formation is deposited in the littoral to neritic zone. This formation is conformity or interfingering with the Wainukendi Formation and conformably overlies by Napisendi Formation.
- The Napisendi (Tmn) Formation consists of layered limestone, tuff clastic limestone with smoothly to roughly grained and a little solid limestone; conglomerate, breccia, sandy limestone and marl as well as smoothly to roughly grained sandy limestone intercalations. The results of the foraminifers analysis shows the Middle Miocene age of the top section. This formation is deposited in the littoral to neritic zones. Conformity located on top of the Wainukendi and Wafordori Formation and unconformably overlies by the younger formation.
- The Korem Formation (Tmk) consists of marl and marl limestone, localized with sandy marl and marl limestone intercalations; white and brown colored, solid. The results of the foraminifer fossil analysis show the Late Miocene age. Located conformably on top of the older formations and the top of this formation is interfingering with Wardo Formation.
- The Wardo (Tmpw) formation consists entirely of marl limestone and sandy limestone, localized at the top mainly composed of limestone. The limestone is brownish-white colored and contains small foraminifer fossils showing the Late Miocene to the Pliocene. This formation is

conformably overlies by the Mokmer Formation while the bottom part is interfingering with the Korem Formation.

- The Mokmer Formation (Qm) consists of coralline limestone at the top and limestone at the bottom. The limestone contains planktonic foraminifer fossils, coral fragments, algae and mollusks. The bottom of this formation may shift into the Wardo Formation. The results of the fossil analysis show the age as Pleistocene.
- Terrestrial sediment, composed of sand, loose gravel.
- The coastal sediments are composed of mud, sand and gravel and; the overlay are limited on both sides of the Sorendidori Strait.

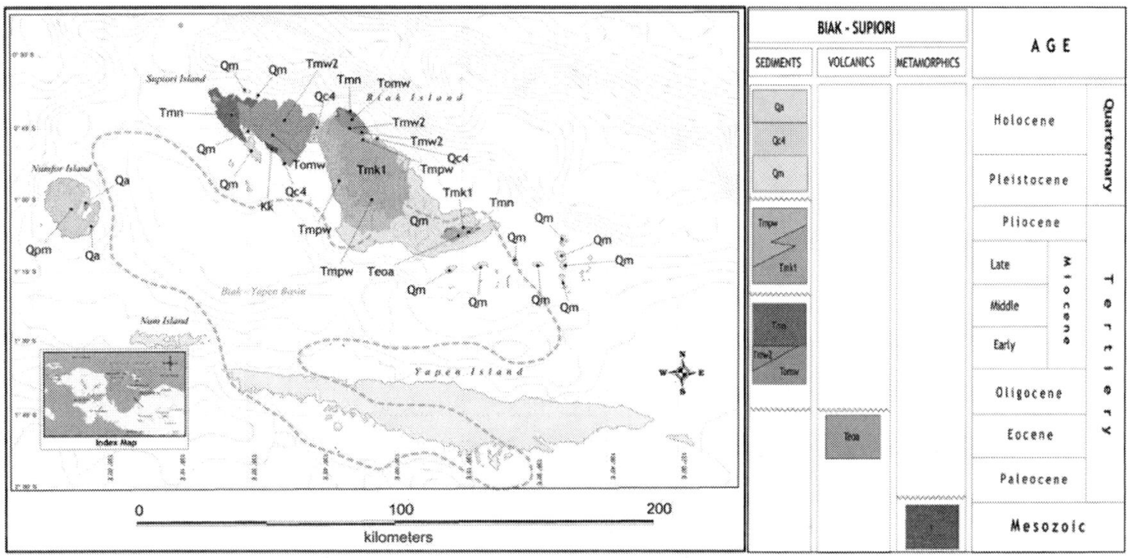

Figure 3. Regional stratigraphy of Biak Island and Supiori Island in [8] and [9].

2.2. Regional Structural Lineament

The lineament structure of the geological map in the study area was made based on residual gravity [10]. As for the subsurface geology was supported by magnetotelluric cross-sections [11]. From the results of residual gravity and magnetotelluric can be reconstructed the differences in the structural lineament type between Biak and Supiori Island (Figure 4). The structural lineament segments in the eastern part of Biak Island are more likely to be dominated by reverse fault structures with N - S and NNW – SSE relative direction, resulting in uplift due to compressional activities of the Pleistocene to Recent. This region has distribution of carbonate sedimentary rocks at the age of Late Miocene to Pleistocene. Meanwhile the Supiori Island has extensional fault segments with NW - SE relative direction generating fault blocks which tends to form graben. The extensional activity of the Supiori Island is estimated formed at the age of Miocene which is marked by the presence of carbonate rock formations of the Early Miocene to the Late Miocene.

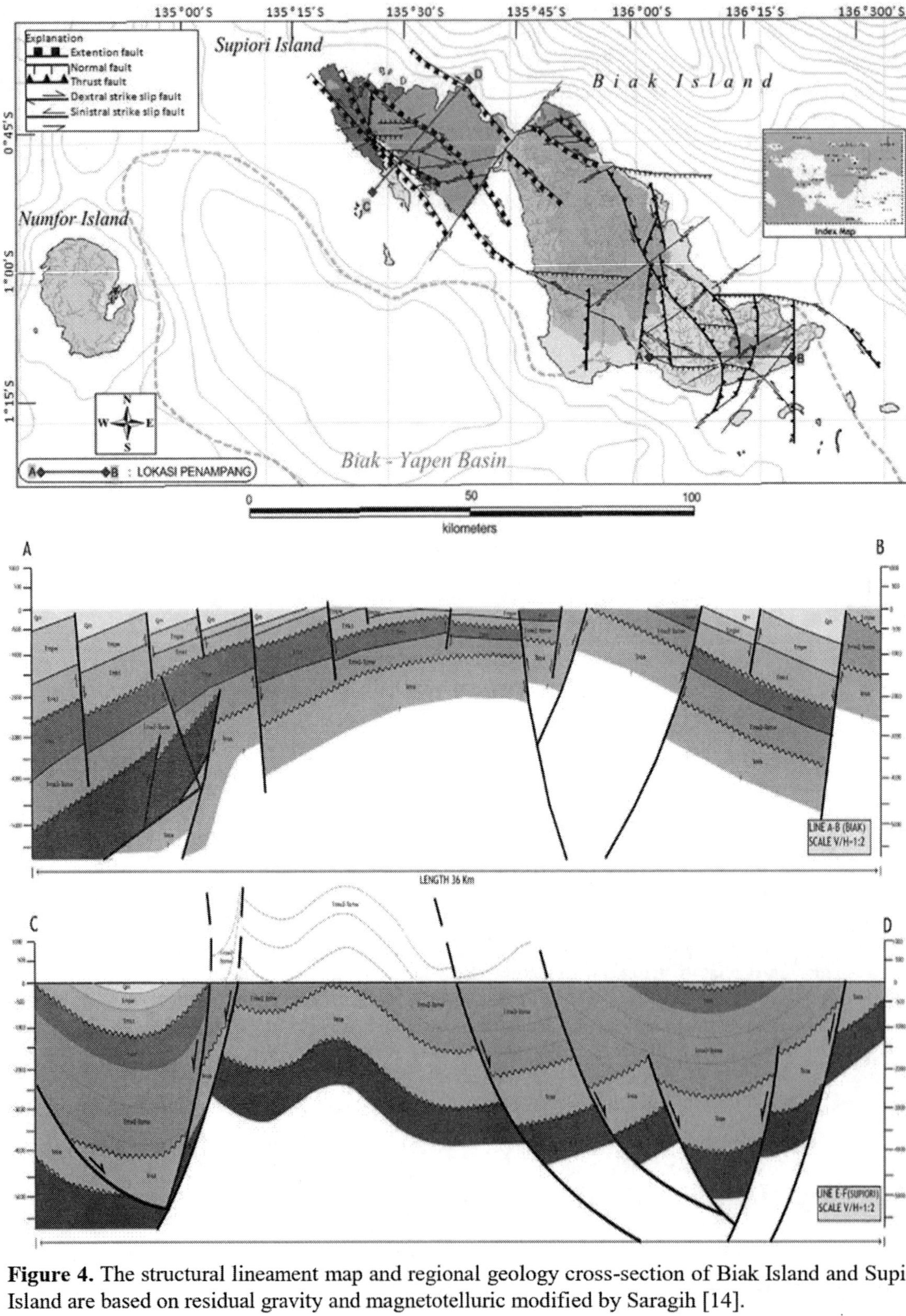

Figure 4. The structural lineament map and regional geology cross-section of Biak Island and Supiori Island are based on residual gravity and magnetotelluric modified by Saragih [14].

3. Methods

The structural data processing used Delvaux concept [12] to obtain regime strees along with the track and Zoback concept to obtain vertical stress (Sv), maximum horizontal stress (Shmax), minimum horizontal stress (Shmin). The Zoback concept aims to analyze a rifting or lifting that was formed by the activity of the fault lineament in the research area [13].

Generally, the methodology of this research is divided into three stages, namely; field survey, analysis data, and map modeling (Figure 5). The field surveys activity includes collecting data on the field in the form of geological structure, stratigraphy, and geophysics which are gravity residuals and magnetotelluric. The analysis data is clustering geological structure data to determine rifting and rifting phases based on field observation such as folds, joints and faults. Rifting and lifting phase map modelling is generated from simple statistical calculations from the cumulative measurement of field observation on the Biak Island and Supiori Island. Statistically, it shows the direction of the rifting and lifting phase as a percentage.

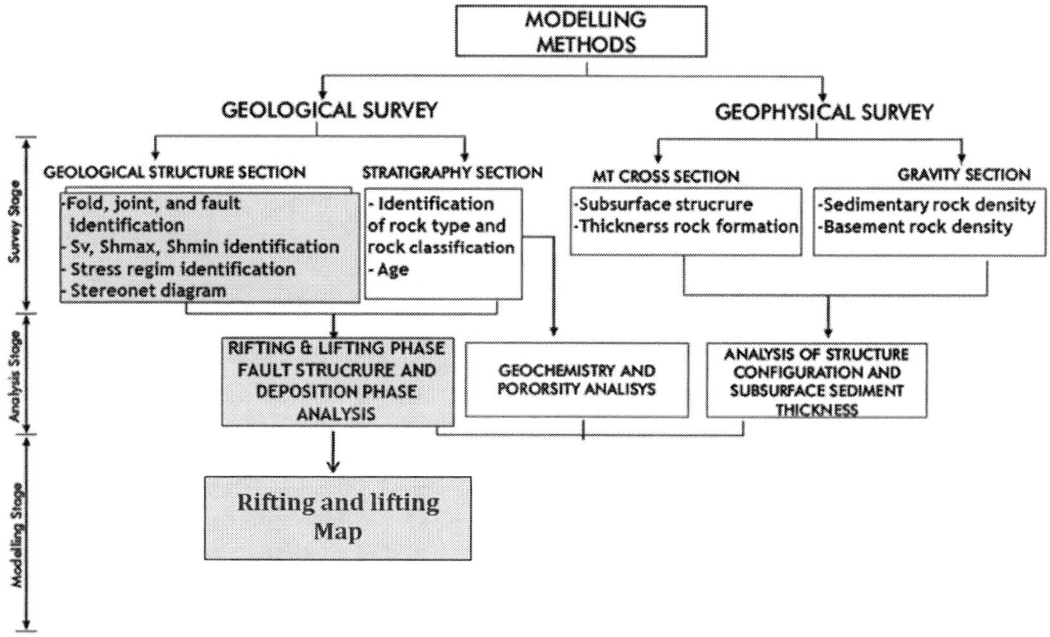

Figure 5. The research diagram.

4. Result

4.1. Trajectory Measurement

The local structural measurement trajectory on Biak Island consists of 49 measurement locations (Figure 6), across the northern, southern, central, and eastern part of Biak Island. Generally the rocks present are dominated by grainstone, boundstone, and calcareous siltstone. These rocks belongs to the members of Korem and Wardo Formations. These rock formations have conformable relationship, which is deposited at the Late Miocene to Early Pliocene [9]. The present of local structure shows two type of joints namely shear joint and hybrid joint and also present the fault fields of extensional and compressional groups.

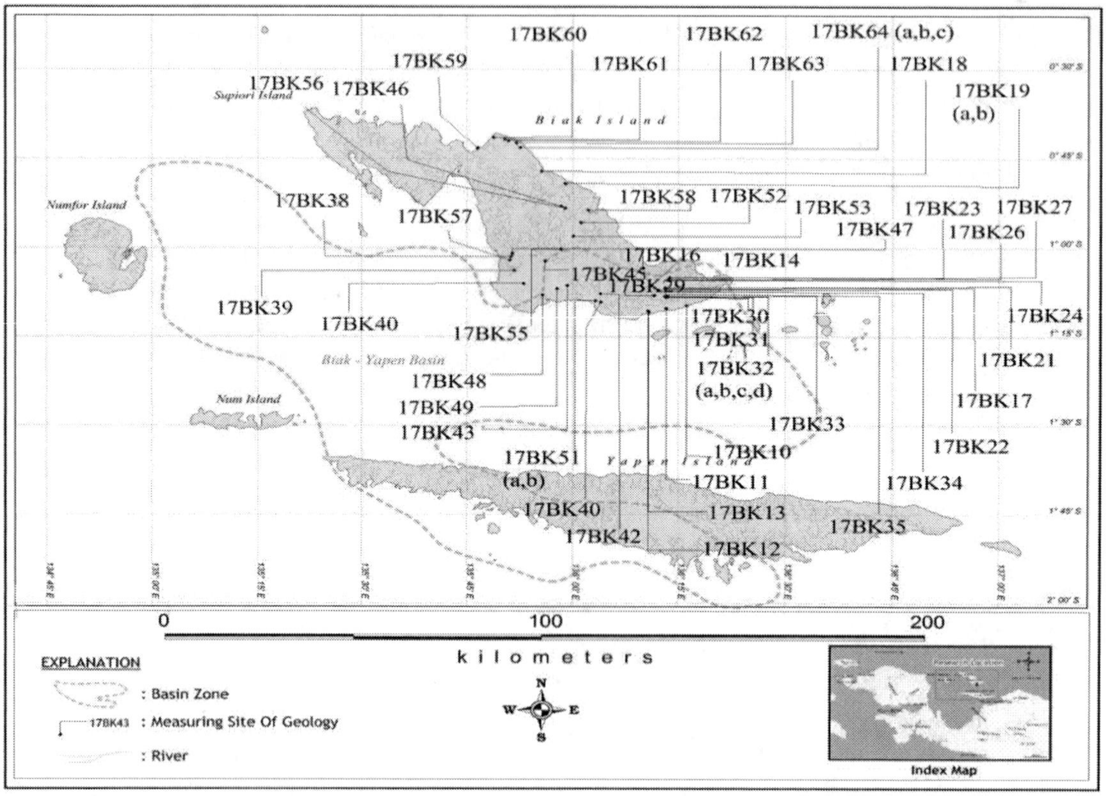

Figure 6. The trajectory map of local structure measurement in Biak Island [14].

The local structural measurement trajectory on Supiori Island consists of 29 measurement locations (Figure 7), across the northern and southern part of Supiori Island. In both regions were found calcareous sandstone, limestone, boundstone-grainstone. The bioclast grains in the limestone are dominated by forams, gastropods, fragments of mollusc shell and some parts of algae and coral. This rock group belongs to the members of Wainukendi, Wafordori and Napisendi Formations. These rock formations have conformable relationship, the lithological changes are influenced by sea level rise and are deposited at Early to Middle Miocene [9]. In the southern part, there are also metamorphic rocks in the form of schist, gneiss, and meta-limestone of Mesozoic age which is included as member of Korido Formation [9]. These rocks show the existence of structural fields from compressional to transpressional groups. In the northern part of Supiori Island, there are limestone of the Early Miocene to the Late Miocene showing the structural fields extensional and transpressional groups.

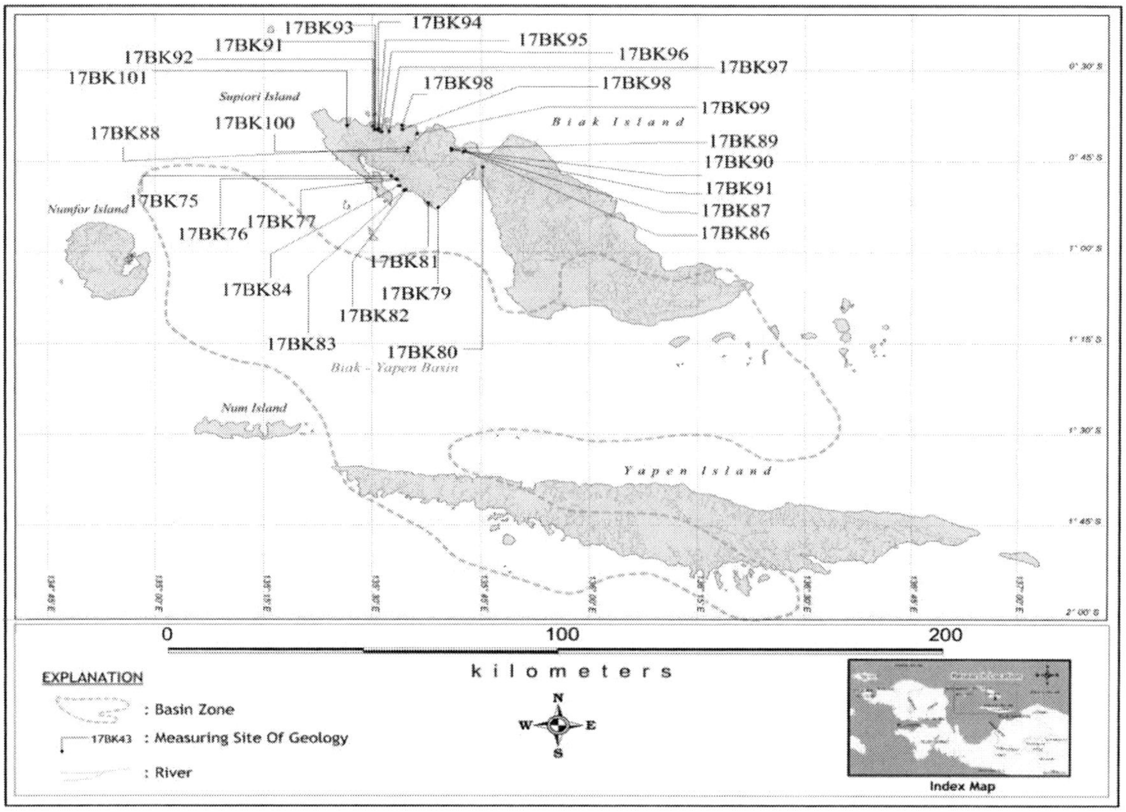

Figure 7. The trajectory map of local structure measurements in Supiori Island [14].

The local structure measurements of the in Biak and Supiori Island are projected in the stereonet diagram producing the position and the intensity of structural fields in their relative trends. The Biak Island trajectory has 430 structural fields with maximum field intensity showing a relative NNW-SSE trend of 18.82% and minimum field intensity with a relative W-E trend of 5.58% (Figure 8). In addition, the Supiori Island trajectory has 254 structural fields with maximum field intensity showing a relative NE-SW trend of 18.9% and a minimum field intensity with a NW-SE relative trend of 9.1% (Figure 9).
In these two trajectories also shows the slope angle quantity of dip right and the slope angle quantity of plunge, where Biak Island trajectory shows the maximum slope angle quantity of dip right with values ranging from 37.2° to 68.5° and the maximum slope angle quantity of plunge with values ranging from 25.4° to 50.7°. Whereas in the Supiori Island trajectory shows the maximum slope angle quantity of dip right with values ranging from 38° to 46° and the maximum slope angle quantity of plunge range with values ranging from 44° to 52° (Figure 10).

4.2. Structural Lineament.
The lineament drawing is based on the gravity map interpretation of the research area in which the structural lineament developed in the research area, in term of intensity (%) consists of lineament direction: WNW – ESE of 28%, NW – SE of 24%, W – E of 18%, N – S of 13%, NNE – SSW of 10%, and NE – SW of 7% (Figure 11).

4.3. Stress Regime Structure.

Structural data processing in the stress regimes study results stress ratio (R) and stress index (R') by using Delvaux concept [6]. The results of stress regime developed from 78 measurement locations along the trajectory in Biak Island and Supiori Island (Figure 12) can be seen in Table 1.

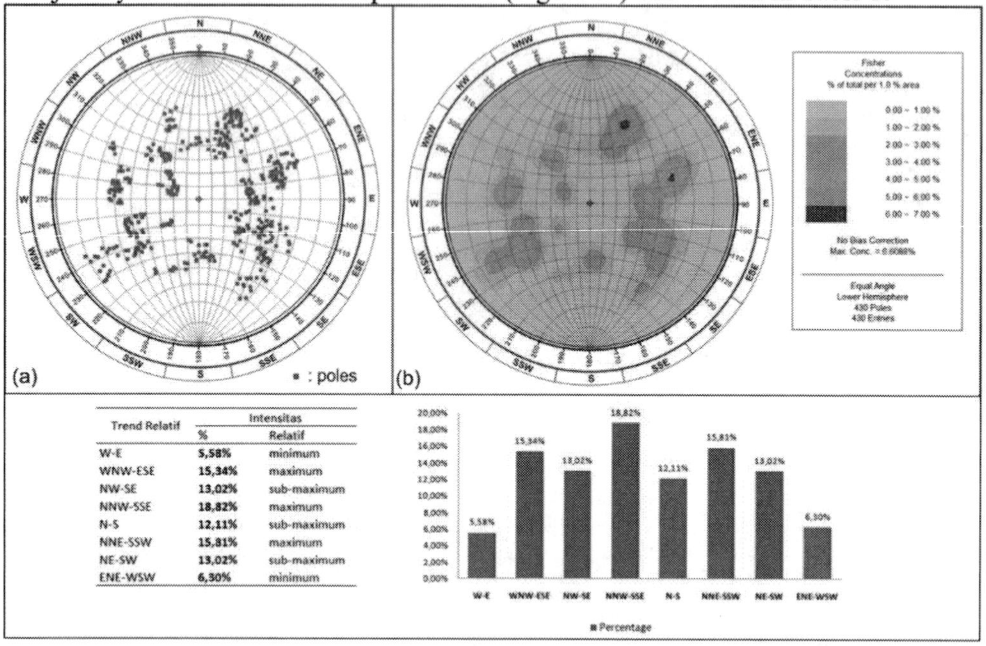

Trend Relatif	Intensitas %	Intensitas Relatif
W-E	5,58%	minimum
WNW-ESE	15,34%	maximum
NW-SE	13,02%	sub-maximum
NNW-SSE	18,82%	maximum
N-S	12,11%	sub-maximum
NNE-SSW	15,81%	maximum
NE-SW	13,02%	sub-maximum
ENE-WSW	6,30%	minimum

Figure 8. (a) Pole position and (b) field intensity of the local structure measurements (430 fields) on the Biak Island trajectory, shown in the stereonet diagram.

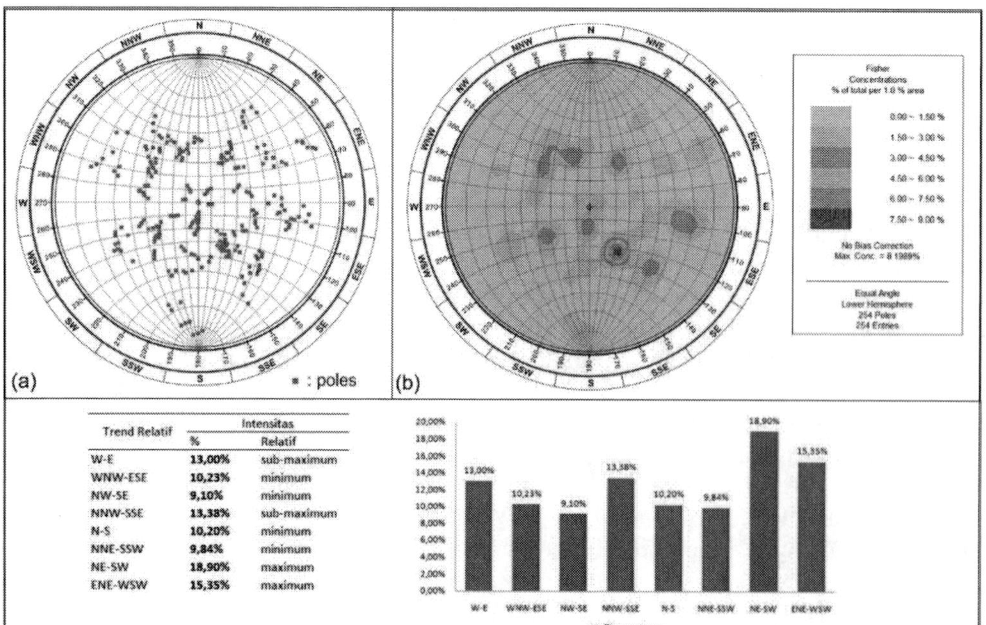

Trend Relatif	Intensitas %	Intensitas Relatif
W-E	13,00%	sub-maximum
WNW-ESE	10,23%	minimum
NW-SE	9,10%	minimum
NNW-SSE	13,38%	sub-maximum
N-S	10,20%	minimum
NNE-SSW	9,84%	minimum
NE-SW	18,90%	maximum
ENE-WSW	15,35%	maximum

Figure 9. (a) Pole position and (b) field intensity of the local structure measurements (254 fields) on the Supiori Island trajectory, shown in the stereonet diagram.

IOP Conf. Series: Materials Science and Engineering **982** (2020) 012042 doi:10.1088/1757-899X/982/1/012042

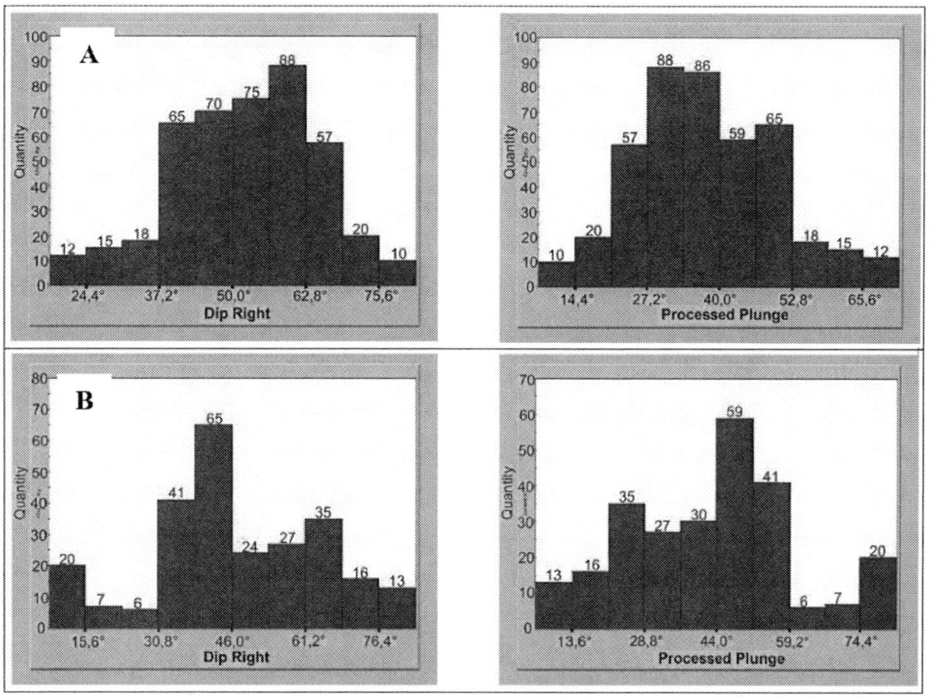

Figure 10. The graphic shows the slope angle quantity of dip right and plunge from the local structure measurements about 430 fields in Biak Island (A) and about 254 fields Supiori Island trajectory (B).

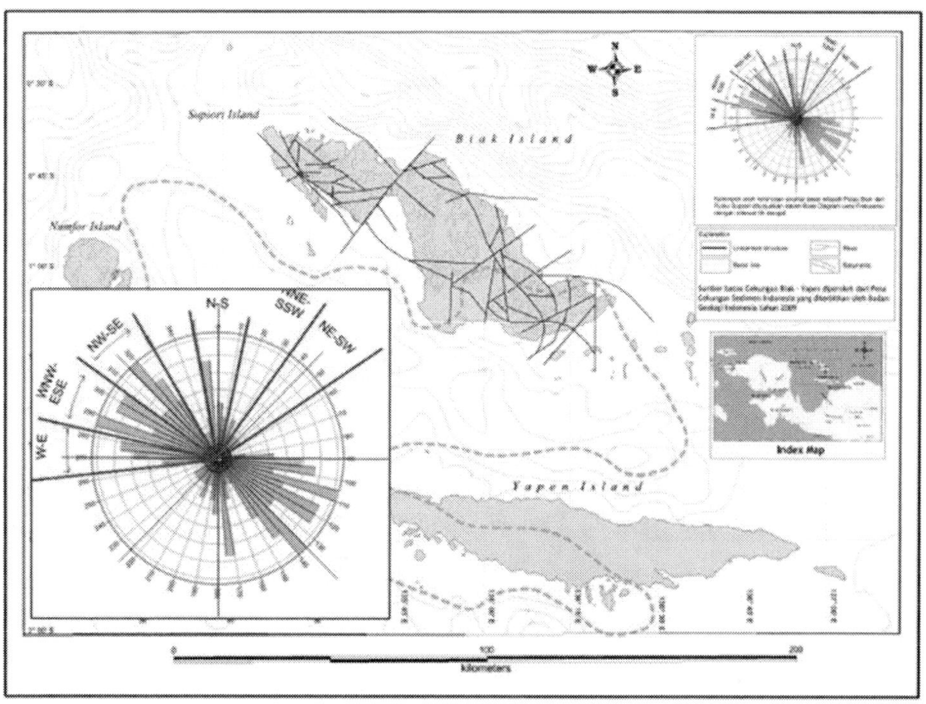

Figure 11. The map of lineament structure direction shown in rosette diagram [14].

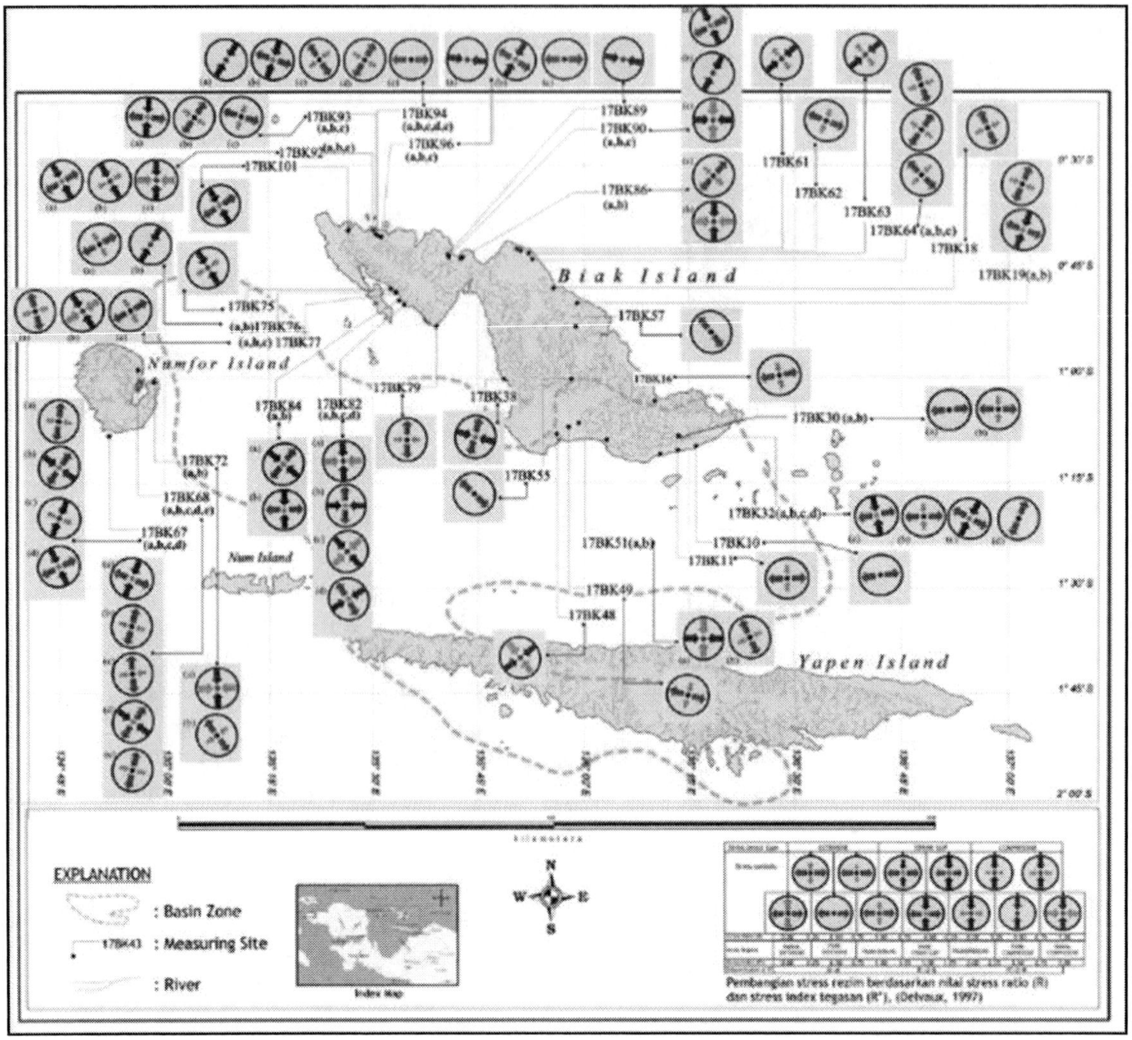

Figure 12. The map of stress regimes location in Biak Island and Supiori Islands [14].

4.4. Rifting and Lifting.

The rifting and lifting analysis was based on the minimum horizontal stress of σ3 relatively horizontal will form an extensive to transtensive regime and the maximum horizontal stress σ1 relatively horizontal will form a compressive or transpressive regime.

4.4.1. Rifting phase directed at NE – SW. The Shmin candidate relatively directed at NE-SW was identified in 7 stations or 194 local structures field with a stress ratio value of 0.75 – 0.25, stress index between 0.75 – 1.75, included in the transtensive regime to strike slip regime [12] and the average value of Shmin from σ3 is formed in 04°/ N42°E and 04°/ N220°E, then it is identified as forming the rifting zone (Figure 13).

This phase is estimated to occur at the age of the final Middle Miocene to the Late Miocene causing the formation of rifting zone in the area of Biak Island and Supiori Island. The structural lineament is generally directed at NW-SE, NNW-SSE forms a transtensive fault to pure extensive and the other is relatively NE-SW, N-S, and W-E strike slip faults. The rifting activity forming subsidence caused an opening of the first phase sub-basins indicated to 9 sub-basins (see Figure 13). The subsidence is simultaneously followed by sedimentation of the carbonate from the Napisendi Formation and Korem

Formation occurring at the final Middle Miocene to Late Miocene and the carbonate sediment of the lower Wardo Formation in the Late Miocene and is widespread on Biak Island. While on Supiori Island, rifting activity also forms a subsidence in the southern and northern parts of the island, and it is also estimated that it is followed by carbonate sediment deposition of Napisendi Formation and Korem Formation.

Table 1. The result of stress regime analysis in the Biak Island and Supiori Island trajectory using the Delvaux concept [14].

STATION	QUANTITY	σ_1		σ_2		σ_3		STRESS MAGNITUDE			STRESS HORIZONTAL		STRESS RATIO (R)	STRESS INDEX (R')	TREND RELATIVE Shmax	STRESS REZIM	TRAJECTORY
	N	°	N°E	°	N°E	°	N°E	σ_1	σ_2	σ_3	Shmax	Shmin					
BK-10	10	50	1	40	171	5	265	Sv	Shmax	Shmin	176	76	0,55	0,55	N-S	Pure Extensive	Biak Island
BK-11	30	82	72	8	243	1	333	Sv	Shmax	Shmin	63	333	1	1	ENE-WSW	Transtensive	Biak Island
BK-16	20	58	112	12	2	29	266	Sv	Shmax	Shmin	176	86	1	1	N-S	Transtensive	Biak Island
BK-18	30	48	87	37	235	16	338	Sv	Shmax	Shmin	68	338	0,97	0,97	ENE-WSW	Transtensive	Biak Island
BK-19A	14	24	24	58	160	20	285	Shmax	Sv	Shmin	21	291	0,3	1,7	NNE-SSW	Pure Strike-Slip	Biak Island
BK-19B	24	75	334	13	124	7	216	Sv	Shmax	Shmin	126	216	0,95	0,95	NW-SE	Transtensive	Biak Island
BK-21A	28	85	45	5	225	0	315	Sv	Shmax	Shmin	45	315	0,95	0,95	NE-SW	Transtensive	Biak Island
BK-21B	24	59	187	31	4	1	95	Sv	Shmax	Shmin	5	95	0,95	0,95	N-S	Transtensive	Biak Island
BK-21C	32	69	144	21	336	4	244	Sv	Shmax	Shmin	153	243	0,5	0,5	NW-SE	Pure Extensive	Biak Island
BK-21D	22	39	46	51	228	1	137	Shmax	Sv	Shmin	47	137	0,5	1,5	NE-SW	Pure Strike-Slip	Biak Island
BK-30A	26	66	353	24	173	0	83	Sv	Shmax	Shmin	173	93	0,5	0,5	N-S	Pure Extensive	Biak Island
BK-30B	12	37	200	43	334	25	90	Shmax	Sv	Shmin	0	90	0,95	1,05	N-S	Transtensive	Biak Island
BK-32A	24	35	356	54	161	7	261	Shmax	Sv	Shmin	172	262	0,62	1,35	N-S	Pure Strike-Slip	Biak Island
BK-32B	12	70	165	19	5	6	273	Sv	Shmax	Shmin	2	272	0,8	0,8	N-S	Transtensive	Biak Island
BK-32C	24	63	204	27	24	0	294	Sv	Shmax	Shmin	24	294	0,5	0,5	NNE-SSW	Pure Strike-Slip	Biak Island
BK-32D	30	70	92	20	280	3	189	Sv	Shmax	Shmin	99	9	0,65	0,65	W-E	Pure Extensive	Biak Island
BK-38	20	31	107	59	289	1	198	Shmax	Sv	Shmin	107	197	0,5	1,5	WNW-ESE	Pure Strike-Slip	Biak Island
BK-48	12	31	57	31	168	43	292	Shmax	Shmin	Sv	24	14	0,95	2,95	NNE-SSW&WNW-ESE	Radial Compressive	Biak Island
BK-50	10	52	72	16	183	34	284	Sv	Shmax	Shmin	14	284	1	1	NNE-SSW	Transtensive	Biak Island
BK-51A	10	40	90	33	327	33	212	Sv	Shmax	Shmin	90	0	0	0	W-E&N-S	Radial Extensive	Biak Island
BK51B	10	46	22	23	266	35	159	Sv	Shmax	Shmin	65	155	0,85	0,85	ENE-WSW	Transtensive	Biak Island
BK-55	32	68	206	22	35	3	304	Sv	Shmax	Shmin	33	303	0,55	0,55	NNE-SSW	Pure Extensive	Biak Island
BK-57	12	14	235	76	46	2	144	Shmax	Sv	Shmin	55	325	0,5	1,5	NE-SW	Pure Strike-Slip	Biak Island
BK-61	12	26	234	31	341	47	122	Shmax	Shmin	Sv	52	142	0,1	2,1	NE-SW	Transpressive	Biak Island
BK-62	24	64	126	6	24	25	291	Sv	Shmax	Shmin	18	288	0,55	0,55	NNE-SSW	Pure Extensive	Biak Island
BK-63	12	25	234	40	347	40	121	Shmax	Shmin	Sv	54	144	0	2	NE-SW	Transpressive	Biak Island
BK-64A	12	60	304	16	63	25	160	Sv	Shmax	Shmin	70	160	1	1	ENE-WSW	Transtensive	Biak Island
BK-64B	10	45	181	18	290	39	36	Sv	Shmax	Shmin	127	37	0,95	1	NW-SE	Transtensive	Biak Island
BK-64C	10	46	258	23	234	35	135	Sv	Shmax	Shmin	41	131	0,83	0,83	NE-SW	Transtensive	Biak Island
BK-76A	14	44	307	33	177	27	67	Sv	Shmax	Shmin	153	63	0,8	0,8	NNW-SSE	Transtensive	Supiori Island
BK-76B	16	25	22	14	285	61	169	Shmax	Shmin	Sv	30	120	0,5	2,5	NNE-SSW	Pure Compressive	Supiori Island
BK-77A	12	50	90	37	245	13	345	Sv	Shmax	Shmin	75	165	1	1	ENE-WSW	Transtensive	Supiori Island
BK-77B	14	39	141	11	42	49	299	Shmax	Shmin	Sv	29	119	1	3	NNE-SSW&WNW-ESE	Radial Compressive	Supiori Island
BK-77C	18	53	165	35	323	11	61	Sv	Shmax	Shmin	151	61	1	1	NNW-SSE	Transtensive	Supiori Island
BK-79A	12	59	184	30	16	5	283	Sv	Shmax	Shmin	13	103	1	1	NNE-SSW	Transtensive	Supiori Island
BK-79B	16	71	217	19	47	3	316	Sv	Shmax	Shmin	46	136	1	1	NE-SW	Transtensive	Supiori Island
BK-82A	22	0	0	10	90	80	270	Shmax	Shmin	Sv	0	90	1	3	N-S&W-E	Radial Compressive	Supiori Island
BK-82B	16	33	3	57	191	4	95	Shmax	Sv	Shmin	4	94	0,5	1,5	N-S	Pure Strike-Slip	Supiori Island
BK-82C	12	16	315	17	50	67	185	Shmax	Shmin	Sv	93	183	1	3	N-S&W-E	Radial Compressive	Supiori Island
BK-82D	18	23	248	57	19	22	148	Shmax	Sv	Shmin	62	152	0,6	1,4	ENE-WSW	Pure Strike-Slip	Supiori Island
BK-84A	12	26	121	43	95	36	323	Shmax	Sv	Shmin	41	131	0,5	1,5	NE-SW	Pure Strike-Slip	Supiori Island
BK-84B	16	2	4	60	271	30	95	Shmax	Sv	Shmin	4	94	0,5	1,5	N-S	Pure Strike-Slip	Supiori Island
BK-85A	12	62	317	28	131	2	222	Sv	Shmax	Shmin	132	42	1	1	NW-SE	Transtensive	Supiori Island
BK-85B	20	11	0	0	270	79	180	Shmax	Shmin	Sv	90	0	1	3	N-S&W-E	Radial Compressive	Supiori Island
BK-89	14	15	100	3	190	75	290	Shmax	Shmin	Sv	98	8	0,5	2,5	W-E	Pure Compressive	Supiori Island
BK-87A	10	34	175	23	69	47	312	Shmax	Shmin	Sv	12	102	0,5	2,5	NNE-SSW	Pure Compressive	Supiori Island
BK-87B	14	12	238	1	148	78	53	Shmax	Shmin	Sv	58	148	0,5	2,5	ENE-WSW	Pure Compressive	Supiori Island
BK-88A	12	43	195	15	90	43	346	Sv	Shmax	Shmin	75	165	0,95	0,95	ENE-WSW	Transtensive	Supiori Island
BK-88B	10	56	311	34	133	1	42	Sv	Shmax	Shmin	132	42	1	1	NW-SE	Transtensive	Supiori Island
BK-90A	16	29	329	60	168	8	64	Shmax	Sv	Shmin	152	242	0,5	1,5	NNW-SSE	Pure Strike-Slip	Supiori Island
BK-90B	16	25	18	25	276	53	147	Sv	Shmax	Shmin	29	119	0,5	2,5	NNE-SSW	Pure Compressive	Supiori Island
BK-90C	12	0	180	5	90	85	270	Shmax	Shmin	Sv	360	270	1	3	N-S&W-E	Radial Compressive	Supiori Island
BK-92A	10	1	338	62	70	28	247	Shmax	Sv	Shmin	158	248	0,5	1,5	NNW-SSE	Pure Strike-Slip	Supiori Island
BK-92B	18	17	153	13	59	68	293	Shmax	Shmin	Sv	154	144	0.15	2,15	NNW-SSE	Transpressive	Supiori Island
BK-92C	12	0	180	5	90	85	270	Shmax	Shmin	Sv	360	270	1	3	N-S&W-E	Radial Compressive	Supiori Island
BK-93A	14	25	356	58	217	18	95	Shmax	Sv	Shmin	1	91	0,5	1,5	N-S	Pure Strike-Slip	Supiori Island
BK-93B	10	83	293	7	127	1	35	Sv	Shmax	Shmin	125	35	1	1	NW-SE	Transtensive	Supiori Island
BK-93C	16	76	9	14	197	2	107	Sv	Shmax	Shmin	17	107	1	1	NNE-SSW	Transtensive	Supiori Island
BK-94A	18	26	46	33	154	46	286	Shmax	Shmin	Sv	35	125	0,5	2,5	NE-SW	Pure Compressive	Supiori Island
BK-94B	14	7	19	55	279	34	114	Shmax	Sv	Shmin	21	111	0,5	1,5	NNE-SSW	Pure Strike-Slip	Supiori Island
BK-94C	14	47	243	41	42	11	141	Sv	Shmax	Shmin	51	141	0,95	0,95	NE-SW	Transtensive	Supiori Island
BK-94D	18	79	322	11	131	2	221	Sv	Shmax	Shmin	131	221	0,95	0,95	NW-SE	Transtensive	Supiori Island
BK-94E	12	42	47	29	287	34	175	Sv	Shmax	Shmin	84	354	0,95	0,95	W-E	Transtensive	Supiori Island
BK-96A	22	2	184	30	275	60	91	Shmax	Shmin	Sv	3	93	0,7	2,7	N-S	Pure Compressive	Supiori Island
BK-96B	14	17	31	56	274	29	131	Shmax	Sv	Shmin	35	125	0,45	1,55	NE-SW	Pure Strike-Slip	Supiori Island
BK-96C	20	45	180	45	0	0	270	Sv	Shmax	Shmin	360	270	0,5	0,5	W-E	Pure Extensive	Supiori Island
BK-101	18	40	310	44	166	19	57	Shmax	Sv	Shmin	143	53	0,65	1,35	NW-SE	Pure Strike-Slip	Supiori Island

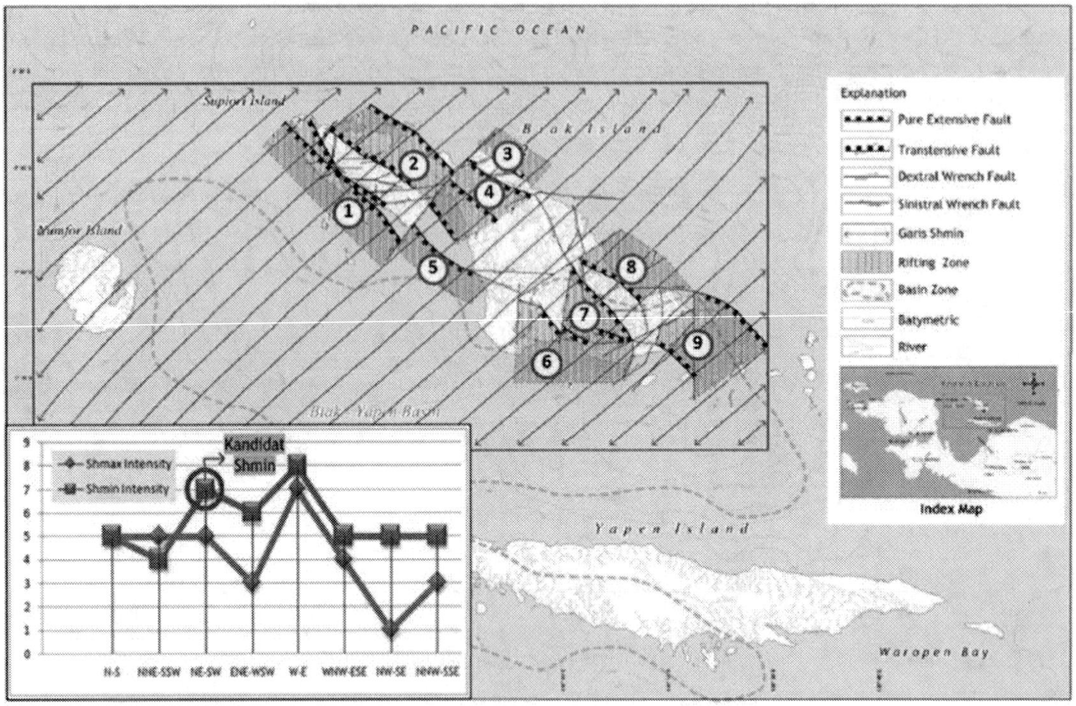

Figure 13. The rifting phase map at the age of the Middle Miocene to Late Miocene, showing the existence of sub-basin openings 1, 2, 3, 4, 5, 6, 7, 8, and 9.

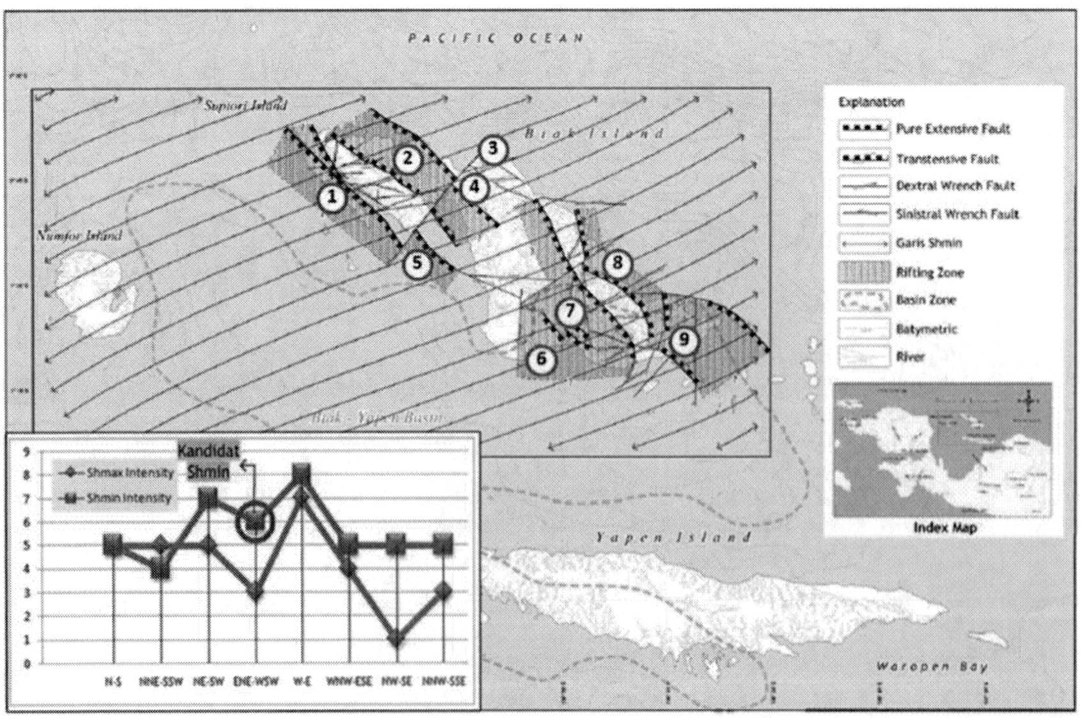

Figure 14. The rifting phase map at the age of the Late Miocene to Pleistocene showing the existence of sub-basin openings 1, 2, 4, 5, 6, 7, 8, and 9.

4.4.2. Rifting phase directed at ENE – WSW. The Shmin candidate relatively directed at ENE-WSW was identified in 6 stations or 132 local structures field with a stress ratio value of 0.75 – 0.25, stress index between 0.75 - 1.75, included in the transtensive regime to strike slip regime [12] and the average value of Shmin from σ3 is formed in 16°/ 62°E and 08°/ N251°E, thus, it is identified as forming the rifting zone (Figure 14).

This phase is estimated to occur at the age of the Late Miocene to Pliocene causing the formation of rifting zone in the area of Biak Island and Supiori Island. The structural lineament is generally directed at NW-SE, NNW-SSE forms a transtensive fault to pure extensive and the other is relatively NE-SW, N-S, and W-E strike slip faults. The rifting activity forming subsidence caused an increase in the opening from the sub-basin of the first phase to the second phase. The indicated phase two continues the first phase can be seen in reactivity of sub-basin deepening 1, 2, 4, 5, 6, 7, 8, 9 (see Figure 14), while sub-basin 3 does not. This rifting was followed by deposition of carbonate sediments from the upper Wardo Formation occurring on the Pliocene and spreading widely on Biak Island to the south. While on Supiori Island, rifting activitiy also forms subsidence occurring in the south and northern onshore parts of the island, and it is also estimated that it is followed by carbonate sediment deposition of the Wardo Formation.

4.4.3. Rifting phase directed at W – E. The Shmin candidate relatively directed at W-E was identified in 8 stations or 185 local structures field with a stress ratio value of 0.75 – 1.00, stress index between 0.75 – 1.25, included in the transtensive regime [12] and the average value of Shmin from σ3 is formed in 09°/ N89°E and 10°/ N269°E, thus, it is identified as forming the rifting zone (Figure 15).

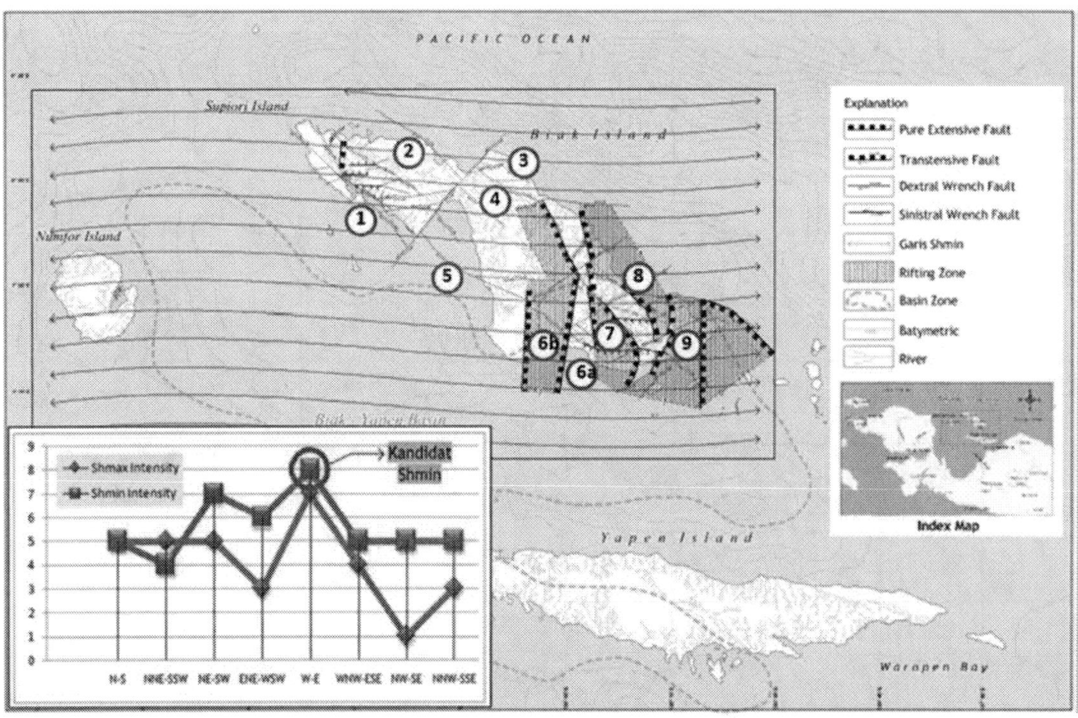

Figure 15. The rifting phase map at the age of the Pleistocene showing the existence of sub-basin openings 6, 7, 8, and 9.

This phase is estimated to occur at the age of Pleistocene causing the formation of rifting zone in the eastern part of Biak Island. The rifting activity forms a subsidence resulting an increase in the opening from the sub-basin of the second phase to the third phase. The indicated phase three continues the phase two can be seen in reactivity of sub-basin deepening 6, 7, 8, 9 (see Figure 15), while sub-basin 1, 2, 3, 4, and 5 do not. The rifting activity in sub-basin is followed by sedimentation of carbonates from the Mokmer Formation that occurs in Pleistocene and is widespread in the eastern part of the island. In Supiori Island, the rifting activity does not occur, so generally the sedimentation of the Pleistocene age is not well developed. This island region is only dominated by reactivity of strike slip faults relatively directed at NW-SE and NE-SW.

4.4.4. Lifting phase directed at W – E. The Shmax candidate relatively directed at W-E was identified in 7 stations or 158 local structures field with a stress ratio value of 0.25 - 1.00, stress index between 2.25 – 3.00, included in the pure compressive to radial compressive regime [12] and the average value of Shmax from $\sigma 1$ is formed in 06°/ N96°E and 03°/ N274°E, thus, it is identified as forming the lifting zone (Figure 16).

This phase is estimated to occur at the age of the Late Pleistocene - Recent causing the formation of a compressive lifting zone in the eastern region of Biak Island. The structure lineament generally forms reactivity of reverse fault relatively N-S and the strike slip faults of relative NW-SE and NE-SW. The compressive lifting activity is indicated to cause lifting and compression in sub-basin 6, 7, 8, and 9 (see Figure 16), while in sub-basin 1, 2, 3, 4, and 5 did not experience full compression, but experienced transpressional, presenting the uplifting of the Late Miocene to Recent Formation as in the Korem Formation, Wardo Formation and Mokmer Formation becoming the mainland on Biak Island. In Supiori Island, the activities of the compressive lifting do not occur significantly so that the Korem Formation, Wardo Formation and Mokmer Formation is thought to be located on the offshore section mainly in the southern part and a small part in the north.

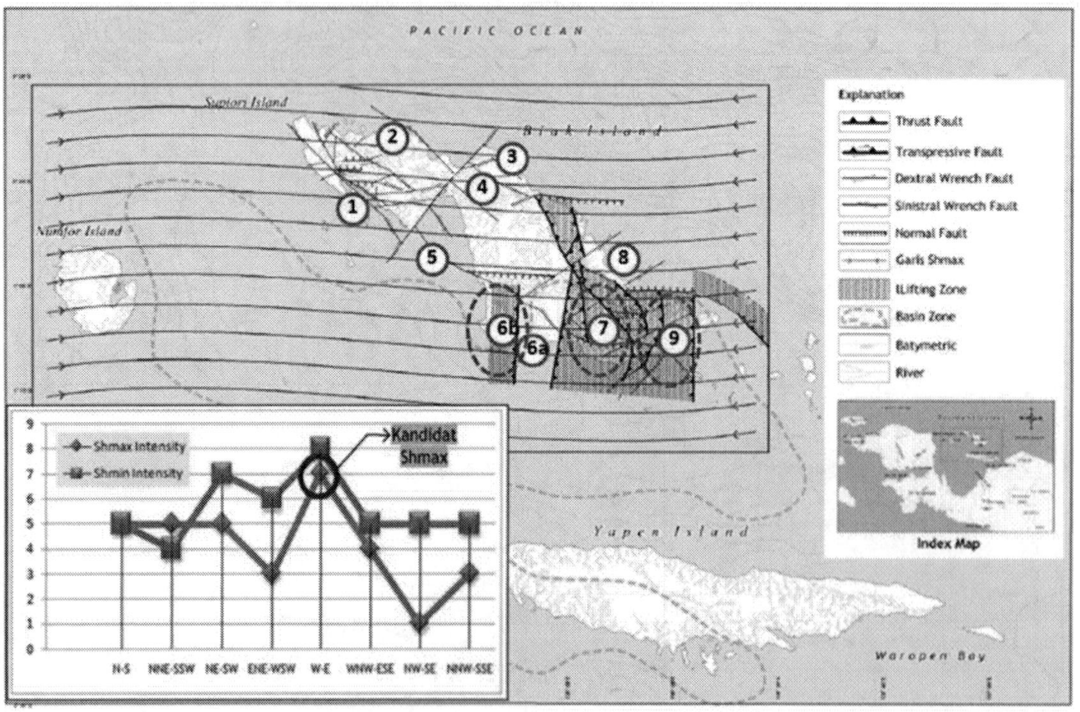

Figure 16. The compressional lifting phase map at the age of the Late Pleistocene to Recent showing the existence of sub-basin openings 6, 7, 8, and 9.

5. Conclusion

Field observations on Neogene-aged rocks obtained shows high structural intensity. The calculation of geological structure contained in these rocks based on the Delvaux and Zoback results the Biak Island and Supiori Islands underwent four tectonic phases. Rifting activities consisted of 3 phases, namely phase 1 at the Middle Miocene to Late Miocene, phase 2 at the Late Miocene to the Pliocene, and phase 3 at the Pleistocene. While the lifting phase occurred within the Pleistocene to the Recent. Rifting and lifting phase activities potentially to create structural geology which become trap zone for the hydrocarbon presence in the subsurface.

Of the three rifting phases in the Biak Island and Supiori Islands, it produced 9 sub-basins where sub-basins 6,7, 8, 9 experienced a deepening reactivity that occured as many as 3 times from the Middle Miocene to Pliocene in the eastern and south-eastern parts of the Islands Biak. Sub-basin 6, 7, 8, 9 is predicted having a fairly good sediment thickness in the southern, especially offshore.

6. Recommendation

Sub-basin areas 6, 7, 8, 9, are interesting enough to be followed up to further exploration stages to find out the potential sources of oil and gas, especially in the offshore (Figure 17). The existence of sediment thickness is quite good because it is supported by 3 times rifting activities. At the subsurface, the 3 times rifting activity will produce an increased rock pressure and also temperature as well, especially in rock formations aged Eocene, Oligocene to Early Miocene, then further studies is needed in the form of subsurface analysis in detail with the seismic cross-section method and is correlated to stratigraphic in the surface.

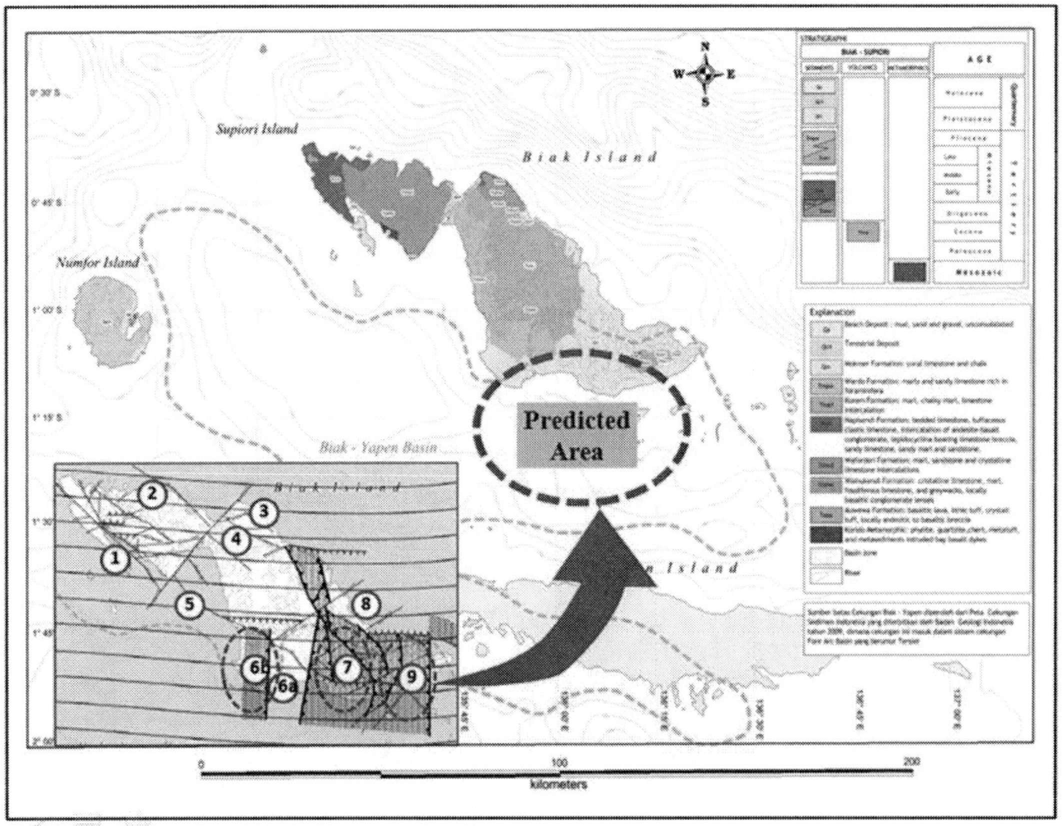

Figure 17. The map of increasing temperature and pressure based three rifting phase and a lifting phase once in the Biak Island.

Re-evaluation of oil and gas blocks presence around the study area is needed because this is necessary for adjustment to preliminary results of research on the Biak Island and Supiori Island trajectory as a target area of oil and gas potentials (Figure 18).

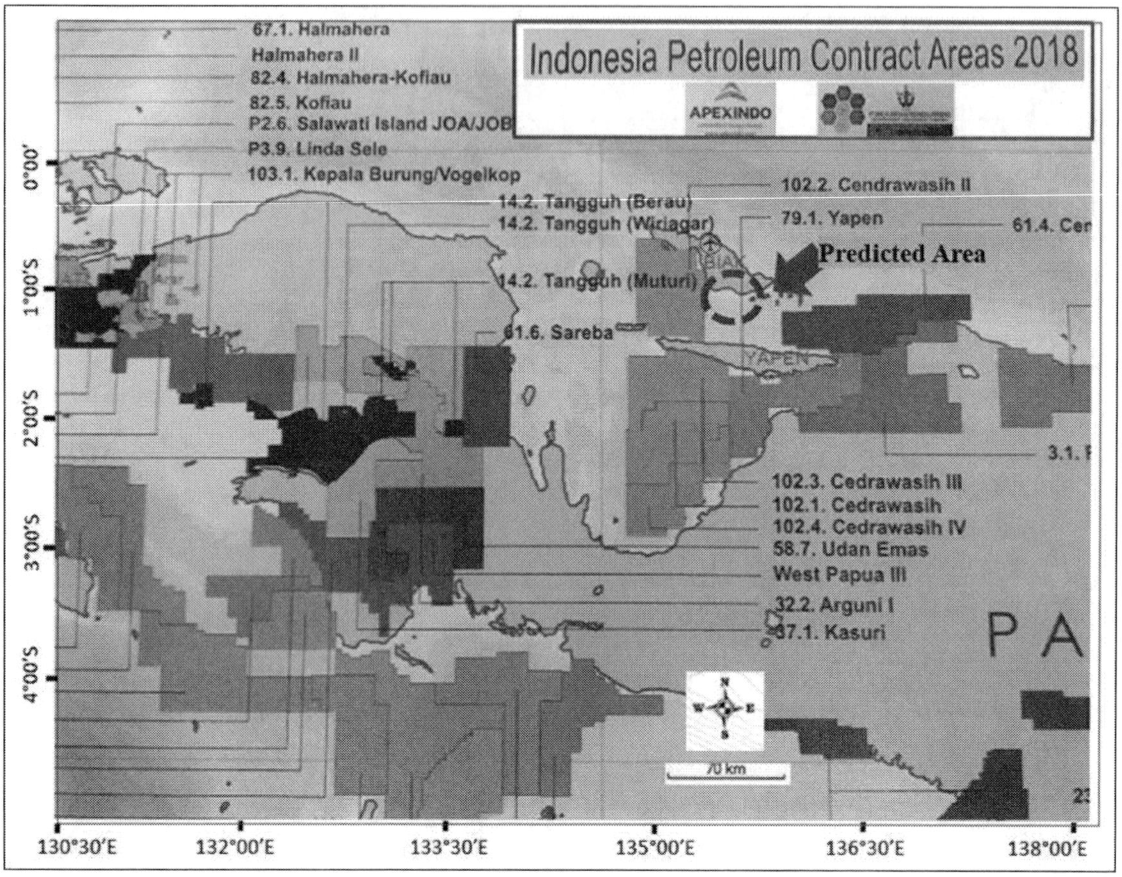

Figure 18. The evaluation map of oil and gas locations in the form of pressure and temperature increasing in the Biak Island.

Acknowledgments

We would like to acknowledge the support from Geological Survey Center, Bandung. We also wish thanks to Biak-Yapen *WK-Migas* 2017 team for the cooperation during fieldwork in the Biak and Supiori Island.

References

[1] Dewey, J. F., and Bird, J. M., 1970, Mountain Belts and The New Global Tectonics, *J. Geophys. Research* **75** p 2625-2647 (Journal)

[2] Milsom, J., Masson, D., Nichols, G., Sikumbang, N., Dwiyanto, B., Parson, L., and Kallagher, H., 1992, The Manokwari Trough and The Western End of The New Guinea Trench, *Tectonics* **11** p 145-153 (Journal)

[3] Hall, R., 2002, Cenozoic Geological and Plate Tectonic Evolution of SE Asia and The SW Pacific: Computer-based Reconstruction Model and Animations, *J. Asian Earth Sci* **20** p 353-431 (Journal)

[4] Cloos, M., Sapiie, B., Ufford, A. Q.,Weiland, R. J., Warren, P. Q., and McMahon, T., 2005, Collisional Delamination in New Guinea: The Geotectonics of Subducting Slab Breakoff.

The Geological Soc. of America Spec.Pap. p 400 (Journal)

[5] Hill, K. C., and Hall, R., 2003, Mesozoic-Cenozoic Evolution of Australia's New Guinea Margin in a West Pacific Context, *Geol. Soc. Australia Spec. Publ.* **22**, and *Geol. Soc. America Spec. Pap.* **372** p 265-290 (Journal)

[6] Gold, D. P., Hall, R., Burgess, P., and White, L., 2014, Neogene Structural History of Biak and The Biak Basin, Eastern Indonesia, *South East Asia Research Group Poster* (Journal)

[7] Anonim, 2009, Indonesia Sedimentary Basin Map, scale 1: 5,000,000, Badan Geologi, Bandung (Journal)

[8] Masria, M., Ratman, N., and Suwitodirdjo, 1981, Geological Map of Biak, Irian Jaya, Scale 1:250.000, *Center for Geological Research and Development*, Bandung (Map)

[9] Alviyanda, Junursyah, G. M. L., and Saragih, R, Y., 2019, Reservoir Potential of Neogene Rocks in the Biak Island and Its Surrounding Areas, *Indon. J. of Sedimentary Geol*, **42** ed Minarwan and Darman H, p 33-42 (Journal)

[10] Panjaitan, S., and Subagio, 2015, Prospect Of Energy Resources Based On Analysis Of Gravity Anomaly Pattern at Biak And Its Surrounding, Papua, *Jurnal Geologi Kelautan* vol **13** p 87 (Journal)

[11] Junursyah, G. M. L., 2013, Final Rreport of Magnetotelluric Survei in Biak-Yapen and surrounding area, Papua, *Center for Geological Research and Development,* Bandung (Book)

[12] Delvaux, D., Moeys, R., and Stapel, G., 1997, Paleostress Reconstructions and Geodynamics of The Baikal Region, Central Asia, Part II : Cenozoic Tectonic Stress and Fault Kinematics, *Tectonophysics* (Journal)

[13] Zoback, M., 2007, The Tectonic Stress Field, *In Reservoir Geomechanics,* Cambridge University Press p 3-26 (Book)

[14] Saragih, R. Y., Junursyah, G. M. L., Badaruddin, F., Alviyanda, 2018, Structural Trap Modelling of The Biak-Yapen Basin as a Neogene Frontier Basin in North Papua, *Indonesian Petroleum Association Proc* 42nd, Jakarta (Journal)

The Genesis of Gemstone based on Pillow Lava Primary Structure: A Case Study from Kumbang Formation, Limbasari

Huzaely Latief Sunan[1,4], Siswandi[1], Akhmad Khahlil Gibran[1], Maulana Rizki Aditama[1], Rahmat Yantono Saragih[2], Widhiatmoko Herry Purnomo[3], FX Anjar Tri Laksono[1]

[1]Geological Engineering Department, Universitas Jenderal Soedirman, Indonesia
[2]Geological Survey Center, Indonesia
[3]Electrical Engineering Department, Universitas Jenderal Soedirman, Indonesia

[4]huzaely.sunan@unsoed.ac.id

Abstract. Gemstone/Jasper is an ornamental rock included in the quartz (SiO2) microcrystalline chalcedony variety. Jasper is translucent, has a microcrystal structure, and consists of several mixes of other minerals resulting in appealing color and patterns. In general, Jasper forms as a result of hydrothermal activity controlled by geological structure (secondary structure). On the other hand, Jasper is formed in Limbasari, controlled by a primary structure on pillow lava. The methods used are field data collection, petrography, and XRF (X-Ray Fluorescence). From the data collected in the field and results from petrography analysis and genesis geochemistry. We can infer that the Jasper found in the research site can be called Peperite. The jasper there is a result of unconsolidated tuff deposits still in the form of wet deposits rolled up by the movement of pillow lava and sealed in the gaps of the pillow lava primary structure, starting a partial melting process, and then cooled swiftly, resulting in glassy texture.

1. Introduction

The term 'peperino' is used by [1] to describe clastic rocks from central France which consisted of a mixture of limestone and basalt lacustrine. [2] interpreted the rocks as having originated by a 'violent and intimate union of volcanic fragmentary matter with limestone while yet in a soft state'. Gemstone or commonly known as Jasper lately become one of the highly sought-after gemstones. As one of the industrial minerals for determining the prospect area, the geological research is required because of the genesis controlled by geological processes. The initial outcrop evidence such as fractures, tensions, and breccias, indicate the structural geology involvement [3]. [4] research revealed the genesis of gemstones on the mountain range on the southern part of java in Pacitan is the result of a hydrothermal process, supergent enrichment, and being an ancient river deposit. Typically gemstone can be found spread out at the tertiary volcanic area, an area with complex geological structures, and alteration, especially at the Serayu Utara mountain range. Magmatic activity occurred along the volcanic arc and parallel to the tertiary subduction zone of the island of Java. Magmatic activity that occurred along the volcanic arc and parallel to the tertiary subduction zone of the island of Java is a potential area for the formation of ore or non-ore mineral deposits. The formation of gemstone in this zone has a high possibility of controlled by tertiary magmatic activity. Tectonism that happened as a result of a collision between oceanic crust and Jawa island arc on Oligo-Miocene resulted in magmatism and volcanic activities in this arc with the marine-land environment. This research aims to gain clarity about the genesis of gemstone in Kumbang formation, Limbasari. (Fig. 1)

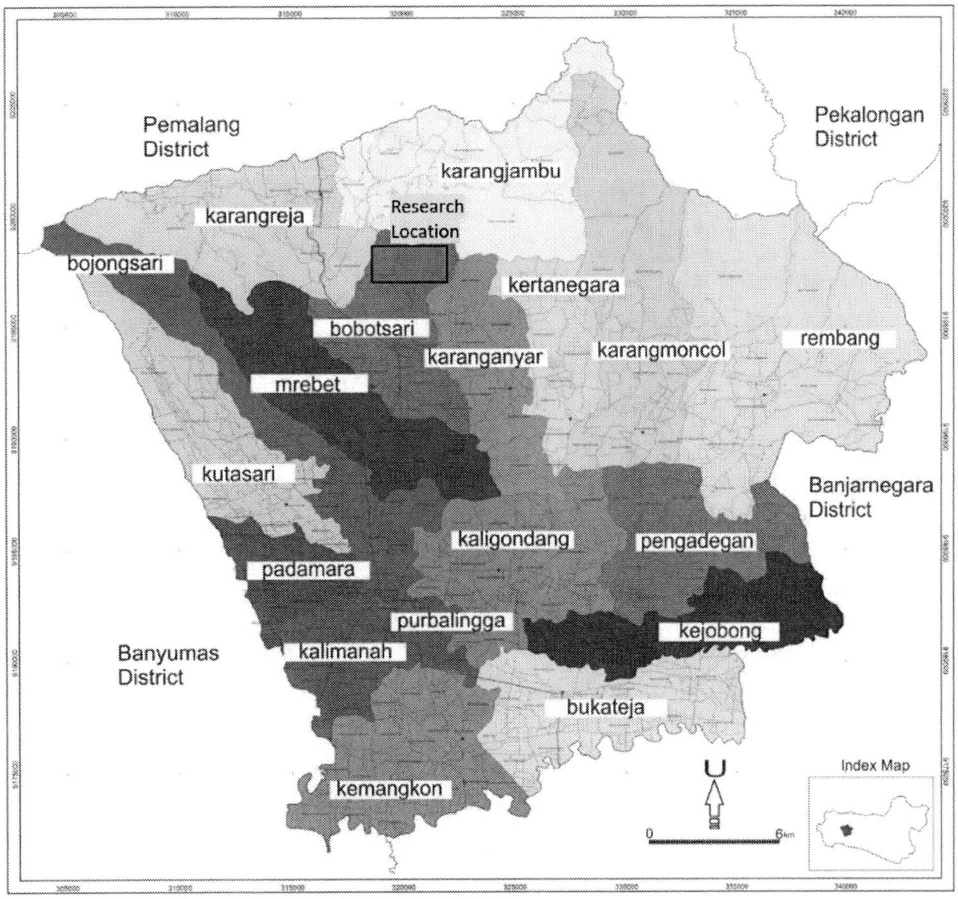

Figure 1. Research Location [5]

Stratigraphy of the area based upon [6] composed of Tertiary-Quaternary rock, which consists of Pemali Formation, Rambatan Formation, Halang Formation, Rock Intrusion, Kumbang Formation, Tapak Formation, Kalibiuk Formation, Ligung Formation, Mengger Formation, Gintung Formation, Linggopodo Formation, terrace deposit, Quaternary Volcanic Sediment, Andesitic Lava dan and the youngest are Slamet mountain laharic deposit and alluvial deposits. The research conducted in this area is focused only on Kumbang formation Gemstone/Jasper. The lower part of the formation consists of breccias with an angular fragment and an andesitic lava deposit. The upper part consists of alternating layers between tuff and breccias and tuff sandstone. This formation is middle-late Miocene in age [7] and has a thickness of 2000 m with a good outcrop at Mount Kumbang.

2. Methods

This research is conducted with field observation and laboratory/studio approach. The approach in the field is carried out to retrieve geologic information of the field (location, lithology, stratigraphy, structure, and rock structures). Afterward, petrography analysis is carried out in Geological Engineering UNSOED, and X-RD analysis.

IOP Conf. Series: Materials Science and Engineering **982** (2020) 012043 doi:10.1088/1757-899X/982/1/012043

3. Result and Interpretation

The spreading of gemstones or jasper in the research site, in general, is highly related to the geologic process happening in the past/present in the area. It is showed that the jasper reserve in the area is an in-situ rock that associates with pillow lava (Fig. 2).

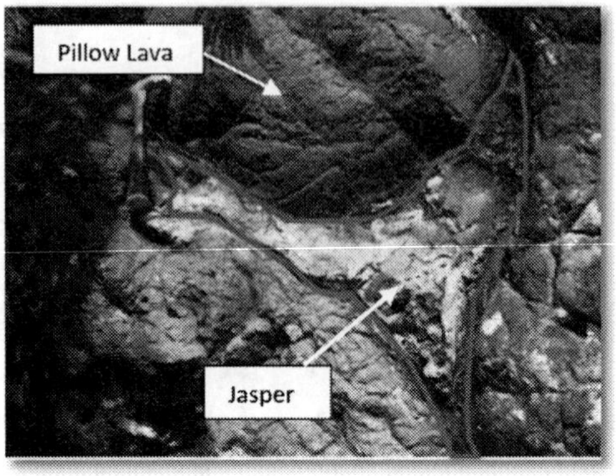

Figure 2. Reserve of gemstones/jasper as an in-situ rock on the gaps of pillow lava primary structure.

3.1 Characteristic of gemstones or jasper

Gemstone or green jasper inside of pillow lava can be found in the field as an insertion between gaps in spheres of pillow lava. In general, it takes a form as an irregular pattern following the gaps between the spheres of pillow lava (Fig. 2). A few meters from the location of the gemstone/jasper reserve found an outcrop resembling a breccia with basalt igneous rock fragment and a matrix of whitish mudrock, with a hardness level comparable to the gemstone/jasper, described as peperite (Fig. 3). Pepperite is a rock naming determination according to the genesis of the rock, which is the process of mixing between a magma intrusion or a lava flow with an unconsolidated deposit or weakly consolidated deposit [8].

Figure 3. Pepperite outcrop as an association for the gemstones/jasper in the research site.

3.2 Petrography Characteristic

Petrography analysis is done to discern the mineralogy characteristic of the research site gemstone/jasper. Petrography observation result of the gemstone/jasper (Fig. 4). Thin section SM-04 consists of 5% white-colored or transparent quartz with low relief and 3% black opaque minerals. Embed on 77% brownish glass fragment matrix, transformed into chlorite. In petrography known as vitric tuff [9]. Thin section SPT 09 consists of 2 % of black opaque minerals and 2 % of low relief quartz. Embed on 94% of fine glass fragment matrix transformed into chlorite clay minerals. In petrography known as vitric tuff [10].

IOP Conf. Series: Materials Science and Engineering **982** (2020) 012043 doi:10.1088/1757-899X/982/1/012043

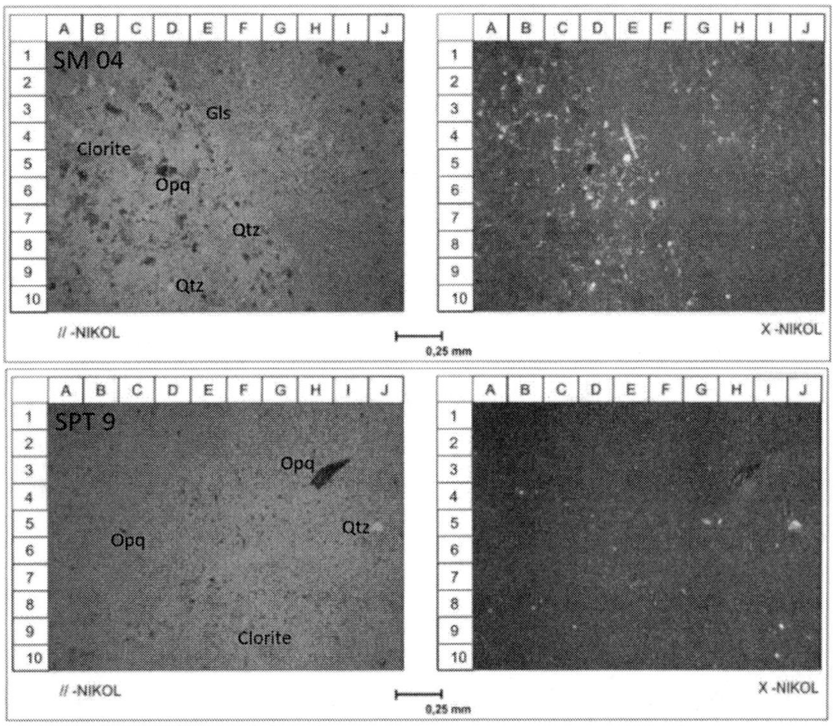

Figure 4. thin section of the petrography samples. In the sample SM04 the chlorite, quartz, Opaque, Glass are seen. While in the SPT9 sample there are Opaque, quartz, and chlorite.

3.3 Geochemistry Analysis

Geochemistry analysis was done through the XRF method to comprehend the chemical element content of gemstone/jasper as an altered rock. Pillow lava acts as the alteration medium and tuff has not been altered into becoming a gemstone/jasper.

Table 1. XRF analysis result of gemstone/jasper (SYJ01), basalt (SYB02), and tuff (SYL03)

No	Sample code	SYJ01	SYB02	SYL03
1	SiO_2	83	50.1	72.4
2	TiO_2	0.21	1.41	0.43
3	Al_2O_3	5.44	17.9	9.86
4	Fe_3O_2	5.56	13.9	5.19
5	MgO	1.19	3.11	0.95
6	MnO	0.02	0.25	0.05
7	CaO	1.19	11.5	4.69
8	K_2O	3.32	1.27	6.09
9	P_2O_5	0	0.22	0
10	TOTAL	99.93	99.66	99.63

The result of the analysis showed a difference in silica compound (SiO2) content on the unaltered to become jasper tuff (SYL03) and gemstone/jasper (SYJ01). It indicates a silicification process as a result of the interaction between tuff and pillow lava (SYB02). The green color found in gemstone/jasper is because of the dominant existence of chlorite minerals [11].

4. Discussion

Gemstone/jasper genesis and primary structure as the controller

In megascopic, gemstone/jasper on the research site has a dark green color but in some places, a gradation from the inside to the outside respectively showed the dark green color to the yellowish-green. (Fig. 5)

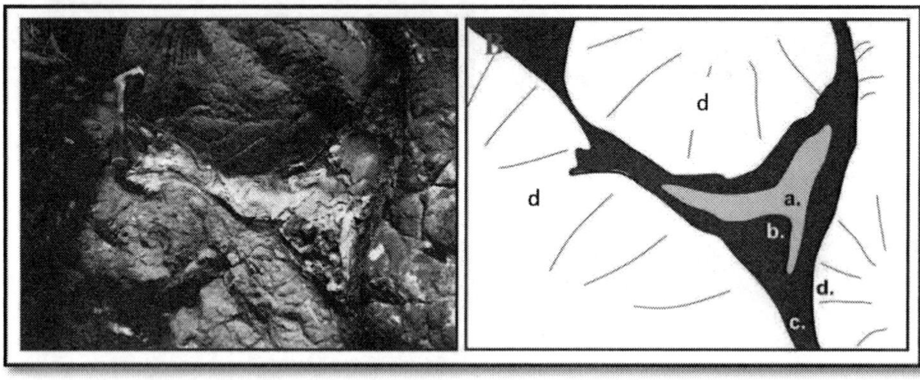

Figure 5. (A) Picture of gemstone/jasper controlled by primary structure outcrop taken on the research site. (B) Sketch showing changing color gradation a). Yellowish-green color part of jasper indicating the imperfect maturity of it. b). Dark green color part of the jasper indicating it is mature enough. c). Interpreted as a backing effect with a reddish-brown color, d). Pillow lava.

[10] mentioned that Jasper is the metamorphic process within the lower degree, and its explained as the product of greenschist metamorphism from metasomatism and contact metamorphism. We concur with [11], that the term peperite is best used in a genetic sense. [12] defined the term as follows: Peperite (n): a genetic term applied to a rock formed essentially in situ by the disintegration of magma intruding and mingling with unconsolidated or poorly consolidated, typically wet sediments. The term also refers to similar mixtures generated by the same processes operating at the contacts of lavas and other hot volcaniclastic deposits with such sediments. Evidence of unconsolidated tuff still wet deposit contact with basaltic lava, shown in the field on pepertite breccias outcrop as an unaltered deposit located in the gaps and joints igneous rock. (Fig. 6).

IOP Conf. Series: Materials Science and Engineering **982** (2020) 012043 doi:10.1088/1757-899X/982/1/012043

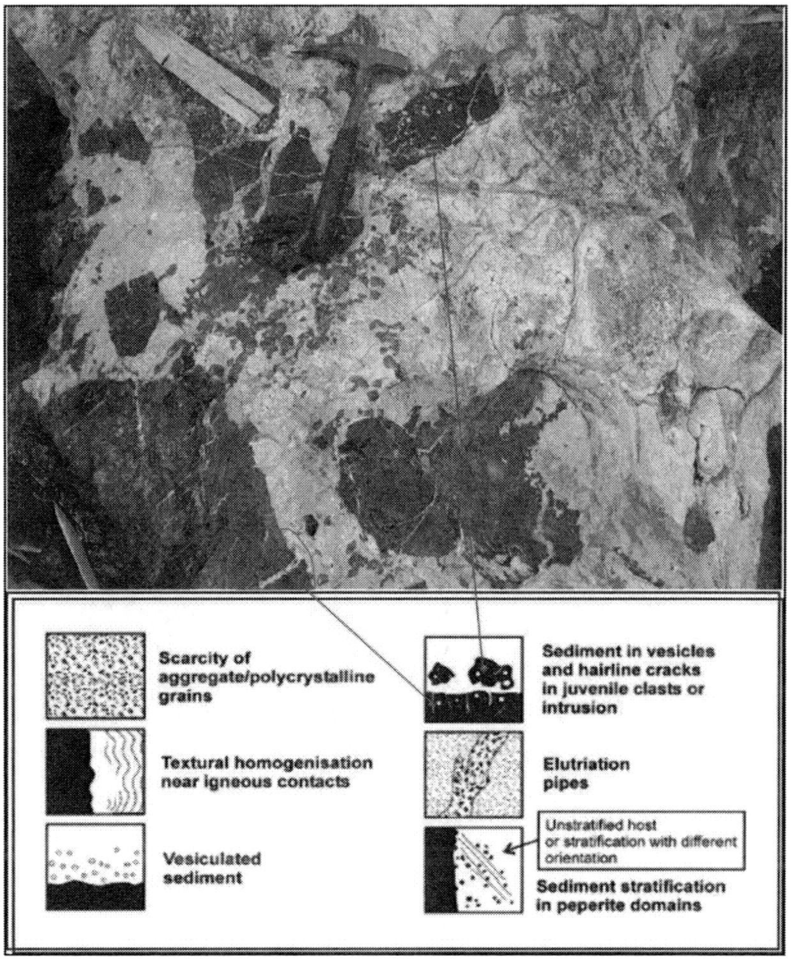

Figure 6. Shown evidence in the field of unconsolidated or weakly consolidated deposit upon contact with lava,(A) tuff deposit turning into igneous rock fragment. (B) tuff deposit in the gaps of igneous rock [8].

According to the appearance data and gemstone/jasper association in the field backed by data from petrology, petrography, structural geology, and geochemistry analysis aspects, it is concluded that genesis process of gemstone/jasper in the research site is highly related to volcanism at that time, as described in peperite formation terminology. As a result, gemstone/jasper in the site as seen from the genesis itself can be called peperite. The original characteristic of it is unconsolidated tuff deposits still in the form of wet deposits, contact with lava, rolled up and sealed in the gaps of the pillow lava primary structure, starting a partial melting process, and then cooled swiftly, resulting in glassy texture (Fig. 7). The role of structural geology in this is the primary structure on pillow lava as a medium for the formation of peperite in the Kumbang Formation in Limbasari.

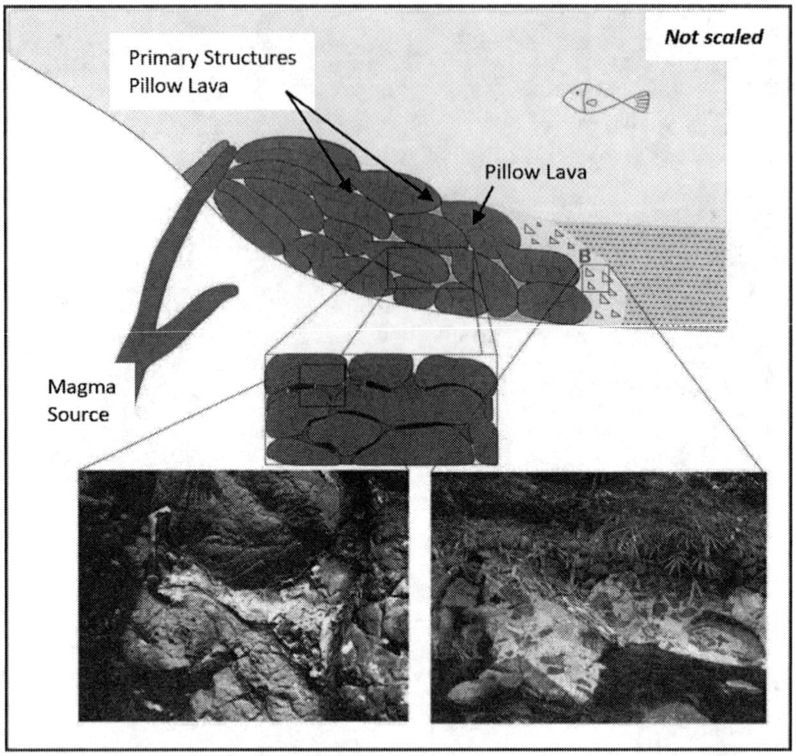

Figure 7. Formation of peperite controlled by primary structure modeling, (A) wet deposits rolled up by the movement of pillow lava and sealed in the gaps of the pillow lava primary structure, starting a partial melting process, (B) peperite formed by fragmentation of pillow lava as a result of some processes one of them is fragmentation because of cooling and autobreccia by a mechanical process [8].

5. Conclusion

Pillow lava's primary structure becomes one of the controlling factors in the formation process in the research site. Kumbang formation gemstones/jasper found in Limbasari can be called as peperite, as a result of unconsolidated tuff deposits still in the form of wet deposits rolled up by the movement of pillow lava and sealed in the gaps of the pillow lava primary structure, starting a partial melting process, and then cooled swiftly, resulting in glassy texture.

References

[1] Scrope G P 1827 *Memoir on the Geology of Central France Including the Volcanic Formations of Auvergne the Velay and the Vivarais* (London: Longman Rees Orme Brown and Green) p 79

[2] Scrope G P 1858 *The Geology and Extinct Volcanoes of Central France* (London: John Murray) p 258.

[3] Sunan H L, Pratomo S U 2020 *Potential Gold Deposits Associated with the Geological Structure of the East Lio Region Ende of East Nusa Tenggara* Dinamika Rekayasa J. DRJ16 (2020)001

[4] Ansori C, Defry H, Edi H 2007 *Proses Pembentukan Batumulia Pada Rangkaian Pegunungan Selatan Jawa Di Pacitan* Prosiding Seminar Geokronologi Kontribusi Ilmu Kebumian DalamPembangunan Berkelanjutan **3** L12

[5] Laksono F X A T, Purwanto B S, Indra P 2020 *Peningkatan Nilai Tambah Curug Bandung Desa Sumingkir Kecamatan Kutasari Kabupaten Purbalingga sebagai Destinasi Geowisata Baru* Dinamika J. DJ02(2020)001

[6] Djuri M 1996 *Peta Geologi Lembar Purwokerto dan Tegal, Jawa* (Jakarta: Geological Agency of the Ministry of Energy and Mineral Resources) p 16

[7] Mcphie J, Doyle M G and Allen R 1993 *Volcanic Textures: A guide to the interpretation of textures in volcanic rocks* (Hobarth: CODES Key Centre,University of Tasmania) p 196

[8] Schmid R 1981 *Descriptive Nomenclature And Classification Of Pyroclstic Deposits And Fragments: Recommendations Of The IUGS Subcommission On The Systematics Of Igneous Rocks Geology* (New York: The Geological Society of America) p 18

[9] Kievlenko E Y A 1980 *Prospecting and Evaluation of Deposits of Gem and Decorative Stones* (Moscow: Nedra) p 34.

[10] Saggerson E P, Turner L M 1982 *General Comments On The Identification Of Chlorites In Thin Section* (Durban: Department Of Geology) p 17

[11] Brooks E R 1995 *Palaeozoic£uidization, folding and peperite formation, northern Sierra Nevada, California* Canadian Journal of Earth Sciences, 32(3), pp.314-324.

[12] White J D L, McPhie J, Skilling I P 2000 *Peperite: a useful genetic term* Bull. Volcanol. BV62(2000)065

IOP Conf. Series: Materials Science and Engineering **982** (2020) 012044 doi:10.1088/1757-899X/982/1/012044

Limestone Facies Change of Jonggrangan To Sentolo Formation in The Western Part of Yogyakarta-Central Java Basin

Rachmad Setijadi, Asmoro Widagdo, and Januar Aziz Zaenurrohman[1)]
Department of Geological Engineering, Universitas Jenderal Soedirman, Purwokerto
e-mail :rachmad.setijadi@unsoed.ac.id

Abstract. The research location is in the western of Yogyakarta-Central Java Basin, where limestone of Jonggrangan and Sentolo Formation are found with the same age but have different depositional facies characteristics. These differences will be interesting to study the historical process of its formation.The research was conducted through a series of methods. Starting from geological mapping to determine the order of rock stratigraphy, the depiction of outcrop profiles, and limestone sampling for petrographic analysis. Eventually, the depositional environment of the two rock formations was discovered. Field observation results and laboratory analysis were further integrated to determine the rock facies that make up the two formations.The rock units in the study area from old to young are the pyroxene andesite breccia unit of Upper Oligocene age and limestone. The bioclastic limestone and layered limestone units are deposited by interfingering, unconformity over the pyroxene andesite breccia unit. The Middle Miocene bioclastic limestone unit is equivalent to Jonggrangan Formation in the inner shelf environment (Restricted Lagoon and Open Lagoon facies environment). Whereas layered limestone of Middle Miocene age is equivalent to the Sentolo Formation in the lower slope environment (Restricted Lagoon, Open Lagoon, and Open Sea Shelf) facies.
Keywords: Yogyakarta basin, limestone, facies, diagenesis, petrography.

1. Introduction

The study area is in the Kalibawang area, the western part of Yogyakarta (Figure 1). In this area limestone sedimentary rocks are found above volcanic rocks. These rocks are the Jonggrangan and Sentolo Formations which are Middle Miocene aged rocks [1]. [2] describe the stratigraphy of the Jonggrangan-Sentolo Formation which are in the Middle Miocene to the Pliocene age and have an interfingering relationship between the two.

[3] researched the facies and depositional environment of the Jonggrangan Formation in the south of the study area. [4] have conducted biostratigraphic studies and facies of the Sentolo Formation by comparing locations to the west and east of the Progo River. [5] conducted a study on the development of limestone sedimentology based on petrographic data on the Sentolo Formation in the Pengasih area. [5] conducted limestone sedimentology research in the Jonggrangan Formation along the Kiskendo Cave pass in the Girimulyo area.

Conducted a study of the contact limits of the Old Andesite Formation and Sentolo Formation in the north of the study area. Conducted a biostratigraphic study of the planktonic foraminifera and sedimentation environment of the Sentolo Formation along the Niten River, Serang River and Ngramang Village, KulonProgo [6].

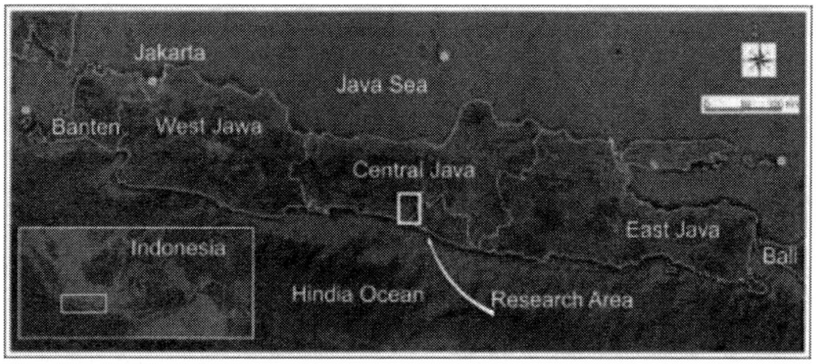

Figure 1. Location of the study area

The Jonggrangan limestone formation is mainly a massive bioclastic limestone while the Sentolo Formation is a layered bioclastic limestone. Parts of the Sentolo Formation are locally found with coral limestone, showing the same age as the Jonggrangan Formation, but in some places, the Sentolo Formation is younger [7].

2. Regional stratigraphy

[4] states that volcanic groups or the Old Andesite Formation [8] or in the eastern part of Gajah Mountain [9] in the western part of Yogyakarta are covered unconformably by shallow marine deposits of the Jonggrangan and Sentolo Formation. [1] stated that the first carbonate sedimentation occurred at the top of the Early Miocene and occupied the highlands to produce the Sentolo and Jonggrangan Formations which were of early of Middle Miocene age. Reefs are built on the heights of former volcanic bodies. [10] conducted research aimed at identifying and investigating coral fossils from the Jonggrangan Formation and analyzing the role of paleoecology in the structure of ancient coral reefs containing layered limestone and reef limestone.

On top of the Old Andesite Formation, besides the Jonggrangan Formation, it was also deposited unconformably with the Sentolo Formation. The relationship between the Sentolo Formation and the Jonggrangan Formation is interfingering. The Sentolo Formation consists of limestone and calcareous limestone. The lower part consists of conglomerates stacked by calcareous tuff with a tuff insertion. This rock upward gradually turns into a fine layered limestone rich in foraminifera.

3. Research method

The research methods used include data collection by mapping the distribution of limestone, measuring stratigraphic detail in several pathways, taking rock samples, and conducting the petrographic analysis. Mapping is carried out on the distribution of the Jonggrangan and Sentolo Formation, both of which are separated by Old Andesite Formation rocks or East Gunung Gajah rocks which are stratigraphically older. Limestone classifications are used according to [11]. The facies obtained will be compared with the carbonate facies model [12]. Some rock samples were collected for paleontological analysis (10 samples) and petrography (10 samples) in the laboratory. Field observations and laboratory analyses are then integrated to find out the rock facies that make up the two formations.

4. Research stratigraphy

4.1. Reef limestone unit

Based on field observations, the compiler of this unit has the characteristics of bioclastic limestone, reefs, and crystalline, reddish-white. Some areas are clastic, medium-coarse sand, porosity is quite good because of the dissolution process. At the bottom, there are limestone fragments of igneous gravel, there are many foraminiferas, shell fragments, elongated algae, and calcite crystals. The thickness of this unit based on the different height contours from the topographic map is ± 325m.
Age determination based on planktonic fossils in this limestone unit, Orbulina Universa (D'Orbigny), the age for this unit is not older than N 9 [13] or Middle Miocene. In a petrographic incision found large foraminifera of Lepidocyclina. This foram has an age range from the Upper Oligocene to Middle Miocene [14]. Based on these data, this unit is interpreted to be deposited not older than N 9 and not younger than N 14, and the age of this unit is equivalent to the Jonggrangan Formation with the age of the Middle Miocene. Determination of the depositional environment of this unit using the calculation P / B ratio obtained values of 0% - 5.88%. From these data, the unit's settling environment is in the inner shelf environment [15].

4.2. Layered Limestone Unit
Based on field observations the compiler of this unit has the characteristics of either layered limestone between calcarenite (dominant) and calculative. The general color is gray-brown - dark brown, bioclastic, compact, parallel lamination and burrows sedimentary structures, good sorting, good porosity due to the dissolution process, there are many foraminiferas, broken mollusk shells. The thickness of this unit based on the geological cross-section reconstruction is ± 330 m.
The micropaleontological analysis shows fossils of planktonic foraminifera, benthonic foraminifera, and large foraminifera in these rock units. Planktonic index fossils, as interval zones for determining depositional age, are present in layered limestone units taken from locations at Location F and Location H ie Globorotalia praemenardii (D'Orbigny) as the beginning of the deposition process, Globorotalia peripheroronda (Blow & Banner) as the end from the deposition process. From the fossil data so that the age for this unit is not older than N 10 and not younger than N 10 [13] or Middle Miocene age. This unit's age is equivalent to the Sentolo Formation with the age of the Middle Miocene.

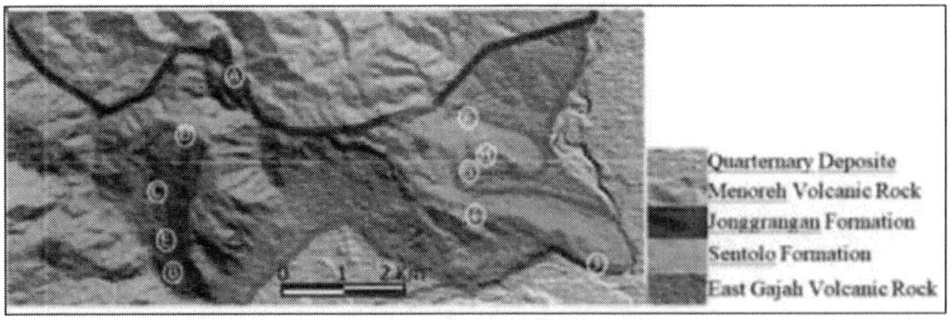

Figure2. Geological map and location for taking rock samples.

Determination of the depositional environment of this unit using the calculation P / B ratio obtained values of 89.93% - 93.92%. From these data, the unit settling environment is in the lower slope environment [15]. A layered limestone unit is a unit that is deposited together (interfingering) with a bioclastic limestone unit deposited in the study area.

5. Batugamping faces

IOP Conf. Series: Materials Science and Engineering **982** (2020) 012044 doi:10.1088/1757-899X/982/1/012044

5.1. Bioclastic limestone facies

5.1.1. Location A
The results of the thin-section analysis of bioclastic limestone at Location A (figure 4), show that this facies has the characteristics of light brown color, grain supported, poor sorting, irregular arrangement of grains. The condition of the grains is 90% incomplete, arranged by allochemintraclast (4%), allochem bioclast (miliolidae = 15%, red algae = 5%, coral = 7%, planktonic foram = 6%, discocyclina = 5%, nummulites = 2%) in the percentage of 40%, micrite with a percentage of 10%, and sparite with a percentage of 30% (in the form of blocky).
According to [16], the name of this rock is packstone and according to [11] is rudstone. The porosity that develops is fracture stylolites, vuggy, and moldic. These limestone characteristics are generally following Standard microfacies types (SMF): grainstone or packstone with abundant foraminifera or algae according to [12]. This facies is interpreted as being deposited on restricted platform areas (lagoon) characterized by an abundance of milliolidae as the main characters in the lagoon area.

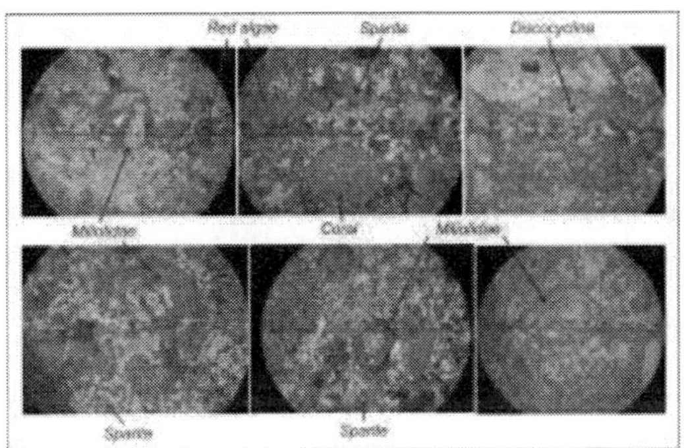

Figure 3. Milliolid appearance (Facies at Location A)

5.1.2. Location B
The results of the analysis of the bioclastic limestone thin section at Location B (figure 5), show that this facies has the characteristics of a light brown, skeleton type foraminifera, filled with a dominant matrix of sparite calcite, grain supported, poor sorting, random arrangement. The condition of the item 80% is not intact, composed by allochemintraclast = 13%, allochem bioclast (corals = 16%, red algae = 8%, planktonic foraminifera = 5%, lepidocyclina = 14%, nummulites = 6%) in the percentage of 52%, sparite with a percentage of 11% (blocky form), micrit with a percentage of 7%.

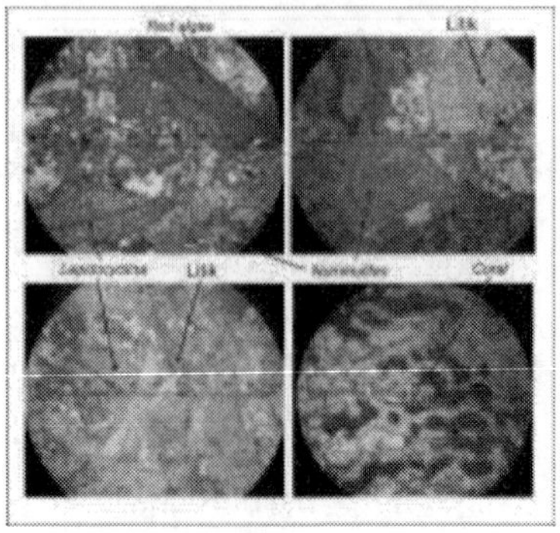

Figure 4. The appearance of Coral - Large Foraminifera grainstone (Location B)

Naming according to [16] is grainstone and according to [11] is rudstone. The developing porosity isvuggy, moldic, and intraparticle types.These limestone characteristics are generally following Standard microfacies types (SMF) is grainstone or packstone with abundant foraminifera or algae according to [12]. This facies is interpreted to be deposited in an open lagoon indicated by the appearance of red algae, it is estimated that this facies was deposited on a medium-high current flow which causes the constitution of large foraminifera (lepidocyclina) to be present in this facies. The presence of coral is an indication of detritus in the Organic Build Up part of the current, so it is estimated that these facies grows near the reef. Then the presence of plankton and land clastic sediments in these facies reinforces that these facies are deposited close to land or still in the interior platform environment.

5.1.3. Location C

The results of the bioclastic limestone thin-section analysis at Location C (figure 6), show that this facies has the characteristics of a light brown, skeleton type foraminifera, filled with a dominant matrix of calcite sparite, grain supported. The grains show poor sorting, random arrangement, grain conditions 95% are not intact, arranged by allochemintraclast = 5%, allochem bioclast (red algae = 20%, planktonic foraminifera = 4%, lepidocyclina = 8%, broken mollusca 3%) in percentage 35%, sparite with a percentage of 17%, micrite with a percentage of 8%.

According to [16], the name of this rock is packstone and according to [11] is rudstone. The porosity that develops is the type of cavern, vuggy, and moldic. These limestone characteristics are generally following Standard microfacies types (SMF) is coated skeletal packstone or wackestone according to [12]. This facies is interpreted to be deposited in an open lagoon indicated by the appearance of red algae, it is estimated that this facies was deposited on a medium-high current flow which causes the constitution of large foraminifera (lepidocyclina) to be present in this facies.

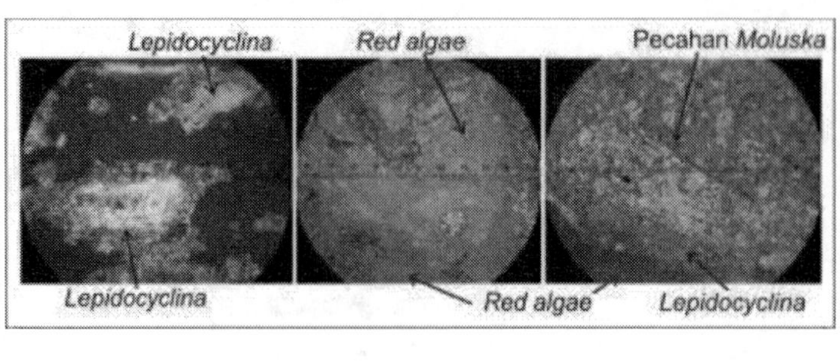

IOP Conf. Series: Materials Science and Engineering **982** (2020) 012044 doi:10.1088/1757-899X/982/1/012044

Figure 5. The appearance of Red algae - Lepidocyclinapackstone (Location Facies C)

5.1.4. Location D

The results of the analysis of the thin-section of the bioclastic limestone at Location D (figure 7), show that this facies has the characteristics of a light brown color, filled with a dominant matrix of calcite sparites, poor sorting, random arrangement. The condition of the item 95% is not intact, arranged by allochem bioclast (red algae = 22%, mollusk fraction 6%) in the percentage of 28%, sparite with a percentage of 37%, micrit with a percentage of 10%.

According to [16], the name of this rock is packstone and according to [11] rudstone. The porosity that develops is the type of vuggy, and moldic. These limestone characteristics are generally following Standard microfacies types (SMF) is worn skeletal packstone or wackestone according to [12]. This facies is interpreted as being deposited on an open lagoon which is indicated by the appearance of red algae.

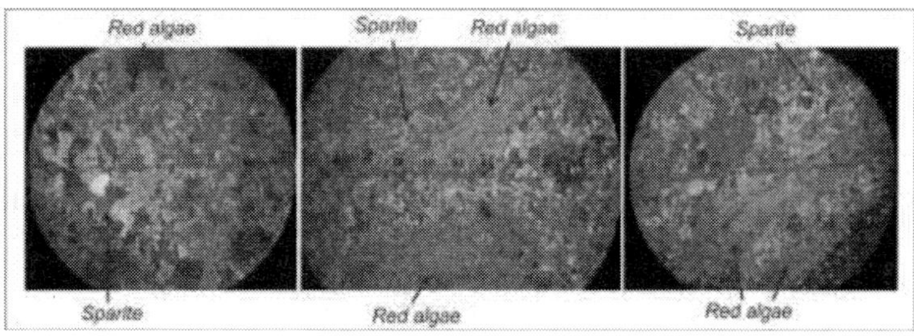

Figure 6. Red algae packstone appearance, facies at location D

5.1.5. Location E

The results of the analysis of the thin-section of the bioclastic limestone at Location E (figure 8), show that this facies has the characteristics of a light brown color, filled by the calcite sparite dominant matrix, poor sorting, random arrangement. The condition of the grains of 95% is not intact, arranged by allochem bioclast (red algae = 15%, green algae = 5%, bivalve = 2%, milliolidae = 3%, mollusk fragments 13%) in the percentage of 38%, sparite with a percentage of 30%, micrit with a percentage of 20%.

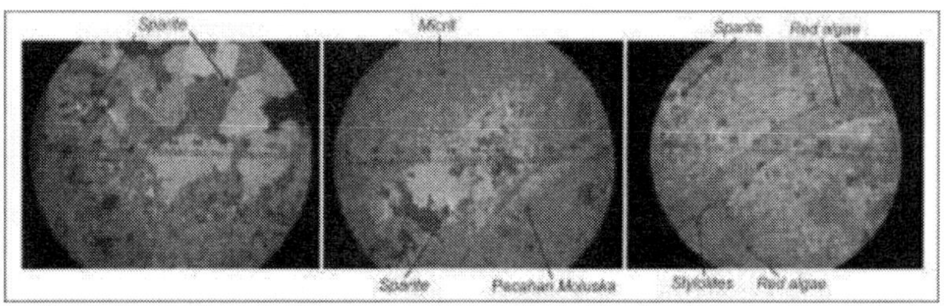

Figure 7. The appearance of Red algae packstone, Facies at Location E

Naming of this rock according to [16] is packstone and according to [11] is rudstone. The porosity that develops is the type of vuggy, and moldic.These limestone characteristics are generally following Standard microfacies types (SMF) is grainstone or packstone with abundant foraminifera or algae

IOP Conf. Series: Materials Science and Engineering **982** (2020) 012044 doi:10.1088/1757-899X/982/1/012044

according to [12]. This facies is interpreted as being deposited on a restricted platform (lagoon) which is indicated by the emergence of miliolidae as the main characters in the lagoon area.

5.2. Layered Limestone Facies

5.2.1. Location F

The results of layered limestone thin-section analysis at Location F (figure 9), show that this facies has the characteristics of light gray, type of foraminifera skeleton, filled with calcite dominant sparite matrix, grain supported, poor sorting, random arrangement. The condition of the item 90% is not intact, composed by allochemintraclast = 16%, allochem bioclast (coral = 5%, red algae = 8%, lanktonic foraminifera = 2%, lanktoni = 3%, milliolidae = 2%, lepidocyclina = 18%, discocyclina = 2%, nummulites = 1%, gastropods = 3%) in the percentage of 44%, sparite with a percentage of 12% (blocky form), micrit with a percentage of 8%.

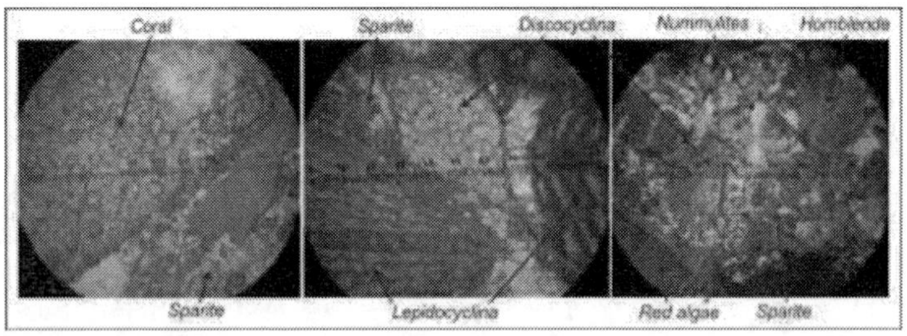

Figure 8. The appearance of Coral - Large Foraminifera grainstone, Location Facies F

The name of this rock according to [16] is grainstone and according to [11] is rudstone. The porosity that develops is the type of vuggy and moldic.These limestone characteristics are generally following Standard microfacies types (SMF) is grainstone or packstone with abundant foraminifera or algae according to [12]. This facies is interpreted to be deposited on a restricted platform (lagoon) which is indicated by the emergence of miliolidae, it is estimated that these facies is deposited on medium-high currents which causes the constitution of large foraminifera (lepidocyclina) and planktonic foram present in this facies.

5.2.2. Location G

The results of layered limestone thin-section analysis at Location G (figure 10), show that this facies has the characteristics of light brown, skeleton type by planktonic foraminifera, grain supported, filled by micritic and sparite matrices, poor sorting, random arrangement, grain condition 80 % incomplete. Composed by allochemintraclast = 14%, allochem bioclast (Coral = 2%, Red algae = 2%, Foram planktonic = 12%, Foramplanktonic = 5%, lepidocyclina = 7%, Nummulites = 2%, broken mollusk = 6%) in the percentage of 36%, sparite with a percentage of 14%, micrit with a percentage of 10%.

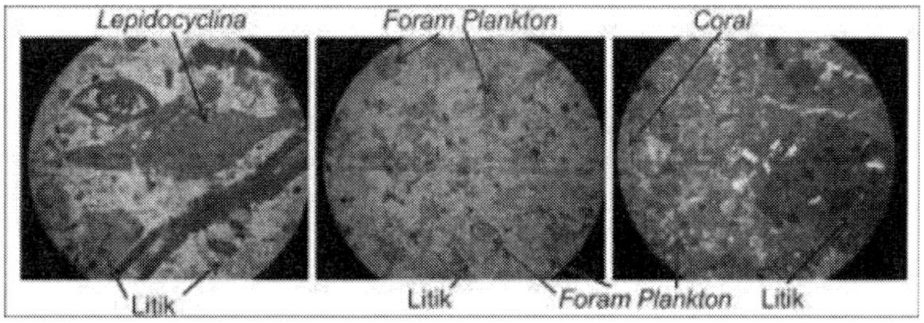

Figure 9. The appearance of planktonic Foraminifera - Lepidocyclina at Location G.

According to [16], the name of this rock is packstone and according to [11] is rudstone. The porosity that develops is intraparticle, vuggy, and moldic types. These limestone characteristics are generally following Standard microfacies types (SMF) is worn skeletal packstone or wackestone according to [12]. This facies is interpreted as being deposited on an open lagoon, which is indicated by the appearance of lepidocyclina and red algae, deposited with a calm - moderate current.

5.2.3. Location H
The results of layered limestone thin-section analysis at Location H (figure 11), show that this facies has brown characteristics. Types of skeletons by planktonic foraminifera, grain supported, filled with micritic and sparite matrices, poor sorting, random arrangement, 80% grain condition is not intact. Composed by allochemintraclast = 7%, allochem bioclast (foramplanktonic = 25%, foramplanktonic = 10% broken mollusk 5%) in the percentage of 40%, sparite with a percentage of 28%, micrit with a percentage of 10%.

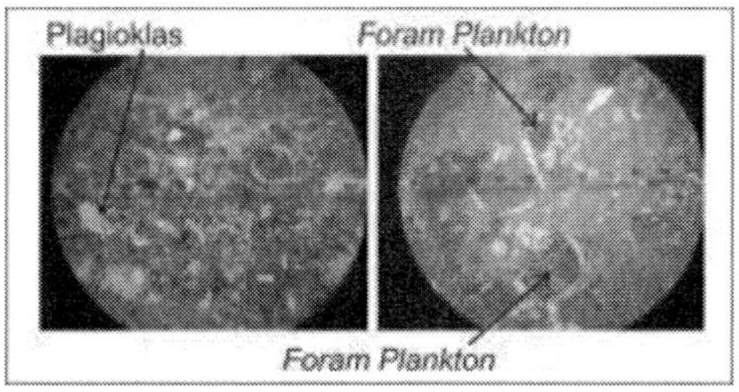

Figure 10. The appearance of planktonic Foraminifera, Facies at Location H

According to [16], the name of this rock is packstone and according to [11] is rudstone. The developing porosity is the type of burrow, vuggy, and moldic.These limestone characteristics are generally following Standard microfacies types (SMF) is worn skeletal packstone or wackestone according to [12]. This facies is interpreted to be on an open sea shelf, which is indicated by the appearance of abundant planktonic foraminifera.

5.2.4. Location I
The results of layered limestone thin-section analysis at Location I (figure 12), show that this facies has the characteristics of light brown, skeleton type by planktonic foraminifera, grain supported, filled by micritic and sparite matrices, sorting quite well, 70% grain condition is not intact. Composed by allochemintraclast = 4%, allochem bioclast (foramplanktonic = 35%, foramplanktonic = 19%,

IOP Conf. Series: Materials Science and Engineering **982** (2020) 012044 doi:10.1088/1757-899X/982/1/012044

lepidocyclina = 1%, broken mollusk 1%) in the percentage of 56%, sparite with a percentage of 15%, micrit with a percentage of 10%.
According to [16], the name of this rock is packstone and according to [11] is rudstone. Experiencing the compacting process, it can be seen from the relationship between the grains showing concavo-convex, the porosity that develops is the type of vuggy and moldic. These limestone characteristics are generally following Standard microfacies types (SMF) is worn skeletal packstone or wackestone according to [12]. This facies is interpreted to be on an open sea shelf, which is indicated by the appearance of abundant planktonic foraminifera, the appearance of lepidocyclina but not so dominant.

Figure 11. The appearance of planktonic Foraminifera (Facies at Location I)

5.2.5. Location J
The results of layered limestone thin-section analysis at Location J (figure 13), show that this facies has the characteristics of light brown, skeleton type by planktonic foraminifera, grain supported, filled with micritic and sparite matrices, poor sorting, random arrangement, grain condition 85 % incomplete. Composed by allochemintraclast = 4%, allochem bioclast (planktonic foram = 30%, planktonic foram = 10%) in the percentage of 40%, sparite with a percentage of 25%, micrit with a percentage of 10%.

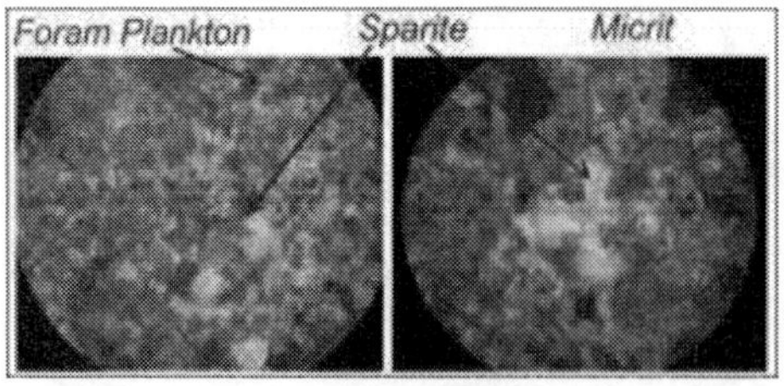

Figure 12. Planktonic Foraminifera Appearance (Facies at Location J)

Naming [16] packstone and according to [11] rudstone. Experiencing the compacting process, it can be seen from the relationship between the grains showing the suture, the porosity that develops is the type of vuggy and moldic. These limestone characteristics are generally following Standard microfacies types (SMF) is worn skeletal packstone or wackestone according to [12]. This facies is interpreted to be on an open sea shelf, which is indicated by the appearance of abundant planktonic foraminifera.

6. Discussion
The facies environment of the study area (figure 19) is divided into bioclastic limestone and layered limestone environments. Bioclastic limestone units with bioclastic limestone lithology occur in restricted

lagoon facies (restricted circulation shelf & tidal flats) and open lagoon (shelf lagoon open circulation) environments. Shown by the emergence of Red algae and Miliolidae, then reinforced by the presence of coral and clastic sediments, which indicates its presence not far from the environment of organic build-up, and still near land.

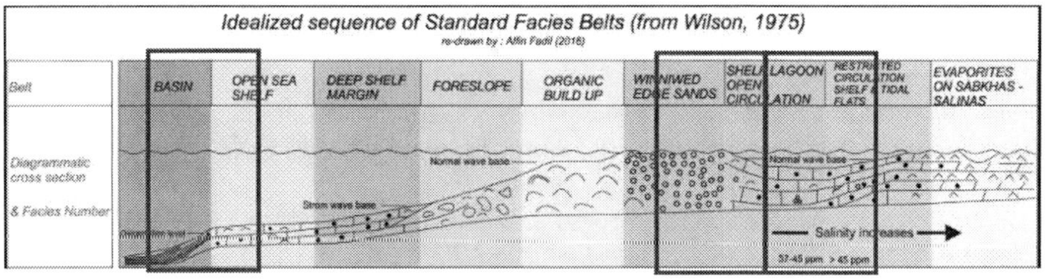

Figure 13. Facies environment according to Wilson (1975)

Layered limestone unit with layered bioclassical limestone lithology, indicating that this limestone is a limestone originating from the deposition, not growing in its place of origin.The facies environment is located in three environments, namely the restricted lagoon, open lagoon, marked by the appearance of red algae and miliolidae, indicating that they are deposited with a calm - moderate current. The appearance of coral fragments indicates that it is still near the organic build-up area, then there are lithic rocks that indicate that it is still deposited near land or exposed old volcanic bodies [17]. The open sea shelf environment is indicated by the abundance of planktonic foraminifera, although there is large foraminifera it is not dominant, this can be indicated that this environment is deposited with calm currents.

7. Conclusion

Bioclastic limestone facies are deposited in restricted lagoon environments (restricted circulation shelf & tidal flats) to open lagoon (shelf lagoon open circulation), characterized by the presence of miliolidae and red algae.

Layered limestone facies are in restricted lagoon environments (open circulation shelves & tidal flats), open lagoon (shelf lagoon open circulation) are characterized by miliolidae to open sea shelf environments, characterized by an abundance of planktonic foraminifera.

References

[1] Satyana AH and Purwaningsih MEM 2003 Oligo-miocenecarbonates of Java: tectonic setting and effects of vulcanism*Proceedings of Joint Convention Jakarta 2003* The 32nd IAGI and 28th HAGI Annual Convention and Exhibition

[2] Rahardjo W,Rosidi HMD, and Sukandarrumidi1995 Geology Map of the Yogyakarta - Java. *Bibliography : Geological Survey Center* p 8 Bandung.

[3] Kusuma HD, Ramadhani H, Novian MI, Furqon M A, Arimawan Y, and Citra IA 2019 Facies and environment deposition of Jonggrangan Formation on Sentul pathway of Jonggol Mountain, eastern KulonProgo Mountains*Journal of Geological and Mining Research*Vol. 29**2** *(141-152)* DOI: 10.14203/risetgeotam2019.v29.1002

[4] Pandita H, Pambudi S, and Winarti 2016 Study of Sentolo Formation stratigraphy and facies in the Gulurejo and Ngaran areas of Bantul Regency to identify the presence of the Progo Fault *Proceedings National Seminar* UniversitasPadjajaran Bandung

[5] Maryanto S 2015 Development of limestone sedimentology based on petrographic data in Sentolo Formation Along Pengasih path Kulonprogo*Journal of Mineral Resources Geology (JGSM)* Vol16 **3**p 129 – 139

[6] Riandari BL and Wijono S 2013 Planktonic foraminifera biostratigraphy and sedimentation environments of the Sentolo Formation along the Niten River, Serang River and Ngramang District *Proceedings National Earth Seminar*6, Geological Engineering UGM, Yogyakarta.

[7] Pringgoprawiro *H* 1968 On the age of SentoloFormaion based on planktonik. *Dept. of Geology*. ITB. Bandung.

[8] Van Bemmelen RW 1970 The Geology Of Indonesia Vol. IA: General Geology Of Indonesia And Adjacent Archipelagoes*Government Printing Office* : Amsterdam.

[9] Widagdo A, Pramumijoyo SP, and Harijoko A 2018 Morphotectono-volcanic of Menoreh-Gajah-Ijovolcanic rock in western side of Yogyakarta-Indonesia *Journal of Geoscience, Engineering, Environment, and Technology (JGEET)* Vol.03 **03**. http://journal.uir.ac.id/index.php/JGEET; DOI: 10.24273/ jgeet.2018. 3.3.1715

[10] Sulistyoningrum D and Rahardjo W 2010 Identification and paleoecology of coraline fossil (cnidaria: anthozoa) from Jonggrangan Limestone Western Slope of Kucir Hill West Progo Area Yogyakarta *Proceedings PIT IAGI Lombok*The 39th IAGI Annual Convention and Exhibition.

[11] Embry AF and Klovan JE1971 A late devonian reef tract on northeastern banks island. *N.W.T. Bull.* Canadian Petrol. **19**.

[12] Wilson and Lee J 1975Carbonat facies in geology history. *Springer-Verlag* New York

[13] Blow W H 1969 Late middle Eocene to Recent planktonic foraminiferal biostratigraphy. *Proceedings of the First International Conference on Planktonic Microfossils*, Geneva 1967 Vol. 1 Eds: Bronnimann, P.Renz, H.H. p. 199-422

[14] Velrk V D and Umbgrove 1927 Distribution of large Foraminifera in the "East Indian" Tertiary.

[15] Grimsdale TF and Van Morkhoven FPCM 1955 The ratio between pelagic and benthonic foraminifera as a means of estimating the depth of deposition of sedimentary rock. *Proc. World Pet.* Congr. 4th (Rome) Sect. 1/D4.

[16] Dunham, R. J., 1962, Classification of carbonate Rocks according to depositional texture. In: Ham, W. E. (ed.), *Classification of carbonate Rocks: American Association of Petroleum Geologists Memoir*, p. 108-121.

[17] Widagdo A, Pramumijoyo SP, and Harijoko A 2019 The structural geology constellation of western boundary of Yogyakarta Basin. *Proceedings National Earth Seminar*12, Geological Engineering UGM, Yogyakarta.

IOP Conf. Series: Materials Science and Engineering **982** (2020) 012045 doi:10.1088/1757-899X/982/1/012045

Reservoir Characteristic of Triassic Sandstone, Eastern Seram, Maluku, Indonesia

Akhmad Khahlil Gibran[1,3], Aries Kusworo[2], Joko Wahyudiono[2], Huzaely Latief Sunan[1], Deventi Nur Aeni[1] and Afif Alghazali[1]

[1]Geological Engineering Department, Universitas Jenderal Soedirman, Jawa Tengah, Indonesia
[2] Geological Survey Center, Geological Agency, Ministry of Energy and Mineral Resources, Indonesia

[3]akgibran@unsoed.ac.id

Abstract. The Triassic aged sandstone from Kanikeh Formation, are spread across Seram, Kesui and Teor Island. The Kanikeh formation has been known as excellent source rock, however, reservoir characteristics of this formation are unknown. This paper aimed at the diagenetic influence on reservoir characteristics was investigated for the Carnian-Norian sandstone sequence in the Eastern Seram. This study comprises two methods which are petrographic analysis and Scanning Electron Microscope-Energy Dispersive Spectrometer (SEM-EDS). The petrographic analysis shows the sandstones of the Carnian sequence in the Eastern Seram are lithic wacke, litharenite, arkose. SEM-EDS observations showed that the minerals belong to the silica group derived from the silica minerals, feldspar, clay, and mica minerals. The silica mineral is a quartz grain. The observed feldspar mineral is K-Feldspar in the form of granules. The most dominant clay minerals are smectite, illite, chlorite, kaolinite, halloysite. The observed mica group is muscovite. Diagenetic identified in these sandstones include compaction, cementation by calcite, quartz, clay minerals, iron oxides, dissolution, and alteration of unstable clastic grains, tectonically induced grain fracturing. Unstable clastic grains like feldspars suffered considerable alteration to clay minerals. Based on characteristics of the Diagenetic, the Carnian-Norian sandstones in the study area have a negligible reservoir characteristic.

1. Introduction

In Bula Basin, could be found 2 active oil and gas blocks, which are Bula Block with sandstones and limestone play aged Quaternary and Oseil Block with limestone play aged Triasss-Jurassic. Based on biomarker analysis, the oil from Bula and Oseil Field come from the same sources[1]. Kanikeh Formation has been known for the long term and has a good characteristic for main source rock[2][3][4]. Carnian-Norian sandstones in Kanikeh Formation could be indicated has a good reservoir[4]. Kanikeh Formation is the oldest sediment that could be found in Bula Basin[5]. This formation arranged by siliciclastic rocks that pressed unconformity by the oldest metamorphic rocks and could be found fault contact with the other oldest rock[6]. Kanikeh Formation consists of the repetition of sandstone and shale which spread across Seram Island to Kesui Island and Teor in Southeast Maluku[7][8].

Study of Kanikeh Formation sequence had been chosen due to a wide resource in Maluku (figure 1) and the newest discovery for oil sources that comes out from Mesozoic rock in the Bula Basin of sediment, likewise in the research area. The aim of the study in Kanikeh Formation clastic sequence is to reveal the diagenetic history with reservoir characteristics using a surface sample as analog subsurface for understanding the reservoir quality using an approach of petrography and Scanning Electron Microscope-Energy Dispersive Spectrometer (SEM-EDS). The research area is by passed by many faults that cut the basement with direction north west-south east and elongated to eastern of Papua Island[9]. The structure along Seram Island represented as a sinistral and dextral strike-slip fault[9] (figure 2). The folds that grow with strong deformation and associated with every rupture in different directions. A fold formed asymmetric anticline that dipping two directions and could be found in the field.

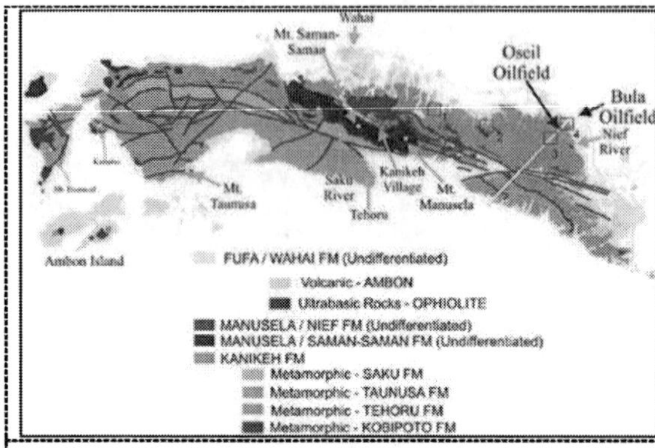

Figure 1. Geology map of Seram Island (Hill,2012). A red box is the research Area, (1) Lofin observation, (2) Niner observation, (3) Oseil observation and (4) Bula observation

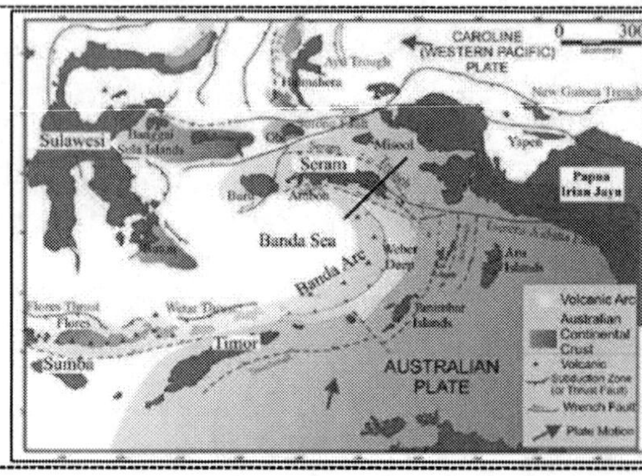

Figure 2. Tectonic setting of East Indonesia, showing that Tarera Aiduna Fault that becomes one of the structures that controlling the development of structural geology in Seram Island[9].

2. Methodology

The method that we use for collecting the stratigraphy sample and data is using the stratigraphy measuring section. The stratigraphy measuring section have been done in Bula, Oseil, Niner, and Lofin area. The literature review is our first stage for the research, which include a literature review that has been published. It used for understanding the major interpretation of geology condition of the area, the location of outcrop observation, and knowing the major issue in these formation, which shows the diagenetic process that affect the reservoir quality. Geology map and to topography map analysis in this research area have been done before fieldwork. After doing the literature review, we do the preliminary survey for knowing the main road access and outcrop condition which could be indicated as our first research target. The map of the geology track that has been chosen is road, river, and cliffs to get a good outcrop. The major data collected include the stratigraphy measuring section, rocks sample, and documentation.

26 samples are used for representing the study of Petrography and SEM-EDS. Major description and Autogenic Mineral from Sandstones based on the 26 samples consist of Petrography samples and SEM-EDS. The result of these petrography laboratory work is a thin section of rocks that produced by Basic Geology Laboratory, Centre For Geological Suvey Bandung. Research under polarization microscope in Petrography Laboratory, Centre For Geological Suvey by a researcher to know the mineral composition of rocks, type, and the diagenetic from each rock. The purpose of SEM analysis to observe the morphology, and rock composition or mineral. Morphology to analyzed shape and dimension from the sample; Composition to analyzed a composition from surface quantitatively and qualitatively. Preparation and testing of the SEM sample have been done in Basic Geology Laboratory, Centre For Geological Suvey, Bandung. The testing is using Scanning Electron Microscope JEOL JSM-6360LA, complete with *Energy Dispersive Spectrometer* (EDS) system JEOL

JED-2300. and could enlarge the object up to 300.000 times and determine the concentration of chemical elements semi-quantitatively.

Diagenetic is a physics, chemical, and biological process which turns sediment into sediment rocks. As a result, mineralogy and/or texture become different[10]. Diagenetic occurred after the deposit to metamorphosis (high temperature until 150°/200°c with/without pressure). Diagenetic occurred where mineralogy of rocks in unstable conditions as the result of transformation in physics, chemical, or biological conditions. The unstable mineralogy usually occurred in grains contact and pores between grains. A change of pressure and temperature causing a new mineral formed or the existing mineral before occurred a change, adjusting on new equilibrium. There are a few diagenesis classifications that could be used for knowing the grade of sandstones diagenetic[11][12][13][14].

The diagenetic process based on the burial effect to early shallow subsurface becomes Group A, late deep subsurface as Group B and Group C[12]. The diagenesis stage becomes eodiagenetic, an immature mesodiagenetic, semi-mature mesodiagenetic, mesodiagenetic mature A, mesodiagenetic mature B, and super mature[11]. The diagenetic stage becomes stage Ia, I, II, III, dan IV[14].

3. Field Study

Outcrop Observation and collecting the sample could be focused on siliciclastic rocks aged Triassic, in this case, sediment rocks that placed in Kanikeh Formation. All of the samples collected based on stratigraphy sequence and geographic location (Figure 1).

4. Results

4.1. Petrography analysis

The purpose of petrography analysis toward 5 samples of siliciclastic rocks such as sandstone of Kanikeh Formation have been done by doing some microscopic analysis to observe the mineral composition (Table 1), texture, and diagenetic stages that occurred in these rocks. Based on the presentation of mineral composition (Quartz (Q), Feldspar (F), Lithic (L)), and the representation of the matrix amount, a rock could be classified into sediment classification[13] (Figure 3).

The number of presentations calculate in every type of mineral and determines the name of rocks show in the Petrography Classification of Siliciclastic Rock (Figure 3). The result of 3 type are Lithic Wacke, Arkose, and Litharenit.

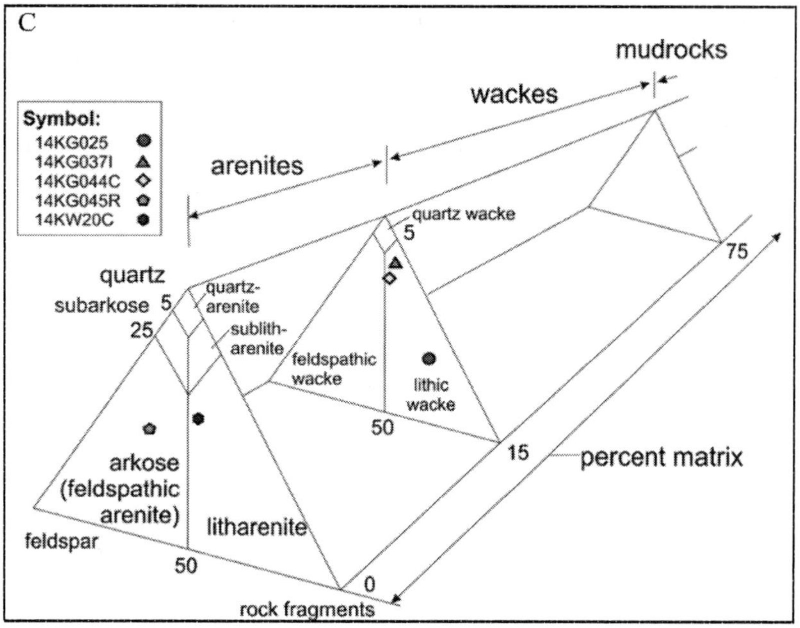

Figure 3. Classification of petrography for siliciclastic in Kanikeh Formation in the research area[13].

Table 1. Composition of all the rocks based on petrography analysis.

Sample Code	Fragment									Cement				Replacement		Porosity			
	Monocrystalline Quartz	Polycrystalline Quartz	Total Quartz	Feldspar	Lithic	Mica	Carbon	Opaque	Matrix	Clay	Carbonate	Oxide	Recrystallization of calcite	Chlorite	Recrystallization of matrix	Intergranular	Intragranular	Total	Rock Type of Pettijohn (1987)
14KG025	20	4	24	7	25	6	2	3	16	10	0	2	-	-	-	5	-	100	Lithic wacke
14KG037I	30	12	42	5	10	3	-	8	19	3	-	-	-	-	-	10	-	100	Lithic wacke
14KG044C	20	10	30	5	7	3	8	5	20	14	-	5	-	-	-	3	-	100	Lithic wacke
14KG045R	20	10	30	12	5	3	5	4	10	-	2	2	5	2	5	5	10	100	Arkose
14KW20C	18	7	25	5	7	4	5	3	11	5	20	3	2	-	-	7	3	100	Litharen ite

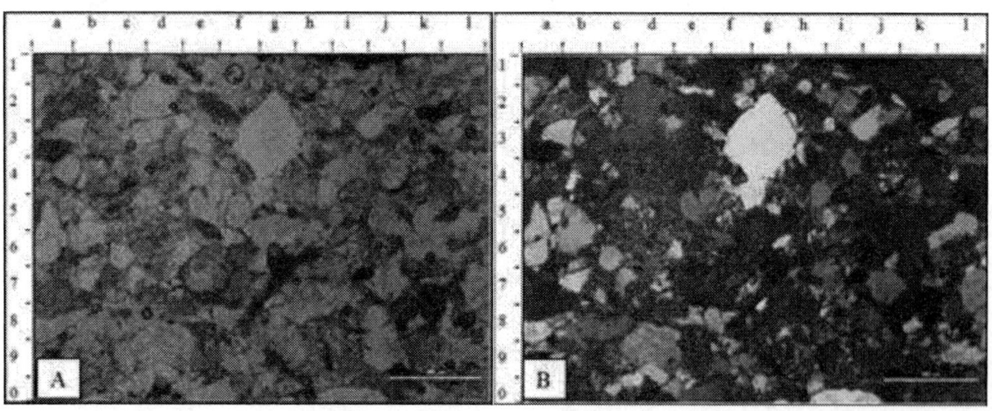

Figure 4.A. Position for plane light; B. Position for cross-polarized light. A thin-section of arenit felspathic, has a good-sorted, grain relation generally long contact, embedded and suture contact, cementation system are iron oxide, matrix recrystallization, and calcite, replacement isan opaque mineral, calcite, and chlorite, pore system is channel and intragranular.

4.2. SEM-EDS analysis

The purpose of SEM-EDS analysis toward 21 samples of rocks it which collected from 4 different field study and type of lithology consist of sandstone and claystone is to observe the topography, morphology, and rock or mineral composition.

SEM-EDS observation that have been done and could be indicated in the silica group that comes from silica, feldspar, clay, and mica mineral group. The silica group consists of quartz grains, feldspar group that observed is K-feldspar such as grains (figure 5). The most dominant clay mineral group appears between smectite, illite, chlorite, kaolinite, halloysite, and mica group that observed is muscovite.

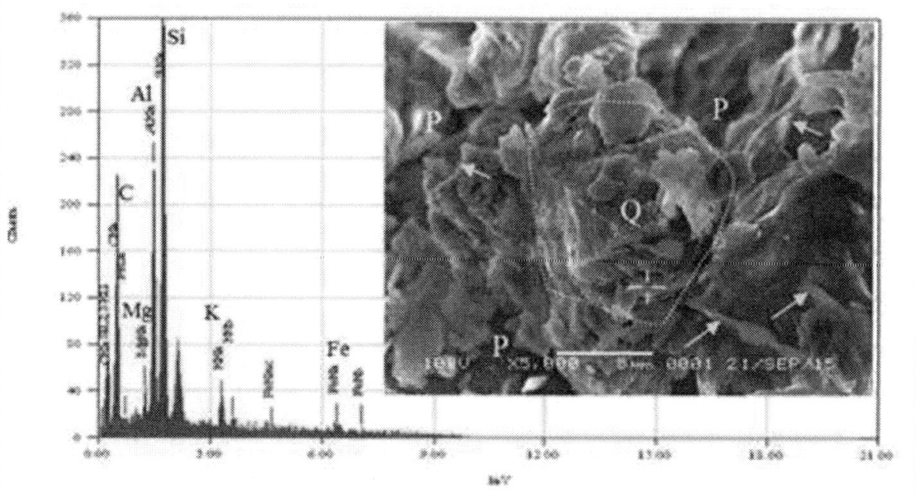

Figure 5.KW23L sample under SEM shows quartz grains (Q) and pores of rock (P) that surround by illite mineral (yellow arrows) and chromatogram showing the spectrum of major elements such as C, Mg, Al, Si, K, Fe.

4.3. The quality of rock porosity

The potential of reservoir rock could be determined for the observation result of lithology and the diagenetic process which occurred in it. The detail of petrography analysis from the thin section shows 5 samples of sandstones from kanikeh formation has an immature texture with a subangular shape and poorly-sorted. The appearance of abundant lithic shows as immature mineralogy. Immature texture, in general, has poor porosity, but has the good permeability, due to the grain shape that angular toward the pores of rock. Characteristic of the reservoir of sandstones depend on the diagenetic process, this process could increase or decrease the rock porosity, the important process is compaction, dissolution, and cementation by calcite, clay mineral and iron oxide.

The process of mechanic compaction that occurred in these 5 samples shows that the rocks getting joint strongly and create the secondary porosity, the secondary porosity consists of many fractures in rocks could be the dissolution media that could increase grade porosity and permeability, but joint that we observe not intensify. Dissolution that occurred for the 5 samples on sandstone shows that the dominant dissolution is feldspar. Dissolution could be occurred near the earth's surface by meteoric water[15] or the deeper depth by water that results from diagenetic process from organic material or clay mineral[16][17]. The dissolution process formed the secondary porosity and be a major factor for pore rocks forming.

The cementation process that occurred is cementation by calcite and iron oxide. All of the cementations could decrease the permeability and porosity grade. The process of authigenic mineral forming also observed as clay mineral forming but couldn’t be identified what kind of clay mineral is. The appearance of clay mineral will form discrete-particle pore bridging and pore lining. Forming of discrete-particle formed by kaolinite pore bridging by chlorite mineral and pore-lining by illite. From three kind of formed which comes due to authigenic mineral could decrease the permeability and porosity grade, but discrete particle didn’t affect the permeability rock. The reservoir characteristic that we get from the research shows the porosity grade around 3%-15%, moreover, it could be classified as denying to enough (porosity classification[18]) so, the reservoir potential could be denied.

5. Discussion

The sandstone has a various composition of quartz and lithic. Classified into *lithic wacke*, *arkose*, dan *litharenit.* The diagenetic process formed sandstones including compaction, cementation, and dissolution. Characteristics of the reservoir could be affected by the sandstone composition, the burial depth, and the uplift massively. The major mineral of diagenetic appears in sandstones, such as calcite, quartz, and iron oxide with a clay mineral. Feldspar has been altered during the diagenetic process and shows form the dissolving and alteration which occurred intensively in clay mineral. Secondary porosity is the result of early and late dissolution, in clastic sandstones, authigenic mineral, and cement. Compaction and cementation of calcite as the main major for decreasing porosity in sandstones. Kaolinite is an authigenic mineral, clay is the most common appear in sandstone, also appear illite and smectite in small occurrence. The porosity is controlled by the compaction and cementation, in this case, the porosity, in general, will be decreased during the compaction and cementation by iron oxide, quartz, calcite, and kaolinite. The secondary porosity formed from the dissolution of feldspar and calcite as the response form burial progressive. The appearance of clay mineral that covered or filled the pores caused decreasing of porosity and permeability. The permeability is caused by the influence of *authentic clays* appearances such as kaolinite, smectite, and illite. Dissolution in calcite cement in a few has a contribution by influencing the amount of secondary porosity. The abundance of authigenic kaolinite also brings the influence towards microporosity. Permeability would be increase by the increase of porosity, but it has weak relation, which means the porosity not give the major contribution towards the permeability grade. The appearance of tiny content and or the difference in the size of pore could be represented as the weak correlation between porosity and permeability. The index of reservoir quality depends on permeability if the permeability getting highest the reservoir quality getting better so in reverse.

The petrography observation shows that Carnian-Norian sandstone has an *immature* texture, based on the number of subangular grains, not compact of grain size has contained so many clay mineral as matrix and contains much of feldspar grain. These conditions caused the porosity grade in Carnian-Norian sandstones poorly (Table 1). Based on compaction process it could be characterized by domination between long contact grain (Figure 4), concave-convex contact (Figure 4), with a tiny expose of mica, the appearance overgrowth quartz cause pressure solution (Figure 4) and change of smectite clay mineral become illite (Figure 5), with the dissolution of feldspar grain (figure 5), and Carnian-Norian sandstones could be into late deep subsurface diagenetic (Group C)[12]. While, it could be interpreted as mature mesogenic B[11], these sandstones interpreted as low part from stage III with a temperature around 95°-120°C with the depth burial around 2700m-4000m[14].

6. Conclusion

Reservoir characteristics of sandstones could be determined by the diagenetic process, these processes could increase or even decrease the rock porosity. The important process is compaction, dissolution, and cementation by calcite, clay mineral, and iron oxide.

The compaction process could be dominant to influence the contact in grains, causes of carnian-norian sandstones could be compacted and only left less intergranular porosity, the compaction process could leave much fracture, so the secondary porosity couldn't be formed dominantly. Due to not dominant secondary porosity, the dissolving process not occurs intensively, the dissolving only occurred in feldspar mineral dissolving that the amount not significant. The decrease of rock porosity getting decreases by cementation activity because of calcite, clay mineral, and iron oxide. The influence of diagenetic formed reservoir quality into the poorly.

Acknowledgments
The writer is feeling grateful to head and all of the geologist from Geological Survey Center, Geological Agency, Ministry of Energy and Mineral Resources, Indonesia could give a best chance to facilitate the writer when fill the field data and laboratory analysis. The writer also being thankful to all of the college lecturer from major Geological Engineering, Universitas Jenderal Soedirman, Purwokerto, Indonesia.

References

[1] Wahyudiono J, Adlan R, Permanadewi S and Gibran A K 2018a Karakteristik minyak bumi di blok bula dan blok oseil, pulau seram, Maluku *Jurnal Geologi dan Sumberdaya Mineral* **19**: 233-241

[2] Kemp G and Mogg W 1992 A re-appraisal of the geology, tectonics, and prospectivity of seram island, eastern Indonesia *Proceeding of Indonesian Petroleum Association* **21**: 521-552

[3] Surjono S S and Wijayanti H D K 2012 Tectono-stratigraphic framework of eastern indonesia and its implication to petroleum systems. *ASEAN Engineering Journal Part C*, **1**: 109-123

[4] Adlan R, Wahyudiono J, Susilo A, Salimudin B, Gibran A K and Wiratmoko E S 2018 Petroleum system potential of lofin and banggoi area, seram island. *Proceedings of Indonesian Petroleum Association* **42** IPA18-250-G

[5] Tjokrosapoetro S and Budhitrisna T 1982 Geology and tectonics of northern banda arc *Bulletin Geological Research and Development Centre* Bandung. **6**: 1–17

[6] Gafoer S, Suwitodirdjo K and Suharsono 1993 *Peta geologi lembar bula dan watubela, maluku, skala 1:250.000* Pusat Penelitian dan Pengembangan Geologi Bandung

[7] Wahyudiono J, Susilo A, Adlan R, Salimudin B, Gibran A K and Wiratmoko E S 2018b Integrated field mapping, organic chemistry and subsurface geological interpretation of kanikeh formation as potential source rock in seram island *Proceedings of Indonesian Petroleum Association* **42** IPA18-247-G

[8] Chandra B Y and Kusworo A 2019 Palynostratigraphy and paleoenvironment of the triassic kanikeh formation, seram island indonesia *Proceedings of Indonesian Petroleum Association* **43** IPA19-G-280

[9] Hill K C 2012 Tectonic and regional structure of seram and the banda Arc *Berita Sedimentologi, Indonesian Journal of Sedimentary Geology* **23**: 6-16

[10] Santy L D 2014 Diagenetic batupasir eosen di cekungan ketungau dan melawi, kalimantan barat *Jurnal Geologi dan Sumberdaya Mineral* **15(3)**: 117-131

[11] Schmidt V and McDonald D A 1979 The role of secondary porosity in the course of sandstone Diagenetic, In: Schole P A and Schluger P R (Eds), Aspect of Diagenetic *Society of Economic Paleontologist and Mineralogist* Special Publication **26**:175-207

[12] Helmod K P and Van De Kamp P C 1984 Diagenetic mineralogy and controls on a albitization and laumontite formation in Paleogene Arkose, Santa Yenz Mountains, California, In: McDonald D A and Surdam R C (Eds), Clastic Diagenetic *American Association of Petroleum Geologist* Memoir **37**:239-276

[13] Pettijohn F J, Potter P E, and Siever R 1987 *Sand and Sandstone, 2ndeds* SpringerVerlag New York NY

[14] Burley S D and Kantorowicz J D 1987 Clastic Diagenetic *American Association of Petroleum Geologist, Treatise of Petroleum Geology Reprint Series* 408-455

[15] Mathisen M E 1984 Diagenetic of plio-pleistocene non-marine sandstones, cagayan basin, philippines, early development of secondary porosity in volcanic sandstones *Association of Petroleum Geologists Memoir* **37**: 177–193

[16] Seibert R M, Moncure G K and Lahann R W 1984 A theory of framework grain dissolution in sandstones *Association of Petroleum Geologists Memoir* **37**: 136–175

[17] Surdam R C, Boese S W and Crossey L J 1984 The chemistry of the secondary porosity *Association of Petroleum Geologists Memoir* **37**: 127–149

IOP Conf. Series: Materials Science and Engineering **982** (2020) 012045 doi:10.1088/1757-899X/982/1/012045

[18] Koesoemadinata R P 1980 *Geologi Minyak Dan Gas bumi* Edisi Kedua Jilid **2** ITB Bandung.

IOP Conf. Series: Materials Science and Engineering **982** (2020) 012046 doi:10.1088/1757-899X/982/1/012046

The Control Of Geological Structure At Songgo River Dyke Intrusion In Nanggulan Formation, Western Part Of Yogyakarta-Central Java

Asmoro Widagdo[1,3], Rachmad Setijadi[1] and Eko Bayu Purwasatriya[1]

[1] Geological Engineering Department, Jenderal Soedirman University (UNSOED), Purwokerto, Indonesia

[3]asmoro.widagdo@unsoed.ac.id

Abstract. This study to explain the presence of intrusive rocks in Naggulan Formation was carried out with a series of geological research methods. The method used in this research is through reference studies from previous research, observing the image of the study area and its surroundings, direct observation of Paleogenic rock bodies, and measurement of fault structures, joints, and folds in the field. Measurement of the elements of geological structure is used to determine the main structure and stress forming structure in Nanggulan Formation rock which is of Paleogene age and surrounding rocks which are Neogene age. The igneous rocks found in the study area is a dyke that cut the sedimentary rock layers of Nanggulan Formation. Dyke igneous rock distribution in the study area forms a northwest-southeast (NW-SE) direction. The distribution of igneous rocks is controlled by geological structures with the main stress direction of northwest-southeast which creates extensional vertical fractures in the direction of the main compressional stress source.

1. Introduction

The study area is in the Nanggulan area, the western part of the Yogyakarta Special Region-Central java (Figure 1). This area is very geologically interesting because it is one of the oldest sedimentary rocks found on the island of Java. This rock is the Nanggulan Formation which is an Eocene age rock according to [1] and [2]. In the western part of Yogyakarta, this rock forms a base for Oligocene - Miocene volcanic and sedimentary rocks at the top.

The presence of Nanggulan Formation rocks in this area has attracted many researchers including Van [3], [4], [5], [6], [7], [8] and [9]. Research by [10] described the presence of fold structure, strike-slip faults, and normal faults in the Songgo River area and surroundings. Research by [11], map the distribution of the Nanggulan Formation, and describe north-south trending anticlines and basalt intrusions in the Songgo River area.

IOP Conf. Series: Materials Science and Engineering **982** (2020) 012046 doi:10.1088/1757-899X/982/1/012046

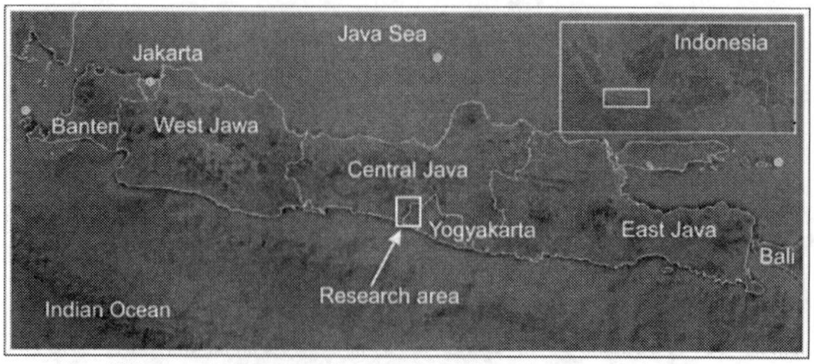

Figure 1. Research area location in Java Island.

2. Regional Stratigraphy

The Nanggulan Formation has a type location on the Songgo River, Nanggulan area. [3], explains that this formation is the oldest rock in the western part of Yogyakarta. The constituent lithology consists of sandstones with lignite, sandy marl, claystone with limonite, marl and limestone concretions, sandstones, tuffs rich in foraminifera and mollusks, estimated thickness up to 350 m. Based on the study of planktonic foraminifera, the Nanggulan Formation has an age range between the Middle Eocene to the Oligocene (Figure 2).

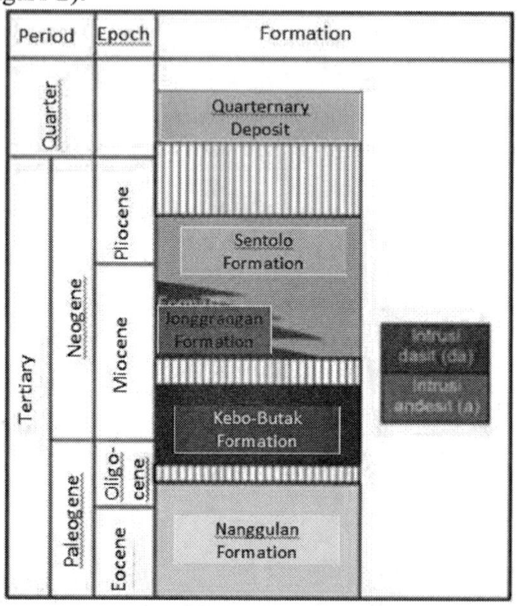

Figure 2. Stratigraphic column of study area according to [13] with modifications.

The volcanic rocks referred to by [3] as the Old Andesite Formation (OAF) are the same as those referred to as the Kebo Butak Formation (Figure 2) by [12] and [13]. [13] mentions the constituent rocks of the Kebo Butak Formation are andesite breccias, tuffs, lapilli tuffs, agglomerates and inserts andesite lava flows. Research of [14] describes the rocks that make up the Old Andesite Formation including volcanic breccias, lava, and intrusion. Volcanic breccias consist of volcanic breccias associated with lava; associate with lapilituf and tuff; with volcanic sandstone lenses; and alternating with finely layered sandstones. Lava is found around the center of the eruption (feeder) of the existing Oligo-Miosen volcano. Intrusion rocks consist of andesite, dacite, and basal intrusion.

Publication of [15] states, this volcanic group are covered unconformably by the shallow marine deposits of the Jonggrangan Formation and Sentolo Formation. Based on [16] states that the first carbonate sedimentation occurred at the top of the Early Miocene and occupied high areas such as Kulon Progo and Karang Bolong (Kebumen) resulting in the Sentolo and Jonggrangan Formations of Early Middle Miocene. Reefs are built on the heights of former volcanic bodies. Result of [17] states that carbonate sedimentation formed at the top of the Early Miocene and took place in high areas such as at the height of the early Middle Miocene Kulon Progo.

3. Research Method

The study was conducted through a series of reference studies from previous research, observation of the image of the study area and its surroundings, direct observation of the rock body, and measurement of fault structures, joints, and folds in the field. Measurement of the elements of geological structure is used to determine the main structure and the main stress of structure formation in the Nanggulan Formation rock which is of Paleogene age and the surrounding rocks which are of Neogene age.

4. Research Result

Nanggulan folds develop on the Nanggulan Formation rocks (Figure 3) in the Girimulyo and Kalibawang areas, Kulon Progo. The anticline fold has a Southwest-Northeast-directed axis. Anticline limb facing to the southeast can be seen in the coal seam on the Watupuru River with dipping 25°. The limb that leads to the northwest has a slope between 7°-17°. Limb in the southwest is found with dipping 35° and 24°.

Figure 3. Outcrops of Nanggulan fold limb. a. The northwestward view of the anticline limb on sandstone was inserted by claystone dipping to the NW. b. Westward view of the southeast anticline limb on claystone intercalated by siltstone.

The results of the stereographic analysis of the Nanggulan Folds data (Figure 4) resulted in the position of the northwest limb is N228°E / 13°. Southeast Limb domiciled N4°E / 29°. The Nanggulan fold is formed by the main stress (T1) 7° / N105°E. The formed axis has the position N196°E / 83° with the axis line 07° / N015°E. Based on the results of the analysis, the Naggulan Fold is an Upright Horizontal Fold type according to Rickard's Classification, 1971. This fold is formed by the maximum main stress of T1: 7° / 105°E or stress from the southeast with an almost horizontal of plunge.

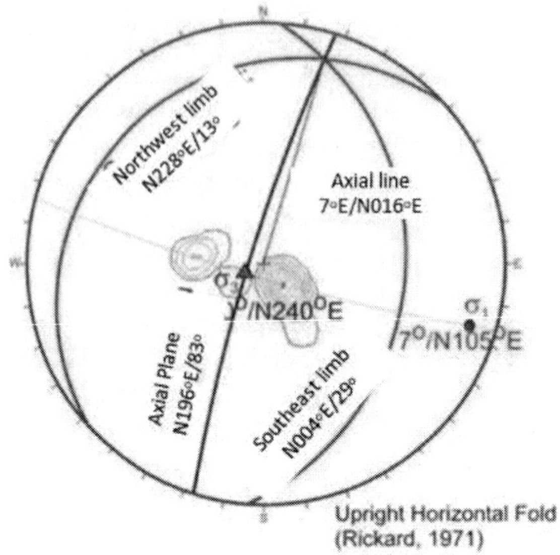

Figure 4. Stereographic analysis of Nanggulan folds.

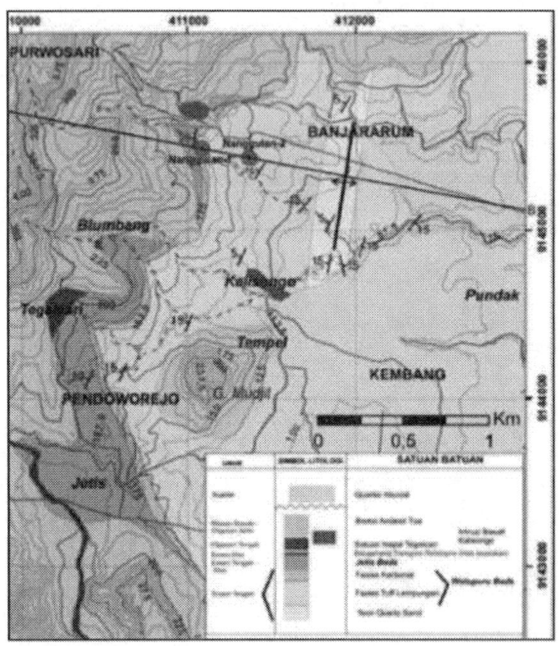

Figure 5. Geological map of the study area according to [15]. Songgo River dyke is in the center of the map.

Field observations show the distribution of igneous rocks as shallow intrusions (Figures 5 and 6). These rocks are slightly light gray to dark gray, fine afanite-porphyritic textures, vesicular structures (the structure of holes in dyke), and scoria structures. The vesicular structure only develops in the middle part of the body of intrusion rock that forms a columnar joint.

Figure 6. The appearance of basalt igneous rock in the study area [7].

5. Discussion

Based on fault kinematics, dyke is possible to form due to maximum extension, which allows magma to flow into the low-pressure zone. The relationship between the dyke pathway and the anticline structure shows its relationship with normal faults or extensional fracture and magmas fill spaces in the strain zone produced by the fault [18]. Magmatism activities are closely related to local area tectonic activities. As long as the area's tectonic activities are still ongoing, it can cause reactivation of magmatism activities [19].

The direction tendency of dyke is likely to be concentrated in the direction of the maximum compressional stress (T1) with the opening towards the minimum compressional stress (T3). The T1 direction is usually horizontal in the contractional tectonic region which is characterized by the presence of a strike-slip fault or thrust fault that is not in line with the vertical dyke intrusion. The trajectory can be controlled by the arrangement of stress-field lithospheric and local stress fields and fractures caused by such as normal faults, thrust faults, and strike-slip faults [20]. According to [21], [22] and [23] mentioned the presence of anticline and thrust faults as controllers for the emergence of the Nanggulan Formation in western Yogyakarta.

Figure 7 shows the movement of magma in the crust along the plane of the thrust fault, illustrating that magma can spread to the surface in a tectonic regime characterized by contractional deformation. If the thrust fault is below the surface with a gentle angle in a flat view, vertical tension fractures can form in the shallow part of the hanging wall subparallel to the imposed shortening. These structures can be formed due to local loading, due to uplifted areas in regional stress fields [24].

Dyke which intrudes into an area can be simplified into a vertical plane, better known as the dyke plane, as the main magma escape route. Normal-tension is perpendicular to the dyke plane. Volumetric changes around the magma sac may be related to the influence of magma overpressure as a source of magma. Active fault zone can cause wider swelling in the magma reservoir which allows volcanic eruptions [25].

IOP Conf. Series: Materials Science and Engineering **982** (2020) 012046 doi:10.1088/1757-899X/982/1/012046

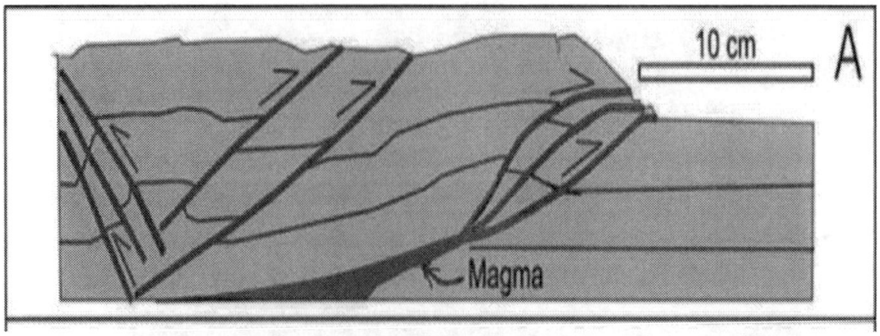

Figure 7. Intrusion models in contractional tectonic settings [24]. Shows the movement of magma under the surface along the plane of the thrust fault.

Magmatism in the study area is of the same age as northwest-southeast-directed local compression. The maximum main stress is horizontal T1 and the minimum main stress is vertical T3 creating a horizontal / almost horizontal thrust fault structure. Magma flows through the composite system of sub-horizontal tension fractures and thrust fault planes.

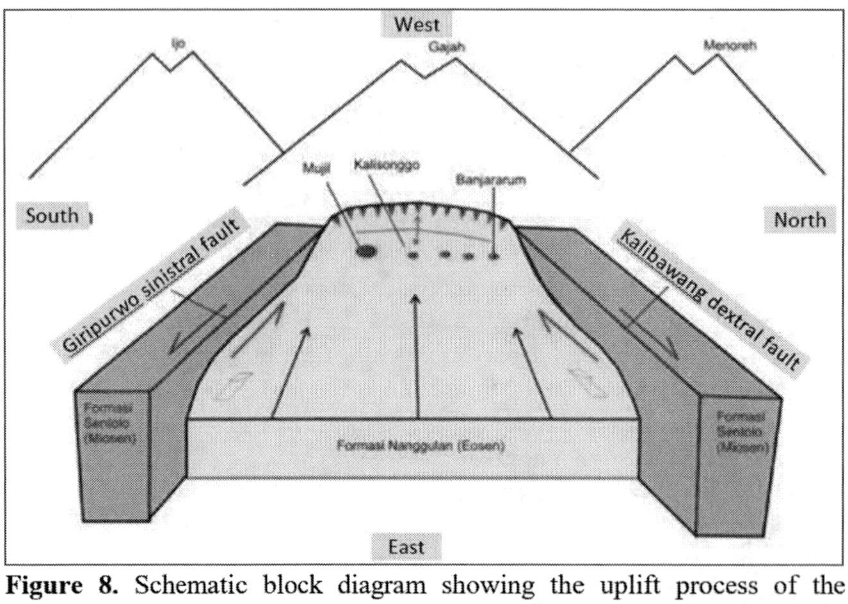

Figure 8. Schematic block diagram showing the uplift process of the Nanggulan Formation due to strike-slip and thrust fault.

The block diagram scheme (Figure 8) shows the thrust fault and anticline that gave rise to the Nanggulan Formation. In this picture, the rising of the Nanggulan (Eocene) Formation rock is drawn up to cut the young rock of the Sentolo Formation (Middle Miocene). To the north and south of the Nanggulan Formation, the Sentolo Formation is found. So it can be said that laterally the Naggulan Formation is flanked by the Sentolo Formation in the north and south. While in the west there is contact with the Old Andesite Formation [3] or the Kebo Butak Formation [13] or the Gajah Volcano rock according to [26] and [15].

6. Conclusion

The igneous rock found in the study area is a dyke that cuts anticline structures of the Nanggulan Formation sedimentary rocks. Dyke igneous rock distribution in the study area forms a northwest-southeast direction. The distribution of igneous rocks is controlled by geological structures with the main direction of stress is southeast-northwest which creates an extensional vertical fracture in the direction of forming stress.

References

[1] Premonowati, 1994. Informasi Paleobatimetri Asosiasi Fosil Moluska. Preceedings Geologi dan Geotektonik Pulau Jawa Sejak Akhir Mesozoik Hingga Kuarter, Yogyakarta.

[2] Harley and Morley 1995. Ultrastructural studies of some fossil and extant palm pollen, and the *reconstruction* of the biogeographical history of subtribes Iguanurinae and Calaminae, Review of Palaeobotany and Palynology **85** 153-182.

[3] Van Bemmelen R W 1970 The Geology of Indonesia. Vol. IA, General Geology of Indonesia and Adjacent Archipelago, Government Printing Office, The Hague.

[4] Prasetyadi C 2008 Provenan Batupasir Eosen Jawa Bagian Timur. Proceeding of annual meeting IAGI **37**.

[5] Rahmad B, Maha M, and Rodhi A 2008 Reflektan Vitrinit dan Komposisi Maseral Seam Batubara Eosen Formasi Nanggulan Daerah Kalisonggo, Kecamatan Girimulyo, Kabupaten Kulon Progo, DIY. Prosiding Pertemuan Ilmiah Tahunan IAGI **37**

[6] Winardi, S, Toha, B, Imron M, and Amijaya D H 2010 The Potency Of Eocene Shale Of Nanggulan Formation As Hydrocarbon Source Rock, Proceedings PIT IAGI Lombok.

[7] Hartono H G and Pambudi S 2015 Gunung Api Purba Mujil, Kulon Progo, Yogyakarta: Suatu Bukti dan Pemikiran. National Seminar ReTII Ke-**10**, STTNAS, Yogyakarta.

[8] Pambudi, S and Sujono 2016 Konfigurasi Cekungan Purba Formasi Nanggulan di Daerah Nanggulan, Kulon Progo, Daerah Istimewa Yogyakarta, Proceeding of **11**

[9] Hartono H G dan Sudradjat A 2017 Nanggulan Formation and Its Problem As a Basement in Kulonprogo Basin, Yogyakarta, Indonesian Journal on Geoscience **4** 71-80.

[10] Astuti B S, Humantoro R, Hidayat M and Kusuma H D 2016 Analisis Struktur Geologi Jalur Kali Watupuru Dan Kali Songgo Daerah Degan Kulonprogo, Dan Implikasinya Terhadap Penyebaran Batupasir Kuarsa Formasi Nanggulan Yang Berpotensi Sebagai Reservoar. National Seminar, UNPAD, Bandung.

[11] Saputra R and Akmaluddin 2015 Biostratigrafi Nannofosil Gampingan Formasi Nanggulan Bagian Bawah Berdasarkan Batuan Inti dari Kec. Girimulyo dan Kec. Nanggulan, Kab. Kulon Progo, D.I. Yogyakarta. Proseding Seminar Nasional Kebumian Ke-**8**, FT-UGM.

[12] Rahardjo W, Sukandarrumidi, and Rosidi H M D 1995 Peta Geologi Lembar Yogyakarta. Pusat Penelitian dan Pengembangan Geologi, Bandung.

[13] Rahardjo W, Sukandarrumidi, and Rosidi H M D 2012 Peta Geologi Lembar Yogyakarta. Pusat Survey Geologi-Badan Geologi-Kementrian ESDM.

[14] Suroso, Rodhi A, dan Sutanto, 1987 Usulan Penyesuaian Tata Nama Litostratigrafi Kulon Progo, Daerah Istimewa Yogyakarta. Kumpulan Makalah Pertemuan Ilmiah Tahunan XV Ikatan Ahli Geologi Indonesia, Volume **1**, IAGI-Yogyakarta.

[15] Harjanto A 2011 Vulkanostratigrafi di Daerah Kulon Progo dan Sekitarnya, Daerah Istimewa Yogyakarta. Jurnal Ilmiah MTG, Vol. **4** Yogyakarta.

[16] Satyana A H and Purwaningsih M E M 2003 Oligo-Miocene Carbonates of Java: Tectonic setting and effects of Vulcanism. Proceedings of Joint Convention Jakarta 2003, The **32**nd IAGI and **28**th HAGI Annual Convention and Exhibition.

[17] Satyana, A H 2005 Oligo-Miocene Carbonates of Java, Indonesia : Tectonic-Volcanic Setting and Petroleum Implications. Proceedings Indonesian Petroleum Association.

[18] Alvarez S A, Samaniego A F N, Zenteno D J M, and Aldave L A 2002 Rhyolitic volcanism in extension zone associated with strike-slip tectonics in the Taxco region, southern Mexico. Journal of Volcanology and Geothermal Research **118** 1-14.
[19] Bronto S 2010 Geologi Gunung Api Purba. Special publication, Badan Geologi, Kementrian Energi dan Sumberdaya Mineral, Bandung.
[20] Thouret 1999 Volcanic geomorphology-an overview. Earth-Science Reviews **47** 95–131.
[21] Widagdo A, Pramumijoyo, S P, Harijoko A, Setiawan A 2016 Kajian Pendahuluan Kontrol Struktur Geologi Terhadap Sebaran Batuan-Batuan Di Daerah Pegunungan Kulonprogo-Yogyakarta, Proceeding Seminar Nasional Kebumian ke-**9** TG FT-UGM, Yogyakarta.
[22] Widagdo A, Pramumijoyo S P, and Harijoko A 2018 Morphotectono-Volcanic of MenorehGajah-Ijo Volcanic Rock In Western Side of Yogyakarta-Indonesia, Journal of Geoscience, Engineering, Environment, and Technology (JGEET) **03**. http://journal.uir.ac.id/index.php/JGEET; DOI: 10.24273/jgeet.2018.3.3.1715
[23] Widagdo A, Pramumijoyo S P, and Harijoko A 2019 The Structural Geology Constellation of Western Boundary of Yogyakarta Basin. Proceeding Seminar Nasional Kebumian **12**, 5-6
[24] Tibaldi A, Pasquare F, and Tormey D 2010 Volcanism in Reverse and Strike-Slip Fault Settings. In: S. Cloetingh, J. Negendank (eds.), *New Frontiers in Integrated Solid Earth Sciences*, International Year of Planet **315**, Earth, DOI 10.1007/978-90-481-2737-5_9
[25] Hosono M, Mitsui Y, Ishibashi H, and Kataoka J, 2016 Elastostatic Effects Around A magma reservoir and pathway Due To Historic Earthquake : A Case Study Of Mt. Fuji, Japan, Progress In Earth and Planetary Science, **3** 33, DOI 10.1186/s40645-016-0110-9 .
[26] Barianto D H, Kuncoro P, and Watanabe K, 2010 The Use of Foraminifera Fossils for Reconstructing the Yogyakarta Graben, Yogyakarta, Indonesia. Journal of South East Asian Applied Geology **2** 138-143.

IOP Conf. Series: Materials Science and Engineering **982** (2020) 012047 doi:10.1088/1757-899X/982/1/012047

Run-up Height and Flow Depth Simulation of the 2006 South Java Tsunami Using COMCOT on Widarapayung Beach

FX Anjar Tri Laksono[1,2], Maulana Rizki Aditama[1], Rachmad Setijadi[1] and Gumilar Ramadhan[1]
[1] Geological Engineering Department, Universitas Jenderal Soedirman, Jawa Tengah, Indonesia

[2] anjar.trilaksono@unsoed.ac.id

Abstract. The destructive force of the tsunami could induce a considerable amount of casualties, infrastructures, and properties. One of the large tsunamis in Indonesia is 2006 South of Java Tsunami. There were 664 fatalities, 498 injured, 55 million dollars in losses, and 1623 homes damaged. Simulations of flow depth and run-up of the 2006 tsunami on Widarapayung Beach have never been conducted. Although they can be used as a basis for determining tsunami hazard zones. The method that might be used for tsunami impact analysis is tsunami waves modeling by considering the focal mechanism. The aims of this research are to determine the run-up height and flow depth of the 2006 tsunami on Widarapayung Beach. The calculations of modeling were performed using COMCOT numerical model. It operates calculations by solving shallow water equations in both linear and non-linear equations. Using this model, the generation and propagation of tsunami in a multi-grid system could be simulated. The result of the simulations on three different models depicts that the maximum run-up height on Widarapayung Beach is 3 to 3.8 m. The flow depth result from the three different models also suggests vulnerable areas up to 1 km from the shoreline.

1. Introduction

The destructive force of tsunami can cause a considerable amount of fatality and infrastructure damage. One of such tsunami is the 2006 South Java Tsunami (Fig. 1). This event occurred at 225 km from Pangandaran Beach (9.222°S, 107.320°E) with 7.7 on the Richter scale [1]. Based on a survey carried out by [2], this tsunami resulted in 664 fatalities. While based on the [3] survey in South Java 802 fatalities, 498 injuries, 55 million dollars in losses, and 1623 homes destroyed. On Pengandaran Beach, the intensity of the tsunami in the Papadopoulos and Imamura scale reached values IX and X (destructive to very destructive) [4]. The tsunami is a series of rapid long waves caused by large disturbance of a water body. To cause large scale disturbance, the event needs to release a large amount of energy in a short period of time. A tsunami usually triggered by fault displacement deep inside the convergence plate boundary. Most of the time it was caused by an earthquake but a big enough landslide that releases large amounts of energy could also trigger tsunami [5]. The July 17, 2006 tsunami in South Java was triggered by a small-scale earthquakes[6]. The local resident at the time barely felt the tremor of the earthquake [7]. In 2006, [8] conducted research using a long-period body and Rayleigh waves. The research focused on the mechanism of deformation fault and obtained an estimation of 1.0-1.5 km/s rupture velocity. With the low velocity and a short duration of 185 s, this earthquake is considered to have a slow breaking speed. These are in accordance with the characteristics of a tsunami earthquake [8].

A **B**

Figure 1. A). Rest area building destroyed after the tsunami at Widarapayung Beach. **B)**. Tsunami monument in Widarapayung, Cilacap [2].

As disaster countermeasures, numerical modeling can be used to obtain a mapping of inundation areas. The inundation map then can be used to identify areas prone to tsunami flooding and to take preventative measures [9]. The propagation and run-up data obtained from modeling are verified by observations in the field so that we can get a reliable map of vulnerable areas [10]. Widarapayung Beach is a beach tourist attraction located in Widarapayung Village, Binangun District, Cilacap Regency, Central Java, Indonesia. The location is about 35 km to the east of Cilacap. This beach was hit by the 2006 tsunami in South Java. This event occurred on July 17, 2006. It caused many fatalities and infrastructure damages. Based on a field survey by [11] run-up height on Widarapayung Beach is around 4 meters. Meanwhile, according to [12], the height of the tsunami run-up reached 4.26 meters.

Cornell Multi-grid Coupled Tsunami Model (COMCOT) is used to study the entire tsunami process, from generation, propagation to inundation. Using a modified leap-frog finite difference numerical method, it can solve shallow water equations as the basis of the simulation [5]. Although the 2006 Java tsunami has been studied by many researchers, there is no research on the tsunami hazard zones around the southern Cilacap coastal area. Using tsunami modeling, we simulate the 2006 tsunami to understand the physical impact of the tsunami and delineate the area vulnerable to future tsunami. The coast between Cilacap city and Kebumen is known to be a tourism destination [2]. Besides many tourists, the coast is also occupied as residential areas [10]. Therefore, this research is important to determine the tsunami inundation in the southern of Cilacap. We believe that Widarapayung Beach could represent the tsunami risk in the southern of Cilacap.

2. Method

Tsunami modeling is used to simulate the whole of a tsunami process beginning from generation, propagation, and inundation at the coast using sets of mathematical formulas. With the simulated result, we can understand physical characteristics like flow depths, wave speeds, amplitude, and inundation range. There are two types of tsunami modeling: numerical models and empirical models [12]. In this research, we use a numerical model which is a simulation of the mathematical models using computers. Numerical tsunami modeling involves three stages: source modeling, propagation modeling, and inundation modeling. Source modeling is the simulation on the generation of the tsunami, either by an earthquake, a submarine landslide, a volcanic eruption, or a bolide impact. Propagation modeling is the simulation of the movement and dispersal of the waves from the initial deformation area. Inundation modeling is the simulation of the water flow over the shoreline [13]. One of the models that are currently used by researchers for tsunami modeling is COMCOT.

COMCOT tsunami source model is based on [5] formula. The initial seafloor deformation for the propagation is calculated with the [5] formula. This formula assumes the earthquake represents a finite dislocation within an elastic body. Tsunami model COMCOT implements shallow water equations, either in Spherical or Cartesian Coordinates to simulate the propagation and run-up of Tsunami. For this study, we used Spherical Coordinates both the linear and nonlinear equations. Linear shallow water equations in spherical coordinates used in COMCOT is as below:

$$\frac{\partial\eta}{\partial t} + \frac{1}{R\cos\varphi}\left\{\frac{\partial P}{\partial\psi}(cos\varphi Q)\right\} = -\frac{\partial h}{\partial t} \tag{1}$$

$$\frac{\partial P}{\partial t} + \frac{gh}{Rcoscos\,\psi}\frac{\partial\eta}{\partial\psi} - fQ = 0 \tag{2}$$

$$\frac{\partial P}{\partial t} + \frac{gh\partial\eta}{\partial\psi} + fP = 0 \tag{3}$$

Water surface elevation is denoted with η. The volume flux components in x and y directions are denoted with P and Q respectively. Latitude and longitude of the Earth are denoted with φ and ψ. The radius of the earth is denoted with R. Gravitational acceleration is denoted with g. Water depth is denoted with h. Due to the rotation of the Earth Coriolis force coefficient is used and represented with f and lastly rotation of the Earth is denoted with Ω [5].

$$f = \Omega\sin\varphi \tag{4}$$

As tsunami wave propagates over the continental shelf and into the coastal area the wave-length of the incident tsunami becomes shorter and the amplitude becomes larger therefore linear shallow water equations become invalid. In coastal areas as waves become nearer to shore nonlinear convective inertia force and bottom frictions, terms are getting more important while Coriolis force and the frequency dispersion become less important. Thus the nonlinear shallow water equations including bottom frictions effect are adequate [14]. The non-linear shallow water equations used in COMCOT for spherical coordinates are as below:

$$\frac{\partial\eta}{\partial t} + \frac{1}{R\cos\varphi}\left\{\frac{\partial P}{\partial\psi}(cos\varphi Q)\right\} = -\frac{\partial h}{\partial t} \tag{5}$$

$$\frac{\partial P}{\partial t} + \frac{1}{R\cos\varphi}\frac{\partial}{\partial\psi}\left\{\frac{P^2}{H}\right\} + \frac{1}{R}\frac{\partial}{\partial\varphi}\left\{\frac{PQ}{H}\right\} + \frac{gH}{R\cos\varphi}\frac{\partial\eta}{\partial\psi} - fQ + F_x = 0 \tag{6}$$

$$\frac{\partial Q}{\partial t} + \frac{1}{R\cos\varphi}\frac{\partial}{\partial\psi}\left\{\frac{PQ}{H}\right\} + \frac{1}{R}\frac{\partial}{\partial\varphi}\left\{\frac{Q^2}{H}\right\} + \frac{gH}{R}\frac{\partial\eta}{\partial\varphi} + fQ + F_x = 0 \tag{7}$$

Water depth is denoted with H. Bottom friction in x and y are denoted with Fx and Fy, respectively. Fx and Fy are also evaluated using Manning's formula. Manning's roughness coefficient is denoted with n.

$$F_x = \frac{gn^2}{H^{\frac{7}{3}}}P(P^2 + Q^2)^{1/2} \tag{8}$$

$$F_y = \frac{gn^2}{H^{\frac{7}{3}}}Q(P^2 + Q^2)^{1/2} \tag{9}$$

The parameters used for the fault model are modified data compiled from [14] and [15]. The epicenter location we used for the fault model is from field observation by [16]. For Model 1 we used data extracted from [17]. In their research, they used GSN broadband waveforms to obtain the slip history using a finite fault inverse algorithm. For Model 2, we used data from research by [18]. Using inversion of tsunami waveforms, they concluded the tsunami source was 200 km long with a slip of about 2.5 meters. For Model 3, we used data from research by [19]. The slip dislocation varies

IOP Conf. Series: Materials Science and Engineering **982** (2020) 012047 doi:10.1088/1757-899X/982/1/012047

between 8-15 meters. In this model, we used the average value of 12 meters instead. [18] found that the slip distribution at the source is distributed unevenly with five to six sharp differences near the epicenter. However, in this research, we only used a single uniform fault in our source (Table 1).

Table 1. Compiled fault parameters

	Fault Parameter		
	Model 1 [17]	Model 2 [18]	Model 3 [19]
Slip Angle (degree)	93	95	95
Strike (degree)	297	289	289
Dip (degree)	6	10	10
Depth (km)	10	10	12
Dislocation (meter)	5	2.5	12
Fault Width (km)	40	40	40
Fault Length (km)	200	200	200
Epicenter (long)	107.310997	107.310997	107.310997
Epicenter (lat)	-9.22999954	-9.22999954	-9.22999954

3. Bathymetry and topography

The data read by COMCOT need to be written in 3 columns for longitude, latitude, and water depth. On reading bathymetry and land topography data, COMCOT has a particular way of reading data from the file format. Water depth in ETOPO file format the data will be read based on the sign of the number, negative sign means the data is below the water surface and positive sign means the data is above water surface [15]. Thus COMCOT can read data directly from the ETOPO file format. Other than ETOPO file format, COMCOT can also read data from the XYZ file format or its old propriety file format, just with the slight change of sign. In ETOPO data. The bathymetry data we used, however, is ETOPO1 data from [14]. ETOPO1 is a 1 arc-minute global relief model of Earth's surface that integrates land topography and ocean bathymetry. In Figure 2 the deep blue color shows the deeper water depth and green color show the shallower water depth. The negative sign illustrates under mean sea level. On the other hand, a positive sign shows above mean sea level.

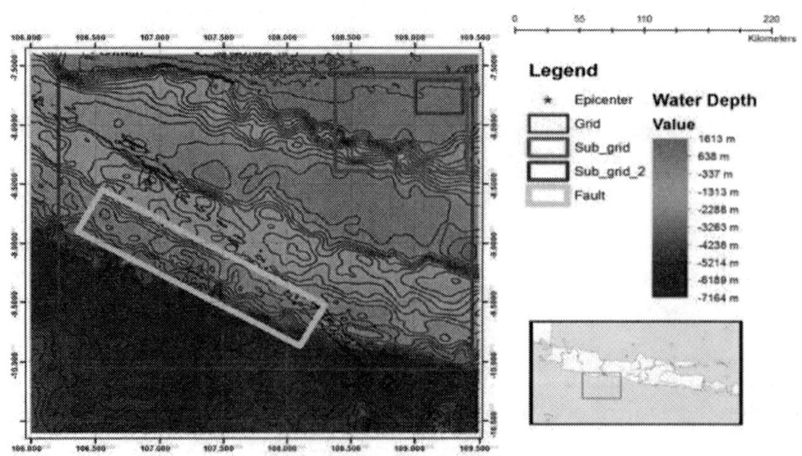

Figure 2. the bathymetry data and model domain. A red box displays the parent layer grid area. A pink box depicts the sub layer 1 grid area. A blue box shows the sub layer 2 grid area. A green box illustrates the fault dimension and direction.

4. Numerical domain and setup

In this research, we used a nested grid form with three different grid levels and grid resolution. The first grid or layer encompass big regions below the south coast of Java with 106.208504 to 109.433746 longitude and -7.55985355 to -10.0633001 latitude. This layer uses linear shallow water equations. with a grid dimension of 707x549 with a grid size of 0.27 arch minutes and a step size of 0.1366. The second encompasses the coast of Pangandaran to coast of Ayah with 108.381889 to 109.393776 longitude and -7.59982634 to -8.39243221 latitude. This layer uses non-linear shallow water equations with a grid size ratio of 2 to the parent layer. The third layer encompasses the area around Widarapayung Beach with 109.006989 to 109.364838 longitude and -7.62647533 to -7.90666676 latitude. In this layer, we use non-linear equations and enabled bottom friction with constant roughness of 0.013.

We compile run-up height data from several sources (Fig.3). From these data, we compared the simulated results and the observed data. As a comparison between simulated wave height and observed wave height, we choose four locations on the southern coasts of Java. The observed run-up height data and their respective sources are presented in Table 2.

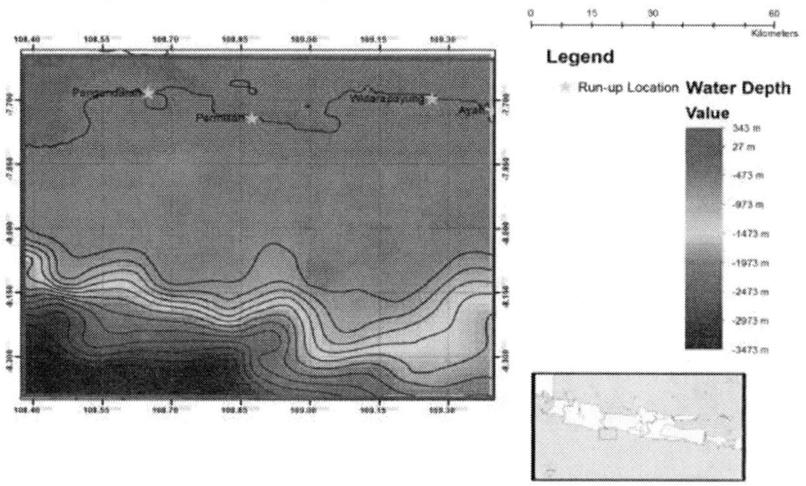

Figure 3. Bathymetry data and studied run-up location.

Table 2. Compiled observed run-up height.

No	Locations	Coordinates		Run-up Height (m)		
		Longitude	Latitiude	[2]	[11]	[17]
1	Ayah	-7.72356	109.3949	1	6.7	1.9
2	Widarapayung	-7.69597	109.2643	2-5	5	4.26
3	Pangandaran	-7.6833	108.65	3-5	7.8	5
4	Permisan	-7.74184	108.8749	20	5-8	10-20

5. Result and discussion

Based on the result of the simulation we compare the observed run-height on four locations on the southern coasts of Java and the maximum height from the models. The maximum height resulted from the simulation on those locations are presented in Table 3.

Table 3. Compiled simulated run-up height.

No	Locations	Coordinates		Simulated Maximum Run-up Height (m)		
		Longitude	Latitiude	[17]	[18]	[19]
1	Ayah	-7.72356	109.3949	1.3 - 2	0.6 - 1	2.7 - 3
2	Widarapayung	-7.69597	109.2643	2 - 2.4	1 - 1.5	3.5 - 3.8
3	Pangandaran	-7.6833	108.65	1 - 1.5	1.2 - 1.6	3 - 3.5
4	Permisan	-7.74184	108.8749	4.3 - 5	1.5 - 2	7.5 - 8

5.1.Model 1

Figures 4, 5, 6 show the tsunami propagation for the three models every 900s from 0 to 5400s with Model 1. With this model, using parameters from [17], tsunami waves first reach land at 1699s on Parang Kakapa Beach with a wave height of 7 meters. The Tsunami then hit Permisan at 2799s, Pangandaran at 3199, Ayah at 3259, and Widarapayung at 3600. The maximum run-height on Ayah is 1.3 to 2 m, 2 to 2.4 m on Widarapayung, 1 to 1.5 m on Pangandaran, and 4.3 to 5 m on Permisan. Figure x shows the maximum run-height on sub-grid 1. Compared to the observed run-up height, maximum run-up height from Model 1 fit with the data from [2] the most.

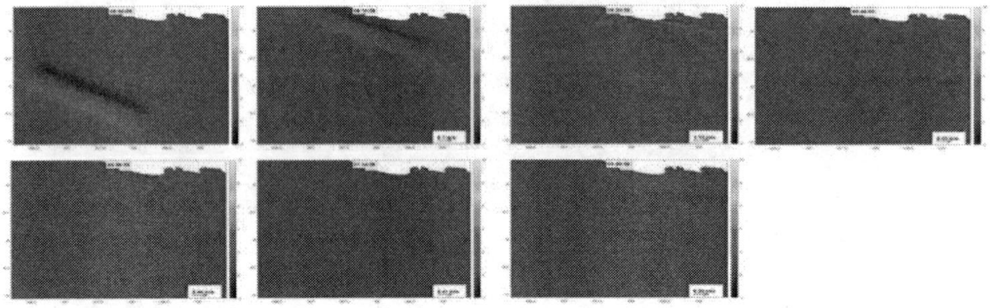

Figure 4. Tsunami propagations with Model 1 on parent layer grid every 15 minutes for 90 minutes

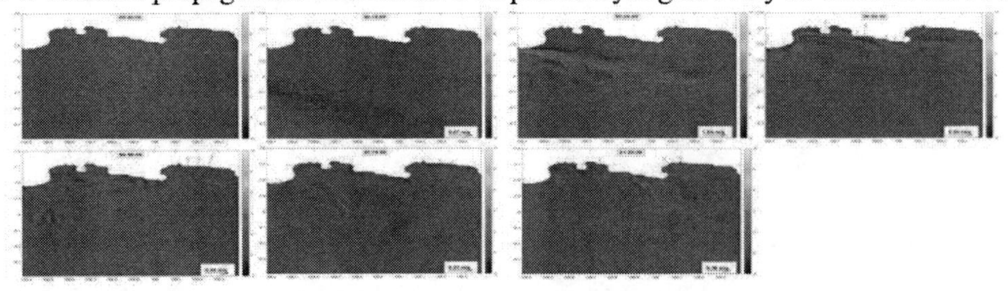

Figure 5. Tsunami propagations with Model 1 on sub layer 1 grid every 15 minutes for 90 minutes

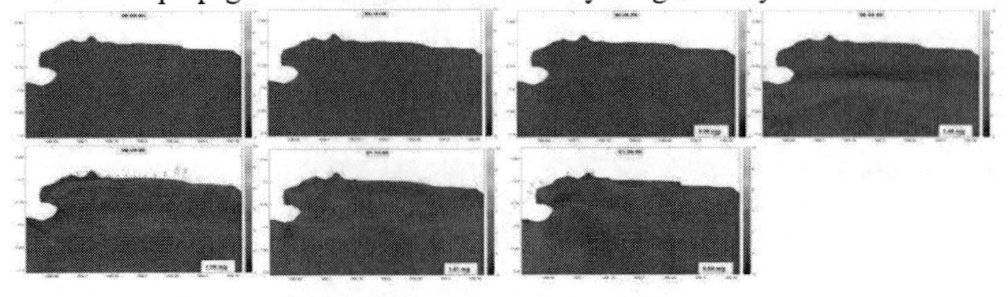

Figure 6. Tsunami propagations with Model 1 on sub layer 2 grid every 15 minutes for 90 minutes

5.2.Model 2

Figures 7, 8, 9 show the tsunami propagation for the three models every 900s from 0 to 5400s with Model 2. With this model, using parameters from research by [18], tsunami waves first reach land at 1699s on Parang Kakapa Beach with a wave height of 3 meters. The Tsunami then hit Permisan at 2799s, Pangandaran at 3199, Ayah at 3259, and Widarapayung at 3600. The maximum run-height on Ayah is 0.6 to 1 m, 1 to 1.5 m on Widarapayung, 1.2 to 1.6 m on Pangandaran, and 1.5 to 2 m on Permisan. Model 2 compared to the other 3 models generated the smallest tsunami waves and run-up height on the 4 study locations. Compared to the observed run-up height, the maximum run-up height from Model 2 is much smaller than any of the observed data. This may be caused by the small slip dislocation of only 2.5 m.

Figure 7. Tsunami propagations with Model 2 on parent layer grid every 15 minutes for 90 minutes

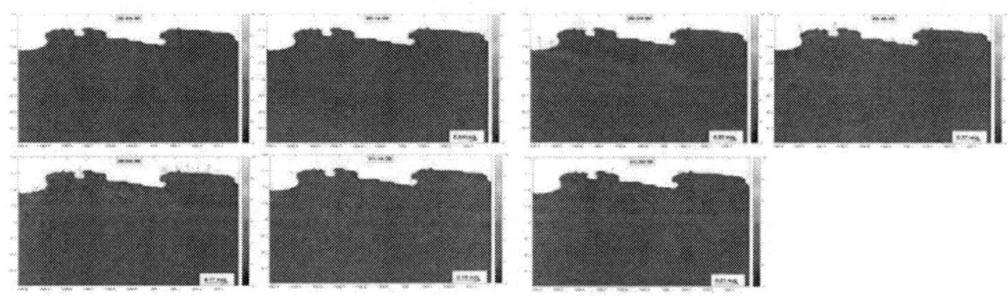

Figure 8. Tsunami propagations with Model 1 on sub layer 1 grid every 15 minutes for 90 minutes

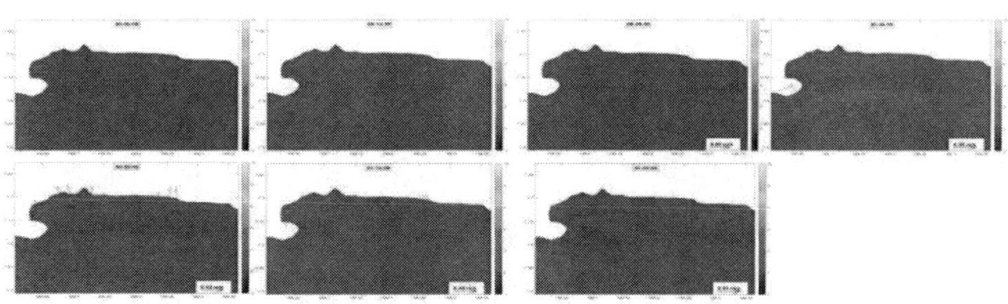

Figure 9. Tsunami propagations with Model 1 on sub layer 2 grid every 15 minutes for 90 minutes

5.3. Model 3

Figures 10, 11, and 12 show the tsunami propagation for the three models every 900s from 0 to 5400s with Model 1. With this model, using parameters from research by [19], tsunami waves first reach land at 1699s on Parang Kakapa Beach with a wave height of 12 meters. The Tsunami hit Permisan at 2799s, Pangandaran at 3199, Ayah at 3259, and Widarapayung at 3600. The maximum run-height on Ayah is 2.7 to 3 m, 3 to 3.8 m on Widarapayung, 3 to 3.5 m on Pangandaran, and 7.5 to 8 m on Permisan. Model 3 compared to the other 3 models generated the biggest tsunami waves and run-up height on the 4 study locations. Compared to the observed run-height, maximum run-up height from Model 3 fit closer with the data from USGS the most.

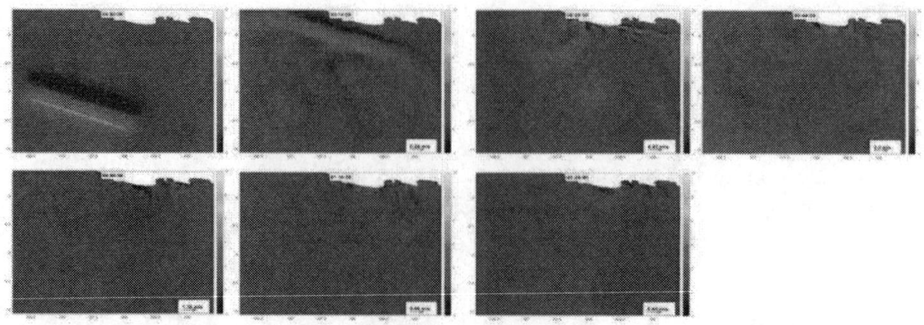

Figure 10. Tsunami propagations with Model 2 on parent layer grid every 15 minutes for 90 minutes

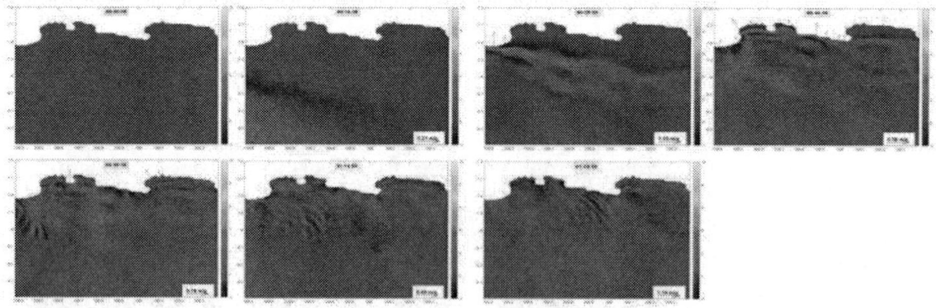

Figure 11. Tsunami propagations with Model 1 on sub layer 1 grid every 15 minutes for 90 minutes

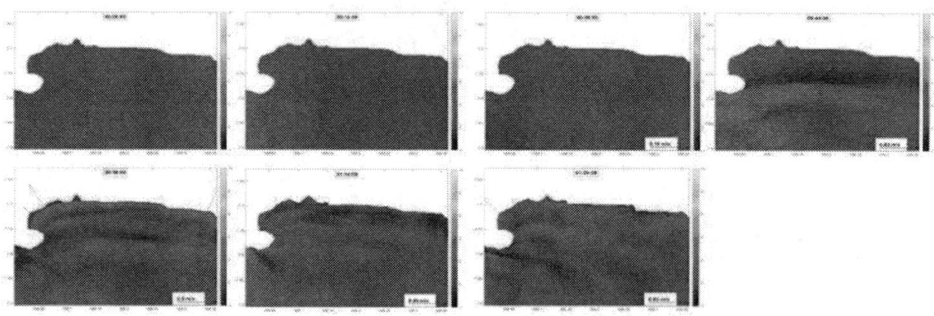

Figure 12. Tsunami propagations with Model 3 on sub layer 2 grid every 15 minutes for 90 minutes

5.4.Flow depth at Widarapayung

In this study, we also calculated the flow depth at Widarapayung Beach. With this, we intend to find out how far water inundated inland with all three models. This information will help in determining area vulnerable areas for future tsunami hazard mitigation. For all three models, we use non-linear shallow equations and take account of the bottom friction. A new grid was made around Widarapayung Beach with a grid size ratio of 3 to its parent layer. we also overlay satellite images taken in 2020 of Widarapayung Beach into the grid. Based on the simulated flow depth, in model 1 the tsunami wave inundated about 600-750 meters inland. In model 2 the tsunami wave inundated about 300-450 meters inland. While in model 3 the tsunami waves inundated about 750 to 1000 meter inland. Out of all three models, 2 inundated the shortest, and model 3 inundated the furthest. Figure 13 shows the flow depth of all three models and their wave height. However, data from [3] illustrate only 76-meter horizontal inundation at Widarapayung.

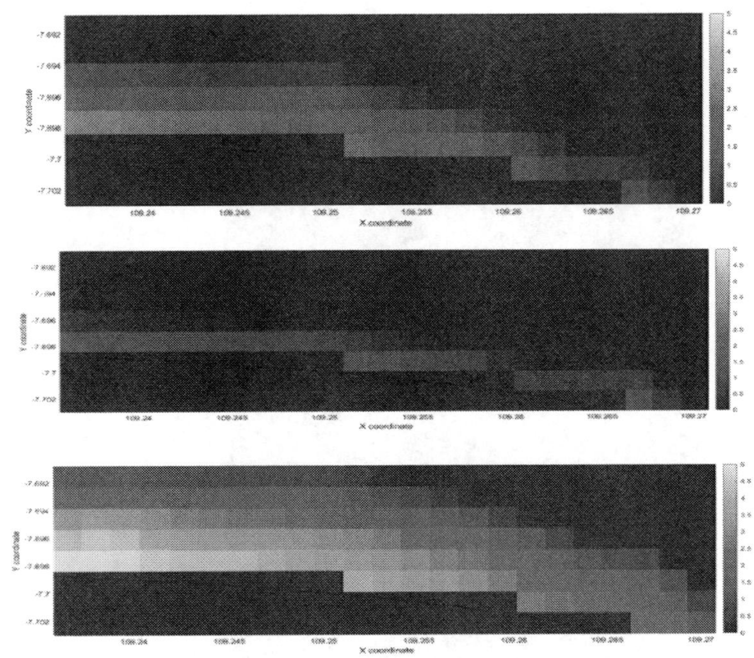

Figure 13. Simulated flow depth, black line refer to the coastline based on the ETOPO1 bathymetry data a). Model 1 b). Model 2 c). Model 3

6. Conclusion

Based on the result of the simulation (Fig. 14 and 15), model 1 and 2 generated too small tsunami waves and run-up height at the study location. This might be because of an insufficient amount of slip and focal depth. We concluded model 3 using the fault parameter compiled from research by [19] fit the observed run-up height better. Based on the simulation, the tsunami hit Permisan at 2799s, Pangandaran at 3199, Ayah at 3259, and Widarapayung at 3600. The maximum run-height on Ayah is 2.7 to 3 m, 3 to 3.8 m on Widarapayung, 3 to 3.5 m on Pangandaran, and 7.5 to 8 m on Permisan. The huge difference between observed run-up height and simulated run-up height at Permisan presumably because the data collected are from the inundated wave that hit the cliffsides around Permisan instead of the beach. Between the three models, it can be inferred that the slip amount and the focal depth had a big role in generating the maximum tsunami height. The horizontal inundation based on Model 3 is 750 to 1000 at Widarapayung.

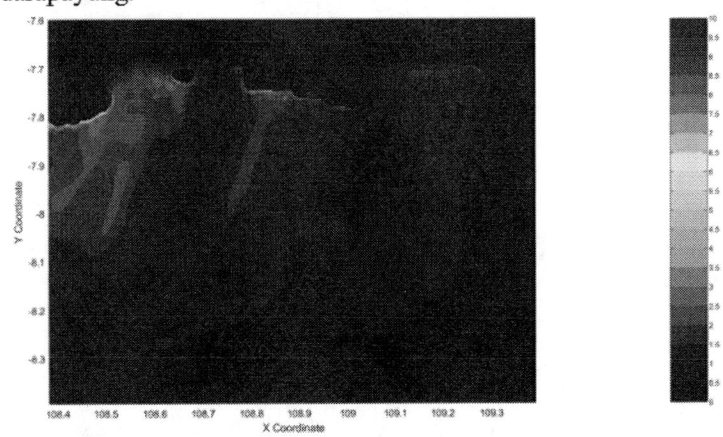

Figure 14. Maximum simulated wave height of model 1.

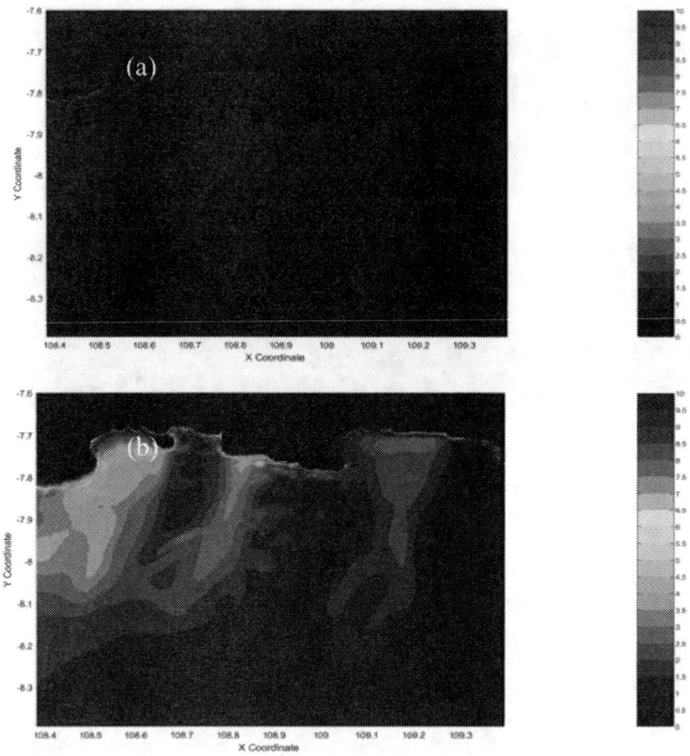

Figure 15. Maximum simulated wave height of model 2 (a) and model 3 (b)

7. References

[1] Fan Wenyuan, Basset D, Jiang J, Shearer P M and Ji C 2017 Rupture Evolution of the 2006 Java Tsunami Earthquake and the Possible Role of Splay Faults *Tectonophysics* 721: 143-150.

[2] Hanifa R, Meilano I, Sagiya S, Kimata F and Abidin H 2007 Numerical Modeling of the 2006 Java Tsunami Earthquake *Advances in Geosciences* 13: 231-248.

[3] Hartoko A, Helmi M, Sukarno M and Hariyadi 2016 Spatial Tsunami Wave Modelling for the South Java Coastal Area, Indonesia *International Journal of GEOMATE* 11(25): 2455-2460.

[4] Motoaki S, Ittetsu I, Ichiro A, Hiroyuki S and Junji K 2014 Numerical Simulations and Experiments on Tsunami for the Design of Coastal and Offshore Structures *Engineering Review* 46(2): 21-25.

[5] X Wang and W Power 2011 COMCOT: a Tsunami Generation Propagation and Run-up Model *GNS Science Report* 2011/43 129 p

[6] Y Fujii and K Satake 2006 *Geophysics Research Letter* 33 L24317.

[7] C J Ammon, H Kanamori, T Lay and A A Velasco 2006 *Geophysics Research Letter* 33 L24308.

[8] Samaras A G, Karambas Th V and Archetti R 2015 Simulation of Tsunami Generation, Propagation and Coastal Inundation in the Eastern Mediterranean *Ocean Science* 11:643-655.

[9] Selvan S C and Kankara R S 2016 Tsunami Model Simulation for 26 December 2006 and Its Effect on Koodankulam Region of Tamil Nadu Coast *International Journal of Ocean and Climate Systems* 7(2): 62-69.

[10] Rizal Y, Aswan, Zaim Y, Santoso W D, Rochim N, Daryono Anugrah S D, Wijayanto, Gunawan I, Yatimantoro T, Hidayanti, Rahayu R H and Priyobudi 2017 Tsunami Evidence

in South Coast Java, Case Study: Tsunami Deposit along South Coast of Cilacap *Earth and Environmental Science* 71:1-12.

[11] F Lavigne, C Gomez, M Giffo, P Wassmer, C Hoebreck, D Mardiatno, J Prioyono and R Paris 2006 Field observations of the 17 July 2006 Tsunami in Java *Natural Hazards Earth System Sciences* 7:177–183

[12] Bazykina A Y, Mikhailichenko S Y and Fomin V V 2018 Numerical Simulation of Tsunami in the Black Sea Caused by the Earthquake on September 12, 1927 *Physical Oceanography* 25(4):295-304.

[13] Kulikov E A, Gusiakov V K, Ivanova A A and Baranov B V 2016 Numerical Tsunami Modeling and the Bottom Relief *Physics of Earth, Atmosphere, and Hydrosphere* 71(6): 527-536.

[14] Kowalik Z 2012 Introduction to Numerical Modeling of Tsunami Waves *Institute of Marine Science University of Alaska* Fairbanks.

[15] Syamsidik Al'ala M, Fritz H M, Fahmi M and Hafli T M 2019 Numerical Simulations of the 2004 Indian Ocean Tsunami Deposits Thicknesses and Emplacements *Natural Hazard Earth System Science* 19:1265-1280.

[16] Amouzgar R, Liang Q and Clarke P J 2016 Computationally Efficient Tsunami Modelling on Graphics Processing Unit (GPU) *International Society of Offshore and Polar Engineers* 26(2):1-6.

[17] Raharja R, Gunawan E, Meilano, I, Abidin H Z and Efendi J 2016 Long Aseismic Slip Duration of the 2006 Java Tsunami Earthquake Based on GPS Data *Earthquake Science* 29(5): 291-298.

[18] Reid A 2012 Historical Evidence for Major Tsunamis in the Java Subduction Zone *Asia Research Institute Working Paper Series* 178:1-9.

[19] Prasetya T and Imamura F 2010 Numerical Simulation For The Assessment Of Tsunami Scenarios in Southern Yogyakarta *Bulletin of International Institute of Seismology and Earthquake Engineering* 44:121-126.

A Combination of Landsat-8 Satellite Imaging with Geothermal Manifestation Point at Wayratai Geothermal Prospect Area to identify *"the zone of interest"* on Geo-track and Geo-tourism planning

Taufiq[1,3] and Maharani[2]

[1] Geothermal Research, Research & Technology Center of Pertamina, South Jakarta, Indonesia.
[2] Graduated Student, University of Lampung, Bandar Lampung, Indonesia.

[3]E-mail: taufiq.tg10@gmail.com

Abstract. In this paper, we will identify the zone of interest on Wayratai geothermal prospect area based on the Satellite Imagery Map and Geothermal Manifestation Point to make a geo-track and geo-tourism planning in the future. Based on Satellite Imagery Map, we composite several bands, such as band (6,4,2) to describe an alteration rock and composite band (4,3,2) to describe a vegetation variation. After that, we process several bands to get the land surface temperature in this area, then we correlated the satellite data with surface measurement (with X-Ray Thermogun) on several geothermal manifestations. Based on both correlation, it shows a positive correlation, such as hot spring on the East Area with temperature 70-80°C is correlated by a high value of LST (indicated by dark-red zone), fumarole with temperature 74-78°C, and steaming ground on the West Area with temperature 26-59°C. Based on this data, then we make a geo-tourism planning on this area with make geo-track to connect all of "zone of interest" on satellite imagery with regional geology spot with a circular route, so in the future, Wayratai geothermal prospect area can be one of a geo-tourism spot on Lampung Province.

1. Introduction

Wayratai geothermal prospect area is located in the administrative area of Pesawaran, Lampung Province. It has a huge potential for tourism locations such as mountain, coastal, island, river, and waterfall that have not been explored and managed properly. This potential occurs because the geographical of Wayratai is directly related to Mt. Ratai in the West, Lampung Bay, and Pahawang Island in the East, Betung Mountain in the North, and Shark Teeth Geo-park in the South. Based on the tectonic setting, this area is the result of divergences between the two plates that meet in the southwest, It is the Indo-Australian plate (ocean plate) which has a basaltic composition (base) and Eurasian plate (continental plate) which has a granitic (acid) composition. The divergences of two plates and volcanic activities[3] in the past made this area have complex geology, it is dominated by lava, breccia and tuff rocks from young volcanic sediments, and we can found an alluvial composition such as cobble, pebble, sand, clays, and peat. Based on a previous preliminary survey and research result[6][12], we obtained several geothermal manifestations in the Wayratai area such as hot spring, solfatara, and the steaming ground that did not utilize and developed immediately. Then, in this research, we try to extract and calculate the land surface temperature from Satellite Imagery Landsat-8 data that combined with geothermal manifestation point from previous research[6][12] to get "the zone of interest" or we call it "spot with many geological or geothermal studies". After that, we made geo-tracking to connect all "the zones" on a circular route, and in the future, Wayratai can become one of geo-tourism place on Lampung.

2. Theory

In this research, we used two methods to collect data and analyze the data :

2.1. Collect Information from Regional Geology Map

Way Ratai geological map has a complex several formation. The first formation is dominated by young volcanic sediment from the eruption of complex Ratai volcano (Qhv-R) containing andesite basalt lava, breccia, and tuff. Then Menanga Formation (Km) contains alternating shale and claystone with basalt, intercalations of chert, and lenses of limestone. Then Alluvium contains a cobble, pebble, sand, clay, and peats. Then Hulusimpang Formation (Tomh) contains andesite basalt lava, tuff, and volcanic breccia with lenses of limestone. Then Sabu Formation (Tpos) contains alternating conglomerate breccia with sandstone. Then Dacite Piabung (Tmda) contains dacite and last is Tarahan Formation (Tpot) contain welded tuff, breccia with intercalations of chert[8].

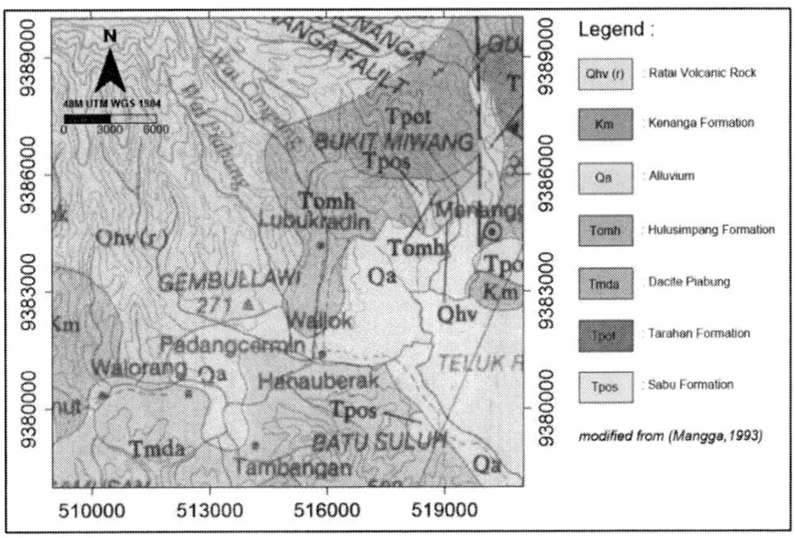

Figure 1. Geological Map of Wayratai Geothermal Prospect Area (Modified from Mangga, 1993).

2.2. Collect Information from using Landsat-8: Composite Band

Satellite Imagery Landsat-8 has several combinations to describe a variation on the surface such as vegetation, alteration of rock, hot spot, etc. In this case, we will use three Combination of Band (Composite Band), there are Natural View (Band 4,3,2), Atmospheric Penetration (Band 7,5,3), and Alteration Rock (Band 6,4,2). The First is Natural Band (Band 4,3,2), which is a typical combination of false color[5] cables as in aerial photographs. Channel 4 is to detect a part of vegetation and differentiates of vegetation, apart from separating the soil and conservation. This combination displays red, bright red vegetation indicating larger vegetation. Soil with less or zero vegetation between white (sand or salt) to green or brown depending on moisture and organic content. Waters is appeared a blue color, at clearwater will look dark blue or black, while on shallow water or high sediment concentration water will appear a light blue. And look a brown-blue style in a settlement.

Second is the Atmospheric Penetration[9] (Bands 7,5,3) give colors like a natural and also give a penetration ability of atmospheric particles, smoke, and fog vegetation looks blackish and light green when growing season. A settlement has appeared in white, gray, cyan, or purple. Sand, soil, and minerals show in a variety of colors. Almost all absorption in the middle of the infrared (IR) is in water, ice, and snow giving us a clear boundary of coastlines and sea. Snow and ice look dark blue, and water shows a black or dark blue. Hot surfaces such as forest fires and volcanic caldera absorb medium infrared (IR) and show in red or yellow. The application for this combination is forest fire monitoring.

Third, is Alteration Rock (Bands 6,4,2) will show like a "natural-like" as well as combinations of 4-3-2 but the difference is Band 6,4,2 can be used to detect a difference of vegetation and containing rock at this case is alteration rock for geothermal manifestation. Where the vegetation will appear dark red (because of IR penetration), the ground will appear brown-red, the water will appear blue, and last the alteration can show green. This combination is very useful in geology, agriculture, and wetlands.

2.3. Processing and analysis of Land Surface Temperature Map using Landsat-8

Land Surface Temperature is obtained by using satellite imagery map from Landsat-8 especially using Band 3,4,10 and 11. First step is radiometric calibration[1][2], is to process or convert a satellite imagery data with Digital Number (DN) format to radiance or reflectance format, as well as to brightness temperature (for Infra-Red Thermal Channels). For this case, note to attention a radiometric resolution which shows how many bits are used in a single pixel. In this research we will used gain offset to process satellite imagery map, with using radiance or reflectance multiple rescale factor (Gain) and additive rescale factor (Offset) with equation :

$$L_\lambda = M_L\, Q_{cal} + A_L \tag{1}$$

Where:

L_λ : TOA spectral radiance (Watts/(m2 * srad * μm))
Q_{cal} : Quantized and calibrated standard product pixel values (DN)
M_L : Band-specific multiplicative rescaling factor from the metadata (RADIANCE_MULT_BAND_x, where x is the band number)
A_L : Band-specific additive rescaling factor from the metadata (RADIANCE_ADD_BAND_x, where x is the band number)

Then we do the At-Satellite Brightness Temperature [10] with equation:

$$T = \frac{K_2}{\left(In\left(\frac{K_1}{L_\lambda + 1}\right)\right)} \tag{2}$$

Where:

T : At-satellite brightness temperature (K)
L_λ : ToA Spectral radiance (Watts/(m2 * srad * μm))
K_1 : Band-specific thermal conversion constant from the metadata (K1_CONSTANT_BAND_x, where x is the band number, 10 or 11)
K_2 : Band-specific thermal conversion constant from the metadata (K2_CONSTANT_BAND_x, where x is the band number, 10 or 11).

Brightness Temperature (T) is not a surface temperature on the earth, so to obtain a real temperature, At-Satellite Brightness Temperature must be changed to Surface Temperature by utilizing Emmisivity data (derived from NDVI data) by the formula[4]:

$$\mathrm{NDVI} = (\mathrm{NIR} - \mathrm{VIS})/(\mathrm{NIR} + \mathrm{VIS}) \tag{3}$$

After get NDVI value, we continue to find LSE value to get Land Surface Temperature with equation:

$$\mathrm{LSE} = \mathrm{NDVI} - (\mathrm{NDVI_{min}} \,/\, \mathrm{NDVI_{max}} + \mathrm{NDVI_{min}}) \tag{4}$$

$$LST = \frac{T_B}{\left[\left(1 + \left(\lambda * \frac{T_B}{C_2}\right) * In(e)\right)\right]} \tag{5}$$

Where:

LST	: Land Surface Temperature (°K)
λ	: Wavelength of Emitted Radiance
C_2	: 14388 µm.K
T_B	: Band-specific brightness temperature conversion from the metadata
e	: LSE value

3. Result and Discussion

Wayratai geothermal prospect area has a huge potential of tourism location such as mountain, coastal, island, river, and waterfall that have not been explored and managed properly. In this research, we will use the satellite imagery map using Landsat-8 to know and correlate the geothermal manifestation with a land surface temperature that made a geo-track for geo-tourism.

Based on Satellite Imagery Map, we used a composite band or combination band with the band (4,3,2) to know the area as a natural like[5][7] (see **Figure 2**). On the composite map, we will look at the penetration of forest or trees vegetation with black-green color, then the soil, land, and flow of sediment river can be seen in brown color, and last the water or sea shows with dark-blue color. This view of the composite band can be used to track the flow of the sediment river from west to east the research location, so we can know the heat-flow water from Mt. Ratai as a source (in the West direction).

Second composite map, combined (band 7,5,3) (see **Figure 3**) as an Atmospheric Penetration[9], we will look the penetration of forest or trees vegetation with green color, then the soil or land is indicated with brown color, and river or water object can be seen in blue until dark-blue color. This view of the composite band can be used to track vegetation, soil, and stagnant water and flow of water indicated from the area. So based on this data, we can know the vegetation and water content of Wayratai area.

Third composite map using (band 6,4,2) (see **Figure 4**) as an Alteration Rock Penetration, we will look the penetration of forest or trees vegetation with dark-red color, then the soil, land, and river are indicated with brown-red color, and alteration rock object can be seen in dark-green color. This view of the composite band can be used to track vegetation, a soil, and an alteration rock caused by geothermal manifestation on these locations. So based on this data, we can know the alteration of rock in Wayratai.

Then we used Band 10 and Band 11 to process Land Surface Temperature Map[11] (see **Figure 5**), after the process, we can look at the difference of temperature (**Figure 5** on the right side), on this map, temperature move from low until high temperature. The low temperature indicated by green

IOP Conf. Series: Materials Science and Engineering **982** (2020) 012048 doi:10.1088/1757-899X/982/1/012048

until yellow color (cloud effect value) and high temperature indicated by orange to red color. Based on this map, we obtained the red zone is indicated by hot spot or geothermal manifestation because the red zone has a high temperature, is equal to 20-30 °C on the rescaling of satellite imagery.

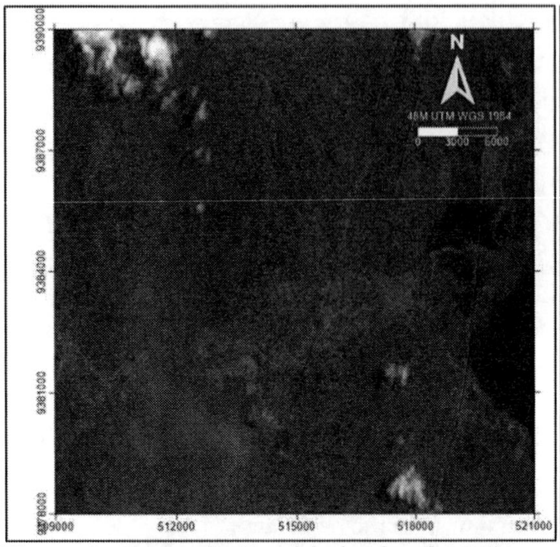

Figure 2. Composite Band (using band 4,3,2) as a Natural View

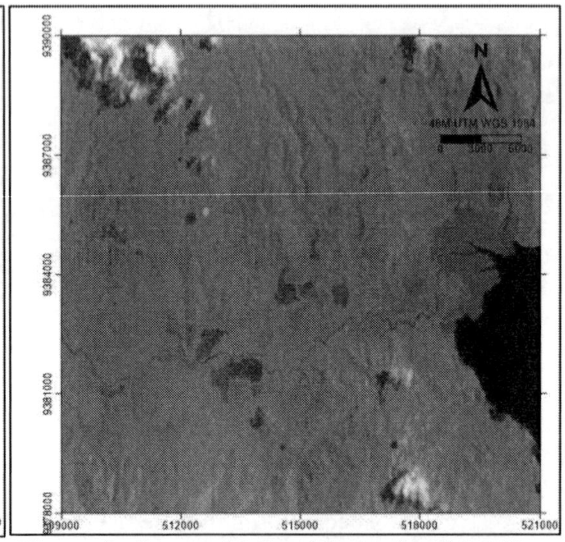

Figure 3. Composite Band (using band 7,5,3) as *an* Atmospheric Penetration

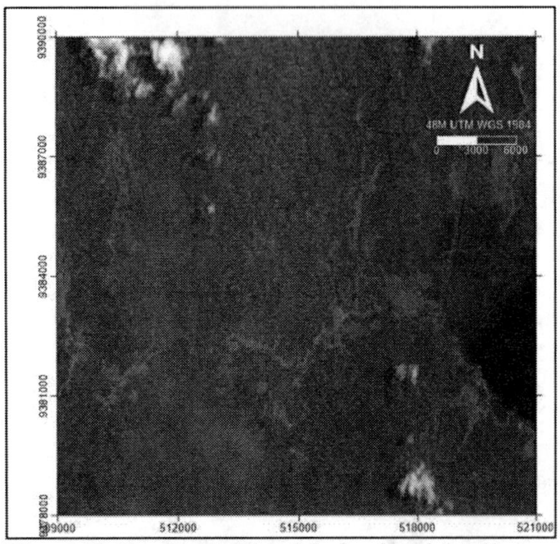

Figure 4. Composite Band (using band 6,4,2) as an Alteration of Rock Penetration

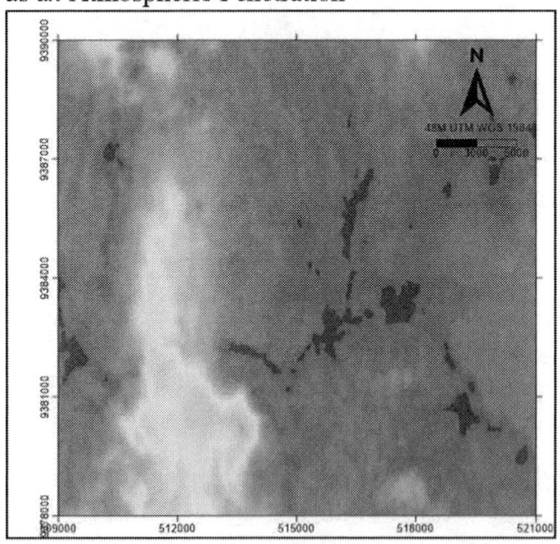

Figure 5. The result of Landsat-8 process to get Land Surface Temperature Map

Table 1. The temperature of Geothermal Manifestation on Wayratai Geothermal Prospect Area

Geothermal Manifestation Point	Temperature
Area 1 (*Steaming Ground*)	26-59 °C
Area 1 *(Hot Spring)*	70-80 °C

Area 1 (Fumarol)	74-78 °C
Area 2 (Solfatara)	78-81 °C

Based on Satellite Imagery Map especially Land Surface Temperature, then we do the tracking on geothermal manifestation in this area based on previous results [6][12]. First, we track on near settlement at the east and then we follow the sedimentation from the river, trace the manifestation and heat from the hotspot at this map, then we get a geothermal manifestation on this area such as solfatara, hot spring, mud pool, and steaming ground. After that, we used X-Ray Temperature Detector to measure surface temperature on the geothermal manifestation (see **Table 1**).

After we plot a manifestation point on the top of Land Surface Temperature Map (**Figure 6**), we look a positive correlation between the high temperature on LST Map and Geothermal Manifestation on the red-box (**Figure 6**). This result can be plan to make a geo-tracking in the next research. Geo-tracking is a geo-travel journey that follows a trail that have been designed to be tracked either by foot or by vehicle, this method is combined with satellite imagery model to facilitate geo-tourism travel in the area of Wayratai. Therefore, the challenges is not unavailability track to the surface manifestation, second is geothermal manifestation is located on the slopes of Mt. Ratai, and last is the geothermal manifestation have a steaming ground as a weak ground so we will carefully to step in this area.

From **Figure 6.** Many areas have not been roamed directly due to limited access to the location, but from the land surface temperature map, we can track the area in knowing the existence of the hotspot zones to make Wayratai become geo-tourism area. To develop geo-tourism on the geothermal manifestation of Wayratai area, we studied in two aspects that is destination and market, but the concept of geo-tourism can not be separated from the conservation area hence in the form of a group driving nature tour which is managed by the local population. In the aspect of the destination, we can utilize the surface manifestation to be studied and developed as a hot pool or bathpool area. Wheres in market aspect we can utilize yields like coffee, cocoa, some type of fruit and handicrafts typical Wayratai, as self-managed to advance BUMDES (Business owned by the village) at Wayratai area so from that to be used for the management of geo-tourism area of the sight further.

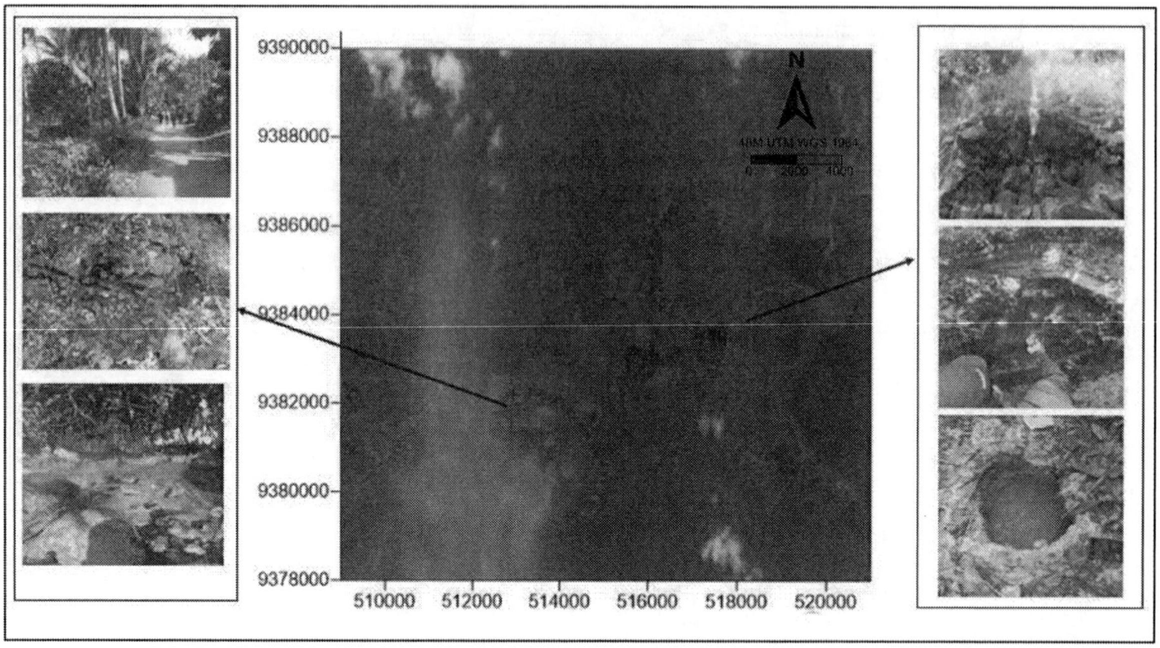

Figure 6. A Correlation of Geothermal Manifestation and Land Surface Temperature (LST) Map from Landsat-8 to make a geo-tourism track on geothermal manifestation point

5. Conclusion

The conclusions derived from this paper are:

1. Based on Geological Map, we conclude the Wayratai area is dominated by young volcanic sediment from the eruption of Ratai volcano (Qhv-R) containing andesite-basalt lava, breccia, and tuff.
2. Based on Composite Band, we conclude the alteration rock can be described with a band (6,4,2) that shows the dark-green color and the vegetation can be described with the band (4,3,2) as a green form.
3. Based on Land Surface Temperature Map, the high temperature is indicated by dark-red color with value is 20-30 °C.
4. Based on X-Ray Temperature Measurement from Geothermal Manifestation, we obtained the area 1 (east) have a steaming ground with values 26-59 °C, Hot spring with values 70-80 °C, and Sulfatara with values 74-78 °C on Area 1 and 74-78 °C on Area 2.
5. Based on the land surface temperature map and geothermal manifestation point, we can correlate the temperature and the geothermal manifestation point in the next research as a geo-tracking circular on the Wayratai geothermal prospect area.

Acknowledgment

We would thank to University Lampung to support this plan to make the geo-tracking concept a geo-tourism on Lampung Province.

References

[1] Bakruddin, Utama W, and Warnana D D 2016 Application of Landsat-8 Satellite to Identified Structure at Arjuno-Weilerang Geothermal Area East Java *Proceedings APTW IX, **I**,* 37-44.

IOP Conf. Series: Materials Science and Engineering **982** (2020) 012048 doi:10.1088/1757-899X/982/1/012048

[2] Chander G, Markham B, and Helder D L 2009 Summary of Current Radiometric Calibration Coefficient for Landsat MSS, TM, ETM+ and EO-1 ALI sensors *Remote Sensing of Environtment*
[3] Goff F and Cathy J J 2000 Encyclopedia of Volcanoes : Geothermal system *Academic Press* 817-834.
[4] Jaelani L M 2010 Land Surface Temperature Mapping at Medium Scale Using Landsat-8 – TIRS Satellite Image Geomatics (Institute of Sepuluh November Surabaya).
[5] Lopes F C, Pereira A J S C, and Mantas V M 2012 Mapping of Salt Structure and Related Fault Lineament Based on Remote Sensing and Gravimetric Data: The Case of Monte Real Salt Wall (Onshore West-Central Portugal) *AAPG Bulletin,* **96** 615-634.
[6] Karyanto and Haerudin N 2013 Kualitatif Interpretation of surface temperature at Wayratai Geothermal Prospect Lampung *National Sains and Technology V Course* (Research Institution of University of Lampung).
[7] Hu T, Yang J, Li X, and Gong P 2016 Mapping Urban Land Use by using Landsat Imaging and Open Social Data *Remote Sensing,* **8**, 151.
[8] Mangga S A, Amirudin, Suwarti T, Gafoer S, and Sidarto 1993 Geological Map of the Tanjungkarang Quadrangle Sumatera *Geological Research and Development Center*.
[9] Seldak P 2002 Using Landsat TM Data for mapping of the Quatenary Deposits in Central Sweden *Geographica, 37*.
[10] Ridwan M A, Radzi N A M, Ahmad W S H M W, Mustafa I S, Din N M, Jalil Y E, Isa A M, Othman N S, and Zaki W M D W 2018 Application of Landsat-8 Data: a Survey *International Journal of Engineering & Technology,* **7** 436-441.
[11] Solanky V, Singh S, Katiyar S K 2018 Land Surface Temperature Esrimation using Remote Sensing Data *Hydrogeologic Modelling,* 343-351.
[12] Suharno, Amukti R, Hidayatika A, and Putri M K 2015 Geothermal Prospect of Padang Cermin Pesawaran Lampung Province Indonesia *Proceedings World Geothermal Congress 2015*.

IOP Conf. Series: Materials Science and Engineering **982** (2020) 012049 doi:10.1088/1757-899X/982/1/012049

Gravity Modelling And Second Vertical Derivative Calculation To Analysis Subsurface Structure At Wayratai Geothermal Prospect Area, Lampung

Taufiq[1]

[1]Geothermal Research, Research & Technology Center of Pertamina, South Jakarta, Indonesia.

E-mail: taufiq.tg10@gmail.com

Abstract. Wayratai Geothermal Prospect Area, Lampung is near a cluster of Mount Ratai with geological condition in the form of young volcanic sediment such as lava, breccias, and tuff obtained from the eruption of Mount Ratai. To determine geothermal potential in Wayratai area, we needed some preliminary surveys such as geological, geochemistry, and geophysical survey. Based on geological mapping, we obtained several point of manifestations such as 5 hotspring points (with a temperature 86-90°C), mudpool, steaming ground, and altered rock. Afterthat, performed with a geophysical survey using the gravity method to determine subsurface structure and describe the initial conditions of the geothermal system. Gravity method is used because this method is sensitive to analyze the vertical changes, and can describe the subsurface conditions based on density variation of rock. Based on data acquisition and literature studies performed on Wayratai geothermal prospects area, obtained 2D and 3D subsurface model based on density variations. From the model, obtained area with a density value 2.3-2.6 gr/cm^3 and depth from 400 to 2000 m were assumed to be a reservoir zone. Then, the area with a density value 2.8-3.2 gr/cm^3 and depth from 2000-2500 m assumed to be a heat source. With these density, we can assume that the rock in reservoir area is calcareous-breccia and in the heat source is an andesitic-basaltic rocks. Based on the analysis of second vertical derivative calculation, obtained value of SVD maximum is greater than SVD minimum which indicated that the area is a basin structure or normal fault, and its strenghtening of the assumption that the modelling of the geothermal reservoir is graben form.

1. Introduction

Wayratai geothermal prospect area is located at Pesawaran district, 40 km from the capital city of Lampung Province with coordinates 5°36.7'S and 105°10.38'E. Wayratai geothermal prospect area bordered by two clusters of young volcanic, namely Mt. Ratai and Mt. Pesawaran are located in the north, Mt. Menangis in the west and Ratai Bay in the east. With this conditions, Wayratai area is attracted to do an preliminary study on the geothermal potential. Some preliminary studies were conducted to determine geothermal potential such as geological, geochemistry, and geophysical surveys. Based on the geological mapping performed in Wayratai area, we obtained several point of manifestations such as 5 hotspring points (with an average temperature of 85-98°C)[4], mudpool, steaming ground, and altered rock (silica sinter) spreading at the prospect area. From the results of geological and geochemical surveys, geophysical surveys then performed with using the gravity method to determine the subsurface conditions.

2. Theory

In this research, we collect the data information from:

2.1. Regional Geology

The condition of the surface geology at Wayratai geothermal prospect area (Figure 1) is composed by the composer deposition of swamp deposits (Qs), alluvial (Qa), young volcanic deposit (Qhv),

Lampung Formation (QTl), Hulusimpang Formation (Tomh), Sabu Formation (Tpos), Tarahan Formation (Tpot), Menanga Formation (Km), unconsolidated sediment from Mt.Kasih, Way Galih Schist (Pzgs), Piabung Dacite, and Dulan Granodiorite[8].

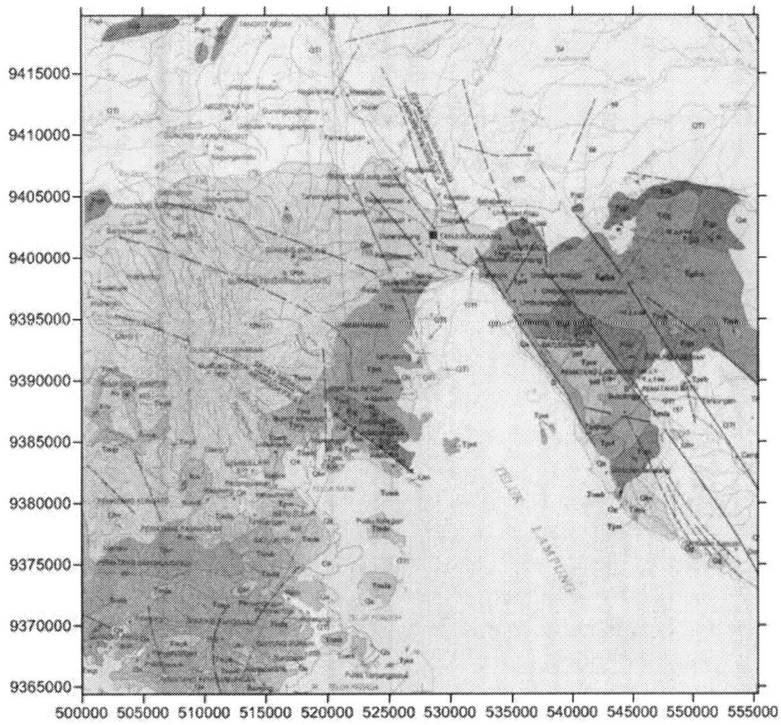

Figure 1. Geological Map of Wayratai Geothermal Prospect (Modifier from[5]).

Based on structural geology, Sumatera island is located along the southwestern edge of the Sunda Shelf, on the extension of the Eurasian Plate to mainland Southeast Asia is part of the Sunda Arc. Oceanic crust that have lay a portion of the Indian Ocean and the Indian-Australian Plate, have subducted tilted along the Sunda Trench in the west coast of Sumatra. Lane slope meeting is included in the Sunda Arc trench system that stretched more than 5,000 miles from Burma to eastern Indonesia[5].

The location of the arc and trench are contained now that possible since the Miocene. The pressure caused by the oblique subduction periodically mirrored by parallel faults to the edge of the plate and proved in the Sumatra Fault System which runs along the island arc sequence and Merentas sequence. With connected to the arc of magma, from west to east, Sumatra can be divided into four tectonic, namely: Mentawai lane, Bengkulu Fore Arc lane, Magma Arc lane, and Barisan Rear Arc lane or Jambi-Palembang lane[8].

2.2. Geothermal System

Geothermal energy is the heat stored in the rock beneath on the earth's surface and the fluid contained therein. Geothermal system consist of elements that made up the system. Essential elements of constituent geothermal system is separated in three aspects, contained a heat source, a permeable reservoir rock, and fluid that bring the heat flow[2].

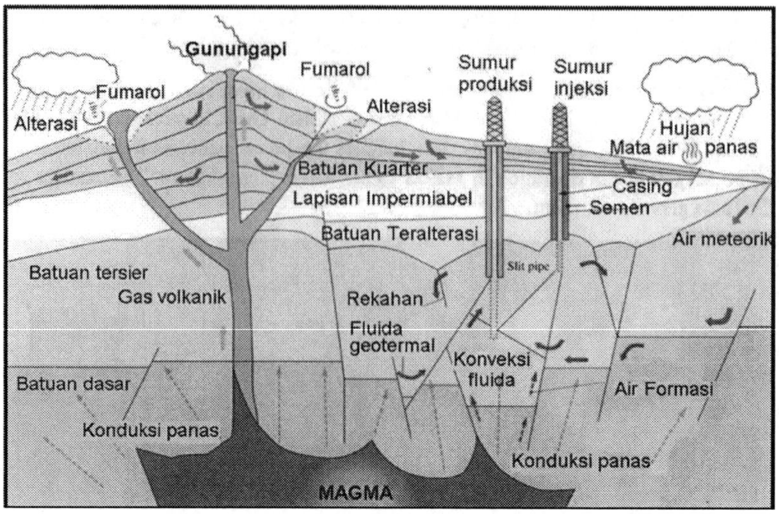

Figure 2. Geothermal System.

White[9] was assumed that the geothermal fluids contained in hydrothermal reservoir come from surface water, such as rainwater (meteoric water) which penetrated the subsurface and heated by a heat source (Figure 2). Afterthat, water will enter the cracks into permeable rocks. The heat will be propagated through the rock (by conduction) and through the fluid (by convection). Heat flow then transfered by convection occurs because buoyancy. Gravity makes the water always has a tendency to move down. But, if the water is in contact with a heat source, the heat transfer will occur so that the water temperature becomes higher and the water becomes lighter. This causes more hot water moves upward and cooler water moves downward to the bottom reservoir, make the water in circulation or convection system.

The temperature of reservoir depend on the intensity of the heat that goes from the heat source rock, characteristic of rock thermal, such as the ability to spread out the hot rock and store heat (conductivity and heat capacity of rock), the ability of the fluid in the rock (permeability). Because of the diversity of rock properties, of course, the temperature from one place to another isn't same, unique, within the reservoir temperature is not homogeneous, and different from one reservoir to another reservoir. Based on the magnitude of temperature, Hochstein[3] classifies geothermal system into three, namely:

- The high-temperature reservoir, is a system with containing fluid in reservoir with temperature above 225°C.
- The medium-temperature reservoir, is a system with containing fluid in reservoir with temperature between 125°C and 225°C.
- The low-temperature reservoir, is a system with containing fluid in reservoir with a smaller temperature below 125°C.

2.3. Gravity Method

The theory of gravity is applied at Newton's Law which state that the force of attraction between two particle depends on the distance of the mass of each particle, which formula :

$$\vec{F}(r) = -G\,\frac{m_1\,m_2}{r^2} \qquad (1)$$

Where :

F (r) : Force of attraction (N)

M_1, m_2 : Mass of object 1 and 2 (kg)
r : Distance beetwen two object (m)
G : Universal Gravitation constant (6.67×10^{-11} m^3 kg^{-1} s^{-2}).

Gravity anomaly read on the surface or called Complete Bouguer Anomaly (CBA) is the value of gravity observations that have been reduced with elevation and terrain correction. The equation is:

$$\Delta g(x,y,z) = g_{obs} - (g_{\varphi} - FAC + BC - TC) \quad (2)$$

Where :
Δg (x,y,z) : Complete Bouguer Anomaly (mGall)
g_{obs} : Observed Gravity
g_{φ} : Latitude Correction
FAC : Free Air Correction
BC : Bouguer Correction
TC : Terrain Correction

2.4. Spectrum Analysis & Moving Average Filter

Spectrum Analysis is used to determine windows lenght in filtering process, can be used to separate regional and residual anomaly. Spectrum analysis using fourier transformation to change anomaly to be sinusoidal signal in various of frequency. The result of fourier transformation are formed amplitudo spectrum and phase spectrum. With this transformation, we can obtain the information about depth with the gradient, can estimate the value of wave number (k) and amplitude (A) to calculate windows length, this value is used to filtering process to separate about regional and residual anomaly. Blackely[1] makes the derivative of spectrum from gravity potential that observed from horizontal area, afterthat, he do the logaritmic from result of fourier transformation, and get relationship between amplitudo (A) with wave number (k) and depth (z_0-z') :

$$In\, A = (z_0 - z')|k| \quad (3)$$

Anomaly then separate with using moving average filter which have a character to pass anomaly with high frequency (High-Pass Filtering), there are not replace phase and rectangular filter, so can conform to process gravity data. Moving Average Filter equation in 1D and 1 windows is[6] :

$$\Delta g_n(x_i) = \Delta g(x_i) - \frac{1}{N}\sum_{k=-n/2}^{k=n/2} \Delta g(x_i - k) \quad (4)$$

Whereas moving average filter in 2D for windows width NxN is:

$$\Delta g_{reg}\left(\frac{N+1}{2}, \frac{N+1}{2}\right) = \sum_{i=1}^{N}\sum_{i=1}^{N} \frac{\Delta g(i,j)}{N^2} \quad (5)$$

After that the residual anomali is:

$$\Delta g_{res}(i,j) = \Delta g(i,j) - \Delta g_{reg}(i,j) \quad (6)$$

2.5. Second Vertical Derivative

Second Vertical Derivative method is used to show local or regional anomaly. This method is good to know discontinuity from subsurface structure, especially fault structure in a survey area. Based on theory, this method obtained from derivative of Laplace Equation for gravity anomaly in surface with equation[7] :

$$\nabla^2.\Delta g = 0 \tag{7}$$

or :

$$\frac{\partial^2 \Delta g}{\partial x^2} + \frac{\partial^2 \Delta g}{\partial y^2} + \frac{\partial^2 \Delta g}{\partial z^2} = 0 \tag{8}$$

so second vertical derivative can be:

$$\frac{\partial^2 \Delta g}{\partial z^2} = -(\frac{\partial^2 \Delta g}{\partial y^2} + \frac{\partial^2 \Delta g}{\partial x^2}) \tag{9}$$

For 1-D data (graphical data) the equation is:

$$\frac{\partial^2 \Delta g}{\partial z^2} = -\frac{\partial^2 \Delta g}{\partial x^2} \tag{10}$$

Second vertical derivative equation and 1-D data show the second vertical derivative from the surface gravity anomaly is equal from negative value of second horizontal derivative. It means that second vertical derivative anomaly can through horizontal derivative which practically more easy to do it. From calculation, we can obtain the characteristic:

- For basin/normal fault applied :

$$\left(\frac{\partial^2 \Delta g}{\partial z^2}\right)_{maks} > \left|\left(\frac{\partial^2 \Delta g}{\partial z^2}\right)_{\min}\right| \tag{11}$$

- For intrusion/reverse fault applied :

$$\left(\frac{\partial^2 \Delta g}{\partial z^2}\right)_{maks} < \left|\left(\frac{\partial^2 \Delta g}{\partial z^2}\right)_{\min}\right| \tag{12}$$

3. Result and Discussion

Figure 3. Wayratai Geothermal Prospect Area view in Google Earth

Wayratai geothermal prospect area is located at the coordinates 5° 36.7' S and 105° 10.38' E, 40 km from Bandar Lampung which can be reached in 2 hours drive (see Figure 3). Gravity measurements performed by gravity base points at the Wayratai Brigid 3 base that calibrated by gravity base points at Physics Building University of Lampung. Gravity measurements were made from 10 until 12 June 2015 as many as 30 point measurement points with grid spacing 200 meters.

Based on measurement results show the Wayratai geothermal prospect area has an altitude from 142 m to 548 m (see **Figure 4 left**). From gravity Observation data (see **Figure 4 right**), show the area with high anomaly in the Southeast direction with value 977985-978015 mGall and the area with low anomaly in the Northwest direction with a value 977910-977945 mGall. By using topographic data and g.Obs data, we can conclude that both of data have a reverse value (correlation data is -1), its indicated that the data acquired is good.

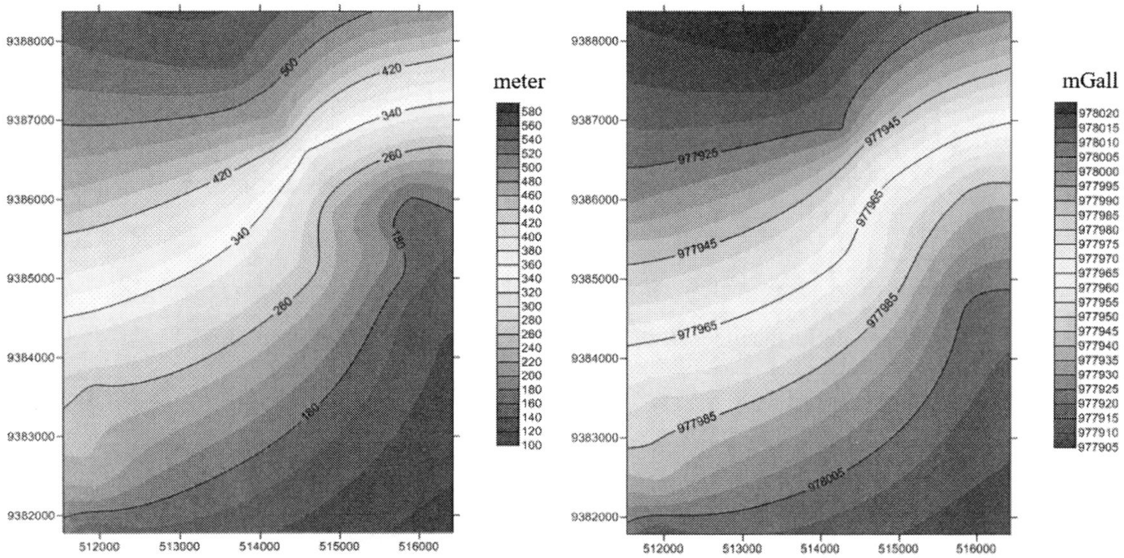

Figure 4. Topographic Map (left) and Gravity Observation (right) Map

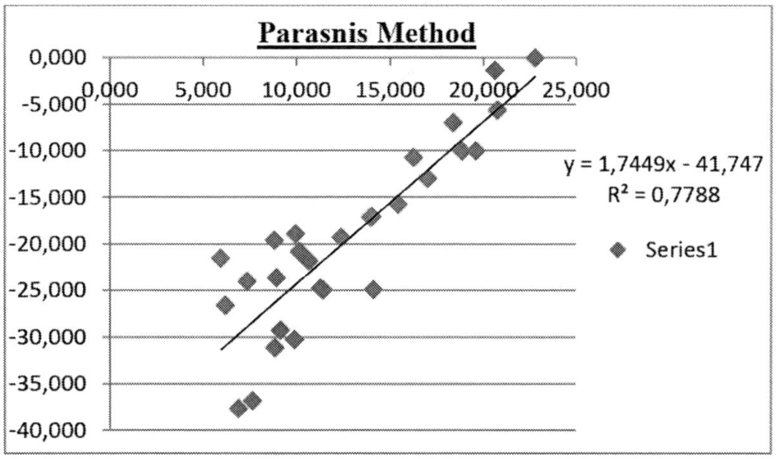

Figure 5. Surface Density Estimation Chart with Parasnis Method

In this research, we used Parasnis and Nettletton method to analyze estimation of surface density that use in Bouguer and Terrain correction. With Parasnis method, obtained the surface density estimation in research area is 1.75 gr/cm^3 (see **Figure 5**). With Nettleton method, obtained the surface density estimation is 1.75 g/cc (see **Figure 6**). Based on both results, we conclude to use that value (1.75 gr/cm^3) as a surface density in correction. G.Obs data have corrected for latitude, free air, Bouguer and terrain called Complete Bouguer Anomaly (CBA).

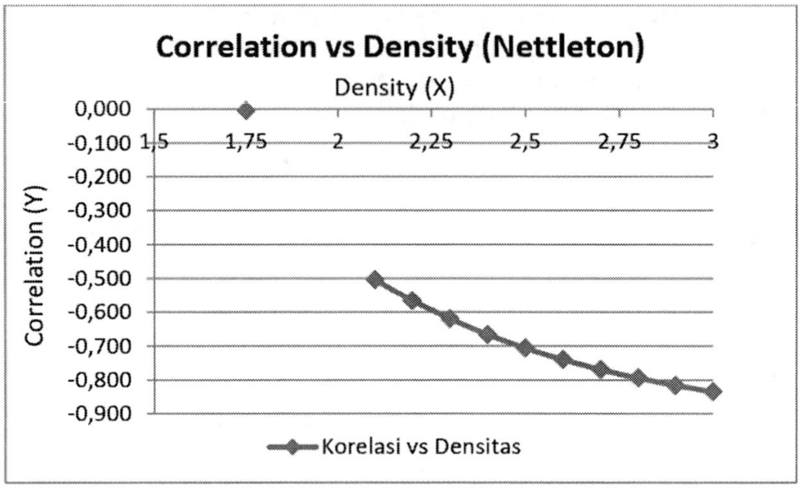

Figure 6. Surface Density Estimation Chart with Nettleton Method

Based on the CBA map (see **Figure 7**), showed the area with low anomaly from -45 to -49 mGall in the middle of research area, and the area with high anomaly from -32 to -38 mGall is located in the northern and southern of research area.

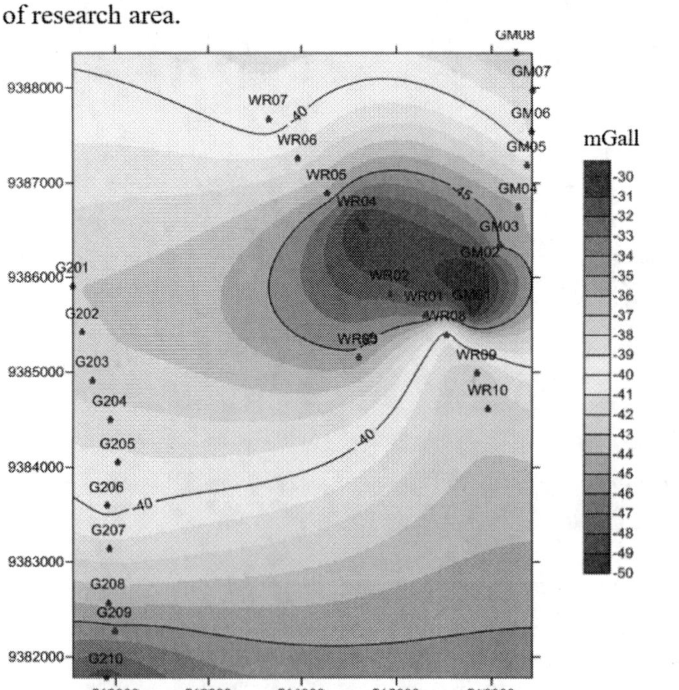

Figure 7. Complete Bouguer Anomaly Map

In this research, we performed the 2D Forward Modelling and 3D Inverse Modelling to describe subsurface structure. The first is Forward Modeling, in this method we require the spectrum analysis (with Fourier transformation) to determine the window length in filter to know boundary between regional zones, residual zones, and noise. Based on spectrum analysis, obtained the crossplot graph between In A and K (see Figure 8) that shows linear regression between the zones which gained the window width is 25x25, with depth classification: the surface zone is 33 m, the residual zone is 307 m and the regional zone is 1809 m. From this data, can be a reference to make a 2D Forward model of the subsurface structure.

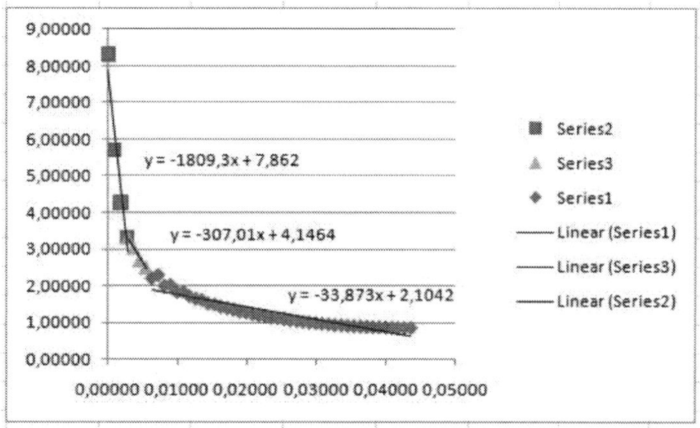

Figure 8. Spectrum Analysis (windows width 25x25)

Afterthat, filtering process is requiring to separate CBA data into difference frequence. In this research, we use Moving Average Filter with window width 25x25 to get regional map (see **Figure 9 left**) and residual map (see **Figure 9 right**). If we see at residual map, we can get a different anomaly between low and high anomaly in center of the map. By slicing from point A to point B on residuals map, we can obtain a data to make the 2D Forward Modelling.

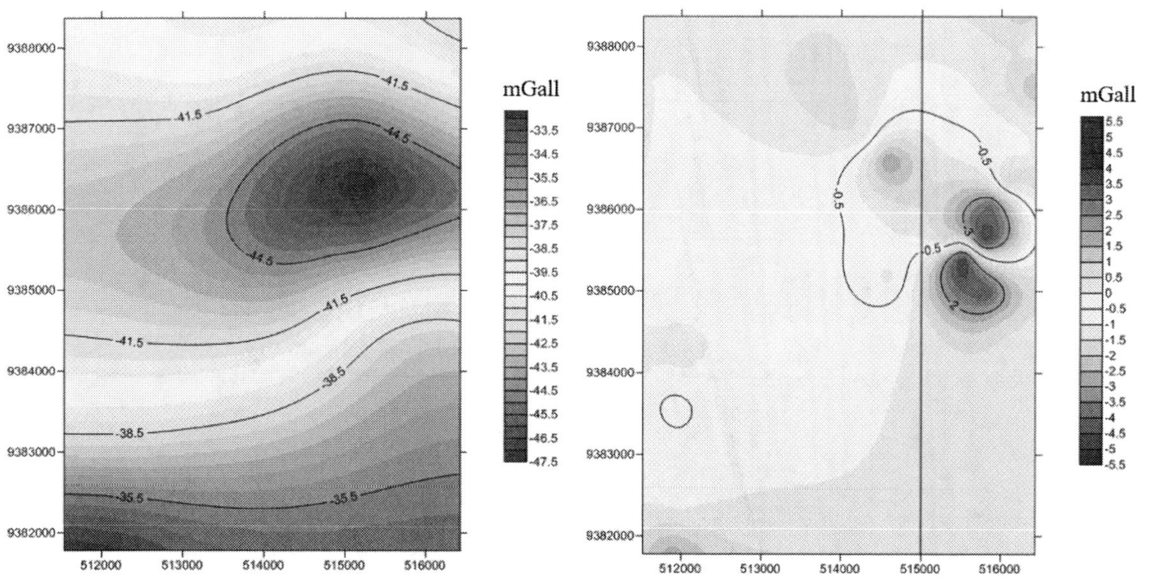

Figure 10. Regional Anomaly Map with filter 25x25 (left) and Residual Map (right)

IOP Conf. Series: Materials Science and Engineering **982** (2020) 012049 doi:10.1088/1757-899X/982/1/012049

To identify the structure boundary in research area, we analysis the gradient in the form of horizontal gradient (used to know the structure boundary in horizontal) and vertical gradient (used to identify the structure boundary in vertical). In this research, SVD analysis is conducted with Elkins filter to get the structure of research area, and with calculation of gradient analysis from 1-D SVD graph obtained the structure boundary and type of structure to reference the 2D modelling.

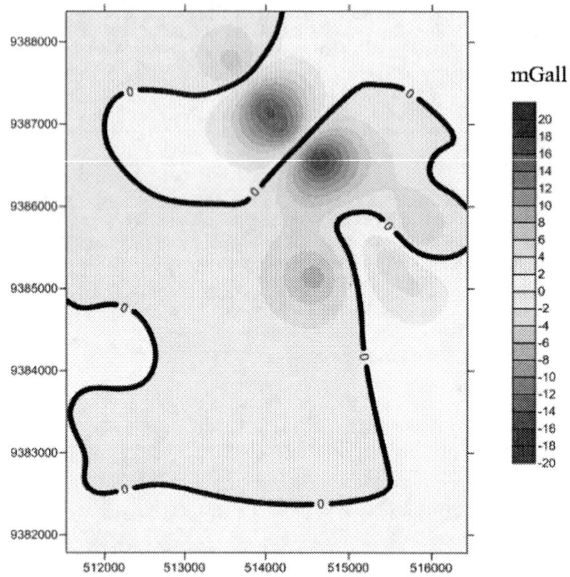

Figure 11. Second Vertical Derivative (Elkins) Mapping

From SVD mapping (see **Figure 11**), obtained the area with SVD value equal than 0, this indicated the fault zone or the structure boundary in the center of the map and around of negative anomaly. After analysis, slicing area in the north to south (N-S) in the prospect area, obtained gradient linear of SVD is showed the value of SVD maximum is greater than |SVD minimum| (see **Figure 12**), so we can conclude in the area have a basin structure (graben reservoir).

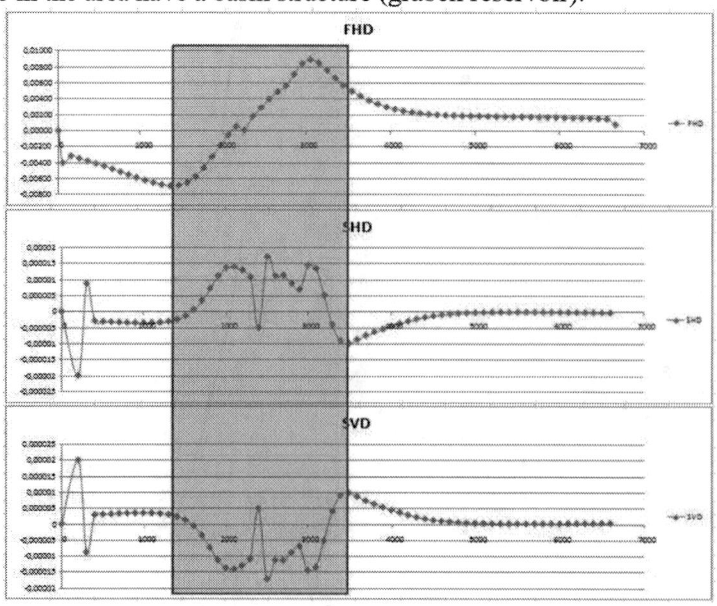

Figure 12. Second Vertical Derivative Calculation

Based on results of 2D Forward Modeling, obtained subsurface cross-section as shown in Figure 14, where the surface layer with the density 2.4 gr/cc is andesite, the second layer consisted of 3 types of rocks with a density value 2.8 gr/cc is andesite-basaltic, the density is 2.3 gr/cc is a calcareous-breccia rock that is assumed to be a reservoir zone (graben), and the density 2.5 gr/cc is breccia-limestone. Whereas the bottom layer with a density value of 3.2 gr/cc identified andesite-basaltic. This identification is based on the (Telford, 1990) and Mangga[5].

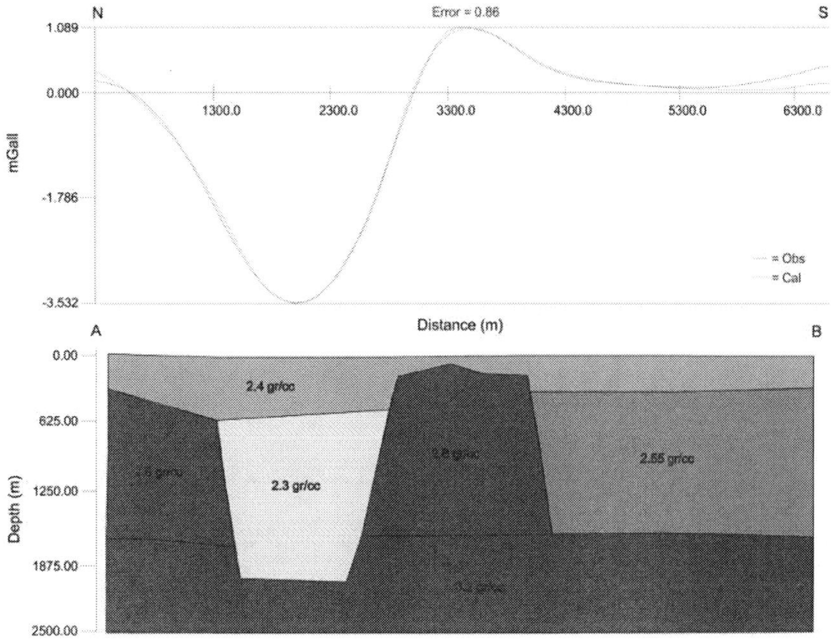

Figure 13. 2D Forward Modelling Wayratai Geothermal Prospect Area Line A-B in direction N-S

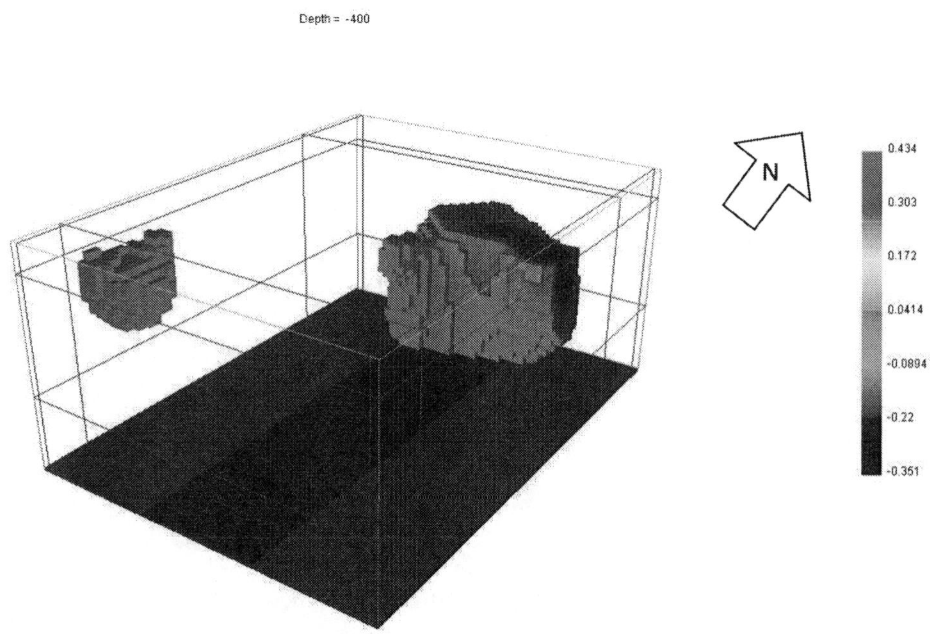

Figure 14. 3D Inverse Modelling Wayratai Geothermal Prospect area (Northeast View)

Based on results of the 3D Inverse Modelling, obtained subsurface cross section (see **Figure 14**) which is focused on reservoir conditions similar to the conditions in 2D Forward Modeling (shape of syncline/graben). In the 3D modeling is known that the reservoir has a density value between 2.35 to 2.57 g/cc are identified as andesite-calcareous with depth 400 to 2000 meter.

5. Conclusion

The conclusions derived from this paper are:

1. Based on Geology and Geochemistry mapping, obtained 5 hotspring points (with average temperature is 85-98°C), mudpool, steaming ground, and alterated rock (silica sinter).
2. From SVD mapping and gradient analysis, obtained SVD maximum > |SVD minimum|, this indicated the normal fault structure or basin structure in this area (graben reservoir).
3. Based on 2D Forward modelling and 3D Inverse modelling, obtained subsurface cross-section with top layer with density 2.4 to 2.8 gr/cc is andesite rock, reservoir layer with density 2.35 to 2.57 gr/cc is calcareous-breccia (graben reservoir structure), and basement layer with density 2.8 to 3.2 gr/cc is andesite-basaltic rock.
4. From the gravity modelling, with additional second vertical derivative calculation, we can conclude that the Wayratai Geothermal Prospect Area will be one of the future geothermal power plant in Lampung Province.

Aknowledgement

We would like to thanks Geophysical Engineering UNILA to join data acquisition in this area.

References

[1] Blakely R J 1996 Potential Theory in Gravity and Magnetic Applications *Cambridge University Press Cambridge, p. 439.*

[2] Goff F and Cathy J J 2000 Encyclopedia of Volcanoes : Geothermal System *Academic Press* **817-834**.

[3] Hochstein 1990 Introduction of Geothermal Prospecting *Geothermal Institute University of Auckland New Zealand.*

[4] Karyanto and Haerudin N 2013 Kualitatif Interpretation of surface temperature at Wayratai Geothermal Prospect Lampung *National Sains and Technology V Course* Research Institution of University of Lampung.

[5] Mangga S A, Amirudin, Suwarti T, Gafoer S, and Sidarto 1993 Geological Map of the Tanjungkarang Quadrangle Sumatera *Geological Research and Development Center.*

[6] Suhadiyatno 2008 Pemodelan Metode Gravitasi Tiga Dimensi dengan menggunakan Matlab Final Project of Geophysics Department UI.

[7] Reynolds J M 1997 An Introduction to Applied and Environtmental Geophysics *Baffios Lane.*

[8] Suharno, Amukti R, Hidayatika A, and Putri M K 2015 Geothermal Prospect of Padang Cermin Pesawaran Lampung Province Indonesia *Proceedings World Geothermal Congress 2015.*

[9] White D E 1967 Some Principle of Geyser activity, mainly from steambot spring Nevada *Am. J. Sci* **265** 641-684.

IOP Conf. Series: Materials Science and Engineering **982** (2020) 012050 doi:10.1088/1757-899X/982/1/012050

Comparison of Preprint Publication Server Using Analytical Hierarchy Process

Rio Aurachman[1,2]

[1]Telkom University, Bandung, Indonesia

[2]rioaurachman@telkomuniversity.ac.id

Abstract. Publishing a scientific paper has many alternatives. One of them is to publish in preprint servers. In this work, we aim to compare available preprint servers by using several assessment criteria such as ease of citation and online accessibility. Based on the comparison, we recommend osf.io as a more convenient place to publish preprint articles. Osf.io meets the criteria of ease of citation and index on Google Scholar. Osf.io also fulfills several other assessment criteria.

1. Introduction

To make publications that are open and easily accessible, some researchers are encouraged to start publishing their work on several Preprint Servers. The preprint also serves as an initial stage for publishing scientific papers before they are published in journals or proceedings. This early-stage publication allows the author to get input and suggestions for improvements to the paper before the paper is published in a reputable journal.

There are various types of server preprint alternatives. Comparison and comparison between the preprint servers are needed so that researchers can choose which preprint server is suitable for them. This comparison, for preprint server managers, can be input and improvement. From these improvements will provide benefits for stakeholders

Several studies have been conducted that focus on examining the preprint. Some of them are exploring the new model of preprint [1], studying the relationship between preprint and some field of research [2] [3], exploring the importance of preprint [4], studying the relationship between preprint and plagiarism [5], and several other topics. However, we have not succeeded in finding a paper that tries to compare the preprint servers to compare one with a certain criterion. This needs to be done for the reasons as stated in the previous paragraph.

Through this article, we will review the quality of those preprint server. Through this paper, we will examine whether it has the features needed by researchers. Specifically, researchers who are interested in publicly and freely publishing their work for the advancement of science

2. Method

The method to be used, as shown in Figure 1, is using comparison and analytical Hierarchy Process [6] [7] [8] . In the first phase, we will do a comparison between several preprint servers. From this comparison, several preprint servers will be studied in more depth. After that, the evaluation criteria will be made from each of the preprint servers. Each assessment criteria has a way to measure it. The final step is to build an additional guide to using the selected preprint server.

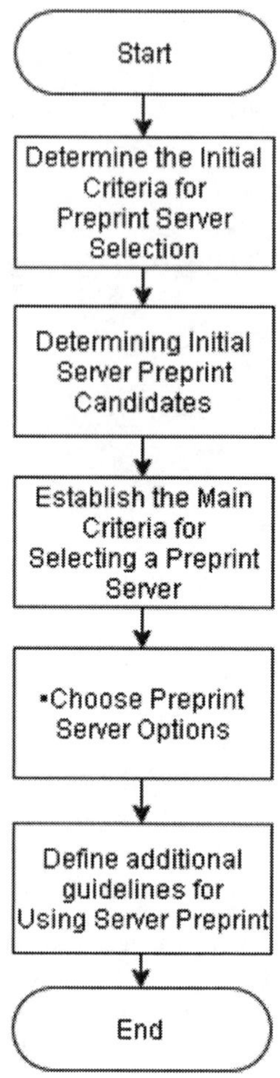

Figure 1. Server Preprint Determination Method

3. Result and Discussion

The first step is to establish preliminary criteria for the preferred preprint server. The criteria that we use is the preprint server which is quite general and not specific to certain scientific fields. This is because it will give the author the freedom to give thoughts and results of any research that he feels needs to be published without worrying about restricted topics. Some web pages have featured discussions and presentations on preprint alternatives. One comprehensive discussion process is on the Q&A menu of researchgate. Some of the list presented on that page, including the following:

1. Research Preprint Server List
2. Rising tide Preprint Server

Based on the alternatives presented there, we made a selection of preprint servers that were in accordance with the initial criteria of having a broad, general, and general range of knowledge. Some of the preprint servers are

1. PeerJ Preprints
2. Preprints (MDPI)
3. Figshare
4. Zenodo
5. F1000Research
6. Authorea / Winnower

7. HaL
8. Researchgate
9. Osf.io

Of the several alternatives, it is necessary to choose a server that has the expected features. Previously will be determined in advance the criteria used to choose. The main criteria in selecting a server are as follows

1. Cover general knowledge
2. Indexed to Google Scholar
3. The indexing process on Google Scholar is quite fast
4. Allows writing in several Languages
5. Articles uploaded in the preprint can be cited
6. Citation done on the preprint paper is expected to be counted as a sum of citations
7. Have DOI
8. Have a link
9. There is sufficient freedom to make publications without being limited by long moderation and review processes

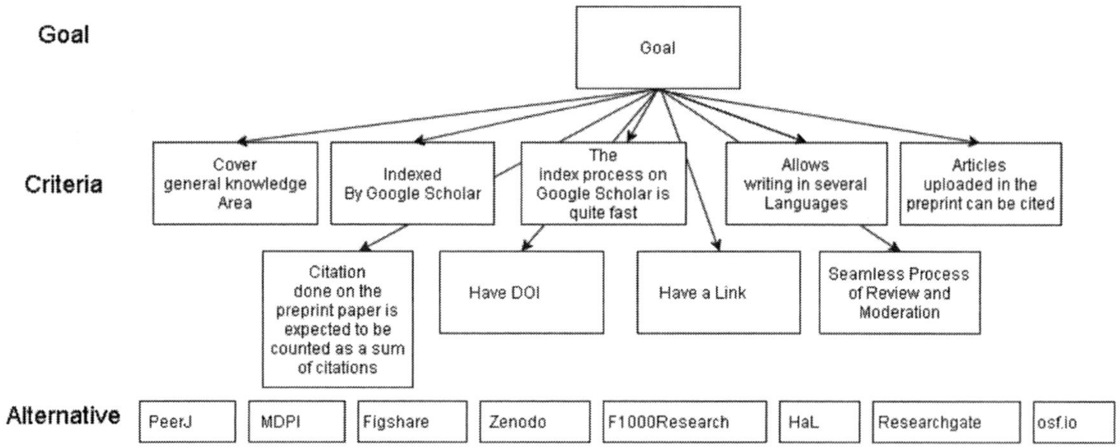

Figure 2. Hierarchy of Analytical Hierarchy Process

These criteria are measured by directly testing each preprint server. We use the hierarchy of analytical hierarchy process as shown in Figure 2. One method used is to test one piece of paper in the preprint and then see if there is also a paper on Google Scholar. Some servers apparently cannot fulfill these criteria. Citation criteria are also seen on Google Scholar. Some preprints also cannot be cited. Researchgate, for example, preprints can cite other papers outside Researchgate, but it turns out that citing researchgate preprints itself will not be counted. We conducted this trial in our paper [9] [10] One analysis of the causes we found was possible because the researchgate paper on Google Scholar did not contain information on the year of publication. We evaluate the index speed criteria by seeing whether the paper published on today's date has been indexed on Google Scholar. Some preprints actually take longer, even up to several weeks. The moderation and review process is evaluated by looking at the submission procedures raised by the preprint website.

Based on these criteria, the best preprint server so far is the osf.io. preprint server. Osf.io has several important features that are reviewed in the criteria list. Paper published on Osf.io with a fast time soon appeared on Google Scholar. For example, the main paper takes five days from the first upload on osf.io until then recorded in Google Scholar [11]. The written upload date is May 11, 2020, on osf.io. when we checked Google Scholar on May 16, 2020, the paper was already listed in Google Scholar. Paper uploaded on osf.io can also be cited and the citation is recorded on Google Scholar. For example, the paper titled "FBXL8 is differentially expressed and transcriptionally induced in MERS coronavirus infection" [12] published only on osf.io. The paper was cited by another paper which was also published in osf.io titled "IL6 and IL1RN is differentially expressed and transcriptionally induced in models of SARS coronavirus infection" [13] By the same author, in the same year. This allows the development of research networks more quickly. The citation can also be done not only by fellow osf.io but by papers from other journals, for example, a paper entitled The Scholarly Commons-principles and practices to guide research communication [14], also cited by a paper in another journal. The citations are counted and accumulated in the paper and also in the author's profile. The citation needs to write a link paper in the bibliography. For example, it can be written in an alternative form among the following or all three at once:

1. DOI code "doi: 10.31219/osf.io/g564w"
2. Access link that contains DOI https://doi.org/10.3897/rio.2.e9340
3. Access link to website https://osf.io/preprints/weqsf/

So far we have not been able to find a citation process that displays without a link or DOI. It also needs to be tested on researchgate preprints using links. Osf.io also does not require a moderation that takes a long time during the upload process of the paper. Osf.io also allows publishing processes using any language [11]. We found several osf.io papers using Arabic and Italian.

4. Conclusion

Based on the process that has been carried out, we get the conclusion that we recommend OSF.IO to be used as a means to publish writing. Osf.io meets the established criteria. Further analysis can be performed after uploading experiments on the osf.io. server. In addition, further analysis can be done using other criteria

References

[1] C. Hu, Y. Zhang and G. Chen, "Exploring a new model for preprint server: A case study of CSPO," *The Journal of Academic Librarianship,* vol. 36, no. 3, pp. 257-262, 2010.

[2] J. T. Rutka, "Preprint servers and neurosurgical publications," *Journal of Neurosurgery,* vol. 1, no. aop, pp. 1-3, 2020.

[3] C. Rawlinson and T. Bloom, *New preprint server for medical research,* British Medical Journal Publishing Group, 2019.

[4] P. Chaddah, "The importance of a preprint repository," *Current Science,* vol. 111, no. 6, pp. 979-980, 2016.

[5] J. Giles, *Preprint server seeks way to halt plagiarists,* Nature Publishing Group, 2003.

[6] T. L. Saaty, "Theory and applications of the analytic network process: decision making with benefits, opportunities, costs, and risks," *RWS publications,* 2005.

[7] T. L. Saaty, "'Decision making with the analytic hierarchy process', Int. J," . *Services Sciences, Vol. 1, No. 1,* pp. 83-98, 2008.

[8] T. Saaty and K. Kułakowski, "Axioms of the Analytic Hierarchy Process (AHP) and its Generalization to Dependence and Feedback: The Analytic Network Process (ANP)," *arXiv preprint arXiv:1605.05777,* 2016.

[9] R. Aurachman, "Analytical Hierarchy Process Sebagai Tools untuk Menentukan Pasangan Hidup," 2020.

[10] R. Aurachman, "Konsep Probabilitas untuk Memodelkan Resolusi Antara Ikhtiar, Tawakal, dan Doa dalam Menghadapi Wabah Covid19," 2020.

[11] W. S. Utami, "Perangkat Lunak (Software) Pada Komputer," 2020.

IOP Conf. Series: Materials Science and Engineering **982** (2020) 012050 doi:10.1088/1757-899X/982/1/012050

[12] S. Mamoor, "FBXL8 is differentially expressed and transcriptionally induced in MERS coronavirus infection.," 2020.

[13] S. Mamoor, "IL6 and IL1RN is differentially expressed and transcriptionally induced in models of SARS coronavirus infection.," 2020.

[14] J. Bosman, I. Bruno, C. Chapman, B. G. Tzovaras, N. Jacobs, B. Kramer, M. E. Martone, F. Murphy, D. P. O'Donnell, M. Bar-Sinai and others, "The Scholarly Commons-principles and practices to guide research communication," 2017.

Work Mental Load Analysis on Car Repair Mechanics in Yogyakarta

Atyanti Dyah Prabaswari[1,3], Bagus Wahyu Utomo[2]
[1]Industrial Engineering, Universitas Islam Indonesia, Yogyakarta, Indonesia
[2]Industrial Engineering, Sekolah Tinggi Teknologi Adisutjipto, Yogyakarta, Indonesia

[13]atyanti.dyah@uii.ac.id

Abstract. Mechanics must work with precision, excellent and consistent speed, and other work that drains both the mechanic's physical and mental energy. Official motorbike mechanics experience complaints of work stress due to high work demands. Regarding the measurement of mental workload, a subjective workload is easily given and has a high assessment ability because the measurement is not dependent on the task. The NASA-TLX (Task Load Index) scale is the most widely used subjective scale by asking participants to rank separately on the subscale of mental demand, physical demand, temporal demand, performance, effort, and frustration level. This will be useful to be able to distinguish between tasks in terms of the specific subscale. Mental demand has the highest result (305) because the car workshop is demanded to find the cause of damage to the car, then remember the car components when dismantled and re-installed. Based on statistical analysis, between mental demand and working hours indicates a positive relationship. Heavy mental activities can be reduced by training workshop workers relating to administration, expertise in each type of car, and comfortable and safe work areas.

1. Introduction

Work fatigue is a part of general fatigue usually characterized by a reduced willingness to work due to monotony, intensity and duration of physical work, environmental conditions, mental causes, health status, and nutritional status [2]. Mechanics must work with precision, excellent and consistent speed, and other work that drains both the mechanic's physical and mental energy. Some researchers hypothesized that the indirect costs of poor working conditions, in terms of reduced performance and increased error rates, significantly impact a system's financial performance than the direct costs [3]. Official motorbike mechanics experience complaints of work stress due to high work demands [3]. Regarding the measurement of mental workload, a subjective workload is easily given and has a high assessment ability because the measurement is not dependent on the task [4]. The NASA-TLX (Task Load Index) scale is the most widely used subjective scale by asking participants to rank separately on the subscale of mental demand, physical demand, temporal demand, performance, effort, and frustration level. NASA-TLX ratings were obtained quickly (it took less than one minute to obtain the six ratings after each experimental condition). It took no more than two minutes to obtain the weights for each different type of task. This suggests that the proposed multidimensional rating scale would be a practical tool to apply in operational environments (which the nine-factor scale was not). Data analysis is substantially easier to accomplish than with SWAT, requiring a specialized conjoint analysis program. The weighted combination of factors provides a sensitive indicator of the overall workload between different tasks and different levels of each task. In contrast, the weights and the magnitude of the individual scales' ratings provide crucial diagnostic information about the specific loading source within the task [5].This will be useful to distinguish between tasks in terms of the specific subscale [6].

2. Literature Review

The staff has many tasks that not only the primary task but an additional task. The additional task is from superior in the same place or the other place. The SOP for duties and distribution of tasks is not clear, so that fatigue can affect productivity. The method to measure mental workload is NASA-TLX. NASA TLX has six indicators: mental demand, physical demand, temporal demand, performance, frustration, and effort. The analysis is not only based on the result of NASA TLX but also using statistic test. The result is that the performance indicator is high that described the staff not satisfied with their job has done. The staff not satisfy but the effort has been taken out is very high. The result of Mann Whitney is not different from two groups of the period long term of working, which is divided in < 7 years and >7 years. The correlation test result is if the staff make more effort, their tasks' satisfaction is more significant [7].

The workers will face working hours, loads, remember many patterns, and frustration when doing an outstanding task. Solar street lighting (PJU TS) installation project is affected by various risks. NASA TLX is a tool to measure the workloads. The Mental Demand (MD) has a score of 325 (the highest score) due to considerable mental activity. The Performance (P) has a score close to zero that is categorized quite well. Regression and correlation analysis between mental demand and duration of work show the direction of positive relationships. More experience makes more MD scores. That is because every PJU installation project has different criteria. The solution is bookkeeping and mind mapping regarding the type of PJU TS. Each PJU TS has the book as a guide in the field. [8].

The central control room is one department in the oil processing company; employees are tasked with monitoring the processing unit for 24 hours nonstop with a combination of 3 shifts in 8 hours. NASA-TLX is one of the subjective mental workload measurement using six factors, namely the Mental demand (MD), Physical demand (PD), Temporal demand (TD), Performance (OP), Effort (EF), frustration levels (FR). Measurement of a subjective mental workload is most widely used because it has a high degree of validity. Based on the mental workload calculation, there are 5 units (DTU, NPU, HTU, DIST, and OPS) at the control chamber (94; 83.33; 94.67; 81, 33, and 94.67 respectively) that categorize as very high mental workload. The high level of mental workload on the operator in the Central Control Room requires high accuracy, alertness, and can make decisions quickly [9].

Gojek is a service that helps people's lives by opening jobs for the community and making it easier for the community itself to live their lives. Many users of the GOJEK application also do not rule out the possibility of excessive passengers during rush hour. The needs of passengers who want to be fast and the company's needs cause gojek drivers to raise workloads on the driver itself. Therefore a study was conducted by analyzing the relationship between mental workload and work stress experienced by gojek drivers using the NASA-Task Load Index (TLX) method. Thirty gojek drivers were respondents in this study. By observing the indicators on NASA TLX, the results obtained are as many as three respondents categorized as having a very high mental workload, 26 respondents having a high category, and 1 respondent having a relatively high category. It is necessary to improve immediately to make the work become not hampered. The workload is a significant stressor that raises a type of destructive psychological reaction [10].

The literature review results show that research on work mental load is considered important to improve the work productivity of car repair mechanics. Increased productivity is useful for the car repair shop's progress and safety for the mechanic who works in the workshop, so this research is considered important to see the mental workload of the auto mechanic who is working in the car repair service.

IOP Conf. Series: Materials Science and Engineering **982** (2020) 012051 doi:10.1088/1757-899X/982/1/012051

3. Research Methodology

3.1. The Concept

NASA TLX questionnaire was distributed to car repair mechanics with a working-age between 1 year and 8 years. The NASA TLX method is used to analyze the mental workload faced by workers. Hancock and Meshkati (1988) explain that there are 6 indicators in conducting the NASA TLX method: Mental Demand (MD) is the amount of mental and perceptual activity needed to see, remember, and search. Does the work include difficult, simple or complex; Physical Demand (PD) is the amount of physical activity required by the worker (e.g. pulling, pushing, turning, and so on); Temporal Demand (TD) is time pressure felt during the work in progress. Is the work can be done slowly or quickly so that it feels tiring; Performance (P) is the success of workers in carrying out their duties and how satisfaction with the results of their work; Frustration Level (FL) is measuring how many workers feel insecure, hopeless, offended, or disturbed when doing their job; Effort (EF) is measuring a lot of hard work required by workers to achieve the required level of performance.

3.2. Data Processing

Data processing was performed to determine the value of each laboratory assistant and study program administrator's mental workload with the NASA TLX method and statistical tests. Data processing using NASA TLX calculations is as follows (Hancock and Meshkati, 1988):

3.2.1 Weighting

Respondents were asked to choose one of two more dominant indicators, causing mental workload. The selection of indicators uses a pairwise comparison.

3.2.2 Rating

Respondents were asked to give a rating of the six mental load indicators. Rating given is subjective, depending on the choice of respondents following what they feel.

3.2.3 Value

Multiply the rating by the weight factor for each indicator.

$$\text{Product} = \text{rating x weighting factor} \tag{1}$$

3.2.4 Weighted Workload (WWL)

WWL is obtained by adding up the six product values.

3.2.5 WWL Average

Divide the number of WWL values by the value 15. The value is the total weight of the pairwise comparison.

3.2.6 Score Interpretation

[5] In [11] explain that the workload scores obtained are divided into Table 1.

Table 1. Interpretation

Workload Rank	Value
Low	0-9
Medium	10-29
Quite High	30-49
High	50-79
Very High	80-100

3.3. The Result

The data that has been processed is analyzed how the categories of each workload mental and interpretations of values. The analysis was also assisted with statistical tests, namely the different regression and correlation tests.

A different test was carried out to see the old group working. The test is to see whether the group working for a long time affects the score. At the same time, the correlation test to find out the highest value on the indicator of age. Conclusions are made to summarize all the results in the study, besides the advice to be given is expected to be useful for further research.

4. Result

4.1. Respondent Profile

Demographic to knowing long of work and working hours show in Tabel 2. This data to analyze using correlation testing.

Table 2. Respondent Profile

Responden	Sex	Age (year)	Long of Work (year)	Working Hours
R1	M	17	8	8
R2	M	17	0,2	6
R3	M	20	0,2	6
R4	F	20	1,6	9
R5	F	19	1	9
R6	M	21	1	8
R7	M	27	3	8
R8	M	27	4	8
R9	M	20	3	8
R10	M	24	4	8

4.2. NASA TLX

Based on the value of NASA TLX in Table 3 and the recapitulation of workload scores in Table 4, it was found that the Technicians had an average WWL of 83.47 and included in the very high category and the dominant indicator affecting the amount of mental admin workload was Mental Demand (MD) with an average of 305.

Table 3. Value

	MD	PD	TD	P	EF	FR
R1	270	160	180	270	450	0
R2	100	300	450	180	320	0
R3	500	200	200	100	300	200
R4	350	120	360	100	160	0
R5	350	120	360	100	160	0
R6	500	200	300	100	400	0
R7	180	180	80	360	400	80
R8	320	180	210	180	320	0
R9	320	180	240	140	30	100
R10	160	270	160	70	320	180

Table 4. Interpretation

Category	Total	Average
MD	3050	305
PD	1910	191
TD	2540	254
P	1600	160
EF	2860	286
FR	560	56
Average Total	1252	

Score	83.47
Interpretation	Very High

4.3. Correlation Testing

Values of Sig. (2-tailed) between Work Hours and Mental Demand 0.808> 0.05. Based on the r count (Pearson Correlations), the r count value for the relationship of Work Experience with Mental Demand is -0.089 <0.549, so it can be concluded that there is no significant correlation between Working Hours and Mental Demand variables. Based on the positive Pearson correlation, the relationship between the two variables is positive. This means that if working hours get higher then Mental Demand will be higher too.

5. Discussion and Conclusion

The MD value (305) has the highest value because the work in car repair shops is demanded to find the cause of damage to the car and then to remember the car's components when it is dismantled and re-installed. The EF value (286) has the second-highest value because the workshop worker must work hard to complete the customer's car repair. The TD value (254) has the third-highest value because a workshop worker is required to repair a customer's damaged car as quickly and as well as possible.

PD value (191) has a high enough value because workers in the workshop must do work that requires a large amount of physical force, such as jacking, removing bolts, and lifting large components from the car. P-value (160) has a high enough value because workers in the workshop feel not entirely satisfied with their work results. After all, the car's completion is often late, some customers complain that the car is damaged again after being repaired, and sometimes fails to repair the damaged car. The FR value (56) has the smallest value among the other categories. This shows that workers are not disturbed by the existing working conditions.

6. Conclusion

Analysis of the correlation between mental demand and working hours shows the direction of a positive relationship. Heavy mental activities in the workshop area can be reduced by conducting training to workshop workers relating to administration, expertise in each type of car, and comfortable and safe work areas. After training or after having experienced working in a better workplace, the workers choose the improvements themselves. This is also essential for successful implementation. Since workers have selected the improvement themselves, this stage can be reduced to the realization by promotion, training, and instruction [12].

Acknowledgment

Researchers thank the support of the Industrial Engineering Department, Faculty of Industrial Technology, Universitas Islam Indonesia, and the Industrial Engineering Department, Sekolah Tinggi Teknologi Adisutjipto.

References

[1] Tarwaka, B. SHA and Sudiajeng, L 2004 *Ergonomics for Safety, Occupational Health and Productivity.*

[2] Pertiwi, L.I. 2018 Relationship between Noise and Mental Workload Against Job Stress at Official Motorcycle Workshop Mechanic in Tembalang Semarang *Skripsi UNDIP.*

[3] Rose, L. M., Orrenius, U. E., & Neumann, W. P 2013 Work environment and the bottom line: Survey of tools relating work environment to business results *Human Factors and Ergonomics in Manufacturing & Service Industries* **23**(5) 368-381.

[4] Casali, J.G. and Wierwille, W.W 1983 A comparison of rating scale, secondary-task, physiological, and primary-task workload estimation techniques in a simulated flight task emphasizing communications load *Human Factors* **25**(6) 623-641.

[5] Hart, S.G. and Staveland, L.E 1988 Development of NASA-TLX (Task Load Index): Results of empirical and theoretical research *Advances in psychology* **52** 139-183.

[6] Tsang, P.S. and Velazquez, V.L 1996 Diagnosticity and multidimensional subjective workload ratings *Ergonomics* **39**(3) 358-381.

[7] Prabaswari, A. D., Basumerda, C., & Utomo, B. W 2019 Analysis of Staff in Study Program of Private Educational Organization *IOP Conference Series: Materials Science and Engineering* **528**(1) p. 012018.

[8] Prabaswari, A. D., Utomo, B. W., & Tauhida, D 2020 Mental Analysis Workload on Solar Street Lighting Workers *Journal of Physics: Conference Series* **1430**(1) p. 012041.

[9] Sugarindra, M., Suryoputro, M. R., & Permana, A. I 2017 Mental workload measurement in operator control room using NASA-TLX *IOP Conference Series: Materials Science and Engineering* **277**(1) p. 012022.

[10] Prabaswari, A. D., Hamid, A. N., & Purnomo, H 2020 The mental workload analysis of gojek drivers. *IOP Conference Series: Materials Science and Engineering* **722**(1) p. 012008.

[11] Hancock, P. A. & Meshkati, N., 1988 *Human Mental Workload* Elsevier.

[12] Vink, P., Peeters, M., Gründemann, R. W. M., Smulders, P. G. W., Kompier, M. A. J., & Dul, J 1995 A participatory ergonomics approach to reduce mental and physical workload *International Journal of Industrial Ergonomics* **15**(5) 389-396.

Analysis of quality control of chippendale furniture products using seven tools approach (case study of PT. Bothwell Indonesia)

Atyanti Dyah Prabaswari[1], Ahmed Joko Susilo[2]

[12]Industrial Engineering, Universitas Islam Indonesia, Yogyakarta, Indonesia

[1]atyanti.dyah@uii.ac.id

Abstract. One of the products produced at PT. Bothwell Indonesia is a small shelf. This product requires a high level of accuracy because the product has a screw and has separate small parts which are then put together. This can be found on the Chippendale Lacquered Library Side Table product. On this product, according to the staff section production and marketing, these products often face problems. Therefore, in this study, an approach with quality control is carried out to conduct further analysis, and it can be seen how the quality control that occurs in the product. The method used in this research is to use seven quality control tools. Based on the results of the product defect analysis, which has the highest value is the type of uneven paint defects with the number of products of 7 pieces. Next followed by the type of gross defect by five pieces, broken/cracked by three pieces. Based on the results of the analysis and discussion of Chippendale products included in controlled quality control. This is because when viewed statistically the number of products that have defects does not exceed or be less than the control limit.

1. Introduction

Business to business is one of the business models that can be applied to company manufacture. Business to business itself is a trade interaction by using internet networks or between business people, i.e. buyers of products sold by non-sellers the final consumer but the fabrication or the individual business at which the product purchased will be processed again or sold directly to end consumers [1]. The business model generally has the definition of a business that provides a product or service to give to other businesses or companies.

A company that runs this business model will have a strong relationship with his companies. That is because every product was given to the company others, always provide the best quality to meet client desires. This is to maintain trust between these companies. The quality that the company gives to customers is proof that they can provide high quality following expectations customers [2]. Products that have not been able to satisfy the customer can be said to have a lower quality product. This can affect the decision customers to buy products [3].

PT. Bothwell Indonesia is one of the companies engaged in the field furniture making with business models used is business to business (B2B) and the production system used is make to order, the MTO system is one type of type production used by a company, where products are made by this company based on the wishes of the buyer/client directly. Companies based on B2B business models and MTO production systems, companies are required to provide services and quality products best to the buyer/client with many product variations. Therefore, controlling the quality of a product becomes a challenge that must be met by companies with many types of products produced.

One of the products produced by PT. Bothwell Indonesia is a small shelf. This product requires a high level of accuracy because the product has a screw and has separate small parts which are then put together. This can be found on the Chippendale Lacquered Library Side Table product. On this product, according to the staff section production and marketing, these products often face problems. Based on data obtained on products using the Checksheet table, there are still defective products.

Table 1. Defective Products

Product	Defect Type				Total
	Dirty	Crack/Break	Uneven paint	Rough	
Chippendale 1	-	-	-	-	0
Chippendale 2	-	√	-	-	1
Chippendale 3	-	√	√		2
Chippendale 4	-	-	-	-	0
Chippendale 5	√	√	-	-	2
Chippendale 6	-	-	-	-	0
Chippendale 7	-	-	√	-	1
Chippendale 8	√	-	-	-	1
Chippendale 9	-	-	-	-	0
Chippendale 10	-	-	-	-	0
Chippendale 11	√	-	-	-	1
Chippendale 12	-	-	√	-	1
Chippendale 13	√	-	√	-	2
Chippendale 14	-	-	-	-	0
Chippendale 15	-	-	√	-	1
Chippendale 16	-	-	-	-	0
Chippendale 17	√	-	-	-	1
Chippendale 18	-	-	-	-	0
Chippendale 19	-	-	√	-	1
Chippendale 20	-	-	√	-	1
Total	5	3	7	0	15

Therefore, in this study, an approach with quality control is carried out to conduct further analysis, and it can be seen how the quality control that occurs in the product. The method used in this research is to use seven quality control tools. This method is an online type of quality control, which is quality control when the product runs during the production process.

2. Literature Review

2.1. Quality

Quality is the degree or level of perfection; in this case, quality is a relative measure of goodness [4]. Operationally, a quality product or service, is one that meets the quality attributes or is often referred to as the quality dimension. In other studies, quality is a dynamic statement related to products, services, people, processes, and environments that can meet or exceed expectations [5].

Quality also has a part in the decision to purchase a product. Based on the calculation results of multiple linear regression tests, it is seen that the quality of the product has a low level of significance compared to the significance of service quality [6]. For this reason, the quality of products in CV Jati Endah Lodoyo, Blitar must be improved so that the quality of these products can attract more consumers to buy furniture at CV Jati Endah Lodoyo, Blitar [6]. This is seen from the t-test of service quality variables having the most significant regression coefficient compared to product quality variables [6].

Several factors need to be considered by the company so that consumers can be attracted to the products produced, namely product quality, product diversity, product design, and price [7]. Product quality is the physical condition, function, and nature of the product concerned that can satisfy the tastes and needs of consumers satisfactorily according to the value of money that has been spent [7].

2.2. Quality Control

Quality is the overall characteristics contained in a product in the form of goods or services that can provide satisfaction to consumers who use the product. Products with good quality will undoubtedly be very in demand by consumers; therefore, it is necessary to have quality control because it is one way to improve the quality of a product [8]. Quality control is a series of 37 activities carried out following the plan, and if deviations occur can be immediately corrected so that something will be achieved [9]. To produce quality products that can compete in the market and can be accepted by the community [10].

There are more tools for achieving operative quality management targets, and the most frequent method is probably measurement and evaluation of the capability of processes through capability indexes [11]. In addition to other histogram and Ishikawa diagram are the following frequently used tools for quality improvement processes [11].

2.3. Statistical Processing Control (SPC)

SPC is statistical quality control that has 7 (seven) main statistical tools that can be used as tools to control quality [12]. These tools include check sheets, histograms, control charts, diagrams, Pareto, causal diagrams, scatter diagrams, and process diagrams.

2.4. Control Chart

A control chart is one of the tools (tools) to control statistical processes (SPC). A control chart is used to analyze the output in a process. Analysis of the control chart carried out can be visualized graphically to monitor and evaluate whether the activity/process is in statistical quality control or not so that it can solve problems and produce quality improvements [13]. In this tool shows a change from time to time, but cannot know the cause of the deviation occurred [14]. The control chart is divided into two based on the object being measured, namely the variable control map and the attribute control map. A variable control map is a tool to measure quality control by taking measurements on objects obtained directly from the measurement results. On the attribute control map is a measure of quality control carried out based on the characteristics of an object.

2.5. P Control Chart

One of the measuring instruments on the attribute control chart is the control chart p. This control chart is to calculate the proportion of characteristics that have been determined. The following are the steps in carrying out quality control with the P control chart [15]:

Calculate damage percentage $$p = \frac{np}{n} \quad (1)$$

Center line (CL) $$CL = \bar{p} = \frac{\sum np}{\sum n} \quad (2)$$

Upper control limit (UCL) $$UCL = \bar{p} + 3\left(\sqrt{\frac{\bar{p}(1-\bar{p})}{n}}\right) \quad (3)$$

Lower control limit (LCL) $$LCL = \bar{p} - 3\left(\sqrt{\frac{\bar{p}(1-\bar{p})}{n}}\right) \quad (4)$$

3. Research Methodology

In this study, the method of analysis carried out is a seven tools approach. The subject used in the study was one of the products produced at PT. Bothwell Indonesia. The product is Chipandell. The object used as research is the quality of the product.

4. Result

In this study, based on observations and interviews. Can be known the types of defects found in a product. The following types of defects have been identified:

Table 2. Type of Defect

Picture	Type
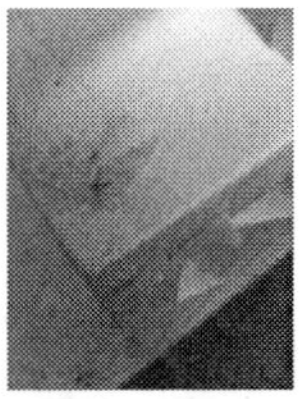	Dirty
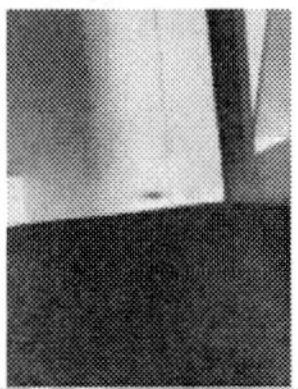	Uneven paint
	Crack/Break

4.1. Checksheet

Making a table (Check sheet) is useful to facilitate the process of data collection and analysis. Because making this table is the first step taken to analyze quality control statistically. The following is a defect data for each product in the Chippendale Lacquered Library Side Table type in February 2020 at Table 1.

4.2. Control Chart

4.2.1. Percentage of Product Defects

The percentage of product defects is used to see the percentage of product defects in each sub-group (observation). The data is processed to find the percentage of damage from each subgroup (observation). The following is a table of data processing results:

Table 3. Percentage of Product Defect

Observation	Product	Defective Product	Defective Proportion
1	3	2	0.67
2	4	2	0.5
3	5	3	0.6
4	3	2	0.67
5	5	3	0.6

In one product can have more than one type of product defect, in these conditions the product is still counted as one defective product.

4.2.2. Center Line (CL)

Central line / Central Line is the middle line that is between the upper control limit (UCL) and the lower control limit (LCL) or is the average value of the proportion of defective products. Based on formula (2) we get Central Line (CL).

$$\sum np = 12$$

$$\sum n = 20$$

$$CL = \bar{p} = \frac{\sum np}{\sum n} = \frac{12}{20} = 0.6 \quad (5)$$

4.2.3. UCL and LCL

The upper control limit and the lower control limit are statistically measured indicators that a process can be said to be distorted or not. The upper control limit (UCL) can be calculated using formula (3); the lower control limit can be obtained as follows:

$$\bar{p} = 0.6$$
$$n = 5$$

$$UCL = \bar{p} + 3\left(\sqrt{\frac{\bar{p}(1-\bar{p})}{n}}\right) = 0.6 + 3\left(\sqrt{\frac{0.6\,(1-0.6)}{5}}\right) = 1.257 \quad (6)$$

$$LCL = \bar{p} - 3\left(\sqrt{\frac{\bar{p}(1-\bar{p})}{n}}\right) = 0.6 - 3\left(\sqrt{\frac{0.6\,(1-0.6)}{5}}\right) = -0.057 \quad (7)$$

4.2.4. Control Chart

After all the required values have been calculated, the next step is to create a p-chart. The control chart p is created using a spreadsheet. The products produced are still within the control limits. That is because for the 5 points in the observation are between the control limits (UCL) and (LCL). Besides that, the value of the proportion of defects is close to the average value of the proportion of defects.

IOP Conf. Series: Materials Science and Engineering **982** (2020) 012052 doi:10.1088/1757-899X/982/1/012052

5. Discussion

5.1. Control Chart

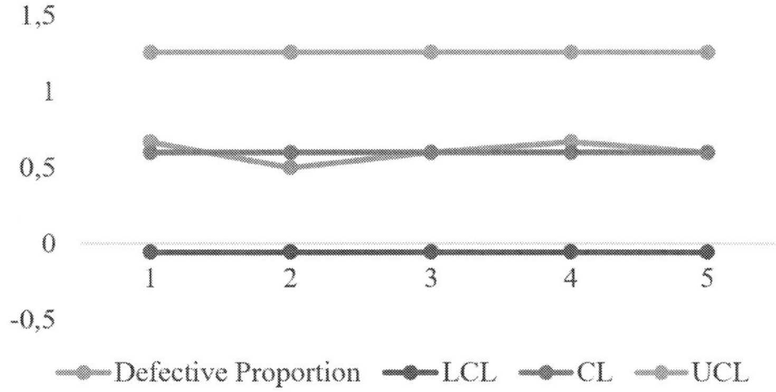

Figure 1. Control Chart

Based on the picture above, it can be seen that the products produced are still within the control limits. That is because for the 5 points in the observation are between the control limits (UCL) and (LCL). Besides that, the value of the proportion of defects is close to the average value of the proportion of defects.

5.2. Fishbone

5.2.1. Fishbone (uneven paint)

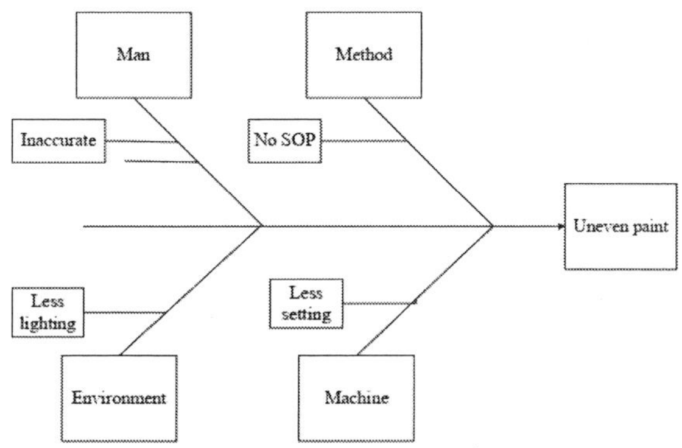

Figure 2. Uneven Paint

Uneven paint results due to poor lighting conditions. This causes the spray section officer to be less careful in giving the paint to the product so that there are parts that are not exposed to colour. Also, because the ability of the expert very much needs the spray process, so the products included in the spray process are not the same result. At the corner of the product cannot be reached by a large spray because it needs to be replaced with a small size spray to improve accuracy in colouring.

IOP Conf. Series: Materials Science and Engineering **982** (2020) 012052 doi:10.1088/1757-899X/982/1/012052

5.2.2. Fishbone (Crack/break)

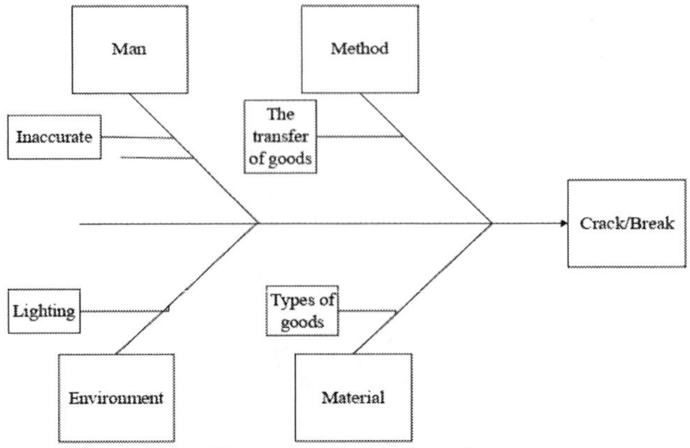

Figure 3. Crack/break

Product breakage or cracking results from poor lighting conditions. This causes the putty section staff to be less careful in giving putty on every connection that is on the product so that there are parts that are not putty. Also, in every process of moving goods, especially at close distances, the transfer is carried out by sliding or pushing the product so that it causes vibrations to the product. The final cause of this type of defect is the type of raw material used because each type of wood used has different physical resistance.

5.2.3. Fishbone (Dirty)

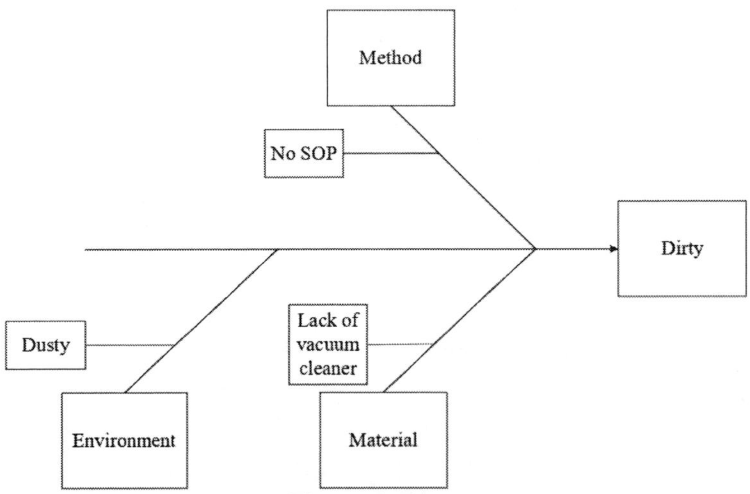

Figure 4. Dirty

The results of dirty products are caused by dusty environmental conditions resulting from the remnants of the refinement process that are not cleaned regularly. This causes the wind can easily carry the remaining refinement. The lack of a vacuum cleaner cause products to become dirty because the intensity of the refinement process is high enough to make the environment full of dust. Then the next cause is the absence of SOPs in moving goods. As with the use of gloves that are required in every transfer of the finished product. This is to reduce employees who do not maintain cleanliness, which is due to environmental conditions and every production process.

IOP Conf. Series: Materials Science and Engineering **982** (2020) 012052 doi:10.1088/1757-899X/982/1/012052

6. Conclusion and Recommendation

Based on the results of the product defect analysis, which has the highest value is the type of uneven paint defects with the number of products of 7 pieces. Next followed by the type of gross defect by five pieces, broken/cracked by three pieces.

In the type of paint defect that is not evenly caused by tools that have not been rearranged, lack of lighting around the spray room, lack of operator accuracy in spraying, and lack of written Standard Operating Procedures for tool use, and in spraying. In the type of defect cracked/broken caused by lack of lighting on the retarding. Resulting in less careful employees in doing putty on every part or connections on the product. Next is the removal of goods without written SOP and types of wood raw materials used in production. In the type of gross defect caused by environmental conditions that are different due to the remnants of the refinement process. Then the lack of a vacuum cleaner and SOP in moving goods. Based on the results of the analysis and discussion of Chippendale products included in controlled quality control. This is because when viewed statistically, the number of products that have defects does not exceed or be less than the control limit.

Acknowledgement

Researchers thank the support of the Industrial Engineering Department, Faculty of Industrial Technology, Universitas Islam Indonesia.

References

[1] Rayport, J. F., & Bernard, J. J., 2001. Introduction to e-commerce. Boston: McGraw Hill.

[2] Mulyono, B., Yoestini, N. R., & Kamal, M., 2007. Analisis Pengaruh Kualitas Produk dan Kualitas Layanan Terhadap Kepuasan Konsumen (Studi Kasus Pada Perumahan Puri Mediterania Semarang). Semarang: Pasca Sarjana Universitas Diponegoro.

[3] Kodu, S., 2013. Harga, Kualitas Produk, dan Kualitas Pelayanan Pengaruhnya Terhadap Keputusan Pembelian Mobil Toyota Avanza. EMBA, 1251-1259.

[4] Hansen, D. R., & Mowen, M. M., 2009. Akuntansi Manajemen, Terjemahan Dewi Fitriasari dan Deny Armor Kwary. Jakarta: Salemba Empat.

[5] Goetsch, D. D., & Davis, S. B., 2006. Quality Management; Introduction Total Quality Management for Production, Processing, and Service. Prentice Hall: Pearson.

[6] Mukti, M.Y.D., 2015. Pengaruh Kualitas Produk Dan Kualitas Layanan Terhadap Keputusan Pembelian (Studi Kasus Pada Mebel Cv Jati Endah Lodoyo Blitar). *Riset Mahasiswa Ekonomi (RITMIK)*, *2*(1).

[7] Basyir, A., 2012. *Pengaruh Kualitas Produk, Keragaman Produk, Desain Produk dan Harga Terhadap Keputusan Pembelian Konsumen pada Jati Jepara Furniture Kabupaten Jepara* (Doctoral dissertation, Universitas Muria kudus).

[8] Susetyo, J., Winarni, & Hartanto, C., 2011. Aplikasi Six Sigma dan Kaizen Sebagai Metode Pengendalian dan Perbaikan Kualitas Produk. Jurnal Teknologi, 53-61.

[9] Assauri, S., 1998. Manajemen Operasi dan Produksi. Jakarta: LP FE UI.

[10] Montgomery, D. C., 1990. Pengantar Pengendalian Kualitas Statistik. Yogyakarta: Gajah Mada Universitas Press.

[11] Simanová, Ľ. and Gejdoš, P., 2015. The use of statistical quality control tools to quality improving in the furniture business. *Procedia Economics and Finance*, *34*, pp.276-283.

[12] Heizer, J., & Barry, R., 2006. Manajemen Operasi. Jakarta: Salemba Empat.

[13] Arsyad, A. G., Ferdinant, P. F., & Ekawati, R., 2017. Analisis Peta Kendali p yang Distandarisasi Dalam Proses Produksi Regulator Set Fujiyama (Studi Kasus : PT. XYZ). Jurnal Teknik Industri, 86-92.
[14] Ariani, D. W., 2004. Pengendalian Kualitas Statistik. Yogyakarta: Andi Offset.
[15] Ilham, M. N., 2012. Analisis Pengendalian Kualitas Produk Dengan Menggunakan Statistical Processing Control (SPC) Pada PT. Bosowa Media Grafika (Tribun Timur). Makassar: Fakultas Ekonomi dan Bisnis Unvirsitas Hasanuddin.

IOP Conf. Series: Materials Science and Engineering **982** (2020) 012053 doi:10.1088/1757-899X/982/1/012053

Work Hazard Risk Analysis and Control in Grey Finishing Department Using HIRARC (Hazard Identification, Risk Assessment and Risk Control)

Atyanti Dyah Prabaswari[1], Dyah Ari Susanti[2], Bagus Wahyu Utomo[3] and Bebie Rizka Shintira[4]

[1,4]Industrial Engineering Department, Universitas Islam Indonesia, Jalan Kaliurang Km. 14,5, Sleman, Yogyakarta 55584, Indonesia

[2]Industrial Engineering Department, Universitas Sarjanawiyata Tamansiswa, Jalan Miliran No 16, Muja Muju, Umbul Harjo, Yogyakarta 55165, Indonesia

[3]Industrial Engineering Department, Sekolah Tinggi Teknologi Adisutjipto, Jalan Janti Blok. R, Lanud Adisutjipto, Bantul, Yogyakarta 55198, Indonesia

[1]atyanti.dyah@uii.ac.id

Abstract. PT. Primissima (persero) engaged in textile manufacturing. PT. Primissima (persero) processes raw materials cone thread become a grey or calico cloth through weaving process. As a manufacturing industry, workers at PT. Primissima (persero) are integrated with the plant facilities such as the production floor, machines, equipment, and warehouse. It potentially carry hazard risk in the form of work accident and occupational illness. Grey finishing department at PT. Primissima (persero) has a role to ensure product that produced has good quality through inspection and folding activities. Based on observation, there are several potential significant hazard such as air pollution, scrap exposure, and also physiological hazards such as aches to environmental hazards, electrical hazards and fire hazards. So, is necessary to apply a control through hierarchy of control. The results of this research contribute to solve the problem of the and give solutions of the company regarding risk controls.

1. Introduction

In manufacturing industry, high productivity is obtained by considering the activity of transforming raw materials into finished goods. Human labor or workers as the main pillar of the running of the industry. As one of the main resource, the workers often experiencing the risk of occupational health and hazards.

Health, Safety, and Environment (HSE) is important because it has direct and indirect impacts on the company's productivity. HSE discusses the relationship of workers with equipment, machinery and the environment that have the potential risks to occupational health and safety.

The company is responsible for implementing safety strategy, even if it is simple, it uses as first precaution [1]. By using this concept, the level of safety of employees increases because they dont have to feel worry about occupational illness or accidents that could potentially occur.

PT. Primissima (Persero) is state-owned company engaged in manufacturing fine textiles. PT. Primissima (Persero) processes cone thread become a grey cloth through weaving process. The production of this cloth uses machinery, equipment and environment that have risk of occupational

illness and work accidents. One of the causes is due to lack of awareness and not optimal of supervision and implementation of HSE in the company. Therefore, real efforts are needed to reduce the risk of accidents and occupational illness.
This research was conducted by identifying the potential risks of occupational hazards and work accidents at the grey finishing department of the company. The results of this research contribute to solve the problem of the and give solutions of the company regarding risk controls.

2. Literature Review

2.1 Definition of Hazard Identification and Risk Assessment (HIRA)
Hazard Identification and Risk Assessment (HIRA) is one method of identification of occupational accidents with an assessment of risk level as one of the important points to implement the Occupational Health and Safety Management System [2]. HIRA method aims to identify the potential hazards in a company to assess the chances of an accident or losses. Hazard identification and risk assessment and control must be carried out throughout the company's activities, whether the work is carried out by direct employees or contract employees, suppliers and contractors, as well as facilities or personal activities that enter the workplace.
Hazard identification means defining the potential that cause a loss impact. The potential is in the form of event or exposure to a hazardous situation. Hazard identification involves systematic recognition of all aspects of the project that have potential hazards that affect employee performance [3]. Risk evaluation can be determined into two factors, namely severity of harm and likelihood of occurrence [4]. The risk level is in the form multiplication of severity and likelihood. Risk assessment uses semi-quantitative measurement techniques that provide probabilistic and objective results on potential risk events. Semi-quantitative techniques provide values as a range of levels that use numerical values as impacts and probabilities and combine them to produce a level of risk. Figure 1. shows level of risks. The green color means a low level of risk, yellow means medium, orange means high and red means extremely high. The level of risk should affect the existing hazard control.

		Severity				
		Negligible (1)	Minor (2)	Moderate (3)	Major (4)	Extreme (5)
Likelyhood	Almost Certain (5)	M	H	H	E	E
	Likely (4)	M	M	H	H	E
	Possible (3)	L	M	M	H	H
	Unlikely (2)	L	M	M	M	H
	Rare (1)	L	L	L	M	M

Figure 1. Risk Level Adapted from the AS/NZ 4360 Standard Risk Matrix and NHS QIS Risk Matrix[5]

2.2 Hierarchy of Hazard Controls
Generally, Hierarchy of Controls is used to determine how control can be carried out by implementing feasible and effective solutions. The National Institute for Occupational Safety and Health (NIOSH) is one part of the U.S Department of Health & Human Services developing the Hierarchy of Controls Research Method to become the Hierarchy of Hazard Controls [6].

3. Research Methodology
Data collected from observation of production process flow, identification of hazards and risks, and interviews of accident experiences in the grey finishing department. The data were processed using the Hazard Identification and Risk Assessment (HIRA) method, the results of data collection were used to identify hazards that might occur. After the hazard risks have been identified, a risk assessment was carried out to determine the level of the risks that. After the hazard risk has been assessed, the recomendations were made in accordance with the hazard control analysis using the Hierarchy of Hazard

Control method. Through this method, the existing controls should be eliminated, need improvement or need a new control.

3.1 Production Process Flow

There are two types of products produced, grey cloth and cambrics. Where the process of making grey cloth are the weaving and spinning activities. The flow of the production process of making grey cloth and cambrics can be seen in the Figure 2. below.

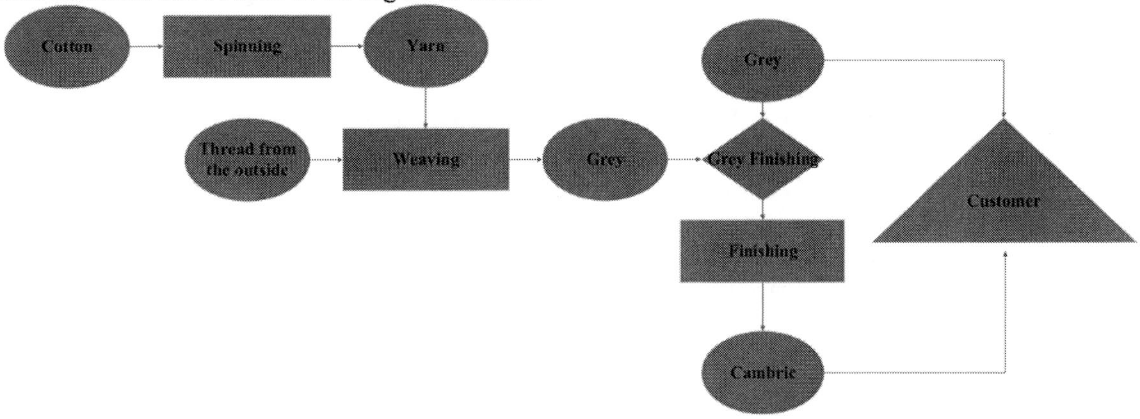

Figure 2. Production Process Flow of Grey Cloth and Cambrics

3.2 Potential Hazards and Risks

The potential hazards in this observation were reviewed based on direct observation and the results of interviews with two department supervisors and three machine operators. Here is the results of the identification of hazards and risks at Table 1.

Table 1. Identification of Hazard and Risk

No	Work Station	Task	Hazard Identification	Risk
1	Inspecting Repair Machines Line	Semi-finished grey cloth inspection	Noise Hazard	Hearing loss and headache
			Hazard of cotton scrap	Sneezing, shortness of breath, allergies
			Physiological hazard	Aches, varicose veins
			Machine hazard	Scratched, pinched, broken bones, broken fingers
2	Folding Machine Line	Folding inspected grey cloth	Machine hazard	Scratched, pinched, broken bones, broken fingers
			Noise Hazard	Hearing loss and headache
			Hazard of cotton scrap	Sneezing, shortness of breath, allergies
			Physiological hazard	Aches, varicose veins
3	Grey Finishing Department Office	Administration and data storage of defective goods	Electrical Hazard	Electricity Shocked
4	Temporary Storage	storage of semi-finished and finished grey cloth	Falling Objects Hazard	Broken bones, bruises, injuries

3.3 Accident Experience

Based on interviews with the head of the grey finishing department, information was obtained that almost all employees had suffered minor injuries such as scratches But there was once an incident where an Inspecting Repair machine operator suffered a fracture due to procedural errors when installing cloth rollers on an Inspecting Repair machine. Then there was also the incident of a folding operator whose nail was broken and even had to be amputated on one of his fingers due to procedural errors when operating the machine.

4. Result

4.1. Hazard Identification

The results of hazard identification at the grey finishing department are as follows:

1. Equipments and locations: inspecting repair, folding machines, passageways, grey finishing department offices, and temporary storage areas.
2. The following is a description of the existing activities:
   a. Inspecting Repair Machine
   It is a machine to inspect the defects. The machine automatically curls the fabric in front of a bright screen, then the operator inspects whether there is a defective cloth roll. The operator uses the Inspecting Repair machine in a standing position.
   b. Folding Machine
   Folding machine is a machine to fold fabric that has been completed by inspecting repair machines. Folding machines have rotational drive with chains. The operator works in standing position.
   c. Grey Finishing Office Department
   The department office is the administrative part where there are important documents related to the process of finishing grey cloth.
   d. Temporary storage is a storage place for grey cloth that has been inspected.
3. The method/approach used in this study is the Hazard Identification and Risk Assessment (HIRA) with a descriptive analysis involving workstations in the Grey Finishing Department. Subject of this study are the heads of departments, operators and machine technicians.
   Data analysis starts from process of identifying potential sources of hazards that cause occupational accidents and occupational illness. Result of hazard identification are carried out a risk assessment to determine the level of risk and review of the control efforts that have been made. Data analysis refers to Undang-undang No. 1 tahun 1970 concerning Work Safety and Peraturan Pemerintah No. 50 Tahun 2012 concerning the Implementation of the Occupational Health and Safety Management System.
4. Person affected of the research as follow:
   a. Factory Production Employee (production line operator and mechanic)
   b. Staff (grey finishing department staff and head of grey finishing department)

4.2 Potential Significant Hazards

Potential Significant Hazards were obtained after hazard identification has been carried out as Table 2.

Table 2. Potential significant hazards

No	Work Station	Task	Hazard	Significant Hazards
1	Inspecting Repair Machines Line	Semi-finished grey cloth inspection	Noise	Exposed to noise without using adequate earplugs resulting in hearing loss and headaches
			Hazard of cotton scrap	Exposed to factory waste so that it can experience respiratory distress (sneezing, shortness of breath, itching due to allergies and eye infections due to exposure to the remaining scrap spinning and inspection activities)
			Physiological hazard	Experiencing aches and even varicose veins due to prolonged standing
			Machine hazard	Pinched roll roller calico cloth, pricked by a needle
2	Folding Machine Line	Folding inspected Grey cloth	Machine hazard	Pinched chain machine
			Noise	Exposed to noise without using adequate earplugs resulting in hearing loss and headaches
			Hazard of cotton	Exposed to factory waste so that it can experience

No	Work Station	Task	Hazard	Significant Hazards
			scrap	respiratory distress (sneezing, shortness of breath, itching due to allergies and eye infections due to exposure to the remaining scrap spinning and inspection activities)
			Physiological hazard	Experiencing aches and even varicose veins due to prolonged standing
3	Grey Finishing Department Office	Administration and data storage of defective goods	Environment Hazard	Difficult to emergency evacuation, flammable
			Electrical Hazard	Electricity shocked
4	Temporary Storage	Storage of semi-finished and finished grey cloth	Falling Objects Hazard	Falling piles of grey cloth causes bruises and even broken bones
			fire Hazard	Flammable goods

4.3 Risk Assessment

Risk Assessment is a systematic method and analytical technique used to define a variety of business, trade and industry needs in the context of occupational health and safety, the application is focused on identifying, assessing and controlling risks to minimize all hazards that can occur in the workplace. The following Table 3. a risk assessment of the hazards were identified in the grey finishing department.

Table 3. Risk assessment

No	Work Station	Task	Hazard Identification	Risk	Risk Value				Risk Value	Risk Level
					Likelyhood		Severity			
					Criteria	Score	Criteria	Score		
1	Inspecting Repair Machines Line	Semi-finished grey cloth inspection	Noise Hazard	Hearing loss and headache	Almost Certain	5	Minor	2	10	High
2	Inspecting Repair Machines Line	Semi-finished grey cloth inspection	Hazard of cotton scrap	Sneezing, shortness of breath, allergies	Almost Certain	5	Minor	2	10	High
			Physiological hazard	Aches, varicose veins	Unlikely	2	Minor	1	2	Low
			Machine hazard	Scratched, pinched, broken bones, broken fingers	Possible	3	Moderate	3	9	Medium
3	Folding Machine Line	Folding inspected Grey cloth	Machine hazard	Scratched, pinched, broken bones, broken fingers	Unlikely	2	Extreme	5	10	High
			Noise Hazard	Hearing loss and headache	Almost Certain	5	Minor	1	5	Medium
			Hazard of cotton scrap	Sneezing, shortness of breath, allergies	Almost Certain	5	Minor	1	5	Medium
			Physiological hazard	Aches, varicose veins	Unlikely	2	Minor	1	2	Low
4	Grey Finishing Department Office	Administration and data storage of defective goods	Electrical Hazard	Electricity Shocked	Unlikely	2	Negligible	1	2	Low
5	Temporary Storage	storage of semi-finished and finished grey cloth	Falling Objects Hazard	Broken bones, bruises, injuries	Occasional	3	Moderate	3	9	Medium

4.4 Control Measure

Every identified hazard was observed based on what controls exist in the work station can be seen in Table 4.

Table 4. Control Measure

No	Work Station	Task	Hazard	Existing Controls
1	Inspecting Repair Machine Line	Semi-finished grey cloth inspection	Noise Hazard of cotton scrap Physiological hazard Machine Hazard	(+) SOP regarding the use of PPE (Earmuffs, disposable masks, special glasses, closed shoes) is complete (+) The availability of SOP) for safe work processes (-) No first aid in this work station (-) There are no strict regulations on the use of PPE so workers have not used PPE properly (-) No national standardized PPE from the company
2	Folding Machine Line	Folding inspection of grey cloth	Machine Hazard Noise Hazard of cotton scrap Physiological hazard	(+) SOP regarding the use of PPE (Earmuffs, disposable masks, special glasses, closed shoes) is complete (+) The availability of SOP for safe work processes (-) No first aid in this work station (-) There are no strict regulations on the use of PPE so workers have not used PPE properly (-) No national standardized PPE from the company
3	Grey Finishing Department Office	Administration and data storage of defective goods	Environment Hazard Electrical Hazard	(+) SOP regarding the use of PPE (Earmuffs, disposable masks, special glasses, closed shoes) is complete (+) The availability of SOP for safe work processes (+) The availability first aid in this work station (+) There is already a safety switch and adequate electrical equipment (+) Fire Extinguisher is available (-) There are no strict regulations on the use of PPE so workers have not used PPE properly (-) No national standardized PPE from the company (-) There are no danger sign and evacuation routes
4	Temporary Storage	Storage of semi-finished and finished grey cloth	Falling Objects Hazard Fire Hazard	(+) The availability of SOP for safe work processes (-) Fire Extinguisher is available (-) No first aid in this work station (-) There are no strict regulations on the use of PPE so workers have not used PPE properly (-) The incomplete SOP regarding the use of PPE headgear and closed shoes (-) No national standardized PPE from the company

4.5 Notification of Changes

The following were the notification of changes during the observation:

1. Description of changes to activity/equipment, there is no change in activity or the addition of equipment.
2. New potential hazard associated with changed activity/equipment, because there has been no change in activity or the addition of new tools, so there is no potential for new hazards.
3. Control measures to be implemented, no new controls are needed yet.

4.6 Hierarchy of Control

Controls that must be carried out in Table 5.

Table 5. Hierarchy of Control

No	Hazard	New Control	Regulation
1	Noise	Apply 5 minutes for safety before work (administrative control)	[7]
		Applying inspection to certain times related to the use of PPE (administrative control)	[8]
		The company prepares earplugs / earmuffs according to SNI (substitution/engineering control)	
2	Hazard of cotton scrap	Apply 5 minutes for safety before work (administrative control)	[7]
		Applying inspection to certain times related to the use of PPE (administrative control)	[8]
		The company prepares earplugs / earmuffs according to SNI (substitution/engineering control)	
3	Machine hazard	Apply 5 minutes for safety before work (administrative control)	[7]
		The use of tools according to the standard (Engineering Control)	[8]
		Ensure workers are in good condition so workers more focus to the objects around	
4	Falling objects hazard	Apply 5 minutes for safety before work (administrative control)	[7]
		The use of tools according to the standard (Engineering Control)	[8]
		Ensure workers are in good condition so workers more focus to the objects around	
5	Fire hazard	Apply 5 minutes for safety before work (administrative control)	[7]
		Workers must fill in the permit entry form before do the work.	[8]
		The use of tools according to the standard (Engineering Control)	
		Ensure workers are in good condition so workers more focus to the objects around	

5. Conclusion and Recommendation

The conclusions obtained from research related to workplace hazard risk analysis and hazard control at the grey finishing department of PT. Primissima (Persero) as follows:

1. Potential significant hazards in Grey Finishing Department are exposed to noise pollution that can interfere with hearing and headaches. In addition, exposure to cotton waste has respiratory problems such as sneezing, shortness of breath and allergies [9]. Then the physiological danger that can cause low back and lower extremity symptoms [10], musculoskeletal disorders of the lower limbs [11], negative effects for ankle/foot discomfort [12] due to prolonged standing. The machines also have the potential cause of injury due to

being pinched. Other potential hazards are environmental hazards [13], electrical hazards [14][15] and fire [16] that cause fatal injuries. And also hazard of failing objects that cause injuries such as bruises and broken bones.

2. Based on observations, the existing controls are Standard Operating Procedures (SOP) including the use of PPE, the availability of fire extinguisher, first aid kit and electrical safety equipment which are good but still inadequate. The company has not provided first aid boxes, PPE according to SNI and signs for emergencies according to specifications provided by the Government. That is because there are no special departments for Occupational Health and Safety Management System and high costs of procurement and maintenance of safety equipment.
3. The Risk Assessment show that the highest hazard are noise hazard and hazard of cotton scrap in inspecting repair machines line, and machine hazard in folding machine line.
4. The controls that can be done by the company are administrative controls, subtitution/engineering controls according to the hazards that have been identified.

6. Acknowledgement

Researchers thank to the support of Industrial Engineering Department, Universitas Islam Indonesia; Industrial Engineering Department, Universitas Sarjanawiyata Tamansiswa; and Industrial Engineering Department, Sekolah Tinggi Teknologi Adisutjipto.

References

[1] Salguero-Caparrós F, Pardo-Ferreira M C, Martínez-Rojas M, and Rubio-Romero J C 2020 Management of Legal Compliance in Occupational Health and Safety. A literature review. Safety Science, 121(August 2019), 111–118.

[2] Roehan K R A, Yuniar Y, and Desrianty A 2014 Usulan Perbaikan Sistem Manajemen Keselamatan dan Kesehatan Kerja (SMK3) Menggunakan Metode Hazard Identifikasi and Risk Assessment (HIRA).

[3] Griffith A and Howarth T 2001 Construction Health and Safety Management. London: Pearson Education Limited.

[4] Odimabo O and Oduoza C F 2018 Guidelines to Aid Project Managers in Conceptualising and Implementing Risk Management in Building Projects. Procedia Manufacturing, 17, 515–522.

[5] Whitfield J 2005 Risk Management Guidelines Companion to AS/NZS 4360:2004 (Vol. 428).

[6] NIOSH 2015 Hierarchy of Controls. Retrieved March 10, 2020, from The National Institute of Occupational Safety and Health website : https://www.cdc.gov/niosh/topics/hierarchy/default.html

[7] Undang Undang No.1. 1970 Hak dan Kewajiban Tenaga Kerja. Bab VIII, Indonesia

[8] Peraturan Menteri 2010 APD sesuai dengan peralatan dan mesin yang digunakan. PER.08/MEN/VII/2010, Indonesia.

[9] Hinson A V, Lokossou V K, Schlünssen V, Agodokpessi G, Sigsgaard T, and Fayomi B 2016 Cotton Dust Exposure and Respiratory Disorders Among Textile Workers at a Textile Company in the Southern Part of Benin. International Journal of Environmental Research and Public Health, 13(9), 895.

[10] Coenen P, Parry S, Willenberg L, Shi J W, Romero L, Blackwood D M, and Straker L M 2017 Associations of Prolonged Standing with Musculoskeletal Symptoms—a Systematic Review of Laboratory Studies. Gait & Posture, 58, 310-318.

[11] Capodaglio E M 2017 Occupational Risk and Prolonged Standing Work in Apparel Sales Assistants. International Journal of Industrial Ergonomics, 60, 53-59.

[12] Baker R, Coenen P, Howie E, Lee J, Williamson A, and Straker L 2018 Musculoskeletal and Cognitive Effects of a Movement Intervention During Prolonged Standing for Office Work. Human Factors, 60(7), 947-961.

[13] Bick R, Halsey E, and Ekenga C C 2018 The Global Environmental Injustice of Fast Fashion. *Environmental Health*, *17*(1), 92.

[14] Shaikh M A, Weiguo S, Shahid M U, Ayaz H, and Ali M 2018 An Assessment of Hazards and

Occupational Health & Safety Practices for Workers in the Textile Industry: A Case Study. Journal of Academic Research in Business and Social Sciences, 8(12), 333-347.
[15] Thaker M N, Phadke B N, and Patel P D 2013 The effects of Electrical Hazards. Perception, 1, 369-374.
[16] Venkataramanan P, Prathap P, Sivaprakash P, and Sivaprakash K 2019 Fire Safety in Textile Industries-A Review. Industria Textila, 70(6), 523-526.

IOP Conf. Series: Materials Science and Engineering **982** (2020) 012054 doi:10.1088/1757-899X/982/1/012054

City-TSP with Objective Minimizing Distance and Noise

Rio Aurachman

Telkom Univesity, Bandung, Indonesia

rioaurachman@telkomuniversity.ac.id

Abstract. Urbanization has encouraged people to have a home in the city. The process of transportation and logistics is also forced to pass through the city area. There is a need to formulate solutions to transportation, supply chain, and logistics problems for cities. One of the commonly used transportation models is the TSP, the Traveling Salesman Problem. Through this paper, we propose a City-TSP, which is a TSP that considers objectives, parameters, and constraints in urban areas. One of the parameters considered is the noise generated from a truck trip. This paper presents the City-TSP model that simultaneously minimizes distance and noise.

1. Introduction

The transportation system supports the economic processes carried out by humans. Human needs are different. Humans also have different abilities from one another. Based on these different abilities humans provide and provide different benefits. These differences give rise to the need to barter or exchange goods and services. Trade processes and economic transactions occur. The transaction is assisted by the transportation system.

Transportation itself is a waste. Waste must be eliminated or at least reduced. The transportation system is one of the objects of study in the field of Supply Chain Management and Logistics. One of the models used is the VRP (Vehicle Routing Problem)[1][2]. The Vehicle Routing Problem has a simple version, namely the TSP (Traveling Salesman Problem)[3]

With increasing urbanization, people shift their activities to urban areas. Production and consumption activities are carried out in the city area. This causes the transportation process, which initially took place outside the city, also took place in the city center. This situation raises the need for knowledge of urban logistics [4]. The urban logistics system has different stakeholders than other systems [5][6]. The existence of these different stakeholders raises different needs and issues. These differences in needs and issues cause the Objective and Constraints of the model to be different. One of the issues of concern is the noise generated by the transportation process. There is one paper that reviews all literature that researching about city VRP [7]

Each route has different noise implications. A route can produce the closest distance. However, these routes may produce noise that is too high. A model and algorithm are needed to help the decision-making process. The model and algorithm are expected to be able to simultaneously minimize mileage and at the same time minimize noise produced

2. Method and Algorithm

In addition to the distance matrix that is commonly used in the Traveling Salesman Problem, this research also involves the Noise matrix. Matrix noise has the same principle as the distance matrix. This matrix has dimensions according to the destinations involved in the Traveling Salesman Problem. For this case study using 7 destinations the matrix noise also has a dimension of 7 x 7. Each row shows the origin and each column shows the destination. The number that becomes the intersection between the row and column is the noise number for when traveling from the row point to the column point. Based on this matrix, the system must find which route is chosen so as to produce the minimum distance and noise. Following in a row are distance matrix and noise matrix

```
distance=([1000,6,4,4,20,12,8],
        [6,1000,4,6,7,4,12],
        [4,4,1000,12,8,9,4],
        [4,6,12,1000,3,5,12],
        [20,7,8,3,1000,8,8],
        [12,4,9,5,8,1000,5],
        [8,12,4,12,8,5,1000])

noise=([1000,26,41,42,20,85,18],
        [60,1000,41,16,56,41,12],
        [14,41,1000,12,18,91,41],
        [41,15,12,1000,31,51,12],
        [20,72,16,13,1000,18,18],
        [12,24,92,35,18,1000,15],
        [81,12,42,12,38,25,1000])
```

The system will work based on an algorithm to get the optimal solution. The algorithm can be illustrated as a process. The process can be described using a flowchart. Figure 1 shows the flowchart of the main process of this algorithm. As for each process described can be detailed to explain the sub-process. Each sub-process can be modeled in the form of a separate flowchart.

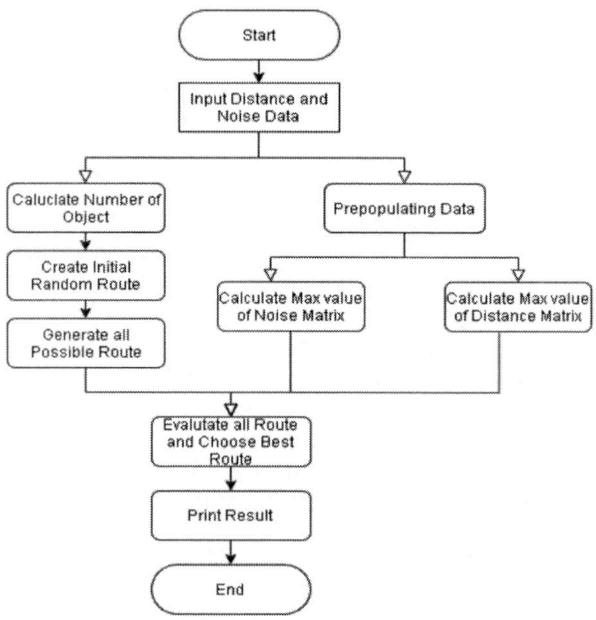

Figure 1. Flowchart of Algorithm

The process in the flowchart is divided into approximately three major groups. The first group is responsible for generating all possible tours. There are many algorithm alternatives to solve VRP or TSP [8][9] [10], but for this research, we used the Brute Force algorithm. All tour possibilities are considered because the algorithm used is brute force. The brute force evaluates all possible alternatives to choose the one that produces the best performance.

The first group is the iteration process to find all the results of route permutations. This process requires the longest time compared to other processes. For example, a route with position 1,2,3,4,5,6,7 can be built permutations 3,6,1,5,2,4,7. All possibilities are counted and registered in an array. The route that has been obtained, then evaluated and found the best route in accordance with the objectives that have been determined.

The second group in charge is the process that functions to get the greatest value from the distance and matrix noise. The distance matrix and noise matrix have values with different scales. However, at the end of the evaluation process, distance and noise are summed. Then it is necessary to normalize the distance and noise based on the maximum value. Without normalization, it will cause one of the variables of high value to be prioritized. Though both of these objectives need to be achieved. Decision-makers can have a preference to prioritize one of the objectives to be optimized, whether it's distance or noise. However, the system must standardize and normalize the two objective variables. Here is the programming code for calculating maximum noise and maximum distance from the matrix

```
def maxvalue(array):
    jar=np.array(array)
    jar2=np.where(jar==1000, 0, jar)
    max=np.amax(jar2)
    #min=np.amin(distance)
    return max                                                    (1)
```

In getting the maximum value from a matrix, you can use the numpy package, namely numpy.amax. However, in the distance matrix and noise matrix, there is a Big M parameter that functions as a constraint. So that the route does not depart and go to the same destination, the diagonal value of the matrix needs to be set as a large number. The number will automatically be selected when using numpy.amax to get the maximum value. Then it is necessary to eliminate 1000 values. This is done specifically for the process of calculating the maximum value. The process is carried out with the numpy. where command which temporarily changes the number 1000 to 0.

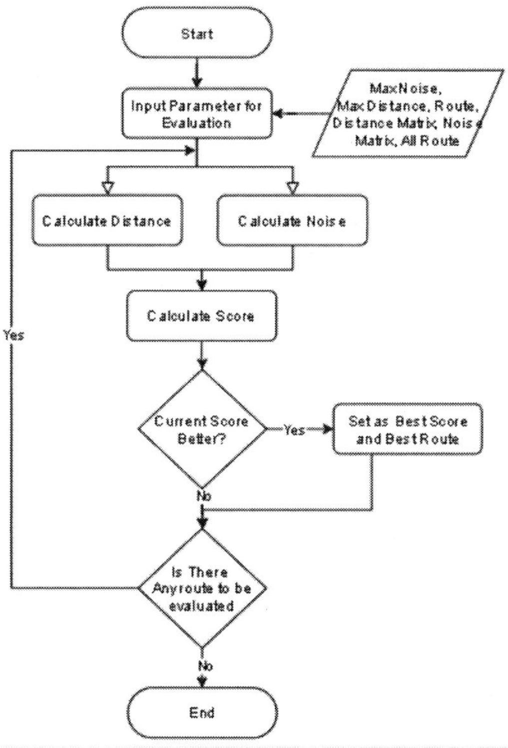

Figure 2. Flowchart of Optimum Solution Algorithm

IOP Conf. Series: Materials Science and Engineering **982** (2020) 012054 doi:10.1088/1757-899X/982/1/012054

The third group is a process that functions to evaluate all routes, as shown in figure 2. Route evaluation is based on the measurement of the total distance and total noise of a route. After that, the two distance and noise calculations are combined into the final score. After that, the system will evaluate whether the resulting score is better than the previous solution score. If yes, then the score and the route being processed will be used as the best route and best temporary score. The next step is to continue the route evaluation process if there are still routes that have not been evaluated.

```
for i in range(len(allpossibletour)):

l=a*totaldistancetour2(allpossibletour[i])/mj+b*totalnoisetour2(allpossibletour[i])
/mn
        if l<lbest:
            dbest=copy.copy(totaldistancetour2(allpossibletour[i]))
            nbest=copy.copy(totalnoisetour2(allpossibletour[i]))
            lbest=copy.copy(l)
          ibest=copy.copy(i)

    def totaldistancetour2(tour):
        d=0
        for i in range(0,len(tour)-1):
            d=d+distance[tour[i]][tour[i+1]]
        d=d+distance[tour[len(tour)-1]][tour[0]]
        return d
```
(2)

3. Result and Discussion

After the development of algorithms and mathematical models, the model of the algorithm and model is tested. To produce the final score, a combination of distance score and noise score is combined. The two scores are summed by giving the weighting of the two scores first. Weights are determined based on decision-maker preferences.

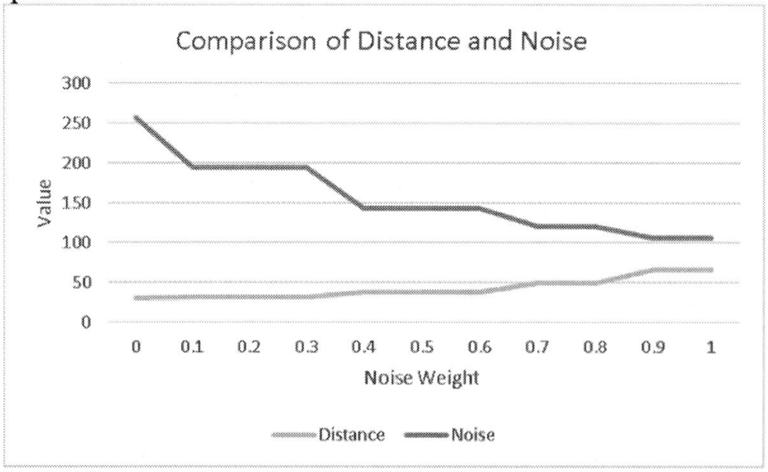

Figure 3. Comparison of Distance and Noise based on Weight

As seen in Figure 3, if we run the model with the minimum coefficient of noise, the optimum solution is obtained, the noise generated is high and the distance calculated is low. However, if the chosen noise coefficient is high, and of course results in the distance coefficient being low, the model will recommend an optimum solution with a high distance, but low noise.

Table 1. Total Score Calculation

a	b	Distance	Noise	Normalized Distance	Normalized Noise	Score
1	0	31	257	0.221428571	0.39906832	0.221428571
0.9	0.1	32	195	0.228571429	0.30279503	0.235993789
0.8	0.2	32	195	0.228571429	0.30279503	0.243416149
0.7	0.3	32	195	0.228571429	0.30279503	0.250838509
0.6	0.4	38	144	0.271428571	0.22360248	0.252298137
0.5	0.5	38	144	0.271428571	0.22360248	0.247515528
0.4	0.6	38	144	0.271428571	0.22360248	0.242732919
0.3	0.7	50	120	0.357142857	0.1863354	0.23757764
0.2	0.8	50	120	0.357142857	0.1863354	0.220496894
0.1	0.9	66	106	0.471428571	0.16459627	0.195279503
0	1	66	106	0.471428571	0.16459627	0.164596273

As shown in Table 1, distance and noise have different values, so that if the two scores are added directly, the system will prioritize the variables with the highest maximum value. To prevent this, normalization is done first for the value of distance and noise. Each distance and noise value is made to be a maximum of 1 and a minimum of zero. After the normalization, the next step is running the code to get the optimum solution. Can be seen in figure 4, the movement of the distance, noise, and total score by changing the noise weight.

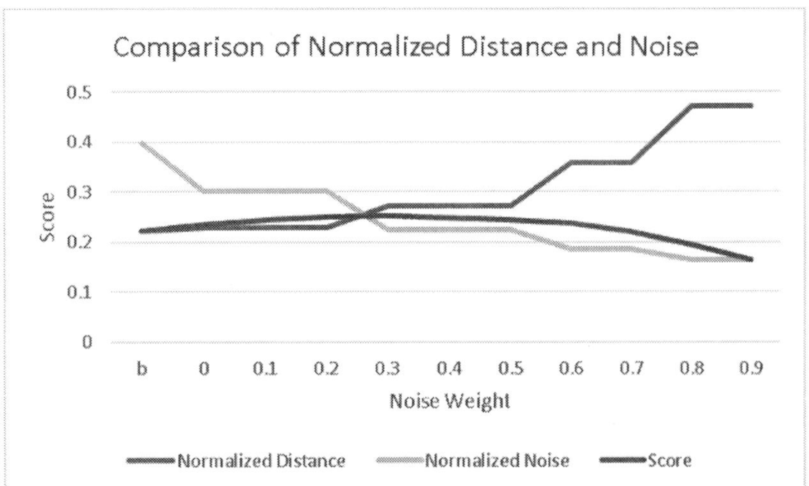

Figure 4. Comparison of Normalized Distance and Noise based on Weight

4. Conclusion

Based on research that has been done, models and algorithms have been found to get the optimum traveling salesman problem solution by considering objective distance minimization and noise minimization. Both objectives are achieved simultaneously.

Before the model and system work, the decision-maker first needs to determine how much weight is set for the two variables. If according to the decision-maker, noise priority is higher than distance, the noise weight is set higher. If the distance priority is higher than noise, the distance weight is set higher.

To make the resulting solution not biased, then the value of the distance and noise needs to be normalized first. The two parameters have different maximum and minimum values so that when calculated together it will cause an imbalance from the optimum solution.

Some further research can be done by considering other constraints on TSPs with an Urban context. City TSP can continue to grow with increasing urbanization. Some constraints that can be used are to consider congestion in the city and consider the comfort of the city dwellers.

References

[1] Eksioglu B, Vural A V, Reisman A 2009 *The vehicle routing problem: A taxonomic review* Computers & Industrial Engineering 57(4) pp. 1472-1483.

[2] Laporte G 2009 *Fifty years of vehicle routing Transportation science* 43(4) pp.408-416.

[3] Näher S 2011 *The travelling salesman problem* In Algorithms unplugged Springer Berlin Heidelberg pp. 383-391

[4] Jiang G, Lam S K, Ning F, He P and Xie J 2019 *Peak-Hour Vehicle Routing for First-Mile Transportation: Problem Formulation and Algorithms* IEEE Transactions on Intelligent Transportation Systems.

[5] Kim G 2018 *Optimal Vehicle Routing for Parcel Delivery with Considering Two Time Periods In 2018* IEEE International Conference on Industrial Engineering and Engineering Management (IEEM) pp. 918-922.

[6] Erdoğan S and Miller-Hooks E 2012 *A green vehicle routing problem* Transportation research part E logistics and transportation review 48(1) pp.100-114.

[7] Kim G, Ong Y S, Heng C K, Tan P S, and Zhang N A 2015 *City vehicle routing problem (city VRP): A review* IEEE Transactions on Intelligent Transportation Systems 16(4) pp.1654-1666.

[8] Kaewman S and Akararungruangkul R 2018 *Heuristics Algorithms for a Heterogeneous Fleets VRP with Excessive Demand for the Vehicle at the Pickup Points, and the Longest Traveling Time Constraint: A Case Study in Prasitsuksa Songkloe* Ubonratchathani Thailand Logistics 2(3) p.15.

[9] GONG J L and CHENG Z M 2006 *The Realization of Genetic Algorithm of VRP Based on the Matlab [J]* Logistics Sci-Tech, 7.

[10] James J Q, Yu W and Gu J 2019 *Online vehicle routing with neural combinatorial optimization and deep reinforcement learning* IEEE Transactions on Intelligent Transportation Systems 20(10) pp.3806-3817.

Application Design of Farmbot Mobile App Using Internet of Things on Android

Firman Setia Nugraha[1,2], Haris Rachmat[1], Denny Sukma Eka Atmaja[1] and Murni Dwi Astuti[1]

[1]Department of Industrial Engineering, School of Industrial and System Engineering, Telkom University, Indonesia

[2]firman.decalvata@gmail.com

Abstract. Food is essential for country development. It can impact both sustainability and the succession of the process itself. As for a developing country, Indonesia's main food supplies are coming from the agriculture sector. It was estimated at least 49.81 million tons of rice must be produced every year. However, this sector has faced some serious threats, the land conversion is one of the major issue, almost 4,000 agriculture hectares is getting convert every year resulting in a mass reduction of food supply. Furthermore, urbanization also decreasing the farmers with a ratio of 0.7 million people annually. These two factors alone will give a huge impact which leading to food shortage. Gladly, the improvements of Internet of Things give a sight at the technological solution. Robotics, combined with IoT, and machinery can produce an agriculture CNC based robotic called Farmbot which functioned to help urban people grow and maintain plants in a small indoor or outdoor area automatically. Farmbot can be controlled by the user using an installed interface on a connected device through internet which allows it to receive command and execute it with great accuracy. In this study, the automation system to operate Farmbot from smartphone devices will be created.

1. Introduction

The need for food is fundamental for all kinds of life, especially in human life. Not only on the individual scale, but it also made a huge impact on country sustainability. As the global population keeps rising, food will become more vital to maintain country development. Indonesia which was stated as the 4th most populated country with more than 268 million people, was expected to produce 49.81 million tons of rice in 2019 according to "Portal Epublika Pertanian" [1]–[3]. This number was indicating a huge demand for food must be fulfilled. Agriculture is the most important and crucial sector in terms of food fulfillment especially in Indonesia as one of the agrarian country [3]–[5]. However, based on current observations, Food fulfillment was facing several threats such as urbanization, which caused the reducing numbers of land and clean water, and so do the farmers.

According to Djauhari's and Nurchamidah's research, it was stated that 4000 hectares of agricultural land annually were converted into other sector lands which are industrial area, housing, roads, and other facilities and infrastructures [2]. Also, on the other hand, the Ministry of Agriculture of Indonesia mentioned that every year from 2012 until 2017, the number of people who works as farmers were reducing as much as 0.7 million residents [6]. The bad news of this phenomenon is that the food fulfillment challenges are not just happened in a particular country or half side of the earth, but it is happening in a whole globe. As no one can prevent the urbanization, then a new grant solution must be given and put into a realize in order to provide a great amount of food supply. One of the solutions is the merger between the agricultural sector with IoT which can lead to smart agriculture.

Smart agriculture can provide modern agriculture using the automation system and IoT. The main concept of IoT's technology needed to help solving the fulfillment food issue is simply the integration of various components and processes to generate highly intelligent and practical tools [7]–[9]. The agricultural sector must embrace the IoT for this advancement and develop more technology to sustain food fulfillment. Also, based on the main factor which is urbanization, the solution which makes agriculture smarter should follow up on the condition in the urban area which only has a small land area for agriculture. This makes the agricultural sector have to transform by utilizing modest resources. By using a modern agricultural concept, smart agriculture will be able to be carried out in urban areas with a high level of population, both on vacant land and in the yard [10].

The basic integration of IoT-based technology (Internet of Things) was accompanied by Artificial Intelligent, Robotic, Sensor, and Human Machine Interface [7], [11]. Smart Agriculture which is a modern farming method will be more efficient using IoT technology, and will also increase the level of productivity and food quality. The IoT system will facilitate the automated implementation of Smart Agriculture on both a large scale and a small scale. There are also some examples of the use of Smart Agriculture using IoT-based technology, namely, (a) agricultural drone which is the use of technology in the form of drones in the agricultural sector, (b) smart greenhouses which are smart greenhouse technology, (c) livestock monitoring or in other words supervision livestock, and (d) precision agriculture which is precision agriculture [12]. Also considering the development of smartphone technology, it is possible to integrate the agricultural tech into the mobile phone to make it easier to use.

Farmbot will be a perfect solution for urban smart farming, by connecting it into a smartphone device, a specific user can control and monitoring the plant from anywhere and harvest it at the correct time. Along with that, by considering the easiness factor, and user satisfactory point of view, it is important to create the UI to be easy to use, understandable, useful, and comfortable to use. All of those factors were wrapped up on this research which leading the research as into designing a mobile app that can be connected to Farmbot and send a specific command to execute watering feature, and spindle movement with high accuracy and the intensity of the app to be complacent to use [5], [13].

2. Materials and Method

2.1. Materials

2.1.1. User Interface. The user interface (UI) is one part of Human-Computer Interaction (HCI), which is a study, planning, and design of interactions between humans and computers [14]. UI in a nutshell can mean a display on a screen, which can be seen or controlled by the user by displaying information in an interactive system [15]–[17]. In the current Industry 4.0 era, a user interface can mean a display that connects a function in the form of a program in it, to do what is desired by one or more users using a computer or smartphone to allow interaction between humans and computers and smartphones. The UI must also prioritize the layout arrangement of the application or website display design which is simple, easily recognizable and consistent [18]

2.1.2. Farmbot. Farmbot is a small agricultural robot that resembles an open-source CNC that moves on a special track that navigates a plot of land with a plot of the Cartesian field [13]. There are several inseparable parts of a farmbot, including tracks, tools, servo, and containers of tools or tools bay. Farmbot conducts all activities from seeding, watering, scheduling, and monitoring (including monitoring plant moisture conditions) by being on the track that has been available on a plot of land using the right tool heads in accordance with its functions which are stored in the tool bay located in one side of the plot. Farmbot is controlled using electronic components such as Arduino / RAMPS and Raspberry Pi 3 [13].

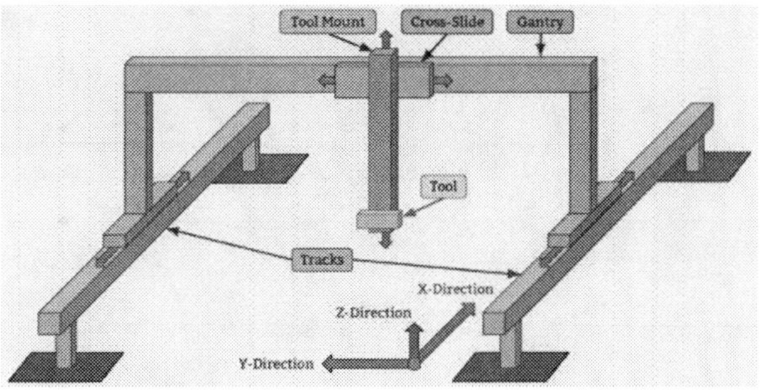

Figure 1. High Level Cartesian FarmBot Design

Figure 1 above shows the high level of Farmbot Cartesian design which consist of 3 axis direction X, Y, and Z. This allowing better movement with precision to execute seeding, watering, and monitoring command. Farmbot is also identified by Web application that enables access from anywhere in which user located and using any internet-connected device. Figure 2 below showing the Farmbot web application which represents the garden design as one of the features available.

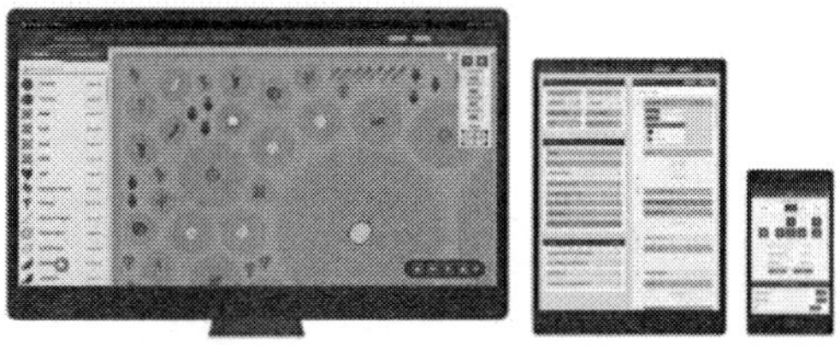

Figure 2. Farmbot Web-Based App

2.1.3. Usability. Usability can be interpreted as a convenience that can be learned and understood by users in the process of operating, preparing, and interpreting the output of a system or component. Usability can also mean as a form of achieving a specific goal related to the satisfaction, effectivity, and efficiency of an application user [14], [19].

In determining the usability value, there is an interval scale of values that become a reference to determine whether the usability value obtained satisfies the user or not. According to Arikunto in a book entitled "Research Procedure for a Practical Approach", it was stated that if the usability results obtained ranged from 0 - 52% then it can be said that the interface design was not successful or could not meet user expectations. Then, if the range of usability values ranges from 53 - 68%, then it is said that the interface design is sufficient to meet user expectations and if the usability value ranges between 60-100%, then the interface design can be said to be successful. The following table 1 shows the scale interval of the qualification value from the usability test to be performed.

2.2. Method

Figure 3 below describes a frame of mind that consists of a step by step sequence of problem solving that will make it easier for writers to carry out the research process from the beginning to completion. The research was divided into 3 main sequences that began with the observation of the existing system of Farmbot hardware and software in research planning. Then it was processed with the software design or the app generation starting from gathering the user needs until testing it out using the Usability test. The last process was the integration phase which concludes the final step of the research.

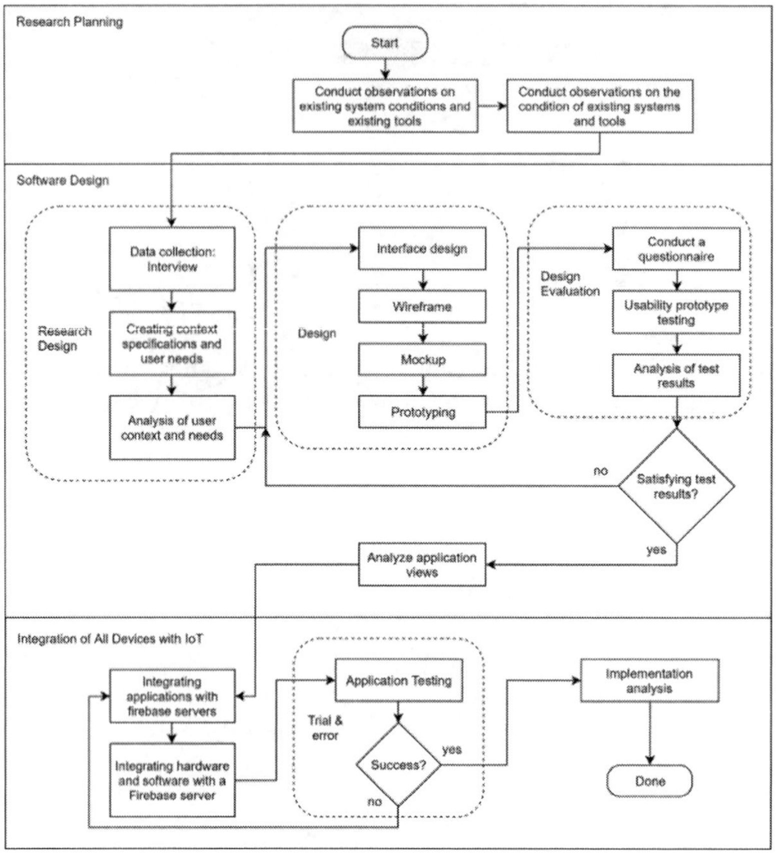

Figure 3. Conceptual Model

2.2.1. Research Planning

The author began the research by do some benchmarking on existing systems of the Farmbot app and the system software to be used during the research. By doing research planning, the author found Android Studio will be selected for the app creation and Firebase Realtime-Database as the server.

2.2.2. Software Design Phase

Conduct 3 main sub-phase which are part of the User-Centered Design method, research design, design, and design evaluation [14], [20]. The phase began by interviewing the previous researchers of the Farmbot Windows-based app in order to generate a User Persona. The user persona is used as guidance for the app making process. Starting from creating the wireframe, mockup, and finalize it as a complete installable app. Just before the interface is being finalized, the fixed user interface is reviewed using usability test to assure it meets the user expectance. The usability test is begun by sending some questionnaires to some respondents while showing the interface designed, also some of the respondents got the chance to try the prototype. Author using Likert Scale Value as shown in table 1 for scaling the questionnaire result.

Table 1. Likert Scale Value

Category	Score
Strongly Disagree	1
Disagree	2

Neutral	3
Agree	4
Strongly Agree	5

The results are measured using the formula as follow:

$$Usability\ (\%) = \frac{A+B+C+D}{4} x100\% \quad (1)$$

The usability test will be judge by the author using Qualification Usability Percentage by Arikunto as shown in table 2 below to determine whether the app interface has met the users' expectations or not.

Table 2. Qualification Usability Percentage [21]

Interval (%)	Qualification	Result
85 - 100	Very good	Succeed
69 – 84	Good	Succeed
53 – 68	Enough	Not Successful
37 – 52	Poor	Not Successful
20 - 36	Very Poor	Not Successful

2.2.3. Integration System with IoT.
Internet of things is the most important setup in this research. It is used to transfer the information it needed from the app onto the Farmbot hardware. The server which used in this research was Firebase system from google, it is connected to Android studio by giving certain code to send and retrieve information as the command for Farmbot to act. The command then received by Farmbot hardware through an embedded WIFI module inside Raspberry Pi3 used. Then Raspberry Pi3 forwarded it to Arduino Mega 2560 to utilize which hardware parts should move. It's connected to all parts of Farmbot such as motor driver, soil moisture sensor, water pump, camera, and motion sensor. Once the command is given using the app, the information will remain saved inside Firebase Realtime-Database and it can be changed by giving new commands. The integration phase is finished when the trial and error of the whole system succeed.

3. Result and Discussion

3.1. User Persona
Farmbot mobile app is an app that will be able to help farmbot user to operate it within such a long or short distance using a mobile platform. The app itself is meant for the user conformance thus make it has to satisfy the user needs. User Persona helps the author to map and determine the layout of User Interface and the structural function of the app. Table 3 below shows the User Persona gained by the author.

Table 3. User Persona

Needs	Requirement
Want ease in controlling the Farmbot device	• Farmbot application based on Android so that it can be used on smartphones • Firebase server that connects applications with farmbot devices to enable interactions using IoT so that they can be controlled remotely

Want ease in managing watering events on the Farmbot	• Scheduling feature that can be controlled via an Android-based application • Plant coordinate feature that makes it easy for users to determine the coordinates of the farmbot watering
Want to display applications that are simple, fun and easy to understand	• The interface is made as simple as possible using the appropriate color so that it is fun to look at

Each feature that was made in the app are mapped using scenario context as shown in Table 4 and 5:

Table 4. Scenario Context

No	Task	Sub-Task	User	Goals	System
1	Scheduling	Show Schedule	Students and further farmbot researcher	Users can view information related to events that are taking place on the Farmbot device. Some indicators can be seen such as: action (what activity is carried out by the farmbot), repeat (when the event will repeat), time (when the event will be executed), and moisture (indicator of soil moisture).	Flow in Show Schedule: Open the application then select the scheduling menu.
		Edit Schedule		Users can input events in the form of watering plant schedules and plant scanning schedules.	Flow in Edit Schedule: Open the Scheduling menu then touch the Edit button.

Table 5. Scenario Context (Continued)

No	Task	Sub-Task	User	Goals	System
1		Activate Schedule		Users can activate events that have been input before. The number of events that can be activated is 4.	Flow on activate schedule: Open the Scheduling menu then touch the red switch button until the color turns green.
		Deactivate Schedule		Users can deactivate events that have been previously activated.	Flow on activate schedule: Open the Scheduling menu then touch the green switch button until the color turns red.

		Start Point	Students and further farmbot researcher	The user can set the first Start Point from the watering location.	Flow on Start Point: Open the Control menu then fill in the start point coordinates then touch the Submit button.
2	Watering Coordinate	Watering Step	Students and further farmbot researcher	Users can adjust the number of crop location loops.	Flow on Start Point: Open the Control menu then fill in the number of watering steps then touch the Submit button.
		Interval	Students and further farmbot researcher	Users can adjust the distance between watering locations.	Flow on Start Point: Open the Control menu then fill in the interval coordinates then touch the Submit button.

3.1.1. Scheduling
The application to be designed is planned to have 4 scheduling slots that can be inputted by the user both for watering and scanning the planting area. Task scheduling that will manage all the functions of the event or scheduling feature that consists of:

3.1.1.1 Show Schedule
Shows event data that has been inputted previously. Users can see the minimum moisture content before the Farmbot device does the watering and other data including indicators whether the schedule is active or not.

3.1.1.2 Edit Schedule
Allows users to change the schedule for watering and scanning of planting land.

3.1.1.3 Activate Schedule
The four schedules that have been inputted can be activated in each event show interface using the button switch.

3.1.1.4 Deactivate Schedule.
Provides the ability for users to deactivate events that have been inputted.

3.1.2. Watering Coordinate
There are 3 components that determine the location of watering, the number of watering locations, and the distance between watering locations, including:

3.1.2.1 Start Point
Users can specify the location of watering manually by changing the start point in the Firebase database using the interface control.

3.1.2.2 Watering Step
Watering step determines the number of looping watering locations in the x and y-axis cartesian format. Users can set the number of watering locations manually through this function.

3.1.2.3 Interval
The interval determines the distance between watering locations. Users can set it manually using this function.

3.2. Interface Architecture
Figure 4 below shows the arrangement of the wireframe UI app which was generated based on the user persona and scenario context. This application has 4 main menus shown in the interface architecture below which are the Scheduling menu, the Control menu, the Monitoring menu, and the About menu. However, in this study, there will be only 2 menus for discussion, which are scheduling and control menu. From the main menu, users can access scheduling and control menu. On the Scheduling menu, the user will immediately be directed to Schedule 1 Show menu which gives information about the first watering schedule. Users can navigate the schedule menus 2, 3, and 4 via the arrow buttons on each interface. Also included is a feature to activate or deactivate schedules that have been inputted. From the home menu, users can also access the Control menu which can display the coordinates of the temporary spindle of the Farmbot device. At the bottom of the interface, there is a watering coordinate feature. In this feature, the user can manually adjust the watering coordinates and specify the number of locations that must be watered. More details about each feature will be discussed in the prototype figures below.

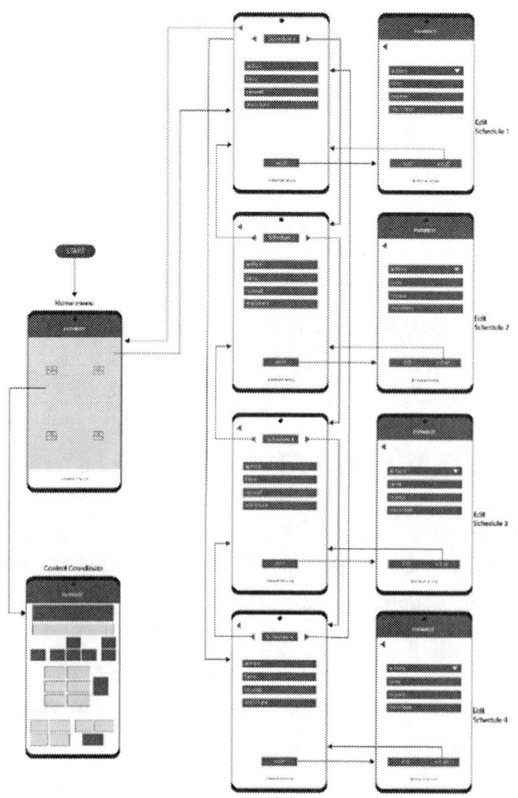

Figure 4. Interface Architecture Farmbot Mobile App

4. Prototypes

In this application, there are several views to perform the previously mentioned scenario of the app which accessible from the main menu view which can be seen in Figure 5.

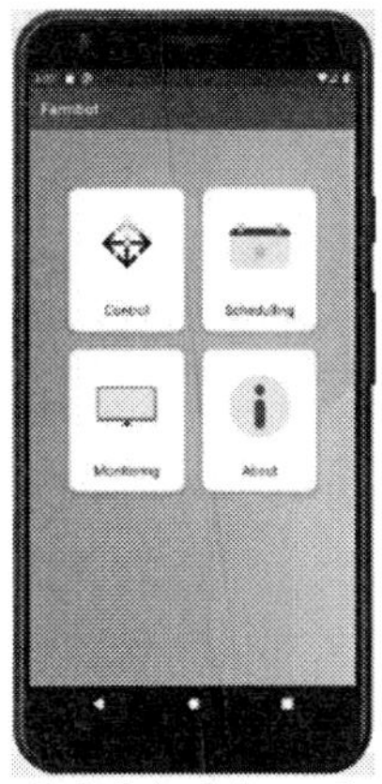

Figure 5. Main Menu Interface

Figure 5 above is the interface of the main menu in this application which has 4 main sub-menus that described as follows:

4.1. Scheduling

Figure 6 below which is the Scheduling menu which directs the user to an interface full of the Farmbot watering activity information. Its main functions are to provide information to the user regarding when it will perform watering to the plant, when it should do the watering again (it could be every day, every week, or every one-hour), and even in this menu the user could either activating or deactivating the watering schedule manually. It also providing some information regarding the moisture indicator to determine whether it should perform watering or not.

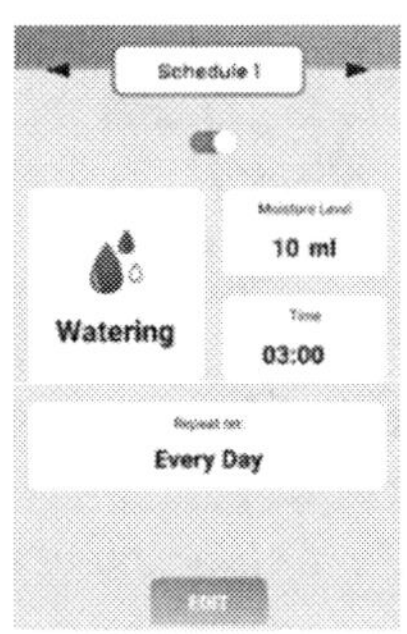

Figure 6. Scheduling Menu Interface

4.2. Schedule Edit

It's the sub-menu from Scheduling menu which can be accessed by touching the EDIT button at the bottom interface from the previous menu. Its main function is to edit the scheduled event that has been inputted before and change it with different parameters as the user's wills. Figure 7 below is the Schedule Edit interface.

IOP Conf. Series: Materials Science and Engineering **982** (2020) 012055 doi:10.1088/1757-899X/982/1/012055

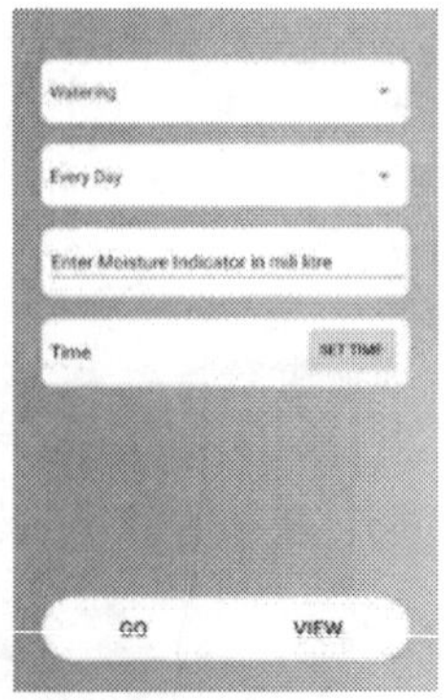

Figure 7. Scheduling Edit Interface

4.3. Control

Control is the sub-menu which gives the user full control of farmbot movement by using the *x,y, z*-axis. Control menu is also divided into several functions which are described as follows.

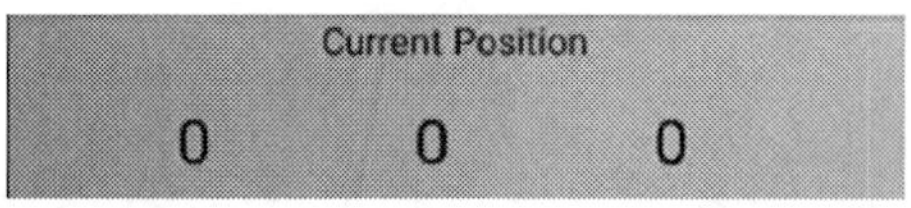

Figure 8. Current Position Function Interface

Figure 8 above shows the current position of Spindle Farmbot or the main tools that will perform watering, monitoring, or seeding. It will give information to Users on which coordinate the spindle is.

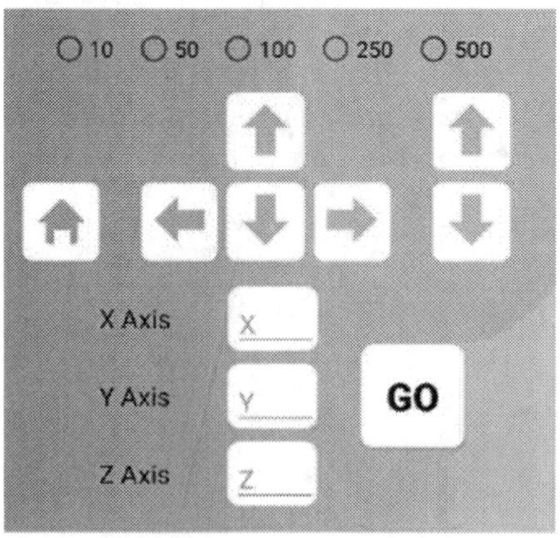

Figure 9. Move Coordinate Function Interface

Figure 9 above shows a function interface that can be used to move the spindle manually to a specific location.

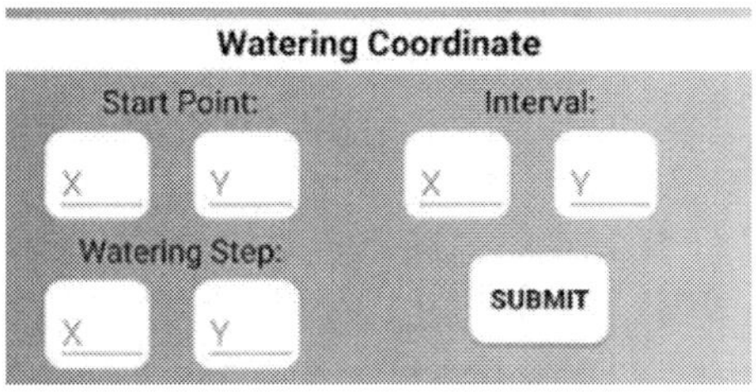

Figure 10. Watering Coordinate Function Interface

In figure 10, it is shown the Watering Coordinate function interface that can be used to change the watering coordinates manually as wills.

5. Usability Test

The usability test in this research is meant to determine either the interface has met user expectations or not. The idea is to create a Farmbot mobile app which more understandable, useful, easy to use, and easy to learn. Several questions were spread in a questionnaire into 50 respondents whim given a sight of the app and tried to use it. The questionnaire was using Likert Scale from 1 to 5, and the result obtained was calculate using the usability formula as follows.

$$Usability\ (\%) = \frac{A+B+C+D}{4} x100\% \qquad (2)$$

- usefulness score = A
- easy to use score = B
- easy to learn score = C
- satisfaction score = D

Questionnaire result calculated as follows in Table 6.

Table 6. Questionnaire Result in USE Result

Usefulness	Easy of Use	Easy of Learn	Satisfaction
84.9%	81.6%	76.9%	82.4%

Using the equation above, the usability result obtained as showed in the calculation below.

$$Usability\ (\%) = \frac{84{,}9 + 81{,}6 + 76{,}9 + 83{,}4}{4} x100\%$$

$$Usability\ (\%) = 81{,}45\%$$

6. Integration System

This application is connected to a firebase realtime database server to store data or commands from the user using through the app and transfer it into existing farmbot hardware to perform a specific action.

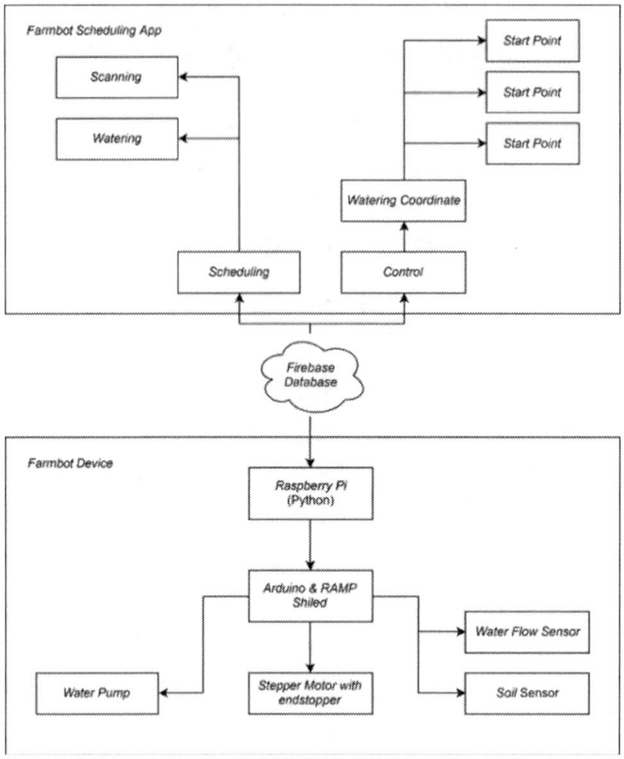

Figure 11. Hardware and Software Integration

Figure 11 explains that the Farmbot system consists of two main sub-systems consisting of a Farmbot schedule app and a Farmbot device. The two sub-systems are connected using a Firebase server which is an intermediary in sending and retrieving data from the Farmbot app to the Farmbot device and vice versa. Farmbot devices connect various farmbot devices such as water pumps, stepper motors, soil sensors, and water flow sensors. The Farmbot device is controlled by Arduino Mega which is connected using Raspberry Pi to the Firebase realtime database. Then the Farmbot scheduling application will retrieve information on firebase and store various scheduling commands in the database.

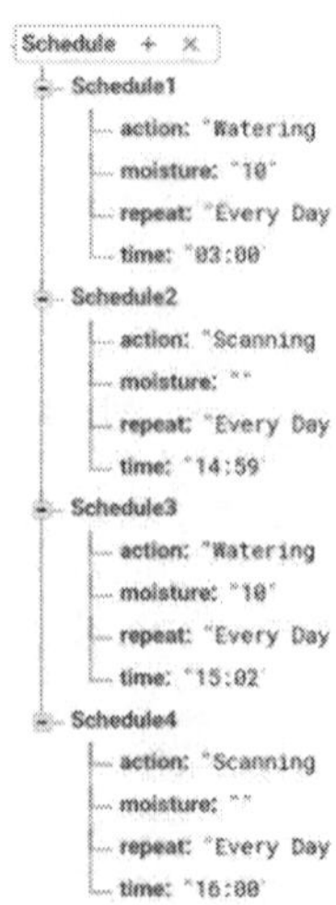

Figure 12. Firebase Scheduling Database

Figure 12 and 13 give brief information about the information stored in the Realtime Database which also can be transferred into Farmbot Hardware to do a specific command.

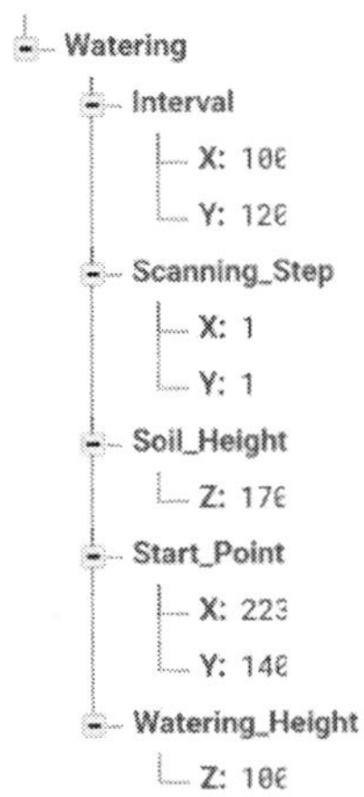

Figure 13. Firebase Scheduling Database (Continued)

7. Conclusion

Farmbot is a CNC based robotic device which functions are to help people in the urban area to grow plants and enable them to produce plant-based food supply without consuming huge areas. Thus, the operation of Farmbot that was previously using computer which requires short distance between users and Farmbot hardware, can also be operated using user's smartphone with unlimited distance as far as the internet signal can reach it. Firebase holds an important role in this research as to store the information given by the user. The Raspberry Pi at Farmbot device helps it to get the information from Firebase as it has an embedded wifi module to access internet. Farmbot also detects area which information is transferred back to the Firebase, and the application which connected to Firebase can obtain it and shows it to the user. By using this app, users can control the position of Farmbot spindle manually, set scheduling events to do watering or monitoring the land area, and set the new watering coordinates by using mobile smartphone at any distance. The interface also user-oriented by the results of the usability test had shown. The advantage of this android based Farmbot app are the user-friendliness and its borderless, no matter where the user is giving the command to Farmbot, will be delivered by the Firebase server.

References

[1] Bapenas, "Jumlah Penduduk Indonesia 2019 Mencapai 267 Juta Jiwa," *Katadata*, 2019. [Online]. Available: Https://Databoks.Katadata.Co.Id/Datapublish/2019/01/04/Jumlah-Penduduk-Indonesia-2019-Mencapai-267-Juta-Jiwa. [Accessed: 18-Sep-2019].

[2] L. Nurchamidah And Djauhari, "Pengalih Fungsian Lahan Pertanian Ke Non Pertanian Di Kabupaten Tegal," *Akta*, Vol. 4, No. 4, Pp. 699–706, 2017.

[3] Sabarella *Et Al.*, *Buletin Konsumsi Pangan*, Vol. 10. Jakarta: Epublika.Setjen.Pertanian, 2019.

[4] B. Murdyantoro, "Perancangan Aplikasi Farmbot Berbasis Internet Of Things (Iot)," Telkom University, 2019.

[5] R. Sukmana, "Perancangan Tools Watering Dan Soil Sensor Pada Spindel Farmbot Menggunakan Metode Kansei Engineering," Telkom University, 2019.

[6] Kementrian Pertanian, *Rencana Strategis Kementrian Pertanian Tahun 2015-2019.* Kementrian Pertanian, 2015.

[7] A. Hartarto, "Making Indonesia 4.0," *Kementrian Perindustrian*, Pp. 1–8, 2019.
[8] A. Kamilaris, F. Gao, And F. X. Prenafeta-Bold, "Agri-Iot : A Semantic Framework For Internet Of Things-Enabled Smart Farming Applications."
[9] B. Murdyantoro, D. Sukma, E. Atmaja, And H. Rachmat, "Application Design Of Farmbot Based On Internet Of Things (Iot)," Vol. 9, No. 4, Pp. 1163–1170, 2019.
[10] M. Zainal And S. Raba, "Urban Agriculture : The Role Of Knowledge Among Farmer In Malaysia," Vol. 7, No. 2003, Pp. 77–85, 2017.
[11] M. Wollschlaeger, T. Sauter, And J. Jasperneite, "The Future Of Industrial Communication: Automation Networks In The Era Of The Internet Of Things And Industry 4.0," *Ieee Ind. Electron. Mag.*, Vol. 11, No. 1, Pp. 17–27, 2017.
[12] L. Lipper *Et Al.*, "Climate Smart Agriculture For Food Security," *Nat. Clim. Chang.*, Vol. 1072, No. November 2014, Pp. 1068–1072, 2014.
[13] L. Swart, "Farmbot Rfid Integration Senior Project Final Report," No. June, 2017.
[14] A. Al Ghiffari, E. Darwiyanto, And D. Junaedi, "Perancangan Ulang User Interface Website Politeknik Kesehatan Makassar Menggunakan Metode User-Centered Design," *E-Proceeding Eng.*, Vol. 6, No. 1, Pp. 2291–2341, 2019.
[15] A. Nugroho, M. Imrona, And E. R. Kaburuan, "Perancangan User Interface Aplikasi Pesan Antar Warung Makan Dengan Metode User Centered Design Untuk Operator Lemah Didik (Studi Kasus Warung Makan Pesan Antar Wilayah Kampus Universitas Telkom)," 2016.
[16] R. Hilmi And Ardiansyah, "Pengembangan User Ecperience Dan User Interface Aplikasi Igometer Taksi," *Sarj. Tek. Inform.*, Vol. 1, No. Agustus, Pp. 1–10, 2015.
[17] A. I. Yunus, "Perancangan Desain User Interface Dan User Experience Pada Aplikasi Siakad Dengan Menggunakan Metode User Centered Design (Ucd) Pada Universitas Islam Negeri Sunan Ampel Surabaya," 2018.
[18] A. F. Patulak, J. Purwadi, And T. H. R, "Implementasi Metode User Centered Design Dalam Perancangan Aplikasi Pelbelajaran Vlan & Inter-Vlan Routing," 2008.
[19] E. Saputra *Et Al.*, "Usability Testing Untuk Mengukur Penggunaan Website," No. 12, 2014.
[20] M. Lubas, J. Mitchell, And G. De Leo, "User-Centered Design And Augmentative And Alternative Communication Apps For Children With Autism Spectrum Disorders," 2014
[21] Arikunto, S., "Prosedur Penelitian Suatu Pendekatan Praktik", Edisi Revisi 6. Jakarta : Rineka Cipta, 2009.

IOP Conf. Series: Materials Science and Engineering **982** (2020) 012056 doi:10.1088/1757-899X/982/1/012056

Maritime Inventory Routing Problem: Application on Discharge the Load of the Ship in Cement Companies to Minimize the Total Transportation Cost

Febryan Khoirun Yusuf[1,3], Ari Yanuar Ridwan[1] and Hardian Kokoh Pambudi[2]

[1]Industrial Engineering Department, School of Industrial Engineering, Telkom University, Bandung, Indonesia
[2]Logistics Engineering Department, School of Industrial Engineering, Telkom University, Bandung, Indonesia

[3]febryanky@student.telkomuniversity.ac.id

Abstract. Distribution activities are an important part and very considered in the world of logistics because distribution is one of the key drivers of profits earned by companies. One that is related to distribution is transportation. Transportation refers to moving products from one location to another where the product moves from the beginning of the supply chain to consumers where this transportation will incur costs and is one of the costs that affect the price of a product. This research aims to schedule ship transportation from 3 production ports to 6 consumption ports with a heterogeneous fleet of ships to minimize the total transportation costs in the cement industry companies. Maritime Inventory Routing Problem (MIRP) is a problem of ship scheduling which is not only related to the distribution of products from production ports to consumption ports, it also manages the inventory at these ports and is usually used for bulk industrial products. The method used in this research is MIRP with Mixed-Integer Linear Programming (MILP) approach where this method can minimize the total transportation costs. The results show that the method used can reduce the total waiting time so that the total transportation costs are also reduced.

1. Introduction

Nowadays companies in competing change their focus, which at first the focus of competition between the companies turned into a competition between supply chains[1]. Supply chain management is the integration of business processes from end-users through initial suppliers that provide products, services, and information that add value to customers[2]. Included in supply chain management is good coordination and collaboration between suppliers and customers. In other words, supply chain management integrates negotiation and demand management in the company[3].

One of the things related to the supply chain is distribution activities. Distribution is taking steps to move and store products from the supplier stage to the consumer stage in the supply chain[4]. Distribution in the world of logistics has become an important part and very considered, where distribution is the process of moving goods from suppliers to the end customers involved in the supply chain, basically distribution is one of the main drivers of the overall profit earned by the company because it affects the cost of the supply chain[5].

Meanwhile, transportation refers to moving products from one location to another where the product moves from the beginning of the supply chain to the consumers[4]. Transportation costs can be categorized into two, namely outbound transportation costs and inbound transportation costs. Outbound transportation costs for shipping to customers are usually included in sales, general, and administrative costs. while inbound transportation costs are usually included in the cost of goods sold, therefore this transportation cost is one that affects the price of a product[4].

There are various types of modes of transportation, one of which is sea mode. In general, vehicles used in transportation by sea are ships[4]. The research conducted is in the company engaged in the cement industry in Indonesia. This company in meeting the demands of its customers in Indonesia, the distribution of its products uses several modes of transportation, one of which is the sea mode using vessels. The vessels owned by this company are chartered vessels of different types and capacities or heterogeneous fleets.

The company has 3 production ports (Tarjun, Tanjung Priok and Cirebon Port) and 6 consumption ports (Samarinda, Pontianak, Surabaya, Lombok, Palembang, and Lampung Port) and will distribute its cement by vessel. The vessel loads cement at the production port and will sail to the consumption port followed by discharging the cement at the consumption port. Vessels can discharge their cargo when the inventory storage from the port of consumption is not full. However, in reality, many vessels when arriving at the port of consumption cannot discharge the cargo, due to the storage of cement supplies still full. So, the vessels must wait until the cement they have is sold to consumers.

Table 1. Total Idle Time of Vessel in 2018

No	Month	Total Idle Time (Hours)
I	Jan	163.1
II	Feb	171.3
III	Mar	310.3
IV	Apr	12.7
V	May	0
VI	Jun	281.8
VII	Jul	21.3
VIII	Aug	66
IX	Sep	173.6
X	Oct	215.4
XI	Nov	34.7
XII	Dec	210.5
Total		1660.7

In 2018, the total idle time that occurred was 1660.7 hours in sending cement to all consumption ports. A large amount of idle time due to the vessel being unable to discharge cement. One of the causes of a large idle time is improper scheduling and it impacts a large total transportation cost.

Table 2. Existing Total Transportation Cost in 2018 (a)

No	Month	Total Transportation Cost
I	Jan	Rp 24,392,140,341
II	Feb	Rp 14,284,236,389
III	Mar	Rp 21,230,310,242
IV	Apr	Rp 8,961,013,584
V	May	Rp 19,310,321,060
VI	Jun	Rp 15,103,512,463
VII	Jul	Rp 20,387,184,523
VIII	Aug	Rp 23,685,842,976

Table 3. Existing Total Transportation Cost in 2018 (b)

No	Month	Total Transportation Cost
IX	Sep	Rp 24,441,641,005
X	Oct	Rp 25,580,885,240
XI	Nov	Rp 27,039,256,182
XII	Dec	Rp 28,385,171,692
Total		Rp 252,801,515,696

Maritime Inventory Routing Problem is a problem method used for a company that must be responsible for carrying products from the supplier port or loading port to the consumer port or unloading port and also responsible for managing inventory levels at the port, MIRP is usually often used for bulk industrial products[6]. Thus, scheduling cement distribution is solved using the MIRP problem method with a Mixed Integer Linear Programming mathematical formulation to minimize the total transportation cost.

The focus in this paper is the use of the Maritime inventory routing problem method with a mixed-integer linear programming approach that has been widely used in several cases. As in the research conducted by Agra[7], which is formulating a mathematical model that is used by solving the problem using the software Xpress Optimizer version 21.01.00 with Xpress Mosel version 3.2.0 and also research conducted by Diz[6] for solving problems in the case of petroleum in Brazil using the platform commercial optimization called AIMMS 3.13 and also GUROBI 5.5 Solver. So that this research will use the same methods and mathematical models to solve a case in one of the cement industry companies in Indonesia, where the difference with previous research is code that was built using the Python Programming Language with the library PuLP, which is easy to understand for humans and a powerful library for modeling in Python with open-source packages and solved using the GUROBI 9.0.1 solver version of the free academic license. Besides, this research will provide optimal results for a full year by providing sequential monthly results. So, this research is expected to be able to enrich the use of the MIRP method with different cases.

2. Literature Review

Maritime Inventory Routing Problem is defined as a combination of ship routing and ship scheduling problems so finding routes and scheduling vessels with minimal transportation costs and inventory management problems so that scheduling does not interrupt production or consumption at the storages[8].

MIRP is a problem method used for a company that must be responsible for carrying products from the supplier port or loading port to the consumer port or unloading port and also responsible for managing inventory levels at the port, MIRP is usually often used for bulk industrial products[6] such as cement, rice, liquid chemicals, and oil.

Pioneering this problem was the work of Dantzig and Fulkerson in 1954[9]. The problem solved was minimizing the number of tankers needed to carry out a certain set of schedules. This can be considered as a matter of fleet size where there is only one type of ship. This shows that the problem is a transportation problem[10]. Next, the MIR problem is to address the ammonia supply chain problem. To overcome this problem, Christiansen in 1999[11] proposed a formulation of mixed-integer linear programming (MILP).

Furthermore, researchers and industry pay attention to the monetary benefits that can be achieved by coordinating inventory management planning activities and ship schedules. Besides, coordination can also increase the robustness and flexibility of operations as done by Christiansen et al. in 2013[12]. This perception has made the focus on MIR issues increase over the past decade[6].

Although researchers and industry recognize that the optimization model can improve efficiency in the maritime inventory routing problem, they also realize that solving this type of problem until optimality for real-world problems is very complicated. Although it is complex, Agra, Andersson, et

IOP Conf. Series: Materials Science and Engineering **982** (2020) 012056 doi:10.1088/1757-899X/982/1/012056

al., 2013[7] claims that the MIR problem is usually very focused on capital, and a simple improvement in fleet scheduling and cargo quantity will mean large economic benefits.

In the last few decades, due to complex problem solving it requires the approach used. most approaches in the literature use MILP techniques to deal with these problems. Regarding appropriate methods such as MILP, researchers have investigated decomposition techniques and the development of reformulations to improve computational performance. In 1999 Christiansen[11] proposed decomposing the MIR problem into two types of subproblems: ship routing and inventory management. Agra, Andersson, et al., 2013[7] also proposed a reformulation called the fixed load flow model (FCNF) and has implemented several important valid inequalities that provide more stringent formulations in terms of a linear relaxation. Furthermore, with the formulation carried out by Agra, Andersson, et al.[7], In 2017 Diz[6] used the formulation to resolve the petroleum case in Brazil concerning crude oil offloading and supply problems (COSP).

3. Mathematical Model

The mathematical model used is a model for Maritime Inventory Routing Problem with Mixed Integer Linear Programming (MILP) approach. MILP is a mixture of Linear Program (LP) and Integer Linear Programming (ILP), i.e., a set of variables can be partitioned into two subsets where the first subset contains variables with integer domains and variables in the second subset have real value domains which has a linear objective function[13]. MILP is a linear integer programming model that has a function to optimize the goal of the model[14].

Mathematical formulation using the MILP approach is a mathematical model that was formulated by Agostinho Agra[7] in 2013 which later was also implemented by Gustavo Souto dos Santos Diz[6] in 2017 which was applied to the case of a petroleum company in Brazil.

3.1. Sets

The set used in the formulation of this mathematical model as follows.

N	Set of all ports indexed by p and q;
T	Set of time periods indexed by t;
V	Set of vessels or ships indexed by v;
NP	Set of loading or production ports indexed by p and q;
ND	Set of discharge or consumption ports indexed by p and q;

3.2. Parameters

As for the parameters used in the formulation of this mathematical model as follows.

${C^T}_{pqv}$	Sailing or traveling cost from ports p to q using vessel v;
${C^W}_v$	Waiting for costs due to vessels v for each time period;
${C^P}_{pv}$	Port cost for vessel v operating in port p;
$o(v)$	The initial position of the vessel v at the beginning of the planning horizon;
$d(v)$	Artificial destination node for each vessel;
T_{pqv}	Time is taken by vessels v to sail from port p to q;
B_{pt}	Number of vessels that can berth at port p during time period t;
Q_v	Maximum number of products that can be (un)loaded by vessel v in one time period;
${L^0}_v$	Initial inventory in the vessel v on the planning horizon;
K_v	The capacity of vessel v;
D_{pt}	Amount of demand for each time period t at the discharge or consumption port p;
P_{pt}	The capability of production for each time period t at the loading or production port p;

SMX_{pt} Upper limit or the maximum of inventory level on each time period *t* in port *p*;
SMN_{pt} Lower limit or the minimum of inventory level on each time period *t* in port *p*;
$S^0{}_p$ Initial inventory level at the beginning of the planning horizon at port *p*.

3.3. Variables

As for the variables used in the formulation of this mathematical model as follows.

o_{pvt} For $o_{pvt} \in \{0,1\}$
Equal to 1 if port *p* used by vessel *v* to loading or unloading during time period *t* and 0 otherwise,
$o^A{}_{pvt}$ For $o^A{}_{pvt} \in \{0,1\}$
Shows whether vessel *v* starts operating in port *p* in the period *t*,
$o^B{}_{pvt}$ For $o^B{}_{pvt} \in \{0,1\}$
Shows whether vessel *v* continue operating in port *p* in the period *t*,
x_{pqvt} For $x_{pqvt} \in \{0,1\}$
Equal to 1 if vessel *v* leave from port *p* and sailing to port *q* during the time period *t* and 0 otherwise,
w_{pvt} Where $w_{pvt} \in \{0,1\}$
Equal to 1 if during time period *t* vessel *v* is waiting at port *p* and 0 otherwise,
$f^X{}_{pqvt}$ That shows the load onboard vessel *v* when sailing from port *p* to port *q*, leaving port *p* in period *t*.
$f^{OA}{}_{pvt}$ That shows the load onboard vessel *v* when starts operating at port *p* in period *t* and in the time period *t*-1 has not yet operated.
$f^{OB}{}_{pvt}$ That shows the load onboard vessel *v* before continue operating at port *p* in period *t* and in the time period *t*-1 has been operating.
$f^W{}_{pvt}$ That shows the load onboard vessel *v* when waiting at port *p* during time period *t*.
q_{pvt} For $q_{pvt} \in$ R
The quantity (un)loaded from or to vessel *v* during the time period *t* on port *p*,
s_{pt} For $s_{pt} \in$ R
Inventory level during time period *t* at port *p*.

3.4. Assumption

As for getting results that can be optimized there are assumptions used in this research as follows.

a. Weather along the time horizon is always good or bad weather does not occur.
b. The speed of the ship during the trip is always considered constant.
c. Inventories of bulk cement at loading ports or loading ports are always considered to be available for distribution to unloading ports.
d. Ships that have been rented are considered always in good condition.
e. The vessel's path in carrying products from the loading port to the loading port is always the same as the distance between ports.
f. The vessel's initial position is at the point the ship is carrying out the first operation and the vessel's final position is at the point where the ship is carrying out the last operation.
g. Ships that have finished operating in that month are immediately off-hired and will be leased again the following month.

3.5. Objective Function

The objective functions are functions that describe the goals or things to be achieved in the problem model[15]. The objective function used in the formulation of this mathematical model as follows.

$$min\sum_{v\in V}\sum_{p\in N\cup\{o(v)\}}\sum_{q\in N\cup\{d(v)\}}\sum_{t\in T}C^{T}_{pqv}x_{pqvt}+\sum_{v\in V}\sum_{p\in N}\sum_{t\in T}C^{P}_{pv}o_{pvt}+\sum_{v\in V}\sum_{p\in N}\sum_{t\in T}C^{W}_{v}w_{pvt} \quad (1)$$

The minimization function (25) sailing or traveling costs, operating or port costs, and waiting costs.

3.6. Constraint Function

Constraint function is a form of mathematical representation of the available capacity limits for optimal allocation[15]. The constraint function used in the formulation of this mathematical model as follows.

a. Routing Constraints

$$\sum_{q\in N\cup\{d(v)\}}\sum_{t\in T}x_{o(v)qvt}=1,\forall v\in V, \quad (2)$$

$$\sum_{p\in N\cup\{o(v)\}}\sum_{t\in T}x_{pd(v)vt}=1,\forall v\in V. \quad (3)$$

Constraints (2) and (3) ensure that each vessel departs from the initial artificial port and completes the voyages at the artificial destination port.

b. Sailing, Waiting, and Operating Time Constraints

$$\sum_{q\in N\cup\{o(v)\}}x_{qpv,t-T_{qpv}}+w_{pv,t-1}=w_{pvt}+o^{A}_{pvt},\forall v\in V,p\in N,t\in T, \quad (4)$$

$$o^{A}_{pvt-1}+o^{B}_{pvt-1}=o^{B}_{pvt}+\sum_{q\in N\cup\{d(v)\}}x_{pqvt}\,,\forall v\in V,p\in N,t\in T, \quad (5)$$

$$o^{A}_{pvt},o^{B}_{pvt}\in\{0,1\},\forall v\in V,p\in N,t\in T, \quad (6)$$

$$o^{A}_{pvt}+o^{B}_{pvt}=o_{pvt},\forall v\in V,p\in N,t\in T, \quad (7)$$

$$\sum_{v\in V}o_{pvt}\le B_{pt},\forall p\in N,t\in T. \quad (8)$$

Constraint (4) represents the vessel arrives at a port and (5) represents when the vessel is sailing from one of the port. Constraint (6) represents the variables ${o^{A}}_{pvt}$ and ${o^{B}}_{pvt}$ are binary. Constraint (7) presents the relation between the old operating variable and the new operating variable. Constraint (8) provides berth limitation at each node and in consequence waiting time.

c. Loading Constraints

$$0\le q_{pvt}\le Q_{v}o_{pvt},\forall v\in V,p\in N,t\in T, \quad (9)$$

$$\sum_{q\in N\cup\{o(v)\}}f^{X}_{qpv,t-T_{qpv}}+f^{W}_{pv,t-1}=f^{W}_{pvt}+f^{OA}_{pvt}\ \forall v\in V,p\in N,t\in T, \quad (10)$$

$$f^{OA}_{pv,t-1}+f^{OB}_{pv,t-1}+q_{pv,t-1}=f_{pvt}OB+\sum_{q\in N\cup\{d(v)\}}f^{X}_{pqvt}\,, \quad (11)$$

$$\forall v\in V,p\in NP\cup^{\{o(v)\}},t\in T,$$

$$f_{pv,t-1}^{OA} + f_{pv,t-1}^{OB} - q_{pv,t-1} = f_{pvt}OB + \sum_{q \in N \cup \{d(v)\}} f_{pqvt}^{X}, \quad \forall v \in V, p \in ND \cup \{o(v)\}, t \in T, \tag{12}$$

$$f_{o(v)qvt}^{X} = L_v^0 x_{o(v)qvt}, \forall v \in V, q \in N \cup \{d(v)\}, t \in T. \tag{13}$$

The quantity loaded or unloaded (q_{pvt}) must not exceed the maximum number of products loaded or unloaded from vessel v (constraint 9). The flow conservation constraint (10) - (13) ensure a balanced load on board along every arc of the structure. The variables upper bound and non-negative constraints are shown in (14) - (18).

$$0 \leq f_{pqvt}^{X} \leq K_v x_{pqvt}, \forall v \in V, p \in N \cup \{o(v)\}, q \in N \cup \{d(v)\}, t \in T \tag{14}$$

$$0 \leq f_{pvt}^{OA} \leq K_v o_{pvt}^{A}, \forall v \in V, p \in N, t \in T, \tag{15}$$

$$0 \leq f_{pvt}^{OB} \leq K_v o_{pvt}^{B}, \forall v \in V, p \in N, t \in T, \tag{16}$$

$$0 \leq q_{pvt} \leq Q_v o_{pvt}, \forall v \in V, p \in N, t \in T, \tag{17}$$

$$0 \leq f_{pvt}^{W} \leq K_v w_{pvt}, \forall v \in V, p \in N, t \in T. \tag{18}$$

d. Inventory Control Constraints

$$s_{p,t-1} + \sum_{v \in V} q_{pvt} = D_{pt} + s_{pt}, \forall p \in ND, t \in T, \tag{19}$$

$$s_{p,t-1} + P_{pt} = \sum_{v \in V} q_{pvt} + s_{pt}, \forall p \in NP, t \in T, \tag{20}$$

$$SMN_{pt} \leq s_{pt} \leq SMX_{pt}, \forall p \in N, t \in T, \tag{21}$$

$$s_{pt} = S_p^0, p \in N, \tag{22}$$

$$x_{pqvt} \in \{0,1\}, \forall v \in V, p \in N \cup \{o(v)\}, q \in N \cup \{d(v)\}, t \in T, \tag{23}$$

$$o_{pvt}, w_{pvt} \in \{0,1\}, \forall v \in V, p \in N, t \in T, \tag{24}$$

$$s_{pt}, q_{pvt} \in \mathbb{R} \, \forall v \in V, p \in N, t \in T. \tag{25}$$

The inventory level at every port is controlled during every time unit of the planning horizon at the loading and unloading port (constraints 19 and 20). Constraint (21) provides operational ranges that must be at the inventory level, and constraint (22) provide initial inventory at each port. All binary variables are declared in constraints (23) and (24) and constraint (25) gives a continuous variable.

4. Research Method

The research methodology in this research uses systematic problem solving, so the problems that occur can be solved appropriately. Systematic problem solving is divided by four-stage such as the first stage named initial identification. At this stage, a field study is carried out, which is to look directly at the existing conditions on the research object in the field. Based on these field conditions and associated with literature studies and various other sources such as books, journals, and previous studies and know the formulation of the problems that occur so that the objectives and benefits of the research conducted. second stage named data collecting and processing, third stage named analysis stage, and the fourth stage named conclusion and suggestion. The second stage to the last stage will be explained later. This following figure is all stage of systematic problem-solving in this research.

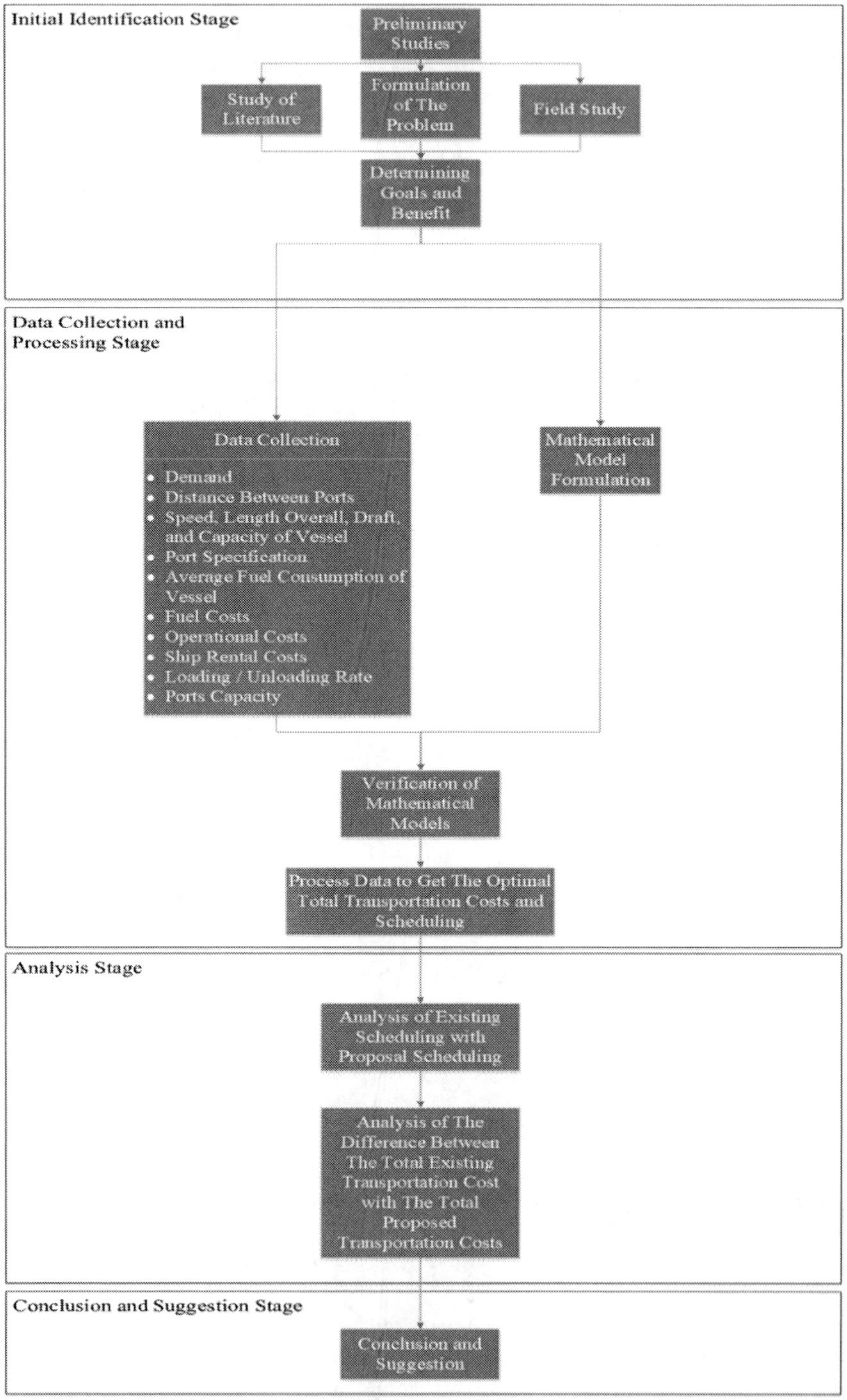

Figure 1. Systematic Problem Solving

Next, the second stage is the data collection and processing stage. At this stage, Collecting all the data needed and using the mathematical formulation proposed by Agra[7] and also used by Diz[6]. After that, verify mathematical models using left and right side comparisons to equate units of both left and right sides and also verify mathematical models built-in Python Programming Language version 2.7 with PuLP library. Next, by all the data and source code that built-in Python, solve the problem using Gurobi 9.0.1 solver with academic license and will produce an output of the optimal total transportation cost and schedule.

The third stage is the analysis stage, which is conducting an analysis of the existing scheduling by scheduling proposals and also analysis of the total existing transportation cost by the total proposed transportation cost that obtained from the output results.

The last stage is the conclusion and suggestion stage. This final stage is concluded from the results of the analysis conducted and provides suggestions for further research.

5. Data Processing and Analysis Result

5.1. Mathematical Model Formulation

The mathematical model used is a mathematical model that has been developed by Agostinho Agra[7] and also has been implemented by Gustavo Souto dos Santos Diz[6] with the assumptions previously described, the difference is in the assumptions used. Because the research conducted has a similar case, so the mathematical model used is also the same as the mathematical model formulated by Agra and implemented by Diz without developing the model.

5.2. Scheduling Result

As for the data after it is complete, then the data is processed using the Python programming language with PuLP library and Gurobi Solver version 9.0 with an academic version license. The results provided from the code that was built in the form of scheduling ships from January to December in 2018.

Table 4. Scheduling Results for January 2018

	Ports		
Vessel's Name	From	To	Days to-
Bellini 7	Tarjun	Lombok	8
Sophia	Tarjun	Samarinda	5
	Samarinda	Tarjun	8
	Tarjun	Samarinda	10
	Samarinda	Tarjun	13
	Tarjun	Lombok	15
	Lombok	Tarjun	18
	Tarjun	Samarinda	21
	Samarinda	Tarjun	24
	Tarjun	Samarinda	26
Sari Andalas	Cirebon	Surabaya	7
	Surabaya	Tarjun	11
	Tarjun	Samarinda	15
	Samarinda	Tarjun	19
	Tarjun	Lombok	22
Meutia Andalas	Tarjun	Lombok	6
Tahta	Tanjung Priok	Pontianak	6
	Pontianak	Tanjung Priok	11
	Tanjung Priok	Pontianak	15
Tiro	Tarjun	Surabaya	5
Total Transportation Cost		Rp10,174,278,466	

This result is the optimal results obtained for January. The total cost of transportation for January is Rp10,174,278.46. This result can save as much as 58% of the total existing transportation costs in January 2018. The output provided by the code built provides total transportation costs from January to December in 2018 which results from each month are added up and will be total transportation costs for 2018.

IOP Conf. Series: Materials Science and Engineering **982** (2020) 012056 doi:10.1088/1757-899X/982/1/012056

5.3. Inventory Level Result

The result of scheduling needs to be considered the inventory level of each consumption port, this needs to be considered to ensure that the level of inventory of each port does not exceed or less than its limits or capacity.

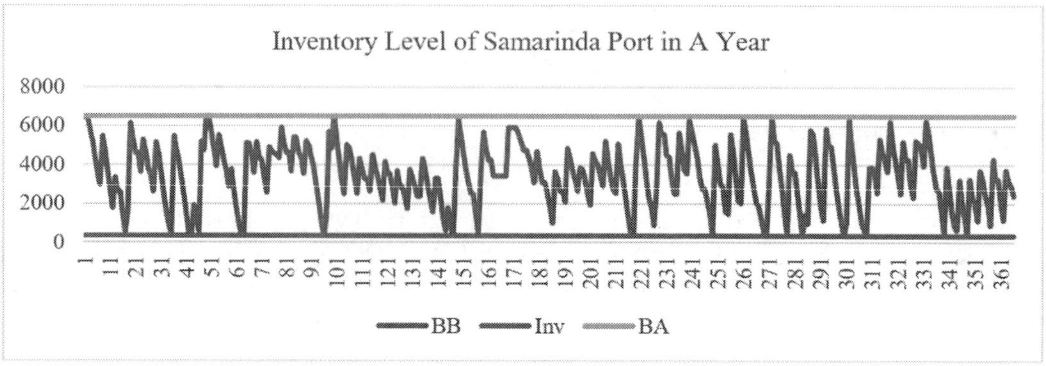

Figure 2. Inventory Level of Samarinda Port in A Year

Based on the picture above, it can be seen that the inventory level (Inv) for Samarinda Port during 2018 using scheduling proposals is never less than the lower limit (BB) or exceeds the upper limit (BA) of the inventory storage capacity at Samarinda Port. Likewise, for the other consumption ports that have never violated that limits

5.4. Waiting / Idle Time Result

In addition to producing scheduling with the optimal total transportation costs for each month, the code that was built also provides results related to waiting time that occurs every month in 2018. The total waiting time that occurs is as follows.

Table 5. Total Waiting Time with Proposal Scheduling

No.	Month	Waiting Time (Hours)
I	Mar	24
II	May	48
III	Jun	72
IV	Sep	48
V	Nov	48
VI	Dec	48
	Total Waiting Time	288

The result of the total waiting time by using proposed scheduling is 288 hours. The waiting time that occurs in the months that are not listed in the table indicates that in that month there is no waiting time. In general, this waiting time occurs at the production port where the ship is better to do waiting time in the selected production port compared to other production ports because it allows for a longer time to travel and long distances which can increase sailing costs and thus resulting a total transportation cost will be increased.

5.5. Total Transportation Cost Result

The result of proposal scheduling also contains total transportation costs carried out each month in 2018 which for the total transportation costs each month are summed up and will be the total transportation costs for 2018. The total transportation costs are as follows.

Table 6. Total Transportation with Proposal Scheduling

No	Month	Transportation Cost
I	Jan	Rp 10,174,278,466
II	Feb	Rp 10,970,001,219
III	Mar	Rp 12,407,927,742
IV	Apr	Rp 8,656,906,556
V	May	Rp 8,635,742,296
VI	Jun	Rp 9,042,789,991
VII	Jul	Rp 14,286,639,970
VIII	Aug	Rp 13,847,001,809
IX	Sep	Rp 14,533,750,020
X	Oct	Rp 16,850,955,124
XI	Nov	Rp 14,966,809,435
XII	Dec	Rp 16,451,731,416
Total Transportation Cost		Rp 150,824,534,044

The total transportation cost to deliver cement to all consumption ports owned by the company using the scheduling proposal in 2018 is Rp 150,824,534,044. This can reduce 40% or Rp 101,976,981,652 from the existing transportation costs of Rp 252,801,515,696.

6. Conclusion

Based on research conducted using the Maritime Inventory Routing Problem (MIRP) Method and Mixed-Integer Linear Programming (MILP) approach, it can be concluded that MIRP method with MILP approach can be applied in the case of cement distribution with three production ports and six consumption ports, providing ship scheduling which provides an increase for ships to carry out operations at the Port of Cirebon which allows providing more minimal both sailing time and operational activities. In addition, the proposal scheduling gives a total waiting time reduced by 83% of the existing total waiting time and also the proper scheduling of ships provides lower ship travel costs, fast operating activities, and does not result in ships not being able to lower cargo. Thus, providing a minimum total transportation cost of Rp150,824,534,044 which can reduce total transportation costs by Rp101,976,981,652 or 40% of the total existing transportation costs with the inventory level of all consumption ports never lower than the lower limit or exceeds the upper limit. However, this makes the utilization of some ships low.

References

[1] Simons S 1991 Minimax theorems with staircases *Arch. der Math.* vol **57** pp 169–179 doi: 10.1007/BF01190004

[2] Waaly A N, Ridwan A Y and Akbar M D 2019 Development of sustainable procurement monitoring system performance based on Supply Chain Reference Operation (SCOR) and Analytical Hierarchy Process (AHP) on leather tanning industry *MATEC Web Conf.* vol **204** doi: 10.1051/matecconf/201820401008

[3] Sari D I, Ridwan A Y and Santosa B Design of Risk Management Monitoring System Based on Supply Chain Operations Reference (SCOR): A Study Case at Dairy Industry in Indonesia vol **5** pp 104–115

[4] Chopra S and Meindl P 2012 *Supply Chain Management: Global Edition* (London: Pearson Education)

[5] Deviani S, Ridwan A Y and Santosa B 2015 Usulan Perencanaan dan Penjadwalan Aktivitas Distribusi Air Minum Dalam Kemasan Gallon Untuk Meningkatkan Kemampuan Pemenuhan Permintaan Di Pabrik MKS Menggunakan Metode Distribution Requirement Planning (DRP) vol **2** pp 7550–7556

[6] Diz G S, Oliveira F and Hamacher S 2017 Improving maritime inventory routing: application to a Brazilian petroleum case *Marit. Policy Manag.* vol **44** pp 42–61 doi:

10.1080/03088839.2016.1216622
[7] Agra L W A, Andersson H and Christiansen M 2013 A Maritime Inventory Routing Problem: Discrete Time Formulations and Valid Inequalities *Networks* vol **60** pp 45–58 doi: 10.1002/net
[8] Christiansen M and Fagerholt K 2008 *Maritime Inventory Routing Problems in Encyclopedia of Optimization* (US: Springer) pp 1947–1955
[9] Dantzig G B and Fulkerson D R 1954 Minimizing the number of tankers to meet a fixed schedule *Nav. Res. Logist. Q.* vol **1** pp 217–222 doi: 10.1002/nav.3800010309
[10] Christiansen M, Fagerholt K and Ronen D 2004 Ship Routing and Scheduling: Status and Perspectives *Transp. Sci.* vol **38** pp 1–18 doi: 10.1287/trsc.1030.0036
[11] Christiansen M 1999 Decomposition of a combined inventory and Time Constrained Ship Routing ttp://Dx.Doi.Org/10.1287/Trsc.33.1.3
[12] Christiansen M, Fagerholt K, Nygreen B and Ronen D 2013 Ship routing and scheduling in the new millennium *Eur. J. Oper. Res.* vol **228** pp 467–483 doi: 10.1016/j.ejor.2012.12.002
[13] Mehlbeer F and Vien N A 2014 Mixed-Integer Linear Programming Applied to Temporal Planning of Concurrent Actions
[14] Amelia P, Ridwan A Y and Santosa B 2019 Designing of Raw Material Scheduling Supply Multi on Supplier Strategies with Price, Lead time, and Stochastic Demand Variations. Case Study: Electricity Manufacturer *IOP Conf. Ser. Mater. Sci. Eng.* vol **598** doi: 10.1088/1757- 899X/598/1/012111.
[15] Subagyo P, Asri M and Handoko T H 2013 *Dasar-Dasar Operation Research* (Yogyakarta: BPFE)

Feasibility Study of Determination a New Distribution Warehouse Location Using P-Median and Analytical Network Process Methods in One of the Cement Industries

Nadya Saniyya Ramadhanti[1,2], Ari Yanuar Ridwan[1] and Hardian Kokoh Pambudi[1]

[1]Industrial Engineering Department, Telkom University, Jawa Barat, Indonesia

[2]nadyasaniyya@student.telkomuniversity.ac.id

Abstract. This research was carried out at one of the companies engaged in the cement industry, which distributed its products throughout Indonesia. However, based on audit results for the Central Java region, there are still a number of minor and major statuses in warehouse management perspectives and the percentage of warehousing cost components to the total distribution costs that exceed the maximum limit set at the warehouses rented from the distributor. Therefore, it's necessary to determine a new distribution warehouse location owned by the company by considering qualitative and quantitative aspects, then stop the operation of a rented warehouse from a distributor because of the high subsidy costs. The method used in this research is investment feasibility analysis using Net Present Value and Benefit Cost Ratio calculations, determining optimal locations using the P-Median method, and determining location based on multi-criteria using the Analytical Network Process method. The results showed that the location with the highest weight based on multi-criteria decision making and declared feasible based on investment feasibility analysis was located in Cilacap Regency with the optimal location in North Cilacap District. If this decision is carried out, the company can reduce the total distribution costs that must be incurred by the company.

1. Introduction

Along with the development of the times, human needs today are constantly increasing and experiencing significant changes and it is a reason to support the emergence of industries that can meet human needs. Therefore, increasingly intense competition in the industry requires companies to have a strategy to win the market compared to competitors. By having sustainable strategic competitiveness, the company will have a competitive advantage because it is able to create value for its buyers[1].

One key to the success of a company to be able to win market competition lies in good supply chain management. Supply Chain Management includes the planning of all activities[2]. The purpose of the Supply Chain is to satisfy customer needs and increase company profits[3]. So, to achieve that success, one of the things that need to be considered in the supply chain is the product distribution process, where one of the costs incurred in the product distribution process is warehousing costs. Warehousing costs usually dominate about 20-30% of the total distribution costs[4].

A good distribution process is a process that can run efficiently and effectively so that the volume of production, warehouse capacity and product allocation in each region in the distribution area reaches a good level of balance. In distribution planning, it is necessary to consider the distribution location that is considered the most potential and related to the area that is the area of distribution[3].

IOP Conf. Series: Materials Science and Engineering **982** (2020) 012057 doi:10.1088/1757-899X/982/1/012057

This research was carried out at one of the companies engaged in the cement industry, which distributed its products throughout Indonesia. However, based on audit result for Central Java region, there are still a number of minor and major statuses in warehouse management perspectives at the rented warehouse from the distributor, so that need to be improved. Then, the percentage of warehousing cost components at the rented warehouses from distributors on the total distribution costs exceeds the maximum limit set, at 37% while the maximum limit set by the company is 25-30%. One improvement step that can be done is determining a new distribution warehouse location owned by the company by considering qualitative and quantitative aspects, then stop the operation of the rented warehouse from a distributor because of the high subsidy costs.

The method used in this research is investment feasibility analysis using Net Present Value and Benefit Cost Ratio calculations, determining optimal locations using the P-Median method, and determining location based on multi-criteria decision making using Analytical Network Process methods. Therefore, by conducting this research, which is determining the right location for a new warehouse and stop the operation in one of the rented warehouses from the distributor is expected to reduce distribution costs, reduce warehousing costs, and increase company competitiveness.

2. Basic Theory

2.1. Warehouse Location

In a logistical flow, of course, there are many activities that support the logistics mission, there are the right place, the right goods, and the right time. Activities that play a role are related to one another and cannot be separated. The warehouse is one of the important elements that support logistics activities. The warehouse is usually used to store product inventor. For companies, this is useful for storing products that have not yet reached the hands of consumers. As for consumers, the warehouse serves to speed up the arrival of the product in their hands[5]. Then, in determining the location of a warehouse, a company should consider several things, including transportation costs, demand, product distribution networks, infrastructure, expansion, supporting facilities, geographical location, and climate[5].

2.2. P-Median

In the P-Median problem, there are divisions of several nodes, which within each node have the same opportunity to become a median or center the cluster. The main goal in solving problems with the P-Median is to place the P (median) facility closest to a group of consumers (nodes) around it so that it has the shortest distance between its facilities. In the settlement using P-Median, at least one optimal solution has to be placed into the network[5]. The mathematical model of the P-Median method is as follows[6].

$$\text{Minimize: } \sum_i \sum_j hi\ dij \tag{1}$$

Description:
hi = Request at node - i
dij = Distance between the request point of node i and node j, i.e. the candidate built. dij is zero if i = j

2.3. Feasibility Study

The feasibility of investment or business is a study of whether a business or investment can be carried out successfully. The main objective is to avoid the risk of loss that will be suffered by the company if the business plan is failed or does not meet expectations[7]. In the feasibility study, there are several aspects that must be considered, namely legal aspects, market and marketing aspects, financial or financial aspects, technical or operational aspects, management or organizational aspects, economic and social aspects, and environmental impact aspects[7]. Like a business that has been planned if viewed through financial aspects or financial aspects can be known through the calculation of Net Present Value (NPV) and Benefit Cost Ratio (BCR)[8].

2.3.1. Net Present Value. Net Present Value (NPV) is the calculation of the difference between benefits (revenues) and costs (expenses) that have been estimated benefits or benefits from the planned project. Thus, NPV calculations rely on discounted cash flow techniques. The NPV calculation formula is as follows[8].

$$NPV = PV\ Benefit - PV\ Cost \tag{2}$$

NPV Assessment Criteria:
If NPV > 0, the investment proposal is feasible to proceed
If NPV < 0, the investment proposal is not feasible to proceed
If NPV = 0, investment only experiences Break Event Point, which is only to cover production costs

2.3.2. Benefit Cost Ratio. Benefit Cost Ratio (BCR) is a comparison between the present value of net benefits and the present value of net costs. The BCR calculation formula is as follows[8].

$$BCR = \frac{present\ value\ cash\ inflow}{present\ value\ initial\ investment} \tag{3}$$

BCR Assessment Criteria:
B/C Ratio ≥ 1 = Decent
B/C Ratio ≤ 1 = Ineligible

2.4. Analytical Network Process

Analytical Network Process (ANP) is a method that is able to represent the level of importance of various parties by considering the relationship of dependency, both between criteria and between sub-criteria that are points for measuring several predetermined criteria. Analytical Network Process can produce a more accurate approach because the ANP method is able to handle complex problems, which are closely related to dependency and feedback[9].

This ANP method is a development of the Analytical Hierarchy Process (AHP) method, in which the ANP method can correct the deficiencies found in the previous method[9]. There are some differences between ANP and AHP, mainly the AHP there is no relationship between sub-criteria, while in ANP there is a relationship, which is usually called the inner dependences and outer dependencies. Inner dependence is a relationship between elements in the same cluster, so this cluster will connect with itself so that it will form a loop relationship, while outer dependence is a relationship between elements in a different cluster, so that one cluster will be connected to one another cluster[10]. AHP describes complex multi-criteria issues into a hierarchy while ANP describes it into a network[11].

There are four main procedures in completing using the Analytical Network Process method, there are developing a construct a decision model, making a pairwise comparison matrix of interrelated variables based on a 1-9 ratio scale, processing and forming the Super Matrix, and determining the best alternative[12].

2.4.1. Construct a Decision Model. Model construction is based on existing problems, so it is necessary to clearly describe the problem, and form it into the network. In describing the problem, there are two steps that must be done, namely the determination of alternatives used as research objects and determination of criteria and sub criteria[12]. Then, in order to construct a decision network model, it is necessary to identify the dependency relationships between criteria and sub-criteria or better known as inner and outer dependencies[10].

The alternative determination that will be used as the object of research can be done by calculating pairwise comparison, rating calculation, and weighted value calculation[6]. Calculations using this method are suitable if there are many alternatives and only a few alternatives that will be chosen. In

IOP Conf. Series: Materials Science and Engineering **982** (2020) 012057 doi:10.1088/1757-899X/982/1/012057

addition, to be able to do calculations with this method, so certain criteria are needed to choose several alternatives that have the highest priority. The weighted value calculation formula is as follows[6].

$$yn = a.RAn + b.RBn + c.RCn \quad (4)$$

Description:
yn = Weighted value of the alternative-n
a = Priority vector of criteria A
b = Priority vector of criterion B
c = Priority vector of C criteria
RAn = Rating criterion A on alternative-n
RBn = Rating criterion B on alternative-n
RCn = Rating criterion C on alternative-n

Then, This following is a calculation of the rating value for criterion X in alternative-n (RXn):

$$RXn = \frac{RXn}{Total\ RX} \times 100 \quad (5)$$

Description:
n = Alternative-n
RX = Criteria X
RXn = Rating value of criterion X in alternative-n

Other than that, to identify the inner and outer dependencies, it can be done by filling out the questionnaire by experts. This questionnaire aims to determine the relationship of dependence between criteria and sub-criteria as a basic reference to construct the ANP models[10]. To state that the criteria and sub-criteria have a dependence, it is necessary to look at the number of experts who provide an assessment of the dependency between the criteria and sub-criteria with the following formula[10].

$$Q = \frac{N}{2} \quad (6)$$

Description:
N = Number of experts who provide an assessment of the dependency
If $N \geq Q$, so, there is a relationship between the criteria and sub-criteria

2.4.2. Making a Pairwise Comparison Matrix. Pairwise comparisons on ANP method are performed by comparing the importance of each element to its control criteria. The scale used for comparison is the numerical scales 1-9[12]. The following is a further explanation of the numerical scale on ANP.

Table 1. The Numerical Scale on ANP Method

Scale	Definition	Annotation
1	Equally Important	Two elements are equally important to alternatives
3	A Little More Important	One element is slightly more important than the other elements
5	More Important	One element is more important than the other elements
7	Far More Important	One element is far more important than the other elements

9	Absolute More Important	One element is absolutely more important than the other elements
2, 4, 6, 8	Middle Value	Values between two value considerations are close together

To get a pairwise comparison matrix assessment, it can be done by filling out the questionnaire by experts. This questionnaire was made based on the dependency relationship obtained from the previous questionnaire[10]. If the number of respondents is more than one person, a Geometric Mean (GM) calculation is performed. The formula is:

$$GM = (R1 \times R2 \times \ldots \times Rn)^{1/n} \quad (7)$$

Description:
G = Geometric Mean
n = Number of Respondents
R = Respondent Questionnaire Value-n

Then, if the pairwise comparisons have been made entirely, the next is to calculate the priority vector w (called the eigenvector). The eigenvector is the priority weight of the matrix which is then used in the preparation of the supermatrix calculated by the formula[10]:

$$A * w = \lambda max * w \quad (8)$$

Description:
A = Pairwise Comparison Matrix
λmax = Eigen Value

Then, check the inconsistency ratio. Consistency ratio is the ratio that states whether the judgments given by experts are consistent or not. Inconsistency ratios less than 0.1 or 10% have reliable and consistent results[13]. The consistency index (CI) of a comparison matrix is calculated by the formula:

$$CI = \frac{\lambda max - n}{n - 1} \quad (9)$$

The consistency ratio is obtained by comparing the consistency index with the values of the random consistency index (RI) numbers, as follows:

$$CR = \frac{\mathrm{CI}}{RI} \quad (10)$$

Description:
λmax = The largest eigenvalue of the pairwise comparison matrix n x n
n = Number of Items Compared
CI = Consistency Index
RI = Random Consistency Index

The mathematical model used to find the Ratio Index (RI) is as follows:

$$RI = \frac{1{,}98(\mathrm{n} - 2)}{n} \quad (11)$$

2.4.3. Construct The Super Matrix. Super matrix is a matrix consisting of sub-matrices that are composed of a set of relationships between the two levels contained in the model. There are three supermatrix stages in the ANP model, namely the Unweighted Super Matrix, Weighted Super Matrix, and Limit Matrix[12].

IOP Conf. Series: Materials Science and Engineering **982** (2020) 012057 doi:10.1088/1757-899X/982/1/012057

2.4.4. Determination of the Best Alternative. After obtaining the value of each element in the limit matrix, next is to choose the best alternative. The alternative with the highest global priority is the best alternative.

3. Research Method

The initial stages carried out in this research are identifying market potential criteria and doing some calculations to obtain five regencies or cities in the Central Java region which will be used as the focus of research. The next stage is the identification of criteria and sub-criteria. After that, a calculation is made to get the optimal sub-districts of each selected district and city using the P-Median Method. Next, an investment feasibility study was conducted on five alternative locations using NPV and BCR calculations. The final stage is data processing using the ANP method and decision making is made to determine the best and optimal location.

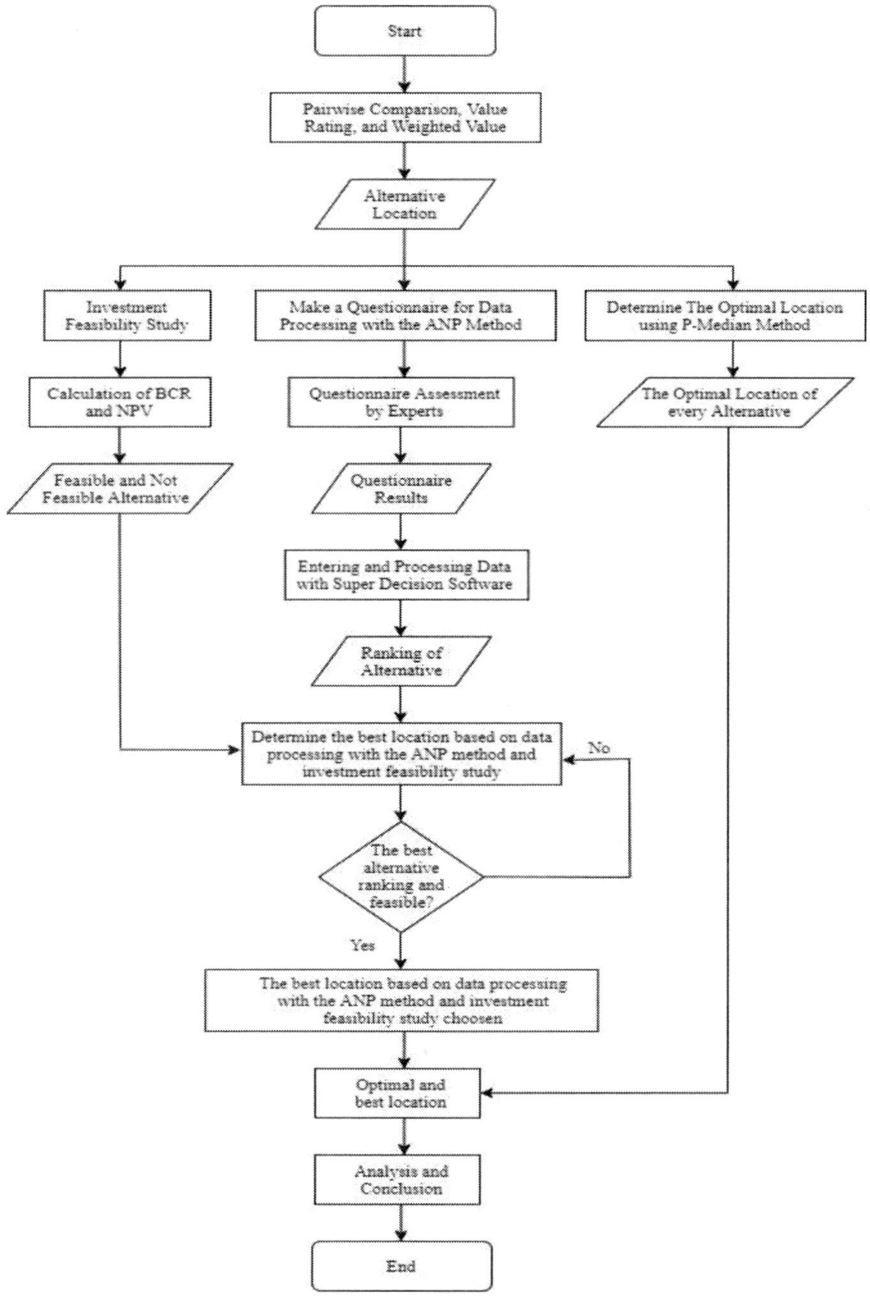

Figure 1. Research Flowchart

4. Analysis and Result

Three criteria are used to determine alternatives based on the market potential set by the company, including the number of retailers, the number of construction service companies, and the number of development projects. The number of alternatives selected as alternative locations is five regencies or cities. The steps taken to determine alternatives are the calculation of pairwise comparison, calculation of the rating value, and calculation of the weighted value. The results are shown in **Table 2**.

Table 2. Weighted Values for each District and City

No	District/City	RR	a	RK	b	RP	c	Weighted Value
1	Banjarnegara	2.86463209	0.7	3.026998	0.2	2.935151	0.1	2.908368852
2	Banyumas	5.9726643	0.7	3.745114	0.2	2.963959	0.1	5.10016593
3	Batang	2.13443175	0.7	1.945278	0.2	3.300045	0.1	2.250027915
4	Blora	2.63995506	0.7	3.526952	0.2	2.355803	0.1	2.789538047
5	Boyolali	2.82718592	0.7	1.681665	0.2	2.978362	0.1	2.605201879
6	Brebes	3.98801723	0.7	3.608763	0.2	1.726842	0.1	3.606026649
7	Cilacap	5.09267927	0.7	6.226707	0.2	3.04878	0.1	5.059597563
8	Demak	3.01441678	0.7	3.981456	0.2	3.067985	0.1	3.225998851
9	Grobogan	3.16420146	0.7	2.663394	0.2	3.090391	0.1	3.048478931
10	Jepara	3.12675529	0.7	2.627034	0.2	2.931951	0.1	2.995112261
11	Karanganyar	2.84590901	0.7	1.472593	0.2	3.063184	0.1	2.584591859
12	Kebumen	3.74461711	0.7	2.917917	0.2	2.984764	0.1	3.468429471
13	Kendal	2.86463209	0.7	1.908917	0.2	3.027975	0.1	2.684396922
14	Klaten	3.66972477	0.7	3.126988	0.2	3.196018	0.1	3.491761236
15	Kudus	3.05186295	0.7	2.599764	0.2	3.208821	0.1	2.977239738
16	Magelang	3.25781689	0.7	1.07263	0.2	3.058383	0.1	2.76924899
17	Magelang City	0.67403108	0.7	1.563494	0.2	1.99251	0.1	1.03805079
18	Pati	4.17524808	0.7	2.23616	0.2	2.826324	0.1	3.585274238
19	Pekalongan	1.92847781	0.7	2.21798	0.2	3.189617	0.1	2.158012344
20	Pekalongan City	1.44167759	0.7	0.963549	0.2	2.011715	0.1	1.416688538
21	Pemalang	3.29526306	0.7	3.126988	0.2	3.114397	0.1	3.235548321
22	Purbalingga	2.95824752	0.7	1.872557	0.2	3.035977	0.1	2.73910804
23	Purworejo	2.52761655	0.7	1.799836	0.2	2.856731	0.1	2.417693494
24	Rembang	2.17187793	0.7	3.572403	0.2	2.952756	0.1	2.572178163
25	Salatiga City	0.86126194	0.7	1.06354	0.2	1.803662	0.1	1.029812688
26	Semarang	1.92847781	0.7	2.072539	0.2	3.019973	0.1	2.104622486
27	Semarang City	4.88672533	0.7	10.8172	0.2	2.643877	0.1	5.841055525
28	Sragen	2.65867815	0.7	3.072448	0.2	3.140004	0.1	2.810395183
29	Sukoharjo	3.18292455	0.7	3.417871	0.2	3.096793	0.1	3.221095516
30	Surakarta City	1.89103164	0.7	2.536133	0.2	3.064785	0.1	2.184075665
31	Tegal	3.18292455	0.7	4.090537	0.2	3.301645	0.1	3.390640159
32	Tegal City	0.91743119	0.7	1.308972	0.2	1.976506	0.1	1.141565585
33	Temanggung	2.02209324	0.7	1.754386	0.2	3.176813	0.1	2.119629296
34	Wonogiri	2.52761655	0.7	3.090628	0.2	2.831125	0.1	2.687148594
35	Wonosobo	2.50889347	0.7	3.29061	0.2	3.026375	0.1	2.743220281

Based on the result of the weighted value that is shown in Table 2, it is known that five regencies/cities that are selected for the alternative, namely Banyumas Regency, Brebes Regency, Semarang City, Cilacap Regency, and Klaten Regency. After determining the selected alternative next determine the optimal sub-district of each selected district and city by using the P-Median Method and the criteria considered are the smallest demand-weighted distance. The results are shown in Table 3.

Table 3. Optimal Sub-District for each Regency or City

Regency/City	Sub-District Optimal	Demand-Weighted Distance	Average Distance
Banyumas	Purwokerto Timur	3151239.9	16.43324937
Semarang City	Semarang Selatan	446671.8	6.755676215
Cilacap	Cilacap Utara	11825073.8	32.30576719
Brebes	Wanasari	1439020.4	20.77167932
Klaten	Klaten Tengah	54125.2	9.627392387

Next is an investment feasibility analysis of each alternative location that has been determined. This investment feasibility analysis will be processed by calculating the Net Present Value (NPV) and Benefit Cost Ratio (BCR) with the criteria considered are NPV > 0 and BCR > 1. Then. the results of the investment feasibility analysis calculation can be used as consideration for experts in providing an assessment of alternative locations based on consideration of several other aspects which will then be processed using the Analytical Network Process method. The results of calculations for investment feasibility analysis in each selected alternative are shown in Table 4.

Table 4. Calculation Results for Investment Feasibility Study

Feasibility Study	Banyumas Regency	Klaten Regency	Semarang City	Cilacap Regency	Brebes Regency
NPV	-Rp256.544.937.556	Rp22.555.409.371	-Rp645.325.297.905	Rp883.283.157.634	-Rp99.000.175.861
BCR	-1.565449376	1.225554094	-5.453252979	9.832831576	0.009998241
Note	NPV < 0. BCR < 1	NPV > 0. BCR > 1	NPV < 0. BCR < 1	NPV > 0. BCR > 1	NPV < 0. BCR < 1
Conclusion	**Not Feasible**	**Feasible**	**Not Feasible**	**Feasible**	**Not Feasible**

After that. the criteria and sub-criteria will be determined to be used in data processing with the ANP method and identify the dependence between the criteria and sub-criteria which will then be input to construct the ANP network model. So. the criteria and sub-criteria that used in determining the location of the warehouse are:

Table 5. List of Criteria and Sub-Criteria (a)

Criteria	Sub-Criteria	Source
Legal	Government Policy	Ashrafzadeh. et al.. 2012
	Legal Restrictions	Handayani. et al.. 2010
	Political Environment	Handayani. et al.. 2010
Market and Marketing	Lead Time and Responsiveness	Ashrafzadeh. et al.. 2012
	Retailer's reputation	Noviandri. et al.. 2015
	Marketing Scope	Handayani. et al.. 2010
Technical	Water Infrastructure	Turgut. et al.. 2011
	Quality and Reliability of Utilities	Ashrafzadeh. et al.. 2012
	Energy and Electricity Infrastructure	Turgut. et al.. 2011

Table 6. List of Criteria and Sub-Criteria (b)

Criteria	Sub-Criteria	Source
Economy and Social	Level of Social Acceptance and Independent Security System	Gracia. et al.. 2008
	Criminality	Gracia. et al.. 2008
	Proximity to Sources of Labor	Achsin. et al.. 2011

	Environmental Issues Arising	Ashrafzadeh. et al.. 2012
Environmental Impact	Conformity of the Climate Environment	Ashrafzadeh. et al.. 2012
	The health of the surrounding environment	Degener. et al.. 2013

The identification of dependency between criteria and sub-criteria is done by filling out questionnaires by three experts. The results of the identification are as follows:

AFFECTING \ AFFECTED	GP	LR	PE	LTR	RR	MS	WI	QRU	EEI	TPS	TK	KTK	IL	KIL	KLS
GP				3		3	1		2	1					
LR				1		2									
PE						3				3	3	3			
LTR	3	1				1							1		
RR						1		3							3
MS	3	2	3	1	1										
WI	1											3	3	3	3
QRU					3				2				3		
EEI	2							2				1	3		
TPS	1		3								3				
TK			3							3					
KTK			3				3		1					1	3
IL				1			3	3	3						
KIL							3					1			2
KLS					3		3					3		2	

Figure 2. Result of Identification of dependencies between Criteria and Sub-Criteria

Based on the results of the identification of dependencies between the criteria and sub-criteria that have been attached in Figure 2. it can be seen that there are several sub-criteria that have Outer Dependences and Inner Dependences. Inner Dependences can form a loop. After all, inputs have been completed. an ANP network model can be constructed as shown in Figure 3.

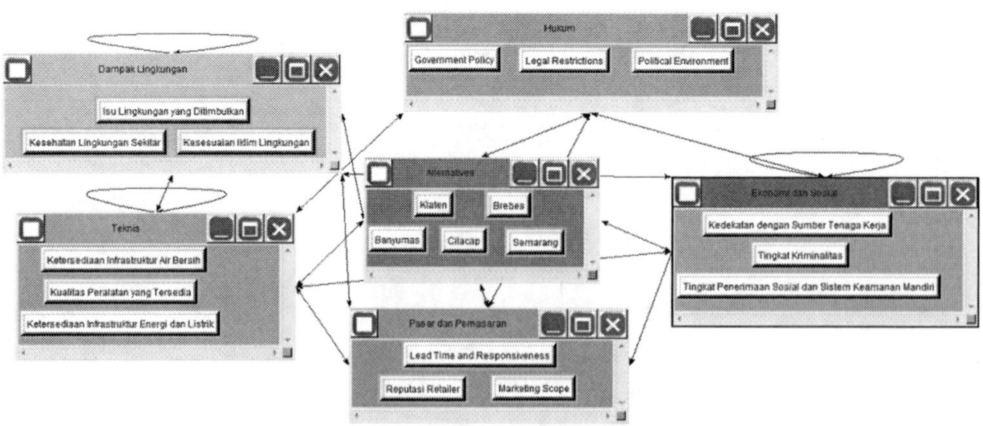

Figure 3. ANP Network Model

Based on data processing using the Analytical Network Process method. it can be seen that the alternative with the first priority sequence is Cilacap Regency. The weight of the value obtained by each alternative can be seen in Table 7.

Table 7. Result of ANP

Name	Normalized By Cluster	Limiting
Cilacap Regency	0.32809	0.087194
Klaten Regency	0.22155	0.058880
Semarang City	0.20615	0.054788
Banyumas Regency	0.14791	0.039308
Brebes Regency	0.09630	0.025594

Based on data processing that has been done. it can be concluded that the alternative is chosen as the location for the construction of a new distribution warehouse. namely Cilacap Regency because it is the first priority sequence based on processing by the ANP method and also Cilacap Regency is feasible based on the results of an investment feasibility study. Next. because the work on this data processing is parallel. then the optimal location for Cilacap Regency can be known based on calculations using the P-Median method that has been done previously. The optimal location for Cilacap Regency is located in the North Cilacap District with a demand-weighted distance of 11825073.8 and an average distance of all demand points of at least 32.305 km. Therefore. further analysis will be carried out on the construction of a new distribution warehouse owned by the company if it is carried out in the North Cilacap District.

The proposed improvement that can be done is to build a new distribution warehouse owned by a company located in North Cilacap Sub-District and stop the operation of the rented warehouse from the distributor located in Cilacap Regency because of the high cost of subsidies. With the proposed improvements. it can reduce the warehousing cost component rented from the distributor in the total distribution cost by 2% or decrease in cost components warehousing of Rp 3.243 for each ton of cement. Then. there was a decrease in total distribution costs of Rp 3.505 for each ton of cement. Thus. the total cost savings that can be achieved by the company is Rp 9.731.066.748.

5. Conclusion

Based on research that has been done. then there are several conclusions that can be drawn. the first is the steps taken to determine the location of the new distribution warehouse in this research are determining alternatives. determining the optimal location using the P-Median Method. analyzing the investment feasibility of each alternative using Net Present Value and Benefit Cost Ratio calculations. and multi-criteria decision making using Analytical Network Process Method. Second. the criteria to be considered at each stage are the number of retailers. the number of construction service companies. and the number of development projects for the alternative determination. the smallest demand-weighted distance for optimal location determination. NPV > 0 and BCR > 1 for investment feasibility analysis. and several other criteria. namely legal. market and marketing. technical. economic and social. and environmental impact to determine the location in multi-criteria decision making.

Based on data processing that has been done. it can be concluded that the alternatives were chosen as the location for the construction of a new distribution warehouse. namely Cilacap Regency because it is the first priority sequence with the highest weight and declared feasible. Next. because the work on this data processing is parallel. then the optimal location for Cilacap Regency is located in North Cilacap District. With the proposed improvements. it can decrease the cost components warehousing rented from the distributor of Rp 3.243 for each ton of cement. Then. there was a decrease in total distribution costs of Rp 3.505 for each ton of cement. Thus. the total cost savings that can be achieved by the company is Rp 9.731.066.748.

References

[1] Cahyadi B and Aulia B 2018 *Penentuan Titik Pusat Warehouse dan Sistem Distribusi Portland Composite Cement di Wilayah Wonogiri Sem. Nasional Sains dan Teknologi* vol **1** pp 1–9

[2] Sari D I. Ridwan A Y and Santosa B 2019 *Design of Risk Management Monitoring System Based on Supply Chain Operations Reference (SCOR): A Study Case at Dairy Industry in*

Indonesia ICORE vol **5** pp 104–115

[3] Chopra S and Meindl P 2003 *Supply Chain Management: Strategy. Planning. and Operation* (London: Pearson Education. Inc)

[4] Parji 2018 *Model Peningkatan Kinerja Warehouse 5. Services and Warehousing (S&W)-Procurement PT. Pertamina (PERSERO) Refinery Unit V Balikpapan* (Yogyakarta: Indonesian Islamic University Press)

[5] Siswanto N 2016 *Penentuan Lokasi Gudang Penyangga Regional PT 'X' Wilayah Jawa Timur* vol **3** pp 184–194

[6] Handayani S and Singgih M 2016 *Penentuan Lokasi Cabang Baru Laboratorium Klinik X dengan menggunakan Analytical Network Process (ANP) dan Kelayakan Investasi*

[7] Hidayati N and Warnana D D 2017 *Analisis Kelayakan Finansial Pengembangan Kelas Alam Terbuka Kebumian dan Lingkungan Berkonsep Rekreasi dan Inspirasi untuk Anak di Surabaya* vol **3** pp 650–656

[8] Suryanto M H 2017 *Metode Riset dan Analisis Saluran Distribusi* (Jakarta: PT Grasindo)

[9] Amalia C 2012 *Perancangan dan Pengukuran Kinerja Rantai Pasokan Sayuran dan Perusahaan dengan Pendekatan Analytical Network Process Serta Data Envelopment Analysis (Studi Kasus : PT Saung Mirwan . Bogor)* (Bogor: Bogor Agricultural Institute Press)

[10] Arvianto A. Sari D P and Olivia G 2014 *Pemilihan Strategi Pemasaran Pada PT Nyonya Meneer Dengan Menggunakan Pendekatan Metode Analytical Network Process (ANP) dan Technique for Order Preference By Similarity to an Ideal Solution* (TOPSIS) *J@Ti Undip J. Tek. Ind.* vol **9** pp 35–44

[11] Waaly A N. Ridwan A Y and Akbar M D 2018 *Development of sustainable procurement monitoring system performance based on Supply Chain Reference Operation (SCOR) and Analytical Hierarchy Process on leather tanning industry* MATEC Web Conf. vol **204**

[12] Ekawati R. Trenggonowati D L and Aditya V D 2018 P*enilaian Performa Supplier Menggunakan Pendekatan Analytic Network Process (ANP)* J. Ind. Serv. vol **3** pp 152–158

[13] Novar M F. Ridwan A Y and Santosa B 2018 *SCOR and AHP based monitoring dashboard to measure rice sourcing performance at Indonesian Bureau of Logistics* Proceeding 2018 12th Int. Conf. Telecommun. Syst. Serv. Appl. TSSA pp 1–6

IOP Conf. Series: Materials Science and Engineering **982** (2020) 012058 doi:10.1088/1757-899X/982/1/012058

Design Mitigation and Monitoring System of Blood Supply Chain Using SCOR (Supply Chain Operational Reference) and HOR (House of Risk)

Maria Francisca Raras Dewantari[1,2], Ari Yanuar Ridwan[1] and Hardian Kokoh Pambudi[1]

[1]Faculty of Industrial Engineering, Telkom University, Bandung, Indonesia
[2]mfrarasd@gmail.com

Abstract. Blood supply chain manages the flow of blood product from donors to patients. One of the service provider is PMI which contributes 92% of blood national donation. In fact, demands and supplies of in a blood supply chain are often unpredictable. They contribute to the occurrence of risks which have direct impacts on human life. Thus need a risk management to mitigate such impacts. One way to do so is by using Supply Chain Operational Reference (SCOR) model for mapping the activities of blood chain. Thus, it can facilitate the identification of risk events and risk agents. Further risk events and risk agents are processed using House of Risk (HOR). While the aim of HOR1 is to identify priority of the risk events, the HOR2 arranges necessary mitigation strategies. Moreover, in this work, there are 9 risk agents chosen from HOR1 and 8 preventive actions for the mitigation. Additionally, this research develop a monitoring system that may assist to monitor the occurring risks.

1. Introduction

The blood supply chain manages the flow of blood products from donors to the patients through 5 stages: donors, mobile collection sites, blood centre, demand nodes, and patients [1]. The concept of blood supply chain is to distribute blood products safely, inexpensively, and accessible to the patients [2]. One of the blood service providers in Indonesia is Unit Donor Darah (UDD). UDD PMI (Unit Donor Darah Palang Merah Indonesia) takes a significant role in blood service in Indonesia since it contributed 92% of national blood donation in 2016 [3].

The demand and supply of blood supply chain are hard to be predicted. That uncertainty causes risks which have direct impacts on human life [2]. A few examples of risks in blood supply chain are the sudden change in blood demands, the delay in product delivery from the supplier, errors in the process of taking blood, and damaged blood [4]. The potential risks of blood supply chain may affect the overall performance of the healthcare network. Therefore, it is important for a risk management to be conducted in order to generate the risk mitigation [2].

One of the methods that can be applied to the risk management is House of Risk (HOR). HOR1 can identify the risk and the cause of it, as well as setting up the mitigation strategy and decreasing the possibility of a risk using HOR2. The identification of risks is done using a Supply Chain Operational Reference (SCOR) model [4]. HOR model has been used in managing the blood supply chain in Regional Blood Centre 5, Thailand [2]. It has also been used to set up the risk mitigation of blood supply chain in PMI of Sleman regency [4].

The risks integrated in the monitoring system will be useful to observe the actual condition that is compared to the target so that it is easier in making a decision [5]. The purpose of this research is to set up the mitigation plan and design a monitoring system in order to minimize the risks of blood supply chain in UDD PMI of Klaten regency.

2. Theoretical Basis

2.1. Supply Chain

Supply chain consists of all parties that contribute directly and indirectly in meeting the demands of the consumer, including manufacturers, suppliers, transporter, warehouse, retailer, and consumer. The main object of every blood supply chain is to fulfill the consumer demand [6]. According to Council of Supply Chain Management, the management of supply chain consist of planning, management, and a good coordination on stocking, manufacturing up to distributing process [7].

According to Xanthopoulos and friends's point of view, main principle of supply chain risk management is to identify, to assess, and to prioritize supply chain risks in order to minimize, to monitor, and to control the occurrences and the impacts of uncertain risk events [2]. Three phases on supply chain risk management are identification phase, risk assessment phase and risk mitigation [8].

Supply Chain Operational Reference (SCOR) is a reference model in operating supply chain [9]. It is useful for descripting and analyzing the supply chain [10]. SCOR model has a hierarchal characteristic. There are three levels in SCOR, namely the type of process, process categories, and process elements [11].

2.2. House of Risk (HOR)

HOR is a modification model of Failure Mode and Effect Analysis (FMEA) to quantify risks and adapting House of Quality (HOQ) model in order to choose which risk agent is more influential. Quantifying process is done to define the process of supply chain based on Supply Chain Operation Reference (SCOR) [12].

2.2.1. HOR1. HOR1 has the aim to decide which risk agent is more influential. The followings are the steps done in HOR1 [12].

- Map the blood supply chain activity based on the SCOR model that is categorized into 5 categories: plan, source, deliver, make, dan return. The HOR1 Framework is displayed in Figure 1.

Business processes	Risk event (E_i)	Risk agents (A_j) A_1	A_2	A_3	A_4	A_5	A_6	A_7	Severity of risk event i (S_i)
Plan	E_1	R_{11}	R_{12}	R_{13}					S_1
	E_2	R_{21}	R_{22}						S_2
Source	E_3	R_{31}							S_3
	E_4	R_{41}							S_4
Make	E_5								S_5
	E_6								S_6
Deliver	E_7								S_7
	E_8								S_8
Return	E_9								S_9
Occurrence of agent j		O_1	O_2	O_3	O_4	O_5	O_6	O_7	
Aggregate risk potential j		ARP_1	ARP_2	ARP_3	ARP_4	ARP_5	ARP_6	ARP_7	
Priority rank of agent j									

Figure 1. HOR1 Framework

- Identify the risk event (E_i) and the risk agent (A_j). Risk event is a negative unexpected event. Risk agent is the cause of risk event occurrence.
- Assess the severity (S_i) of every risk event. This assessment shows how big the effect of risk event. The assessment uses scale of 1 – 10. The bigger the number of the scale is, the bigger the effect of risk event it will have.

- Assess the possibility of risk agent (O_j) occurrence. The assessment uses scale of 1 – 10. Scale 1 shows that a risk agent never happens, while scale of 10 shows that a risk agent almost always happens.
- The next step is to examine relationship matrix in order to assess the correlation between a risk agent and a risk event. Correlation assessment (R_{ij}) has 4 scales: 0,1,3, and 9. Scale of 0 indicates that there is no correlation, scale of 1 reveals a low correlation, scale of 3 shows an average correlation, and scale of 9 indicates a high correlation.
- Calculate Aggregate Risk Potential (ARP_j) of every risk agent using the following formula:

$$ARPj = O_j \sum_i S_i R_{ij} \quad (1)$$

2.2.2. HOR2. HOR2 aims to decide which action (preventive action) that needs to be done first by considering the effectiveness of an action and the difficulty level of the action. The followings are the steps in HOR2 [12].

- Choose which risk agent has a priority rank that can be assisted using pareto chart to analyze the ARP_j value.
- Identify preventive action (PA_k) or the preventive action against risk agent.
- Determine the correlation between every preventive action and every risk agent. The correlation assessment (E_{jk}) has 4 scales: 0,1,3, and 9. The scales display the effectiveness of preventive action k to decrease the possibility of risk agent j occurrence.
- Calculate the total effectiveness of every preventive action using the formula:

$$T_{ek} = \sum_j ARP_j E_{jk} \ \forall k \quad (2)$$

- Assess the degree of difficulties (D_k). There are three scales: scale of 3 means easy, scales of 4 means average, and scale of 5 means difficult.
- Calculate the total effectiveness to difficulty ratio using the formula:

$$ETD_k = Te_k / D_k \quad (3)$$

- Classify the rank of every preventive action. The one with the biggest ETD_k becomes the first in rank. The HOR2 framework is displayed in Figure 2

To be treated risk agent (A_j)	Preventive action (PA_k)					Aggregate risk potentials (ARP_j)
	PA_1	PA_2	PA_3	PA_4	PA_5	
A_1	E_{11}					ARP1
A_2						ARP2
A_3						ARP3
A_4						ARP4
Total effectiveness of action k	TE_1	TE_2	TE_3	TE_4	TE_5	
Degree of difficulty performing action k	D_1	D_2	D_3	D_4	D_5	
Effectiveness to difficulty ratio	ETD_1	ETD_2	ETD_3	ETD_4	ETD_5	
Rank of priority	R_1	R_2	R_3	R_4	R_5	

Figure 2. HOR2 Framework

2.3. Monitoring System

According to Council of Supply Chain Management, monitoring is an activity to observe the reality and the targeted condition. Monitoring can be applied in a system to help in the decision-making process since a monitoring process shows the report, for example figures of data chart [5].

3. Discussion

3.1. Risk event and Risk Agent Identification

In order to identify risks, the mapping of blood supply chain is conducted previously using SCOR model. Supply chain activity mapping is conducted by discussing the matter with the head of blood service in UDD PMI of Klaten regency. There are 28 risks in the supply chain that are categorized in plan, source, make, deliver, and return. Table 1 is the summary of blood supply chain risk event.

Table 1. Risk Event Blood Supply Chain

Code	Risk Event	SCOR Stage	Code	Risk Event	SCOR Stage
E1	Error in demand forecasting	Plan	E15	Error in blood-typing diagnosis	Make
E2	Excessive blood stocks	Plan	E16	Serology test failure	Make
E3	Shortage in blood stocks	Plan	E17	Production failure	Make
E4	Blood donation failed to meet requirement	Plan	E18	Bloods are damaged during the production process	Make
E5	Supporting material shortage	Plan	E19	Data input error	Make
E6	The sudden change in production schedule	Plan	E20	No matching blood type	Make
E7	Blood demand cancellation	Plan	E21	Blood damaged in the storage	Make
E8	Failed negotiation	Plan	E22	Damaged blood and good blood mixed up	Make
E9	Order specification failure	Source	E23	Failed in meeting blood demand	Deliver
E10	Delayed delivery from supplier	Source	E24	Difference in between the inventory system and the actual condition	Deliver
E11	Damaged supporting material	Source	E25	Blood product labelling error	Deliver
E12	Damaged supporting material passed the selection process	Source	E26	Blood damaged during delivery process	Deliver
E13	Supporting material damaged in the storage	Source	E27	Blood product failed to be delivered to the customer	Deliver
E14	Error in blood-taking process	Make	E28	Late in filling complaints to suppliers	Return

After identifying risk event, risk agent is identified. Risk agent is the cause risk event. There are 23 risk event identified in the blood supply chain in UDD PMI Klaten regency. Table 2 is the summary of risk agent in blood supply chain. The process of identifying risk event and risk agent are conducted by applying literature study which verified by expert(s) afterwards.

Table 2. Risk Agent Blood supply chain

Code	Risk Agent	Code	Risk Agent
A1	Fluctuating demand	A13	Difficulty in detecting blood vessels
A2	Demand forecasting error	A14	Production employee's inaccuracy in reading the result
A3	Uncertainty in blood donor numbers	A15	Damaged machine
A4	Staff shortage	A16	Unsuitable temperature of the production site
A5	Sudden huge demands	A17	Blood mishandling
A6	The hospital has gotten blood product from another blood service	A18	Negligence of production employee
A7	Logistic workers' negligence	A19	IT system disruption
A8	Natural disaster	A20	Empty blood stocks
A9	Lack of communication with supplier	A21	Delays in updating stock data
A10	Carelessness of logistic officer	A22	Negligence of blood service employee
A11	Improper handling of supporting material	A23	Accidents in delivering blood products
A12	Uncertainty in blood donor's health		

3.2. *Aggregate Risk Potential Calculation*

In order to calculate the value of Aggregate Risk Potential (ARP), the value of severity (S_i), occurrence (O_j), and correlation (R_{ij}) are required. The assessment of those value as well as the calculation of ARP are provided in Table 3.

Table 3. Risk Agent Blood Supply Chain

Bus. Process	Risk Event (Ei)	Risk Agents (Aj)																							
		A1	A2	A3	A4	A5	A6	A7	A8	A9	A10	A11	A12	A13	A14	A15	A16	A17	A18	A19	A20	A21	A22	A23	Si
Plan	E1	3	3	0	0	1	0	0	0	0	0	0	0	0	0	0	0	0	0	0	0	0	0	0	3
Plan	E2	3	9	9	0	9	0	0	1	0	0	0	0	0	0	0	0	0	0	0	0	0	0	0	1
Plan	E3	9	9	9	0	9	0	9	9	0	0	0	0	0	0	3	0	0	0	0	9	0	0	0	7
Plan	E4	0	0	0	3	3	0	0	3	0	0	0	0	0	0	0	0	0	0	0	0	0	0	0	2
Plan	E5	3	3	3	0	3	0	1	0	0	0	0	0	0	0	0	0	0	0	0	0	0	0	0	5
Plan	E6	3	0	0	0	9	0	0	0	0	0	0	0	0	0	0	0	0	0	0	0	0	0	0	2
Plan	E7	0	0	0	0	0	3	0	0	0	0	0	0	0	0	0	0	0	0	0	0	0	0	0	2
Plan	E8	0	0	0	0	0	0	3	0	3	0	0	0	0	0	0	0	0	0	0	0	0	0	0	2
Source	E9	0	0	0	0	0	0	3	0	3	0	0	0	0	0	0	0	0	0	0	0	0	0	0	2
Source	E10	0	0	0	0	0	0	3	0	3	0	0	0	0	0	0	0	0	0	0	0	0	0	0	7
Source	E11	0	0	0	0	0	0	3	0	0	0	9	0	0	0	0	0	0	0	0	0	0	0	0	1
Source	E12	0	0	0	0	0	0	0	0	0	9	0	0	0	0	0	0	0	0	0	0	0	0	0	2
Source	E13	0	0	0	0	0	0	9	0	0	0	9	0	0	0	0	0	0	0	0	0	0	0	0	2
Make	E14	0	0	0	0	0	0	0	0	0	0	0	9	9	0	0	0	0	9	0	0	0	0	0	2
Make	E15	0	0	0	0	0	0	0	0	0	0	0	0	0	3	3	0	0	3	0	0	0	0	0	5
Make	E16	0	0	0	0	0	0	0	0	0	0	0	0	0	0	9	0	0	9	0	0	0	0	0	2
Make	E17	0	0	0	0	0	0	0	0	0	0	0	0	0	0	9	0	9	9	0	0	0	0	0	4
Make	E18	0	0	0	0	0	0	0	0	0	0	0	0	0	0	0	9	9	9	0	0	0	0	0	2
Make	E19	0	0	0	0	0	0	0	0	0	0	0	0	0	0	0	0	0	0	3	0	0	9	0	6
Make	E20	1	0	0	0	0	0	0	0	0	0	0	0	0	0	0	0	0	0	0	9	0	0	0	2
Make	E21	0	0	0	0	0	0	0	0	0	0	0	0	0	0	0	0	9	0	0	0	0	0	0	3
Make	E22	0	0	0	0	0	0	0	0	0	0	0	0	0	0	0	0	0	3	0	0	0	0	0	2
Deliver	E23	0	3	0	0	0	0	0	0	0	0	0	0	0	0	0	0	0	0	0	9	0	0	0	2
Deliver	E24	0	0	0	0	0	0	0	0	0	0	0	0	0	0	0	0	0	0	3	0	3	0	0	6
Deliver	E25	0	0	0	0	0	0	0	0	0	0	0	0	0	0	0	0	0	0	0	0	0	9	0	3
Deliver	E26	0	0	0	0	0	0	0	0	0	0	0	0	0	0	0	0	9	0	0	0	0	0	3	4
Deliver	E27	0	0	0	0	0	0	0	0	0	0	0	0	0	0	0	0	0	0	0	0	0	3	3	4
Return	E28	0	0	0	0	0	0	3	0	0	0	0	0	0	0	0	0	0	0	0	0	0	0	0	2
Occurrence of agent j		6	3	3	4	7	3	7	6	5	3	3	3	3	3	6	3	3	3	5	7	3	3	3	
Aggregate risk potential j		588	306	261	24	798	18	896	420	165	54	81	54	54	45	540	54	351	333	180	693	54	279	72	
Priority rank of agent		4	9	11	22	2	23	1	6	13	16	14	16	16	21	5	16	7	8	12	3	16	10	15	

The assessment is conducted by handing over/distributing questioner to the expert. In this case, the expert is the head of blood service of UDD PMI in Klaten regency. The table provide the ranks of ARP. It can be seen that rank 1 is a *risk agent* which has the highest ARP.

3.3. *Choosing Risk Agent*

In order to choose risk agent, the pareto chart is applied. The basic concept of pareto, which is 80/20, is applied in choosing a risk agent. Therefore, the chosen one is *risk agent* that has ARP cumulative percentage until 80%. The chosen risk agents are: (a) the negligence of logistic employee (896), (b) sudden huge demands (798), (c) empty blood stocks (693), (d) fluctuating demands (588), (e) damaged machine (540), (f) natural disaster (420), (g) blood mishandling (351), (h) the negligence of production employee (333), (i) and forecasting error (306). The pareto chart is displayed in Figure 3.

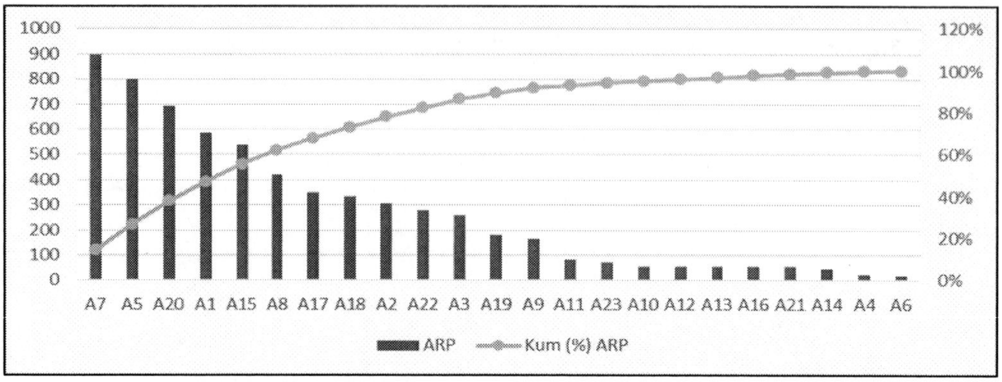

Figure 3. Pareto Chart

3.4. Preventive Action Classification

The preventive action is organized based on the risk agent. There are eight preventive actions as shown in Table 4 below.

Table 4. Preventive Action

Code	Preventive Action
PA1	Punishment for PMI employee's negligence
PA2	Regular training for PMI employees
PA3	Cooperation with other PMIs
PA4	Increasing the number of blood donation socialization
PA5	Analysing blood demand statistically
PA6	Increase blood storage capacity
PA7	Checking the blood processing machine regularly
PA8	Improve communication with hospitals

Preventive action classification is conducted by calculating ETD. ETD indicates the effectiveness degree of an action to minimalize a risk cause. In order to classify preventive action, the value of ARP_j, correlation (E_{jk}), total effectiveness (Te_k), degree of difficulties (D_k), and total effectiveness to difficulty ratio (ETD_k) are required. The assessment of correlation and degree of difficulties are conducted by distributing questioner to the head of UDD blood service in Klaten regency. The result of preventive action classification is displayed in Table 5 as follows.

Table 5. HOR 2

	Preventive Action (PA_k)								
Risk Agent (A_j)	PA1	PA2	PA3	PA4	PA5	PA6	PA7	PA8	ARP_j
A7	9	0	0	0	0	1	0	0	896
A5	0	0	9	1	1	0	0	3	798
A20	0	0	9	9	9	0	0	3	693
A1	0	0	9	1	1	0	0	1	588
A15	0	0	0	0	0	0	9	0	540
A8	0	0	3	0	1	0	0	0	420
A17	3	1	0	0	3	0	0	0	351
A18	3	1	0	0	3	0	0	0	333
A2	0	0	1	0	0	0	0	0	306

Total effectiveness of Action (TE_k)	10116	684	20277	7623	11013	798	4860	5061
Degree of difficulty preforming action (D_k)	3	3	3	4	3	3	3	4
Effectiveness to difficulty ratio (ETD)	3372	228	6759	1906	3671	266	1620	1265
Rank	3	8	1	4	2	7	5	6

Based on the table above, the order of preventive action which can be done by PMI in Klaten regency in sequence: (1) Cooperate with other PMIs, (2) analyze blood demands statistically, (3) give punishment for every negligence of employee in PMI, (4) increase the number of blood donation socialization, (5) check blood processor machine regularly, (6) improve communication with hospitals, (7) increase blood storage capacity, and (8) give a regular training routine to every employee in PMI.

3.5. Monitoring System

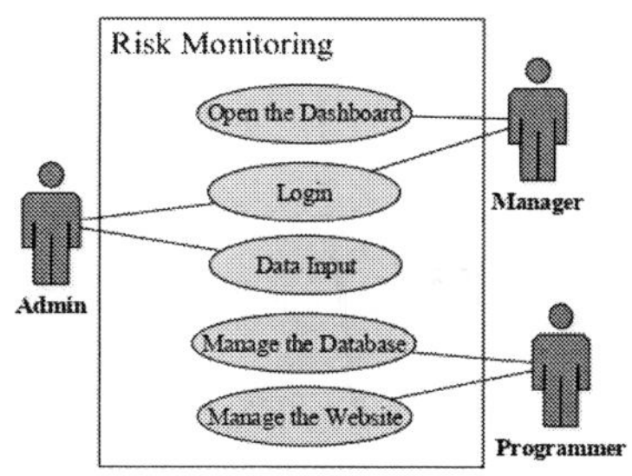

Figure 4. Use Case Diagram

Monitoring system can be used in the risks observing process. Use case diagram used to understand the system behavior and the system function [8]. There are three acts in the monitoring system: an admin, a manager, and a programmer, such as displayed in Figure 4 above. the Manager has access to login and check the dashboard. Meanwhile, the admin may login and input the data. In this case, the programmer takes role in managing the database and the website.

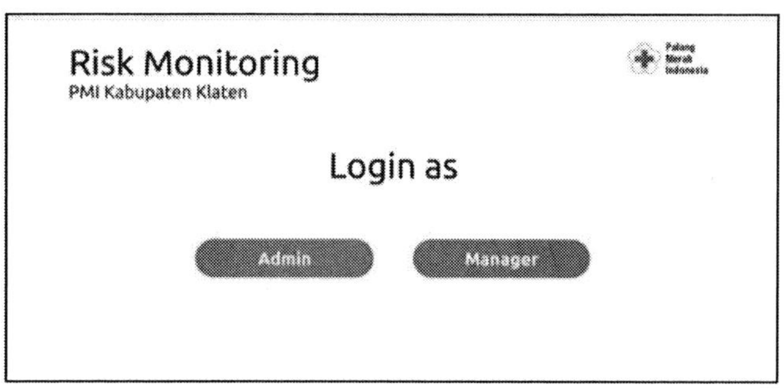

Figure 5. User Interface Login As

When logging in, a user should choose to login whether as a leader or as an admin as displayed in Figure 5. Then the user should enter username and password as illustrated in Figure 6.

IOP Conf. Series: Materials Science and Engineering **982** (2020) 012058 doi:10.1088/1757-899X/982/1/012058

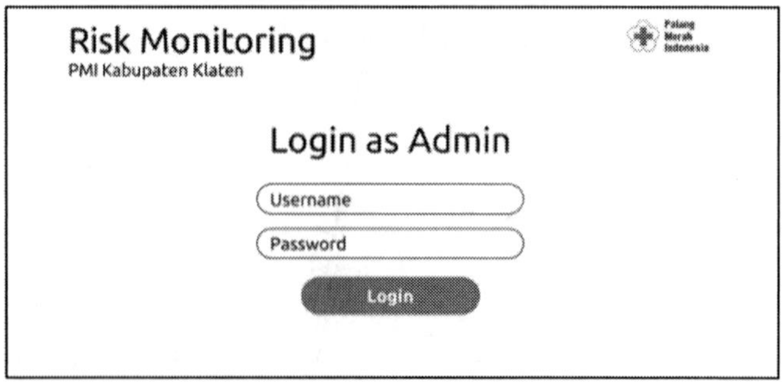

Figure 6. User Interface Login

If the user chooses to log in as an admin, then he may decide whether to input the risks or to do the threshold input as shown in the Figure 7. The risk input is used to add the date, risks, and the causes. While the threshold input is used to input the limit of every risk agent.

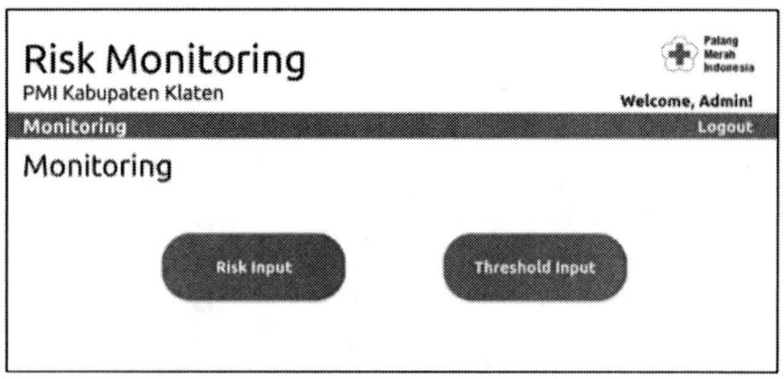

Figure 7. User Interface Menu Admin

Figure 8 shows us what the system be like if the admin chooses risk input. Admin needs to add the risk date, risk event, and risk agent that may cause the risk event happen. Then the admin needs to click "add". The result then can be seen on the table under the input form.

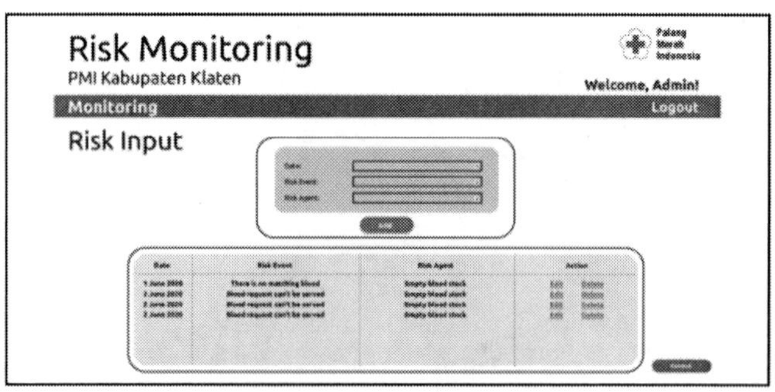

Figure 8. User Interface Risk Input

Figure 9 shows us what the system be like if the admin chooses threshold input. Admin needs to add the risk agent and the lower limit, middle limit and the upper limit. Then the admin needs to click "add". The result then can be seen on the table under the input form.

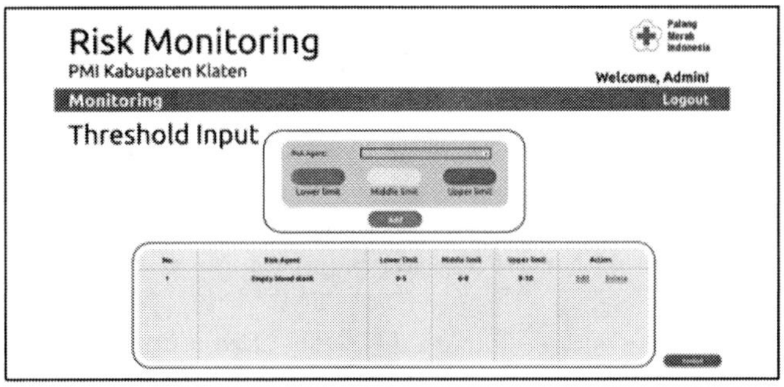

Figure 9. User Interface Threshold Input

If the user chooses to login as the Manager, the system will open the dashboard as Figure 10. The manager needs to choose one of the risk agents to present, and then click show button. As the result, he may see how many times the risk happens. Then, it will be compared to the limit/the threshold.

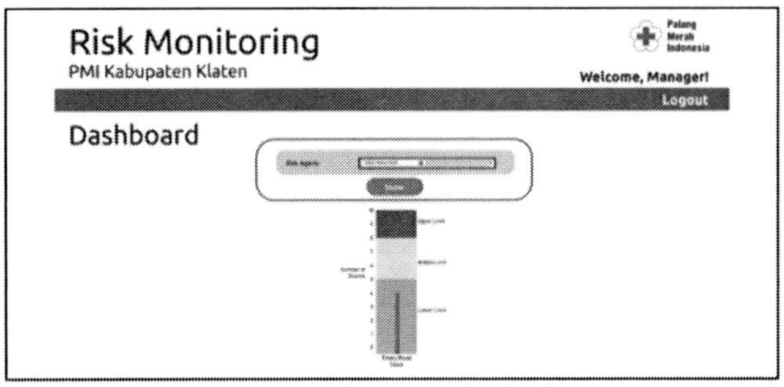

Figure 10. User Interface Dashboard

4. Conclusion

There are 28 risk events and 23 risk agents that have been identified in the blood supply chain of UDD PMI of Klaten regency. Those risks are tabulated using HOR method, resulting the existence of nine chosen risk agents. Those risk agents become an input in arranging mitigation strategy. In arranging the mitigation strategy, there are eight preventive actions that can be applied by UDD PMI of Klaten regency which are expected to decrease the possibility of risk. Monitoring system can be used in observing risks. Monitoring system can observe the occurrence of risks that happen by comparing it to maximum amount that is tolerated by UDD PMI in Klaten regency.

References

[1] Pirabán A, Guerrero W J and Labadie N 2019 Survey on Blood Supply Chain Management: Models and Methods Comput. Oper. Res. 112 104756
[2] Boonyanusith W and Jittamai P 2019 Blood Supply Chain Risk Management using House of Risk Model Walailak J. Sci. Technol. 16 573–591
[3] Pusdatin Kemenkes 2018 Situasi Pelayanan Darah di Indonesia 7
[4] Achmadi R E and Mansur A 2018 Design Mitigation of Blood Supply Chain using Supply Chain Risk Management Approach Proc. Int. Conf. Ind. Eng. Oper. Manag. March 1763–1772
[5] Permana R A Ridwan A Y and Yulianti F 2019 Perancangan Sistem Monitoring Ketahanan Pangan dan Mitigasi Risiko Distribusi Beras Menggunakan Metode FMEA dan AHP pada Bulog Subdivre Bandung e-Procceeding Eng. 6 7137–7145
[6] Chopra M P and Sunil 2012 Supply Chain Management: Global Edition

[7] Ismadhia A S Ridwan A Y and Hadi R M E 2018 A Scor-based Model For Green Sales And Distribution Performance Measurement In The Leather Tanning Industry J. Rekayasa Sist. Ind. 5 1
[8] Sari D I, Ridwan A Y and Santosa B 2019 Design of Risk Management Monitoring System Based on Supply Chain Operations Reference (SCOR): A Study Case at Dairy Industry in Indonesia International Conference On Rural Development And Enterpreneurship 5 104–115
[9] Sutrisno A 2018 A Conceptual Model on Integrating Supply Chain Operation Reference (SCOR) into Disaster Response Service Supply Chain FMEA Int. J. Innov. Enterp. Syst. 2
[10] Waaly A N, Ridwan A Y and Akbar M D 2018 Development of Sustainable Procurement Monitoring System Performance Based on Supply Chain Reference Operation (SCOR) and Analytical Hierarchy Process (AHP) on Leather Tanning Industry MATEC Web Conf. 204
[11] Paul J 2014 Transformasi Rantai Suplai Dengan Model SCOR Jakarta Penerbit PPM
[12] Pujawan I N and Geraldin L H 2009 House of risk: a Model for Proactive Supply Chain Risk Management Bus. Process Manag. J. 15 953–967

IOP Conf. Series: Materials Science and Engineering **982** (2020) 012059 doi:10.1088/1757-899X/982/1/012059

Mental workload analysis on fruit truck suppliers using NASA-TLX method in giwangan market area

Bagus Wahyu Utomo[1], Atyanti Dyah Prabaswari[2], Riani Nurdin[3] dan Calvin Hasiolan Sinaga[4]

[134]Industrial Engineering, Sekolah Tinggi Teknologi Adisutjipto, Yogyakarta, Indonesia
[2]Industrial Engineering, Universitas Islam Indonesia, Yogyakarta, Indonesia

[2]atyanti.dyah@uii.ac.id

Abstract. Several times the truck driver received complaints from consumers (merchants) related to the inaccuracy in delivering the goods, and the quality of the goods declined. The driver stated that most of this happened because of the unfavorable market conditions so that it was not possible for drivers who arrived at the Giwangan market to carry out the loading and unloading process directly. That is because drivers are forced to wait up to 1-5 hours in the Giwangan market area to unload goods. These things are thought to add to the mental workload of truck drivers. The method used to measure mental workload using a questionnaire on the NASA-TLX method. The number of respondents sampled in this study were 20 (twenty) people, and all male. The respondents' age is divided into two categories, namely early adulthood (18-40) and middle adulthood (40-60). The NASA-TLX calculation results stated that the workload of fruit truck drivers on the market during normal day conditions is classified as "High" with a score obtained of 70.26. The correlation test has resulted in the more age of the fruit truck driver, the lower the workload's value.

1. Introduction

All human activities must have a light, moderate, or heavy workload. The workload is an effort that must be expended by someone to fulfill the purpose of the work. The workload is also defined as a person's limited capacity to carry out their duties [1]. In work, someone gets physical and mental workload that directly or indirectly affects work productivity, where the heavier workload will reduce worker productivity. The physical workload can be measured using medical devices, so it tends to be easy to see how heavy and large the level of fatigue is caused by these workers and can even be seen in plain view even though not using medical devices. This is different from mental workloads, where mental workloads expended relatively less energy than physical workloads. However, roles and mental workload responsibilities are clear and more massive than the physical workload. This is because a heavy mental workload will have an impact on work stress. Therefore, so that workers can work and produce an optimal output, it will be essential to consider various aspects of humans and their work [2].

Giwangan Market is one of the leading markets for fruit and vegetables in the Special Region of Yogyakarta (DIY). The market is a fruit and vegetable primary market located at Jalan Imogori No. 212, Yogyakarta City. This market has a land area of 24,594 m2 and a building area of 18,984 m2 utilized by 981 vegetable and fruit merchants who originally came from 3 traditional markets, namely: Sri Wedani Market, Shopping Center, and Jalan Pabringan. This market has a hangar, a place for loading and unloading fruits and vegetables from and in goods transport vehicles such as pick-up

(L300), diesel trucks, and others.

In the beginning, the hangar function's planning could run well by the layout arranged by the City Government, but over time the market conditions became more dense and irregular. This happened because merchants traded in parking lots, loading and unloading goods, and roads that were supposed to cross vehicles, and many vehicles parked in places that were not supposed. This happened because merchants already occupied their parking spaces. The current condition around the giwangan market has become dense and irregular, making it difficult for vehicles to get in and out. This makes the market participants feel pressured, especially fruit truck drivers. The truck driver is responsible for delivering the fruit to the hands of consumers (merchants). Some consumers ask to deliver their goods on time and ask to maintain the quality of goods ordered. The truck driver received complaints from consumers (merchants) several times related to the inaccuracy in delivering the goods, and the quality of the goods declined. The driver stated that most of this happened because of the unfavorable market conditions so that it was not possible for drivers who arrived at the Giwangan market to carry out the loading and unloading process directly. Several places should be designated for loading and unloading of goods but it has been occupied by merchants who sell using their vehicles such as tosa or open shanties so that there is less space for loading and unloading for truck drivers. This is because drivers are forced to wait up to 1-5 hours in the Giwangan market area to unload goods. These things are thought to add to the mental workload of truck drivers. Meanwhile, a worker certainly has different capacities in accepting workloads.

One factor in the difference in the capacity of workers is age. The age of truck drivers who carry out the loading and unloading process in the Giwangan market also varies from 22-60 years. The age of adulthood is divided into three namely early adulthood is 18-40 years, middle adulthood is 4-60 years, and advanced adults> 60 years [3]. Each age group has its capacity for receiving workload, so it is necessary to look at the relationship between age and mental workload score.

So that in solving these problems, it is necessary to measure the mental workload of the drivers of fruit transport trucks in the Giwangan market. Several methods can do measurement of mental workload, one of which is the National Administration Task Load Index (NASA-TLX) developed by Sandra G. Hart of NASA-Ames Research Center and Lowel E. Staveland from San Jose State University in 1981 This method is in the form of a questionnaire developed based on the emergence of subjective measurement needs that are easier but more sensitive to workload measurement [4]. This method was developed based on the need for subjective measurements consisting of a scale of nine factors (task difficulty, time pressure, type of activity, physical effort, mental effort, performance, frustration, stress, and fatigue).

By knowing the level of mental workload of fruit truck drivers, these results can be used to consider further decision-making, whether re-structuring the access road in the Giwangan market hangar and strategies for implementing workers for drivers that the driver's mental workload score can be reduced. Through measurements can also be known as indicators that dominate the mental workload of the driver. So that it can be made a recommendation for market managers to improve accessibility to reduce the mental workload of fruit truck drivers in the Giwangan market.

2. Literature Review

Analysis of work mental load on the study program admin at the industrial technology faculties of Staff's mental workload analysis in the study program of the private educational organization was conducted by [5]. Study program worker in managing the management contained in the study program often gets high enough pressure so that the admin's mental workload increases. This study will determine the mental workload of the questionnaire distribution to study program administrators because remembering this research is subjective. The questionnaire was then analyzed using the NASA-TLX method and statistics such as different tests, correlations, regression.

In the case study of PT. PAHALA KENCANA Bandung Branch is one of the inter-city intercity bus transportation service companies serving routes to 35 destinations on Sumatra, Java, Madura, and Bali [6]. One of the routes investigated is the driver activity of the Bandung-Denpasar route. This route is the longest to be sure that the workload on this route is very high. Therefore the bus driver needs high physical and concentration needs to control the bus properly so that the safety and comfort of passengers can be maintained. This research is also motivated by the number of accidents that have increased from year to year. Therefore the mental workload of bus drivers needs to be done to reduce the risk of accidents.

The workload is a term used to refer to the price or cost of achieving an activity target. PT. ABC is a company that manufactures mobile phones [7]. This research is motivated by the high demand every day, making workers in the logistics department unable to complete the work according to the specified schedule, so workers must work overtime to complete the work. Frequent overtime can have an impact on workers, both physically and mentally. To determine the amount of mental workload of employees in the logistics department, measurements were made using the NASA-TLX (National Aeronautical and Space Administration Task Load Index) method. Based on the scores obtained, it is known that 38.10% of employees have high metal workloads, 57.41% have moderate mental workloads, and 4.76% have mild mental workloads. The solution given in this research is to add workers, especially helper and scanner, who have the highest mental workload.

Analysis of mental workload on PT. DBM cargo operators were once conducted by [1]. In carrying out its activities the cargo operator PT. DBM is not uncommon to get high enough pressure so that the mental workload of operators increases. Data collection is carried out by distributing questionnaires in the data uniformity test and the data's adequacy to ensure that the data taken is uniform and sufficient. After that data, it is used to calculate NASA-TLX scores to see the operator's classification of workload. The NASA-TLX calculation results obtained mental workload values for all workers at the level of "medium" with the largest mental workload indicator on the operator of PT. DBM is the scale of PD (Physical Demand) and MD (Mental Demand). Proposed improvements are given, namely dividing the work and rotating work.

Mental workload as an evaluation of the operator of marginal load (the difference between his motivational capacity and the tasks are given) when carrying out work quite well in motivating conditions [4]. The concept emphasized here is the marginal workload which is the difference between workload demands of a task with maximum capacity (motivated) one's mental load in motivated conditions. The mental workload of someone in handling work is influenced by work situation and type of activity, response time and completion time available, individual factors such as motivation level, expertise, fatigue/boredom, and allowable performance tolerance. The physical workload measurement can be done by measuring energy consumption and / or oxygen consumption during the activity [8]. However, the subjective workload measurement can be done and is a measure of mental workload based on the respondent/worker's subjective perception.

3. Research Methodology

3.1. Research Object
The study object was the mental workload of fruit transport truck drivers at the Giwangan market from outside the city of Yogyakarta.

3.2. Data Collection Methods
Interviews and questionnaires are active primary data collection by questioning respondents. Interviews and questionnaires, conducted by giving NASA-TLX questionnaires that have been made to truck drivers who are willing to provide responses relating to the mental workload they receive while working [9].

3.3. NASA TLX
The method used to measure mental workload using a questionnaire on the NASA-TLX method. This method is done with a subjective approach to statements representing the conditions of work in the field [10].

The factors namely Mental Demand (MD), Physical Demand (PD), Temporal Demand (TD), Performance (P), Effort (EF), and Frustration Level (FR) [4]. Mental Demand (MD) with its low - high rating. MD is how much mental and perceptual activity it takes to see, remember, and search. Is the job simple or complex? Physical Demand (PD) with a low - high rating. PD is how workers need much

IOP Conf. Series: Materials Science and Engineering **982** (2020) 012059 doi:10.1088/1757-899X/982/1/012059

physical activity (for example attracting, encouraging, etc.). Temporal Demand (TD) with a low - high rating. TD is how much time pressure is felt during the work takes place. Is the work can be done slowly, or quickly so that it feels tiring. Own Performance (OP) with an incorrect rating - perfect. OP is how much the success rate of workers in carrying out their duties and how satisfaction with their work results. Effort (E) uses a rating from low to high. E is how much effort or hard work is needed to achieve the required level of performance. Frustration (FR) uses a low-high rating. FR is how many workers feel insecure, hopeless, offended, or disturbed while doing their job.

NASA-TLX measurements have five stages [4]. The weighting phase is carried out by filling in by selecting one of the two most dominant indicators felt during work. After the respondent successfully fills out the 15 comparison indicators, a weighting is done by adding each selected indicator's total. After that, add up each mental workload indicator that the respondent has chosen to get the indicator's weight.

In the rating phase, respondents are asked to choose a rating value from the range of scores from 100 to 100 in each mental workload indicator statement. The product rating phase is multiplying the rating by the indicator weights for each scale of mental workload. calculating the average weighted workload (WWL) phase is done by adding up the product value of each indicator and dividing the results of the sum of the indicators by the number of respondents. After obtaining the average WWL from each factor, the next is to determine the average value of the total WWL by dividing the total value of the average WWL by a value of 15 (derived from the sum of the total weights from the pairwise comparison of the NASA-TLX workload mental scale on the questionnaire). Score interpretation is divided into several categories [4].

Table 1. Score Interpretation

Workload Group	Score
Low	0-9
Middle	10-29
Rather High	30-49
High	50-79
Very high	80-100

3.4. Statistic Test

Mann-Whitney difference test use between early adult and intermediate age groups. There are testing to find differences in mental workload scores between early adult and middle adulthood. The division of adulthood is divided into three [3]. First, early adulthood begins at the age of 18 years to approximately age 40 years. When the physical and psychological changes that accompany a decrease in reproductive capacity. Middle adulthood begins at the age of 40 years until the age of 60, which is when both physical and psychological abilities are declining, which is see in everyone. Advanced adulthood begins at the age of 60 years until death. At this time, both physical and psychological abilities are rapidly declining, but modern medical techniques and efforts in dressing and grooming enable men and women to look, act and feel like they were when they were young.

Besides that, using the correlation test to test the relationship between age and mental workload scores obtained. Before conducting the correlation test, the data must be tested for normality first. This is done to be able to determine what correlation method is suitable to use.

4. Result

4.1. Population Fruit Truck Driver

In determining the number of drivers of fruit transport truck drivers from outside the Yogyakarta city, observations were made based on expert judgment by taking as many as eight respondents who were market participants who had worked for a very long time in the area around the Giwangan market hangar. Market participants who are take questionnaires were often in direct contact with fruit transport truck drivers such as ticket attendants, parking attendants, fruit merchants, and truck drivers. Based on Table 2, the minimum respondent sample is seven people. However, the number of respondents sampled in this study were 20 (twenty) people.

Table 2. *The population of Fruit Truck Drivers Based on Expert Judgement*

Respondent Expert	Age (year)	Job	Started Working at Giwangan Market (year)	Number of tracks originating from outside the city/day (unit)		*Most Likely* (per day)
				Lowest	Highest	
E1	58	Ticket Attendant	2005	5	7	6
E2	52	Fruit Merchant	2004	5	7	6
E3	50	Fruit Merchant	2004	5	9	7
E4	33	Fruit Merchant	2000	5	8	6,5
E5	70	Parking Attendant	2004	4	10	7
E6	46	Fruit Merchant	2004	6	8	7
E7	56	Fruit Merchant	2001	5	10	7,5
E8	58	Truck Driver	2005	5	8	6,5
		Poplation Average				6,68 ≈ 8

Table 3. *Respondent Profile*

Respondent	Sex	Age (year)	Place of Delivery
R1	Male	35	Banyuwangi
R2	Male	29	Kediri
R3	Male	31	Lampung
R4	Male	38	Malang
R5	Male	22	Jember
R6	Male	51	Palembang
R7	Male	59	Madura
R8	Male	46	Malang
R9	Male	39	Banyuwangi
R10	Male	30	Jember
R11	Male	44	Palembang
R12	Male	38	Kediri

R13	Male	35	Jember
R14	Male	25	Jember
R15	Male	32	Lampung
R16	Male	50	Madura
R17	Male	46	Malang
R18	Male	54	Malang
R19	Male	42	Palembang
R20	Male	55	Banyuwangi

The respondents' age is divided into two categories, namely early adulthood (18-40) and middle adulthood (40-60). The following are the results of age acquisition based on age categories shown in Table 4.

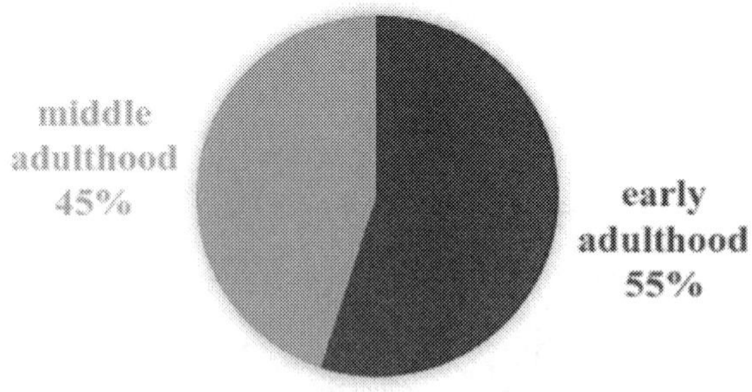

Figure 1. Percentage of Driver Age Group

4.2. *Score Interpretation*

WWL average calculation results are obtain from the sum of each indicator's average scores divided by 15, which is the total initial weighting of each indicator. Before calculating the average, WWL must calculate the average score of the indicator workload by dividing the total number of indicator product values by the number of respondents. The average score of workload (weighted workload) and the interpretation of these values can be seen in Table 4.

Table 4. *Interpretation Score*

Category	Truck Driver	
	Total	Average
MD	5760	288
PD	1180	59
TD	5420	271
OP	4260	213
E	3470	173.5
FR	990	49.5
Average WWL	70.26	
Interpretation	High	

4.3. Mann-Whitney Test in Workload Score Between Early Age Groups and Intermediate Adults

This test is conducted to determine whether there is a difference in NASA-TLX scores between the early adult (18-40 years) and middle adulthood (41-60 years) groups. To prove this, the nonparametric difference test is used, the Mann-Whitney test. The average WWL score for each respondent is obtained by adding up the respondent's product scores and divided by 15 (total comparison of NASA-TLX indicators). Based on these tests, it can be seen that asymp.Sig. (2-tailed) for the two-tailed test is 0.381, or the probability is above 0.05 (0.381> 0.05). So H0 is accepted or indeed, the mental workload received by the first adult trucker group is the same as the mental workload received by the middle adult trucker group.

4.4. Correlation Test Between Truck Driver Mental Workload Scores and Truck Driver Age

A correlation test is done between age and mental workload acquisition. Furthermore, the correlation test with the Spearman method is performed using SPSS 23 software because of the two data that is not normally distributed [11].

Table 5. *Correlation Test*

			Age	Mental Workload Score
Spearman's rho	Age	Correlation Coefficient	1,000	-0.321
		Sig. (2-tailed)		,167
		N	20	20
	Mental Workload Score	Correlation Coefficient	-0.321	1,000
		Sig. (2-tailed)	,167	
		N	20	20

Based on Table 5, it can be seen that the age with a mental workload value has -0,321. This figure shows the weak relationship between age and the value of mental workload (under 0.5), while the negative sign (-) indicates that the more age of the fruit truck driver, the lower the value of the workload.

5. Discussion

Based on the calculation of each NASA-TLX indicator's average workload, the Mental Demand score obtained or the greatest mental needs of the truck drivers are 288. So it can be concluded that the work as a fruit truck driver has high mental needs even though it is not so complex or difficult. The score not only obtained from the job activities as a driver, but external factors at work can also influence the high level of mental needs of truck drivers, such as the current market conditions in Giwangan.

Based on the calculation of each NASA-TLX indicator's average workload, the Temporal Demand score or time requirement is quite large, namely 251. This is triggered because in delivering fruit, the driver must be on time and work as a driver will require hours of work. long. Truck drivers are sometimes required to carry out the loading and unloading process immediately, but there are still long enough queues before arriving at the loading and unloading location.

Based on the calculation of each NASA-TLX indicator's average workload, the Own Performance score for truck drivers is 213. This shows that truck drivers are not very satisfied with their business. Drivers delivering fruit to the hands of consumers (merchants) are often not on time. One of the drivers' causes is not on time because the market conditions are too crowded and irregular that does not allow drivers to be able to directly complete the work of delivering goods to consumers so that the time for loading and unloading takes a long time. Consumers (merchants) complaints related to the quality of goods to truck drivers. For example, such as chili and tomatoes that almost rot because of too long a fall, due to waiting for the turn of loading and unloading goods.

Based on the calculation of the average workload of each NASA-TLX indicator, it obtained an Effort or effort score of 173.5. This shows that the driver's business is good at delivering fruit to the Giwangan market but cannot reach the consumers (merchants) due to crowded market conditions and reduced loading and unloading places due to being occupied by merchants who make drivers forced to have to queue to do the process loading and unloading so that consumers sometimes complain because of old items that reach their hands.

Based on the calculation of each NASA-TLX indicator's average workload, the Physical Demand score or physical needs for truck drivers is obtained, which is 59. This shows that work as a fruit transport truck driver does not have a job that requires much physical activity. The work done is not heavy because the driver just sits driving the truck. The loading and unloading process is carried out by the porters in the market so that the truck drivers just sit and wait for the process to finish. Based on the results of interviews with the driver, the activity that requires the driver to do physical activity is to open a tarpaulin installed above the tub when he wants to load and unload goods that require the driver to expel his physical exertion.

Based on the calculation of each NASA-TLX indicator's average workload, the Frustration Level score for the truck driver is 49.5. This shows that the driver's work does not cause the driver to feel insecure, hopeless, offended, disturbed. Based on interviews, because the average driver feels that their work has become a hobby, it is not too stressful when carried out, although several factors make truck drivers feel stressed, such as market conditions. That thing might make them stress, but the stress is not prolonged because they realize they are making a living. So the Frustration Level (FR) value is at the lowest position among the six indicators.

The early adult age group, referred to as early adulthood, is starting at the age of 18 years to approximately 40 years. The category is when the physical and psychological changes accompany a decrease in reproductive capacity (Hurlock, 1980). At this age, the truck driver's motivation and enthusiasm is still high. During this time, truck drivers will learn a lot and adjust to the work done. This can make the new truck driver sometimes overwhelmed when faced with several situations at work. Giwangan market conditions are currently not good. Some merchants trade in places that are not supposed to be like parking lots and loading and unloading processes. As a result of this situation, locations to carry out the loading and unloading of goods will be increasingly reduced and some vehicles parked in places that are not supposed to be. This condition will certainly cause congestion in the Giwangan market so that when a truck driver arrives at the Giwangan market, he cannot immediately carry out the loading and unloading process but must wait his turn. This makes the driver at an early age adult will be depressed because of the difficulty of parking their vehicles, and also remember they have responsibility for the goods they carry that must be immediately delivered to the merchant.

The middle adulthood group is at the age of 40 to the age of 60, namely when both the decline in physical and psychological abilities are visible in everyone (Hurlock, 1980). There is a decline in physical and psychological abilities at this age, so drivers become easily tired and stressed when doing work, especially when facing the dense and irregular market conditions in Giwangan. At this age, drivers often experience physical pain such as muscle or bone when parking trucks in crowded market conditions. Of the two age groups, both early adulthood, and middle adulthood, they did not significantly differ in mental workload scores even though the two groups differed physically and psychologically. This shows that no matter how old a fruit truck driver is from out of town, the mental workload score caused by market conditions will be the same.

6. Conclusion and Recommendation

The NASA-TLX calculation results stated that the workload of fruit truck drivers on the market during normal day conditions is classified as "High" with a score obtained of 70.26. It can be concluded that current market conditions affect the high mental workload of fruit truck drivers from outside Yogyakarta. This is due to unfavorable market conditions where traders trade in places where they are not supposed to, so loading and unloading places are limited to fruit trucks. Therefore, the fruit truck driver who arrives at the Giwangan market is forced to wait his turn to carry out the loading and unloading of goods, making the time consumers (merchants) receive their goods for a long time. It is then necessary to improve the layout and order in the Giwangan market area related to the layout of the facilities and market participants so that the mental workload of the fruit truck drivers can decrease.

7. Acknowledgment

Researchers thank the support of the Industrial Engineering Department, Sekolah Tinggi Teknologi Adisutjipto and Industrial Engineering Department, Faculty of Industrial Technology, Universitas Islam Indonesia

References

[1] Arasyandi M and Bakhtiar A 2016 Analisa beban kerja mental dengan metode nasa tlx pada operator kargo di PT. Dharma Bandar Mandala (PT. Dbm). *Industrial Engineering Online Journal*, *5*(4)

[2] Okitasari H and Pujotomo D 2016 Analisis beban kerja mental dengan metode nasa tlx pada divisi distribusi produk PT. Paragon Technology and Innovation. *Industrial Engineering Online Journal*, *5*(3)

[3] Hurlock E, Istiwidayanti, Sijabat R, and Soedjarwo 1990 *Psikologi perkembangan: Suatu pendekatan sepanjang rentang kehidupan*. Erlangga, Jakarta

[4] Hancock P and Meshkati N 1988 *Human mental workload*. Amsterdam: North-Holland, pp. 139-183

[5] Prabaswari A D, Basumerda C, and Utomo, B. W 2019 The mental workload analysis of staff in study program of private educational organization. In *IOP Conference Series: Materials Science and Engineering* Vol. 528, No. 1 p. 012018 IOP Publishing.

[6] Akbar H 2017 Analisis beban kerja terhadap pengemudi bus jurusan bandung-denpasar dengan menggunakan metode NASA-TLX (Studi Kasus di PT. PAHALA KENCANA Cabang Bandung).

[7] Putri, U and Handayani N 2017Analisis beban kerja mental dengan metode NASA TLX pada departemen logistik PT ABC. *Industrial Engineering Online Journal*, *6*(2)

[8] Widyanti A, Johnson A, and Waard, D 2010 Pengukuran beban kerja mental dalam searching task dengan metode rating scale mental effort (RSME). *J@ Ti Undip*, (1), 1-6

[9] Mudrajad K 2013 Metode riset untuk bisnis dan ekonomi edisi 4. *Erlangga, Jakarta*

[10] Hariyanti, N and Muslimah I 2017 *Perbandingan dan pengukuran beban kerja mental dengan pendekatan NASA-TLX (Studi Kasus: IKM Maria Jaya dan IKM Mukti Abadi)* Doctoral dissertation, Universitas Muhammadiyah Surakarta

[11] Sugiyono D 2001 Statistik non parametris untuk penelitian. *Bandung: Alfabeta*

IOP Conf. Series: Materials Science and Engineering **982** (2020) 012060 doi:10.1088/1757-899X/982/1/012060

Facility Layout Design with Corelap Algorithm for Educational Tour

Nayoko Prasetyo Jati[1], Apsari Dita Indah Rahayu[1], Shelly Elvina Salsabila[1] dan Abdullah 'Azzam[1]
[1]Department of Industrial Engineering, Faculty of Industrial Technology, Univrersitas Islam Indonesia, D. I. Yogyakarta

abdullah.azzam@uii.ac.id

Abstract. Yogyakarta has potential in the field of tourism. Many places can be chosen to be visited, including tourism education village. This village is interesting because not only a vacation spot, but we can also increase our knowledge there. Currently, Widodomartani villagers are planning to create a tourism education village inspired by the tourism education village in Kebumen. To make a tourism education village there must be good planning. The purpose of this study is to assist Widodomartani villagers in planning the facility layout of an educational village. The analysis that will be carried out includes making business processes, identifying facilities, and establish facility layouts that can be used to plan the layout of facilities. Layouting facilities are done by the Corelap algorithm based on the relationship between the facilities. Based on TCR (Total Closeness Rating) for each department, where TCR is the number of value calculated based on the rating of a systematic approach. Determination of the layout of facilities for 24 facilities is done using Activity Relationship Diagrams by considering the Total Closeness Rating between existing facilities as a basis for determining the layout of facilities. From the results of the closeness relationship, layout optimization is done using CORELAP (Computerized Relationship Layout Planning) software to get the optimal layout which is then actualized described in Widodomartani tourism education village land.

1. Introduction

Yogyakarta is one of the tourist destinations in Indonesia. The number of tourists visiting Yogyakarta has increased every year (Figure 1), in 2017 the number of tourists reached 5,229,298, with a proportion of 8% of foreign tourists and 92% of domestic tourists [1]. The high level of tourist visits has become an opportunity for the development of new tourist destinations, especially tourist destinations that offer uniqueness for tourists.

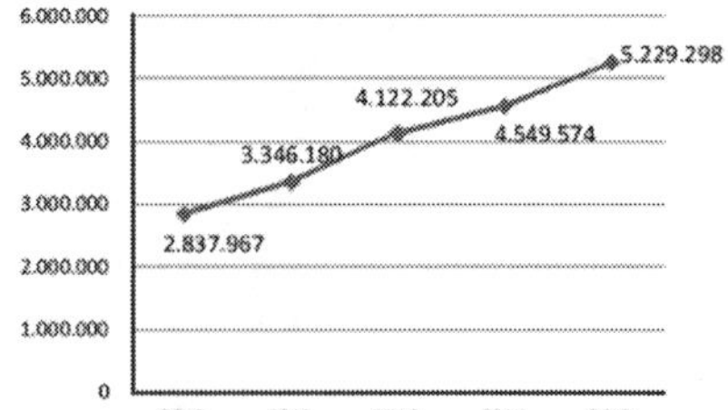

Figure 1. Tourist Visits in Yogyakarta

The tourism village is one of the developing tourist destinations because it offers uniqueness for tourists. Yogyakarta is a province that has a high level of success in developing tourist villages, where Sleman Regency has 19 Tourism Villages with a visit rate of 206,934 tourists in 2017 [1]. Each tourism village has its uniqueness, which distinguishes one village from another.

Widodomartani Village in 2015 formed a "Forum Kelompok Usaha Bersama (Kube)" which consisted of Kube representatives from each region. Furthermore, in 2018 "Forum Kube" Widodomartani Village, after visiting a tourist village in Kebumen, they decided to establish an educational destination in the form of "Tourism education" in Widodomartani Village. This idea requires well planning to be realized and requires collaboration with several agencies.

To initiate the idea, careful development planning is needed, one of which is planning the layout of facilities. It aims to maximize existing facilities seen from several aspects such as the distance between facilities, the closeness between facilities, etc. Because this is still an idea, a method of formation or commonly called Craft is used, which can be used to design a facility layout that has not yet been built. One of the supporting tools for the Craft method is to use the CORELAP application by adding activity relationship chart (ARC), activity relationship diagram (ARD), and area allocation diagram (AAD) based on field studies that will be conducted next. So by using the method of forming by using CORELAP, it is expected that the laying of the facilities to be built will be right with a long life span.

Based on the background above, it can be formulated that the problem that will be resolved is how to design an optimal layout of integrated agricultural and farming facilities using the CORELAP method. The purpose of this study is to produce an optimal layout of agricultural and animal husbandry educational facilities using the CORELAP method. In this study, the results of the layout are used as a basis for the construction of integrated animal husbandry and agriculture tours.

2. Literature Review

2.1. Tourism education Village

Criteria for a village can be developed into a tourist village, if it has several supporting factors, including; (1) has product potential and attractiveness, (2) has the support of human resources (HR), (3) strong motivation from the community, (4) has adequate infrastructure and support, (5) has facilities to support tourism activities, (6) has institutions that regulate tourism activities and (7) availability of land/areas that are possible to be developed into tourist destinations [2]

The model of an educational village for sustainable rural development must be creative and can prioritize the characteristics of a village. The formation of an educational village can build the economy of the people who live and can help build knowledge for visitors who come.

2.2. Facility Layout Design

The facility layout design is a technique and steps to change or improve the layout of an industry to be able to obtain layout changes to handle facilities and material handling so that the process activities are more optimal.

ARC is a technique for planning the layout of a facility or department based on the degree of activity relations of each facility or department. ARC Qualitative, using letter codes indicating the degree of activity relations and also numbers to explain the reasons for determining the letter code [3].

There are three main parts to the relationship between layout design activities that can be broken down as follows [4]:

- Identifying activities that have been defined as factory facilities.
- Prepare the ARC sheet and fill it in with the names of the facilities specified in step 1.
- Formulate reasons that can be used as a basis that facilities can be brought closer or must be kept away.

2.3. CORELAP

The CORELAP algorithm uses the proximity relationship rating stated in TCR in the selection of work station placement. This algorithm is a development algorithm (construction algorithm), which is an algorithm used to produce a new layout design that does not depend or does not require an initial layout. Based on TCR for each department, where TCR is the number of value calculated based on the rating of a systematic approach. [5]

CORELAP algorithm is a construction algorithm that can change qualitative data into quantitative data [6]. To determine the initial facilities that can be placed in a layout, data related to the activity relationship is needed. In research at PT. X also uses CORELAP algorithm and the results of the proposed layout are better as well as saving electricity [7]. CORELAP algorithm reduces the total distance difference on the production floor and reduces the total cost of moving material compared to the initial layout [8]. The application of the CORELAP algorithm is also used in the design of facility layouts at an oven manufacturing plant and an area savings of 8.83% is obtained [9].

According to (Tompkins, et al., 1996) there are CORELAP algorithm steps as follows:

- Calculate the TCR for each department.
  To calculate the TCR value, add up the weighting of the proximity values for each department, where the proximity values for each department are obtained from the ARC, and the FTC. From the ARC and FTC above, we can calculate TCR for each department from the TCR.
- Select one of the departments with maximum TCR, then place it first in the layout center.
- If there is the same TCR, first select the area that has a larger area then if the area is the same, then choose the department with the smallest number.
- Place the department with the link A, with the one chosen, then the linkages E, I, O, U, and X. If there are several of the same, the criteria are used the same as the previous step.
- If a department has been selected, determine its placement based on the Placing Rating, if the number of Weighted Closeness Ratings between departments that have entered with who will enter. If the Placing Rating is the same, then the length of the boundary or the number of square units adjacent to adjacent ones is compared.
- Layout evaluation through layout score:

$$\sum_{n} \textit{Closeness Rating Numeric x Shortest}\ \text{Track Length}$$

n = all the department

2.4. State of The Art

Based on the previous research [10], to obtain the layout of production facilities that have a minimum total transfer moment. In this study, using the BLOCPLAN (Black Layout Overview with Layout Planning) and CORELAP (Computerized Relationship Layout Planning) analysis was performed by comparing the total moment of movement between the actual layout and the layout of the proposal. Moment of displacement in the initial layout is 7,593,352 meters displacement/year. The results of the study found that the layout using the CORELAP algorithm was chosen as the layout of the proposal because it has a material transfer efficiency of 19.52% and a total moment of material transfer of 6,111,172 meters per year. While the layout using the BLOCPLAN algorithm has a moment of displacement of 7,449,682 meters/year displacement with material handling efficiency of 89%.

Subsequent research [11] is a company that processes plastic bottles into plastic pellets ready for processing. This processing requires several processes using machines that have been prearranged without specific calculations. As a result, the production process is hampered by seeing the condition of the movement of goods that pass through the pedestrian path and the top of employees who are working on the production machine. There is a need to change the layout of the new production facility so that it is expected to reduce the distance of material handling. So it is processed by CORELAP algorithm. The results of his research in the form of CV. Robbani, there are 2 material transfer flows from the raw material warehouse which is far apart, namely raw material warehouse 1 and raw material warehouse 2. So the distance of material handling warehouse raw material 1 is 27.79 m and distance of material handling warehouse raw material 2 is 31.57, so that the total length of material handling distance in the initial conditions is 59.36 m. Based on the Corelap Algorithm method, the Proposed Layout obtained with a distance of material transfer as far as 30.27 meters from the Initial Layout with a material transfer distance of 59.36 meters, the difference saves Material Handling distance by 49%.

The latest research [12] that is engaged in manufacturing rubber seal gripper manufacturers. This company has a problem with the layout of its production floor that is the occurrence of cross-movement and the distance at some stations is also too far which causes the flow of material disrupted. This problem can be solved by improving the layout of the production floor using the CORELAP and ALDEP methods then simulating it with Flexsim software. This study aims to design the layout of proposed facilities that can minimize material transfer distance by comparing the efficiency of the actual layout displacement moment with the proposed layout. The results of this study indicate a decrease in the total moment of displacement on the production floor of PT. Tourism education from 14,495.08 meters/month to 5930.19 meters/month using the CORELAP algorithm and 7,369.7 meters/month on the ALDEP algorithm. The distance efficiency in the proposed layout also increased from 53.67% to 93.74% in the CORELAP algorithm and 78.18% in the ALDEP algorithm. After doing a simulation to find the best method, the layout of the selected proposals is the layout of the results of the CORELAP algorithm with kilometers traveled per day 1.9 km/day.

3. Research Methodology

This research was conducted at the Widodo Martani tourism education located on Jalan Cangkringan, Umbul Martani village, Sleman, Yogyakarta Special Region. The implementation time was carried out in April 2020 until the beginning of May 2020. The formulation of the problem that was designed was the making of a good facility layout to turn Widodomartani Village into an educational village. Data collection is done by field study. Field studies were carried out in Widodomartani Village in the area where the Tourism education development was carried out. Observation of visitors flow and facility layout at similar Tourism education namely Sinatria Farm Tourism education Farm and Smesta Tourism education located in Pakem. Data processing is carried out to obtain business processes that will be applied to the Widodomartani Tourism education village and to search for secondary data to support research. After obtaining the right business processes, grouping the closeness of relations between each facility using the ARC (Activity Relationship Chart). After knowing the relationship between the facilities, layout planning is done with the CORELAP application. After processing the data, an analysis of the results has been obtained. The analysis is the distance between the buildings, the location of the building based on CORELAP, and the feasibility of the layout if applied to the place that has been prepared. Conclusions based on research that has been done. And provide suggestions for further development of subsequent research.

4. Result

4.1. Business Process

Widodomartani Tourism education Village is a newly developed tourism site that does not yet have a patent business process. So that researchers do business processes based on data obtained from surveys on tourism education sites that have been built and have business processes, that is Tourism education Sinatria and Tourism education Smesta.*Convergent Validity*

IOP Conf. Series: Materials Science and Engineering **982** (2020) 012060 doi:10.1088/1757-899X/982/1/012060

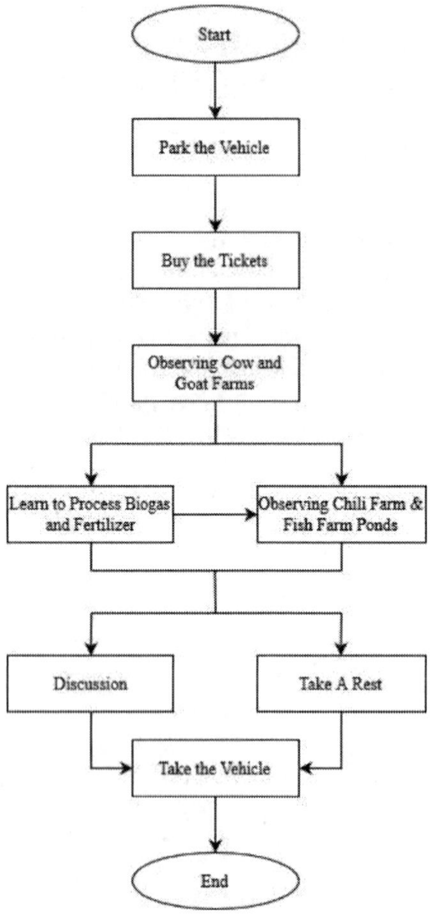

Figure 2. Research flow

Based on the business process data above, the next step is to determine what facilities can support the business process. So the researchers determined 24 types of facilities that would be used in the Widodomartani education and the area and amount needed, the determination of this data based on the needs taken from the field survey. Determination of what facilities to be built are as follows:

Table 1 Extensive Facilities

Facilities	Length	Width	Number of Facilities	Total Area
Security	3	3	1	9
Parking Park	30	18	1	540
Ticket office	3	2	1	6
Playground	10	18	1	180
Office	12	7	1	84
Toilet	2	2	1	4
Library	8	8	1	64
Mushola	5	6	1	30
Goat Pen	2	1,5	2	6

Facilities	**Length**	**Width**	**Number of Facilities**	**Total Area**
Cowshed	2	2	2	8
Cow Field	6	6	1	36
Goat Field	5	5	1	25
Feed Warehouse	2	2	1	4
Feed Land	4	4	1	16
Chili Farm	2	2	3	12
Fish Farm Pond	2	3	1	6
Biogas Processing	2	2	1	4
Gazebo	4	4	1	16
Canteen and Souvenir Store	4	3	1	12
Fertilizer Processing	3	3	1	9
Water Tower	1	1	1	1
Garbage Bank	2	2	1	4
Draw Well	1	1	1	1
Laboratory	3	4	1	12
Total Area				1089

4.2. Proposed Layout

Based on the research method described earlier, the initial data processing is by making an ARC, after which a numerical value is given for each proximity of the facility so that the TCR is obtained and continued with the allocation of TCR values to form a layout the new one.

4.2.1. ARC is based on existing facilities.

ARC application is obtained from facility data that will be linked in pairs to find out how close the two facilities are to be paired. Determination of the level of closeness between the facilities in terms of several aspects among them is the interrelationships between the production process, material flow, equipment used, humans, and others. In this study, the facilities used are based on table 1 above.

ARC is a map of the relationship of activities in the form of a rhombus, consisting of 2 parts. The upper part shows the symbol of the degree of interrelation between two departments/rooms, while the lower part is the reason used to measure the degree of interrelation [13]. Based on the degree of relationship between activities and their reasons, the ARC for 24 facilities at the Widodomartani Tourism education can be found in Figure 3. In the ARC it has been described the relationship between the pairs of facilities that are equipped with the degree of closeness A, E, I, O, U, and X along with the reasons for their closeness. The results of making this ARC are used to calculate the TCR. The following is an ARC for the entire production process and its supporting facilities, can be seen in Figure 3:

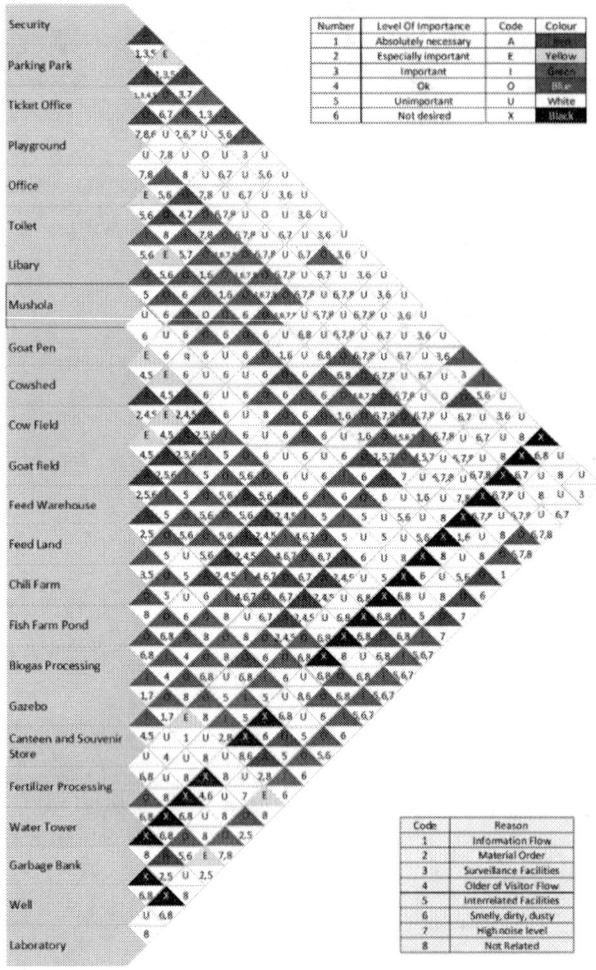

Figure 3. Activity Relationship Chart

4.2.2. TCR

The establishment of ARC is the basis for calculating TCR which will be used in allocating facilities. TCR calculation is done by converting each degree of closeness into a rating value as follows A = 5, E = 4, I = 3, O = 2, U = 1 and X = 0.

As for determining the awarding of value in each facility by looking at the interests of the paired facilities. The higher the level of closeness, the TCR value will be higher. Facilities that have a high TCR value will be used as the main reference placed in the layout allocation when computed using the next CORELAP application.

Based on table 2, the facility that has the largest TCR value is the Goat Umbaran facility, because the facility is a source of biogas processing. Followed by cow slap. After doing TCR calculations, it can be seen which facilities have the highest proximity value with other facilities and will be the main area for further analysis.

Table 1. TCR

Facilities	A	E	I	O	U	X	Total	TCR
	5	4	3	2	1	0		
Security	1	1	3	3	14	1	23	38

Table 1. TCR

Facilities	A	E	I	O	U	X	Total	TCR
	5	4	3	2	1	0		
Parking Park	2	0	0	4	16	1	23	34
Ticket office	1	1	0	1	19	1	23	30
Playground	0	0	2	14	6	1	23	40
Office	0	1	2	12	7	1	23	41
Toilet	0	2	2	12	6	1	23	44
Library	0	0	2	12	8	1	23	38
Mushola	0	1	3	3	15	1	23	34
Goat Pen	4	2	3	8	5	1	23	58
Cowshed	4	2	3	8	5	1	23	58
Cow Field	4	2	3	7	6	1	23	57
Goat Field	4	2	3	9	4	1	23	59
Feed Warehouse	6	0	2	5	10	0	23	56
Feed Land	1	0	6	7	9	0	23	46
Chili Farm	0	0	2	15	5	1	23	41
Fish Farm Pond	1	0	3	13	5	1	23	45
Biogas Processing	5	2	1	5	10	0	23	56
Gazebo	0	0	11	6	5	1	23	50
Canteen and Souvenir Store	0	0	4	13	5	1	23	43
Fertilizer Processing	4	2	0	4	12	1	23	48
Water Tower	1	0	3	3	15	1	23	35
Garbage Bank	0	0	0	0	3	20	23	3
Draw Well	2	0	0	6	14	1	23	36
Laboratory	0	2	6	9	5	1	23	49

4.2.3. Corelap Application

Data processing using the CORELAP method is done using the help of CORELAP 1.0 software. The input used is the area of the number of departments, the area of each department, and the ARC used to calculate TCR. In the use of the CORELAP method, only one proposal layout will be produced, which is the best layout. Here are departmental allocations using CORELAP Algorithm based on TCR values, Facility 1 would be the center of the place allocation.

- Iteration 1

  Location 1 = 1 x 5 = 5
  Location 2 = 0,5 x 5 = 2,5
  Location 3 = 1 x 5 = 5

IOP Conf. Series: Materials Science and Engineering **982** (2020) 012060 doi:10.1088/1757-899X/982/1/012060

Location 1 showed the highest score in the earliest sequence, so we can conclude that location 1 is the best location for Facility II.

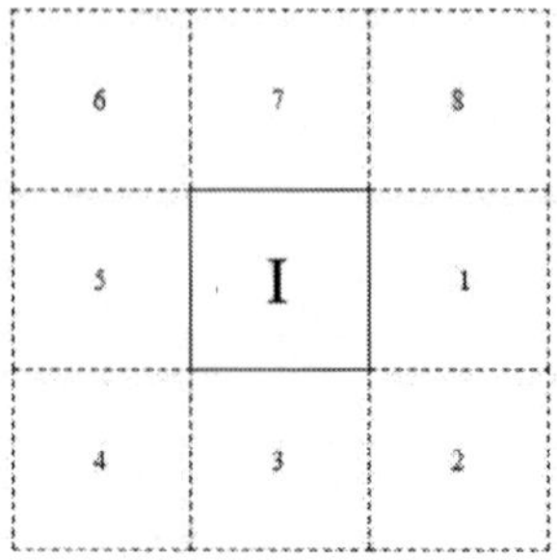

Figure 4. Iteration 1

- Iteration 2

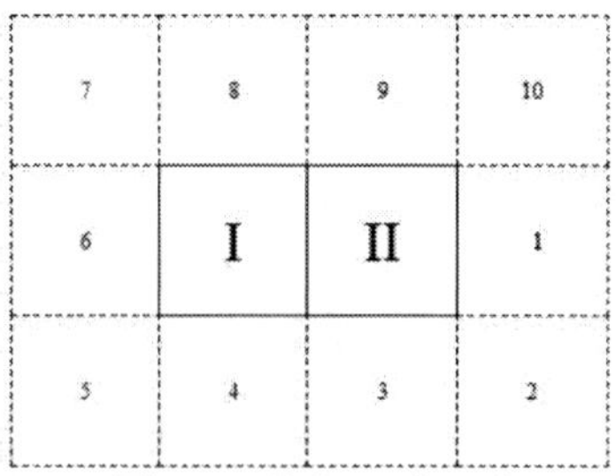

Figure 5. Iteration 2

Location 1 = 1 x 5 = 5
Location 2 = 0,5 x 5 = 2,5
Location 3 = (1 x 5) + (0,5 x 4) = 7

Location 3 showed the highest score in the earliest sequence, so we can conclude that Location 3 is the best location for facility III. This calculation continues until all facilities have been allocated. After the calculation is done with CORELAP, the final layout result that is produced is like Figure 5.

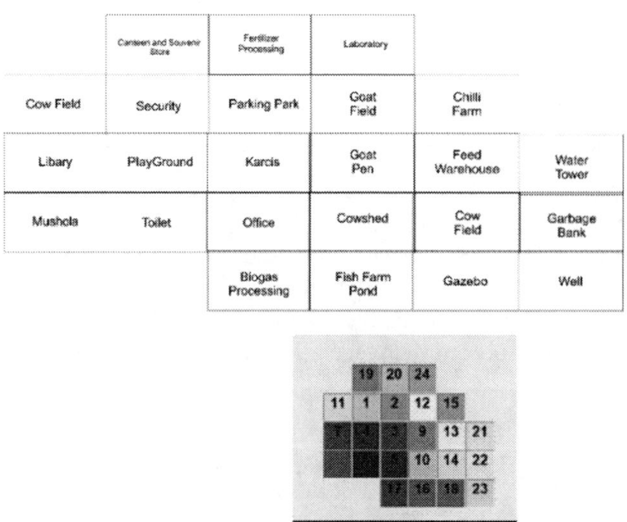

Figure 6. CORELAP Results Layout

4.3. Layout Overview

After getting the layout from the CORELAP method, the next step is to adjust the layout to the field conditions that will be used. On the Umbulmartani Tourism education land which has an area of 2,400 m2, several facilities have been built before doing this research, namely the Office and the mosque. This, of course, makes the layout that has been produced requires adjustments to the buildings that have been built, so with the above problems, a map with a ratio of 1: 500 is made as follows:

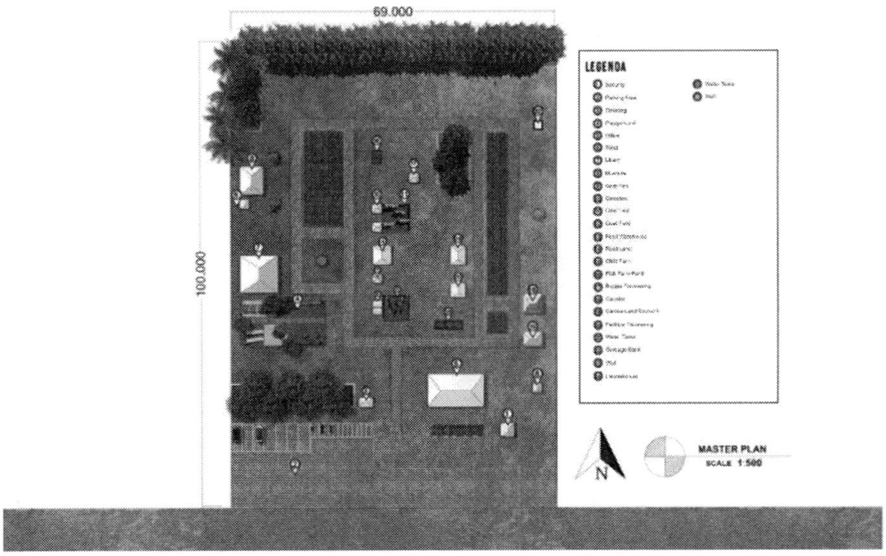

Figure 7. Layout Overview

5. Conclusion and Recommendation

5.1. Conclusion

The construction of the Widodomartani Tourism education village requires 24 kinds of facilities, the determination of this amount was obtained from the needs of the Widodomartani Tourism education village as well as the benchmarking process in several tourism education villages that had been surveyed. Thus, the total land area needed for development is 1089 m2, this area is the total facility area and does not include vacant land that can be used for other facilities. Determination of the layout of facilities for 24 facilities is done by considering the TCR between existing facilities as a basis for determining the layout of facilities. From the results of the closeness relationship, layout optimization is done using CORELAP software to get the optimal layout which is then actualized described in Widodomartani tourism education village land.

5.2. Acknowledge

The proposed layout generated in this study considers the level of closeness between facilities, to better describe the business processes that occur in the Widodomartani Tourism education village. Further business process analysis can be carried out which is then simulated to find out the effectiveness of the layout.

References

[1] Kepariwistaan S, *Yogyakarta BPS* Maret 2017 [Online] Available Yogyakarta.bps.go.id..
[2] Utomo J. S. *Strategi Pengembangan Desa Wisata Di Kecamatan Karangploso Kabupaten Malang* Neo.Bis. Vol. **11** Number 2 2017
[3] Pamulasih T. *Usulan Rancangan Tata Letak Fasilitas dengan Menggunakan Metode Automated Layout Design Program (ALDEP) di Edem Ceramic* 2017

[4] Safitri D. *Analisis perancangan tataletak fasilitas produksi menggunakan metode activity relationship chart (ARC)* pp. 38-47 2017

[5] Rahmadani W. I. *Perancangan Ulang Tata Letak Gudang Menggunakan Metode Konvensional, Corelap dan Simulasi Promodel* Jurnal Optimasi Teknik Industri 2020

[6] Dannang and Dalliya *Improvement of Production Facility Layout of Fried Soybean using BLOCPLAN and CORELAP Method (A Case Study in UKM MMM Gading Kulon,Malang)* Jurnal Teknologi and Manajemen Agroindustri 2017

[7] Agustina Maukar and Dewi *Perencanaan Produksi dan Perbaikan Tata Letak di PT. Berkat Anugrah Alam Cemerlang* Widya Teknik 2007

[8] Fitriani A. Azis G. P. and Mulyono A. *Facility Layout Improvement: Based on Safety and Health at Work* 2015

[9] Rajesh N. K. *Plant Layout Optimization of Oven Manufacturing UNIT Using CORELAP Algorithm* International Journal of Current Engineering and Technology 2016

[10] Siregar R. M. Sukatendel D. and Tarigan U. *Perancangan Ulang Tataletak Fasilitas Produksi Dengan Menerapkan Algoritma Blocplan Dan Algoritma Corelap Pada PT. Xyz* Jurnal Teknik Industri USU 1 2013

[11] Putra Y. P. *Meracang Tata Letak Fasilitas Pabrik dengan Metode Algoritma Corelap di CV. Robbani Singosari* Jurnal Valtech 1 num. 1 pp. 65-70 2018

[12] Tarigan U. Simbolon R. Sembiring M. Tarigan T. U. P. P. Sembiring N. and Tarigan I. R. *Perancangan Ulang dan Simulasi Tata Letak Fasilitas Produksi Gripper Rubber Seal dengan Menggunakan Algoritma Corelap, Aldep, dan Flexsim* Jurnal Sistem Teknik Industri 21 num. 1 2019

[13] Siregar R. M. Sukatendel D. and Tarigan U. *Perancangan Ulang Tataletak Fasilitas Produksi Dengan Menerapkan Algoritma Blocplan Dan Algoritma Corelap Pada PT. Xyz* Jurnal Teknik Industri USU 1 2013

IOP Conf. Series: Materials Science and Engineering **982** (2020) 012061 doi:10.1088/1757-899X/982/1/012061

African Buffalo Optimization for Solving Flow Shop Scheduling Problem to Minimize Makespan

M Krisnawati[1,3], A A Sibarani[1], A Mustikasari[2], D Aulia[1]

[1]Department of Industrial Engineering, Universitas Jenderal Soedirman, Central Java, Indonesia

[2]Department of Management, Universitas Negeri Yogyakarta, Yogyakarta, Indonesia

[3]maria.krisnawati@unsoed.ac.id

Abstract. In this paper, African Buffalo Optimization is proposed to solve the flow-shop Scheduling Problem (FSP). The FSP involves n-job that was processed in m-machine. The aim is to reduce the makespan of the whole process. In this study, we also compare the African Buffalo Optimization with an exact solution and other meta-heuristic methods, such as Particle Swarm Optimization (PSO), Hybrid Genetics Algorithm, and also Crow-search Algorithm (CSA) to know the performance of the methods in solution quality and computational time. Friedman test showing African Buffalo Optimization gives an optimal solution in solution quality, the same result with the exact solution, PSO, and hybrid GA.

1. Introduction

In the manufacturing and service industries, scheduling becomes an important role in the decision-making process. Also, scheduling and effective sequencing become necessary in surviving in those industries [1]. In the scope of deterministic machine scheduling, optimization, approximation algorithms, and complexity results are used to overcome problems that involving parallel or single machine, job shop, open shops, and flow shops [2].

Flow shop is a type of scheduling in which the duty sequence of each job is set and every job comes to the work machines or stations for the same processing order [3]. In the Flow-shop scheduling problem (FSP) m different machines $\{M_1,..., M_m\}$ must process n independent jobs $\{J_1,...,J_n\}$. Each job is consists of m operations where each operation needs a distinct machine. The operation on the machine i of job j is denoted O_{ij}. Each operation also needs processing time p_{ij} [4].

Meta-heuristic and heuristic is also kind of exact methods that are used to solve FSP. The exact method gives the optimal problem solution. Yet, the exact method becomes impractical for large flow shop problems because of the complexity raises. Large size FSPs are included in the NP-hard problem class. Hence, heuristic and meta-heuristics methods are used to overcome this kind of problem [5]. Meta-heuristics methods are a set of strategies to increase the heuristic procedures efficiency [6]. Meta-heuristics could give a better result that comes from heuristics at first. It also has an objective to give efficient and detailed optimized output for an optimized schedule of jobs, which decreases the computational time [7].

Many meta-heuristics methods have been used by researchers to solve FSP, such as Iterated Greedy Algorithms [8]–[10], and Hybrid Genetic Algorithm [11], [12]. Other methods for solving FSP are Hybrid Ant Colony algorithm (ACO) [13], Cuckoo Search algorithm [14], Particle Swarm Optimization (PSO) [15]–[17] and Crow search algorithm (CSA) [18]. But, our research proposed to

Improve African Buffalo Optimization (IABO) for solving the flow shop scheduling problem. African Buffalo Optimization (ABO) was introduced by [19], this algorithm shows better performance on the Traveling Salesman Problem [20], Asymmetric TSP [21] and the design of PID Controller in Automatic voltage regulator system [22], when compared with SA, GA, PSO, ACO, and Randomized Insertion Algorithms. BFO has also been used by [23] in choosing a biodiversity conservation scope with a budget that is limited and [24] in the Low carbon Flexible job shop scheduling problem.

In our research, we use the basic ABO method that has been promoted by [25]. We compared IABO with an exact solution and other meta-heuristics methods, such as PSO, CSA, and Hybrid Genetic Algorithm, to find out the performance of the proposed method in solving the Flow shop scheduling problem. The choice of that methods is because PSO, CSA, Hybrid Cross entropy, and Genetic Algorithm have better performance better results in terms of time calculation and optimum solution compare with other algorithms for solving scheduling problems [12], [17], [18].

2. Research Methodology

2.1. Data collecting.

Data is the real problem of a muffler industry in Purbalingga. Muffler industry which is the object produces several types of car muffler products and motorcycle muffler (n-products) that are processed on the m-machine. The process of production of mufflers was done semi-manually, with some processes using the machine of production [25]. The process of production includes rolling plates with beatings, cutting plates, cutting with cutting grinders, welding, assembly, and finishing. The data needed is the processing time of each job on each machine and the number of jobs. Data is collected in 24 days for the type and number of jobs. Retrieval of data at the beginning of the processing time is repeated 10 times to get the standard time in each process. In calculating the standard time, the adequacy test and uniformity test are carried out first. Retrieval of processing time data is taken if the data are not sufficient. The standard time calculated will be the job processing time on each machine used in the experiment.

2.2. Data Processing

2.2.1 Flow Shop Scheduling Model. This is a flow shop scheduling model to minimize makespan. Objective function:

$$\text{Min}\left(\sum_{i=1}^{n}\sum_{j=1}^{m}\sum_{p=1}^{k} Y_{(j,p)} \times W_{(i,p)}\right) \quad (1)$$

Subject to:

$$\sum_{p=1}^{k} Y_{(j,p)} = 1, \quad j\forall 1,2,\dots,m \quad (2)$$
$$\sum_{j=1}^{m} Y_{(j,p)} = 1, \quad p\forall 1,2,\dots,k \quad (3)$$
$$C_{(1,1)} = \sum_{j=1}^{n} W_{(1,j)} \times Y_{(j,1)} \quad (4)$$
$$C_{(1,p)} \geq C_{(1,p-1)} + \sum_{j=1}^{n} W_{(1,j)} \times Y_{(j,p)} \quad (5)$$
$$C_{(i,1)} \geq C_{(i-1,1)} + \sum_{j=1}^{n} W_{(i,j)} \times Y_{(j,p)} \quad (6)$$
$$C_{(i,p)} \geq C_{(i-1,p)} + \sum_{j=1}^{n} W_{(i,j)} \times Y_{(j,p)} \quad (7)$$
$$C_{(i,p)} \geq C_{(i,p-1)} + \sum_{j=1}^{n} W_{(i,j)} \times Y_{(j,p)} \quad (8)$$
$$Y_{(j,p)}\forall(0,1) \quad (9)$$

Our model has a binary solution with Eq. 1 is our objective function to minimize makespan. The value of $Y_{(j,\ p)}$ is the decision variable of the model, where the value of $Y_{(j,\ p)}$ is 1 if job j is assigned to position p. Whereas $W_{(i,\ j)}$ is the time processing of job j on machine i. Equation (2) until (9) are constraints of the model. Equation (2) ensures that only 1 job is processed at each position p and equation (3) ensures that each position gets a job. Equation (4) ensures job in position p cannot start until a job in position p-1 finishes. Equation (5) ensures jobs on machine 1 have no predecessors. Equation (6) ensures a job in position on the machine i cannot start until it finishes on i-1. Equation (7) ensures job in position p cannot start on m until the previous job finishes on i. Equation (8) ensures a job in position on the machine i cannot start until it finishes on i-1. And equation (9) shows $Y_{(j, p)}$ has a value of 0 or 1.

2.2.2 Proposed Algorithm. African Buffalo Optimization (ABO) is included in swarm intelligence optimization. This algorithm emulated migration from African Buffalo in search of lush grazing land. [26] These are the two basic sounds of the African Buffalos with which they are able to organize themselves to search for food and defend themselves whenever they are attacked. The "maaa" sound tells the buffalos to stay on to exploit their environment since it is safe and has sufficient pastures and "waaa" sounds used to mobilize the buffalos to move on to explore the search space. Each African buffalo will choose the best position in relation to the target solution and compare it with the previous position that has been stored in memory. The animals move to explore other locations depending on the democratic equation (eq.11) [25]. The ABO algorithm for flow shop scheduling problems in this study is as follows:

1. Initialization: randomly place buffalos to nodes at the solution space by generating a random number (r0) of m x k for the solution of matrix Y a number of N ($Y = [Y_1, Y_2, \ldots, Y_N]^T$). For the FSP problem in our research, solution taken binary value, where 1 denotes job i processed in position k, 0 otherwise. How to transform fraction into zero-one value can be done by :
$$Y_{ik} = \begin{cases} 1, if\ r0\ \leq p0 \\ 0, otherwise \end{cases} \quad (9)$$
Where p0 is the probability of success in the initial matrix [27].
2. Update the buffalo's fitness value using equation (10)
$$w_{t+1} = w_t + lp_1 r_1 (bg_{max} - m_t) + lp_2 r_2 (bg_{max,t} - m_t) \quad (10)$$
where m_t and w_t describe exploitation and exploration during iteration t. Moves respectively from b^{th} buffalo (b = 1, 2, ..., N), depending on lp_1 and lp_2 which are learning factors and r_1 and r_2 are random numbers between [0,1]. bg_{max} is the best fitness and bp_{max} is the best position for individual buffalos.
3. Update the location buffalo b ($bp_{max,b}$ dan $bg_{max,b}$) using equation (11)
$$m_{t+1} = \lambda(w_t + m_t) \text{ where } \lambda \text{ is a unit time} \quad (11)$$
4. Judge whether bg_{max} is updating. If yes, go to step 5, otherwise, go to step 1
5. If the termination condition is met. If yes, go to step 6, otherwise, go to step 2
6. End the procedure. Output best solution

2.2.3 Experiment. The optimization model was solved using the proposed algorithm (ABO). Here, we applied the model to the real case of muffler production. To assess the performance of the proposed algorithm, we divided data into 3 types, which is 8 days 16 days and 24 days. Experiment with other meta-heuristics methods using parameters as shown in Table 1. To achieve an optimal solution, ABO uses the parameters lp_1 0.4 and lp_2 0.2, PSO uses the parameters of the value range inertia $\rho_{max} =$ 0.9 and $\rho_{min} = 0.4$, CSA uses the parameters of awareness probability $AP = 0.1$ and flight length $fl =$ 0.2. The third metaheuristics method using 2 parameters setting, but hybrid GA uses 3 parameters that are smoothing coefficient $\alpha = 0.9$, mutation parameter $c_m = 0.1$ and cross over parameter $c_r = 0.8$. Experiment with African Based Optimization algorithm and other meta-heuristic methods using population size 100 and the number of iteration 40. To ensure proposed algorithm performance and to prove its ability in FSP, this section compares the solution of the proposed method (ABO) with several well-known meta-heuristics methods: Hybrid GA, PSO, and CSA. We also give a statistical test for experimental results and computational time. In this research, the proposed method (ABO) also will be

IOP Conf. Series: Materials Science and Engineering **982** (2020) 012061 doi:10.1088/1757-899X/982/1/012061

compared with the exact solution from LINGO. The comparison is done with the Friedman test statistical test.

3. Result and Discussion

In 24 days, a muffler production in Purbalingga has 47 jobs that processed in 6 machine departments, cutting plates, rolling plates with beatings, cutting with cutting grinders, assembly, welding, and finishing. This is the number of data set:

Table 1. The number of data set

No.	Data Type	Number of Job
1	Type 1 (8 days)	17
2	Type 2 (16 days)	39
3	Type 3 (24 days)	47

Experiments were carried out for the exact method with LINGO, ABO, Hybrid GA, PSO, and CSA. Table 2 shown the experiment results in the best minimum makespan of 30 times models testing, and the average computational time is shown in Table 3.

Table 2. Minimize makespan (in a second)

Data type	Exact solution	ABO	Hybrid GA	PSO	CSA
Type 1	200530.5	200530.5	200530.5	200530.5	201498.5
Type 2	447510.4	447510.4	447510.4	447510.4	447510.4
Type 3	530493.7	530493.7	530493.7	530493.7	530493.7

Table 3. Computational time (in a second)

Data type	ABO	Hybrid GA	PSO	CSA
Type 1	67.059	13.469	2.511	2.575
Type 2	637.551	27.710	7.506	11.951
Type 3	1087.717	28.379	10.236	15.209

To find out the performance of ABO, we will do Friedman test on experiment result. Table 4 shows the results of the Friedman test on the objective function (makespan) and the computational time.

Table 4. Friedman test for makespan and computational time

Method	Makespan	Computational time
Mean rank		
Exact Solution	2.83	-
ABO	2.83	4.00
Hybrid GA	2.83	3.00
PSO	2.83	1.00
CSA	3.67	2.00
N	3	3
Chi-Square	4.000	9.000
df	4	3
Asymp. Sig.	0.406	0.029

Friedman test result has shown a significant difference in the solutions produced by ABO and comparison methods. ABO has the optimal solution with the same mean rank as the exact solution,

Hybrid GA and PSO. But in computational time, Friedman test showed ABO has the longest computational time.

4. Conclusions

For solving the Flow Shop Scheduling Problem (FSP) we can use the exact method, heuristic or meta-heuristic method. African Buffalo Optimization (ABO) algorithm has shown better results in minimizing makespan than the Crow search algorithm (CSA). ABO will be an alternative for solving FSP, but the computation time of ABO is slower than other meta-heuristic methods. So for the future, the models can be extended in the NP-hard problem. But to speed up the calculation time, it is necessary to improve the ABO algorithm, for example by hybridizing with other meta-heuristics methods or improvement in the ABO step.

References

[1] Pinedo M L 2016 *Scheduling Theory, Algorithms, and Systems*. Springer Berlin Heidelberg
[2] Lawler E L, Lenstra J K, Rinnooy Kan A H G, and Shmoys D B 1989 *Sequencing and Scheduling: Algorithms and Complexity* Handbooks Oper. Res. Manag. Sci. vol 4 no C pp 445–522
[3] Hamilton E and Vairaktarakis G 2012 *Flow Shop Scheduling Theoretical Results, Algorithms, and Applications* vol 182. Springer Science & Business Media
[4] Ruiz R, Maroto C, and Alcaraz J 2006 *Two New Robust Genetic Algorithms for The Flowshop Scheduling Problem* Omega vol 34 no 5 pp 461–476
[5] Yenisey M M and Yagmahan B 2014 *Multi-Objective Permutation Flow Shop Scheduling Problem: Literature Review, Classification and Current Trends* Omega (United Kingdom) vol 45 pp 119–135
[6] Tian Z and Fong S 2016 *Survey of Meta-Heuristic Algorithms for Deep Learning Training Optim* Algorithms - Methods Appl
[7] Arora D and Agarwal G 2016 *Meta-Heuristic Approaches for Flowshop Scheduling Problems : A Review* Int. J. Adv. Oper. Manag vol 8 no 1 pp 1–16
[8] Tasgetiren M F, Kizilay D, Pan Q K, and Suganthan P N 2017 *Iterated Greedy Algorithms for The Blocking Flowshop Scheduling Problem with Makespan Criterion* Comput. Oper. Res. vol 77 pp 111–126
[9] Shao W, Pi D, and Shao Z 2017 *Optimization of Makespan for The Distributed No-Wait Flow Shop Scheduling Problem with Iterated Greedy Algorithms* Knowledge-Based Syst. vol 137 pp 163–181
[10] Ribas I, Companys R, and Tort-Martorell X 2019 *An Iterated Greedy Algorithm for Solving The Total Tardiness Parallel Blocking Flow Shop Scheduling Problem* Expert Syst. Appl. vol 121 pp 347–361
[11] Utama D M, Ardiansyah L R, and Garside A K 2019 *Penjadwalan Flow Shop untuk Meminimasi Total Tardiness Menggunakan Algoritma Cross Entropy–Algoritma Genetika* J. Optimasi Sist. Ind. vol 18 no 2 p. 133
[12] Widodo D S and Kamaliyah N L 2017 *Pengembangan Cross Entropy-Genetic Algorithm (Cega) Pada Penjadwalan Model Flow Shop Untuk Meminimalkan Makespan* Teknoterap, no 1 pp 1–11
[13] Engin O and Güçlü A 2018 *A New Hybrid Ant Colony Optimization Algorithm for Solving The No-Wait Flow Shop Scheduling Problems* Appl. Soft Comput. J. vol 72 pp 166–176, 2018
[14] Wang H, Wang W, Sun H, Cui Z, Rahnamayan S, and Zeng S 2017 *A New Cuckoo Search Algorithm with Hybrid Strategies for Flow Shop Scheduling Problems* Soft Comput. vol 21 no 15 pp 4297–4307
[15] Tseng C T and Liao C J 2008 *A Particle Swarm Optimization Algorithm for Hybrid Flow-Shop Scheduling with Multiprocessor Tasks* Int. J. Prod. Res. vol 46 no 17 pp 4655–4670

[16] Allahverdi A and Al-Anzi F S 2006 *A PSO And A Tabu Search Heuristics for The Assembly Scheduling Problem of The Two-Stage Distributed Database Application* Comput. Oper. Res. vol 33 no 4 pp 1056–1080
[17] Singh M R and Mahapatra S S 2012 *A Swarm Optimization Approach for Flexible Flow Shop Scheduling with Multiprocessor Tasks* Int. J. Adv. Manuf. Technol. vol 62 no 1–4 pp 267–277
[18] Adhi A, Santosa B, and Siswanto N 2018 *A Meta-Heuristic Method for Solving Scheduling Problem: Crow Search Algorithm* IOP Conf. Ser. Mater. Sci. Eng. vol 337 no 1
[19] Odili J B and Kahar M N M 2015 *African Buffalo Optimization (ABO): A New Meta-Heuristic Algorithm* Journal of Advanced & Applied Sciences 3(03) pp 101-106
[20] Odili J B and Kahar M N M 2016 *Solving Traveling Salesman's Problem Using African Buffalo Optimization* Comput. Intell. Neurosci pp 1-12
[21] Odili J B, Kahar M N M, Noraziah A, and Odili E A 2016 *African Buffalo Optimization and The Randomized Insertion Algorithm for The Asymmetric Travelling Salesman's Problems* J. Theor. Appl. Inf. Technol. vol 87 no 3 pp 356–364
[22] Odili J B and Kahar M N M 2016 *African Buffalo Optimization Approach to The Design of PID Controller in Automatic Voltage Regulator System* Natl. Conf. Postgrad. Res. Univ. Malaysia Pahang pp 641–648
[23] Almonacid B, Reyes-Hagemann J, Campos-Nazer J, and Ramos-Aguilar J 2018 *Selecting a Biodiversity Conservation Area with A Limited Budget Using The Binary African Buffalo Optimisation Algorithm* IET Softw. vol 12 no 2 pp 96–111
[24] Zhu H and Jiang T 2020 *Discrete African Buffalo Optimization Algorithm for the Low-carbon Flexible Job Shop Scheduling Problem* J. Adv. Manuf. Syst.
[25] Odili J and Kahar M 2015 *Numerical Function Optimization Solutions Using the African Buffalo Optimization Algorithm (ABO)* Br. J. Math. Comput. Sci. vol 10 no 1 pp 1–12
[26] Jiang T, Zhu H, Deng G 2020 *Improved African Buffalo Optimization Algorithm for The Green Flexible Job Shop Scheduling Problem Considering Energy Consumption* J. Intell. Fuzzy Syst. vol 38 no 4 pp 4573–4589, 2020
[27] Krisnawati M, Santosa B, and Rusdiansyah A 2011 *Comparison of Cross Entropy And Differential Evolution To Solve Crew Rostering Problem* Pros. Semin. Nas. Manaj. Teknol. XIV

3. Model Development

In this paper, disruption results in changes in customer preferences so that demand on each channel also changes. Therefore, we develop the demand portions in the retail channel where demand portions channel is depending on customer preferences in the channel. The portion of customers who buy products in the retail channel, according to [14] and [15] is :

$$\theta = \frac{E(X_1 + X_2 + \cdots + X_{a_m})}{a_m} = \frac{\sum_{j=1}^{a_m} P(p_r - p_d \leq \xi_j)}{a_m} = \frac{a_m\, P(p_r - p_d \leq \xi_j)}{a_m} = P(p_r - p_d \leq \xi_j) \quad (1)$$

Parameter ξ_j describe the product brand loyalty. If the value of $p_r - p_d < \xi_j$customer will consider buying the product in the retail channel, but if the value of $p_r - p_d \geq \xi_j$ Customers will consider buying the product in an online channel.

Because $\xi_j$is uniformly distributed on [a,b] in which $-1 \leq a \leq 0$ and $0 \leq b \leq 1$, therefore

$$\theta = \begin{cases} 1, & p_r - p_d \leq a \\ \frac{b-(p_r-p_d)}{b-a}, & a < p_r - p_d < b \\ 0, & p_r - p_d \geq b \end{cases} \quad (2)$$

The customer only can buy the product in either retail channel or online channel in the system, therefore the consumer portion in retail channel become:

$$\theta = \frac{b-(p_r-p_d)}{b-a} \quad (3)$$

The customer portion then develops by considering disruptions that affect customer preferences. The disruptions parameter using parameter by [16]. The disruption probability denoted by $\emptyset$ and the disruption proportions denoted by λ. According to [13], disruption can be caused by various reasons such as a breakdown in channels, customers rushing to e-channel (and therefore sales website going out of reach due to overload), legal impositions, terrorist attacks, and so forth, channels may go disrupted. Therefore, the number of demands in the retail and online channels will also change. The difference between initial customer portions and customer portions affected with disruption will be the new value of customer portions. Therefore, the customer portions with disruptions model in the retail channel:

$$\theta_a = \theta_o - \left(\left(\frac{b-(p_r-p_d)}{b-a}\right)\lambda\emptyset\right) \quad (4)$$

The next step is to formulate the demand function in the DCSC system. The demand model consists of two demand model. First is the traditional retail channel demand model and online channel demand model. The demand model reference using a demand model by [17]. Each demand function will be explained below:

3.1. Demand Model in Retail Channel

$$D_r = \theta_a D_{max} - \alpha_1 p_r + \beta_1 p_o \quad (5)$$

$$D_r = \left(\theta_o - \left(\left(\frac{b-(p_r-p_o)}{b-a}\right)\lambda\emptyset\right)\right) D_{max} - \alpha_1 p_r + \beta_1 p_o \quad (6)$$

3.2. Demand Model in Online Channel

$$D_o = (1-\theta_a)D_{max} - \alpha_2 p_o + \beta_2 p_r \quad (7)$$

$$D_o = \left(1-\left(\theta_o - \left(\frac{b-(p_r-p_d)}{b-a}\right)\lambda\emptyset\right)\right)D_{max} - \alpha_2 p_o + \beta_2 p_r \quad (8)$$

In the function (5), (6), (7), and (8) the retail channel and online channel denoted by subscript r and o, respectively. The demand function (5) is the retail channel demand function, and (7) is the online channel demand function in DCSC structure. This model is already considering the customer portions and the interplay between retailers and online channels. The share demand or demand portions in the retail channel is denoted by θ. Customer portions with disruption (4) were then substituted in (6) and (8). Therefore, demand function (6) and (8) is a demand function that already considers channel disruption in which the disruptions affect the customer preferences and customer portions. D_{max} is a potential customer if the product is free of charge. The share demand or demand portions for a retail channel is denoted by θ_a, and the rest $1-\theta_a$ goes to the online channel, when p_s and p_o are zeros. α_1 and α_2 are the coefficients of the self-price elasticity of D_r and D_o, respectively. The cross-price sensitivities $\beta_1$and β_2 reflect the degree to which the products sold via the two channels are substitutes.

4. Numerical Example

We present some numerical examples to illustrate the effect of disruption on customers' preferences. When the customers' preferences changes, the demand portions each channel will changes too so that demand in each channel will be affected too. The initial parameter is used in-demand model using parameter by [14], [15], and [16]. The initial parameters are D_{max} = 600, $a = -0.2$, b = 0.2, β_1=0.35, β_2=0.75, $\alpha_1 = 2$, $\alpha_2 = 2.5$, λ = 0.3, ∅ = 0.5, $p_r = 0.772$, $p_o = 0.701$. Numerical example to find the optimum demand by maximizing the demand for both channels with considering the dual channel system. In this numerical example, optimization is carried out separately between offline channel and online channel. The initial solution is $p_r = 0.972$ with demand total in retail channel $D_r = 359$ and for online channel $p_r = 0.901$ with demand total in online channel $D_o = 329$.

The relationship of customers' demands in DCSC and customers demand portions according to customers preferences after channel disruptions shown in Figure 2 and Figure 3,

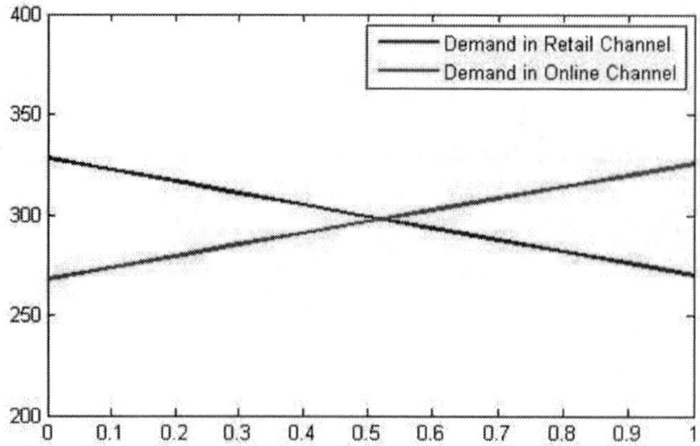

Figure 2. Comparison between demand in online channel and demand

From figure 2 we can see that there is a trade-off between demand in online channel and demand in the offline channel when $\emptyset = \pm 0.5$. When the probability of disruption below ±0.5, the retail channel generates higher demand than the online channel. But when the probability of disruption is high, more than ±0.5, online channel generates higher demand than the retail channel.

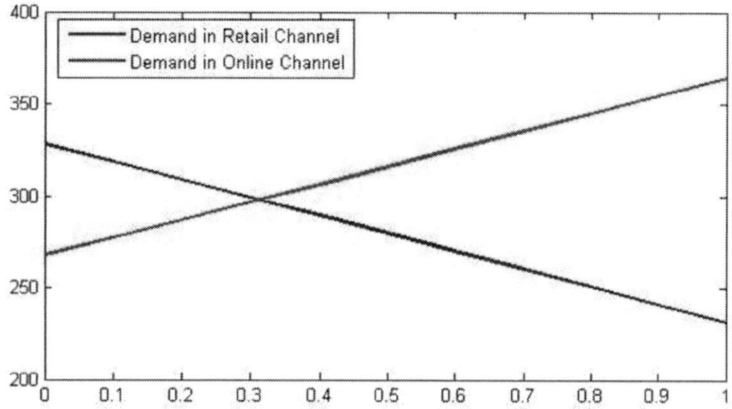

Figure 3. Comparison between demand in online channel and demand

From figure 3, we can see that there is a trade-off between demand in online channel and demand in the offline channel when $\lambda = \pm 0.3$. When disruption proportion below ±0.3, the retail channel generates higher demand than the online channel. But when disruption proportion is high, more than ±0.3, online channel generates higher demand than the retail channel.

5. Conclusion

This paper is considered a dual-channel supply chain system with a single product, one manufacturer, and one independent retailer. The manufacturer owns and manages an online channel. Customers can purchase the product either from the retail channel or online channel, according to their preferences. We study the condition when channel disruption happens, customers will change their preferences depends on disruptions probability and proportions. Therefore, the demand in the retail channel and online channel will change. From a numerical example, it is shown that the retail channel will be lost demand more than an online channel if the probability and proportions of disruption high.

For further studies, this model needs to develop further so that we can see the effect of channel disruption in customer preferences in the retail channel and online channel profit. Secondly, this paper can consider a different scenario in maximizing profit each channel and DCSC profit as a whole system. The retail and online channels can choose the most appropriate strategy in their channel.

References

[1] Saha S, S P Sarmah, and N M. Modak 2017 *Asia Pacific Manag. Rev.* **23** 148-160
[2] Chun S H, B D Rhee, S Y Park, and J C Kim 2011 *Int. Rev. Econ. Financ.* **20** 812-825
[3] Takahashi K, T. Aoi, D. Hirotani, and K. Morikawa 2011 *International Journal of Production Economics* **133** 403-415
[4] Chen J, H. Zhang, and Y. Sun 2012 *Omega* **40** 571-583.
[5] Tsay A A and N Agrawal 2004 *Prod. Oper. Manag.* **13** 93-110
[6] Goldkhul L, "Multiple Marketing Channel Conflict with a Focus on the Internet," *Thesis*, 2005.
[7] Craighead C W, J. Blackhurst, M. J. Rungtusanatham, and R. B. Handfield 2007 *Decis. Sci.* **38** 131-156
[8] Blos M F, R. M. Da Silva, and P. E. Miyagi 2015 *IFAC-PapersOnLine.* **48** 640-645

[9] Li B, M. Zhu, Y. Jiang, and Z. Li 2016 *J. Clean. Prod.* **112** 2029-2042
[10] Huang H, Y He, and D. Li 2018 *J. Clean. Prod.* **180** 280-296
[11] Chen K Y, M Kaya, and Ö. Özer 2008 *Manuf. Serv. Oper. Manag.* **10** 563-691.
[12] Ramanathan R 2011 *Int. J. Prod. Econ.* **130** 255-261
[13] Aslani A and J Heydari 2019 *J. Clean. Prod.* **223** 596-609.
[14] Khouja M, S. Park, and G. (George) Cai 2010 *Int. J. Prod. Econ* **125** 84-95
[15] ZHOU Y, H. FAN, and Y. LI 2012 *Int. J. u- e- Serv. Sci. Technol.* **9** 49-58.
[16] Pi Z, W. Fang, and B. Zhang 2019 *Comput. Ind. Eng.* **138** 106130.
[17] Huang S, C. Yang, and X. Zhang 2012 *Comput. Ind. Eng.* **62** 70-83

IOP Conf. Series: Materials Science and Engineering **982** (2020) 012063 doi:10.1088/1757-899X/982/1/012063

The Contributions of Artificial Intelligence in Achieving Sustainable Development Goals: Indonesia Case

Arwin Datumaya Wahyudi Sumari[1,2,3]

[1]Abdulrachman Air Force Base, 2nd Operation Command, Indonesian Air Force, Malang, East Java

[2]Cognitive Artificial Intelligence Research Group (CAIRG), Department of Electrical Engineering, State Polytechnic of Malang, Jl. Soekarno-Hatta No. 9, Malang 65141, East Java

[3]Faculty of Defense Technology, Indonesia Defense University, Indonesia Peace and Security Center (IPSC), Sentul, Bogor, West Java

*arwin.sumari@tni-au.mil.id, arwin.sumari@polinema.ac.id

Abstract. Sustainable Development Goals (SDG) have been adopted by all United Nations Member states in 2015. There are 17 Global Goals where one goal to another is interrelated and they have to go hand-in-hand. The essential of the goals is to increase the people's quality of life all over the world. Every country has been pacing to achieve all goals by conscripting all national resources as well as all capabilities and masteries of the technology such as Artificial Intelligence (AI). To achieve SDG's goals, there has to be a guidance that guide all resources to have the same perspective and mind-set to what should have to be carried out not only in the short time but also in the medium and the long one. AI is believed can deliver solutions to most of Global Goals such as food security including life below water, water security, and energy security, increase the quality of health service, support the establishment of smart nation, and the most of them is AI has become a leveraging factor of a country's economy. Therefore, many countries have determined the action in utilizing AI for the good to their country and people by developing AI national strategy. In this paper, Indonesian initiative in form of a National Strategy for Artificial Intelligence will be delivered in a concise manner. The deliverables include four focus areas as the basis for AI ecosystem as well as the priorities of the AI usage for supporting Indonesia to achieve its national interests as well as the world goals through SDGs. We also deliver Indonesia's innovation in utilizing AI technology to cope with global pandemic which one of the technologies used is an Indonesian-original AI technology called Knowledge Growing System (KGS).

1. Introduction

Sustainable Development Goals (SDGs) are the continuation of Millennium Development Goals (MDGs). In its core, MDGs which were initiated by United Nations (UN), have eight goals that cover most of the aspects of human life. MDGs put the eradication of extreme poverty and hunger as the first goal followed by universal education, women empowerment and gender equality, the reduction of child mortality, the improvement of maternal health, the combat to severe diseases such as malaria, the establishment of environmental sustainability, and the development of a global partnership [1]. MDGs were declared in 2000 and all goals should have been achieved by all nations in 2015 as reported in [2].

MDGs were a huge task that was carried hand-in-hand among national governments, private sectors, civil societies, and national universities, as well as international communities. The achievement of MDGs' goals is a proof of the results of a collaboration among four helices, or simply it is called as quadruple-helix (quad-helix). It is very understandable that MDGs' goals have not been yet achieved in 2015 because of various factors and each country has its own challenges and obstacles. The continue what have been achieved by MDGs, then the UN created SDGs in 2015 with a collective of goals that have to be achieved by 2030, or 10 years from now. There are 17 goals grafted in the SDGs document and SDGs themselves are the extension of MDGs [3]. Therefore, they are some goals in SDGs which are intersected with MDGs' goals. In general, MDGs' ultimate objective is to enhance the quality of life of the people all over the world by ensuring the health of the earth. The key is the planet. Without a healthy planet then the peace and prosperity will be impossible to achieve.

In the effort to achieve all those goals, various resources will be required and the key is the people themselves. With the availability of advanced technologies brought by a new wave of Industrial Revolution a couple years ago as well as financial supports, what they need to do just creativity and know-how. Those two words represent today's important terminology, namely innovation. One technology which offers huge opportunities of innovation is Artificial Intelligence (AI). AI is not new. This technology has been around us since in the middle of 1990s. Since its reintroduction along with other 4.0 technologies, AI has been a belle or *primadona* because of its very wide applications in many aspects of human life as well as its contributions to increase the quality of life. The mastery of AI technology has been a goal of most of countries in the world. Therefore, there has to be a guidance to be obeyed in those countries to be able to master AI technology along with its complexities.

In this keynote article, we will show the contributions of AI technology in achieving SDGs especially in Indonesia. Before getting into Indonesia's effort in utilizing AI technology to cope with the pandemic situation in Indonesia that has been happening since last February 2020, a brief on AI will be delivered in Section 2. In Section 3, the state-of-the-art of AI National Strategy will be delivered including the recent status in Indonesia. This keynote article will be concluded in Section 5 with some concluding remarks.

2. Artificial Intelligence

AI technology actually was already invented several years before the terminology "Artificial Intelligence" was coined in 1956. The effort to have an intelligent machine has been starting since 19th century or even before that time. Today's most used method in AI is deep learning which is the derivation of one of machine learning methods called as neural networks or simply, NNs. Such networks are derived from the smallest element of human brain called neuron which are interconnected and interacted one to others to form a large network of neurons. This is the most widely used machine learning method in AI applications world. The neuron and NNs were firstly modeled by a Spanish neuroscientist, Santiago Ramón y Cajal in 1888 with the first illustration of human nervous system [4]. His works then were followed by his neuroscientist colleague, Warren McCulloch who worked together with Walter Pitts created a mathematical model of a neuron in 1943, called as McCulloch-Pitts (MCP) neuron [5]. This model of neuron then became the basis for any kind of NN architectures today.

Meanwhile from different perspective, a team of researchers called as Cognitive Artificial Intelligence Research Group (CAIRG) introduced a new perspective in AI named as Knowledge Growing System (KGS) in 2009 [6]. Unlike machine learning's NNs and deep learning that we call as neuron-based AI, KGS is cognitive psychology-based AI. The development of KGS is based on Jean Piaget's constructivism theory or theory of knowledge generation [7]. He observed and made notes of human cognitive development from a new-born infant until 11-year old boy and forth. Based on this concept, KGS is aimed to develop an ever-learning intelligent system. Therefore, we defined KGS as a system that is capable of growing its own knowledge as the accretion of the information as the time passes. Like human, this system starts from having no knowledge up to achieving ultimate knowledge regarding a phenomenon, and can use its knowledge to deliver solutions to given problems. To make KGS able to grow its own knowledge, we have developed a new information-inferencing fusion method called ASSA2010 (Arwin Sumari-Suwandi Ahmad 2010). This method has been registered to Directorate of Intellectual Property, Ministry of Law and Human Rights of Indonesia for getting a national patent since 2019 [8].

Intelligence is not only shown by humans but also by other living things such as bees, fishes, birds, and ants. Humans can learn collective intelligence or swarm intelligence from those insects and birds. Bees are able to fly in a group to a certain location and return to their nest without lose in flight, and a group of fishes that is swimming together can automatically split into several small groups to avoid attacks from predators and return to a big one after the predators are gone. A bunch of ants leave their nest to find food and they are able to return home without getting lost, and a flock of storks flying can show a kind of discipline in a formation flight where other storks will not be flying in front of their leader. If the leader is tired, another stork will replace the leader and continue the formation flight. From these we can see that other living things have their own intelligences, the intelligence in their world, Such as humans, they have their intelligence in their world, human world.

Based on our research, we have categorized AI in three groups based on how the God creations' intelligence is emulated. The first group is neuron-based AI, the second group is cognitive psychology-based AI, and the last group is living thing behavior-based AI. Basically, AI is not created to replace humans. There are two purposes of AI after Herbert Simon. The first one is to give more power to human thinking and the second one is to apprehend the mechanism occurred in human brain when thinking. Collecting all definitions of AI and also its purposes, we define AI as a technology that emulates natural intelligence (human and living things) through mathematical models that are translated into programming codes which are then implemented into a computer system. A computer-based system with such codes embedded within it will show an intelligence as human's or living thing's one. In many cases, AI is artificial human intelligence. Therefore, anything that can done by human is considered can be done by AI. With the advancement of technologies such as electronics, informatics, and materials as well as sciences such as cognitive science, computer science, and neuroscience, and pushed with the need of assistants that are flexible and able to work 24/7, AI has become a new support. Brought back again by 4.0 Industrial Revolution, AI has penetrated many aspects of human life from assisting house works to critical works such as deep sea and other planets explorations, forest and high-rise building firefighting, mine sweeping and detonations, and other tasks that can bring high-risk to human agents. In essentials, AI is a promising technology to leverage a country's economy as well as increase the standard of living of a country's people. AI is an enabler to make a country able to bring welfare to its people. Therefore, AI is believed can give big contributions to the achievement of the SDGs to all countries in the world. There is one condition that has to become a primary concern. The utilization of AI surely has to have a guidance at a national scale because it will impact to all aspects of human life and has to be in line with national interests. This guidance is called as Artificial Intelligence National Strategy.

3. Artificial Intelligence National Strategy

National strategy can be defined as an art to achieve national objectives by making the maximum use of all national resources from all aspects such as ideological, political, economic, socio-culture, science and technology as well as defense and security. Strategy is used mostly in military domain but it has been a while that strategy has also been used in non-military organizations. Strategy defines various things that have to be prepared to achieve the defined objectives. Strategy details the policy with how-to and who-does the steps to implement the policy or simply who does what with what (method).

AI can be thought as a strategic technology that will give huge impact to a nation. The impact is not only to the nation but also to individual people of the nation. But there is a thin line between good impact and the bad one. Unfortunately, AI is a technology that seems easy to be misused by any actor for various purposes. Therefore, a national strategy in utilizing AI is a mandatory. In one side, a nation can make the maximum use of it for uplifting its people well-being and raising the national economics. In another side, the misuse of AI can drown a nation into a ruin. National strategy is the guardian of the misuse of AI. The protection to AI usages has to go hand-in-hand with its mastery and utilizations. According a survey has been carried out by Holon IQ [9], Oxford Insights [10], and Observatory of Public Sector Innovation (OPSI) [11], some countries already have their own AI national strategy.

According to Holon IQ that did the survey dated February 20, 2020 there are 50 countries already have AI strategy in 2020. Indonesia has not been mentioned yet because it just started creating AI National Strategy in March 2020 and it is still ongoing up to now.

In 2020, Oxford Insights released Government Artificial Intelligence Readiness Index 2019 which includes 194 countries in the world. This survey uses 4 cluster parameters namely, governance, infrastructure and data, skills and education, and government and public services. If in [9] Indonesia is not in the list but in [10] Indonesia is in the 57th rank after Thailand. [11] that did the last survey on January 23, 2020 mentioned that Indonesia's Agency for Assessment and Application of Technology or simply BPPT, was charged to develop Indonesia AI National Strategy. The development of Indonesia's AI strategy was started in the beginning of March 2020 after some preliminary meetings in January and February 2020. This a big step for Indonesia in anticipating the fast advancement of AI technology as well as its various applications for humankind. Indonesia will not want to be left behind in mastering and utilizing this technology for the national interests and accelerating the achievement of the SDGs goals.

Indonesia's AI National Strategy has been released for public assessment since the end of July 2020 up to now [12]. In developing this document, the development team was divided into five groups and each group has its own responsibility. As depicted in Figure 1, Group 1 was responsible in developing the strategy for providing human talents in AI starting from the most basic education to the highest one and also the strategy to cope with the brain drain. Group 2 was responsible to create borders for the use of AI technology and protect it to be misused, while Group 3 prepared the strategy for providing the best infrastructure and data to AI utilization. Group 4 designed the strategy to build an AI research and innovation ecosystem which accommodates the quadruple-helix namely, all national resources for AI development and utilizations. The quadruple-helix consists of Academician, Business, Community, and Government (ABCG). The last group, Group 5 was responsible to define the quick wins and the priorities of the applications of AI in Indonesia in a short, medium, and long terms.

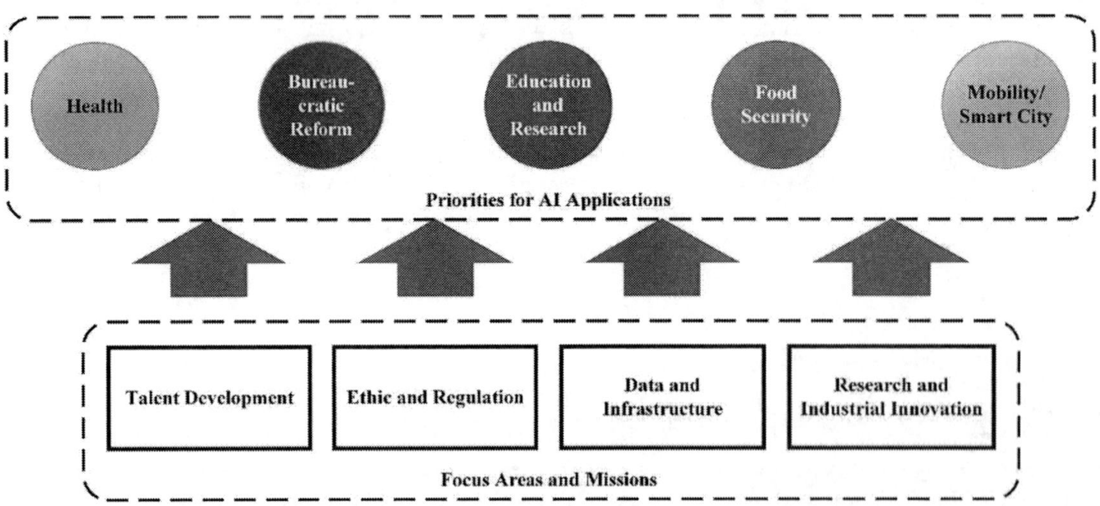

Figure 1 A simple illustration of Indonesia's AI National Strategy.

Indonesia has defined five priority areas in AI National Strategy. They are Healthcare, Bureaucratic Reform, Education and Research, Food Security, and Mobility/Smart City. Healthcare has been an important application area for AI. With the advancement of various supporting technologies such as image processing, voice processing, feature extraction, and also decision-making techniques, AI can deliver many benefits to health sector especially in expediting the diagnostic process to obtain the results. AI also will give many benefits for assisting humans especially in routine works for accelerating public services. Bureaucratic reform can get many helps from AI, especially in new capital of Indonesia which has been preparing by the Indonesian Government since last year. The new capital of Indonesia needs better services and facilities such as intelligent transportation system and intelligent public servants. Education and Research is like two-side of coin that have to go together especially AI, a technology that has to be mastered. AI gives so much help in the situation such as the pandemic that has been occurring since March 2020. Intelligent tutoring systems help lecturers and teachers to deliver the studying materials to students wherever they are. In research side, by using AI we can formulate the most correct composition of nutrition for a specific plant that is needed by most of Indonesia people to guarantee its sustainability. By doing that, we can ensure Food Security in Indonesia. In essentials, AI can deliver many contributions to all stated national priorities as well as quick wins that have to be achieved in a short term.

If we take a look back to 17 goals of SDGs, the above five priorities have already covered eight goals namely, No Poverty, No Hunger, Good Health, Quality Education, Good Jobs and Economic Growth, Innovation and Infrastructure, Sustainable Cities and Communities, and Partnership for the Goals. Actually, the detail of Indonesia's AI National Strategy has already considered SDGs and we have developed the AI strategies for other 9 SDGs' goals. The ultimate goal of Indonesia's strategy is the achievement of 2045 Indonesia Vision or 100 years of Indonesia Independence. That will be the year when Indonesia is in its golden era. We dream that Indonesia becomes 10 countries in the world in economic achievement with Gross National Product (GNP) more than 20% from the utilization of AI technology in all sectors. The other one, we also dream that Indonesia becomes one of top 5 countries in world that masters the AI technology especially for Agro-Maritime applications.

4. Artificial Intelligence Innovations to Cope With the Pandemic Situation: Indonesia Case

Healthcare has become one of quick wins of Indonesia's AI National Strategy knowing that AI can give much help in giving solutions to healthcare issues, especially in pandemic situation that has been lasted for five months. In the effort to find a solution in detecting a person is infected by covid-19 or not, a team called Task Force for Research and Innovation for Handling Covid-19, abbreviated as TFRIC-19 was formed in March 2020 by Indonesia's Agency for Assessment and Application of Technology or BPPT. TFRIC-19 is a collaboration among quad-helix [13]. It has five subtask forces with its own mission and products. One of the subtask force called as subtask force for strengthening information system and AI application, is charged to develop a new imaging system based on AI. Its mission is to develop a Medical Imaging System based on AI by utilizing the most appropriate methods and techniques in AI to process various data needed to assist doctor in establishing medical diagnosis on covid-19.

The requirement is the system that will be developed should have capability to not only process and analyze images data, but also non-images data such as clinical symptoms. For this purpose, the system utilizes three kinds of AI method namely, machine learning, deep learning, and Indonesian-original AI technology, KGS, as already mentioned previously. The illustration of the system is depicted in Figure 2. Each method has different task but delivers mutual support one to another. Machine learning and deep learning are tasked to process and analyses images data such as x-ray and CT scan images, while KGS is tasked to process and analyses clinical symptoms data. The results of each method will be used by the authorized doctor to establish the medical diagnostic. A similar AI-based system is also developed by [14] but it does not use x-ray images.

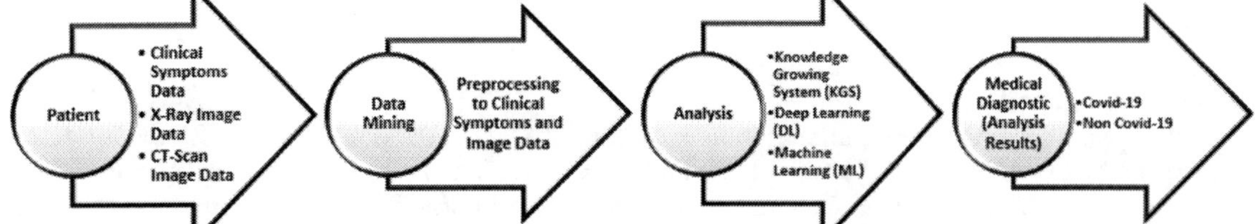

Figure 2 The illustration of Medical Imaging System based AI which utilizes two AI method, namely machine learning and deep learning, and one Cognitive AI method, namely KGS (adapted from [15]).

By using this system, covid-19 indication in a person under observation can be detected faster and more accurate, and more over it can expedite the handling of the person if the system's result indicates that person is positively infected by covid-19. The system's results are knowledge to the authorized doctor in making medical diagnostic decision. This AI-based system does not replace doctor but it gives ever-lasting assistance to the doctor. The decision is on the doctor or human, and not on the AI-based system. Last month, the system had just passed an operational test in a national hospital using Indonesian data and now is waiting for the next tests to ensure its performance. What has been done by TFRIC-19 is the national effort in the research and technology innovation to get the most out of AI technology for Indonesian people especially in health sector. In this research and technology innovation to cope with covid-19, we are glad to inform that the Indonesia government decided to use Indonesian-original AI technology, KGS that was invented by Arwin Datumaya Wahyudi Sumari and Adang Suwandi Ahmad in 2009, as part of Medical Imaging System based on AI.

Different from machine learning and deep learning, in growing or generating the knowledge KGS learns by interaction. Meaning, KGS will grow the knowledge based on the data it is perceiving during interactions with the phenomenon under observation. It does not learn from past data such as AI methods in general especially machine learning and deep learning. KGS does not need data labeling and also does not need many data to understand the phenomenon surrounds it. What it needs just interactions. During the interactions, KGS will use all of its sensory organs to get as many as possible data from the observed phenomenon. KGS' brain will combine all data from all sensory organs to generate new knowledge by a mechanism called as information-inferencing fusion. There is a cognitive characteristic shown by KGS when fusing all data to obtain information, and then fusing all information to obtain inferencing. The inferencing itself is the new knowledge of the system [16]. This mechanism shows cognitive characteristic such the one occurred within human brain. This is the primary reason that KGS is a method in Cognitive AI [17]. KGS actually has been applied in various fields even though some of them are in a research scale. But most of all, the happiest thing is the Indonesian-original AI technology named KGS, has given its real contribution to the achievement of one of SDGs namely, Good Health, for better Indonesia.

5. Concluding Remarks

SDGs is still 10 years to go but if we do not have a good technology, the SDGs' goals may not be achieved as hoped. The reemergence of AI technology brought by 4.0 Industrial Revolution wave has brought a new hope. AI, with its various methods and techniques, is believed can deliver big contributions in any country in pursuing SDGs' goals in time. Because of AI is very prone to be misused, then there has to be a strategy at a national scale in order that AI can give the best to the country. Indonesia has just finished developing its AI National Strategy which the dictated five priorities have already covered eight goals of SDGs. Indonesia is already inline its AI National Strategy with SDGs' goals.

Pandemic situation has accelerated Indonesia to achieve Good Health through Innovation and Infrastructure with the development of an AI-based medical imaging system to help doctors in establishing medical diagnostic if a person is infected by covid-19 or not. This AI-based system is one of quick wins in AI National Strategy in health sector. This system uses Indonesian-original AI technology called KGS, a cognitive psychology-based AI or simply called as Cognitive AI.

6. Acknowledgment

The author would like to express high gratitude to Chief of Staff of the Indonesian Air Force for giving permission to join AI National Strategy team and TFRIC-19. The gratitude is also delivered to Head of Indonesia's Agency for Assessment and Application of Technology for giving permission to deliver some parts of Indonesia's AI National Strategy as part of the author's keynote speech in the international conference hosted by University of General Soedirman, Indonesia. The author also would like to express gratitude to all members of Cognitive Artificial Intelligence Research Group (CAIRG) as well as all colleagues in Indonesia AI National Strategy team and TFRIC-19.

IOP Conf. Series: Materials Science and Engineering **982** (2020) 012063 doi:10.1088/1757-899X/982/1/012063

7. References

[1] United Nations 2001 Road map towards the implementation of the United Nations Millennium Declaration, *Report of the Secretary-General*, A/56/326

[2] United Nations 2015 *The Millennium Development Goals Report 2015* https://www.un.org/millenniumgoals/2015_MDG_Report/pdf/MDG%202015%20rev%20(July%201).pdf.

[3] Kumar S, Kumar N and Vivekadhish S 2016 Millennium Development Goals (MDGs) to Sustainable Development Goals (SDGs): Addressing Unfinished Agenda and Strengthening Sustainable Development and Partnership *Indian Journal of Community Medicine* pp 1-4

[4] DeFelipe J 2013 Cajal and the discovery of a new artistic world: the neuronal forest *Progress in Brain Research* Volume 203 pp 201-220

[5] McCulloch W S and Pitts W 1943 A logical calculus on the ideas immanent in nervous activity *Bulletin of Mathematical Physics* Vol. 5 pp 115-133

[6] Sumari A D W and Ahmad A S 2009 The development of Knowledge Growing System as a new perspective in Artificial Intelligence *Proceedings of Conference on Information Technology and Electrical Engineering 2009 (CITEE2009)* No. 3 pp 46-51

[7] Sumari A D W, Ahmad A S, Wuryandari A I and Sembiring J 2010 Constructing brain-inspired knowledge-growing system: a review and a design concept, *the Second International Conference on Distributed Framework and Applications 2010 (DFmA2010)*

[8] Sumari A D W and Ahmad A S 2019 *Metode komputasi untuk sistem berpengetahuan tumbuh terinspirasi cara manusia berpikir* Registered Patent Number P00201202101 12 March 2019 http://lpik.itb.ac.id/hki/detail/patent/308

[9] The Global AI Strategy Landscape 2019 https://www.holoniq.com/notes/the-global-ai-strategy-landscape/

[10] Oxford Insights 2019 Government Artificial Intelligence Readiness Index 2019 https://ai4d.ai/wp-content/uploads/2019/05/ai-gov-readiness-report_v08.pdf

[11] OECD 2020 AI Strategies & Public Sector Components https://oecd-opsi.org/projects/ai/strategies/

[12] National Secretary for Indonesia Artificial Intelligence 2020 KA Menuju Visi Indonesia 2045 https://ai-innovation.id/

[13] Task Force for Research and Innovation for Handling Covid-19 (TFRIC-19) https://tfric-19.id/

[14] Mei X, Lee H, Diao K, Huang M, Lin B, Liu C, Xie Z, Ma Y, Robson P M, Chung M, Bernheim A, Mani V, Calcagno C, Li K, Li S, Shan H, Lv J, Zhao T, Xia J, Long Q, Steinberger S, Jacobi A, Deyer T, Luksza M, Liu F, Little B P, Fayad Z A and Yang Y 2020 Artificial intelligence–enabled rapid diagnosis of patients with COVID-19 *Nature Medicine* Volume 26 pp 224-1228

[15] Sumari A D W 2020 The utilization of Knowledge Growing System for enhancing telepathology capability *Neurabot Deep Science Webinar 2020* p. 29 https://www.researchgate.net/publication/344781075_The_Utilization_of_Knowledge_Growing_System_for_Enhancing_Telepathology_Capability

[16] Sumari A D W, Ahmad A S, Wuryandari A I and Sembiring J 2009 A mathematical model of Knowledge-Growing System: a novel perspective in Artificial Intelligence *Proceedings of IndoMS International Conference on Mathematics and Its Applications 2009* (IICMA2009) pp 229-240

[17] Sumari A D W and Ahmad A S 2018 Cognitive artificial intelligence: concept and applications for humankind *Intelligent Systems* ed C Wongchoosuk (London: INTECH) chapter 2 pp 23-55

The effect of deposition time on the properties of titanium dioxide thin film prepared using CVD

Jais Lias[1,2], Muhammad Harith Fadhilah Mohd Fauzi[1], Mohd Zainizan Sahdan[1,2] and Nafarizal Nayan[1,2]

[1] Faculty of Electrical and Electronic Engineering, Universiti Tun Hussein Onn Malaysia, Parit Raja, 86400 Batu Pahat, Johor, Malaysia
[2] Microelectronics and Nanotechnology - Shamsuddin Research Centre (MiNT-SRC), Universiti Tun Hussein Onn Malaysia, Parit Raja, 86400 Batu Pahat, Johor, Malaysia

jaisl@uthm.edu.my

Abstract. This research aimed to investigate effect of deposition time on the structural, morphological properties and optical properties of the titanium dioxide (TiO_2) thin film prepared using Chemical Vapour Deposition (CVD). This research involved two processes which are samples preparation process and characterisation process to fulfil the aim. The samples preparation process was done by synthesising TiO_2 on indium thin oxide (ITO) substrates heated at 60℃ substrate temperature. Titanium butoxide used as the precursor for this chemical reaction was volatilised at 210℃. Oxygen gas was flown at 1 litre per minute as the carrier gas. In order to study the effect of deposition time, the synthesise of the thin films were varied to 15, 30, 45 and 60 minutes. After synthesising process, the samples underwent thermal treatment via annealing process for 1 hour at 500℃. For the characterisation process, Raman Spectroscopy technique was employed to investigate the structural properties of the samples. Apart from that, Field Emission Microscopy (FEM) technique, which was performed via FE-SEM, was employed to investigate the morphologies of the samples. Other than that, UV-Vis spectrometry was employed to analyse the optical properties of the samples. Analysis of data from Raman spectroscopy displayed four Raman Shifts for each sample which confirms that the samples exhibit TiO_2 of anatase phases. Whereas, images from FE-SEM displayed reduction of nanoparticle clusters on the samples as the rate of deposition time increases. Meanwhile, UV-Vis analysis displayed transmittance of the samples ranging between 50% to 80% transmittance and each sample exhibits the same optical band gap at 3.25 eV.

1. Introduction

The invention of thin film has marked significant changes in modern days technology especially in electronics and photonics field. The bulk size of transistors was able to scale down while the thinning of solar cells was made possible through this invention. Despite the importance of scaling down devices, economic and safety concern were heavily highlighted in fabrication. Therefore, researchers nowadays tend to look for alternative safe materials and method to fabricating thin film.

Recently, there have been a significant attention worldwide towards titanium dioxide (TiO_2) as thin film materials because of the safe and economic properties of the materials. Anderson *et al.* stated that, the economic and non-toxic characteristics of TiO_2 combines with the high photocatalytic efficiency and photochemical stability has prompted an extensive research interest globally [1]. The authors added that, nanoscale research has been going on for TiO_2 where the characteristics of TiO_2

nanoparticle has been controlling to enhance the materials to suits specific applications. Khalifa & Mahmood shared similar opinions on the economic and non-toxic properties of TiO_2 [2]. The authors claimed that TiO_2 is preferable metal oxide in photocatalytic applications to support their opinion.

At the meantime, the fabrication methods for thin film have also been the subject of interest among researchers to produce high quality products. Thin films can be deposited through variety of methods which have two major branches which are the physical deposition and chemical deposition method. Prior literatures had extensively discussed various technique for depositing thin films. Kääriäinen *et al.* experimented on depositing TiO_2 thin films using atomic layer deposition, ALD [3]. Meanwhile, Evtushenko *et al.* employed sol-gel method for depositing their TiO_2 thin films [4]. Magnetron sputtering on the other hand, was employed by Pansila *et al.* to deposit TiO_2 thin films [5]. On the other hand, Rahim *et al.* [6], Lias *et al.* [7], Maruyama & Arai [8] and Khalifa & Mahmood [2] used CVD to deposit TiO_2 thin films.

CVD is the most prominent in term of economic term amongst all the other methods due to simplicity of the CVD device and process which able to yield good quality end products. Gupta described CVD processes are often used over other deposition technique because the high-purity deposits, substantial variety of chemical composition, moderately fine film adherence and excellent economy and process control [9]. The process for CVD was explained by Elshabini & Barlow who emphasised CVD technique depends on reactive carrier gas to deliver selected precursor to substrate where stable reaction product is produced through decomposition or gaseous reaction [10].

Another factor which influence the fabrication of thin film is time. A research by Shahadan *et al.* researcher identified that the deposition time influenced the properties of TiO_2 thin films deposited using solid precursor, 99.9% pure titanium powder through CVD technique [7]. The used of solid precursor had been analysed to promotes certain drawbacks especially in terms of time and energy consumption. Because of the high boiling point of solid, using solid as precursor has increases the deposition time which might not suit certain application which requires fast manufacturing process. Therefore, the main goal of this research is to seek the effects of deposition time on the structural, morphological and optical properties of TiO_2 thin films deposited on ITO glass using CVD technique with titanium (iv) butoxide as a precursor.

2. Methodology

In order to achieve the goal, the research was split into three processes which are the substrate preparation, thin films deposition and thin films characterisation.

2.1. Substrate preparation

Prior to depositing the TiO_2 thin film, the substrate was prepared by cutting the ITO substrate into 1.5 cm x 2.0 cm. The substrates were then washed with soap to remove impurities. Next, the substrates were immersed in deionised (DI) water and bathed in ultrasonics for 10 minutes at 60°C. Next, the substrates were immersed in acetone and again bathed in ultrasonic for 10 minutes. After that, isopropyl alcohol (IPA) was used to remove the acetone residue through immersing the substrates in IPA and bathed in ultrasonic for another 10 minutes. The substrates were then let to dry in dry oven for 5 minutes before proceeding to the CVD step. The setup for deposition process using the double zone CVD is clearly illustrated in the Figure 1 to prepare the thin film samples. Additionally, since this research aimed to study the effect of deposition time, the deposition time were varied at 15, 30, 45 and 60 minutes.

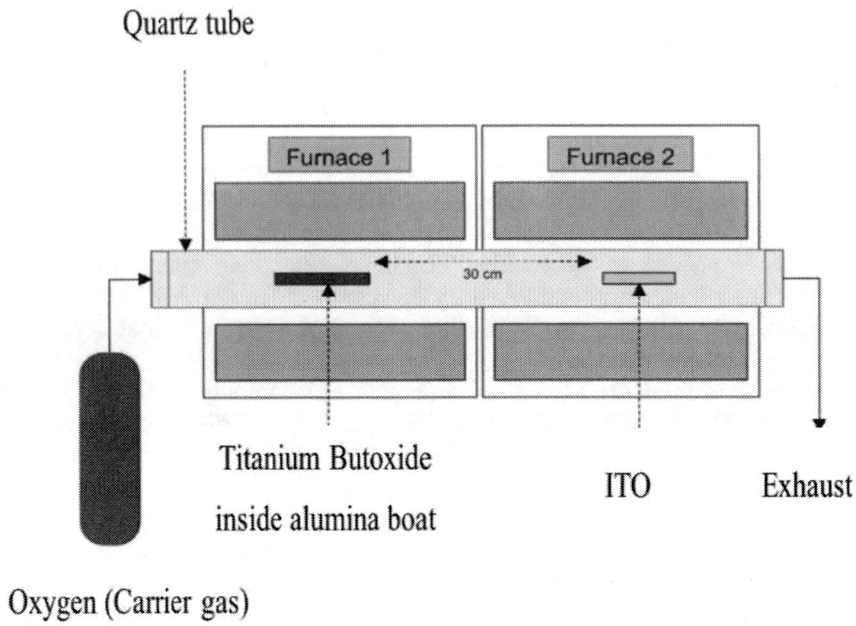

Figure 1. The illustration for the setup of the double zone CVD which was used to deposit TiO_2 thin films

2.2. *Thin films deposition*

The deposition of the thin film was done by first, heating the substrate in the second zone of the CVD to 60°C. The current supply for the second zone should be set to 50 mA. This process should take approximately around 15 to 20 minutes because the CVD will need to stabilise the temperature. When the substrate reaches 60°C, 1 ml of titanium (iv) butoxide was dispensed using pipette inside alumina boat was inserted into first furnace and was heated to 210°C. The distance between the substrate and precursor should be kept constant at 30 cm. After the titanium butoxide was inserted inside CVD, the ends of the quartz tube were plugged with carrier gas holder at the first-zone end while exhaust holder at the second zone end. The current supply for the first zone should be set to 50 mA. The heating of precursor should take around 15 to 20 minutes as the CVD need to stabilise. After the temperature reached 210°C, oxygen gas was flown at the rate of 1 litre per minutes. The flow rate was adjusted using oxygen gas flow rate. The deposition time should start as soon as the oxygen was flown inside the CVD. When the desired deposition time was reached, the deposition process should be stopped in the reverse order. Oxygen supply was first halted. This is followed by reducing the temperature of both zones to 27 °C. Meantime, the current should be set to 0 mA. The substrate and alumina boat can be removed after the temperature for both zones reach 27°C. After that, the samples were annealed for 1 hour at 500°C.

2.3. *Thin films characterisation*

In order to study the effect of deposition time on the structural, morphological and optical properties of TiO_2 thin film prepared using CVD, the thin films were characterised via multiple instruments. Raman spectroscopy was used to analyse the structural properties of the TiO_2 thin film. Field emission scanning electron microscope (FE-SEM) was used to analyse the surface morphology of the TiO_2 thin films. The optical properties of the TiO_2 thin films, the transmittances were measured using ultraviolet visible (UV-Vis) spectrometer. The transmittances were also used for determining the band gap of the thin films.

3. Results and discussion

The findings from all characterisation process are discussed briefly in this section.

3.1. Structural properties

Analysis from Raman Spectroscopy found that there are 4 Raman shifts being presented. The Figure 2 shows the Raman shifts for every sample of TiO_2 thin films meanwhile, the Table 1 recorded the details of the Raman shifts. Based on the Table 1, For the 15 minutes-deposited thin film, the Raman shift for the first E1g band was observed to be 154.339 cm^{-1} this was followed by the first B1g band, A1g + B1g band and E1g bands which are observed to be 414.165, 526.166 and 651.525 cm^{-1} respectively.

As displayed by the Table 1, for the 30 minutes-deposited thin film, the Raman shift for the first E1g band was observed to be 151.197 cm^{-1}. This was followed by the first B1g band, A1g + B1g band and E1g bands which were observed to be 408.070, 523.158 and 642.632 cm^{-1} respectively.

Referring Table 1, the 45 minutes-deposited thin film, the Raman shift for the first E1g band was observed to be 151.197 cm^{-1}. This was followed by the first B1g band, A1g + B1g band and E1g bands which were observed to be 405.019, 514.128 and 639.666 cm^{-1} respectively.

According to the Table 1, 60 minutes-deposited thin film, the Raman shift for the first E1g band was observed to be 148.055 cm^{-1}. This was followed by the first B1g band, A1g + B1g band and E1g bands which are observed to be 401.969, 511.117 and 636.698 cm^{-1} respectively.

Tuschel, states that the anatase TiO_2 should occupies 6 Raman shifts which indicated tetragonal crystal system of anatase TiO_2 which belongs to the group D194h [11]. However, the typical Raman discussed in literature by Chu *et al.* stated that the typical anatase phase of TiO_2 are peaking at 144, 394, 514 and 636 cm^{-1} of Raman [12].

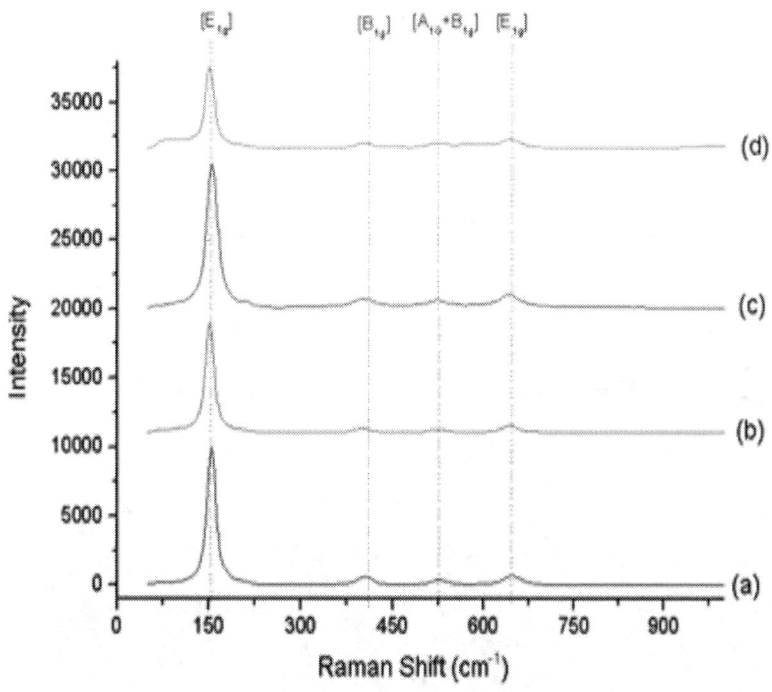

Figure 2. The Raman spectra for the synthesised thin films. The label (a), (b), (c), and (d) are for the 15, 30, 45 and 60 minutes deposited TiO_2 thin films respectively

Table 1. The Raman spectra for the synthesised thin films.

Deposition time (minutes)	**Raman Shifts (cm^{-1})**			
	E_{1g}	B_{1g}	$A_{1g} + B_{1g}$	E_{1g}
15	154.339	414.165	526.166	651.525
30	151.197	408.07	523.158	642.632
45	151.197	405.019	514.128	639.666
60	148.055	401.969	511.117	636.698

3.2. Morphological properties

The Figure 3 shows the FE-SEM images for the TiO_2 thin films under 5000 magnification at 5.0 kV.

According to the images taken from FE-SEM analysis, the 15 minutes deposited TiO_2 thin films produced the highest clusters of TiO_2 nanoparticles. This is followed by 30, 45, and 60 minutes deposited TiO_2 thin films respectively.

Since the deposition of the TiO_2 thin films using CVD included annealing process, which is known for crystal modification, it can be said that the crystal growth process is influenced by the deposition time.

Based on the findings, the shortest deposition time reveals more cluster growth of nanoparticles. Meanwhile, the nanoparticles clusters were observed to be decreasing as deposition time increases.

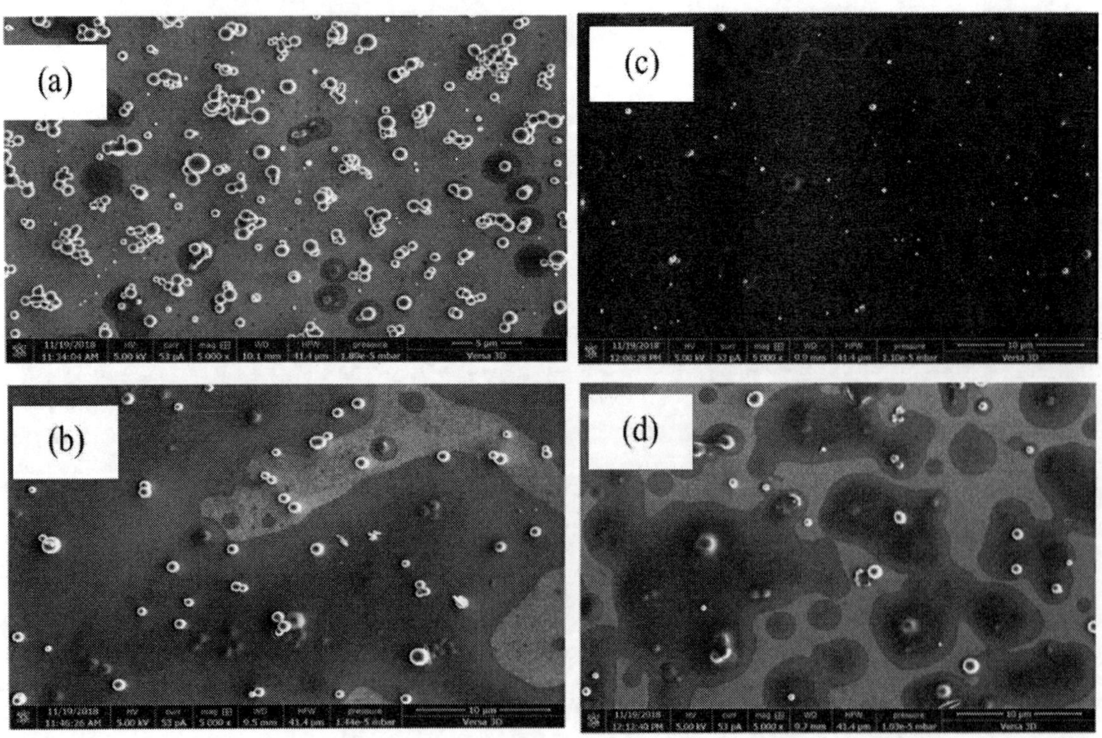

Figure 3. The FE-SEM images for the TiO_2 thin films under 5000 magnification the labels (a), (b), (c) and (d) indicates 15, 30, 45 and 60 minutes deposited film respectively

3.3. Optical properties and band gap

The Figure 4 shows the shows the transmittance spectra for TiO_2 thin films deposited using CVD for 15, 30, 45 and 60 minutes. In general, all the deposited TiO_2 thin films show a lesser than 80% and greater than 50% transmittance. Hence, all the TiO_2 thin films deposited appears translucent.

The 15 minutes deposited TiO_2 thin films shows the highest percentage of transmittance which exceeds 70%. On the other hands, the 60 minutes deposited TiO_2 thin films shows the lowest percentage of transmittance reaching not more than 65%. This suggest that the deposition time influence the transmittance of TiO_2 thin films. The longer the deposition time lower the transmittance of TiO_2 thin films.

Meanwhile, the band gap for all films were determined via Tauc Plot. The band gap of the TiO_2 thin films show a constant band gap trend as the TiO_2 thin films being deposited from 15 minutes to 60 minutes. The 15, 30, 45 and 60 minutes deposited TiO_2 thin films show the same band gap values which are at 3.25 eV. The fundamental band gap recorded in various literature on TiO_2 thin films is 3.2 eV. The 0.5 eV difference from the findings can be assumed mainly because of several defects in the samples.

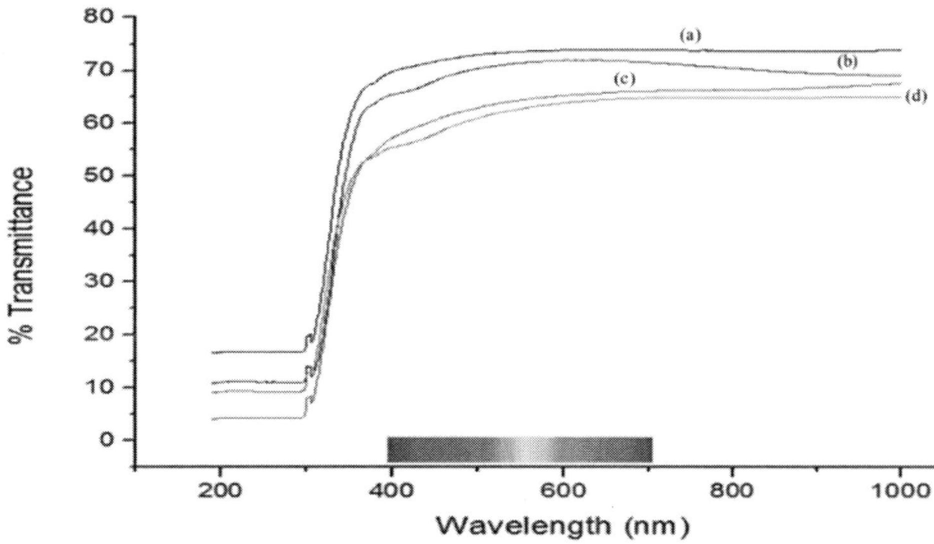

Figure 4. The UV-vis spectra for the TiO_2 thin films labelled (a), (b), (c) and (d) for 15, 30, 45 and 60 minutes deposition time

4. Conclusion

According to the analysis of structural, morphological and optical analysis from Raman Spectroscopy, FE-SEM imaging and UV-vis spectrometry, the relationship between deposition time and these properties can be configured. On structural level, the deposition time influence the stress of TiO_2 thin films. Increasing deposition time reduces the stress of TiO_2 crystalline while reducing the deposition time increases the stress of crystalline in TiO_2. Besides that, on morphological scale, it can be said that, longer deposition time allows less clustering of nanoparticles of TiO_2 thin films growing on ITO substrate after annealing vice versa. On the other hand, on optical scale, the translucency of TiO_2 thin films are declining over time when deposited using CVD while the correlation between band gap and deposition time is that the band gap of TiO_2 thin films deposited using CVD are unaffected by the deposition time.

IOP Conf. Series: Materials Science and Engineering **982** (2020) 012064 doi:10.1088/1757-899X/982/1/012064

Acknowledgement
The author would like to express their thanks to the Microelectronics and Nanotechnology-Shamsuddin Research Centre (MiNT-SRC), Universiti Tun Hussein Onn Malaysia (UTHM) for providing facilities to run the research.

References
[1] A. L. Anderson and R. Binions 2016 A preferential precursor for photocatalytically active titanium dioxide thin films: Titanium bis-ammonium lactato dihydroxide as an alternative to titanium tetra iso-propoxide *Polyhedron* vol. 118, pp. 81–90
[2] Z. S. Khalifa and S. A. Mahmoud 2017 Photocatalytic and optical properties of titanium dioxide thin films prepared by metalorganic chemical vapor deposition *Phys. E Low-Dimensional Syst. Nanostructures* vol. 91, no. February, pp. 60-64
[3] M. L. Kääriäinen, T. O. Kääriäinen, and D. C. Cameron 2009 Titanium dioxide thin films, their structure and its effect on their photoactivity and photocatalytic properties *Thin Solid Films* vol. 517, no. 24, pp. 6666-6670
[4] Y. M. Evtushenko, S. V. Romashkin, N. S. Trofimov, and T. K. Chekhlova 2015 Optical Properties of TiO_2 Thin films *Phys. Procedia* vol. 73, pp. 100-107
[5] P. Pansila, N. Witit-Anun, and S. Chaiyakun 2012 Influence of sputtering power on structure and photocatalyst properties of DC magnetron sputtered TiO_2 thin films *Procedia Eng.* vol. 32, pp. 862–867
[6] M. S. Rahim, M. Z. Sahdan, A. S. Bakri, S. H. A. Yunus, and J. Lias 2016 Low Temperature Growth of High Quality Crystal of Anatase Based-TiO_2 Thin films Grown Using Double Zone CVD Technique *Proc. - 2016 World Symp. Comput. Appl. Res. WSCAR 2016* pp. 126-129
[7] J. Lias, S. A. Shahadan, M. S. A. Rahim, N. Nayan, M. K. Ahmad, and M. Z. Sahdan 2016 Influences of deposition time on TiO_2 thin films properties prepared by CVD technique *Jurnal Teknologi* vol. 78, no. 10–3, pp. 1–5
[8] R. T. Maruyama and S. Arai 1992 Titanium Dioxide Thin films by Chemical Vapor Deposition," *Sol. Energy Mat. Sol. Cells*, vol. 26, pp. 323-329
[9] T. K. Gupta 2003 *Handbook of thick- and thin-film hybrid microelectronics* (NJ: Wiley-Interscience)
[10] A. Elshabini and F. D. Barlow 1998 *Thin film technology handbook*. (NY: McGraw Hill)
[11] D. Tuschel 2015 The Correlation Method for the Determination of Spectroscopically Active Vibrational Modes in Crystals *Spectroscopy*
[12] Chu, Z. Qin, J. Yang, and X. Li 2015 Anatase TiO_2 Nanoparticles with Exposed {001} Facets for Efficient Dye-Sensitized Solar Cells *Sci. Rep.* vol. 5, no. 1, p. 12143

IOP Conf. Series: Materials Science and Engineering **982** (2020) 012065 doi:10.1088/1757-899X/982/1/012065

Comparison between Viscoelastic Mechanical and Chemical Kinetic Aging Model of Asphalt Concrete

Shih-Hsien Yang[1,a], Susanto Hery Awan[2,b,*], Atiya Arief Fath[3,c]
[1]Associate Professor, Department of Civil Engineering, National Cheng Kung University, No. 1 University Rd., Tainan, Taiwan.
[2]Graduate Research Assistant, Department of Civil Engineering, National Cheng Kung University, No. 1 University Rd., Tainan, Taiwan (*corresponding author).
[2]Lecturer, Department of Civil Engineering, Faculty of Engineering, Jenderal Soedirman University, Jl. Mayjen Sungkono Km.5 Purbalingga
[3]Graduate Research Assistant, Department of Civil Engineering, National Cheng Kung University, No. 1 University Rd., Tainan, Taiwan.
[a]shyang@mail.ncku.edu.tw, [b]hery.susanto@unsoed.ac.id, [c]aarieffath@gmail.com

Abstract. The objective of this study was to fully characterize and validate the mechanical-based aging model parameters and compared with the chemical kinetic model. This study found that level of aging applied to both asphalt binder and asphalt concrete showed differences, as verified by all of the p-values from regression analysis test, were greater than 0.05. The aging variable value decreased with the aggregate involved. The aging history dependent (k2) could be simulated the actual aging behavior as a polynomial curve. The Pearson correlation test showed that there is a strong relationship between the carbonyl content and the aging model variable.
Keywords: Aging, Asphalt concrete, Asphalt binder, Carbonyl content, Mechanistic model, Chemical kinetic model

1. Introduction

Asphalt aging is one of the significant factors influencing the performance of asphalt pavements [1]. Aging could result in two major effects: first, load-bearing capacity and permanent deformation resistance may increase due to the pavement becoming stiffer; and second, it can lead to the formation of cracks in the pavement due to reduced pavement flexibility [2].

The most fundamental issue affecting binder aging in pavements is the binder oxidation chemistry. The cause of the oxidative aging was due to the formation of carbonyl compound that transform the light molecular weight compound into large molecular weight within the asphalt binder [3]. Asphalt's carbonyl compound and formation rate depend on the temperature and oxygen pressure [4]. The chemical reactions have resulted in enhanced stiffness, viscosity, and brittleness of asphalt mixture. It will cause more vulnerability to fatigue cracking.

Another way to characterize binder oxidative aging is through mechanical based modeling of asphalt material behavior [5, 6]. Michalica et al. (2008) modeled the impact of aging using loading time and creep compliance as a function of time; Masad et al. (2008) modeled the impact of aging by introducing aging time shift factor in the nonlinear viscoelastic model. On the other hand, a mechanical based oxidative aging modelling by introducing oxidative aging variable which is conceptually model the carbonyl formation rate of asphalt materials during the long-term aging process [7].

Numerous studies conducted to understand constitutive modelling of oxidative aging and its effect on the mechanical response of asphalt concrete. There are three methods can use to determine relationship between oxidative aging and mechanical response, specifically: aging-time shift factor, creep compliance/relaxation moduli, and continuum-based modeling of asphalt concrete [8]. Among the three methods that are very widely used now is continuum-based modeling of asphalt concrete. Because this method can show physically-based oxidative aging constitutive relationship coupled with the viscoelastic response of asphalt concrete due to the process of aging.
Although the constitutive model proposed by Al-Rub et al. (2013) incorporated an aging variable to capture the effect of long-term aging on asphalt concrete's viscoelastic response, the fully characterization and validation of such aging model is not available yet. Several studies have attempted to characterize the mechanical-based aging model parameters, but their work has only limited on characterizing the internal aging variable and temperature-dependent parameter [8, 9]. Therefore, it is necessary to fully characterize and validate the mechanical-based aging model parameters and compared with the chemical kinetic model.
The main objective of this study was to characterize and validate the mechanical-based aging model parameters and compared with the chemical kinetic model. The asphalt concrete and the asphalt binder aged to five different level of aging. Dynamic modulus test, linear frequency sweep test was used to characterize the mechanical response of asphalt concrete and binder respectively. Fourier transform infrared spectroscopy (FTIR) test was employed to investigate the change of chemical compound in the asphalt binder [10]. The result then used to characterize and validate the linear viscoelastic mechanical-based aging model parameters.

2. Material and Methods

Plant mixed polymer modified asphalt mixes were obtained from asphalt plant and materials were aged to various level in the laboratory. Mechanical test of asphalt concrete, as well as rheological and chemical tests of asphalt binder, have been conducted in the study to characterize the evolution of viscoelastic and chemical kinetic properties of the material. The aging variable models derived from the viscoelastic test result were compared to the carbonyl formation rate model from the chemical kinetic tests.

2.1 Material

A plant mix 19 mm polymer modified coarse well grade mix was used in this study. The asphalt binder used is a Polymer Modified Asphalt Type III (PMAIII) binder. Table 1 shows the viscosity of asphalt binder in various temperature. Also from the test results obtained that SG = 1.033 and ρbinder = 1033 kg/cm^3.

Table 1. Viscosity of Asphalt Binder

Temp (°C)	Rotor Model	Rotation Speed (rpm)	Torque Value (%)	Viscosity	
				CP	Poise
160	SC4-18	75	20.2	689	6.89
180			9.3	317	3.17
200			4.9	167	1.67

Aggregate used in the study was crushed river gravel with 19 mm maximum aggregate size. The physical properties tests of aggregate have been conducted as seen in Table 2. The designs result were found to be suitable for Taiwan specifications as illustrated in
Table **3**.

Table 2. Physical Properties of the Aggregate

Test Item	Test Value	Specification Value	Test Standard
Aggregate Abrasion Value at 500 revolutions (%)	21	≤ 40	CNS 490 (2009) ASTM C 131
Fractured Particle in Course Aggregate at one face (%)	92	≥ 90	CNS 15312 (2010)
Fractured Particle in Course Aggregate at one face (%)	90	≥ 75	ASTM D5821-13
Sand Equivalent Value	69	≥ 45	CNS 15346 (2010) ASTM D2419-14
Ratio of Width to Thickness of Aggregate Retained In #4 Larger Than 3 Times of The Ratio (%)	4.7	≤ 15	CNS 15171 (2008) ASTM D4791-10
Soundness of Aggregates (%)	1.2	-	CNS 1167 (1995)

Table 3. Asphalt Mixture Properties Used in the Mixture Design

Term	Experimental Value	Standard Value
Binder Content (%)	4.7	-
Maximum Specific Gravity (Gmm)	2.467	-
Bulk Specific Gravity (Gmb)	2.379	-
Density (kg/m3)	2379	-
Stability (lbf)	5120	≥2700 lbf
Air Void (%)	4	3~5
Voids in Mineral Aggregate (VMA, %)	13.1	≥12.7
Voids Filled with Asphalt (VFA, %)		

2.2 Methods

The BRRC aging method was followed for both binder and asphalt mixes to simulate the long-term aging of materials [11]. Both binder and loose mixes were conditioned in temperature (60 and 95 °C) and time (5 and 10 days) to simulate different levels of aging.

The prony series model was used to fit the dynamic modulus test results to obtain the viscoelastic parameter of asphalt concrete. The SGC specimen was cored and cut into a cylindrical specimen with 101.6 mm in diameter and 152.4 mm in height. The test was conducted at four temperatures (15°C, 25°C, 40°C, 55°C) and six loading frequencies (25, 10, 5, 1, 0.5, 0.1 Hz) at each temperature.

The dynamic shear rheometer (DSR) was employed for frequency sweep test of asphalt binder to characterize the rheological behavior of asphalt binder at intermediate and high service temperature. The frequency sweep tests were performed over a range of temperature 22°C to 34°C representing the intermediate service temperatures and 40°C to 76°C to represent high service temperature with 6°C intervals.

In FTIR test, a sample subjected to infrared light of varying frequencies to determine how well it absorbs the light at each wavelength. A Thermo-Nicolet 6700 FTIR spectrometer was employed to produce the spectra. The background spectra and sample spectra are scanned using wavelength range of 400 to 4000 cm^{-1} and a scan resolution of 4 cm^{-1}. Data analysis then produced using the Omnic 8 Software.

3. Characterization of Asphalt Aging Material

3.1 Viscoelastic Model Parameters for Unaged Materials

This section describes an organized procedure to obtain time-temperature shift factor and the linear viscoelastic model parameters and aging variables. Results obtained from the dynamic modulus test were dynamic modulus (E*) and phase angle (θ) at different frequencies and temperatures. Then, complex compliance (D*) calculated as Eq. (1).

$$|D^*| = \frac{1}{|E^*|} \quad (1)$$

The sigmoidal function regarding complex compliance then utilized to fit experimental data and obtain the time-temperature shift factor as shown in Eq.(2)- (3) respectively. Horizontal shifting then applied to create a dynamic modulus master curve.

$$|D^*| = \delta + \frac{\alpha}{1+\exp[\beta+\gamma\log(\omega_r)]} \quad (2)$$

Where, D*| is complex creep compliance, δ is maximum value of the dynamic compliance, δ + α= minimum value of the dynamic compliance, β + γ= shape parameters, and ω_r = reduced frequency.

$$a_T = \frac{\omega_r}{\omega} \quad (3)$$

Where ω is the frequency and αt is the time-temperature shift factor.

This study employed Schapery's integral form to calculate the strain response due to an applied stress [12]. In general, the total strain of an asphalt mixture decoupled into storage compliance (D'_exp) and loss compliance (D"_exp). Both values computed by utilizing the creep compliance and phase angle. The storage and loss compliance (D'_model and D"_model) that related to Prony series then calculated by minimizing the error between the experimental and model data in Eq. (4)- (5). The error function used is shown in Eq. (6). The Prony series then formulated to represent the creep compliance ΔD as a function of time. It should be mentioned that the result of the Linear Frequency Sweep (LFS) test of binder (complex modulus G* and phase angle δ) also followed the same procedures to get the time-temperature shift factors and linear viscoelastic parameters.

$$D'_{\_exp} = |D^*|\cos\theta\ ; \qquad D''_{\_exp} = |D^*|\sin\theta, \quad (4)$$

$$D'_{\_model} = D_0 + \sum_{i=1}^{N}\frac{D_i}{1+\omega^2\tau_i^2}\ ; \qquad D''_{\_model} = \sum_{i=1}^{N}\frac{D_i\omega\tau_i}{1+\omega^2\tau_i^2} \quad (5)$$

$$\text{error} = \frac{1}{M}\sum_{i=1}^{N}\left[\left(\frac{D'_{fit}}{D'_{exp}}-1\right)\right]^2 + \left[\left(\frac{D''_{fit}}{D''_{exp}}-1\right)\right]^2 \quad (6)$$

Where N is the number of prony series coefficient that could be assumed between 5 to 9, M is the total data set, τn is 1/λ, Dn and λn are the nth coefficients of Prony series and retardation time respectively.

3.2 Viscoelastic Model Parameters for Aged Materials

The term of the aging variable (A) was referred as the oxidative aging variable that increases as the level of aging increases. To identify the aging variable (A), the aged asphalt mixture followed the same procedure to get time-temperature shift factors (αT), storage compliance (D'_exp) and loss compliance (D''_exp). To relate the aged and unaged viscoelastic properties, the loss and storage compliance model (D'_model and D"_model) of the unaged Eq. (5) was recalled and incorporated with Eq.(7), so that a new relationship [7] was created as shown in Eq.(8).

$$D_n^A = (1-A)^{k_2}D_n\ ; \qquad \lambda_n^A = (1-A)^{k_2}\lambda_n, \quad (7)$$

$$D'^A_{\_model} = D_0 + \sum_1^N\frac{(1-A)^{k_2}D_n}{1+\omega^2(1-A)^{k_2}\tau_n^2}\ ; \qquad D''^A_{\_model} = \sum_1^N\frac{(1-A^{k_2}D_n\omega(1-A)^{k_2}\tau_n}{1+\omega^2(1-A)^{k_2}\tau_n^2} \quad (8)$$

The new model certainly indicated a relationship between both compliances and retardation times of aged material altered by the aging level. Since the test covered all of the variable needed except A, the use of Microsoft Excel solver was required to minimize the error Eq. (6) to obtain the aging variable A.

3.3 Chemical Properties of Asphalt Binder

In this study, the test conducted to evaluate the evolution of carbonyl formation groups' chemical composition in a different level of aging. To calculate infrared absorption ratio of the carbonyl content and corresponding functional groups in the asphalt binders following to Yao's method [13]. The denominator in this equation is between wavenumber 600 and 2000 cm^{-1} where the aging components in the asphalt binder formerly concentrate.

3.4 Statistical Analysis

To verify the effect of different aging levels on both materials, mixture, and binder, the statistical analysis was conducted for the dynamic and complex modulus test result. The regression analysis was then performed to test the data with 95% confidence level. To evaluate significance, p value lower than 0.05 determined as significant which means between two sets of data measured, it has significantly different. The same observation procedure also conducted to complex modulus data from binder test.

4. Results and Discussion

4.1 Mechanical Properties of Asphalt Mixture

4.1.1 Viscoleastic Properties of Asphalt Concrete

Using time-temperature superposition principle, Figure 1 shows the dynamic modulus master curve of asphalt concrete in a different level of aging. The results demonstrated that the dynamic modulus interval between different aging level was greater in the low frequency rather than in high frequency. In addition, to verify the difference between each dynamic modulus test result at different aging level (AL), regression test was conducted at confidence level 95% and p-value set to evaluate the significance. The results showed that all of the test data had significantly different since their p-value were all lower than 0.05. Next, the dynamic modulus master curve was further fitted with prony series to determine viscoelastic model parameters. Using iteration method to minimizing the error between the experimental data and the prony series model, Prony series parameters can be obtained. The fact that using more prony series coefficients made the error between the experimental data and the prony series model will come to smaller number.

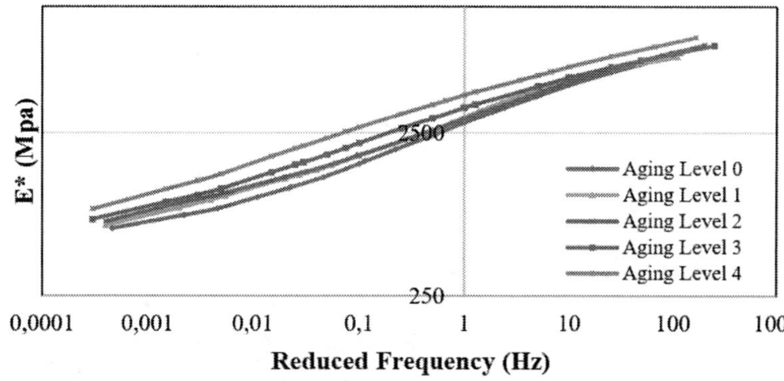

Figure 1. Dynamic modulus master curves at 25 °c for asphalt concrete at various aging level

4.1.2 Rheological Properties of Asphalt Concrete

The dynamic shear modulus master curve of asphalt binder also showed a similar trend to the dynamic modulus master curve of asphalt concrete where the interval between different aging level was greater in the low frequency rather than in high frequency as shown in Figure 2. The result also showed that when the oxygen pressure reduced to 0.8 atm, the dynamic shear modus master curves at all aging level showed lower values than that in the 1 atm oxygen condition.

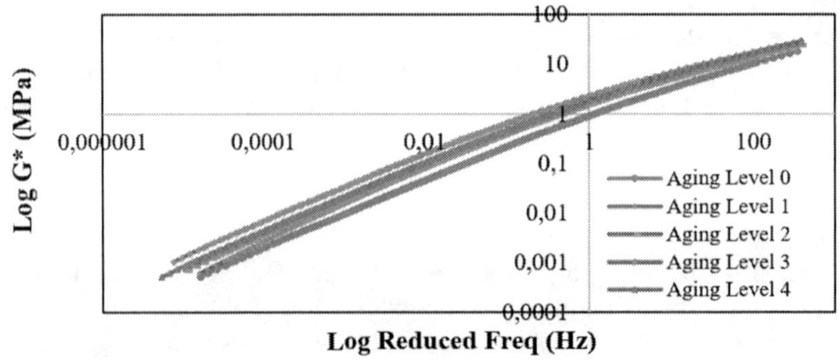

(a)

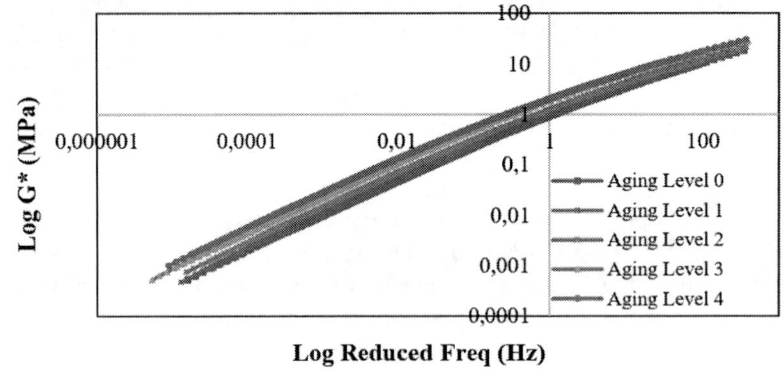

(b)

Figure 2. Complex modulus master curve at 25°c at different aging level (a) with oxygen pressure on 1 atm; (b) with oxygen pressure on 0.8 atm

To verify the difference between each dynamic shear modulus test result at the different aging level and oxygen pressure, regression tests were conducted at confidence level 95% and p-value set to evaluate the significance. The analysis showed that both within same atmospheric pressure and between different atmospheres, there are significant differences between various dynamic modulus master curves. The result of all the tests showed p-value smaller than 0.05. It is noteworthy that the procedure to get time-temperature shift factors, master curves and Prony series also followed the same procedure as derived for asphalt concrete.

4.1.3 Calculation of Aging (A)

Once the dynamic modulus master curves and viscoelastic model parameters of asphalt concrete/binder were obtained, the internal state aging variable (A) can be determined using Eq. (4)-(8). Figure 3 showed the result of aging variable A from asphalt concrete and Figure 4 showed the results from asphalt binder. For instance, in Figure 3, a minimal change of aging variable when asphalt concrete was aged under 60°C. The aging variable changed from 0 (aging level I) to 0.005 (aging level I) to 0.016 (aging level II). In the other hand, when the materials were aged at 95°C, a much higher rate of aging was observed. The aging variable was 0.101 at aging level III and 0.257 at aging level IV. Figure 4 showed that asphalt binder aging variable changed from 0 (aging level I) to 0.214 (aging level I) to 0.271 (aging level II). Otherwise, when the materials were aged at 95°C, a much higher rate of aging was observed.

The aging variable (A) has a limitation range value between 0 and 1 where A=0 stands for unaged material and A=1 means materials have been fully aged so that the carbonyl formation was completely stopped [7]. Practically, the aging variable will never reach 1.0 due to the long-term aging is very slow through the service life of the asphalt pavement [6].

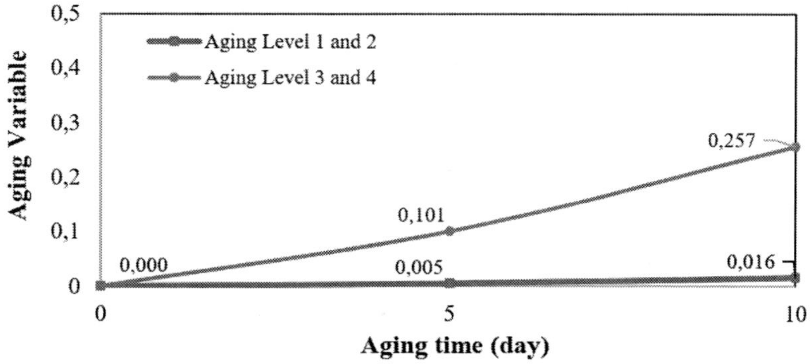

Figure 3. Aging variable of asphalt mixture

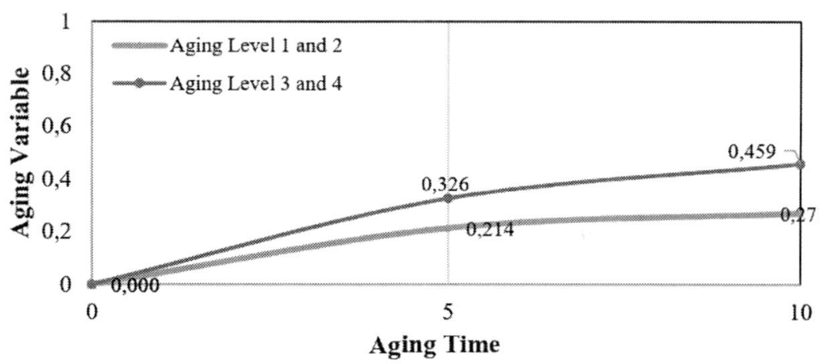

(a)

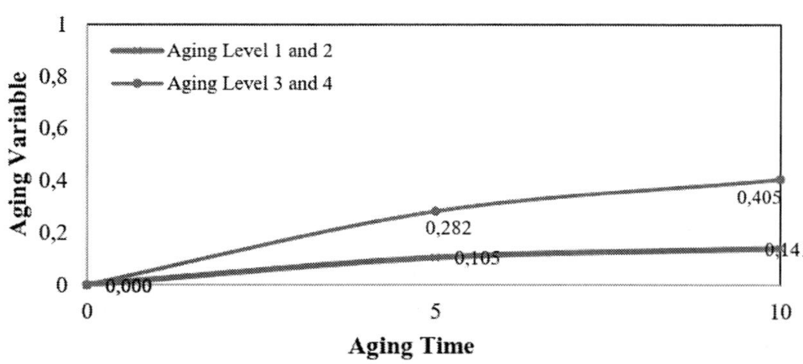

(b)

Figure 4. Aging variable of asphalt binder, (a) with 1 atm, and (b) with 0.8 atm

From the results of the aging variable of asphalt concrete and binder, different trends were observed. The aging variable from the asphalt concrete has lower values than that of the asphalt binder. Also, the increasing rate of an aging variable was higher in the asphalt binder than that in the asphalt concrete.

4.2 Chemical Properties of Asphalt Mixture

The impact of oxidative aging could be determined by evaluating the changes regarding carbonyl and two functional carbonyl groups those are carboxylic acids and ketones that have peak integral value of IR absorbance at 1670-1820 cm^{-1}, 1700-1725 cm^{-1}, 1665-1685 and 1680-1700 cm^{-1} respectively. Figure 5 shows the FTIR test result at all aging levels. Figure 6 to Figure *8* shows the ratio calculations for different bonds with additional 1 atm oxygen pressure. For instance, in Figure 6, a minimal change of aging variable when asphalt binder was aged at 60°C. The aging variable changed from 0.0386 (aging level 0) to 0.044 (aging level I) to 0.048 (aging level II). In the other hand, when the materials were aged at 95°C, a much higher rate of aging was observed.

The figures indicate that the carbonyl increased due to the oxidative aging. Both Carboxylic Acids and Ketones as the main product of oxidation. Yao et al. (2015) indicated that the Carboxylic Acids and Ketones are the main components that correlate well with pavement rutting and fatigue resistance. The chemical reactions between the oxygen and the resins as well as the asphaltenes during the aging of asphalt that would produce the aromatic hydrocarbons and the polar functional group that leads to increasing of carboxylic acid and ketones bonds within the asphalt binder.

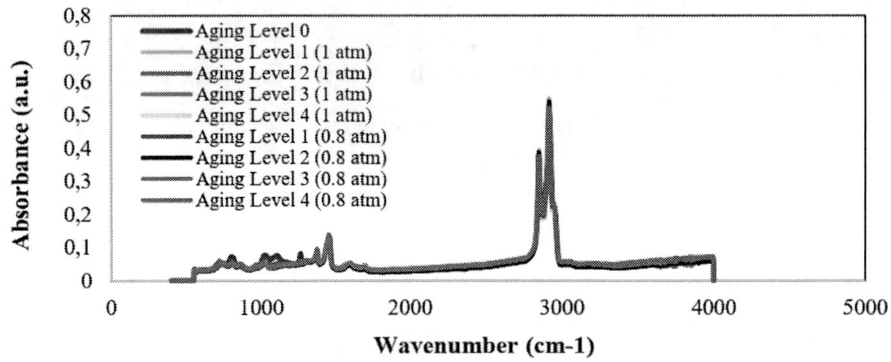

Figure 5. IR spectra of asphalt binder at different aging level

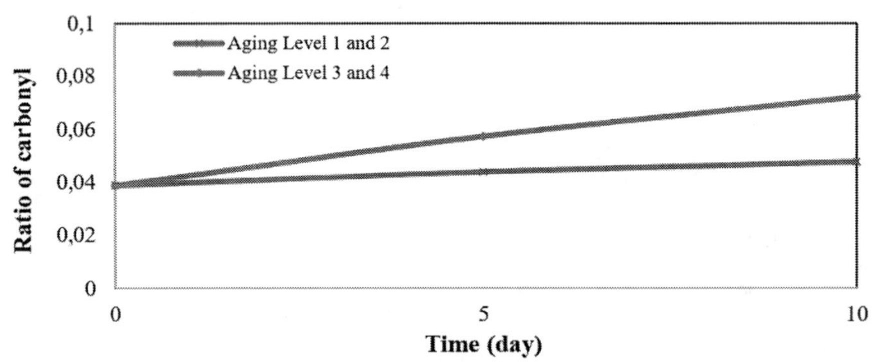

Figure 6. Ratio of carbonyl within the asphalt binder

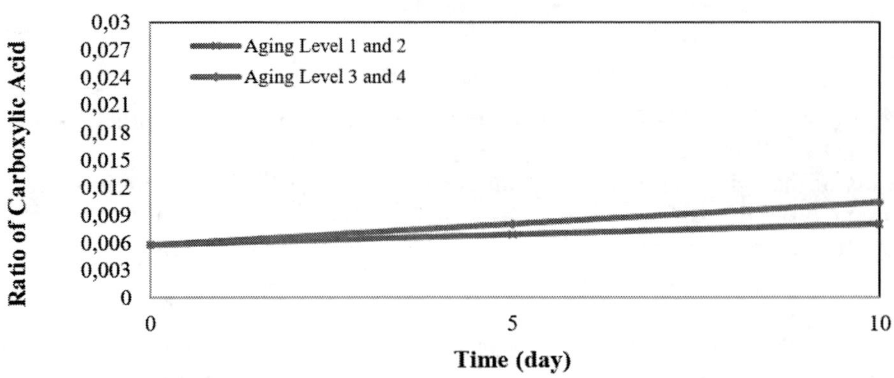

Figure 7. Ratio of carboxylic acid within the asphalt binder

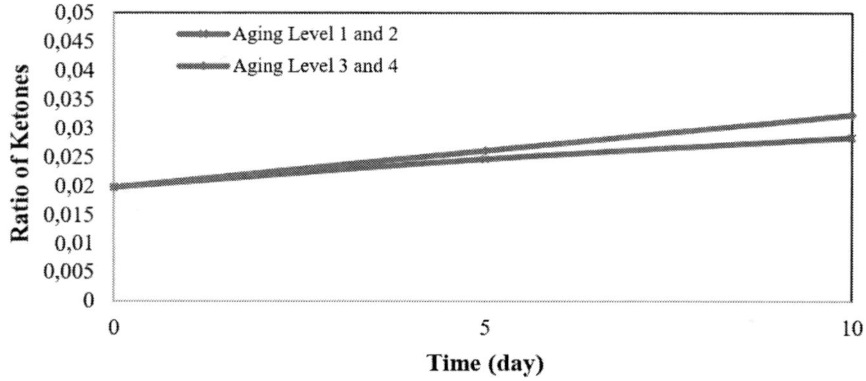

Figure 8. Ratio of ketones within the asphalt binder

4.3 Comparison of Mechanical and Chemical Aging Variable

In this study, to characterize those parameters, two temperatures (60°C and 95°C), and two different days (5 and 10 days) for asphalt mixture and additional two different oxygen content for asphalt binder were applied to be the aging condition in different aging levels. From the aged asphalt binder, with the obtained aging variable values as seen in Figure 4, the k1, k2, k3 and $\Gamma\alpha$ parameters achieved by minimizing the error between aging variable from experiment test result and model prediction. The summary of the model parameters for asphalt binder was presented in Table 4. To capture the evolution of time rate of aging parameter A at different relevancies such as temperature, oxygen content and aging variable, the rate of aging and fitted model parameters were used. The model prediction of Aging variable then calculated and was shown in Figure 9. The aging variable had two rates, which was fast rate at the beginning of the aging period and slower rate by the end of the aging period.

Table 4. Aging Model Parameters for the Asphalt Binder

Γ^{a}	A_0	k_1	k_2	k_3	T_0
0.0269	0	8.57	8.64	1.60	25

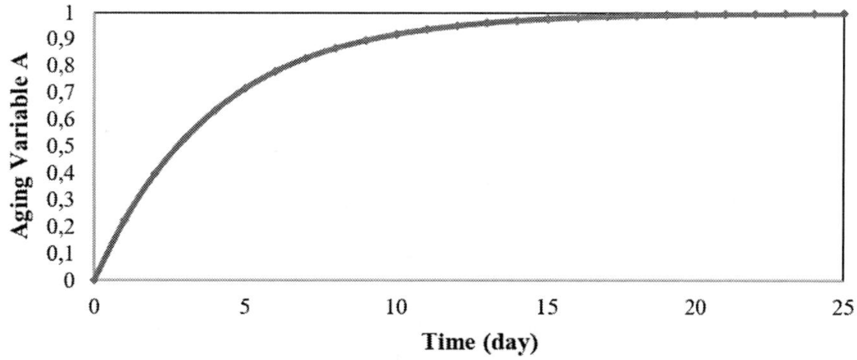

Figure 9. Aging variable model

Besides, using the same procedures of asphalt binder to determine the model parameters and using the k1 value from asphalt binder as the k1 value for the asphalt concrete aging model, the rest of the aging model parameters were calculated and it was shown in Table 5. Figure 10 then showed the evolution of time rate of aging parameter (A) at different relevancies and Figure 11 described the prediction model of aging variable A for asphalt concrete.

Table 5. Aging Model Parameters for the Asphalt Concrete

Γ^{a}	A_0	k_1	k_2	k_3	T_0
0.0013	0	8.57	5.98	0.68	25

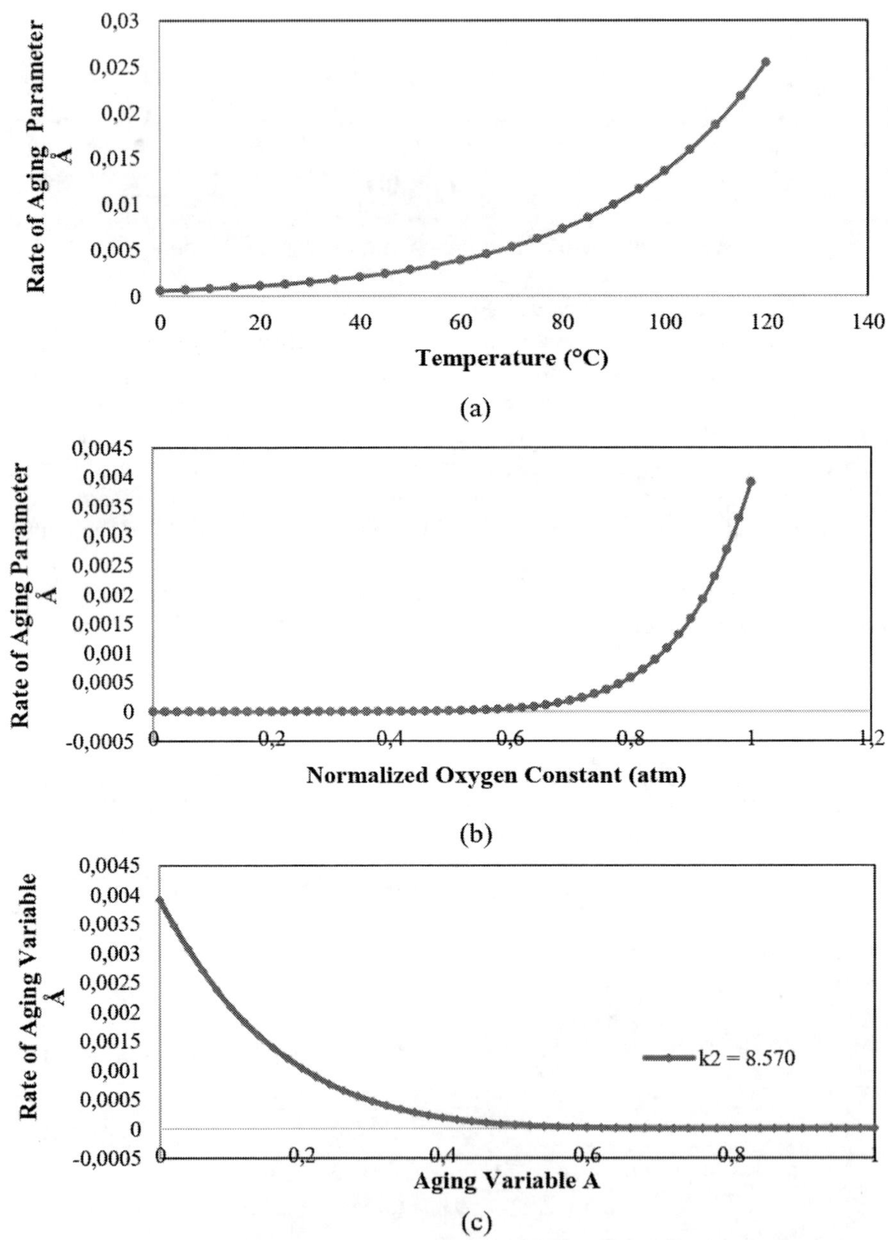

Figure 10. Rate of aging parameters of asphalt concrete, (a) at different temperature, (b) at different oxygen content, (c) at different aging variable

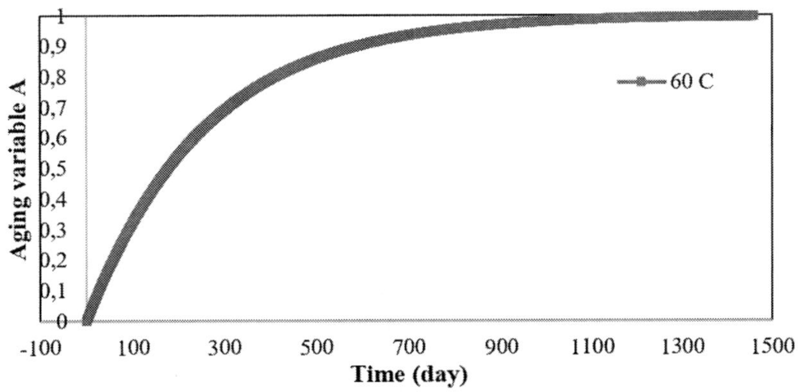

Figure 11. Aging variable model

Figure 10 (a) showed that temperature change has a contribution to the rate of aging variable, especially in higher temperature. It has the same objection with the previous study that said asphalt as a temperature-dependent material and by this study, it was validated that the aging model could predict the aging rate in term of temperature change very well. Previous studies showed the fact that the rate of aging variable was also related to the oxygen content available within the asphalt concrete since the amount of it could lead to higher rate of chemical reaction during the oxidative aging process. That being said, the prediction model of the aging variable also captured the phenomenon was shown in Figure 10 (b) and was validated by the decreasing rate of aging variable during the service life Figure 10 (c). It leads to an interesting finding that by using the assumption of aging history variable k2 = 1, the rate of the aging variable was decreased with the linear curve which is not realistic to the real aging phenomenon since the rate of aging tends to decrease with increasing level of aging and gradually reach a plateau. In this study, the k2 parameter was obtained amount 8.570. Employed the obtained k2 into the rate of aging resulted in a decreasing rate of aging variable by polynomial curve which was a closer relation of the oxidative aging process in reality. Figure 12 showed the comparison between k2=1 and k2=8.570 for the aging variable vs. aging rate.

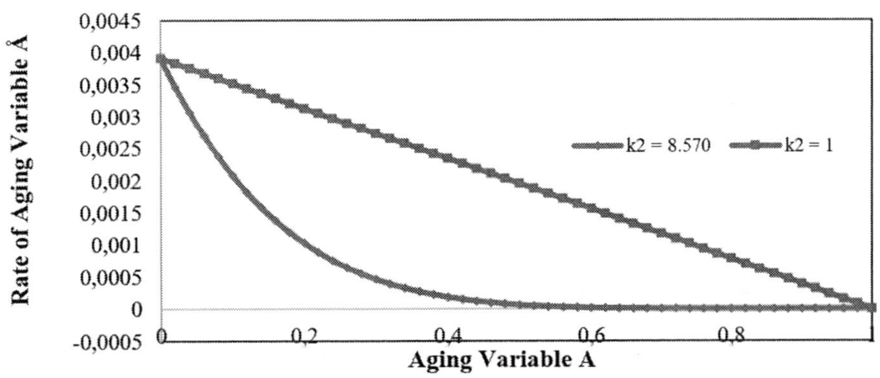

Figure 12. Comparison of aging variable rate with assumption and evaluated aging history parameter k_2 value

It was noteworthy about the aging model that it was more related to simulation of the aging of asphalt concrete rather than asphalt binder. Abu al-Rub et. al. (2013) mentioned that the aging variable value was formulated as a function of the diffused oxygen content and temperature evolution that linked to the mechanical response of aged material during the oxidative aging of asphalt concrete. Babadopulus et. Al. (2014) and Rahmani et. Al. (2017) studies that related to the same aging variable also showed that there is a closer rate of an aging variable due to aged asphalt concrete rather than aged asphalt binder found from this study.

Lastly, FTIR test result was taken to verify the aging model. The Pearson Correlation Analysis applied to obtain the relationship between the aging variable rate of asphalt concrete and the carbonyl rate of aged asphalt binder. Based on the results of this study, aging variable of asphalt concrete was strongly related to the rate of carbonyl content with the result of r = 0.976 and p-value = 0.0046. Figure 13 showed the result of the correlation test.

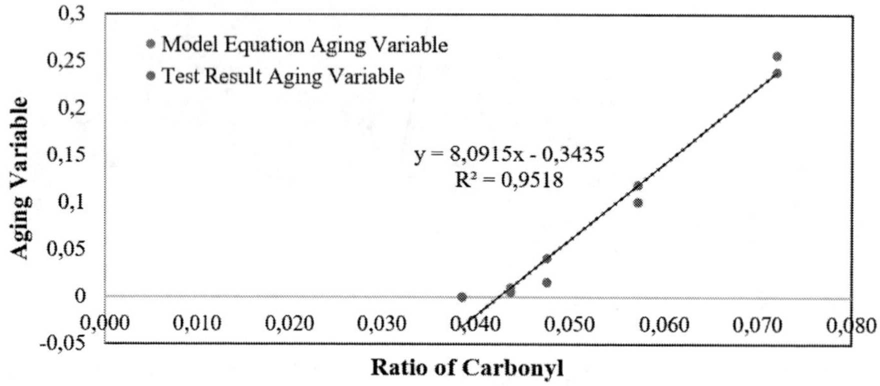

Figure 13. Correlation test result between carbonyl and aging variable

5. Conclusion

The findings of this study summarized as follow:

- The level of aging applied to both asphalt binder and asphalt concrete showed differences, as verified by all of the p-values from the regression analysis test, were greater than 0.05.
- The aging variable values were decreased with the aggregate involved
- The aging model parameters value of asphalt binder were 0.0269, 8.570, 8.640, 1.598 for Γ^{α}, k_1, k_2, k_3 respectively, where asphalt concrete were 0.0013, 8.570, 5.978, 0.780.
- Aging history dependent (k_2) found in this study allow the model to simulate the actual aging behavior as a polynomial curve.
- Carbonyl content was used to validate the aging model variable feasibility using the Pearson correlation test resulting r = 0.976 and p-value = 0.0046 which means there was a strong correlation between both of them.

6. References

[1] Xiaoge T., Jianlong, Z., & Qisen Z. (2004). Effect of aging on viscoleastic performance of asphalt binder. Proceedings of the 23rd Southern African Transport Conference (SATC 2004). ISBN Number: 1-920-01723-2. Pretoria, South Africa.

[2] Nur Izzi, M.Y. (2012). Modelling the linear viscoelastic rheological properties of bituminous binders. Disertation. Department of Civil Engineering. University of Nottingham. United Kingdom.

[3] Mousavi, M., Pahlavan, F., Oldham, D., Hosseinnezhad, S., & Fini, E.H. (2016). Multiscale investigation of oxidative aging in biomodified asphalt binder. The Journal of Physical Chemistry, 120 (31), pp 17224–17233. DOI: 10.1021/acs.jpcc.6b05004.

[4] Glover, C.J., Martin, A.E., Chowdhury, A., Hang, R., Prapaitrakul, N., Jin, X., & Lawrence, J. (2009). Evaluation of Binder Aging and Its Influence in Aging of Hot Mix Asphalt Concrete Evaluation of Binder Aging and Its Influence in Aging of Hot Mix Asphalt Concrete. Project 0-6009, FHWA/TX-08/0-6009-1, Texas Department of Transportation and the Federal Highway Administration.

[5] Michalica, P., Kazatchkov, I. B., Stastna, J., & Zanzotto, L. (2008). Relationship between chemical and rheological properties of two asphalts of different origins. *Fuel*, *87*(15), 3247-3253.

[6] Darabi, M. K., Al-Rub, R. K. A., Masad, E. A., & Little, D. N. (2012). Thermodynamic-based model for coupling temperature-dependent viscoelastic, viscoplastic, and viscodamage constitutive behavior of asphalt mixtures. International Journal for Numerical and Analytical Methods in Geomechanics, 36(7), 817-854. DOI:10.1002/nag.1030.

[7] Al-Rub, R. K. A., Darabi, M. K., Kim, S. M., Little, D. N., & Glover, C. J. (2013). Mechanistic-based constitutive modeling of oxidative aging in aging-susceptible materials and its effect on the damage potential of asphalt concrete. Construction and Building Materials, 41, 439-454. http://dx.doi.org/10.1016/j.conbuildmat.2012.12.044.

[8] Rahmani, E., Darabi, M.K., Little, D.N., & Masad, E.A. (2017). Constitutive modeling of coupled aging viscoelastic response of asphalt concrete. Construction and Building Materials, 131, 1-15. https://doi.org/10.1016/j.conbuildmat.2016.11.014.

[9] Babadopulos, L. F. A. L., Branco, F.C., & Soares, J.B. (2014). Aging and constitutive modeling of asphalt mixtures: research developments in Brazil. Conference paper. DOI: 10.1201/b17219-131.

[10] Bowers, B. F., Huang, B., Shu, X., & Miller, B. C. (2014). Investigation of reclaimed asphalt pavement blending efficiency through GPC and FTIR. Construction and building materials, 50, 517-523.

[11] Piérard, N., & Vanelstraete, A. (2009). Developing a test method for the accelerated ageing of bituminous mixtures in the laboratory. Advanced Testing and Characterization of Bituminous Materials, 163-171.

[12] Schapery, R. A. (1969). On the characterization of nonlinear viscoelastic materials. *Polymer Engineering & Science*, *9*(4), 295-310.

[13] Yao, H., Dai, Q., & You, Z. (2015). Fourier Transform Infrared Spectroscopy characterization of aging-related properties of original and nano-modified asphalt binders. Construction and Building Materials, 101, 1078-1087. DOI: 10.1016/j.conbuildmat.2015.10.085.

AUTHOR INDEX

Institute of Physics
Dirac House, Temple Back
Bristol BS1 6BE UK

ISSN: 1757-8981
ISBN 978-1-7138-4796-0

9 781713 847960

CURRAN ASSOCIATES INC.
proceedings.com